Problem Books in Mathematics

Edited by P. Winkler

Péter Komjáth and Vilmos Totik

Problems and Theorems in Classical Set Theory

Springer

Péter Komjáth
Department of Computer Science
Eotvös Lóránd University, Budapest
Budapest 1117
Hungary

Vilmos Totik
Department of Mathematics
University of South Florida
Tampa, FL 33620
USA
and
Bolyai Institute
University of Szeged
Szeged
Hungary
6720
totik@math.usf.edu

Series Editor:
Peter Winkler
Department of Mathematics
Dartmouth College
Hanover, NH 03755-3551
Peter.winkler@dartmouth.edu

Mathematics Subject Classification (2000): 03Exx, 05-xx, 11Bxx

Library of Congress Control Number: 2005938489

ISBN-10: 0-387-30293-X
ISBN-13: 978-0387-30293-5

springer.com

Komjáth & Totik: Problems and Theorems in Classical Set Theory

First Indian Reprint 2011

ISBN 978-81-8489-840-8

This edition is published by Springer (India) Private Limited, (a part of Springer Science+Business Media), Registered Office: 3rd Floor, Gandharva Mahavidyalaya, 212, Deen Dayal Upadhyaya Marg, New Delhi – 110 002, India

Printed in India by Rashtriya Printers, Delhi.

Dedicated to András Hajnal
and to the memory of
Paul Erdős and Géza Fodor

Contents

Part II Solutions

Part III Appendix

Preface

Although the first decades of the 20th century saw some strong debates on set theory and the foundation of mathematics, afterwards set theory has turned into a solid branch of mathematics, indeed, so solid, that it serves as the foundation of the whole building of mathematics. Later generations, honest to Hilbert's dictum, "No one can chase us out of the paradise that Cantor has created for us" proved countless deep and interesting theorems and also applied the methods of set theory to various problems in algebra, topology, infinitary combinatorics, and real analysis.

The invention of forcing produced a powerful, technically sophisticated tool for solving unsolvable problems. Still, most results of the pre-Cohen era can be digested with just the knowledge of a commonsense introduction to the topic. And it is a worthy effort, here we refer not just to usefulness, but, first and foremost, to mathematical beauty.

In this volume we offer a collection of various problems in set theory. Most of classical set theory is covered, classical in the sense that independence methods are not used, but classical also in the sense that most results come from the period, say, 1920–1970. Many problems are also related to other fields of mathematics such as algebra, combinatorics, topology, and real analysis.

We do not concentrate on the axiomatic framework, although some aspects, such as the axiom of foundation or the rôle of the axiom of choice, are elaborated.

There are no drill exercises, and only a handful can be solved with just understanding the definitions. Most problems require work, wit, and inspiration. Some problems are definitely challenging, actually, several of them are published results.

We have tried to compose the sequence of problems in a way that earlier problems help in the solution of later ones. The same applies to the sequence of chapters. There are a few exceptions (using transfinite methods before their discussion)—those problems are separated at the end of the individual chapters by a line of asterisks.

We have tried to trace the origin of the problems and then to give proper reference at the end of the solution. However, as is the case with any other mathematical discipline, many problems are folklore and tracing their origin was impossible.

The reference to a problem is of the form "Problem x.y" where x denotes the chapter number and y the problem number within Chapter x. However, within Chapter x we omit the chapter number, so in that case the reference is simply "Problem y".

For the convenience of the reader we have collected into an appendix all the basic concepts and notations used throughout the book.

Acknowledgements We thank Péter Varjú and Gergely Ambrus for their careful reading of the manuscript and their suggestions to improve the presentation. Collecting and writing up the problems took many years, during which the authors have been funded by various grants from the Hungarian National Science Foundation for Basic Research and from the National Science Foundation (latest grants are OTKA T046991, T049448 and NSF DMS-040650).

We hope the readers will find as much enjoyment in solving some of the problems as we have found in writing them up.

Péter Komjáth and Vilmos Totik
Budapest and Szeged-Tampa, July 2005

Part I

Problems

1

Operations on sets

Basic operations among sets are union, intersection, and exponentiation. This chapter contains problems related to these basic operations and their relations.

If we are given a family of sets, then (two-term) intersection acts like multiplication. However, from many point of view, the analogue of addition is not union, but forming divided difference: $A\Delta B = (A \setminus B) \cup (B \setminus A)$, and several problems are on this Δ operation.

An interesting feature is that families of sets with appropriate set operations can serve as *canonical models* for structures from other areas of mathematics. In this chapter we shall see that graphs, partially ordered sets, distributive lattices, idempotent rings, and Boolean algebras can be modelled by (i.e., are isomorphic to) families of sets with appropriate operations on them.

1. For finite sets A_i we have

$$|A_1 \cup \cdots \cup A_n| = \sum_i |A_i| - \sum_{i<j} |A_i \cap A_j| + \sum_{i<j<k} |A_i \cap A_j \cap A_k| - \cdots,$$

and

$$|A_1 \cap \cdots \cap A_n| = \sum_i |A_i| - \sum_{i<j} |A_i \cup A_j| + \sum_{i<j<k} |A_i \cup A_j \cup A_k| - \cdots.$$

2. Define the symmetric difference of the sets A and B as

$$A\Delta B = (A \setminus B) \cup (B \setminus A).$$

This is a commutative and associative operation such that $\cap$ is distributive with respect to Δ.

3. The set $A_1 \Delta A_2 \Delta \cdots \Delta A_n$ consists of those elements that belong to an odd number of the A_i's.

4. For finite sets A_i we have

$$|A_1 \Delta A_2 \cdots \Delta A_n| = \sum_i |A_i| - 2\sum_{i<j} |A_i \cap A_j| + 4 \sum_{i<j<k} |A_i \cap A_j \cap A_k| - \cdots.$$

5. Let our sets be subsets of a ground set X, and define the complement of A as $A^c = X \setminus A$. All the operations $\cap$, $\cup$ and $\setminus$ can be expressed by the operation $A \downarrow B = (A \cup B)^c$. The same is also true of $A \mid B = (A \cap B)^c$.

6. For any sets

a)

$$\bigcup_{i\in I} \bigcap_{j\in J_i} A_{i,j} = \bigcap_{f\in\prod_{i\in I} J_i} \bigcup_{i\in I} A_{i,f(i)}$$

b)

$$\bigcap_{i\in I} \bigcup_{j\in J_i} A_{i,j} = \bigcup_{f\in\prod_{i\in I} J_i} \bigcap_{i\in I} A_{i,f(i)}$$

c)

$$\prod_{i\in I} \left(\bigcup_{j\in J_i} A_{i,j} \right) = \bigcup_{f\in\prod_{i\in I} J_i} \left(\prod_{i\in I} A_{i,f(i)} \right)$$

d)

$$\prod_{i\in I} \left(\bigcap_{j\in J_i} A_{i,j} \right) = \bigcap_{f\in\prod_{i\in I} J_i} \left(\prod_{i\in I} A_{i,f(i)} \right)$$

(general distributive laws).

7. Let X be a set and $A_1, A_2, \ldots, A_n \subseteq X$. Using the operations $\cap$, $\cup$ and $\cdot^c$ (complementation relative to X), one can construct at most 2^{2^n} different sets from $A_1, A_2, \ldots, A_n$.

8. Let

$$X = \{(x_1, \ldots, x_n) \ : \ 0 \le x_i < 1, 1 \le i \le n\}$$

be the unit cube of $\mathbf{R}^n$, and set

$$A_k = \{(x_1, \ldots, x_n) \in X \ : \ 1/2 \le x_k < 1\}.$$

Using the operations $\cap$, $\cup$, and $\cdot^c$ (complementation with respect to X), one can construct 2^{2^n} different sets from $A_1, A_2, \ldots, A_n$.

9. Using the operations $\setminus$, $\cap$ and $\cup$ one can construct at most 2^{2^n-1} different sets from a given family $A_1, A_2, \ldots, A_n$ of n sets. This 2^{2^n-1} bound can be achieved for some appropriately chosen $A_1, A_2, \ldots, A_n$.

10. For given A_i, B_i, $i \in I$ solve the system of equations
    (a) $A_i \cap X = B_i, \quad i \in I$,
    (b) $A_i \cup X = B_i, \quad i \in I$,
    (c) $A_i \setminus X = B_i, \quad i \in I$,
    (d) $X \setminus A_i = B_i, \quad i \in I$.
    What are the necessary and sufficient conditions for the existence and uniqueness of the solutions?
11. If $A_0, A_1, \ldots$ is an arbitrary sequence of sets, then there are pairwise disjoint sets $B_i \subseteq A_i$ such that $\cup A_i = \cup B_i$.
12. Let $A_0, A_1, \ldots$ and $B_0, B_1, \ldots$ be sequences of sets. Then the intersection $A_i \cap B_j$ is finite for all i, j if and only if there are disjoint sets C and D such that for all i the sets $A_i \setminus C$ and $B_i \setminus D$ are finite.
13. Let X be a ground set and $\mathcal{A} \subseteq \mathcal{P}(X)$ such that for every $A \in \mathcal{A}$ the complement $X \setminus A$ can be written as a countable intersection of elements of $\mathcal{A}$. Then the σ-algebra generated by $\mathcal{A}$ coincides with the smallest family of sets including $\mathcal{A}$ and closed under countable intersection and countable disjoint union.
14. Define

$$\liminf_{n\to\infty} A_n := \bigcup_{n=1}^{\infty} \bigcap_{m=n}^{\infty} A_m,$$

$$\limsup_{n\to\infty} A_n := \bigcap_{n=1}^{\infty} \bigcup_{m=n}^{\infty} A_m,$$

and we say that the sequence $\{A_n\}$ is convergent if these two sets are the same, say A, in which case we say that the limit of the sets $\{A_n\}$ is A. Then

a) $\liminf_n A_n \subseteq \limsup_n A_n$,

b) $\liminf_n A_n$ consists of those elements that belong to all, but finitely many of the A_n's.

c) $\limsup_n A_n$ consists of those elements that belong to infinitely many A_n's.

15. Let X be a set and for a subset A of X consider its characteristic function

$$\chi_A(x) = \begin{cases} 1 \text{ if } x \in A, \\ 0 \text{ if } x \in X \setminus A. \end{cases}$$

The mapping $A \to \chi_A$ is a bijection between $\mathcal{P}(X)$ and ${}^X\{0,1\}$. Furthermore, if $B = \liminf_{n\to\infty} A_n$, then

$$\chi_B = \liminf_{n\to\infty} \chi_{A_n},$$

and if $C = \limsup_{n\to\infty} A_n$, then

$$\chi_C = \limsup_{n\to\infty} \chi_{A_n}.$$

16. A sequence $\{A_n\}_{n=1}^\infty$ of sets is convergent if and only if for every sequences $\{m_i\}$ and $\{n_i\}$ with $\lim_{i\to\infty} m_i = \lim_{i\to\infty} n_i = \infty$ we have
$$\bigcap_i (A_{m_i} \Delta A_{n_i}) = \emptyset.$$
17. A sequence $\{A_n\}_{n=1}^\infty$ of sets converges if and only if for every sequences $\{m_i\}$ and $\{n_i\}$ with $\lim_{i\to\infty} m_i = \lim_{i\to\infty} n_i = \infty$ we have
$$\lim_{i\to\infty} (A_{m_i} \Delta A_{n_i}) = \emptyset$$
(if we regard Δ as subtraction, then this says that for convergence of sets "Cauchy's criterion" holds).
18. If A_n, $n = 0, 1, \dots$ are subsets of the set of natural numbers, then one can select a convergent subsequence from $\{A_n\}_{n=0}^\infty$.
19. Construct a sequence $\{A_n\}_{n=0}^\infty$ of sets which does not include a convergent subsequence.
20. If $\mathcal{H}$ is any family of sets, then with the inclusion relation $\mathcal{H}$ is a partially ordered set. Every partially ordered set is isomorphic with a family of sets partially ordered by inclusion.
21. Every graph is isomorphic with a graph where the set of vertices is a family of sets, and two such vertices are connected precisely if their intersection is not empty.
22. Let $\mathcal{H}$ be a set that is closed for two-term intersection, union and symmetric difference. Then $\mathcal{H}$ is a ring with Δ as addition and $\cap$ as multiplication, in which every element is idempotent: $A \cap A = A$.
23. If $(A, +, \cdot, 0)$ is a ring in which every element is idempotent $(a \cdot a = a)$, then $(A, +, \cdot, 0)$ is isomorphic with a ring of sets defined in the preceding problem.
24. With the notation of Problem 22 let $\mathcal{H}$ be the set of all subsets of an infinite set X, and let $\mathcal{I}$ be the set of finite subsets of X. Then $\mathcal{I}$ is an ideal in $\mathcal{H}$. If $a \neq 0$ is any element in the quotient ring $\mathcal{H}/\mathcal{I}$, then there is a $b \neq 0, a$ such that $b \cdot a = b$ (in other words, in the quotient ring there are no atoms).
25. If $\mathcal{H}$ is a family of subsets of a given ground set X which is closed for two-term intersection and union, then $\mathcal{H}$ is a distributive lattice with the operations $H \wedge K = H \cap K$, $H \vee K = H \cup K$.
26. Every distributive lattice is isomorphic to one from the preceding problem.
27. If $\mathcal{H}$ is a family of subsets of a given ground set X which is closed under complementation (relative to X) and under two-term union, then $\mathcal{H}$ is a Boolean algebra with the operations $H \cdot K = H \cap K$, $H + K = H \cup K$, $H' = X \setminus H$ and with $1 = X$, $0 = \emptyset$.

28. Every Boolean algebra is isomorphic to one from the preceding problem.

29. $\mathcal{P}(X)$, the family of all subsets of a given set X, is a complete and completely distributive Boolean algebra with the operations $H \cdot K = H \cap K$, $H + K = H \cup K$, $H' = X \setminus H$ and with $1 = X$, $0 = \emptyset$ (in the Boolean algebra set $a \preceq b$ if $a \cdot b = a$, and completeness means that for any set K in the Boolean algebra there is a smallest upper majorant $\sup K$ and a largest lower minorant $\inf K$, and complete distributivity means that

$$\inf_{i \in I} \sup_{j \in J_i} a_{i,j} = \sup_{f \in \prod_{i \in I} J_i} \inf_i a_{i,f(i)}$$

for any elements in the algebra).

30. Every complete and completely distributive Boolean algebra is isomorphic with one from the preceding problem.

31. Let $\mathcal{H}$ be a family of sets such that if $\mathcal{H}^* \subset \mathcal{H}$ is any subfamily, then there is a smallest (with respect to inclusion) set in $\mathcal{H}$ that includes all the sets in $\mathcal{H}^*$, and there is a largest set in $\mathcal{H}$ that is included in all elements of $\mathcal{H}^*$. Then every mapping $f : \mathcal{H} \to \mathcal{H}$ that preserves the relation $\subseteq$ (i.e., for which $f(H) \subseteq f(K)$ whenever $H \subseteq K$) there is a fixed point, i.e., a set $F \in \mathcal{H}$ with $f(F) = F$.

* * *

32. The converse of Problem 31 is also true in the following sense. Suppose that $\mathcal{H}$ is a family of sets closed for two-term union and intersection such that for every mapping $f : \mathcal{H} \to \mathcal{H}$ that preserves $\subseteq$ there is a fixed point. Then if $\mathcal{H}^* \subset \mathcal{H}$ is any subfamily, then there is a smallest set in $\mathcal{H}$ that includes all the sets in $\mathcal{H}^*$, and there is a largest set in $\mathcal{H}$ that is included in all elements of $\mathcal{H}^*$.

33. With the notation of Problem 24 for each $a \neq 0$ there are at least continuum many different $b \neq 0$ such that $b \cdot a = b$.

34. With the notation of Problem 24 let $\mathcal{H}$ be the set of all subsets of a set X of cardinality κ, and let $\mathcal{I}$ be the ideal of subsets of X which have cardinality smaller than κ. Then the quotient ring $\mathcal{H}/\mathcal{I}$ is of cardinality 2^κ.

2

Countability

A set is called *countable* if its elements can be arranged into a finite or infinite sequence. Otherwise it is called *uncountable.* This notion reflects the fact that the set is "small" from the point of view of set theory; sometimes it is negligible. For example, the set Q of rational numbers is countable (Problem 9) while the set $\mathbf{R}$ of real numbers is not (Problem 7), hence "most" reals are irrational. On the other hand, a claim that a certain set is not countable usually means that the set has many elements.

If in an uncountable set A a certain property holds with the exception of elements in a countable subset B, then the property holds for "most" elements of A (in particular $A \setminus B$ is not empty). In this section many problems are related to this principle; in particular many problems claim that a certain set in $\mathbf{R}$ (or $\mathbf{R}^n$) is countable. Actually, the very first "sensational" achievement of set theory was of this sort when G. Cantor proved in 1874 that "most" real numbers are transcendental (and hence there are transcendental numbers), for the algebraic numbers form a countable subset of $\mathbf{R}$ (see Problems 6–8). Other examples when the notion of countability appears in real analysis will be given in Chapters 5 and 13.

The cardinality of countably infinite sets is denoted by ω or $\aleph_0$.

1. The union of countably many countable sets is countable.
2. The (Cartesian) product of finitely many countable sets is countable.
3. The set of k element sequences formed from a countable sets is countable.
4. The set of finite sequences formed from a countable set is countable.
5. The set of polynomials with integer coefficients is countable.
6. The set of algebraic numbers is countable.
7. $\mathbf{R}$ is not countable.
8. There are transcendental real numbers.
9. The following sets are countable:

a) $\mathbf{Q}$;

b) set of those functions that map a finite subset of a given countable set A into a given countable set B;

c) set of convergent sequences of natural numbers.

10. If $A_i \subseteq \mathbf{N}$, $i \in I$ is an arbitrary family of subsets of $\mathbf{N}$, then there is a countable subfamily A_i, $i \in J \subset I$ such that $\cap_{i\in J} A_i = \cap_{i\in I} A_i$ and $\cup_{i\in J} A_i = \cup_{i\in I} A_i$.
11. If A is an uncountable subset of the real line, then there is an $a \in A$ such that each of the sets $A \cap (-\infty, a)$ and $A \cap (a, \infty)$ is uncountable.
12. If k and K are positive integers and $\mathcal{H}$ is a family of subsets of $\mathbf{N}$ with the property that the intersection of every k members of $\mathcal{H}$ has at most K elements, then $\mathcal{H}$ is countable.
13. The set of subintervals of $\mathbf{R}$ with rational endpoints is countable.
14. Any disjoint collection of open intervals (open sets) on $\mathbf{R}$ (in $\mathbf{R}^n$) is countable.
15. Any discrete set in $\mathbf{R}$ (in $\mathbf{R}^n$) is countable.
16. Any open subset of $\mathbf{R}$ is a disjoint union of countably many open intervals.
17. The set of open disks (balls) in $\mathbf{R}^2$ ($\mathbf{R}^n$) with rational radius and rational center, is countable (rational center means that each coordinate of the center is rational).
18. Any open subset of $\mathbf{R}^2$ ($\mathbf{R}^n$) is a union of countably many open disks (balls) with rational radius and rational center.
19. If $\mathcal{H}$ is a family of circles such that for every $x \in \mathbf{R}$ there is a circle in $\mathcal{H}$ that touches the real line at the point x, then there are two intersecting circles in $\mathcal{H}$.
20. Is it true that if $\mathcal{H}$ is a family of circles such that for every $x \in \mathbf{R}$ there is a circle containing x, then there are two intersecting circles in H?
21. Let $\mathcal{C}$ be a family of circles on the plane such that no two cross each other. Then the points where two circles from $\mathcal{C}$ touch each other form a countable set.
22. One can place only countably many disjoint letters of the shape T on the plane.
23. In the plane call a union of three segments with a common endpoint a Y-set. Any disjoint family of Y-sets is countable.
24. If A is a countable set on the plane, then it can be decomposed as $A = B \cup C$ such that B, resp. C has only a finite number of points on every vertical, resp. horizontal line.
25. A is countable if and only if $A \times A$ can be decomposed as $B \cup C$ such that B intersects every "vertical" line $\{(x, y) : x = x_0\}$ in at most finitely many points, and C intersects every "horizontal" line $\{(x, y) : y = y_0\}$ in at most finitely many points.

26. If $A \subset \mathbf{R}$ is countable, then there is a real number a such that $(a+A) \cap A = \emptyset$.

27. If $A \subset \mathbf{R}^2$ is such that all the distances between the points of A are rational, then A is countable. Is there such an infinite bounded set not lying on a straight line?

28. Call a sequence $a_n \to \infty$ faster increasing than $b_n \to \infty$ if $a_n/b_n \to \infty$. If $\{b_n^{(i)}\}$, $i = 0, 1, \ldots$ is a countable family of sequences tending to ∞, then there is a sequence that increases faster than any $\{b_n^{(i)}\}$.

29. If there are given countably many sequences $\{s_n^{(i)}\}_{n=0}^{\infty}$, $i = 0, 1, \ldots$ of natural numbers, then construct a sequence $\{s_n\}_{n=0}^{\infty}$ of natural numbers such that for every i the equality $s_n = s_n^{(i)}$ holds only for finitely many n's.

30. Construct countably many sequences $\{s_n^{(i)}\}_{n=0}^{\infty}$, $i = 0, 1, \ldots$ of natural numbers, with the property that if $\{s_n\}_{n=0}^{\infty}$ is an arbitrary sequence of natural numbers, then the number those n's for which $s_n = s_n^{(i)}$ holds is unbounded as $i \to \infty$.

31. Are there countably many sequences $\{s_n^{(i)}\}_{n=0}^{\infty}$, $i = 0, 1, \ldots$ of natural numbers, with the property that if $\{s_n\}_{n=0}^{\infty}$ is an arbitrary sequence of natural numbers, then the number those n's for which $s_n = s_n^{(i)}$ holds tends to infinity as $i \to \infty$?

32. Let $\{r_k\}$ be a 1–1 enumeration of the rational numbers. Then if $\{x_n\}$ is an arbitrary sequence consisting of rational numbers, there are three permutations π_i, $i = 1, 2, 3$ of the natural numbers for which $x_n = r_{\pi_1(n)} + r_{\pi_2(n)} + r_{\pi_3(n)}$ holds for all n.

33. With the notation of the preceding problem give a sequence $\{x_n\}$ consisting of rational numbers for which there are no permutations π_i, $i = 1, 2$, of the natural numbers for which $x_n = r_{\pi_1(n)} + r_{\pi_2(n)}$ holds for all n.

34. Any two countably infinite Boolean algebras without atoms (i.e., without elements $a \neq 0$ such that $a \cdot b = a$ or $a \cdot b = 0$ for all b) are isomorphic.

35. Let $\mathcal{A} = (A, \ldots)$ be an arbitrary algebraic structure on the countable set A (i.e., $\mathcal{A}$ may have an arbitrary number of finitary operations and relations). Then the following are equivalent:

    a) $\mathcal{A}$ has uncountably many automorphisms;

    b) if B is a finite subset of A then there is a non-identity automorphism of $\mathcal{A}$ which is the identity when restricted to B.

36. Suppose we know that a rabbit is moving along a straight line on the lattice points of the plane by making identical jumps every minute (but we do not know where it is and what kind of jump it is making). If we can place a trap every hour to an arbitrary lattice point of the plane that captures the rabbit if it is there at that moment, then we can capture the rabbit.

37. Let $A \subset [0,1]$ be a set, and two players I and II play the following game: they alternatively select digits (i.e., numbers 0–9) $x_0, x_1, \ldots$ and $y_0, y_1, \ldots$, and I wins if the number $0.x_1y_1x_2y_2\ldots$ is in A, otherwise II wins. In this game if A is countable, then II has a winning strategy.

38. Let $A \subset [0,1]$ be a set, and two players I and II play the following game: I selects infinitely many digits $x_1, x_2, \ldots$ and II makes a permutation $y_1, y_2, \ldots$ of them. I wins if the number $0.y_1y_2\ldots$ is in A, otherwise II wins. For what countable closed sets A does I have a winning strategy?

39. Two players alternately choose uncountable subsets $K_0 \supset K_1 \supset \cdots$ of the real line. Then no matter how the first player plays, the second one can always achieve $\cap_{n=0}^{\infty} K_n = \emptyset$.

* * *

40. Let κ be an infinite cardinal. Then H is of cardinality at most κ if and only if $H \times H$ can be decomposed as $B \cup C$ such that B intersects every "vertical" line $\{(x,y) : x = x_0\}$ in less than κ points, and C intersects every "horizontal" line $\{(x,y) : y = y_0\}$ in less than κ points.

3

Equivalence

Equivalence of sets is the mathematical notion of "being of the same size". Two sets A and B are *equivalent* (in symbol $A \sim B$) if there is a one-to-one correspondence between their elements, i.e., a one-to-one mapping $f : A \to B$ of A onto B. In this case we also say that A and B are of the same *cardinality* without telling what "cardinality" means.

A finite set cannot be equivalent to its proper subset, but things change for infinite sets: any infinite set is equivalent to one of its proper subsets. In fact, quite often seemingly "larger" sets (like a plane) may turn out to be equivalent to much "smaller" sets (like a line on the plane).

The notion of infinity is one of the most intriguing concepts that has been created by mankind. It is with the aid of equivalence that in mathematics we can distinguish between different sorts of infinity, and this makes the theory of infinite sets extremely rich.

This chapter contains some simple exercises on equivalence of sets often encountered in algebra, analysis, and topology. To establish the equivalence of two sets can be quite a challenge, but things are tremendously simplified by the *equivalence theorem* (Problem 2): if each of A and B is equivalent to a subset of the other one, then they are equivalent. The reason for the efficiency of the equivalence theorem lies in the fact that usually it is much easier to find a one-to-one mapping of a set A into B than onto B.

1. Let $f : A \to B$ and $g : B \to A$ be 1-to-1 mappings. Then there is a decomposition $A = A_1 \cup A_2$ and $B = B_1 \cup B_2$ of A and B into disjoint sets such that f maps A_1 onto B_1 and g maps B_2 onto A_2.
2. (Equivalence theorem) If two sets are both equivalent to a subset of the other one, then the two sets are equivalent.
3. There is a 1-to-1 mapping from $A(\neq \emptyset)$ to B if and only if there is a mapping from B onto A.
4. If A is infinite and B is countable, then $A \cup B \sim A$.
5. If A is uncountable and B is countable, then $A \setminus B \sim A$.

6. The set of irrational numbers is equivalent to the set of real numbers.
7. The Cantor set is equivalent to the set of infinite 0–1 sequences.
8. Give a 1-to-1 mapping from the first set into the second one:
   a) $\mathbf{N} \times \mathbf{N}$; $\mathbf{N}$
   b) $(-\infty, \infty)$; $(0, 1)$
   c) $\mathbf{R}$; the set of infinite 0–1 sequences
   d) the set of infinite 0–1 sequences; $[0, 1]$
   e) the infinite sequences of the natural numbers; the set of infinite 0–1 sequences
   f) the set of infinite sequences of the real numbers; the set of infinite 0–1 sequences

   In each of the above cases **a)–f)** the two sets are actually equivalent.
9. Give a mapping from the first set onto the second one:
   a) $\mathbf{N}$; $\mathbf{N} \times \mathbf{N}$
   b) $\mathbf{N}$; $\mathbf{Q}$
   c) Cantor set; $[0, 1]$
   d) set of infinite 0–1 sequences; $[0, 1]$

   In each of the above cases **a)–d)** the two sets are actually equivalent.
10. Give a 1-to-1 correspondence between these pairs of sets:
    a) (a, b); (c, d) (where $a < b$ and $c < d$, and any of these numbers can be $\pm\infty$ as well)
    b) $\mathbf{N}$; $\mathbf{N} \times \mathbf{N}$
    c) $\mathcal{P}(X)$; ${}^X\{0, 1\}$ (X is an arbitrary set)
    d) set of infinite sequences of the numbers $0, 1, 2$; set of infinite 0–1 sequences
    e) $[0, 1)$; $[0, 1) \times [0, 1)$
11. There is a 1-to-1 correspondence between these pairs of sets:
    a) set of infinite 0–1 sequences; $\mathbf{R}$
    b) $\mathbf{R}$; $\mathbf{R}^n$
    c) $\mathbf{R}$; set of infinite real sequences
12. We have
    a) ${}^{B \cup C}A \sim {}^B A \times {}^C A$ provided $B \cap C = \emptyset$,
    b) ${}^C({}^B A) \sim {}^{C \times B} A$,
    c) ${}^C(A \times B) \sim {}^C A \times {}^C B$.
13. Let X be an arbitrary set.
    a) X is similar to a subset of $\mathcal{P}(X)$.
    b) $X \not\sim \mathcal{P}(X)$.

4

Continuum

A set is called *of power continuum* (**c**) if it is equivalent with **R**. Many sets arising in mathematical analysis and topology are of power continuum, and the present chapter lists several of them. For example, the set of Borel subsets of $\mathbf{R}^n$, the set of right continuous real functions, or a Hausdorff topological space with countable basis are all of power continuum.

The continuum is also the cardinality of the set of subsets of **N**, and there are many examples of families of power continuum (i.e., families of maximal cardinality) of subsets of **N** or of a given countable set with a certain prescribed property. In particular, several problems in this chapter deal with almost disjoint sets and their variants: there are continuum many subsets of **N** with pairwise finite intersection (cf. Problems 29–43).

The problem if there is an uncountable subset of **R** which is not of power continuum arose very early during the development of set theory, and the "NO" answer has become known as the continuum hypothesis (CH). Thus, CH means that if $A \subseteq \mathbf{R}$ is infinite, then either $A \sim \mathbf{N}$ or $A \sim \mathbf{R}$ (other formulations are: there is no cardinality κ with $\aleph_0 < \kappa < \mathbf{c}$; $\aleph_1 = 2^{\aleph_0}$). This was the very first problem on Hilbert's famous list on the 1900 Paris congress, and finding the solution had a profound influence on set theory as well as on all of mathematics. Eventually it has turned out that it does not lead to a contradiction if we assume CH (K. Gödel, 1947) and neither leads to a contradiction if we assume CH to be false (P. Cohen, 1963). Therefore, CH is independent of the other standard axioms of set theory.

1. The plane cannot be covered with less than continuum many lines.
2. The set of infinite 0–1 sequences is of power continuum.
3. The set of infinite real sequences is of power continuum.
4. The Cantor set is of power continuum.
5. An infinite countable set has continuum many subsets.
6. An infinite set of cardinality at most continuum has continuum many countable subsets.

7. There are continuum many open (closed) sets in $\mathbf{R}^n$.
8. A Hausdorff topological space with countable base is of cardinality at most continuum.
9. In an infinite Hausdorff topological space there are at least continuum many open sets.
10. If A is countable and B is of cardinality at most continuum, then the set of functions $f : A \to B$ is of cardinality at most continuum.
11. The set of continuous real functions is of power continuum.
12. The product of countably many sets of cardinality at most continuum is of cardinality at most continuum.
13. The union of at most continuum many sets of cardinality at most continuum is of cardinality at most continuum.
14. The following sets are of power continuum.
    a) $\mathbf{R}^n$, $n = 1, 2, \ldots$
    b) $\mathbf{R}^\infty$ (which is the set of infinite real sequences)
    c) the set of continuous curves on the plane
    d) the set of monotone real functions
    e) the set of right-continuous real functions
    f) the set of those real functions that are continuous except for a countable set
    g) the set of lower semi-continuous real functions
    h) the set of permutations of the natural numbers
    i) the set of the (well) orderings of the natural numbers
    j) the set of closed additive subgroups of $\mathbf{R}$ (i.e., the set of additive subgroups of $\mathbf{R}$ that are at the same time closed sets in $\mathbf{R}$)
    k) the set of closed subspaces of $C[0, 1]$
    l) the set of bounded linear transformations of $L^2[0, 1]$
15. $\mathbf{R}$ cannot be represented as the union of countably many sets none of which is equivalent to $\mathbf{R}$.
16. If $A \subset \mathbf{R}^2$ is such that each horizontal line intersects A in finitely many points, then there is a vertical line that intersects the complement $\mathbf{R}^2 \setminus A$ of A in continuum many points.
17. If A is a subset of the real line of power continuum, then there is an $a \in A$ such that each of the sets $A \cap (-\infty, a)$ and $A \cap (a, \infty)$ is of power continuum.
18. Let $\mathcal{A} = (A, \ldots)$ be an arbitrary algebraic structure on the countable set A (i.e., $\mathcal{A}$ may have an arbitrary number of finitary operations and relations). Then the following are equivalent:
    a) $\mathcal{A}$ has uncountably many automorphisms,

b) $\mathcal{A}$ has continuum many automorphisms.

19. A σ-algebra is either finite or of cardinality at least continuum.

20. A σ-algebra generated by a set of cardinality at most continuum is of cardinality at most continuum.

21. There are continuum many Borel sets and Borel functions on the real line (in $\mathbf{R}^n$).

22. There are continuum many Baire functions on $[0,1]$.

23. The power set $\mathcal{P}(X)$ of X is of bigger cardinality than X.

24. If A has at least two elements, then the set ${}^XA$ of mappings from X to A is of bigger cardinality than X.

25. The following sets are of cardinality bigger than continuum.

 a) set of real functions

 b) set of the 1-to-1 correspondences between $\mathbf{R}$ and $\mathbf{R}^2$

 c) set of bases of $\mathbf{R}$ considered as a linear space over $\mathbf{Q}$ (Hamel bases)

 d) set of Riemann integrable functions

 e) set of Jordan measurable subsets of $\mathbf{R}$

 f) set of the additive subgroups of $\mathbf{R}$

 g) set of linear subspaces of $C[0,1]$

 h) set of linear functionals of $L^2[0,1]$

26. Which of the following sets are of power continuum?

 a) the set of real functions that are continuous at every rational point

 b) the set of real functions that are continuous at every irrational point

 c) the set of real functions f that satisfy the Cauchy equation

$$f(x+y) = f(x) + f(y)$$

27. If A is a set of cardinality continuum, then there are countably many functions $f_k : A \to \mathbf{N}$, $k = 0, 1, \ldots$ such that for an arbitrary function $f : A \to \mathbf{N}$ and for an arbitrary finite set $A' \subset A$ there is a k such that f_k agrees with f on A'.

28. The topological product of continuum many separable spaces is separable.

29. There are continuum many sets $A_\gamma \subseteq \mathbf{N}$ such that if $\gamma_1 \neq \gamma_2$, then $A_{\gamma_1} \cap A_{\gamma_2}$ is a finite set (such a collection is called almost disjoint).

30. Let k be a natural number, and suppose that A_γ, $\gamma \in \Gamma$ is a family of subsets of $\mathbf{N}$ such that if $\gamma_1 \neq \gamma_2$, then $A_{\gamma_1} \cap A_{\gamma_2}$ has at most k elements. Then Γ is countable.

31. To every $x \in \mathbf{R}$ one can assign a sequence $\{s_n^{(x)}\}$ of natural numbers such that if $x < y$, then $s_n^{(y)} - s_n^{(x)} \to \infty$ as $n \to \infty$.

32. There are continuum many sequences $\{s^\gamma\}_{n=0}^\infty$ of natural numbers such that if $\gamma_1 \neq \gamma_2$, then $|s_n^{\gamma_1} - s_{k_n}^{\gamma_2}|$ tends to infinity as $n \to \infty$, no matter how we choose the sequence $\{k_n\}$.

33. Let k be a positive integer, and suppose that $\{s_n^\gamma\}_{n=0}^\infty$, $\gamma \in \Gamma$ is a family of sequences of natural numbers such that if $\gamma_1 \neq \gamma_2$ then $s_n^{\gamma_1} = s_n^{\gamma_2}$ holds for at most k indices n. Then Γ is countable.

34. There is an almost disjoint family of cardinality continuum of subsets of $\mathbf{N}$ each with upper density 1.

35. Let $k \geq 2$ be an integer. Then there is a family of cardinality continuum of subsets of $\mathbf{N}$ such that the intersection of any k members of the family is infinite, but the intersection of any $k+1$ members is finite.

36. If $\mathcal{H}$ is an uncountable family of subsets of $\mathbf{N}$ such that the intersection of any finitely many members of the family is infinite, then the intersection of some infinite subfamily of $\mathcal{H}$ is also infinite.

37. There is a family of cardinality continuum of subsets of $\mathbf{N}$ such that the intersection of any finitely many members of the family has positive upper density, but the intersection of any infinitely many members is of density zero.

38. If $\mathcal{H}$ is a family of subsets of $\mathbf{R}$ such that the intersection of any two sets in $\mathcal{H}$ is finite, then $\mathcal{H}$ is of cardinality at most continuum.

39. There is a family $\mathcal{H}$ of cardinality bigger than continuum of subsets of $\mathbf{R}$ such that the intersection of any two sets in $\mathcal{H}$ is of cardinality smaller than continuum.

40. The are continuum many sets $A_\gamma \subset \mathbf{N}$ such that if $\gamma_1 \neq \gamma_2$, then either $A_{\gamma_1} \subset A_{\gamma_2}$ or $A_{\gamma_2} \subset A_{\gamma_1}$.

41. There are continuum many sets $A_\gamma \subset \mathbf{N}$ such that if $\gamma_1 \neq \gamma_2$, then each of the sets $A_{\gamma_1} \setminus A_{\gamma_2}$, $A_{\gamma_2} \setminus A_{\gamma_1}$, and $A_{\gamma_1} \cap A_{\gamma_2}$ is infinite.

42. For every real number x give sets $A_x, B_x \subseteq \mathbf{N}$ such that $A_x \cap B_x = \emptyset$, but for different x and y the set $A_x \cap B_y$ is infinite.

43. There is a family A_x, $x \in \mathbf{R}$ of subsets of the natural numbers such that if $x_1, \ldots, x_n$ are different reals and $\epsilon_1, \ldots, \epsilon_n \in \{0, 1\}$, then the density of the set $A_{x_1}^{\epsilon_1} \cap \cdots \cap A_{x_n}^{\epsilon_n}$ is 2^{-n} (here $A^1 = A$ and $A^0 = \mathbf{N} \setminus A$).

44. There is a function $f : \mathbf{R}^2 \to \mathbf{N}$ such that $f(x, y) = f(y, z)$ implies $x = y = z$.

5

Sets of reals and real functions

This chapter contains various problems from analysis and from the topology of Euclidean spaces that are connected with the notions of "countability" and "continuum". They include problems on exceptional sets (like a monotone real function can have only countably many discontinuities), Lindelöf-type covering theorems and their consequences, Baire properties, Borel sets, and Peano curves.

1. If $A \subset \mathbf{R}$ is such that for every $a \in A$ there is a $\delta_a > 0$ such that either $(a, a+\delta_a) \cap A = \emptyset$ or $(a-\delta_a, a) \cap A = \emptyset$, then A is countable.
2. Any uncountable subset A of the real numbers includes a strictly decreasing sequence converging to a point in A.
3. Every discrete set on $\mathbf{R}$ (in $\mathbf{R}^n$) is countable.
4. A right-continuous real function can have only countably many discontinuities.
5. Let f be a real function such that at every point f is continuous either from the right or from the left. Then f can have only countably many discontinuities.
6. A monotone real function can have only countably many discontinuities.
7. If a real function has right and left derivatives at every point, then it is differentiable at every point with the exception of a countable set.
8. A convex function is differentiable at every point with the exception of a countable set.
9. The set of local maximum *values* of any real function is countable.
10. The set of strict local maximum points of a real function is countable.
11. If every point is a local extremal point for a continuous real function f, then f is constant.
12. If a collection G_γ, $\gamma \in \Gamma$ of open sets in $\mathbf{R}^n$ covers a set E, then there is a countable subcollection G_{γ_i}, $i = 0, 1, \ldots$, that also covers E (this property of subsets of $\mathbf{R}^n$ is called the Lindelöf property).

It is customary to rephrase the problem by saying that in $\mathbf{R}^n$ every open cover of a set includes a countable subcover.

13. If a collection G_γ, $\gamma \in \Gamma$ of semi-open intervals in $\mathbf{R}$ covers a set E, then there is a countable subcollection G_{γ_i}, $i = 0, 1, \ldots$, that also covers E. The same is true if the G_γ's are arbitrary nondegenerated intervals.
14. If a collection G_γ, $\gamma \in \Gamma$, nondegenerated intervals in $\mathbf{R}$ covers a set E, then there is a countable subcollection G_{γ_i}, $i = 0, 1, \ldots$, that also covers E.
15. Let the real function f be differentiable at every point of the set $H \subset \mathbf{R}$. Then the set of those y for which $f^{-1}\{y\} \cap H$ is uncountable is of measure zero.
16. Call a rectangle almost closed if its sides are parallel with the coordinate axes, and it is obtained from a closed rectangle by omitting the four vertices. Show that any union of a family of almost closed rectangles is already a union of a countable subfamily. Is the same true if the rectangles are closed?
17. Call x an accumulation point of a set $A \subset \mathbf{R}$ ($A \subset \mathbf{R}^n$) if every neighborhood of x contains uncountably many points of A. An uncountable set A has an accumulation point that lies in A.
18. For an uncountable $A \subset \mathbf{R}$ let A^* be the set of those $a \in A$ that are accumulation points of both $A \cap (-\infty, a)$ and of $A \cap (a, \infty)$. Then $A \setminus A^*$ is countable, and A^* is densely ordered.
19. The set of accumulation points of any set A is either empty or perfect.
20. Any closed set in $\mathbf{R}$ ($\mathbf{R}^n$) is the union of a perfect and a countable set.
21. A nonempty perfect set in $\mathbf{R}^n$ is of power continuum.
22. A closed set in $\mathbf{R}$ ($\mathbf{R}^n$) is either countable, or of power continuum.
23. Define the distance between two real sequences $\{a_j\}_{j=0}^\infty$ and $\{b_j\}_{j=0}^\infty$ by the formula

$$d\left(\{a_j\}_{j=0}^\infty, \{b_j\}_{j=0}^\infty\right) = \sum_{j=0}^\infty \frac{1}{2^j} \frac{|a_j - b_j|}{1 + |a_j - b_j|}.$$

With this $\mathbf{R}^\infty$ becomes a complete separable metric space.

24. Every closed set in $\mathbf{R}^\infty$ is the union of a perfect and a countable set.
25. Every closed set in $\mathbf{R}^\infty$ is either countable or of cardinality continuum.
26. Every Borel set in $\mathbf{R}^n$ is a (continuous and) one-to-one image of a closed subset of $\mathbf{R}^\infty$.
27. In $\mathbf{R}^n$ every Borel set is either countable or of cardinality continuum.
28. If $a < b$ and $[a, b] = \cup_{i=0}^\infty A_i$, then there is an interval $I \subset [a, b]$ and an i such that the set A_i is dense in I (Baire's theorem).

29. If $a < b$ and $[a, b] = \cup_{i=0}^{\infty} A_i$, then there is an interval $I \subset [a, b]$ and an i such that for any subinterval J of I the intersection $A_i \cap J$ is of power continuum.

30. If $A \subset \mathbf{R}^n$ is a set with nonempty interior, then A cannot be represented as a countable union of nowhere dense sets (Baire's theorem).

31. If $A \subset \mathbf{R}^n$ is a set with nonempty interior and $A = \cup_{i=0}^{\infty} A_i$, then there is a ball $B \subset A$ and an i such that for any ball $B' \subset B$ the intersection $A_i \cap B'$ is of power continuum.

32. There are pairwise disjoint sets $A_x \subset \mathbf{R}$, $x \in \mathbf{R}$ such that for any $x \in \mathbf{R}$ and any open interval $I \subset \mathbf{R}$ the set $I \cap A_x$ is of power continuum.

33. There is a real function that assumes every value in every interval continuum many times.

34. There is a continuous function $f : [0, 1] \to [0, 1]$ that assumes every value $y \in [0, 1]$ continuum many times.

35. There exists a continuous mapping from $[0, 1]$ onto $[0, 1] \times [0, 1]$ (such "curves" are called area filling or Peano curves).

36. There are continuous functions $f_n : [0, 1] \to [0, 1]$, $n = 0, 1, 2, \ldots$ with the property that if $x_0, x_1, \ldots$ is an arbitrary sequence from $[0, 1]$, then there is a $t \in [0, 1]$ such for all n we have $f_n(t) = x_n$ (thus, $F(t) = (f_0(t), f_1(t), \ldots)$ is a continuous mapping from $[0, 1]$ onto the so-called Hilbert cube $[0, 1]^\infty \equiv {}^{\mathbf{N}}[0, 1]$).

* * *

37. If $\{a_\xi\}_{\xi<\omega_1}$ is a transfinite sequence of real numbers which is convergent (i.e., there is an $A \in \mathbf{R}$ such that for every $\epsilon > 0$ there is a $\nu < \omega_1$ for which $\xi > \nu$ implies $|a_\xi - A| \le \epsilon$), then there is a $\tau < \omega_1$ such that $a_\xi = a_\zeta$ for $\xi, \zeta > \tau$.

38. If $\{a_\xi\}_{\xi<\alpha}$ is a (strictly) monotone transfinite sequence of real numbers, then α is countable.

39. For every limit ordinal $\alpha < \omega_1$ there is a convergent, strictly increasing transfinite sequence $\{a_\xi\}_{\xi<\alpha}$ of real numbers (convergence means that there is an $A \in \mathbf{R}$ such that for every $\epsilon > 0$ there is a $\nu < \alpha$ for which $\xi > \nu$ implies have $|a_\xi - A| \le \epsilon$).

6

Ordered sets

Now we equip our sets with a structure by telling which element is larger than the other one. The theory of ordered sets is extremely rich, in fact, this list of problems is the longest one in the book.

This chapter contains problems on ordered sets and mappings between them. The types of ordered sets and arithmetic with types will be discussed in the next chapter. Occasionally later chapters will also discuss problems on ordered sets if the solution requires the methods of those chapters.

Particularly important are the well-ordered sets (see below), for they provide the infinite analogues of natural numbers. Well orderings offer enumeration of the elements of a given set in a transfinite sequence and thereby the possibility of proving results by transfinite induction.

Let A be a set and $\prec$ a binary relation on A. If $a \prec b$ does not hold, then we write $a \not\prec b$. $\langle A, \prec \rangle$ is called an *ordered set* (sometimes called linearly ordered) if

- $\prec$ *irreflexive*: $a \not\prec a$ for any $a \in A$,
- $\prec$ *transitive*: $a \prec b$ and $b \prec c$ imply $a \prec c$,
- $\prec$ *trichotomous*: for every $a, b \in A$ one of $a \prec b$, $a = b$, $b \prec a$ holds.

With every such "smaller than" relation $\prec$ we associate the corresponding "smaller than or equal" relation $\preceq$: $a \preceq b$ if either $a \prec b$ or $a = b$. This $\preceq$ has the following properties:

- *antisymmetric*: $a \preceq b$ and $b \preceq a$ imply $a = b$,
- *transitive*: $a \preceq b$ and $b \preceq c$ imply $a \preceq c$,
- *dichotomous*: for every $a, b \in A$ either of $a \preceq b$ or $b \preceq a$ holds.

If $\langle A, \prec \rangle$ is an ordered set and $B \subset A$ is a subset of A, then for notational simplicity we shall continue to denote the restriction of $\prec$ to $B \times B$ by $\prec$, so $\langle B, \prec \rangle$ is the ordered set with ground set B and with the ordering inherited from $\langle A, \prec \rangle$.

The ordered set $\langle A, \prec \rangle$ is called *well ordered* if every nonempty subset contains a smallest element, i.e., if for every $X \subseteq A$, $X \neq \emptyset$ there is an $a \in X$ such that for every $b \in X$ we have $a \preceq b$.

If $\langle A, \prec \rangle$ is an ordered set, then $X \subseteq A$ is an *initial segment* if $a \in X$ and $b \prec a$ imply $b \in X$ (intuitively, X consists of a starting section of $\langle A, \prec \rangle$), and in a similar fashion $X \subseteq A$ is called an *end segment* if $a \in X$ and $a \prec b$ imply $b \in X$. An initial segment that is not the whole set is called a proper initial segment. The intervals of $\langle A, \prec \rangle$ are its "convex" (or "connected") subsets, i.e., $X \subseteq A$ is an *interval* if $a, b \in X$ and $a \prec c \prec b$ implies $c \in X$. The intervals generate the so-called *interval topology* (also called *order topology*) on A. This is also the topology that is generated by the initial and end segments of $\langle A, \prec \rangle$.

Ordered sets are special algebraic structures (with no operations, and a single binary relation). Isomorphism among ordered sets is called similarity: $\langle A_1, \prec_1 \rangle$ and $\langle A_2, \prec_2 \rangle$ are *similar* if there is an $f : A_1 \to A_2$ 1-to-1 correspondence between the ground sets A_1 and A_2 that also preserves the ordering, i.e., $a \prec_1 b$ implies $f(a) \prec_2 f(b)$. In particular, similarity implies the equivalence of the ground sets. A mapping f from $\langle A_1, \prec_1 \rangle$ into $\langle A_2, \prec_2 \rangle$ (not necessarily onto) is called *monotone* if $a <_1 b$ implies $f(a) <_2 f(b)$. This is just the same as the notion of homomorphism from $\langle A_1, \prec_1 \rangle$ into $\langle A_2, \prec_2 \rangle$.

The *lexicographic product* of $\langle A_1, \prec_1 \rangle$ and $\langle A_2, \prec_2 \rangle$ is the ordered set $\langle A_1 \times A_2, \prec \rangle$ where $(a_1, a_2) \prec (a_1', a_2')$ precisely if $a_1 \prec_1 a_1'$ or if $a_1 = a_1'$ and $a_2 \prec_2 a_2'$ (i.e., in this ordering the first coordinate is decisive). On the other hand, in *antilexicographic ordering* first we compare the second coordinates and only when equality occurs compare the first coordinates. One can define in a similar manner the lexicographic or antilexicographic product of more than two sets. Lexicographic (antilexicographic) ordering is sometimes called ordering according to the first (last) difference.

Let $\langle A_i, <_i \rangle$, $i \in I$ be ordered sets with pairwise disjoint ground sets A_i and let the index set I be also ordered by the relation $<$. The *ordered union* of $\langle A_i, <_i \rangle$, $i \in I$ with respect to the ordered set $\langle I, < \rangle$ is the ordered set $\langle B, \prec \rangle$ in which $B = \cup_{i \in I} A_i$, and for $a \in A_i$ and $b \in A_j$ the relation $a \prec b$ holds if and only if $i < j$ or $i = j$ and $a <_i b$. The antilexicographic product of $\langle A_1, \prec_1 \rangle$ and $\langle A_2, \prec_2 \rangle$ is nothing else than the ordered union of the sets $\langle A_1 \times \{a\}, \prec_a \rangle$, $a \in A_2$ (where $(p, a) \prec_a (q, a)$ if and only if $p \prec_1 q$) with respect to $\langle A_2, \prec_2 \rangle$.

Unless otherwise stated, if A is a subset of the real line, then we regard A to be ordered with respect to the standard $<$ relation between the reals. In this chapter we mean strict monotonicity if we say that a real-valued function on a subset of the reals is monotone.

An important concept related to ordered sets is their cofinality, which will be used many times in later chapters. A theorem of Hausdorff (Problem 44) says that in every ordered set $\langle A, \prec \rangle$ there is a well-ordered *cofinal subset*, i.e., a subset $B \subseteq A$ such that $\langle B, \prec \rangle$ is well ordered and for every $a \in A$ there is a $b \in B$ with $a \preceq b$. Now the *cofinality* $\text{cf}(\langle A, \prec \rangle)$ is defined as the smallest possible order type of such cofinal $\langle B, \prec \rangle$'s.

The solutions of some problems require the following important result of R. Laver (see On Fraïssé's order type conjecture, *Ann. Math.*, **93**(1971), 89–111): If $\langle A_i, <_i\rangle$, $i = 0, 1, 2, \ldots$, are ordered sets such that neither of them includes a densely ordered subset, then there are $i < j$ such that $\langle A_i, <_i\rangle$ is similar to a subset of $\langle A_j, <_j\rangle$. The proof is considerably more complicated than it could be given in this book.

1. Any infinite sequence of different elements in an ordered set includes an infinite monotone subsequence.
2. Any two open subintervals of $\mathbf{R}$ are similar.
3. Give an ordered set with a smallest element, in which every element has a successor and every element but the least has a predecessor, yet the set is not similar to $\mathbf{N}$.
4. Give an ordering on the reals for which every element has a successor, as well as a predecessor.
5. An infinite ordered set $\langle A, \prec\rangle$ is similar to $\mathbf{N}$ if and only if for every $a \in A$ there are only finitely many elements $b \in A$ with $b \prec a$.
6. What are those infinite ordered sets $\langle A, \prec\rangle$ for which it is true that every infinite subset of A is similar to $\langle A, \prec\rangle$?
7. An infinite ordered set $\langle A, \prec\rangle$ is similar to $\mathbf{Z}$ if and only if it has no smallest or largest element, and every interval $\{c : a \prec c \prec b\}$, $a, b \in A$ is finite.
8. What are the infinite ordered sets $\langle A, \prec\rangle$ for which every interval $\{c : a \prec c \prec b\}$, $a, b \in A$ is finite?
9. There is a countable ordered set that has continuum many initial segments.
10. There is an ordered set of cardinality continuum that has more than continuum many initial segments.
11. There are infinitely many pairwise nonsimilar ordered sets such that every one of them is similar to an initial segment of any other one.
12. Let $\langle A, \prec\rangle$ and $\langle A', \prec'\rangle$ be ordered sets such that each of them is similar to a subset of the other one. Then there are disjoint decompositions $A = A_1 \cup A_2$ and $A' = A'_1 \cup A'_2$ such that $\langle A_i, \prec\rangle$ is similar to $\langle A'_i, \prec'\rangle$ for $i = 1, 2$.
13. If $\langle A, <\rangle$ and $\langle B, \prec\rangle$ are ordered sets such that $\langle A, <\rangle$ is similar to an initial segment of $\langle B, \prec\rangle$ and $\langle B, \prec\rangle$ is similar to an end segment of $\langle A, <\rangle$, then $\langle A, <\rangle$ and $\langle B, \prec\rangle$ are similar.
14. If $\langle A, <\rangle$ and $\langle B, \prec\rangle$ are ordered sets such that $\langle A, <\rangle$ is similar to an initial segment and to an end segment of $\langle B, \prec\rangle$ and $\langle B, \prec\rangle$ is similar an interval of $\langle A, <\rangle$, then $\langle A, <\rangle$ and $\langle B, \prec\rangle$ are similar.
15. There are continuum many subsets of $\mathbf{Q}$ no two of them similar.
16. How many subsets A does $\mathbf{R}$ have for which A is similar to $\mathbf{R}$?

17. There are continuum many pairwise disjoint subsets of $\mathbf{R}$ each similar to $\mathbf{R}$.
18. If $A \subseteq \mathbf{R}$, $A \neq \emptyset$, then $\mathbf{R}$ has continuum many subsets similar to A.
19. $\mathbf{R}$ has $2^{\mathbf{c}}$ subsets of cardinality continuum no two of which are similar.
20. If we omit a countable set from the set of irrational numbers, then the set obtained is similar to the set of the irrational numbers.
21. If $\langle A, \prec \rangle$ has a countable subset B that is dense in A (i.e., for every $a_1, a_2 \in A$, $a_1 \prec a_2$ there is $b \in B$ such that $a_1 \preceq b \preceq a_2$), then $\langle A, \prec \rangle$ is similar to a subset of $\mathbf{R}$.
22. Suppose $A, B \subseteq \mathbf{R}$ are two similar subsets of $\mathbf{R}$. Is it true that then their complements $\mathbf{R} \setminus A$ and $\mathbf{R} \setminus B$ are also similar? What if A and B are countable dense subsets of $\mathbf{R}$?
23. Let $\mathcal{M}$ be a set of open subsets of $\mathbf{R}$ ordered with respect to inclusion "$\subset$". Then $\langle \mathcal{M}, \subset \rangle$ is similar to a subset of the reals.
24. There is a family $\mathcal{F}$ of closed and measure zero subsets of $\mathbf{R}$ such that $\langle \mathcal{F}, \subset \rangle$ is similar to $\mathbf{R}$.
25. There is a family of cardinality bigger than continuum of subsets of $\mathbf{R}$ that is ordered with respect to inclusion.
26. Any countable ordered set is similar to a subset of $\mathbf{Q} \cap (0, 1)$.
27. Any countable densely ordered set without smallest and largest elements is similar to $\mathbf{Q}$.
28. Any countable densely ordered set is similar to one of the sets $\mathbf{Q} \cap (0, 1)$, $\mathbf{Q} \cap [0, 1)$, $\mathbf{Q} \cap (0, 1]$, $\mathbf{Q} \cap [0, 1]$ (depending if it has a first or last element).
29. There is an uncountable ordered set such that all of its proper initial segments are similar to $\mathbf{Q}$ or to $\mathbf{Q} \cap (0, 1]$.
30. There is an uncountable ordered set which is similar to each of its uncountable subsets.
31. The antilexicographically ordered set of infinite 0–1 sequences that contain only a finite number of 1's is similar to $\mathbf{N}$.
32. The lexicographically ordered set of infinite 0–1 sequences that contain only a finite number of 1's is similar to $\mathbf{Q} \cap [0, 1)$.
33. The lexicographically ordered set of infinite 0–1 sequences is similar to the Cantor set.
34. The lexicographically ordered set of sequences of natural numbers is similar to $[0, 1)$.
35. Consider the set A of all sequences $n_0, -n_1, n_2, -n_3, \ldots$ where n_i are natural numbers. Then A, with the lexicographic ordering, is similar to the set of irrational numbers.
36. An ordered set is well ordered if and only if it does not include an infinite decreasing sequence.

37. If $A \subseteq \mathbf{R}$ is well ordered, then it is countable.
38. If $\mathcal{U}$ is a family of open (closed) subsets of $\mathbf{R}$ that is well ordered with respect to inclusion, then $\mathcal{U}$ is countable.
39. If $\langle A, \prec \rangle$ is well ordered, then for any $f : A \to A$ monotone mapping and for any $a \in A$ we have $a \preceq f(a)$.
40. There is at most one similarity mapping between two well-ordered sets.
41. A well-ordered set cannot be similar to a subset of one of its proper initial segments.
42. Given two well-ordered sets, one of them is similar to an initial segment of the other.
43. Two well-ordered sets, each of which is similar to a subset of the other one, are similar.
44. (Hausdorff's theorem) For every ordered set $\langle A, \prec \rangle$ there is a subset $B \subseteq A$ such that $\langle B, \prec \rangle$ is well ordered and cofinal (if $a \in A$ is arbitrary, then there is a $b \in B$ with $a \preceq b$). Furthermore, $B \subseteq A$ can also be selected in such a way that the order type of $\langle B, \prec \rangle$ does not exceed $|A|$ (the ordinal, with which the cardinal $|A|$ is identified).
45. If every proper initial segment of an ordered set is the union of countably many well-ordered sets, then so is the whole set itself.
46. If $\langle A, \prec \rangle$ is a nonempty countable well-ordered set, then $A \times [1, 0)$ with the lexicographic ordering is similar to $[0, 1)$.
47. There is an ordered set that is not similar to a subset of $\mathbf{R}$, but all of its proper initial segments are similar to $(0, 1)$ or to $(0, 1]$. Furthermore, this set is unique up to similarity.
48. Call a point $x \in A$ in an ordered set $\langle A, \prec \rangle$ a fixed point if $f(x) = x$ holds for every monotone $f : A \to A$. A point $x \in A$ is not a fixed point of $\langle A, \prec \rangle$ if and only if there is a monotone mapping from $\langle A, \prec \rangle$ into $\langle A \setminus \{x\}, \prec \rangle$.
49. If $x \neq y$ are fixed points of $\langle A, \prec \rangle$, then y is a fixed point of $\langle A \setminus \{x\}, \prec \rangle$.
50. Every countable ordered set has only finitely many fixed points.
51. For each $n < \infty$ give a countably infinite ordered set with exactly n fixed points.
52. If $\langle A, \prec \rangle$ has infinitely many fixed points, then it includes a subset similar to $\mathbf{Q}$.
53. Every ordered set is similar to a set of sets ordered with respect to inclusion.
54. Let $\mathcal{M}$ be a family of subsets of a set X that is ordered with respect to inclusion and which is a maximal family with this property. Define $\prec$ on X as follows: let $x \prec y$ be exactly if there is an $E \in \mathcal{M}$, such that $x \in E$ but $y \notin E$. Then $\langle X, \prec \rangle$ is an ordered set. What are the initial segments in this ordered set?

55. Every ordered set is similar to some $\langle X, \prec \rangle$ constructed in the preceding problem.
56. If $\langle A, \prec \rangle$ is an ordered set, then there is an ordered set $\langle A^*, \prec^* \rangle$ such that if $A^* = B \cup C$ is an arbitrary decomposition, then either B or C includes a subset similar to $\langle A, \prec \rangle$.
57. To every infinite ordered set there is another such that neither one is similar to a subset of the other.
58. To every countably infinite ordered set $\langle A, \prec \rangle$ there is another countably infinite ordered set that does not include a subset similar to $\langle A, \prec \rangle$.
59. For every n show n countable ordered sets such that neither of them is similar to a subset of another one.
60. If $\langle A_i, \prec_i \rangle$, $i = 0, 1, \ldots$, are countable ordered sets, then there are $i < j$ such that $\langle A_i, \prec_i \rangle$ is similar to a subset of $\langle A_j, \prec_j \rangle$.
61. Every countably infinite ordered set is similar to one of its proper subsets.
62. There is an infinite ordered set that is not similar to any one of its proper subsets.
63. In every infinite ordered set the position of one element can be changed in such a way that we get an ordered set that is not similar to the original one.
64. One can add to any ordered set one element so that the ordered set so obtained is not similar to the original one. Is the same true for removing one element?
65. Every ordered set is a subset of a densely ordered set.
66. Every densely ordered set is a dense subset of a continuously ordered set.
67. Any two continuously ordered sets without smallest and largest elements that include similar dense sets are similar.
68. A continuously ordered set containing at least two points includes a subset similar to **R**.
69. If $\langle A, \prec \rangle$ is continuously ordered and $A_n = \{c : a_n \preceq c \preceq b_n\}$ is a sequence of nested closed intervals, i.e., $A_{n+1} \subseteq A_n$ for all $n = 0, 1, \ldots$, then $\cap_{n=0}^{\infty} A_n \neq \emptyset$.
70. There is an infinite ordered set $\langle A, \prec \rangle$ that is not continuously ordered but for every sequence $\{A_n\}_{n=0}^{\infty}$ of nested closed intervals $\cap_{n=0}^{\infty} A_n \neq \emptyset$.
71. Call a subset of an ordered set scattered, if it does not include a subset that is densely ordered. The union of finitely many scattered subsets of an ordered set is scattered.
72. A subset of the real line is scattered if and only if it has a countable closure.
73. A bounded subset A of the real line is scattered if and only if for any sequence $\epsilon_0, \epsilon_1, \ldots$ of positive numbers there exists a natural number N

such that A can be covered with some intervals $I_0, I_1, \ldots, I_N$ of length $|I_i| = \epsilon_i$.

74. If α is an ordinal then let $H(\alpha)$ be the set of all functions $f : \alpha \to \{-1, 0, 1\}$ for which $D(f) = \{\beta < \alpha \ : \ f(\beta) \neq 0\}$ is finite. Order $H(\alpha)$ according to last difference, i.e., for $f, g \in H(\alpha)$ set $f \prec g$ if $f(\beta) \prec g(\beta)$ holds for the largest $\beta < \alpha$ with $f(\beta) \neq g(\beta)$. Then $\langle H(\alpha), \prec\rangle$ is scattered.
75. The product of two scattered ordered sets is scattered.
76. The ordered union of scattered ordered sets with respect to a scattered ordered set is scattered.
77. Every nonempty ordered set is either scattered, or is similar to the ordered union of nonempty scattered sets with respect to a densely ordered set.
78. Let $\mathcal{F}$ be a family of ordered sets with the following properties:
    - if $\langle S, \prec\rangle \in \mathcal{F}$ and $\langle S', \prec'\rangle$ is similar to $\langle S, \prec\rangle$, then $\langle S', \prec'\rangle \in \mathcal{F}$,
    - if $\langle S, \prec\rangle \in \mathcal{F}$ and S' is a subset of S then $\langle S', \prec\rangle \in \mathcal{F}$,
    - $\mathcal{F}$ is closed for well-ordered and reversely well-ordered unions,
    - there is a nonempty $\langle S, \prec\rangle$ in $\mathcal{F}$.

    Then every ordered set is either in $\mathcal{F}$, or it is similar to an ordered union of nonempty sets in $\mathcal{F}$ with respect to a densely ordered set.
79. Let $\mathcal{O}$ be the smallest family of ordered sets that contains $\emptyset$, 1 and is closed for well-ordered and reversely well-ordered unions as well as for similarity. Then $\mathcal{O}$ is precisely the family of scattered sets.
80. An ordered set is scattered if and only it can be embedded into one of the $\langle H(\alpha), \prec\rangle$ defined in Problem 74.
81. We say that an ordered set $\langle A, \prec\rangle$ has countable intervals if for every $a, b \in A$, $a \prec b$ the set $\{c \in A \ : \ a \prec c \prec b\}$ is countable. There is a maximal ordered set $\langle A, \prec\rangle$ with countable intervals in the sense that every ordered set with countable intervals is similar to a subset of $\langle A, \prec\rangle$.
82. Pick a natural number n_1, and for each $i = 1, 2, \ldots$ perform the following two operations to define n_{2i} and n_{2i+1}:

    (i) write n_{2i-1} in base $i+1$, and while keeping the coefficients, replace the base by $i+2$. This gives a number that we call n_{2i};

    (ii) set $n_{2i+1} = n_{2i} - 1$.

    If $n_{2i+1} = 0$ then we stop, otherwise repeat this process. For example, if $n_1 = 23 = 2^4 + 2^2 + 2^1 + 1$, then $n_2 = 3^4 + 3^2 + 3^1 + 1 = 94$, $n_3 = 93$, $n_4 = 4^4 + 4^2 + 4^1 = 276$, $n_5 = 275$, then, since $275 = 4^4 + 4^2 + 3$, we have $n_6 = 5^4 + 5^2 + 3 = 3253$, etc.

    (a) No matter what n_1 is, there is an i such that $n_i = 0$.

    (b) The same conclusion holds if in **(i)** the actual base is changed to any larger base (i.e., when the bases are not $2, 3, \ldots$ but some numbers $b_1 < b_2 < \ldots$).

* * *

83. In every densely ordered set there are two disjoint dense subsets.
84. The elements of any ordered set can be colored by two colors in such a way that in between any two elements of the same color there is another one with a different color.
85. There is an ordered set which is not well ordered, yet no two different initial segments of it are similar.
86. There exists an ordered set that cannot be represented as a countable union of its well-ordered subsets, but in which every uncountable subset includes an uncountable well-ordered subset.
87. There are two subsets $A, B \subset \mathbf{R}$ of power continuum such that any subset of A that is similar to a subset of B is of cardinality smaller than continuum.
88. There is an infinite subset X of $\mathbf{R}$ such that if $f : X \to X$ is any monotone mapping, then f is the identity.
89. To every ordered set $\langle A, \prec \rangle$ of cardinality $\kappa \geq \aleph_0$ there is another ordered set of cardinality κ that does not include a subset similar to $\langle A, \prec \rangle$.
90. For every infinite cardinal κ there is an ordered set of cardinality κ that has more than κ initial segments.
91. In a set of cardinality κ there is a family of subsets of cardinality bigger than κ that is ordered with respect to inclusion.
92. If $\mathcal{H}$ is a family of subsets of an infinite set of cardinality κ that is well ordered with respect to inclusion, then $\mathcal{H}$ is of cardinality at most κ.
93. If κ is an infinite cardinal, then in the lexicographically ordered set ${}^{\kappa}\kappa$ (which is the set of transfinite sequences of type κ of ordinals smaller than κ ordered with respect to first difference) every well-ordered subset is of cardinality at most κ.
94. Let κ be an infinite cardinal and let T be the set ${}^{\kappa}\{0,1\}$ of 0–1 sequences of type κ ordered with the lexicographic ordering. Then
    a) every nonempty subset of T has a least upper bound,
    b) every subset of T has cofinality at most κ,
    c) every well-ordered subset of T is of cardinality at most κ.
95. Every ordered set of cardinality κ is similar to a subset of the lexicographically ordered ${}^{\kappa}\{0,1\}$.
96. Let κ be an infinite cardinal and $\mathcal{F}_\kappa$ the set of those $f : \kappa \to \{0,1\}$ for which there is a last 1, i.e., there is an $\alpha < \kappa$ such that $f(\alpha) = 1$ but for all $\alpha < \beta < \kappa$ we have $f(\beta) = 0$. Every ordered set of cardinality κ is similar to a subset of the lexicographically ordered $\mathcal{F}_\kappa$.

97. If $\langle A, \prec \rangle$ is an ordered set and κ is a cardinal, then there is an ordered set $\langle B, < \rangle$ such that if $B = \cup_{\xi<\kappa} B_\xi$ is an arbitrary decomposition of B into at most κ subsets, then there is a $\xi < \kappa$ such that $\langle B_\xi, < \rangle$ includes a subset similar to $\langle A, \prec \rangle$.

7

Order types

Order types are to ordered sets what cardinals are to sets. To every ordered set we associate its type in such a way that two ordered sets are similar (isomorphic) if and only if they have the same order type. The order types of well-ordered sets have special name: they are called ordinals or ordinal numbers.

This chapter contains problems on order types and on arithmetic with them. Problems on ordinals will be discussed in Chapter 8 and their arithmetic in Chapter 9.

Even though we shall not say exactly what order types are (such a definition is possible within set theory but does not yield any additional information, so we omit it), we shall still speak about their ordered sums and their product as follows.

If θ_i are order types for each $i \in I$ where $\langle I, <\rangle$ is an ordered set, then we define $\sum_{i\in I(<)} \theta_i$ to be the order type of the ordered union of $\langle A_i, \prec_i\rangle$, $i \in I$, with respect to $\langle I, <\rangle$, where $\langle A_i, \prec_i\rangle$ are pairwise disjoint ordered sets with order type θ_i.

If θ_1, θ_2 are order types and $\langle A_1, \prec_1\rangle$, $\langle A_2, \prec_2\rangle$ are ordered sets with types θ_1 and θ_2, respectively, then $\theta_1 \cdot \theta_2$ is defined as the order type of the antilexicographic product of $\langle A_1, \prec_1\rangle$ and $\langle A_2, \prec_2\rangle$ (in this order). There is no particular reason why antilexicographic product is used here; but historically that is what was used and we adhere to this custom.

Note that these operations are order-sensitive, and multiplication is left distributive with respect to addition

$$\theta \cdot \left(\sum_{i\in I(<)} \theta_i \right) = \sum_{i\in I(<)} \theta \cdot \theta_i,$$

but it is not distributive from the right.

In this chapter $\theta, \tau, \ldots$ will always denote order types. ω is the order type of $\mathbf{N}$, η is the order type of $\mathbf{Q}$, and λ is the order type of $\mathbf{R}$ (with the standard $<$ ordering on these sets).

If θ is an order type, say the type of $\langle A, \prec \rangle$, then θ^* denotes its reverse type, i.e., the type of $\langle A, \prec^* \rangle$, where $a \prec^* b$ exactly if $b \prec a$.

1. What are the order types of the initial segments of the set of rational numbers?
2. The following are true:
   a) $\eta + \eta = \eta$,
   b) $\eta + 1 + \eta = \eta$,
   c) $\eta \cdot \eta = (\eta + 1) \cdot \eta = (1 + \eta) \cdot \eta = \eta$.
   d) $\eta + 2 + \eta \neq \eta$,
   e) $\lambda + 1 + \lambda = \lambda$,
   f) $\lambda + \lambda \neq \lambda$,
   g) $\lambda \cdot \lambda \neq \lambda$,
   h) $\eta \cdot \lambda \neq \lambda \cdot \eta$.
3. Solve for θ_1 and θ_2 the equations:
   a) $\theta_1 + \theta_2 = \eta$,
   b) $\theta_1 + \theta_2 = \lambda$.
4. Write η in the form $\theta_1 \cdot \theta_2$ in such a way that $\theta_1 \neq \eta$, $\theta_2 \neq \eta$.
5. Write λ in the form $\theta_1 \cdot \theta_2$ in such a way that $\theta_1 \neq \lambda$, $\theta_2 \neq \lambda$.
6. Every order type different from $0, 1, 2$ can be represented as a sum of two different nonzero order types.
7. Give an infinite order type that cannot be written as a product of two order types different from 1.
8. The order types ω and η cannot be represented as a two-term sum in which each term is different from the given order type.
9. The order types ω, $\eta + 1$, $1 + \eta$ and $1 + \eta + 1$ cannot be represented as a two-term product in which each factor is different from the given order type.
10. Give infinitely many order types that cannot be represented as a two-term sum in which each term is different from the given order type.
11. Give infinitely many order types that cannot be represented as a two-term product in which each factor is different from the given order type.
12. Let A be the set of those 0–1 sequences that contain only finitely many 1's. What is the order type of the lexicographically ordered A? What is the order type if the ordering is antilexicographic?
13. In the preceding problem replace A with the set of all sequences of non-negative integers that contain only finitely many nonzero elements. What are then the answers?

14. The order type of the lexicographically ordered set ${}^{\mathbf{N}}\mathbf{N}$ of infinite sequences of the natural numbers is $1 + \lambda$.

15. Given n, what is the order type of the set

$$\left\{\frac{1}{k_1} + \cdots + \frac{1}{k_n} \ : \ 1 \le k_1, \ldots, k_n < \omega\right\}?$$

16. What is the order type of $\langle \mathbf{N}, \prec \rangle$, where for $m = 2^{p-1}(2q-1) - 1$ and $n = 2^{r-1}(2s-1) - 1$ we set $m \prec n$ if $p/q < r/s$ or $p/q = r/s$ and $m < n$?

17. Every order type can be uniquely represented as $\alpha + \theta$, where α is an ordinal and θ is the order type of an ordered set which does not have a smallest element.

18. If θ_1 and θ_2 are order types such that for some order types τ and ρ we have $\theta_1 = \tau + \theta_2$ and $\theta_2 = \theta_1 + \rho$, then $\theta_1 = \theta_2$.

19. If θ_1 and θ_2 are order types such that for some order types τ, ρ, σ_1 and σ_2 we have $\theta_2 = \theta_1 + \rho$, $\theta_2 = \tau + \theta_1$ and $\theta_1 = \sigma_1 + \theta_2 + \sigma_2$, then $\theta_1 = \theta_2$.

20. If n is a positive integer and for some order types θ_1 and θ_2 one has $n \cdot \theta_1 = n \cdot \theta_2$, then $\theta_1 = \theta_2$.

21. If n is a positive integer and for some order types θ_1 and θ_2 one has $\theta_1 \cdot n = \theta_2 \cdot n$, then $\theta_1 = \theta_2$.

22. Give an order type θ that can be written both in the form $2 \cdot \tau_1 + 1$ and $2 \cdot \tau_2$, and give an order type that has neither of these forms.

23. Give an order type θ that can be written both in the form $\tau_1 \cdot 2 + 1$ and $\tau_2 \cdot 2$, and give an order type that has neither of these forms.

24. Give infinitely many order types θ with the property $\theta \cdot 2 = \theta$.

25. Give infinitely many order types θ with the property $2 \cdot \theta = \theta$.

26. Give infinitely many order types θ with the property $2 \cdot \theta = \theta \cdot 2 = \theta$.

27. Give infinitely many order types θ with the property $\theta \cdot \theta = \theta$.

28. For every order type $\theta \neq 0$ there is an order type τ such that $\tau + \theta = \theta + \tau = \tau$.

29. For every order type $\theta \neq 0$ there is an order type τ such that $\tau \cdot \theta = \theta \cdot \tau = \tau$.

30. Give an infinite sequence of order types such that they form an arithmetic progression, but each one of them is a square (i.e., if $\{\theta_n\}$ is the sequence, then $\theta_n = \tau_n \cdot \tau_n$ for some τ_n, and there is a σ such that $\theta_{n+1} = \theta_n + \sigma$ for all n).

31. For every positive integer n give n order types such that all possible $n!$ sums of them (formed from different rearrangements) are different.

32. For every positive integer n give n order types such that all possible $n!$ products of them (formed from different rearrangements) are different.

33. Give two different order types θ_1 and θ_2 with equal squares: $\theta_1^2 = \theta_2^2$.

34. The order types $\theta_k = (\omega^* + \omega) \cdot (\omega + \omega \cdot \eta + k)$, $k = 0, 1, \ldots$ are all different, but they have the same higher powers: there is an order type τ such that for all k and all $n \geq 2$ we have $\theta_k^n = \tau$.

35. Give order types $\theta_1, \theta_2, \theta_3$ such that for all positive integer n their nth powers are different, but $\theta_1^n + \theta_2^n = \theta_3^n$ is true.

36. Let θ_1 and θ_2 be the order types of the ordered sets $\langle A_1, <_1 \rangle$ and $\langle A_2, <_2 \rangle$, respectively, and say that $\theta_1 \prec \theta_2$ if $\langle A_1, <_1 \rangle$ is similar to a subset of $\langle A_2, <_2 \rangle$, but not vice versa. This $\prec$ is well defined among order types and it is irreflexive and transitive but not trichotomous.

37. If $\langle A_1, <_1 \rangle$ and $\langle A_2, <_2 \rangle$ in the preceding problem are well ordered, then $\theta_1 \prec \theta_2$ if and only $\langle A_1, <_1 \rangle$ is similar to a proper initial segment of $\langle A_2, <_2 \rangle$. As a consequence, $\prec$ is an ordering (irreflexive, transitive, and trichotomous) among ordinals (in which case we write $\theta_1 < \theta_2$ instead of $\theta_1 \prec \theta_2$). In particular, the order type of a subset of a well-ordered set is at most as large as the order type of the whole set.

38. $\theta \preceq \eta$ for every countable order type θ.

39. There are continuum many order types θ with $\theta \prec \eta$.

40. To every infinite order type θ there is another one τ which is incomparable (with respect to $\prec$) with it, i.e., $\tau \neq \theta$, $\theta \not\prec \tau$ and $\tau \not\prec \theta$.

41. To every countably infinite order type θ there is another countably infinite order type τ such that $\theta \not\prec \tau$.

42. For every $n \geq 2$ give n pairwise incomparable (with respect to $\prec$) countable order types.

43. If θ_i, $i = 0, 1, \ldots$, are countable order types, then $\theta_i \preceq \theta_j$ for some $i < j$.

44. For an order type θ we have $1 + \theta = \theta$ if and only if $\theta = \omega + \tau$ for some order type τ.

45. For an infinite order type θ we have $1 + \theta = \theta + 1$ if and only if $\theta = \omega + \tau + \omega^*$ for some τ.

46. For a nonzero order type θ we have $\eta + \theta = \theta + \eta$ if and only if $\theta = \eta + \tau + \eta$ for some order type τ.

47. For an order type θ we have $\lambda + \theta = \theta + \lambda$ if and only if $\theta = \lambda \cdot n$ for some $n = 0, 1, \ldots$ or $\theta = \lambda \cdot \omega + \tau + \lambda \cdot \omega^*$ for some order type τ.

48. For an order type θ we have $\theta = \theta^*$ if and only if θ can be represented either as $\tau + \tau^*$ or as $\tau + 1 + \tau^*$.

* * *

49. For every infinite κ there are 2^κ different order types of cardinality κ.

8

Ordinals

Ordinals are the order types of well-ordered sets. They are the infinite analogues of the natural numbers, and in many respect they behave like the latter ones. In fact, the finite ordinals are the natural numbers, and hence the transfinite class of ordinals can be considered as an endless continuation of the sequence of natural numbers.

This chapter contains various problems on ordinals and on operations on them. The problems specifically related to ordinal arithmetic will be the content of the next chapter.

The von Neumann definition of ordinals is as follows (see below): a set α is called an ordinal if it is transitive and well ordered with respect to $\in$. When we talk about such an α we shall always assume that it is equipped with the $\in$ relation. It can be shown that every well-ordered set $\langle A, \prec \rangle$ is similar to such a unique α. Therefore, we can set α as the order type of $\langle A, \prec \rangle$. In particular, the order type of α is α.

We set $\beta < \alpha$ if $\beta \in \alpha$. It follows that

α is the set of ordinals smaller than α, and among ordinals the relation $\beta < \alpha$ is the same as $\beta \in \alpha$, and $\beta \le \alpha$ is the same as $\beta \subseteq \alpha$.

We shall not explicitly use von Neumann's definition, but we shall use the just-listed boldfaced convention.

In this chapter $\alpha, \beta, \ldots$ always denote ordinals. As always, ω, the smallest infinite ordinal, is the set of natural numbers, i.e., the set of finite ordinals. An ordinal α is called a *successor ordinal* if it is of the form $\beta+1$. The positive ordinals that are not successors are called *limit ordinals.* Thus, α is a limit ordinal if and only if $\beta < \alpha$ implies $\beta + 1 < \alpha$. The first ordinal 0 is neither limit, nor successor.

The first problem deals with the von Neumann definition of ordinals. A set x is called *transitive* if $y \in x$ and $z \in y$ imply $z \in x$ (or equivalently $y \in x \Longrightarrow y \subset x$). We say that $\in$ is a well-ordering on the set x if its restriction to x is a well-ordering on x. Call a set *N-set* (N for Neumann) if it is transitive and well ordered by $\in$. We always consider an N-set with the well-ordering $\in$, and for notational convenience sometimes we write $<_\in$ for $\in$. Part (h) shows that for a well-ordered set $\langle A, \prec\rangle$ we could define its order type as the unique N-set similar to it, and this is exactly the von Neumann definition of ordinals.

1. (a) Every element of an N-set is an N-set.
   (b) If x is an N-set, then $y = x \cup \{x\}$ is an N-set, and if z is an N-set containing x, then $y \subset z$.
   (c) If x is an N-set, $y \in x$, then y is an initial segment of x.
   (d) If x is an N-set and $Y \subset x$ is one of its initial segments, then Y is an N-set, and either $Y = x$ or $Y \in x$.
   (e) If x, y are N-sets, then $x = y$ or $x \in y$ or $y \in x$.
   (f) For N-sets x, y define $x < y$ if $x \in y$. Then this is irreflexive, transitive and trichotomous. Furthermore, if B is a nonempty set of N-sets, then there is a smallest element of B with respect to $<$ ("well order").
   (g) If x, y are different N-sets, then they are not similar.
   (h) Every well-ordered set is similar to a unique N-set.
2. There is no infinite decreasing sequence of ordinals.
3. Arbitrary infinite sequence of ordinals includes an infinite nondecreasing subsequence.
4. The following relations are true:
   a) $1 + \omega = \omega, \quad \omega + 1 \neq \omega,$
   b) $2 \cdot \omega = \omega, \quad \omega \cdot 2 \neq \omega.$
5. If a and b are natural numbers, then what is $(\omega + a) \cdot (\omega + b)$?
6. Solve the following equations for the ordinals ξ and ζ:
   (a) $\omega + \xi = \omega$
   (b) $\xi + \omega = \omega$
   (c) $\xi \cdot \omega = \omega$
   (d) $\omega \cdot \xi = \omega$
   (e) $\xi + \zeta = \omega$
   (f) $\xi \cdot \zeta = \omega$
7. Solve the equation $\xi + \zeta = \omega^2 + 1$ for the ordinals ξ and ζ.
8. Which one is bigger?
   a) $\omega + k$ or $k + \omega$ (k is a positive integer)
   b) $k \cdot \omega$ or $\omega \cdot k$ ($k \geq$ is an integer)
   c) $\omega + \omega_1$ or $\omega_1 + \omega$
   d) $P(\omega) = \omega^n \cdot a_n + \cdots + \omega \cdot a_1 + a_0$ or ω^{n+1}, where $n \geq 1$ and $a_0, \ldots, a_n$ are natural numbers

e) $P(\omega) = \omega^n \cdot a_n + \cdots + \omega \cdot a_1 + a_0$ or $Q(\omega) = \omega^m \cdot a'_m + \cdots + \omega \cdot a'_1 + a'_0$, where $n, m, a_0, a'_0 \ldots, a_n, a'_n$ are natural numbers

9. Addition among ordinals is monotonic in both arguments, and strictly monotonic in the second argument. The same is true of multiplication provided the first factor is nonzero.

10. a) $\gamma + \alpha = \gamma + \beta$ implies $\alpha = \beta$,
 b) $\alpha + \gamma = \beta + \gamma$ does not imply $\alpha = \beta$,
 c) $\gamma \cdot \alpha = \gamma \cdot \beta$, $\gamma > 0$ imply $\alpha = \beta$,
 d) $\alpha \cdot \gamma = \beta \cdot \gamma$, $\gamma > 0$ do not imply $\alpha = \beta$.
 Does the answer change in b) or d) if γ is a natural number?

11. If $\alpha \cdot \gamma = \beta \cdot \gamma$ and γ is a successor ordinal, then $\alpha = \beta$.

12. If k is a positive integer and $\alpha^k = \beta^k$, then $\alpha = \beta$.

13. If ξ is a limit ordinal, then
 a) $\sup_{\eta<\xi}(\alpha + \eta) = \alpha + \xi$,
 b) $\sup_{\eta<\xi}(\alpha \cdot \eta) = \alpha \cdot \xi$.
 Are the analogous relations true if we change the order of the terms in the sums and products?

14. If $\alpha \leq \beta$, then the equation $\alpha + \xi = \beta$ is uniquely solvable for ξ. Is the same true for the equation $\xi + \alpha = \beta$?

15. If $0 < \alpha$, then for any β there are unique ζ and $\xi < \alpha$ such that $\beta = \alpha \cdot \zeta + \xi$.

16. If $\alpha > 0$ is an arbitrary ordinal and β is sufficiently large, then $\alpha + \beta = \beta$.

17. If $\alpha + \beta = \beta + \alpha$ for all ordinals β, then $\alpha = 0$.

18. Every ordinal can be written in a unique manner in the form $\beta + n$ where β is a limit ordinal or zero and n is a natural number.

19. The limit ordinals are the ones that have the form $\omega \cdot \beta$, $\beta \geq 1$.

20. A positive ordinal α is a limit ordinal if and only if $n \cdot \alpha = \alpha$ for all positive integer n.

21. Let n be finite and α a limit ordinal. Then $(\alpha + n) \cdot \beta = \alpha \cdot \beta + n$ if β is a successor ordinal, and $(\alpha + n) \cdot \beta = \alpha \cdot \beta$ if β is 0 or a limit ordinal.

22. If $k \geq 1, n$ are natural numbers and α is a limit ordinal, then $(\alpha \cdot n)^k = \alpha^k \cdot n$.

23. Given $\alpha > 0$, what are those natural numbers n such that α can be written as $\alpha = n \cdot \beta$ for some ordinal β?

24. In each case find all ordinals α that satisfy the given equation.
 a) $\alpha + 1 = 1 + \alpha$
 b) $\alpha + \omega = \omega + \alpha$
 c) $\alpha \cdot \omega = \omega \cdot \alpha$
 d) $\alpha + (\omega + 1) = (\omega + 1) + \alpha$

e) $\alpha \cdot (\omega + 1) = (\omega + 1) \cdot \alpha$

25. If n is a positive integer, then $\sum_{\xi < \omega^n} \xi = \omega^{2n-1}$.

26. For every α there are only finitely many distinct γ such that $\alpha = \xi + \gamma$ with some ξ. Is the analogous statement true for the representation $\alpha = \gamma + \xi$?

27. For every $\alpha \neq 0$ there are only finitely many γ such that $\alpha = \xi \cdot \gamma$ with some ξ. Is the analogous statement true for the representation $\alpha = \gamma \cdot \xi$?

28. Let m be a positive integer. A successor ordinal can be represented as a product with m factors only in finitely many ways.

29. The equation $\xi^2 + \omega = \zeta^2$ has no solution for ξ and ζ.

30. Give infinitely many ξ and ζ such that ξ is infinite, and $\xi^2 + \omega^2 = \zeta^2$.

31. Solve $\alpha^2 \cdot 2 = \beta^2$ for α and β.

32. For every natural number k there is an infinite sequence of ordinals that form an arithmetic progression and in which each term is a kth power.

33. Give ordinals α and β with the property that for no $n = 2, 3, \ldots$ is $\alpha^n \cdot \beta^n$ or $\beta^n \cdot \alpha^n$ an nth power.

34. The sum $\omega + 1 + 2 + \cdots$ does not change if we alter the position of finitely many terms in it.

35. One can get infinitely many different ordinals from the sum $1+2+3+\cdots+\omega$ by changing the position of finitely many terms in it.

36. For every $n \geq 1$ give a sum $\alpha_0 + \alpha_1 + \cdots$ of positive ordinals from which one can get exactly n different sums by taking a permutation of the terms (possibly infinitely many) in the sum.

37. The sum of the $n+1$ ordinals $1, 2, \ldots, 2^{n-1}, \omega$ in all possible orders take 2^n different values.

38. Let $g(n)$ be the maximum number of different ordinals that can be obtained from n ordinals by taking their sums in all possible $n!$ different orders. Then

$$\lim_{n \to \infty} g(n)/n! = 0.$$

39. For every n give n ordinals such that all products of them taken in all possible $n!$ orders are different.

40. Let α be a limit ordinal, and call a set $A \subseteq \alpha$ of ordinals closed in α if the least upper bound of any increasing transfinite subsequence of A is in A or is equal to α. Then A is closed in α if and only if it is a closed subset of the topological space $(\alpha, \mathcal{T})$, where the topology $\mathcal{T}$ is generated by the intervals $\{\xi : \xi < \tau\}$, $\{\xi : \tau < \xi < \alpha\}$, $\tau < \alpha$ (this topology is called the interval topology on α).
    It is also true that A is closed in α if and only if the supremum of every subset $B \subset A$ is in A or is equal to α.

41. With the notation of the preceding problem a function $f : \alpha \to \alpha$ is continuous in the interval topology if and only if $f(\sup A) = \sup_{\xi \in A} f(\xi)$ for any set $A \subset \alpha$ with $\sup A < \alpha$.

42. If $A \subseteq \alpha$ is of cardinality κ, then its closure in the interval topology is also of cardinality κ.

43. If $\{a_\xi\}_{\xi<\omega_1}$ is a transfinite sequence of countable ordinals converging in the topology on ω_1 to a $\sigma \in \omega_1$, then there is a $\nu < \omega_1$ such that $a_\xi = a_\zeta$ for all $\xi, \zeta > \nu$.

44. Assume that $f : \omega_1 \times \omega_1 \to \omega$ has the property that for $\alpha < \omega_1$, $n < \omega$ the set $\{\beta < \alpha : f(\beta, \alpha) \le n\}$ is finite. Then all the sets

$$Z_f(\alpha, n) = \{\beta < \alpha : \text{there are } \beta = \beta_0 < \beta_1 < \cdots < \beta_k = \alpha, \\ \text{with } f(\beta_i, \beta_{i+1}) \le n\}$$

are also finite.

45. There is a function $f : \omega_1 \times \omega_1 \to \omega$ such that for $\alpha < \omega_1$, $n < \omega$ the set $\{\beta < \alpha : f(\beta, \alpha) \le n\}$ is finite and for any $\alpha_0 < \alpha_1 < \cdots$ we have $\sup_{k<\omega} f(\alpha_k, \alpha_{k+1}) = \omega$.

46. Two players, I and II, play the following game of length ω. At round i first I chooses a countable ordinal α_i at least as large as the previous ordinal chosen by him, then II selects a finite subset S_i of α_i. After ω many steps II wins if $S_0 \cup S_1 \cup \cdots = \sup(\{\alpha_i : i < \omega\})$.

    (a) II has a winning strategy.

    (b) II even has a winning strategy that chooses S_i only depending on i, α_{i-1}, and α_i.

47. Two players, I and II, alternatively select countable ordinals. After ω steps they consider the set of all selected ordinals, and II wins if it is an initial segment, otherwise I wins.

    (a) There is a winning strategy for II.

    (b) There is no such winning strategy if the choice of II depends only on the set of ordinals selected before (by the two players).

    (c) Even such a strategy exists if II is allowed to select finitely many ordinals in every step.

* * *

48. Let κ be an infinite cardinal and let two players alternately choose sets $K_0 \supset K_1 \supset \cdots$ of cardinality κ. Then no matter how the first player plays, the second one can always achieve $\cap_{n=0}^{\infty} K_n = \emptyset$.

9

Ordinal arithmetic

This chapter can be regarded as the "infinite analogue" of classical number theory. It contains problems on the arithmetic properties of ordinals such as divisibility, representation in a base, decomposition, primeness, etc.

A special role is played by the so-called normal representation (Problem 16) which is representation in base ω. In fact, many problems simplify considerably if the ordinals are written in normal form.

In this chapter $\alpha, \beta, \ldots$ always denote ordinals.

If $\alpha \cdot \beta = \gamma$, then we say that α (β) is a left (right) divisor of γ, and also that γ is a right (left) multiple of α (β).

1. If A is any set of nonzero ordinals, then there is a largest ordinal γ that divides every element of A from the left (this γ is called the greatest common left divisor of A). Every ordinal that divides every element of A from the left also divides γ from the left.
2. α is a limit ordinal if and only if ω divides α from the left.
3. α is divisible from the left by $\omega+2$ and by $\omega+3$ if and only if it is divisible from the left by ω^2.
4. α is divisible from the right by 2 and 3 if and only if it is divisible from the right by 6. Is the same true for divisibility from the left?
5. α is divisible from the right by $\omega + 2$ and by $\omega + 3$ if and only if it is divisible from the right by $\omega + 6$.
6. Every ordinal α has only a finite number of right divisors. Is the same true of left divisors? What if α is a successor ordinal?
7. If α and β are right divisors of $\gamma \geq 1$, then either
   a) α divides β from the right, or
   b) β divides α from the right, or
   c) $\alpha = \xi+p$, $\beta = \xi+q$, where ξ is a limit ordinal or 0, and p, q are positive natural numbers.

In case c) if $[p,q]$ is the smallest common multiple of p and q, then $\xi+[p,q]$ is the smallest common left multiple of α and β, and $\xi+[p,q]$ also divides γ from the right.

8. Any set of positive ordinals has a greatest common right divisor, and this greatest common right divisor is divisible from the right by any common right divisor.
9. Any set of positive ordinals has a least common (positive) right multiple, and this least common right multiple divides every common right multiple from the left.
10. Exhibit two ordinals that do not have a common (nonzero) left multiple.
11. Define ordinal exponentiation by transfinite recursion in the following way: $\gamma^0 = 1$, $\gamma^{\alpha+1} = \gamma^\alpha \cdot \gamma$, and for limit ordinal α let γ^α be the supremum of the ordinals γ^η, $\eta < \alpha$. For $\gamma > 1$ the following are true:
    (i) $\gamma^\alpha \cdot \gamma^\beta = \gamma^{\alpha+\beta}$,
    (ii) $(\gamma^\alpha)^\beta = \gamma^{\alpha\cdot\beta}$,
    (iii) if $\alpha < \beta$ then $\gamma^\alpha < \gamma^\beta$,
    (iv) $\alpha \le \gamma^\alpha$.
12. Consider the set $\Phi_{\alpha,\gamma}$ of all mappings $f : \alpha \to \gamma$ for which all but finitely many elements are mapped to 0, and for $f, g \in \Phi_{\alpha,\gamma}$, $f \ne g$ let $f \prec g$ if $f(\xi) < g(\xi)$ for the largest $\xi < \alpha$ for which $f(\xi) \ne g(\xi)$. Then $\langle \Phi_{\alpha,\gamma}, \prec\rangle$ is well ordered, and its order type is γ^α.
13. For any integer $n > 1$ we have
    a) $n^{\omega^\omega} = \omega^{\omega^\omega}$,
    b) $(\omega+n)^\omega = \omega^\omega$.
14. If α is a limit ordinal, then $1^\alpha + 2^\alpha = 3^\alpha$.
15. The following are true:
    a) $2^\omega = \omega$,
    b) if α is countable, then so is 2^α,
    c) for any cardinal $\kappa = \omega_\sigma$ we have $2^{\omega_\sigma} = \kappa$,
    d) if α is infinite, then the cardinality of 2^α is equal to the cardinality of α,
    e) every ordinal can be written in a unique manner in the form

$$2^{\xi_n} + 2^{\xi_{n-1}} + \cdots + 2^{\xi_0}, \tag{9.1}$$

    where $\xi_0 < \xi_1 \ldots < \xi_n$.
    What is the form (9.1) of the ordinal $\omega^4 \cdot 6 + \omega^2 \cdot 7 + \omega + 9$?
16. If $\gamma \ge 2$, then every ordinal can be written in a unique way in the form

$$\gamma^{\xi_n} \cdot \eta_n + \cdots + \gamma^{\xi_0} \cdot \eta_0,$$

where $\xi_0 < \xi_1 < \ldots < \xi_n$, and $1 \le \eta_j < \gamma$ for all $1 \le j \le n$. This form is called the representation of the given ordinal in base γ. The representation of an ordinal α in base ω is called the normal form of α.

17. If

$$\alpha = \omega^{\xi_n} \cdot a_n + \cdots + \omega^{\xi_0} \cdot a_0, \tag{9.2}$$

$\xi_0 < \xi_1 < \cdots < \xi_n$, $a_0, a_1 \ldots, a_n \in \mathbf{N}$ is the normal expansion of α, then $\alpha < \omega^{\xi_n+1}$, and for any $\omega^{\xi_n+1} \le \beta$ we have $\alpha + \beta = \beta$.

18. Find the normal form of the sum and product of two ordinals given in normal form.
19. If the normal form (9.2) of α has $(n+1)$ components, then for $m = 1, 2, \ldots$ the normal form of α^m has $(n+1)$ components if α is a limit ordinal and it has $mn+1$ components if α is a successor ordinal.
20. If the normal form of α is (9.2), then every $0 < \beta < \omega^{\xi_0}$ is a left divisor of α, and besides these there are only finitely many left divisors of α.
21. Given $\alpha > 0$, what are those natural numbers k such that α can be written as $\alpha = \beta \cdot k$ for some ordinal β?
22. Given an ordinal α, what is $\sum_{\beta < \omega^\alpha} \beta$?
23. If $\omega^\alpha = A \cup B$, then either A or B is of order type ω^α.
24. For every α there is a natural number N such that if α is decomposed as $\alpha = A_0 \cup \cdots \cup A_N$ into $N+1$ disjoint sets, then there is a j such that $\cup_{i \ne j} A_i$ has order type α.
25. If κ is an infinite cardinal, then every ordinal α of cardinality at most κ can be decomposed as $\alpha = A_0 \cup A_1 \cup \cdots$ such that every A_n is of order type smaller than κ^ω.
26. Call an ordinal $\alpha > 0$ (additively) indecomposable if it cannot be written as a sum of two smaller ordinals. Give the first three infinite indecomposable ordinals.
27. For every ordinal there is a bigger indecomposable ordinal. Also, for every countable ordinal there is a bigger indecomposable countable ordinal.
28. If α is arbitrary, and γ is the smallest ordinal for which there is a β such that $\alpha = \beta + \gamma$, then γ is indecomposable.
29. α is indecomposable if and only if it does not have a right divisor that is a successor ordinal bigger than 1.
30. α is indecomposable if and only if $\xi + \alpha = \alpha$ for every $\xi < \alpha$.
31. The supremum of indecomposable ordinals is indecomposable.
32. If α is indecomposable, then so is every $\beta \cdot \alpha$, $\beta > 0$.
33. If α is indecomposable, then α is divisible from the left by all $1 \le \beta < \alpha$.
34. The smallest indecomposable ordinal bigger than $\alpha \ge 1$ is $\alpha \cdot \omega$.
35. Every positive ordinal can be represented in a unique manner as a sum of a finite sequence of nonincreasing indecomposable ordinals.

36. Let $\alpha = \beta_1 + \beta_2 + \cdots + \beta_n$ be the decomposition of α from the preceding problem. Then $\alpha = \beta + \gamma$ for some $\beta, \gamma \neq 0$ if and only if there are a $1 \leq m \leq n$ such that $\gamma = \beta_m + \beta_{m+1} + \cdots + \beta_n$ and $\beta = \beta_1 + \beta_2 + \cdots + \beta_{m-1} + \delta$, where δ is an arbitrary ordinal smaller than β_m.

37. The indecomposable ordinals are precisely the ordinals of the form ω^α.

38. Call an ordinal $\alpha > 1$ prime if it cannot be written as the product of two smaller ordinals. Give the first three infinite prime ordinals.

39. $\alpha > 1$ is prime if and only if $\alpha = \beta \cdot \gamma$, $\gamma > 1$ imply $\gamma = \alpha$.

40. If α is an indecomposable ordinal, then $\alpha + 1$ is prime.

41. An infinite successor ordinal is prime if and only if it is of the form $\omega^\xi + 1$.

42. A limit ordinal is prime if and only if it is of the form ω^{ω^ξ}.

43. Every ordinal has at most one infinite right divisor that is prime.

44. Every successor ordinal has at most one infinite left divisor that is prime. However, a limit ordinal may have infinitely many infinite left prime divisors.

45. Every ordinal $\alpha > 1$ is the product of finitely many prime ordinals. In general, this representation is not unique even if we require that no factor can be omitted without changing the product.

46. Every $\alpha > 1$ has a unique representation

$$\alpha = a_1 \cdots a_m \cdot b_1 \cdot c_1 \cdot b_2 \cdots b_s \cdot c_s \cdot b_{s+1},$$

where $a_1 \geq \ldots \geq a_m$ are limit primes, $c_1, \ldots, c_s$ are infinite successor primes, and $b_1, \ldots, b_{s+1} > 1$ are natural numbers (some of the terms may be missing).

47. Call two positive ordinals α and β additively commutative if $\alpha + \beta = \beta + \alpha$. If α is additively commutative with both β and γ, then β and γ are also additively commutative.

48. For every positive ordinal α there are only countably many ordinals with which α is additively commutative.

49. Let n, m be given positive integers. Two ordinals α and β are additively commutative if and only if $\alpha \cdot n$ and $\beta \cdot m$ are additively commutative.

50. Two ordinals α and β are additively commutative if and only if there are positive integers n, m such that $\alpha \cdot n = \beta \cdot m$.

51. Two ordinals α and β are additively commutative if and only if there are natural numbers n, m and an ordinal ξ such that $\alpha = \xi \cdot n$, $\beta = \xi \cdot m$.

52. For any α the ordinals that additively commute with α are of the form $\beta \cdot n$, $n = 1, 2, \ldots$, where β is the smallest ordinal additively commutative with α.

53. If the normal form of $\alpha > 0$ is (9.2), then the ordinals additively commutative with α are the ones with normal form

$$\omega^{\xi_n} \cdot c + \omega^{\xi_{n-1}} \cdot a_{n-1} \cdots + \omega^{\xi_0} \cdot a_0$$

where c is an arbitrary positive natural number.

54. The sum of n nonzero ordinals $\alpha_1, \dots, \alpha_n$ is independent of their order if and only if there are positive integers $m_1, \dots, m_n$ and an ordinal ξ such that $\alpha_1 = \xi \cdot m_1$, $\alpha_2 = \xi \cdot m_2$, ..., $\alpha_n = \xi \cdot m_n$.

55. Let $g(n)$ be the maximum number of different ordinals that can be obtained from n ordinals by taking their sums in all possible $n!$ different orders.

 (a) For each n
$$g(n) = \max_{1 \le k \le n-1} (k2^{k-1} + 1) g(n-k).$$

 (b) $g(1) = 1$, $g(2) = 2$, $g(3) = 5$, $g(4) = 13$, $g(5) = 33$, $g(6) = 81$, $g(7) = 193$, $g(8) = 449$, $g(9) = 33^2$, $g(10) = 33 \cdot 81$, $g(11) = 81^2$, $g(12) = 81 \cdot 193$, $g(13) = 193^2$, $g(14) = 33^2 \cdot 81$, $g(15) = 33 \cdot 81^2$.

 (c) For $m \ge 3$ we have $g(5m) = 33 \cdot 81^{m-1}$ $g(5m+1) = 81^m$, $g(5m+2) = 193 \cdot 81^{m-1}$, $g(5m+3) = 193^2 \cdot 81^{m-2}$ and $g(5m+4) = 193^3 \cdot 81^{m-3}$.

 (d) For $n \ge 21$ we have $g(n) = 81 g(n-5)$.

56. Call two ordinals $\alpha > 1$ and $\beta > 1$ multiplicatively commutative if $\alpha \cdot \beta = \beta \cdot \alpha$. If $\gamma > 1$ is multiplicatively commutative with the ordinals β and γ, then β and γ are also multiplicatively commutative.

57. No successor ordinal bigger than 1 is multiplicatively commutative with any limit ordinal, and no finite ordinal bigger than 1 is multiplicatively commutative with any infinite ordinal.

58. For every ordinal $\alpha > 1$ there are only countably many ordinals that are multiplicatively commutative with α.

59. Let m, n be positive integers. Two ordinals α and β are multiplicatively commutative if and only if α^n and β^m are multiplicatively commutative.

60. Two infinite ordinals α, β are multiplicatively commutative if and only if there are natural numbers n, m such that $\alpha^n = \beta^m$.

61. Two limit ordinals $\alpha < \beta$ are multiplicatively commutative if and only if there is a θ and positive integers p, r such that $\beta = \omega^{\theta \cdot r} \cdot \alpha$, and the highest power of ω in the normal representations of α is $\omega^{\theta \cdot p}$.

62. If α is an infinite successor ordinal and $\xi > 1$ is the smallest ordinal multiplicatively commutative with α, then every ordinal that is multiplicatively commutative with α is of the form ξ^n with $n = 0, 1 \dots$.

63. Two infinite successor ordinals α and β are multiplicatively commutative if and only if there is an ordinal ξ and natural numbers n, m with which $\alpha = \xi^n$ and $\beta = \xi^m$.

64. The ordinals $\omega^2+\omega$ and $\omega^3+\omega^2$ are multiplicatively commutative, but there is no ordinal ξ and natural numbers n,m with which $\alpha=\xi^n$ and $\beta=\xi^m$ would be true.

65. The product of n ordinals $\alpha_1,\dots,\alpha_n$, $\alpha_i\geq 2$ is independent of their order if and only if there are positive integers $m_1,\dots,m_n$ for which $\alpha_1^{m_1}=\alpha_2^{m_2}=\cdots=\alpha_n^{m_n}$.

66. For every n give n ordinals such that all products of them taken in all possible $n!$ orders are different.

67. There are no different infinite ordinals that are simultaneously additively and multiplicatively commutative.

68. For infinite α the following statements are pairwise equivalent:

    a) if $\xi<\alpha$ and $\theta<\alpha$, then $\xi\cdot\theta<\alpha$,

    b) if $1\leq\xi<\alpha$ then $\xi\cdot\alpha=\alpha$,

    c) $\alpha=\omega^{\omega^\beta}$ for some β.

69. Call an ordinal α epsilon-ordinal, if $\omega^\alpha=\alpha$. Find the smallest epsilon-ordinal.

70. For every ordinal there is a larger epsilon-ordinal and for every countable ordinal there is a larger countable epsilon-ordinal.

71. If α is an epsilon-ordinal, then

    (i) $\xi+\alpha=\alpha$ for $\xi<\alpha$,

    (ii) $\xi\cdot\alpha=\alpha$ for $1\leq\xi<\alpha$,

    (iii) $\xi^\alpha=\alpha$ for $2\leq\xi<\alpha$.

72. If $\beta\geq\omega$ and $\beta^\alpha=\alpha$, then α is an epsilon-ordinal.

73. α is an epsilon-ordinal if and only if $\omega<\alpha$ and $\beta^\gamma<\alpha$ whenever $\beta,\gamma<\alpha$.

74. For infinite ordinals $\alpha<\beta$ we have $\alpha^\beta=\beta^\alpha$ if and only if α is a limit ordinal and $\beta=\gamma\cdot\alpha$, where $\gamma>\alpha$ is an epsilon ordinal.

75. Define the product $\prod_{\xi<\theta}\alpha_\xi$ of a transfinite sequence $\{\alpha_\xi\}_{\xi<\theta}$ of ordinals, and discuss its properties!

76. If $\alpha_0+\alpha_1+\cdots$ is a sum of a sequence of ordinals of type ω, then by taking a permutation of (possibly infinitely many of) the terms in the sum, one can get only finitely many different ordinals.

77. If $\alpha_0+\alpha_1+\cdots$ is a sum of a sequence of ordinals of type ω, then by deleting finitely many terms and taking a permutation of (possibly infinitely many of) the remaining terms in the sum, one can get only finitely many different ordinals.

78. Given a positive integer n give a sum $\alpha_0+\alpha_1+\cdots$ of a sequence of infinite ordinals of type ω such that one can get exactly n different values by taking a permutation of the terms in the sum.

79. If $\alpha_0 \cdot \alpha_1 \cdots$ is a product of a sequence of ordinals of type ω, then by taking a permutation of (possibly infinitely many of) the terms in the product, one can get only finitely many different ordinals.
80. If $\alpha_0 \cdot \alpha_1 \cdots$ is a product of a sequence of ordinals of type ω, then by deleting finitely many terms and taking a permutation of (possibly infinitely many of) the remaining terms in the product, one can get only finitely many different ordinals.
81. Given a positive integer n give a product $\alpha_0 \cdot \alpha_1 \cdots$ of a sequence of infinite ordinals of type ω such that one can get exactly n different values by taking a permutation of the terms in the product.
82. Permuting finitely many terms in a sum $\sum_{\beta \leq \omega} \alpha_\beta$ (but keeping the permuted sum of type $\omega + 1$), one may get infinitely many different ordinals.
83. If γ is a countable ordinal and $\{\alpha_\beta\}_{\beta<\gamma}$ is a sequence of ordinals, then there are only countably many different sums of the form $\sum_{\beta<\gamma} \alpha_{\pi(\beta)}$, where $\pi : \gamma \to \gamma$ is any mapping.
84. Permuting finitely many terms in a product $\prod_{\beta \leq \omega} \alpha_\beta$ (but keeping the permuted sum of type $\omega + 1$), one may get infinitely many different ordinals.
85. If γ is a countable ordinal and $\{\alpha_\beta\}_{\beta<\gamma}$ is a sequence of ordinals, then there are only countably many different products of the form $\prod_{\beta<\gamma} \alpha_{\pi(\beta)}$, where $\pi : \gamma \to \gamma$ is any mapping.
86. Write $\Gamma(\alpha) = \prod_{\xi<\alpha} \xi$. Calculate $\Gamma(\omega)$, $\Gamma(\omega+1)$, $\Gamma(\omega \cdot 2)$, and $\Gamma(\omega^2)$.
87. Find all operations $\mathcal{F}$ from the ordinals to the ordinals that are continuous in the interval topology and that satisfy the equation $\mathcal{F}(\alpha+\beta) = \mathcal{F}(\alpha) + \mathcal{F}(\beta)$ for all α and β.
88. Is there a not identically zero operation $\mathcal{F}$ from the ordinals to the ordinals that is continuous in the interval topology and that satisfies the equation $\mathcal{F}(\alpha+\beta) = \mathcal{F}(\beta) + \mathcal{F}(\alpha)$ for all α and β?
89. Find all operations $\mathcal{F}$ from the ordinals to the ordinals that are continuous in the interval topology and that satisfy the equation $\mathcal{F}(\alpha+\beta) = \mathcal{F}(\alpha) \cdot \mathcal{F}(\beta)$ for all α and β.
90. Is there a not identically zero and not identically 1 operation $\mathcal{F}$ from the ordinals to the ordinals that is continuous in the interval topology and that satisfies the equation $\mathcal{F}(\alpha+\beta) = \mathcal{F}(\beta) \cdot \mathcal{F}(\alpha)$ for all α and β?
91. Define the Hessenberg sum (or natural sum) $\alpha \oplus \beta$ of ordinals α, β with normal form

$$\alpha = \omega^{\delta_n} \cdot a_n + \cdots + \omega^{\delta_0} \cdot a_0, \qquad \beta = \omega^{\delta_n} \cdot b_n + \cdots + \omega^{\delta_0} \cdot b_0 \tag{9.3}$$

(with possibly $a_i = 0$ or $b_i = 0$) as

$$\alpha \oplus \beta = \omega^{\delta_n} \cdot (a_n + b_n) + \cdots \omega^{\delta_0} \cdot (a_0 + b_0).$$

(a) $\oplus$ is an associative and commutative operation.

(b) If $\beta < \gamma$, then $\alpha \oplus \beta < \alpha \oplus \gamma$.

(c) For a given α how many solutions does the equation $x \oplus y = \alpha$ have?

(d) Is $\mathcal{F}_\alpha(x) = \alpha \oplus x$ continuous?

(e) $\alpha_1 + \cdots + \alpha_n \le \alpha_1 \oplus \cdots \oplus \alpha_n$. When does the equality hold?

(f) $\alpha_1 \oplus \cdots \oplus \alpha_n \le \max\{\alpha_1, \ldots, \alpha_n\} \cdot (n+1)$.

92. $\alpha_1 \oplus \cdots \oplus \alpha_n$ is the largest ordinal that occurs as the order type of $A_1 \cup \cdots \cup A_n$, where $A_1, \ldots, A_n$ are subsets of some ordered set of order types $\alpha_1, \ldots, \alpha_n$, respectively.

93. If $\mathcal{F}(\alpha, \beta)$ is a commutative operation on the ordinals which is strictly increasing in either variable, then $\mathcal{F}(\alpha, \beta) \ge \alpha \oplus \beta$ holds for all α, β.

The "superbase" form of a natural number in base b is obtained by writing the number in base b, and all exponents and exponents of exponents, etc., in base b. For example, if $b = 2$, then $141 = 2^7 + 2^3 + 2^2 + 1 = 2^{2^2+2+1} + 2^{2+1} + 2^2 + 1$, and the latter form is its "superbase" 2 form.

94. Pick a natural number n_1, and for each $i = 1, 2, \ldots$ perform the following two operations to define the numbers n_{2i} and n_{2i-1}:

(i) write n_{2i-1} in "superbase" form in base $i+1$, and while keeping all coefficients, replace the base by $i+2$. This gives a number that we call n_{2i}.

(ii) set $n_{2i+1} = n_{2i} - 1$.

If $n_{2i+1} = 0$, then we stop, otherwise repeat these operations. For example, if $n_1 = 23$, then its "superbase" 2 form is $23 = 2^{2^2} + 2^2 + 2 + 1$, so $n_2 = 3^{3^3} + 3^3 + 3 + 1 = 7625597485018$, $n_3 = 7625597485017$. Since $n_3 = 3^{3^3} + 3^3 + 3$, and here we change the base 3 to base 4, we have $n_4 = 4^{4^4} + 4^4 + 4$, which is the following 155-digit number:

$$
\begin{gathered}
13407807929942597099574024998205846127479365820592393 \\
37772356144372176403007354697680187429816690342769 0 \\
03185818648605085375388281194656994643364900608435 6.
\end{gathered}
$$

(a) No matter what n_1 is, there is an i such that $n_i = 0$.

(b) The same conclusion holds if in **(i)** the actual base is changed to any larger base (i.e., when the bases are not $2, 3, \ldots$ but some numbers $b_1 < b_2 < \ldots$).

10

Cardinals

Cardinals express the size of sets. Saying that two sets are equivalent (are of equal size) is the same as saying that their cardinality is the same. The cardinality of the set A is denoted by $|A|$, and it can be defined as the smallest ordinal equivalent to A: $|A| = \min\{\alpha : \alpha \sim A\}$.

We set $|A| < |B|$ if A is equivalent to a subset of B but not vice versa. It is easy to see that this is the same as $|A|$ being smaller than $|B|$ in the "smaller" relation (i.e., in $\in$) among ordinals. If κ_i, $i \in I$ are cardinals, then their sum $\sum_{i\in I} \kappa_i$ is defined as the cardinality of $\cup_{i\in I} A_i$, where A_i are disjoint sets of cardinality κ_i, and their product $\prod_{i\in I} \kappa_i$ is defined as the cardinality of the product set $\prod_{i\in I} A_i$ (recall that this is the same as the set of choice functions $f : I \to \cup_{i\in I} A_i$, $f(i) \in A_i$ for all i). Finally, we set $|A|^{|B|}$ as the cardinality of the set ${}^{B}A$ (which is the set of functions $f : B \to A$ from B into A).

This chapter contains problems related to cardinal operations. The fundamental theorem of cardinal arithmetic (Problem 2) says that for infinite cardinals κ, λ we have $\kappa + \lambda = \kappa\lambda = \max\{\kappa, \lambda\}$. Quite often this makes questions on cardinal addition and multiplication trivial. The situation is completely different with cardinal exponentiation; it is not trivial at all, and is one of the subtlest question of set theory with problems leading quite often to independence results. For this reason we shall barely touch upon cardinal exponentiation in this book.

An important property of some cardinals is their *regularity*: $\kappa = \mathrm{cf}(\kappa)$. It is equivalent to the fact that κ cannot be reached by (i.e., not the supremum of) less than κ smaller ordinals. Another equivalent formulation is that a set of cardinality κ is not the union of fewer than κ sets of cardinality smaller than κ (see Problems 9, 10). Some properties hold only for regular cardinals, and quite frequently proofs are simpler for regular cardinals than for singular (=nonregular) ones.

The finite cardinals are just the natural numbers. Infinite cardinals are listed in an endless "transfinite sequence" $\omega_0, \omega_1, \ldots, \omega_\alpha, \ldots$, numbered by ordinals α. Here $\omega_0 = \omega$ is the smallest infinite cardinal, and this numbering is done so that $\beta < \alpha$ implies $\omega_\beta < \omega_\alpha$. If $\kappa = \omega_\alpha$, then $\omega_{\alpha+1}$ is the successor cardinal to κ (i.e., the smallest cardinal larger than κ), and is denoted by κ^+. It is always a regular cardinal.

For historical reasons we also write $\aleph_\alpha$ instead of ω_a (note that ω_α has two faces; it is an ordinal and also a cardinal, and we use the aleph notation when we emphasize the cardinal aspect).

CH, the continuum hypothesis (i.e., that there is no cardinal between ω and $\mathbf{c}$) can be expressed as $\mathbf{c} = \aleph_1$ or as $2^{\aleph_0} = \aleph_1$. The generalized continuum hypothesis (GCH) stipulates that for all α we have $2^{\aleph_\alpha} = \aleph_{\alpha+1}$. This is also independent of the axioms of set theory (cf. the introduction to Chapter 4).

1. What is the cardinal $a_0 \cdot a_1 \cdots$ if the a_i's are positive integers?
2. (Fundamental theorem of cardinal arithmetic) For every infinite cardinal κ we have $\kappa^2 = \kappa$.
3. If at least one of $\kappa > 0$ and $\lambda > 0$ is infinite, then
$$\kappa + \lambda = \kappa\lambda = \max\{\kappa, \lambda\}.$$
4. If X is of cardinality $\kappa \geq \aleph_0$, then the following sets are of cardinality κ:
   a) set of finite sequences of elements of X,
   b) set of those functions that map a finite subset of X into X.
5. Let X be a set of infinite cardinality κ, and call a set $Y \subset X$ "small" if there is a decomposition of X into subsets of cardinality κ each of which intersects Y in at most one point. Then X is the union of two of its "small" subsets.
6. The supremum of any set of cardinals (considered as a set of ordinals) is again a cardinal.
7. If $\rho_1 + \rho_2 = \sum_{\xi<\alpha} \lambda_\xi$, then there are cardinals $\lambda_\xi^{(i)}$, $i = 1, 2$, $\xi < \alpha$ such that $\rho_i = \sum_{\xi<\alpha} \lambda_\xi^{(i)}$, $i = 1, 2$, and for all ξ we have $\lambda_\xi = \lambda_\xi^{(1)} + \lambda_\xi^{(2)}$.
8. If α is the cofinality of an ordered set, then α is a regular cardinal.
9. If κ is an infinite cardinal, then $\mathrm{cf}(\kappa)$ coincides with the smallest ordinal α for which there is a transfinite sequence $\{\kappa_\xi\}_{\xi<\alpha}$ of cardinals smaller than κ with the property $\kappa = \sum_{\xi<\alpha} \kappa_\xi$.
10. An infinite cardinal is regular if and only if κ is not the sum of fewer than κ cardinals each of which is less than κ.
11. A successor cardinal is regular.
12. Which are the smallest three singular (i.e., not regular) infinite cardinals?
13. $\mathrm{cf}(\aleph_\alpha) = \aleph_\alpha$ if α is a successor ordinal, and $\mathrm{cf}(\aleph_\alpha) = \mathrm{cf}(\alpha)$ if α is a limit ordinal.

14. Let n be a natural number. The cardinality of a set H is at most $\aleph_n$ if and only if ${}^{n+2}H(\equiv H^{n+2})$ can be represented in the form $A_1 \cup \cdots \cup A_{n+2}$, where A_k is finite "in the direction of the kth coordinate", i.e., if $h_1, \ldots, h_{k-1}, h_{k+1}, \ldots h_{n+2}$ are arbitrary elements from H, then there are only finitely many $h \in H$ such that $(h_1, \ldots, h_{k-1}, h, h_{k+1}, \ldots h_{n+2}) \in A_k$.
15. The cardinality of a set H is at most $\aleph_{\alpha+n}$ if and only if ${}^{n+2}H(\equiv H^{n+2})$ can be represented in the form $A_1 \cup \cdots \cup A_{n+2}$, where the cardinality of A_k "in the direction of the kth coordinate" is smaller than $\aleph_\alpha$, i.e., if $h_1, \ldots, h_{k-1}, h_{k+1}, \ldots h_{n+2}$ are arbitrary elements from H, then there are fewer than $\aleph_\alpha$ elements $h \in H$ such that

$$(h_1, \ldots, h_{k-1}, h, h_{k+1}, \ldots h_{n+2}) \in H_k.$$

16. (Cantor's inequality) For any κ we have $2^\kappa > \kappa$.
17. (König's inequality) If $\rho_i < \kappa_i$ for all $i \in I$, then

$$\sum_{i \in I} \rho_i < \prod_{i \in I} \kappa_i.$$

18. If the set of cardinals $\{\kappa_\xi\}_{\xi<\theta}$, $0 < \kappa_\xi < \kappa$ is cofinal with κ, then $\prod_{\xi<\theta} \kappa_\xi > \kappa$.
19. If κ is infinite, $\kappa = \sum_{\xi<\mathrm{cf}(\kappa)} \kappa_\xi$ where $\kappa > \kappa_\xi > 1$, then

$$\prod_{\xi<\mathrm{cf}(\kappa)} \kappa_\xi = \kappa^{\mathrm{cf}(\kappa)}.$$

20. If κ is infinite, then $\kappa^{\mathrm{cf}(\kappa)} > \kappa$.
21. If $\lambda \geq 2$ and κ is infinite, then $\mathrm{cf}(\lambda^\kappa) > \kappa$.
22. (Bernstein–Hausdorff–Tarski equality) Let κ be an infinite cardinal and λ a cardinal with $0 < \lambda < \mathrm{cf}(\kappa)$. Then

$$\kappa^\lambda = \left(\sum_{\rho<\kappa} \rho^\lambda\right) \kappa.$$

23. If α is a limit ordinal, $\{\kappa_\xi\}_{\xi<\alpha}$ is a strictly increasing sequence of cardinals and $\kappa = \sum_{\xi<\alpha} \kappa_\xi$, then for all $0 < \lambda < \mathrm{cf}(\alpha)$ we have $\kappa^\lambda = \sum_{\xi<\alpha} \kappa_\xi^\lambda$.
24. If λ is singular and there is a cardinal κ such that for some $\mu < \lambda$ for every cardinal τ between μ and λ we have $2^\tau = \kappa$, then $2^\lambda = \kappa$, as well.
25. If there is an ordinal γ such that $2^{\aleph_\alpha} = \aleph_{\alpha+\gamma}$ holds for every infinite cardinal $\aleph_\alpha$, then γ is finite.
26. The operation $\kappa \mapsto \kappa^{\mathrm{cf}(\kappa)}$ on cardinals determines

    (a) the operation $\kappa \mapsto 2^\kappa$,

(b) the operation $(\kappa, \lambda) \mapsto \kappa^\lambda$.

27. If n is finite, then for $\lambda \geq 1$

(a) $\aleph_{\alpha+n}^\lambda = \aleph_\alpha^\lambda \aleph_{\alpha+n}$.

(b) $\aleph_n^\lambda = 2^\lambda \aleph_n$.

28. When does

$$\prod_{n<\omega} \aleph_n = 2^{\aleph_0}$$

hold?

29.

$$\prod_{n<\omega} \aleph_n = \aleph_\omega^{\aleph_0}.$$

30. If for all $n < \omega$ we have $2^{\aleph_n} < \aleph_\omega$, then $2^{\aleph_\omega} = \aleph_\omega^{\aleph_0}$.

31. If $\rho \geq \omega$ is a given cardinal, then there are infinitely many cardinals κ for which $\kappa^\rho = \kappa$, and there are infinitely many for which $\kappa^\rho > \kappa$.

32. There are arbitrarily large cardinals λ with $\lambda^{\aleph_0} < \lambda^{\aleph_1}$.

33. For an infinite cardinal κ let μ be the minimal cardinal with $2^\mu > \kappa$. Then $\{\kappa^\lambda : \lambda < \mu\}$ is finite.

34. For an infinite cardinal κ let $\rho = \rho_\kappa$ be the smallest cardinal such that $\kappa^\rho > \kappa$. Then ρ_κ is a regular cardinal. What is ρ_ω? And ρ_{ω_ω}?

35. The smallest κ for which $2^\kappa > \mathbf{c}$ holds is regular.

36. Let $\kappa_0 = \aleph_0$, and for every natural number n let $\kappa_{n+1} = \aleph_{\kappa_n}$. Then $\kappa = \sup_n \kappa_n$ is the smallest cardinal with the property $\kappa = \aleph_\kappa$.

37. There are infinitely many cardinals κ such that the set of cardinals smaller than κ is of cardinality κ (i.e., $\kappa = \aleph_\kappa$). If we call such cardinals κ "large", then are there cardinals κ such that the set of "large" cardinals smaller than κ is of cardinality κ?

38. Under GCH (generalized continuum hypothesis) find all cardinals κ for which $\kappa^{\aleph_0} < \kappa^{\aleph_1} < \kappa^{\aleph_2}$ hold.

39. Assuming GCH evaluate $\prod_{\beta<\alpha} \aleph_\beta$.

40. Under GCH determine κ^λ.

11

Partially ordered sets

Let A be a set and $\prec$ a binary relation on A. $\langle A, \prec \rangle$ is called a *partially ordered set* if

- $\prec$ *irreflexive*: $a \not\prec a$ for any $a \in A$,
- $\prec$ *transitive*: $a \prec b$ and $b \prec c$ imply $a \prec c$.

Thus, the difference with ordered sets is that here we do not assume trichotomy (comparability of elements).

In a partially ordered set $\langle A, \prec \rangle$ two elements a, b are called *comparable* if (exactly) one of $a = b$, $a \prec b$ or $b \prec a$ holds, otherwise they are *incomparable*. An ordered subset of a partially ordered set is called a *chain* and a set of pairwise incomparable elements an *antichain*.

The main problem that we treat in this chapter is how information on the size of chains and antichains can be related to the structure of the set in question.

1. In an infinite partially ordered set there is an infinite chain or an infinite antichain.
2. If in a partially ordered set all chains have at most $l < \infty$ elements and all antichains have at most $k < \infty$ elements, where k, l are finite numbers, then the set has at most kl elements.
3. If in a partially ordered set all chains have at most $k < \infty$ elements, then the set is the union of k antichains.
4. If in a partially ordered set all antichains have at most $k < \infty$ elements, then the set is the union of k chains.
5. There is a partially ordered set in which all chains are finite, still the set is not the union of countably many antichains.
6. There is a partially ordered set in which all antichains are finite, still the set is not the union of countably many chains.

7. If in a partially ordered set all chains are finite and all antichains are countable, then the set is countable.
8. If in a partially ordered set all antichains are finite and all chains are countable, then the set is countable.
9. There is a partially ordered set of cardinality continuum in which all chains and all antichains are countable.
10. If in a partially ordered set all chains and all antichains have at most κ elements, then the set is of cardinality at most 2^κ.
11. If κ is an infinite cardinal, then there is a partially ordered set of cardinality 2^κ in which all chains and all antichains have at most κ elements.
12. For every cardinal κ there is a partially ordered set $\langle P, \prec \rangle$ in which every interval $[x, y] = \{z : x \preceq z \preceq y\}$ is finite, yet P is not the union of κ antichains.
13. If $\langle P, \prec \rangle$ is a partially ordered set, call two elements strongly incompatible if they have no common lower bound. Let $c(P, \prec)$ be the supremum of $|S|$ where $S \subseteq P$ is a strong antichain, that is, a set of pairwise strongly incompatible elements.
    (a) If $c(P, \prec)$ is an infinite cardinal that is not weakly inaccessible, i.e., it is not a regular limit cardinal, then $c(P, \prec)$ is actually a maximum.
    (b) If κ is a regular limit cardinal, then there is a partially ordered set $\langle P, \prec \rangle$ such that $c(P, \prec) = \kappa$ yet there is no strong antichain of cardinality κ.
14. If $\langle A, \prec \rangle$ is a partially ordered set, then there exists a cofinal subset $B \subseteq A$ such that $\langle B, \prec \rangle$ is well founded (i.e., in every nonempty subset there is a minimal element).
15. If there is no maximal element in the partially ordered set $\langle P, \prec \rangle$, then there are two disjoint cofinal subsets of $\langle P, \prec \rangle$.
16. There is a partially ordered set $\langle P, \prec \rangle$ which is the union of countably many centered sets but not the union of countably many filters. (A subset $Q \subseteq P$ is centered if for any $p_1, \ldots, p_n \in Q$ there is some $q \preceq p_1, \ldots, p_n$ in P. A subset $F \subseteq P$ is a filter, if for any $p_1, \ldots, p_n \in F$ there is some $q \preceq p_1, \ldots, p_n$ with $q \in F$.)
17. For two real functions $f \neq g$ let $f \prec g$ if $f(x) \leq g(x)$ for all $x \in \mathbf{R}$. In this partially ordered set there is an ordered subset of cardinality bigger than continuum. No such subset can be well ordered by $\prec$.

The following problems use two orderings on the set ${}^\omega\omega$ of all functions $f : \omega \to \omega$: let $f \ll g$ if $f(n) < g(n)$ for all large n, and $f \prec g$ if $g(n) - f(n) \to \infty$ as $n \to \infty$.

18. Each of $\langle {}^\omega\omega, \ll \rangle$ and $\langle {}^\omega\omega, \prec \rangle$ has an order-preserving mapping into the other, but they are not isomorphic.

19. For any countable subset $\{f_k\}_k$ of ${}^\omega\omega$ there is an f larger than any f_k with respect to $\prec$.
20. $\langle {}^\omega\omega, \prec\rangle$ includes a subset of order type ω_1.
21. $\langle {}^\omega\omega, \prec\rangle$ includes a subset of order type λ^m for each $m = 1, 2, \ldots$.
22. If θ is an order type and $\langle {}^\omega\omega, \prec\rangle$ includes a subset similar to θ, then it includes such a subset consisting of functions that are smaller than the identity function.
23. If θ_1, θ_2 are order types and $\langle {}^\omega\omega, \prec\rangle$ includes subsets similar to θ_1 and θ_2, respectively, then it includes subsets similar to $\theta_1 + \theta_2$ and $\theta_1 \cdot \theta_2$, respectively. It also includes a subset similar to θ_1^*, where θ_1^* is the reverse type to θ_1.
24. If θ_i, $i \in I$ are order types where $\langle I, <\rangle$ is an ordered set, and $\langle {}^\omega\omega, \prec\rangle$ includes subsets similar θ_i and also a subset similar to $\langle I, <\rangle$, then it includes subsets similar to $\sum_{i\in I(<)} \theta_i$. In particular, $\langle {}^\omega\omega, \prec\rangle$ includes a set of order type α for every $\alpha < \omega_2$.
25. If $\varphi < \omega_1$ is a limit ordinal and

$$f_0 \prec f_1 \prec \cdots \prec f_\alpha \prec \cdots \prec g_\alpha \prec \cdots g_1 \prec g_0, \qquad \alpha < \varphi,$$

then there is an f with $f_\alpha \prec f \prec g_\alpha$ for every $\alpha < \varphi$.
26. There exist functions

$$f_0 \prec f_1 \prec \cdots \prec f_\alpha \prec \cdots \prec g_\alpha \prec \cdots g_1 \prec g_0, \qquad \alpha < \omega_1,$$

such that there is no function f with $f_\alpha \prec f \prec g_\alpha$ for every $\alpha < \omega_1$.

12

Transfinite enumeration

This chapter deals with a fundamental technique based on the well-ordering theorem. Most of the problems in this chapter require the construction of some objects sometimes with quite surprising properties (like Problem 7: there is a set $A \subset \mathbf{R}^2$ intersecting every line in exactly two points). The objects cannot be given at once, but are obtained by a transfinite recursive process. The idea is to have a well ordering of the underlying structure (in the aforementioned example a well-ordering of the lines on $\mathbf{R}^2$ into a transfinite sequence $\{\ell_\alpha\}_{\alpha<\mathbf{c}}$ of type $\mathbf{c}$) and based on that the object is constructed one by one (in the example constructing an increasing sequence $\{A_\alpha\}_{\alpha<\mathbf{c}}$ of sets such that A_α has at most two points on any line, and it has exactly two points on ℓ_α).

Of similar spirit is the transfinite construction of some closure sets such as the set of Borel sets, the set of Baire functions, or the algebraic closures of fields.

This transfinite enumeration technique will be routinely used in later chapters.

1. If A_i, $i \in I$ is an arbitrary family of sets, then there are pairwise disjoint sets $B_i \subset A_i$ such that $\cup_{i\in I} B_i = \cup_{i\in I} A_i$.
2. If there are given $\kappa \geq \aleph_0$ sets X_ξ each of cardinality κ, then there are pairwise disjoint subsets $Y_\xi \subseteq X_\xi$ each of cardinality κ. Further, we can even have $|X_\xi \setminus Y_\xi| = \kappa$ for all $\xi < \kappa$.
3. If there are given $\kappa \geq \aleph_0$ sets X_ξ, $\xi < \kappa$ each of cardinality κ, then there are pairwise disjoint sets Y_α, $\alpha < \kappa$ such that for all $\alpha, \xi < \kappa$ the intersection $Y_\alpha \cap X_\xi$ is of cardinality κ.
4. Let κ be an infinite cardinal, X a set of cardinality κ, and $\mathcal{F}$ a family of cardinality at most κ of mappings with domain X. Then there is a family $\mathcal{H}$ of cardinality 2^κ of subsets of X with the property that if $H_1, H_2 \in \mathcal{H}$ are two different sets and $f \in \mathcal{F}$ is arbitrary, then $f[H_1] \neq H_2$.
5. If X is an infinite set of cardinality κ, then there is an almost disjoint family $\mathcal{H}$ of cardinality bigger than κ of subsets of X each of cardinality

κ (the intersection of any two members of $\mathcal{H}$ is of cardinality smaller than κ).

6. There is a family $\{N_\alpha\}_{\alpha<\omega_1}$ of subsets of $\mathbf{N}$ such that for $\alpha < \beta < \omega_1$ the set $N_\beta \setminus N_\alpha$ is finite, but the set $N_\alpha \setminus N_\beta$ is infinite.
7. There is a subset A of $\mathbf{R}^2$, that has exactly two points on every line.
8. Suppose that to every line ℓ on the plane a cardinal $2 \leq m_\ell \leq \mathbf{c}$ is assigned. Then there is a subset A of the plane such that $|A \cap \ell| = m_\ell$ holds for every ℓ.
9. If L_1 and L_2 are two disjoint sets of lines lying on the plane, then the plane can be divided into two sets $A_1 \cup A_2$ in such a way that every line in L_1 resp. L_2 intersects A_1 resp. A_2 in fewer than continuum many points.
10. $\mathbf{R}$ can be decomposed into continuum many pairwise disjoint sets of power continuum, such that each of these sets intersects every nonempty perfect set.
11. $\mathbf{R}$ can be decomposed into continuum many pairwise disjoint and non-measurable sets.
12. $\mathbf{R}$ can be decomposed into continuum many pairwise disjoint sets each of the second category.
13. There is a subset A of $\mathbf{R}^2$ that has at most two points on every line, but A is not of measure zero (with respect to two-dimensional Lebesgue measure).
14. There is a second category subset A of $\mathbf{R}^2$ that has at most two points on every line.
15. There is a set $A \subset \mathbf{R}$ such that every $x \in \mathbf{R}$ has exactly one representation $x = a + b$ with $a, b \in A$.
16. If $A \subset \mathbf{R}$ is an arbitrary set, then there is a function $f : A \to A$ that assumes every value only countably many times and for which $f(a) < a$ for all $a \in A$, except for the smallest element of A (if there is one).
17. Every real function is the sum of two 1-to-1 functions.
18. There is a real function that is not monotone on any set of cardinality continuum.
19. There is a real function F such that for all continuous real functions f the sum $F + f$ assumes all values $y \in \mathbf{R}$ in every interval.
20. There is a real function f such that if $\{x_n\}_{n=0}^\infty$ is an arbitrary sequence of distinct real numbers and $\{y_n\}_{n=0}^\infty$ is an arbitrary real sequence, then there is an $x \in \mathbf{R}$ such that for all n we have $f(x + x_n) = y_n$.
21. For $X \subseteq \mathbf{R}^n$ let X^L be the set of all limit points of X, and starting from $X_0 = X$ form the sets

$$X_\alpha = \begin{cases} X_\beta^L & \text{if } \alpha = \beta + 1, \\ \cap_{\xi<\alpha} X_\xi & \text{if } \alpha \text{ is a limit ordinal.} \end{cases}$$

Then there is a countable ordinal θ such that $X_\alpha = X_\theta$ for all $\alpha > \theta$, and the set $X \setminus X_\theta$ is countable. Furthermore X_θ is empty or it is perfect.

22. Every closed set in $\mathbf{R}^n$ is the union of a perfect set and a countable set.
23. Starting from an arbitrary set X and a family $\mathcal{H}$ of subsets of X form the families $\mathcal{H}_\alpha$ of sets in the following way: $\mathcal{H}_0 = \mathcal{H}$; for every ordinal α let $\mathcal{H}_{\alpha+1}$ be the family of sets that can be obtained as a countable union of sets in $\mathcal{H}_\alpha$ or that are the complements (with respect to X) of some sets in $\mathcal{H}_\alpha$; and for a limit ordinal α set $\mathcal{H}_\alpha = \cup_{\beta<\alpha}\mathcal{H}_\beta$. Then $\mathcal{H}_{\omega_1} = \mathcal{H}_\alpha$ for every $\alpha > \omega_1$, and $\mathcal{H}_{\omega_1}$ is the σ-algebra generated by $\mathcal{H}$ (this is the intersection of all σ-algebras including $\mathcal{H}$, and is the smallest σ-algebra including $\mathcal{H}$).
24. The σ-algebra generated by at most continuum many sets is of power at most continuum.
25. The family of Borel sets in $\mathbf{R}^n$ is the smallest family of sets containing the open sets and closed under countable intersection and countable disjoint union.
26. Starting from the set $C[0,1]$ of continuous functions on the interval $[0,1]$ form the following families $\mathcal{B}_\alpha$ of functions: $\mathcal{B}_0 = C[0,1]$; for every α let $\mathcal{B}_{\alpha+1}$ be the set of those functions that can be obtained as pointwise limits of a sequence of functions from $\mathcal{B}_\alpha$; and for a limit ordinal α let $\mathcal{B}_\alpha = \cup_{\beta<\alpha}\mathcal{B}_\beta$. Then $\mathcal{B}_{\omega_1} = \mathcal{B}_\alpha$ for all $\alpha > \omega_1$, and $\mathcal{B}_{\omega_1}$ is the smallest set of functions that is closed for pointwise limits and that includes $C[0,1]$ (this is the set of so-called Baire functions on $[0,1]$).
27. Let $\langle \mathcal{A}, \cdots \rangle$ be an algebraic structure with at most ρ finitary operations. Then the subalgebra in $\mathcal{A}$ generated by a subset of $\kappa(\neq 0)$ elements has cardinality at most $\max\{\kappa, \rho, \aleph_0\}$ (the subalgebra generated by a set X of elements is the intersection of all subalgebras that include X).
28. If $\mathcal{F}$ is any field of cardinality κ, then there is an algebraically closed field $\mathcal{F} \subset \mathcal{F}^*$ of cardinality at most $\max\{\kappa, \aleph_0\}$ (a field

$$\mathcal{F}^* = \langle F^*, +, \cdot, 0, 1 \rangle$$

is called algebraically closed if for any polynomial $a_n \cdot x^n + \cdots + a_1 \cdot x + a_0$ with $a_i \in F^*$ there is an $a \in F^*$ such that $a_n \cdot a^n + \cdots + a_1 \cdot a + a_0 = 0$).
29. Every ordered set of cardinality κ is similar to a subset of the lexicographically ordered set ${}^\kappa\{0,1\}$.
30. Every ordered set is a subset of an ordered set no two different initial segments of which are similar.

13

Euclidean spaces

The problems in this section exhibit some interesting sets or interesting properties of sets in Euclidean n space or in their Hilbert space generalizations. Sometimes the set is given by an explicit construction, at other times by the transfinite enumeration technique of the preceding chapter.

1. If $\mathcal{U}$ is a family of open subsets of $\mathbf{R}^n$ that is well ordered with respect to inclusion, then $\mathcal{U}$ is countable.
2. Call a set $A \subset \mathbf{R}^n$ an algebraic variety if there is a non-identically zero polynomial $P(x_1, \ldots, x_n)$ of n variables such that A is its zero set: $A = \{(a_1, \ldots, a_n) \; : \; P(a_1, \ldots, a_n) = 0\}$. Then $\mathbf{R}^n$ cannot be covered by less than continuum many algebraic varieties.
3. There is a set $A \subset \mathbf{R}^3$ of power continuum such that if we connect the different points of A by a segment, then all these segments are disjoint (except perhaps for their endpoints).
4. From any uncountable subset of $\mathbf{R}^n$ ($n = 1, 2, \ldots$) one can select uncountably many points such that all the distances between these points are different.
5. In ℓ_2 there are continuum many points such that all distances between them are rational (hence from this set one cannot select uncountably many points such that all the distances between the selected points are different).
6. If all the distances between the points of a set $H \subset \ell_2$ are the same, then H is countable.
7. If ℓ_2 is decomposed into countably many sets, then one of them includes an infinite subset A such that all the distances between the points in A are the same.
8. There are continuum many points in ℓ_2 of which every triangle is acute.
9. The plane can be colored with countably many colors such that no two points in rational distance get the same color.

10. $\mathbf{R}^n$ can be colored with countably many colors such that no two points in rational distance get the same color.
11. The plane can be decomposed into countably many pieces none containing the three nodes of an equilateral triangle.
12. Call a set $A \subset \mathbf{R}^2$ a "circle" if there is a point $P \in \mathbf{R}^2$ such that each half-line emanating from P intersects A in one point. The plane can be written as a countable union of "circles".
13. $\mathbf{R}^3$ can be decomposed into a disjoint union of circles of radius 1.
14. $\mathbf{R}^3$ can be decomposed into a disjoint union of lines no two of which are parallel.
15. If A, B are any two intervals on the real line (of positive length), then there are disjoint decompositions $A = \bigcup\{A_i : i = 0, 1, \ldots\}$ and $B = \bigcup\{B_i : i = 0, 1, \ldots\}$ such that B_i is a translated copy of A_i.

14

Zorn's lemma

In this chapter we investigate Zorn's lemma, a powerful tool to prove results for infinite structures. Assume $(\mathcal{P}, \leq)$ is a partially ordered set. A chain $L \subseteq \mathcal{P}$ is a subset in which any two elements are comparable, i.e., for $x, y \in L$ either $x \leq y$ or $y \leq x$ holds. Zorn's lemma states that, if in a partially ordered set $(\mathcal{P}, \leq)$ every chain L has an upper bound (an element $p \in \mathcal{P}$ such that $x \leq p$ holds for $x \in L$), then $(\mathcal{P}, \leq)$ has a maximal element, that is, some element $p \in \mathcal{P}$ with the property that for no $x \in \mathcal{P}$ does $p < x$ hold.

Zorn's lemma is equivalent to the axiom of choice as well as to the well-ordering theorem (see Problem 5), in particular it is independent of the other standard axioms of set theory. Still, as is the case with the axiom of choice, in everyday mathematics it is accepted, and it provides a convenient way to establish certain maximal objects. This chapter contains ample examples for that.

1. Deduce Zorn's lemma from the well-ordering theorem.
2. Prove that Zorn's lemma implies the axiom of choice.
3. Give a direct deduction of the well-ordering theorem from Zorn's lemma.
4. Give a direct deduction of Zorn's lemma from the axiom of choice.
5. The axiom of choice, the well-ordering theorem, and Zorn's lemma are pairwise equivalent.
6. With the help of Zorn's lemma, prove the following.
   (a) The set $\mathbf{R}^+$ of positive real numbers is the disjoint union of two nonempty sets, each closed under addition.
   (b) In a ring with unity, every proper ideal can be extended to a maximal ideal.
   (c) Every filter can be extended to an ultrafilter.
   (d) Every vector space has a basis. In fact, every linearly independent system of vectors can be extended to a basis.

(e) Every vector space has a basis. In fact, every generating system of vectors includes a basis.

(f) For Abelian groups the group $D \supseteq A$ is called the *divisible hull* of A if it is divisible and for every $x \in D$ there is some natural number n that $nx \in A$. If D_1, D_2 are divisible hulls of A, then they are isomorphic over A: there is an isomorphism $\varphi : D_1 \to D_2$ which is the identity on A.

(g) Every field can be embedded into an algebraically closed field.

(h) Every algebraically closed field has a transcendence basis.

(i) Assume F is a field in which 0 is not the sum of nonzero square elements. Then F is orderable, that is, there is an ordering $<$ on F in which $x < y$ implies that $x + z < y + z$ holds for every z, and $x < y$, $z > 0$ imply that $xz < yz$.

(k) If G is an Abelian group and A is a divisible subgroup, then A is a direct summand of G.

(l) Every connected graph includes a spanning tree.

(m) If (V, X) is a graph with chromatic number κ then there is a decomposition of V into κ independent (=edgeless) sets such that between any two there goes an edge.

(n) If X is a compact topological space and $+$ is an associative operation on X which is right semi-continuous (i.e., the mapping $x \mapsto p + x$ is continuous for every $p \in X$), then $+$ has a fixed point, that is, an element $p \in X$, that $p + p = p$.

7. Let S be a set, $\mathcal{F} \subseteq \mathcal{P}(S)$ a family of subsets such that every $x \in S$ is contained in only finitely many elements of $\mathcal{F}$ and for every finite $X \subseteq S$ some $\mathcal{G} \subseteq \mathcal{F}$ constitutes an exact cover of X (i.e., every $x \in X$ is contained in one and only one element of $\mathcal{G}$). Then there is an exact cover $\mathcal{G} \subseteq \mathcal{F}$ of S.

8. (a) For any partially ordered set $(P, <)$ there is an ordered set $(P, <')$ on the same ground set that extends $(P, <)$, i.e., $x < y$ implies $x <' y$.

   (b) Prove that actually $x < y$ holds if and only if $x <' y$ for every such extension.

   (c) If, in part (a), $(\mathcal{P}, <)$ is well-founded, then $(\mathcal{P}, <')$ can be made well ordered.

   (d) Why does part (b) imply part (a) ?

9. (Alexander subbase theorem) Assume that X is a topological space with a subbase $\mathcal{S}$ with the finite cover property, i.e., if the union of some subfamily $\mathcal{S}' \subseteq \mathcal{S}$ covers X, then some finitely many members of $\mathcal{S}'$ cover X, as well. Then X is compact.

10. (Tychonoff's theorem) The topological product of compact spaces is compact.

15

Hamel bases

In this chapter we consider *Hamel bases*, i.e., bases of the vector space of the reals ($\mathbf{R}$) over the field of the rationals ($\mathbf{Q}$). To elaborate, such a basis is a set $B = \{b_i : i \in I\}$ such that every real x can be uniquely written in the form $x = \lambda_0 b_0 + \cdots + \lambda_n b_n$ where $\lambda_0, \ldots, \lambda_n$ are nonzero rationals and $b_0, \ldots, b_n$ are distinct elements of B.

Hamel bases can be used in many intriguing constructions involving the reals. This chapter lists some problems on Hamel bases, as well as on their applications.

Let us call a set $H \subset \mathbf{R}$ *rationally independent* if it is an independent set in the vector space $\mathbf{R}$ over the field $\mathbf{Q}$, and let us call H a *generating subset* if the linear hull of H (over $\mathbf{Q}$) is the whole $\mathbf{R}$.

1. If $H \subset \mathbf{R}$ is rationally independent, then there is a Hamel basis including H.
2. If $H \subset \mathbf{R}$ is a generating set, then it includes a Hamel basis.
3. Every Hamel basis has cardinality $\mathbf{c}$.
4. There are $2^{\mathbf{c}}$ distinct Hamel bases.
5. There is an everywhere-dense Hamel basis.
6. There is a nowhere-dense, measure zero Hamel basis.
7. There is a Hamel basis of full outer measure.
8. A Hamel basis, if measurable, is of measure zero.
9. A Hamel basis cannot be an analytic set.
10. If the continuum hypothesis is true, then $\mathbf{R} \setminus \{0\}$ is the union of countably many Hamel bases.
11. (Cont'd) If $\mathbf{R} \setminus \{0\}$ is the union of countably many Hamel bases, then the continuum hypothesis holds.

12. If the continuum hypothesis is true, then there is a Hamel basis $B = \{b_i : i \in I\}$ such that the set B^+ of real numbers x written in the form $x = \sum\{\lambda_i b_i : i \in I\}$ with nonnegative coefficients is a measure zero set.
13. Describe, in terms of Hamel bases, all solutions of the functional equations
    (a) $f(x+y) = f(x) + f(y)$ (additive functions, Cauchy functions);
    (b) $f(x+y) = f(x)f(y)$;
    (c) $f(xy) = f(x)f(y)$;
    (d) $f(\frac{x+y}{2}) = \frac{f(x)+f(y)}{2}$;
    (e) $f(x+y) = f(x) + f(y) + c$ with some fixed constant c;
    (f) $f(x+y) = g(x) + h(y)$;
    (g) $f(x+y) = af(x) + bf(y)$ with some fixed constants a, b.
14. If the real numbers α, β are not commensurable, then for any $A, B \in \mathbf{R}$ there is a function $f : \mathbf{R} \to \mathbf{R}$ for which $f(x+y) = f(x) + f(y)$ always holds and $f(\alpha) = A$, $f(\beta) = B$.
15. The function $F(x) = x$ (for $x \in \mathbf{R}$) is the sum of two periodic functions.
16. (Cont'd) The function $F(x) = x^2$ (for $x \in \mathbf{R}$) is the sum of three periodic functions but not of two.
17. (Cont'd) Let $k \geq 1$ be a natural number. The function $F(x) = x^k$ (for $x \in \mathbf{R}$) is the sum of $(k+1)$ periodic functions but not the sum of k periodic functions.
18. There exists $A \subset \mathbf{R}$ such that there are countably infinitely many subsets of $\mathbf{R}$ congruent to A.
19. There is a set $A \subset \mathbf{R}$ different from $\emptyset$ and $\mathbf{R}$ such that for all $x \in \mathbf{R}$ only finitely many of the sets A, $A + x$, $A + 2x$, $A + 3x, \ldots$ are different.
20. There exists a set $A \subset \mathbf{R}$ with both A, $\mathbf{R} \setminus A$ everywhere dense, which has the property that if a is a real number, then either $A \subseteq A + a$ or $A + a \subseteq A$.
21. There exists a partition of the set $\mathbf{R} \setminus \mathbf{Q}$ of irrational numbers into two sets, both closed under addition.
22. There exists a partition of the set $\mathbf{R}^+ = \{x \in \mathbf{R} : x > 0\}$ of positive real numbers into two nonempty sets, both closed under addition.
23. We are given 17 real numbers with the property that if we remove any one of them then the remaining 16 numbers can be rearranged into two 8-element groups with equal sums. Prove that the numbers are equal.
24. $\mathbf{R}$ is the union of countably many sets, none of which including a (nontrivial) 3-element arithmetic progression.
25. If a rectangle can be decomposed into the union of finitely many rectangles each having commeasurable sides, then the original rectangle also has commeasurable sides.

26. The set of reals carries an ordering $\prec$ such that there are no elements $x \prec y \prec z$, forming a 3-element arithmetic progression (that is, $y = \frac{x+z}{2}$).
27. There is an addition preserving bijection between $\mathbf{R}$ and $\mathbf{C}$.

16

The continuum hypothesis

The continuum hypothesis (CH) claims that every infinite subset of the reals is equivalent either to **N** or to **R**. It is independent of the standard axioms of set theory (see the introduction to Chapter 4), and in general it is not assumed when one deals with set theory or problems related to set theory.

Since the continuum hypothesis says something about the set of the reals, it is no wonder that it has many equivalent formulations involving real functions or sets in Euclidean spaces. This chapter lists several of these reformulations. Also, in the presence of CH the set of reals "looks differently" than otherwise, and this is reflected in the existence of sets (such as Lusin sets or Sierpinski sets) with various properties. The problems below contain several examples of this phenomenon. CH coupled with the enumeration technique of Chapter 12 is particularly powerful, for in a construction only countably many previously constructed objects have to be taken care of.

1. (Sierpinski's decomposition) CH is equivalent to the statement that the plane is the union of two sets, A and B, such that A intersects every horizontal line and B intersects every vertical line in a countable set.
2. CH holds if and only if the plane is the union of the graphs of countably many $x \mapsto y$ and $y \mapsto x$ functions.
3. CH is equivalent to the existence of a decomposition $\mathbf{R}^3 = A_1 \cup A_2 \cup A_3$ such that if L is a line in the direction of the x_i-axis then $A_i \cap L$ is finite.
4. For no natural number m exists a decomposition $\mathbf{R}^3 = A_1 \cup A_2 \cup A_3$ such that if L is a line in the direction of the x_i-axis then $|A_i \cap L| \le m$.
5. $c \le \aleph_n$ if and only if there is a decomposition $\mathbf{R}^{n+2} = A_1 \cup A_2 \cup \cdots \cup A_{n+2}$ such that if L is a line in the direction of the x_i-axis then $A_i \cap L$ is finite.
6. CH holds if and only if there is a surjection from $\mathbf{R}$ onto $\mathbf{R} \times \mathbf{R}$ of the form $x \mapsto (f_1(x), f_2(x))$ with the property that for every $x \in \mathbf{R}$ either $f_1'(x)$ or $f_2'(x)$ exists.
7. CH holds if and only if $\mathbf{R}$ is the union of an increasing chain of countable sets.

8. CH holds if and only if there is a function $f : \mathbf{R} \to \mathcal{P}(\mathbf{R})$ with $f(x)$ countable for every $x \in \mathbf{R}$ and such that $f[X] = \mathbf{R}$ holds for every uncountable set $X \subseteq \mathbf{R}$.

9. CH holds if and only if there exist functions $f_0, f_1, \ldots : \mathbf{R} \to \mathbf{R}$ such that if $a \in \mathbf{R}$ then for all but countably many $x \in \mathbf{R}$ the set $A_{x,a} = \{n < \omega : f_n(x) = a\}$ is infinite.

10. CH holds if and only if there exist functions $f_0, f_1, \ldots : \mathbf{R} \to \mathbf{R}$ such that if $\underline{a} = \{a_0, a_1, \ldots\}$ is an arbitrary real sequence then for all but countably many $x \in \mathbf{R}$ the set $A_{x,\underline{a}} = \{n < \omega : f_n(x) = a_n\}$ is infinite.

11. CH holds if and only if there exist an uncountable family $\mathcal{F}$ of real sequences with the property that if $\{a_0, a_1, \ldots\}$ is an arbitrary real sequence then for all but countably many $\{y_n\} \in \mathcal{F}$ there are infinitely many n with $y_n = a_n$.

12. CH holds if and only if there exist functions $f_0, f_1, \ldots : \mathbf{R} \to \mathbf{R}$ with the property that if $X \subseteq \mathbf{R}$ is uncountable then $f_n[X] = \mathbf{R}$ holds for all but finitely many $n < \omega$.

13. CH holds if and only if there is a family $\{A_\alpha : \alpha < \omega_1\}$ of infinite subsets of ω such that if $X \subseteq \omega$ is infinite then there is some $\alpha < \omega_1$ with $A_\alpha \setminus X$ finite.

14. CH holds if and only if there is a family $\mathcal{H} = \{A_i : i \in I\}$ of subsets of $\mathbf{R}$ with $|I| = c$, $|A_i| = \aleph_0$ such that if $B \subseteq \mathbf{R}$ is infinite then for all but countably many i we have $A_i \cap B \neq \emptyset$.

15. CH holds if and only if $\mathbf{R}$ can be decomposed as $\mathbf{R} = A \cup B$ into uncountable sets in such a way that for every real a the intersection $(A + a) \cap B$ is countable.

16. CH holds if and only if the plane can be decomposed into countably many parts none containing 4 distinct points a, b, c, and d such that $\text{dist}(a, b) = \text{dist}(c, d)$ ("dist" is the Euclidean distance).

17. CH holds if and only if $\mathbf{R}$ can be colored by countably many colors such that the equation $x + y = u + v$ has no solution with different x, y, u, v of the same color.

18. If the continuum hypothesis holds then there is a function $f : \mathbf{R} \to \mathbf{R}$ such that for every $x \in \mathbf{R}$ we have

$$\lim_{h \to 0} \max\left(f(x - h), f(x + h)\right) = \infty.$$

19. CH holds if and only if there exists an uncountable family $\mathcal{F}$ of entire functions (on the complex plane $\mathbf{C}$) such that for every $a \in \mathbf{C}$ the set $\{f(a) : f \in \mathcal{F}\}$ is countable.

20. (a) If CH holds, then there is a set A of reals of cardinality continuum such that A intersects every set of first category in a countable set (such a set is called a Lusin set).

(b) Every Lusin set is of measure zero.

21. CH is equivalent to the statement that there is a Lusin set and every subset of $\mathbf{R}$ of cardinality $< \mathbf{c}$ is of first category.

22. (a) If CH holds, then there is a set A of reals of cardinality continuum such that A intersects every set of measure zero in a countable set (such a set is called a Sierpinski set).

(b) Every Sierpinski set is of first category.

23. CH is equivalent to the statement that there is a Sierpinski set and every subset of $\mathbf{R}$ of cardinality $< \mathbf{c}$ is of measure zero.

24. If CH holds and $A \subseteq [0,1]^2$ is a measurable set of measure one, then there exist sets $B, C \subseteq [0,1]$ of outer measure one with $B \times C \subseteq A$. (Note that there is an $A \subseteq [0,1]^2$ of measure one such that if $B, C \subseteq [0,1]$ are measurable sets with $B \times C \subseteq A$, then they are of measure zero.)

25. If CH holds, then there is an uncountable set $A \subseteq \mathbf{R}$ such that if $G \supseteq \mathbf{Q}$ is an open set then $A \setminus G$ is countable (A is concentrated around $\mathbf{Q}$).

26. If CH holds, then there is an uncountable $A \subset \mathbf{R}$ such that any uncountable $B \subset A$ is dense in some open interval.

27. If CH holds, then there is an uncountable densely ordered set $\langle A, \prec \rangle$ such that any nowhere dense set (in the interval topology) in $\langle A, \prec \rangle$ is countable.

28. If CH holds, then there is an uncountable set $A \subseteq \mathbf{R}$ such that if $\varepsilon_0, \varepsilon_1, \ldots$ are arbitrary positive reals then there is a cover $I_0 \cup I_1 \cup \cdots$ of A such that I_n is an interval of length ε_n.

29. If CH holds, then there is a permutation $\pi : \mathbf{R} \to \mathbf{R}$ of the reals such that $A \subseteq \mathbf{R}$ is of first category if and only if $\pi[A]$ is of measure zero.

17

Ultrafilters on ω

If X is a ground set, then a family $\mathcal{F}$ of subsets of X is called a *filter* if

- $\emptyset \notin \mathcal{F}$,
- $A, B \in \mathcal{F}$ implies $A \cap B \in \mathcal{F}$,
- $A \in \mathcal{F}$ and $A \subseteq B$ imply $B \in \mathcal{F}$.

A filter $\mathcal{F}$ is called *principal* or *trivial* if $\mathcal{F} = \{A \subset X \ : \ A_0 \subset A\}$ for some $A_0 \subset X$.

A filter that is not a proper subset of another filter is called an *ultrafilter.*

The elements of an ultrafilter $\mathcal{F}$ can be considered as "large" subsets of X, and if the set of elements of X for which a property holds belongs to $\mathcal{F}$, then we consider the property to hold for almost all elements of X.

Ultrafilters play important roles in algebra and logic; in particular, the ultraproduct construction is based on them. They also appear in several solutions in this book.

A dual concept to filter is the concept of an *ideal.* If X is a ground set, then a family $\mathcal{I}$ of subsets of X is called an ideal if

- $X \notin \mathcal{I}$,
- $A, B \in \mathcal{I}$ implies $A \cup B \in \mathcal{I}$
- $A \in \mathcal{I}$ and $B \subseteq A$ imply $B \in \mathcal{I}$.

An ideal that is not a proper subset of another ideal is called a *prime ideal.*

It is clear that $\mathcal{F}$ is a filter (ultrafilter) if and only if $\{X \setminus F \ : \ F \in \mathcal{F}\}$ is an ideal (prime ideal).

This chapter contains various problems on, and properties of ultrafilters on the set of natural numbers. Problem 19 gives an application in analysis, it verifies the existence of Banach limits—a limit concept that extends the standard notion of limit to all real sequences.

1. A filter $\mathcal{F}$ on ω is an ultrafilter if and only if for every $A \subset \omega$ exactly one of A or $X \setminus A$ belongs to $\mathcal{F}$.

2. Every filter on ω is included in an ultrafilter.
3. There are $2^{\mathfrak{c}}$ ultrafilters on ω.
4. If $\mathcal{U}_1, \ldots, \mathcal{U}_n$ are nonprincipal ultrafilters on ω, then there is some infinite, co-infinite $A \in \mathcal{U}_1 \cap \ldots \cap \mathcal{U}_n$.
5. If $\mathcal{U}$ is an ultrafilter on ω and $0 = n_0 < n_1 < \cdots$ are arbitrary natural numbers, then there exists an $A \in \mathcal{U}$ with $A \cap [n_i, n_{i+1}) = \emptyset$ for infinitely many $i < \omega$.
6. If $\mathcal{U}$ is an ultrafilter on ω, then $\mathcal{U}$ contains a set $A \subset \omega$ of lower density zero.
7. There is an ultrafilter $\mathcal{U}$ on ω such that every $A \in \mathcal{U}$ has positive upper density.
8. Is there a translation invariant ultrafilter on ω? Is there a translation invariant ultrafilter on $\mathbf{Q}$?
9. Let $\mathcal{U}$ be a nonprincipal ultrafilter on ω. Two players consecutively say natural numbers $0 < n_0 < n_1 < \cdots$ with player I beginning. Player I wins if and only if the set $[0, n_0) \cup [n_1, n_2) \cup \cdots$ is in $\mathcal{U}$. Show that neither player has a winning strategy.
10. (CH) There is nonprincipal ultrafilter $\mathcal{U}$ on ω such that if $A_0 \supseteq A_1 \supseteq A_2 \supseteq \cdots$ are elements of $\mathcal{U}$, then there is an element B of $\mathcal{U}$ such that $B \setminus A_n$ is finite for every n. (Such an ultrafilter is called a *p*-point.)
11. (CH) There is a nonprincipal ultrafilter $\mathcal{U}$ on ω such that if $f : [\omega]^r \to \{1, 2, \ldots, n\}$ is a coloring of all r-element subsets of ω with finitely many colors, then there is a monochromatic element of $\mathcal{U}$. (Such an ultrafilter is called Ramsey ultrafilter).
12. Assume that $(A, \prec)$ is a countable ordered set and $\mathcal{U}$ is a Ramsey ultrafilter on A. Then there is an element $B \in \mathcal{U}$ which is a set of type either ω or ω^*.
13. Let $\mathcal{U}$ be a Ramsey ultrafilter on ω and let $f : \omega \to \omega$ be arbitrary. Then either f is essentially constant (i.e., $\{n < \omega : f(n) = k\} \in \mathcal{U}$ for some $k < \omega$), or f is essentially one-to-one (i.e., $f|_A$ is one-to-one on a set $A \in \mathcal{U}$).
14. Let $\mathcal{U}$ be a Ramsey ultrafilter on ω and $n_0 < n_1 < \cdots$ arbitrary numbers. Then there is a set $A \in \mathcal{U}$ with $|A \cap [n_i, n_{i+1})| = 1$ for all $i = 0, 1, \ldots$.
15. Let $\mathcal{U}$ be a Ramsey ultrafilter on ω, $\{a_n\}$ a positive sequence converging to 0 and $\epsilon > 0$ arbitrary. Then there is an $A \in \mathcal{U}$ with

$$\sum_{n \in A} a_n < \epsilon.$$

16. There are an ultrafilter $\mathcal{U}$ on ω and a positive sequence $\{a_n\}$ converging to 0, such that if $A \in \mathcal{U}$ then $\sum_{n \in A} a_n = \infty$.

17. There is an ultrafilter $\mathcal{U}$ on ω that is not generated by less than continuum many elements, i.e., if $\mathcal{F}$ is a family of subsets of $\mathcal{U}$ of cardinality smaller than continuum, then there is an element $A \in \mathcal{U}$ such that $F \not\subset A$ for $F \in \mathcal{F}$.

18. Associate with every $A \subseteq \omega$ the real number $x_A = 0, \alpha_0\alpha_1 \ldots$ where $\alpha_i = 1$ if and only if $i \in A$. This way to every subset $\mathcal{U}$ of $\mathcal{P}(\omega)$ we associate a subset $X_{\mathcal{U}}$ of $[0, 1]$. Show that if $\mathcal{U}$ is a nonprincipal ultrafilter on ω, then $X_{\mathcal{U}}$ cannot be a Lebesgue measurable set.

19. If D is a nonprincipal ultrafilter and $\{x_n : n < \omega\}$ is a sequence of reals, then set $\lim_D x_n = r$ if and only if $\{n : p < x_n < q\} \in D$ holds whenever $p < r < q$. If this is the case we say that $\{x_n\}$ has a D-limit.
    (a) Every bounded sequence has a unique D-limit.
    (b) The D-limit of a convergent sequence coincides with its ordinary limit.
    (c) $\lim_D cx_n = c \lim_D x_n$.
    (d) $\lim_D(x_n + y_n) = \lim_D x_n + \lim_D y_n$.
    (e) $|\limsup_D x_n| \le \sup_n |x_n|$.
    (f) If the sequences $\{x_n\}$ and $\{y_n\}$ have the property that $x_n - y_n \to 0$, then $\lim_D x_n = \lim_D y_n$.
    (g) If $\lim_D x_n = a$ and f is a real function continuous at the point a, then $\lim_D f(x_n) = f(a)$.
    (h) If $r \in \mathbf{R}$ is a limit point of the set $\{x_n : n < \omega\}$ then there exists a nonprincipal ultrafilter D such that $\lim_D x_n = r$.
    (i) Set $\lim_D x_n = \infty$ if and only if $\{n : p < x_n\} \in D$ holds whenever $p < \infty$, and define $\lim_D x_n = -\infty$ analogously. Then every real sequence has a (possibly infinite) D-limit.

20. Show that there is a function $f : \mathcal{P}(\mathbf{N}) \to [0, 1]$ such that $f(A) = d(A)$ whenever the set $A \subseteq \mathbf{N}$ has density $d(A)$, and f is finitely additive, i.e., $f(A \cup B) = f(A) + f(B)$ when A, B are disjoint.

21. Let there be an infinite sequence of switches, $S_0, S_1, \ldots$ each having three positions $\{0, 1, 2\}$, and a light also with three states $\{L_0, L_1, L_2\}$. They are connected in such a way that if the positions of all switches are simultaneously changed then the state of the light also changes. Let us also suppose that if all the switches are in the ith position then the light is also in the L_i state. Show that there is a (possibly principal) ultrafilter $\mathcal{U}$ that determines the state of the light in the sense that it is L_i precisely when

$$\{j : S_j \text{ is in the } i\text{th position}\} \in \mathcal{U}.$$

22. Suppose that in an election there are $n \ge 3$ candidates and a set of voters I, each of whom makes a ranking of the candidates. There are two rules for the outcome:

- if all the voters enter the same ranking, then this is the outcome,
- if a candidate a precedes candidate b in the outcome depends only on their order on the different ranking lists of the individual voters (and it does not depend on where a and b are on those lists, i.e., on how the voters ranked other candidates).

Then there is an ultrafilter $\mathcal{F}$ on I such that the outcome is an order π if and only if the set F_π of those voters $i \in I$ whose ranking is π belongs to $\mathcal{F}$. In particular, if I is finite, then in every such voting scheme there is a dictator whose ranking gives the outcome.

18

Families of sets

The problems in this chapter discuss various combinatorial properties of families of sets and functions.

1. For every cardinal $\kappa \geq \omega$ there is a family $A_{\xi,\eta}$, $\xi < \kappa$, $\eta < \kappa^+$ of subsets of κ^+ such that for fixed ξ the sets $A_{\xi,\eta}$, $\eta < \kappa^+$ are disjoint, and for each $\eta < \kappa^+$ the set $\kappa^+ \setminus \cup_{\xi<\kappa} A_{\xi,\eta}$ is of cardinality $< \kappa^+$. (Such a family is called an Ulam matrix. The matrix is of size $\kappa \times \kappa^+$, the κ^+ elements in a row are disjoint, and yet the union of the κ elements in every column is κ^+ save a set of size $< \kappa^+$).
2. For every cardinal $\kappa \geq \omega$ there is a family $\mathcal{F}$ of κ^+ *almost disjoint* subsets of κ of cardinality κ, that is, for $A, B \in \mathcal{F}$, $A \neq B$ we have $|A| = |B| = \kappa$ but $|A \cap B| < \kappa$.
3. If X is an infinite set of cardinality κ, then there are 2^κ subsets $A_\gamma \subset X$ such that if $\gamma_1 \neq \gamma_2$, then each of the sets $A_{\gamma_1} \setminus A_{\gamma_2}$, $A_{\gamma_2} \setminus A_{\gamma_1}$, and $A_{\gamma_1} \cap A_{\gamma_2}$ is of cardinality κ.
4. For every cardinal $\kappa \geq \omega$ there are κ^+ subsets of κ so that selecting any two of them, one includes the other.
5. If X is an infinite set of cardinality κ, then there is a family $\mathcal{F}$ of cardinality 2^κ of subsets of A such that no member of $\mathcal{F}$ is a proper subset of another member of $\mathcal{F}$ (such a family is called an *antichain*).
6. Let $\kappa \geq \omega$ be a cardinal. For every S, the set $[S]^\kappa$ is the union of 2^κ antichains.
7. If κ is an infinite cardinal, then there are 2^κ sets A_α, B_α, $\alpha < 2^\kappa$ of cardinality κ such that $A_\alpha \cap B_\beta \neq \emptyset$ if and only if $\alpha \neq \beta$.
8. Let κ be an infinite cardinal and A_i, B_i, $i \in I$ a family of sets with the property $|A_i|, |B_i| \leq \kappa$ and $A_i \cap B_j \neq \emptyset$ if and only if $i \neq j$. Then $|I| \leq 2^\kappa$.
9. There are two disjoint families $\mathcal{F}, \mathcal{G} \subset \mathcal{P}(\mathbf{N})$ of subsets of $\mathbf{N}$ such that every infinite subset $A \subseteq \mathbf{N}$ includes an element of $\mathcal{F}$ and of $\mathcal{G}$.

10. For any infinite set X there are two disjoint families $\mathcal{F}, \mathcal{G} \subset \mathcal{P}(X)$ of countably infinite subsets of X such that every infinite subset $A \subseteq X$ includes an element of $\mathcal{F}$ and of $\mathcal{G}$.

Call a family $\mathcal{F}$ of subsets of a set S *independent* if the following statement is true: if $X_1, \ldots, X_n$ are different members of $\mathcal{F}$, $\varepsilon_1, \ldots, \varepsilon_n < 2$, then

$$X_1^{\varepsilon_1} \cap \cdots \cap X_n^{\varepsilon_n} \neq \emptyset$$

where for a set X we put $X^1 = X$, $X^0 = S \setminus X$.

11. For every $\kappa \geq \omega$ there is an independent family of cardinality 2^κ of subsets of κ.

12. For every $\kappa \geq \omega$ there are 2^{2^κ} ultrafilters on κ.

13. Let A be an infinite set of cardinality κ. Then there is a family $\mathcal{F}$ of cardinality 2^κ of functions $f : A \to \omega$ with the property that if $f_1, \ldots, f_n \in \mathcal{F}$ are finitely many different functions from $\mathcal{F}$, then there is an $a \in A$ where the functions $f_1, \ldots, f_n$ take different values: $f_i(a) \neq f_j(a)$ if $1 \leq i < j \leq n$.

14. Let A be an infinite set of cardinality 2^κ. Then there is a family $\mathcal{G}$ of cardinality κ of functions $f_k : A \to \kappa$ such that for an arbitrary function $f : A \to \kappa$ and for an arbitrary finite set $A' \subset A$ there is a $g \in \mathcal{G}$ such that g agrees with f on A'.

15. Let κ be infinite. If $\mathcal{T}_i$, $i < 2^\kappa$ are 2^κ topological spaces each of which has a dense subset of cardinality at most κ, then the same is true of their product.

16. Let $\mathcal{F}$ be a countable family of infinite sets with $|A \cap B| = 1$ for $A, B \in \mathcal{F}$, $A \neq B$. Then there is a set X with $1 \leq |X \cap A| \leq 2$ for every $A \in \mathcal{F}$.

17. Let $\mathcal{F}$ be a countable family of infinite sets with $|A \cap B| \leq 2$ for $A, B \in \mathcal{F}$, $A \neq B$. Then there are two sets X, Y such that for every $A \in \mathcal{F}$ either $|A \cap X| = 1$ or $|A \cap Y| = 1$.

18. Prove that for every $\aleph_1 \leq \kappa < \aleph_\omega$ there is a family $\mathcal{F} \subseteq [\kappa]^{\aleph_0}$ of cardinality κ such that for every $X \in [\kappa]^{\aleph_0}$ there is some $Y \in \mathcal{F}$ with $X \subseteq Y$. Prove that no such family exists for $\kappa = \aleph_\omega$.

19. If κ, μ are infinite cardinals, then there is an almost disjoint family of μ-element sets which is not κ-colorable. That is, there is $\mathcal{H} \subseteq [V]^\mu$ for some set V with $|H \cap H'| < \mu$ for $H, H' \in \mathcal{H}$, $H \neq H'$, such that if $F : V \to \kappa$ is a coloring then some member of $\mathcal{H}$ is monocolored.

19

The Banach–Tarski paradox

This chapter deals with a surprising consequence of the axiom of choice, namely the so-called Banach–Tarski paradox claiming that any two balls (with possibly different radii) in the space can be decomposed into each other, i.e., if B_1 and B_2 are such balls then there are disjoint decompositions $B_1 = E_1 \cup \cdots \cup E_n$, $B_2 = F_1 \cup \cdots \cup F_n$ such that each E_i is congruent to F_i. Actually, any two bounded sets in $\mathbf{R}^3$ with nonempty interior can be decomposed into each other.

A "common sense" argument against such a decomposition runs as follows: take a nontrivial finitely additive and isometry invariant measure μ on all subsets of $\mathbf{R}^3$ (think of μ as a "volume" associated with each set). Then the μ-measure of B_1 is different from the μ-measure of B_2 if their radii are different, hence the aforementioned decomposition of B_1 into B_2 is impossible, since measure is preserved under isometry. Of course, this argument fails if there is no such measure, and the Banach–Tarski paradox shows precisely that such a measure does not exist in $\mathbf{R}^3$. Hidden behind the Banach–Tarski paradox is the axiom of choice appearing, for example, in the solution of Problem 17,(c).

Let us also note that in $\mathbf{R}$ and $\mathbf{R}^2$ there are finitely additive isometry invariant measures (see Chapter 28), so in $\mathbf{R}$ and $\mathbf{R}^2$ a Banach–Tarski type paradox cannot be established. The difference between $\mathbf{R}, \mathbf{R}^2$, and $\mathbf{R}^3$ (and of course every $\mathbf{R}^n$ with $n \geq 3$) is that the isometry groups of $\mathbf{R}$ and $\mathbf{R}^2$ are relatively simple, while that of $\mathbf{R}^3$ includes a free subgroup generated by two appropriate rotations.

This chapter contains various problems regarding decompositions (via different kinds of transformations on the parts) culminating in Problem 17 containing the Banach–Tarski paradox. We consider the equidecomposability of subsets of some set X, where sets are decomposed into the union of finitely many subsets and are transformed by the elements of Φ, a family of $X \to X$ bijections, containing the identity, closed under composition and taking inverse (i.e., Φ is a group with respect to composition). If $A, B \subseteq X$, then A *is equidecomposable to* B *via* Φ, in symbol $A \sim_\Phi B$, if there are partitions $A = A_1 \cup \cdots \cup A_n$, $B = B_1 \cup \cdots \cup B_n$, such that $B_i = f_i[A_i]$ for some $f_i \in \Phi$.

If there is no danger of confusion we simply write $A \sim B$ instead of $A \sim_{\Phi} B$. $A \preceq B$ if $A \sim B'$ holds for some $B' \subseteq B$. $\{A_1, \dots, A_t\}$ is a *p-cover of* A (is a $\leq$ *p-cover of* A) if $A_1, \dots, A_t \subseteq A$ and every element of A is in exactly p of the A_i's (and every element of A is in $\leq p$ of the A_i's). If $A, B \subseteq X$, then $pA \sim qB$ denotes that there is a p-cover $\{A_1, \dots, A_t\}$ of A such that for appropriate $f_1, \dots, f_t \in \Phi$, the sets $f_1[A_1], \dots, f_t[A_t]$ constitute a q-cover of B. If, on the other hand, $f_1[A_1], \dots, f_t[A_t]$ is just a $\leq q$-cover of B, then we write $pA \preceq qB$. $A \subseteq X$ is *paradoxical* if $A \sim 2A$. Usually it is "obvious" what Φ is, still, in most cases, we indicate it. If $X = \mathbf{S}^n$ (the n-dimensional unit sphere) then Φ is the set of rotations around its center; if $X = \mathbf{R}^n$, then Φ is the set of the congruences; if X is a group, then Φ is the set of left multiplications: $\Phi = \{f_x : x \in X\}$ where $f_x(y) = xy$.

1. $\sim$ is an equivalence relation.
2. If $A \preceq B$ and $B \preceq A$, then $A \sim B$.
3. If $pA \preceq qB$ and $qB \preceq rC$, then $pA \preceq rC$ holds as well (p, q, r are nonzero natural numbers).
4. If $pA \preceq qB$, $qB \preceq pA$ hold for some natural numbers p, q, then $pA \sim qB$.
5. If $pA \sim qB$ and $qB \sim rC$, then $pA \sim rC$ holds as well (p, q, r are nonzero natural numbers).
6. If $kpA \preceq kqB$ holds for some natural numbers k, p, q, $k \geq 1$, then $pA \preceq qB$. Therefore, $kpA \sim kqB$ implies $pA \sim qB$.
7. The following are equivalent.
   (a) $(n+1)A \preceq nA$ for some natural number n;
   (b) A is paradoxical;
   (c) A can be decomposed as $A = A' \cup A''$ with $A' \sim A'' \sim A$;
   (d) For every $k \geq 2$, A can be decomposed as $A = A_1 \cup \dots \cup A_k$ with $A_1 \sim A_2 \sim \dots \sim A_k \sim A$;
   (e) $pA \sim qA$ holds whenever p, q are positive natural numbers.
8. If A is paradoxical and $A \preceq B \preceq nA$ holds for some natural number n, then B is paradoxical as well.
9. (a) There exists a countable, paradoxical planar set.
   (b) There exists a *bounded* paradoxical set on the plane.
10. If $A \subseteq \mathbf{S}^2$, $|A| < \mathbf{c}$ then $\mathbf{S}^2 \sim \mathbf{S}^2 \setminus A$ (via rotations).
11. $[0, 1] \sim (0, 1]$ (with translations).
12. $Q \sim Q \setminus I$, where Q is the unit square, I is one of its (closed) sides (via translations).
13. If P is a (closed) planar polygon, F is its boundary, then $P \sim P \setminus F$ (via translations).
14. If P, Q are planar polygons, equidecomposable in the geometrical sense, then they are equidecomposable (via planar congruences).

15. Assume that $E \sim \mathbf{Z}$ holds (via translations) for some $E \subseteq \mathbf{Z}$. What is E?

16. (a) No nonempty subset of $\mathbf{Z}^n$ is paradoxical (via translations).
(b) No nonempty subset of an Abelian group is paradoxical (via multiplication by group elements).
(c) No nonempty subset of $\mathbf{R}$ is paradoxical (via congruences).

17. (a) For some $A \subseteq F_2$, natural number n, $\aleph_0 A \preceq F_2 = nA$. ($F_2$ is the free group generated by 2 elements.) Notice that this gives that A, therefore F_2 is paradoxical.
(b) There are two independent rotations around the center of $\mathbf{S}^2$.
(c) $\mathbf{S}^2$ is paradoxical (via rotations).
(d) If $A, B \subseteq \mathbf{S}^2$ both have inner points, then $A \sim B$ (via rotations).
(e) $\mathbf{B}^3$, the unit ball of $\mathbf{R}^3$ is paradoxical (via congruences).
(f) **(Banach–Tarski paradox)** If $A, B \subseteq \mathbf{R}^3$ are bounded sets with inner points, then $A \sim B$ (via congruences).

18. If $A, B \subseteq \mathbf{R}^2$ are bounded sets with inner points and $\epsilon > 0$, then A is equidecomposable into B via ϵ-contractions, that is, there are partitions $A = A_1 \cup \cdots \cup A_n$ and $B = B_1 \cup \cdots \cup B_n$ and bijections $f_i : A_i \to B_i$ such that for x, $y \in A_i$ one has $d(f_i(x), f_i(y)) \le \epsilon d(x, y)$ ($d(x, y)$ is the distance of x and y).

20

Stationary sets in ω_1

This chapter deals with two basic notions of infinite combinatorics, namely with the club (closed and unbounded) sets and with stationary sets in ω_1.

First some definitions. We say that a sequence $\{\alpha_n\}_{n=0}^{\infty}$ of ordinals from ω_1 *converges* to α if $\alpha_n \leq \alpha$ for all n and for every $\beta < \alpha$ there is an N such that $\alpha_n > \beta$ for $n > N$. Note that then necessarily $\alpha < \omega_1$. It is easy to see that this is the same as convergence in the *order topology* on ω_1 (generated by sets of the form $\{\alpha : \alpha < \beta\}$ and $\{\alpha : \alpha > \beta\}$). A subset $A \subseteq \omega_1$ is called

- *closed* if $\alpha_n \to \alpha$ and each α_n is in A then $\alpha \in A$,
- *unbounded* if given any $\beta < \omega_1$ there is a $\beta < \alpha \in A$,
- *club set* if it is closed and unbounded.

A set is closed precisely if it is closed in the order topology, and a closed set is unbounded precisely if it is not compact in this topology.

A set $S \subseteq \omega_1$ is *stationary* if it has a nonempty intersection with every club set. Otherwise, it is a *nonstationary* set.

Closed sets play the role of "full measure" sets among subsets of ω_1, while stationary sets play the role of "sets of positive measure". Club sets are very "thick", the intersection of any countable family of club sets is still a club set, while stationary sets are still sufficiently "thick" in the sense that if some property holds for the elements of a stationary set then we consider it to hold for many elements (like elements in a set of positive measure). The analogy with measure theory stops here: there is an uncountable family of disjoint stationary sets.

A function $f : A \to \omega_1$ is a *regressive function* if $f(x) < x$ holds for every $x \in A \setminus \{0\}$. The basic connection between stationary sets and regressive functions is Fodor's theorem (Problem 9): if f is regressive function on a stationary set, then it is constant on a stationary subset.

1. When is a cofinite subset of ω_1 a club?
2. Assume that $A \subseteq B \subseteq \omega_1$.

(a) Does the stationarity of A imply the stationarity of B?

(b) Does the clubness of A imply the clubness of B?

(c) Does the nonstationarity of B imply the nonstationarity of A?

3. The intersection of countably many club sets is a club set again.
4. The union of countably many nonstationary sets is nonstationary.
5. If S is stationary, C is closed, unbounded, then $S \cap C$ is stationary.
6. If C_α are club sets for $\alpha < \omega_1$, then their *diagonal intersection*

$$\nabla\{C_\alpha : \alpha < \omega_1\} = \{\alpha < \omega_1 : \beta < \alpha \longrightarrow \alpha \in C_\beta\}$$

is also a club set.

7. If $f : [\omega_1]^{<\omega} \to \omega_1$ is a function, then the set

$$C(f) = \{\alpha < \omega_1 : \text{if } \beta_1, \ldots, \beta_n < \alpha \text{ then } f(\beta_1, \ldots, \beta_n) < \alpha\}$$

is a closed, unbounded set.

8. If $C \subseteq \omega_1$ is a club set, then there is a function $f : [\omega_1]^{<\omega} \to \omega_1$ such that $C(f) \setminus \{0\} \subseteq C$.
9. A set is closed, unbounded if and only if it is the range of a strictly increasing, continuous $\omega_1 \to \omega_1$ function.
10. If $f, g : \omega_1 \to \omega_1$ are strictly increasing continuous functions, then for club many $\alpha < \omega_1$, $f(\alpha) = g(\alpha)$ holds.
11. The set of countable epsilon numbers, i.e.,

$$\{\epsilon < \omega_1 : \epsilon = \omega^\epsilon\}$$

is a club set.

12. Assume that $f : \omega_1 \to \omega_1$ is a regressive function. Then some value is assumed uncountably many times.
13. Assume $S \subseteq \omega_1$ is a stationary set and $f : S \to \omega_1$ is a regressive function. Then some value is assumed uncountably many times.
14. If $N \subseteq \omega_1$ is nonstationary, then there is a regressive function $f : N \to \omega_1$ that assumes every value countably many times.
15. If $N \subseteq \omega_1$ is nonstationary, then there is a regressive function $f : N \to \omega_1$ that assumes every value at most twice.
16. (Fodor's theorem) If $S \subseteq \omega_1$ is a stationary set and $f : S \to \omega_1$ is a regressive function, then some value is assumed on a stationary set.
17. If $S \subseteq \omega_1$ is a stationary set and $F(\alpha) \subseteq \alpha$ is a finite set for $\alpha \in S$, then for some finite set s the set $\{\alpha \in S : F(\alpha) = s\}$ is stationary.
18. A slot machine returns $\aleph_0$ quarters when a quarter is inserted. Still, no matter what strategy she follows, if somebody starts with a single coin (and plays through a transfinite series of steps), after countably many steps she loses all her money.

19. There are two disjoint stationary sets.
20. If $f : \omega_1 \to \mathbf{R}$ is monotonic, then it is constant from a point onward.
21. If $f : \omega_1 \to \mathbf{R}$ is continuous, then it is constant from a point onward.
22. ω_1, endowed with the order topology, is not metrizable.
23. (a) If $\alpha < \omega_1$, then $\alpha \times \omega_1$ is a normal topological space.
    (b) $\omega_1 \times \omega_1$ is a normal topological space.
24. $(\omega_1 + 1) \times \omega_1$ is not a normal topological space.
25. Assume that we are given $\aleph_1$ disjoint nonstationary sets. Prove that there are $\aleph_1$ of them with nonstationary union.
26. Two players, I and II, play by alternatively selecting elements of a decreasing sequence $A_0 \supseteq A_1 \supseteq \cdots$ of stationary subsets of ω_1. Player II wins if and only if $\bigcap\{A_i : i < \omega\}$ has at most one element. Show that II has a winning strategy.
27. Assume that there are $\aleph_2$ stationary sets with pairwise nonstationary intersection. Show that there are $\aleph_2$ stationary sets with pairwise countable intersection.
28. (CH) Assume that we are given $\aleph_2$ closed, unbounded subsets of ω_1. Prove that the intersection of some $\aleph_1$ of them is a closed, unbounded set.
29. If there are $\aleph_2$ functions from ω_1 into ω such that any two differ on a closed, unbounded set then there are $\aleph_2$ such functions such that any two are eventually different.
30. There exists a regressive function $f : \omega_1 \to \omega_1$ such that for every limit ordinal $\alpha < \omega_1$ there is an increasing sequence α_n, $n < \omega$, converging to α with $f(\alpha_{n+1}) = \alpha_n$ for all n.

21

Stationary sets in larger cardinals

Now we consider the analogues of questions discussed in the preceding chapter but for larger cardinals. In general, the discussion will be given in a regular cardinal (instead of ω_1), but we shall also indicate how everything works in any ordinal of cofinality larger then ω. We shall copy the treatment for ω_1 only to the extent that is necessary; several new features will emerge in the problems. For example, Problem 20 proves the deep result of Solovay: any stationary set in κ can be decomposed into κ disjoint stationary sets.

In this chapter, unless otherwise stated, κ is always an uncountable regular cardinal.

We say that a transfinite sequence $\{\alpha_\tau : \tau < \mu\}$ of elements of κ *converges* to some $\alpha < \kappa$ if $\alpha_\tau \leq \alpha$ for all $\tau < \mu$ and for every $\beta < \alpha$ there is a $\nu < \mu$ such that $\alpha_\tau > \beta$ whenever $\tau > \nu$. A set $C \subseteq \kappa$ is called

- *closed* if whenever a transfinite sequence $\{\alpha_\tau : \tau < \mu\}$ of elements of C converges to some $\alpha < \kappa$ then $\alpha \in C$,
- *unbounded* if for any $\beta < \kappa$ there is an $\alpha \in C$ with $\beta < \alpha < \kappa$,
- a *club set* if it is closed and unbounded.

It is true again that a set $C \subset \kappa$ is closed if and only if it is closed in the order topology on κ, and a closed set is unbounded precisely if it is not compact in this topology.

If something holds for every element of a club set, we sometimes use the lingo *almost everywhere*, or *for almost every*, in short, a.e.

A set $S \subseteq \kappa$ is *stationary* if it has a nonempty intersection with every closed, unbounded set. Otherwise, it is *nonstationary*. For $A \subseteq \kappa$ a function $f : A \to \kappa$ is *regressive* if $f(x) < x$ holds for every $x \in A$, $x \neq 0$.

1. The intersection of less than κ many club sets is a club set again.
2. If $C \subseteq \kappa$ is a club set, then for a.e. α the intersection $C \cap \alpha$ is a cofinal set in α of order type α

3. If $f : [\kappa]^{<\omega} \to [\kappa]^{<\kappa}$ is a function then the set

$$C(f) = \{\alpha < \kappa : \text{if } \beta_1, \ldots, \beta_n < \alpha \text{ then } f(\beta_1, \ldots, \beta_n) \subseteq \alpha\}$$

is a closed, unbounded set. In the other direction, if $C \subseteq \kappa$ is a club set then there is a function $f : \kappa \to \kappa$ such that $C(f) \setminus \{0\} \subseteq C$.

4. Let $\mathcal{A}$ be an algebraic structure on the set A of cardinality κ, with fewer than κ finitary operations, and let $\{a_\gamma : \gamma < \kappa\}$ be an enumeration of A. Then for almost all $\alpha < \kappa$ the set $\{a_\gamma : \gamma < \alpha\}$ is a substructure of $\mathcal{A}$.

5. If C_α are club sets for $\alpha < \kappa$ then their *diagonal intersection*

$$\nabla\{C_\alpha : \alpha < \kappa\} = \{\alpha < \kappa : \beta < \alpha \longrightarrow \alpha \in C_\beta\}$$

is also a club set.

6. The union of less than κ many nonstationary sets is nonstationary.
7. If S is stationary, C is closed, unbounded, then $S \cap C$ is stationary.
8. If $\mu < \kappa$ is regular, then $S = \{\alpha < \kappa : \text{cf}(\alpha) = \mu\}$ is stationary. Is it a club set? What if the condition $\text{cf}(\alpha) = \mu$ is relaxed to $\text{cf}(\alpha) \leq \mu$ or to $\text{cf}(\alpha) \geq \mu$?
9. (Fodor's theorem, pressing down lemma) If $S \subseteq \kappa$ is a stationary set and $f : S \to \kappa$ is a regressive function, then some value is assumed on a stationary set.
10. Assume that $\mu < \kappa$ is such that if $\tau < \kappa$ then $\tau^\mu < \kappa$ (for example, if $\kappa = (2^\mu)^+$). Let $S \subseteq \{\kappa : \text{cf}(\alpha) = \mu^+\}$ be a stationary set and $f(\alpha) \in [\alpha]^{\leq\mu}$ for $\alpha \in S$. Then f is constant on a stationary set.
11. If A_α $(\alpha < \kappa)$ are nonstationary, then so is $\bigcup\{A_\alpha \setminus (\alpha + 1) : \alpha < \kappa\}$.
12. Let $\{A_\alpha : \alpha < \kappa\}$ be disjoint nonstationary sets in κ. Then $A = \bigcup\{A_\alpha : \alpha < \kappa\}$ is stationary if and only if $B = \{\min(A_\alpha) : \alpha < \kappa\}$ is.
13. Out of κ disjoint nonstationary sets the union of some κ is nonstationary.
14. If A, B are subsets of κ define $A \leq B$ if $A \setminus B$ is nonstationary. Set $A < B$ if $A \leq B$ but $B \leq A$ is not true. (This gives a Boolean algebra if we identify two sets when their symmetric difference is nonstationary.) Prove that every family of at most κ sets has a least upper bound.

In Problems 15–19 we extend these notions to subsets of limit ordinals. If α is a limit ordinal, $X \subseteq \alpha$ is *unbounded* if it contains arbitrarily large elements below α. It is *closed* if it contains its limit points smaller than α. For $\text{cf}(\alpha) > \omega$, $S \subseteq \alpha$ is *stationary* if it intersects every closed, unbounded subset of α. If $\text{cf}(\alpha) = \omega$, then we declare α (and all subsets thereof) nonstationary.

15. (a) Every stationary set is unbounded.
    (b) $\text{cf}(\alpha)$ is the minimal cardinality/ordinal of the closed, unbounded sets in α.

(c) If $\mathrm{cf}(\alpha) = \omega$ then there are two disjoint closed, unbounded sets in α.
(d) If $\mathrm{cf}(\alpha) > \omega$ then the intersection of less than $\mathrm{cf}(\alpha)$ closed, unbounded sets is a closed, unbounded set.
(e) If $\mathrm{cf}(\alpha) = \omega$ then $X \subseteq \alpha$ intersects every closed, unbounded set if and only if X includes some end segment of α.

16. Assume that $\kappa = \mathrm{cf}(\alpha) > \omega$. Let $C \subseteq \alpha$ be a closed, unbounded set of order type κ with increasing enumeration $C = \{c_\gamma : \gamma < \kappa\}$.
    (a) If D is closed, unbounded in κ then $\{c_\gamma : \gamma \in D\}$ is closed, unbounded in α.
    (b) If D is closed, unbounded in α then $\{\gamma : c_\gamma \in D\}$ is closed, unbounded in κ.
    (c) $X \subseteq \alpha$ is stationary if and only if $\{\gamma : c_\gamma \in X\}$ is stationary in κ.
17. (a) If $\mathrm{cf}(\alpha) < \alpha$, then there exists a regressive $f : \alpha \setminus \{0\} \to \alpha$ such that $f^{-1}(\xi)$ is bounded for every $\xi < \alpha$.
    (b) If $S \subseteq \alpha$ is stationary, $f : S \to \alpha$ is regressive, then there is a stationary $S' \subseteq S$ such that f is bounded on S'.
18. If $C \subseteq \kappa$ is closed, unbounded, then for a.e. $\alpha < \kappa$ the set $C \cap \alpha$ is a club set in α.
19. If S, $T \subseteq \kappa$ are stationary sets, define $S < T$ if for almost every $\alpha \in T$, $S \cap \alpha$ is stationary in α. Then
    (a) $S < S$ never holds;
    (b) $<$ is transitive;
    (c) $<$ is well founded.
20. (Solovay's theorem) If $S \subseteq \kappa$ is a stationary set, then it is the union of κ disjoint stationary sets. Prove this theorem through the following steps. Assume that S is a counterexample.
    (a) Every stationary subset of S is also a counterexample.
    (b) If $f : S \to \kappa$ is regressive, then it is essentially bounded, i.e., there are an ordinal $\gamma < \kappa$ and a closed, unbounded set $C \subseteq \kappa$ such that $f(\alpha) < \gamma$ holds for $\alpha \in C \cap S$.
    (c) Almost every element of S is a regular cardinal.
    (d) There is a closed, unbounded set $D \subseteq \kappa$ such that if $\alpha \in D \cap S$ then α is an uncountable, regular cardinal and $S \cap \alpha$ is stationary in α.
    (e) Conclude by showing that no set D as in (d) exists.
21. There is a function $f : \kappa \to \kappa$ such that if $X \subseteq \kappa$ has a club subset, then $f[X] = \kappa$.
22. If $S \subseteq \kappa$ is stationary, then there is a family $\mathcal{F}$ of 2^κ stationary subsets of S such that $A \setminus B$, $B \setminus A$ are stationary if A, B are distinct elements of $\mathcal{F}$.

23. Assume that κ, μ are regular cardinals, $\kappa > \mu^+$, $\mu > \omega$. There exists a family $\{C_\alpha : \alpha < \kappa, \operatorname{cf}(\alpha) = \mu\}$ such that C_α is closed, unbounded in α and for every closed unbounded subset $E \subseteq \kappa$, there is some $C_\alpha \subseteq E$.

24. Assume that $\kappa \geq \omega_2$ is a regular cardinal. Then there exists a family $\{C_\alpha : \alpha < \kappa, \operatorname{cf}(\alpha) = \omega\}$ such that C_α is a cofinal subset of α of type ω and for every closed, unbounded subset E of κ, there is some $C_\alpha \subseteq E$.

22

Canonical functions

In this chapter for a regular uncountable cardinal κ we introduce a family of κ^+ functions that possess various canonicity properties. In some sense they are the first κ^+ functions from κ into the ordinals, this makes it possible to use them for various diverse results in set theory.

For $\kappa > \omega$ regular we construct the *canonical functions* $h_\alpha : \kappa \to \kappa$ for $\alpha < \kappa^+$ as follows. $h_0(\gamma) = 0$ for $\gamma < \kappa$. $h_{\alpha+1}(\gamma) = h_\alpha(\gamma) + 1$ $(\gamma < \kappa)$. If $\alpha < \kappa^+$ is limit with $\mu = \mathrm{cf}\,(\alpha) < \kappa$ then fix a sequence $\{\alpha_\tau : \tau < \mu\}$ converging to α and set

$$h_\alpha(\gamma) = \sup\{h_{\alpha_\tau}(\gamma) : \tau < \mu\}$$

for $\gamma < \kappa$.

Finally, if $\mathrm{cf}\,(\alpha) = \kappa$ and $\{\alpha_\tau : \tau < \kappa\}$ converges to α, then let

$$h_\alpha(\gamma) = \sup\{h_{\alpha_\tau}(\gamma) : \tau < \gamma\}.$$

Notice that the values of the functions $h_\alpha(\gamma)$ depend on the above sequences converging to α, as well.

1. Describe h_α for $\alpha \le \kappa \cdot 2$.
2. If $\beta < \alpha < \kappa^+$, then $h_\beta(\gamma) < h_\alpha(\gamma)$ holds for a.e. γ.
3. If $\{f_\alpha : \alpha < \kappa^+\}$ is a system of $\kappa \to \kappa$ functions such that for $\beta < \alpha < \kappa^+$, $f_\beta(\gamma) < f_\alpha(\gamma)$ holds for a.e. γ, then for every $\alpha < \kappa^+$, $f_\alpha(\gamma) \ge h_\alpha(\gamma)$ holds almost everywhere.
4. If $f(\gamma) < h_\alpha(\gamma)$ holds on a stationary set for some function $f : \kappa \to \kappa$, then there is a $\beta < \alpha$ such that $f(\gamma) \le h_\beta(\gamma)$ holds for stationary many γ.
5. If $f(\gamma) < h_\alpha(\gamma)$ holds on a stationary set, then $f(\gamma) = h_\beta(\gamma)$ holds on a stationary set for some $\beta < \alpha$.

6. Assume that $\{f_\alpha : \alpha < \kappa^+\}$ is a family of $\kappa \to \kappa$ functions that
   a) $f_0(\gamma) = 0$ a.e.;
   b) $f_\beta(\gamma) < f_\alpha(\gamma)$ for a.e. γ ($\beta < \alpha < \kappa^+$);
   c) if $f(\gamma) < f_\alpha(\gamma)$ for stationarily many γ then $f(\gamma) \leq f_\beta(\gamma)$ for stationarily many γ, for some $\beta < \alpha$.

   Then $f_\alpha(\gamma) = h_\alpha(\gamma)$ holds for a.e. γ.
7. For every $\alpha < \kappa^+$, $h_\alpha(\gamma) < |\gamma|^+$ holds for a.e. γ. (Here $|\gamma|^+$ is the cardinal successor of $|\gamma|$.)

   In Problems 8–13 we describe an alternative construction of canonical functions. Fix, for every $0 < \alpha < \kappa^+$, a surjection $g_\alpha : \kappa \to \alpha$. Let $f_\alpha(\gamma)$ be the order type of the set $g_\alpha[\gamma]$ (a subset of α). For $\alpha = 0$ set $f_0(\gamma) = 0$ ($\gamma < \kappa$).
8. If $g_\alpha, g'_\alpha : \kappa \to \alpha$ are surjections, then the above derived functions f_α, f'_α agree almost everywhere.
9. If $0 < \beta < \alpha < \kappa^+$ then for a.e. $\gamma < \kappa$, $g_\beta[\gamma] = g_\alpha[\gamma] \cap \beta$ holds.
10. If $\beta < \alpha$ then $f_\beta(\gamma) < f_\alpha(\gamma)$ holds a.e.
11. If $f(\gamma) < f_\alpha(\gamma)$ holds on a stationary set for some function $f : \kappa \to \kappa$, then there is a $\beta < \alpha$ such that $f(\gamma) = f_\beta(\gamma)$ holds for stationary many γ.
12. $f_\alpha(\gamma) = h_\alpha(\gamma)$ almost everywhere.
13. $f_\alpha(\gamma) < |\gamma|^+$ holds for every γ.

23

Infinite graphs

It frequently occurs in mathematics that a relation is visualized by drawing a graph. If the underlying set is infinite, then we get an infinite graph. Formally, a *graph* is a pair $G = (V, X)$ where V is a set (the *vertex set*) and $X \subseteq [V]^2$, i.e., it is a subset of the two element sets of V (the *edge set*). Sometimes we just speak of X, therefore identifying the graph with its edge set. We say that x and y are joined if $\{x, y\} \in X$. The *complement* $(V, \overline{X})$ of a graph (V, X) is $(V, [V]^2 \setminus X)$, that is, it has the same set of vertices and two vertices are joined in $(V, \overline{X})$ if and only if they are not joined in (V, X). The *degree* of a vertex v is the number of edges emanating from v.

We call (V', X') a *subgraph* of (V, X) if $V' \subseteq V$ and $X' \subseteq X$. It is an *induced subgraph* if

$$X' = \{\{x, y\} \, : \, x, y \in V', \; \{x, y\} \in X\},$$

i.e., if two elements in V' are connected precisely if they are connected in (V, X).

A subset $A \subseteq V$ is *independent* if it contains no edges: $X \cap [A]^2 = \emptyset$.

A subset $X' \subseteq X$ is a *matching* if every vertex is an endpoint of precisely one edge in X'.

A *path* in a graph is a (finite, one-way or two-way infinite) sequence $\{\ldots, v_n, v_{n+1}, \ldots\}$ of consecutively joined points (i.e., $\{v_n, v_{n+1}\} \in V$ for all n). A *circuit* is such a finite sequence with the same starting and ending point.

A *forest* is a graph with no circuits.

If (V, X), (W, Y) are graphs, a *topological* (V, X) is given by an injection $f : V \to W$ and a function g that sends every edge $e = \{x, y\}$ in X into a path in (W, Y) connecting $f(x)$ and $f(y)$, the paths $\{g(e) : e \in X\}$ being vertex disjoint except at their extremities.

A *good coloring* or sometimes a *coloring* of a graph (V, X) with a color set C is a mapping $f : V \to C$ such that $f(x) \neq f(y)$ for $\{x, y\} \in X$ (i.e., the vertices are colored in such a way that vertices that are joined get different colors). The *chromatic number* $\mathrm{Chr}(X)$ of a graph (V, X) is the smallest cardinal κ

for which the graph can be colored by κ colors. Therefore, a graph (V, X) has a good coloring with κ colors if and only if $\mathrm{Chr}(X) \leq \kappa$.

More generally, if $\mathcal{F}$ is a set system over a ground set S, then a good coloring of $\mathcal{F}$ is a coloring of S in such a way that for no $F \in \mathcal{F}$ get all points of F the same color (there is no monochromatic F).

One would expect that the chromatic number of a graph is large only if the graph includes a large complete subgraph. Problem 24 shows it otherwise: the chromatic number can be arbitrarily large even if the graph does not contain three pairwise connected points. Still, a large chromatic number does imply the existence of certain types of subgraphs, e.g., every uncountably chromatic graph must include an infinite path, all circuits of even length and all odd circuits of sufficiently large length (Problems 29, 30).

Let K_κ denote the complete graph (i.e., any two different points are joined) on a vertex set of cardinality κ. A graph (V, X) is called *bipartite* if the vertex set can be decomposed as $V = V_1 \cup V_2$ such that all edges go between V_1 and V_2 (in this case V_1 and V_2 are called the *bipartition classes*). $K_{\kappa,\lambda}$ denotes the complete bipartite graph with bipartition classes of cardinality κ and λ, respectively.

We also make the following definition. Given a class $\mathcal{F}$ of graphs, a *universal graph* in $\mathcal{F}$ is a graph $X_0 \in \mathcal{F}$ such that every graph $X \in \mathcal{F}$ is (isomorphic to) a subgraph of X_0. If $X_0 \in \mathcal{F}$ is such that every $X \in \mathcal{F}$ appears as an induced subgraph in X_0 then it is a *strongly universal graph.*

Many problems from this section are used elsewhere in the book. Problem 8 is particularly useful if one wants to deduce a conclusion for infinite sets provided one knows it for all finite subsets. It states the compactness property for graph coloring.

There are some more problems on infinite graphs in Chapter 24.

1. An infinite graph or its complement includes an infinite complete subgraph.
2. The pairs of ω are colored with $k < \omega$ colors. Then there is a partition of ω into k parts such that the ith part is a finite or one-way infinite path in color i.
3. If X is a graph on $\kappa \geq \omega$ vertices then either X or its complement includes a topological K_κ.
4. If the degree of every vertex in a graph is at most $n < \omega$, then the graph can be colored with $n + 1$ colors.
5. If the degree of every vertex in a graph is at most $\kappa \geq \omega$, then the graph can be colored with κ colors.
6. If the vertex set of a graph has a well-ordering in which every vertex is joined to fewer than κ smaller vertices, then the graph is κ-colorable.
7. Let $\kappa \geq \omega$. If the vertex set of a graph has an ordering in which every vertex is joined to fewer than κ smaller vertices, then the graph is κ-colorable.

8. (de Bruijn–Erdős theorem) If, for some $n < \omega$, every finite subgraph of a graph X is n-colorable, then so is X.

9. A graph is finitely chromatic if and only if every countable subgraph is finitely chromatic.

10. Let X be a graph on some well-ordered set. Then X is finitely chromatic if and only if every subset of order type ω is finitely chromatic.

11. Construct a graph X on ω_1^2 such that every subgraph of order type ω_1 is countably chromatic yet X is uncountably chromatic.

12. Given the graphs (V, X) and (W, Y) form their product $X \times Y$ as follows. The vertex set is $V \times W$, and $\langle x, y\rangle$ is joined to $\langle x', y'\rangle$ if and only if $\{x, x'\} \in X$ and $\{y, y'\} \in Y$. If the chromatic number of (V, X) is the finite k and the chromatic number of (W, Y) is infinite, then the chromatic number of $(V \times W, X \times Y)$ is k.

13. **(a)** If the vertices of a graph (V, X) are partitioned as $\{V_i : i \in I\}$ and X_i is the subgraph induced by V_i then $\mathrm{Chr}(X) \leq \sum \mathrm{Chr}(X_i)$.

    (b) If the edges of a graph (V, X) are decomposed into the subgraphs $\{X_i : i \in I\}$, then $\mathrm{Chr}(X) \leq \prod \mathrm{Chr}(X_i)$.

14. Assume that X is a bipartite graph with bipartition classes A and B and for every $x \in A$ the set $\Gamma(x)$ of the neighbors of x is finite. Then there is a matching of A into B in X if and only if for any finite subset $\{x_1, \ldots, x_k\}$ of A the set $\Gamma(x_1) \cup \cdots \cup \Gamma(x_k)$ has at least k elements.

15. Assume that $p, q \geq 1$ are natural numbers and X is a graph as in the preceding problem. There is a function $f : E \to \{0, 1, \ldots, p\}$ on the edge set E such that

$$\sum_{e:\ x \in e} f(e) = p \quad (x \in A),$$

$$\sum_{e:\ y \in e} f(e) \leq q \quad (y \in B)$$

if and only if the following condition holds: for any k-element finite subset $\{x_1, \ldots, x_k\}$ of A, the set $\Gamma(x_1) \cup \cdots \cup \Gamma(x_k)$ has at least pk/q elements.

16. A graph X is planar if and only if

    (a) X includes no topological K_5 or $K_{3,3}$;

    (b) X has only countably many vertices with degree at least 3;

    (c) X has at most continuum many vertices.

    (A graph is planar if it can be drawn in the plane where the vertices are represented by distinct points, the edges by noncrossing Jordan curves.)

17. A graph is spatial (it can be represented as in the previous problem but in the 3-space) if and only if it has at most continuum many vertices.

18. For an infinite cardinal κ the complete graph on κ^+ vertices is the union of κ forests but the complete graph on $(\kappa^+)^+$ vertices is not.

19. The edge set of a graph can be decomposed into countably many bipartite graphs if and only if the chromatic number of the graph is at most $\mathbf{c}$.
20. There exists a strongly universal countable graph.
21. There is no universal countable K_ω-free graph.
22. There is no universal countable locally finite graph (that is, in which every degree is finite).
23. There is no universal $K_{\aleph_1}$-free graph of cardinality $\mathbf{c}$.
24. For every infinite cardinal κ there is a κ-chromatic, triangle-free graph.
25. Define a graph (ω_1^3, X) on the set ω_1^3 in such a way that (α, β, γ) and $(\alpha', \beta', \gamma')$ are connected if and only if $\alpha < \beta < \alpha' < \gamma < \beta' < \gamma'$ or $\alpha' < \beta' < \alpha < \gamma' < \beta < \gamma$. Then $A \subseteq \omega_1^3$ spans a countable chromatic subgraph if and only if its order type (in the lexicographic ordering) is $< \omega_1^3$.
26. If (V, X) is a graph on the ordered set $(V, <)$ we define the following graph (V', X'). The vertex set is $V' = X$. We create the edges X' as follows. The edge $\{x, y\}$ with $x < y$ is joined to the edge $\{z, t\}$ with $z < t$ if and only if either $y = z$ or $x = t$ holds.
    (a) $\mathrm{Chr}(X') \leq \kappa$ if and only if $\mathrm{Chr}(X) \leq 2^\kappa$.
    (b) If (V, X) does not include odd circuits of length $3, 5, \ldots, 2n-1$ then (V', X') does not include odd circuits of length $3, 5, \ldots, 2n+1$.
    (c) For every natural number n and cardinal κ there is a graph with chromatic number greater than κ, and not including odd circuits of length $3, \ldots, 2n+1$.
27. There is an uncountably chromatic graph all whose subgraphs of cardinality at most $\mathbf{c}$ are countably chromatic.
28. If $2^{\aleph_0} = 2^{\aleph_1} = \aleph_2$, $2^{\aleph_2} = \aleph_3$, then there is a graph with chromatic number $\aleph_2$ with no induced subgraph of chromatic number $\aleph_1$.
29. Every uncountably chromatic graph includes $K_{n,\aleph_1}$ for all finite n, the complete bipartite graph with bipartition classes of size n, $\aleph_1$, respectively. In particular, it includes circuits of length $4, 6, \ldots$.
30. Every uncountably chromatic graph includes every sufficiently long odd circuit.
31. Every uncountably chromatic graph includes an infinite path.
32. Assume that X is an $\aleph_1$-chromatic graph on the vertex set V. Then V can be decomposed into the union of $\aleph_1$ disjoint subsets each spanning a subgraph of chromatic number $\aleph_1$.
33. Assume that X is an uncountably chromatic graph on the vertex set V. Then V can be decomposed into the union of two (or even $\aleph_0$) disjoint subsets each spanning a subgraph of uncountable chromatic number.

34. The following graph (V, X) is uncountably chromatic. The vertex set is

$$V = \{f : \alpha \to \omega \text{ injective}, \alpha < \omega_1\},$$

and two functions are joined if one of them extends the other.

35. If the set system $\mathcal{H}$ consists of finite sets with at least two elements and $|A \cap B| \neq 1$ holds for $A, B \in \mathcal{H}$ then $\mathcal{H}$ is 2-chromatic.

36. Assume that the set system $\mathcal{H}$ consists of countably infinite sets such that $|A \cap B| \neq 1$ holds for $A, B \in \mathcal{H}$. Then $\mathcal{H}$ is ω-chromatic but not necessarily finitely chromatic.

37. Assume that $\mathcal{H}$ is a system of $\aleph_1$ three-element sets no two intersecting in two elements. Then $\mathcal{H}$ is ω-colorable.

38. Consider the graph $G_{n,\alpha}$ with vertex set S^n (the unit sphere of $\mathbf{R}^{n+1}$) and two points are connected if their distance is bigger than α. Then $\mathrm{Chr}(G_{n,\alpha}) \geq n + 2$ for all $\alpha < 2$, and $\mathrm{Chr}(G_{n,\alpha}) = n + 2$ for $\alpha < 2$ sufficiently close to 2.

39. For $\alpha < 1/2$ let the vertices of the graph G be those measurable subsets $E \subset [0, 1]$ which have measure α, and let two such subsets be connected if they are disjoint. Then the chromatic number of G is $\aleph_0$.

24

Partition relations

In partition calculus transfinite generalizations are obtained for the (infinite) Ramsey theorem: if $2 \leq k, r < \omega$ and the r-tuples of some infinite set are colored with k colors, then there is an infinite subset, all whose r-element subsets get the same color (Problem 2).

If X is a set and $f : [X]^r \to I$ is a coloring (partition) of its r-tuples, then $Y \subseteq X$ is called *homogeneous* or *monochromatic* with respect to f if there is an $i \in I$ such that $f(\{y_1, \ldots, y_r\}) = i$ holds for all $\{y_1, \ldots, y_r\} \in [Y]^r$. We usually contract the notation $f(\{y_1, \ldots, y_r\})$ to $f(y_1, \ldots, y_r)$. The *partition relation* $\kappa \to (\lambda)^r_\rho$ expresses that if the r-tuples of a set of cardinality κ are colored with ρ colors then there is a monochromatic subset of cardinality λ (Rado's notation). If this statement fails, then we write $\kappa \not\to (\lambda)^r_\rho$. With this notation the infinite Ramsey theorem reads as $\omega \to (\omega)^r_k$ for r, k finite.

This branch of combinatorial set theory investigates how large homogeneous set can be guaranteed for a given coloring. The most important result is the Erdős–Rado theorem stating that $\exp_r(\kappa)^+ \to (\kappa^+)^{r+1}_\kappa$ holds when κ is an infinite cardinal and $1 \leq r < \omega$ (Problem 25). Here $\exp_r$ denotes the r-fold iterated exponential function, i.e., $\exp_0(\kappa) = \kappa, \exp_1(\kappa) = 2^\kappa, \exp_2(\kappa) = 2^{2^\kappa}, \ldots$, etc. These values are sharp.

In this chapter we consider this basic result and various generalizations and variants. We present applications to point set topology, some problems of this chapter will also be used elsewhere in the book.

A *tournament* is a directed graph in which between any two vertices there is an edge in one and only one direction.

1. If $2 \leq k < \omega$, then $\omega \to (\omega)^2_k$; i.e., if we color the edges of an infinite complete graph with finitely many colors, then there is an infinite monochromatic subgraph.
2. (Ramsey's theorem) If $1 \leq r < \omega$, $2 \leq k < \omega$, then $\omega \to (\omega)^r_k$. That is, if we color the r-tuples of an infinite set by finitely many colors, then there is an infinite monochromatic set.

3. Every infinite partially ordered set includes either an infinite chain or an infinite antichain (i.e., either an infinite ordered set or an infinite set of pairwise incomparable elements).
4. Every infinite ordered set includes either an infinite increasing or infinite decreasing sequence.
5. If X is an infinite planar set, then there is an infinite convex subset $Y \subseteq X$, that is, no point in Y lies in the interior of a triangle formed by three other elements of Y.
6. Every infinite tournament includes an infinite transitive subtournament.
7. If X is an infinite directed graph with at most one edge between any two vertices, then either there is an infinite independent set, or there is an infinite, transitively directed subgraph.
8. The edges of a complete directed graph of cardinality continuum can be colored by ω colors so that there are no connected edges of the same color (two edges are connected if the endpoint of one is the starting point of the other).
9. If $f : [\omega]^2 \to \omega$ is a coloring such that for every $i < \omega$ there is a finite set A_i with $f(i,j) \in A_i$ $(i < j < \omega)$, then there is an infinite set $A \subseteq \omega$ which is *endhomogeneous*, that is, in A, $f(i,j)$ only depends on i.
10. If f is a coloring of $[\omega]^2$ with no restriction on the colors, then there is an infinite $H \subseteq$ such that either
    (a) H is homogeneous for f, or
    (b) if $x < y$, $x' < y'$ are from H, then $f(x,y) = f(x',y')$ if and only if $x = x'$, or
    (c) if $x < y$, $x' < y'$ are from H, then $f(x,y) = f(x',y')$ if and only if $y = y'$, or
    (d) the values $\{f(x,y) : \{x,y\} \in [H]^2\}$ are different.
11. Let $f : \omega \to \omega$ be a function with $f(r) \to \infty$ $(r \to \infty)$. Assume that for every $1 \leq r < \infty$ H_r colors $[\omega]^r$ with finitely many colors. Then there is an infinite $X \subseteq \omega$ such that H_r on $[X]^r$ assumes at most $f(r)$ values. The statement fails if $f(r) \not\to \infty$.
12. There is a constant c with the following property. If $f : [\omega]^2 \to 3$ is a coloring, then there is an infinite sequence $a_0 < a_1 < \cdots$ with $a_n < c^n$ for infinitely many n such that f assumes only two values on this sequence.
13. If κ is an uncountable cardinal, then $\kappa \to (\kappa, \aleph_0)^2$. That is, if $f : [\kappa]^2 \to \{0,1\}$, then either there is a set of cardinality κ monochromatic in color 0 or else there is an infinite set monochromatic in color 1. Show this when κ is
    (a) regular,
    (b) singular.

14. For cardinals $\lambda \geq 2$, $\kappa \geq \omega$ order the $\kappa \to \lambda$ functions lexicographically. There is no decreasing sequence of length κ^+. There is no increasing sequence of length $\max(\kappa, \lambda)^+$.
15. If $\langle A, < \rangle$ is an ordered set, $|A| \leq 2^\kappa$, then there is some $f : [A]^2 \to \kappa$ with no $x < y < z$ such that $f(x, y) = f(y, z)$.
16. There is an uncountable tournament with no uncountable transitive subtournament.
17. (Todorcevic) There is a function $F : [\omega_1]^2 \to \omega_1$ such that for every uncountable $X \subseteq \omega_1$ F assumes every element of ω_1 on $[X]^2$.
18. If $\kappa \geq \aleph_0$ is a cardinal, $r \geq 1$ a natural number and f is a coloring of the $(r+1)$-tuples of $(2^\kappa)^+$ with κ colors, then there is a set $X \subseteq (2^\kappa)^+$, $|X| = \kappa^+$ on which f is endhomogeneous, that is, for $x_1 < \cdots < x_r < y < y'$ from X, $f(x_1, \ldots, x_r, y) = f(x_1, \ldots, x_r, y')$ holds.
19. If $\kappa \geq \aleph_0$ is a cardinal, then $(2^\kappa)^+ \to (\kappa^+)^2_\kappa$. That is, if the pairs of $(2^\kappa)^+$ are colored with κ colors, then there is a homogeneous subset of cardinality κ^+.
20. If $\kappa \geq \aleph_0$ is a cardinal, then $(2^\kappa)^+ \to \left((2^\kappa)^+, (\kappa^+)_\kappa\right)^2$. That is, if $f : (2^\kappa)^+ \to \kappa$, then either there is a homogeneous subset in color 0 of cardinality $(2^\kappa)^+$ or else there is a homogeneous subset in some color $0 < \alpha < \kappa$ of cardinality κ^+.
21. If κ is an infinite cardinal and $\{f_\alpha : \alpha < (2^\kappa)^+\}$ is a sequence of ordinal-valued functions defined on κ, then there is a pointwise increasing subsequence of cardinality $(2^\kappa)^+$, that is, there is a set $Z \subseteq (2^\kappa)^+$, $|Z| = (2^\kappa)^+$, such that $f_\alpha(\xi) \leq f_\beta(\xi)$ holds for $\alpha < \beta$, $\alpha, \beta \in Z$, $\xi < \kappa$.
22. If X is a set then $|X| \leq \mathbf{c}$ if and only if there is an "antimetric" on X, i.e., a function $d : X \times X \to [0, \infty)$ which is symmetric, $d(x, y) = 0$ exactly when $x = y$, and for distinct $x, y, z \in X$ for some permutation x', y', z' of them $d(x', z') > d(x', y') + d(y', z')$ holds.
23. $2^\kappa \not\to (\kappa^+)^2_2$. That is, if $|S| = 2^\kappa$, then there is $f : [S]^2 \to \{0, 1\}$ with no monochromatic set of size κ^+.
24. $2^\kappa \not\to (3)^2_\kappa$. That is, if $|S| = 2^\kappa$, then there is $f : [S]^2 \to \kappa$ with no monochromatic triangle.
25. (Erdős–Rado theorem) If κ is an infinite cardinal, set $\exp_0(\kappa) = \kappa$ and then by induction $\exp_{r+1}(\kappa) = 2^{\exp_r(\kappa)}$. If $\kappa \geq \aleph_0$ is a cardinal, then $\exp_r(\kappa)^+ \to (\kappa^+)^{r+1}_\kappa$. That is, if the $(r+1)$-tuples of $\exp_r(\kappa)^+$ are colored with κ colors, then there is a homogeneous subset of cardinality κ^+.
26. If κ is an infinite cardinal, $r < \omega$, there is a function $f : [\exp_r(\kappa)]^{r+1} \to \kappa$ such that if $x_0 < x_1 < \cdots < x_{r+1}$, then $f(x_0, \ldots, x_{r+1}) \neq f(x_1, \ldots, x_{r+2})$, specifically, $\exp_r(\kappa) \not\to (r+2)^{r+1}_\kappa$.

27. Let κ be an infinite cardinal, $|A| = \kappa^+$, $|B| = (\kappa^+)^+$, and k finite. If $f : A \times B \to \kappa$, then there exist $A' \subseteq A$, $B' \subseteq B$, $|A'| = |B'| = k$ such that $A' \times B'$ is monochromatic.
28. If $|A| = \aleph_1$, $|B| = \aleph_0$, k is finite, $f : A \times B \to k$, then there exist $A' \subseteq A$, $B' \subseteq B$, $|A'| = |B'| = \aleph_0$ such that $A' \times B'$ is monochromatic.
29. (Canonization) Assume λ is a strong limit singular cardinal and S, a set of cardinality λ is partitioned as $S = \bigcup\{S_\alpha : \alpha < \mu\}$ where $\mu = \mathrm{cf}\,(\lambda)$ and each S_α is of cardinality $< \lambda$. Assume that $f : [S]^2 \to \kappa$ with $\kappa < \lambda$. Then there is a set $X \subseteq S$, $|X| = \lambda$, on which f is canonical in the sense that if $x, y \in X$ then $f(x, y)$ is fully determined by α, β where α, $\beta < \mu$ are those ordinals with $x \in S_\alpha$, $y \in S_\beta$.
30. If λ is a strong limit singular cardinal with $\mathrm{cf}\,(\lambda) = \omega$, $3 \leq k < \omega$, then $\lambda \to [\lambda]^2_k$ holds, that is, if $f : [\lambda]^2 \to k$ then on some subset of cardinality λ f assumes at most two values.
31. For a set I of indices let the sets $\{A_i, B_i : i \in I\}$ be given with $|A_i|$, $|B_i| \leq \kappa$ and $A_i \cap B_j = \emptyset$ if and only if $i = j$. Then $|I| \leq 2^\kappa$.
32. If $\kappa > \omega$ is regular, then $\kappa \to (\kappa, \omega + 1)^2$. That is, if $f : [\kappa]^2 \to \{0, 1\}$, then either there is a set of order type κ monochromatic in color 0 or else there is a set of order type $\omega + 1$ monochromatic in color 1.
33. For $k < \omega$, $\omega_1 \to (\omega + 1)^2_k$. That is, if we color $[\omega_1]^2$ with k colors, then there is a monochromatic set of order type $\omega + 1$.
34. If $k < \omega$ and λ denotes the order type of the reals, then $\lambda \to (\omega + 1)^2_k$ holds. That is, if $f : [\mathbf{R}]^2 \to k$, then there is a monochromatic set of order type $\omega + 1$.
35. Assume that $\kappa > \omega$ is a cardinal for which $\kappa \to (\kappa)^2_2$ holds. Then κ is
    (a) regular,
    (b) strong limit (i.e., if $\lambda < \kappa$ then $2^\lambda < \kappa$),
    (c) not the least cardinal with (a) and (b).
36. Define, for $k < \omega$, by transfinite recursion on $\alpha < \omega_1$, the notion of semihomogeneous coloring $f : [S]^2 \to k$ for every $\langle S, < \rangle$ of order type ω^α. For $\alpha = 0$, no condition is imposed. For $\alpha = \beta + 1$, f is semihomogeneous if and only if there is a decomposition $S = S_0 \cup S_1 \cup \cdots$ with $S_0 < S_1 < \cdots$, each S_i having order type ω^β, f is semihomogeneous on every S_i, and gets the same value on all pairs between distinct S_i's. For α limit, f is semihomogeneous if and only if there is a decomposition $S = S_0 \cup S_1 \cup \cdots$ where $S_0 < S_1 < \cdots$, with S_i of order type ω^{α_i} where $\alpha_0 < \alpha_1 < \cdots$ converges to α, f is semihomogeneous on every S_i, and gets the same value on all pairs between distinct S_i's. Then given $\beta < \omega_1$, $k < \omega$, there exists $\alpha < \omega_1$, such that every semihomogeneous coloring of $[\omega^\alpha]^2$ with k colors includes a homogeneous set of type β.

37. If V, a vector space over $\mathbf{Q}$ with $|V| \geq \aleph_2$, is colored with countably many colors, then there is a monochromatic solution of $x+y=z+u$ with pairwise distinct x, y, z, u.
38. If V, a vector space over $\mathbf{Q}$ with $|V| \geq \mathbf{c}^+$ is colored with countably many colors, then there is a monochromatic solution of $x+y=z$ with x, y, z different from zero and each other. This is not true for $|V| \leq \mathbf{c}$.
39. If $\langle X, \mathcal{T} \rangle$ is a Hausdorff topological space with a dense set of cardinality κ, then $|X| \leq 2^{2^\kappa}$.
40. If $\langle X, \mathcal{T} \rangle$ is a Hausdorff topological space with $|X| > 2^{2^\kappa}$, then there is a discrete subspace of cardinality κ^+.
41. If $\langle X, \mathcal{T} \rangle$ is a hereditarily Lindelöf Hausdorff topological space, then $|X| \leq \mathbf{c}$ ("hereditarily Lindelöf" means that every open cover of any subspace includes a countable subcover).
42. If $\langle X, \mathcal{T} \rangle$ is a first countable Hausdorff topological space with no uncountable system of pairwise disjoint, nonempty open sets, then $|X| \leq \mathbf{c}$ ("first countable" means that for every point in the space there is a countable family $\{U_i\}_{i<\omega}$ of neighborhoods of x such that every neighborhood of x includes a U_i).
43. If the elements of $\mathcal{P}(\omega)$ are colored with countably many colors, then there is a monocolored nontrivial solution of $X \cup Y = Z$.
44. There is a set S such that if the elements of $\mathcal{P}(S)$ are colored with countably many colors, then there is a monocolored nontrivial solution of $X \cup Y = Z$ with X, Y disjoint.
45. For every set S there is a coloring of $\mathcal{P}(S)$ with countably many colors such that there do not exist pairwise disjoint $X_0, X_1, \ldots \subseteq S$ with all nonempty, finite subunions in the same color class.
46. For every infinite set S there is a coloring $f : [S]^{\aleph_0} \to \{0,1\}$ of the countably infinite subsets of S with two colors that admits no infinite homogeneous subset, i.e., $\kappa \not\to (\aleph_0)_2^{\aleph_0}$ holds for any κ.

25

Δ-systems

Regarding the inclusion relation the simplest possible family is a family of pairwise disjoint sets. Often, from a family of sets one would like to select a subfamily with such a simple structure, however, with pairwise disjoint sets this is not always possible. A possible remedy is the selection of a Δ-system, where $\{A_i : i \in I\}$ is called a *Δ-system* (or a Δ-family) if the pairwise intersections of the members is the same; $A_i \cap A_j = S$ for some set S (for $i \neq j$ in I). Thus, a Δ-system has a simple structure: all sets in it have a common core, and outside this common core the sets are disjoint.

In this chapter we consider the problem how large Δ-systems can be selected from a given family of sets. As an application we shall obtain in Problem 5 that in no power of **R** (regarded as a topological space) can one find an uncountable system of pairwise disjoint open sets.

1. An infinite family of n-element sets ($n < \omega$) includes an infinite Δ-subfamily.
2. An uncountable family of finite sets includes an uncountable Δ-subfamily.
3. Let $\mathcal{F}$ be a family of finite sets, $\kappa = |\mathcal{F}|$ a regular cardinal. Then $\mathcal{F}$ has a Δ-subfamily of cardinality κ. This is not true if κ is singular.
4. Is it true that every family $\mathcal{F}$ of finite sets with $|\mathcal{F}| = \aleph_1$ is the union of countably many Δ-subfamilies ?
5. Let A, B be arbitrary sets, let B be countable, and let $F(A, B)$ be the set of all functions from a finite subset of A into B. Then among uncountably many elements of $F(A, B)$ there are two which possess a common extension.
6. Consider the topological product of an arbitrary number of copies of **R**, regarded as a topological space. In this space there are no uncountably many pairwise disjoint nonempty open subsets.
7. If $\{A_\alpha : \alpha < \omega_1\}$ is a family of finite sets, then $\{A_\alpha : \alpha \in S\}$ is a Δ-subsystem for some stationary set S.

8. (a) Let $\mathcal{F}$ be a family of countable sets, $|\mathcal{F}| = \mathbf{c}^+$. Then $\mathcal{F}$ has a Δ-subfamily $\mathcal{F}' \subseteq \mathcal{F}$ with $|\mathcal{F}'| = \mathbf{c}^+$.

   (b) Let $\mathcal{F}$ be a family of sets of cardinality $\leq \mu$, with $\lambda = |\mathcal{F}|$ regular and with the property that $\kappa < \lambda$ implies $\kappa^\mu < \lambda$ (for example, $\lambda = (2^\mu)^+$). Then $\mathcal{F}$ has a Δ-subfamily of cardinality λ.

9. For μ infinite, there is a set system of cardinality 2^μ, consisting of sets of cardinality μ, with no 3-element Δ-subsystem.

10. For a set I of indices the sets $\{A_i, B_i : i \in I\}$ are given with $|A_i|$, $|B_i| \leq \mu$ and $A_i \cap B_j = \emptyset$ holds if and only if $i = j$. Then $|I| \leq 2^\mu$.

11. Assume that $\lambda > \kappa \geq \omega$ and $\mathcal{F}$ is a family of cardinality λ of sets of cardinality $< \kappa$. Then there is a subfamily $\mathcal{F}' \subseteq \mathcal{F}$ of cardinality λ such that

$$\left| \bigcup_{A \neq B \in \mathcal{F}'} (A \cap B) \right| < \lambda$$

assuming that either

(a) λ is regular or

(b) GCH holds.

26

Set mappings

In the following problems a *set mapping* is a function $f : S \to \mathcal{P}(S)$ for some set S (or, in some cases, $f : [S]^n \to \mathcal{P}(S)$ for some set S and some finite $n \geq 2$) usually with some restriction on the images. We shall always assume, even if we do not explicitly mention it, that $x \notin f(x)$ (or, in the other case, $x_1, \ldots, x_n \notin f(x_1, \ldots, x_n)$). Given a set mapping $f : S \to \mathcal{P}(S)$ a *free set* is some set $X \subseteq S$ with $x \notin f(y)$ for $x, y \in X$. (If $f : [S]^n \to \mathcal{P}(S)$ then the condition is that $y \notin f(x_1, \ldots, x_n)$ for $y, x_1, \ldots, x_n \in X$).

A basic problem for set mappings is how large free set can be guaranteed under a set mapping. In what follows we shall consider both positive and negative results on this problem.

1. Assume that $f : \mathbf{R} \to \mathcal{P}(\mathbf{R})$ is a set mapping with $x \notin \overline{f(x)}$. Then there is a free set that is
   (a) of the second category,
   (b) of cardinality continuum.
2. There is a set mapping $f : \mathbf{R} \to \mathcal{P}(\mathbf{R})$ with $f(x)$ bounded, but with no 2-element free set.
3. There is a set mapping $f : \mathbf{R} \to \mathcal{P}(\mathbf{R})$ with $|f(x)| < \mathbf{c}$ and with no 2-element free sets.
4. If $f : \mathbf{R} \to \mathcal{P}(\mathbf{R})$ is a set mapping with $f(x)$ nowhere dense, then there is always an everywhere dense free set.
5. Assume that $f : \mathbf{R} \to \mathcal{P}(\mathbf{R})$ is a set mapping such that $|f(x)| < \mathbf{c}$, $f(x)$ not everywhere dense in $\mathbf{R}$. Then there is a 2-element free set. Is there a 3-element free set?
6. Assume that $f : \mathbf{R} \to \mathcal{P}(\mathbf{R})$ is a set mapping such that $f(x)$ is always a bounded set with outer measure at most 1. Then for every finite n there is an n-element free set.

7. (CH) There is a set mapping $f : \mathbf{R} \to \mathcal{P}(\mathbf{R})$ such that for every real number $x \in \mathbf{R}$ the image $f(x)$ is a sequence converging to x, yet there is no uncountable free set.
8. Assume $\mu < \kappa$ are infinite cardinals with κ regular. Let $f : \kappa \to [\kappa]^{<\mu}$ be a set mapping. There is a free set of cardinality κ if κ is
   (a) regular (S. Piccard),
   (b) singular (A. Hajnal).
9. Assume that $f : S \to \mathcal{P}(S)$ is a set mapping with $|f(x)| \leq k$ for some natural number k. Then S is the union of at most $2k + 1$ free sets.
10. Assume that $f : S \to \mathcal{P}(S)$ is a set mapping with $|f(x)| < \mu$ for some infinite cardinal μ. Then S is the union of at most μ free sets.
11. Assume that $f : \omega_1 \to \mathcal{P}(\omega_1)$ is a set mapping such that $f(x) \cap f(y)$ is finite for $x \neq y$. Then for every $\alpha < \omega_1$ there is a free subset of type α.
12. Assume that $f : [S]^k \to [S]^{<\omega}$ is a set mapping for some set S where k is finite. If $|S| \geq \aleph_k$ then there is a free set of size $k + 1$, but this is not true if $|S| < \aleph_k$.
13. If $f : [S]^2 \to [S]^{<\omega}$ is a set mapping on a set S of cardinality $\aleph_2$, then for every $n < \omega$ there is a free set of size n.

27

Trees

In this chapter we consider the somewhat technical but important notion of tree. We start with König's lemma, whose easy yet powerful statement can be formulated as: if there will be infinitely many generations, then there is an infinite dynasty. Then we proceed to higher equivalents, that is, to Aronszajn trees and variants.

A *tree* $\langle T, \prec \rangle$ is a partially ordered set in which the set $T_{<x} = \{y : y \prec x\}$ of the elements smaller than x is well ordered for every $x \in T$. The order type $o(x)$ of $T_{<x}$ denotes *how high* the element x is in the tree: those elements with $o(x) = \alpha$ form the *αth level* T_α of T. In order to be reader-friendly, we will occasionally use the nonstandard but self-explanatory notation $T_{>x} = \{y : x \prec y\}$, $T_{<\alpha} = \bigcup\{T_\beta : \beta < \alpha\}$, $T_{>\alpha} = \bigcup\{T_\beta : \alpha < \beta\}$, etc. The *height*, $h(T)$ of T, is the least α with $T_\alpha = \emptyset$. An *α-branch* of a tree $\langle T, \prec \rangle$ is an ordered subset $b \subseteq T_{<\alpha}$ that intersects every level T_β ($\beta < \alpha$) (in exactly one point).

A tree $\langle T, \prec \rangle$ is *normal* if

(A) for every $x \in T$, $T_{>x}$ contains elements arbitrary high below $h(T)$;

(B) if $x \in T$, then there exist distinct y, y' with $x \prec y$, $x \prec y'$, $o(y) = o(y') = o(x) + 1$;

(C) if $\alpha < h(T)$ is a limit ordinal, $x \neq x' \in T_\alpha$, then $T_{<x} \neq T_{<x'}$.

If $s \prec t$, then we call t a *successor* of s, s a *predecessor* of t. If $s \prec t$ or $t \prec s$ holds, then we call s, t *comparable*. If neither $s \prec t$ nor $t \prec s$ holds, then s, t are *incomparable*. If $s \prec t$ and there are no further elements between s and t (i.e., they are on consecutive levels of the tree), then t is an *immediate successor* of s, s is an *immediate predecessor* of t.

If κ is a cardinal, a tree $\langle T, \prec \rangle$ is a *κ-tree* if $h(T) = \kappa$ and $|T_\alpha| < \kappa$ holds for every $\alpha < \kappa$.

An *Aronszajn tree* is an ω_1-tree with no ω_1-branches, and in general, a *κ-Aronszajn tree* is a κ-tree with no κ-branches. If every κ-tree has a κ-branch, that is, there are no κ-Aronszajn trees, then κ is said to have the *tree property*.

In a tree $\langle T, \prec \rangle$ a subset $A \subseteq T$ is an *antichain* if it consists of pairwise incomparable elements. An ω_1-tree is *special* if it is the union of countably many antichains.

A subset $D \subseteq T$ of a tree is *dense* if for every $x \in T$ there is a $y \in D$ with $x \preceq y$. A subset $D \subseteq T$ of a tree is *open* if $x \prec y$, $x \in D$ imply that $y \in D$.

An ω_1-tree is a *Suslin tree* if there is no ω_1-branch or uncountable antichain in it.

Squashing a tree: if $\langle T, \prec \rangle$ is a tree, then we can transform it into an ordered set as follows. Let $<_\alpha$ be an ordering on T_α. If x, y are distinct elements of T, then set $x <_{\text{lex}} y$ if and only if either $x \prec y$ or $T_{\leq x}$ is "lexicographically smaller" than $T_{\leq y}$. That is, if $T_{\leq x} = \{p_\alpha(x) : \alpha \leq o(x)\}$ where $p_\alpha(x)$ is the only element of $T_{\leq x}$ on T_α, and $T_{\leq y} = \{p_\alpha(y) : \alpha \leq o(y)\}$ is the corresponding set for y, then $p_\alpha(x) <_\alpha p_\alpha(y)$ holds for the least α where $p_\alpha(x) \neq p_\alpha(y)$. Notice that if $\langle T, \prec \rangle$ is normal then it suffices to define $<_\alpha$ on T_0 and for every element s of T on the set of immediate successors of s.

A *Specker type* is the order type of an ordered set that does not embed ω_1, ω_1^*, or an uncountable subset of the reals.

A *Countryman type* is the order type of an ordered set $\langle S, \prec \rangle$ if $S \times S$ is the union of countably many chains under the partial order $\langle x, y \rangle \preceq \langle x', y' \rangle$ if and only if $x \preceq x'$ and $y \preceq y'$.

A *Suslin line* is a nonseparable ordered set that is ccc, that is, it does not include a countable dense subset and every family of pairwise disjoint nonempty open intervals is countable.

There are two more notions of trees: in Chapter 31 what we call trees are certain trees of height ω and of course in graph theory the connected, circuitless graphs are called trees.

1. (König's lemma) ω has the tree property, that is, if every level of an infinite tree is finite, then there is an infinite branch.
2. There is a tree T of height ω, with $|T_n| = \aleph_0$ for every $n < \omega$ such that T has no infinite branch.
3. If an infinite connected graph is locally finite (every vertex has finite degree), then it includes an infinite path.
4. Suppose that $\mathcal{H}$ is an infinite set of finite 0–1 sequences closed under restriction, that is, if $a_1 \cdots a_n \in \mathcal{H}$, then $a_1 \cdots a_m \in \mathcal{H}$ holds for every $m < n$. Then there is an infinite 0–1 sequence all whose (finite) initial segments belong to $\mathcal{H}$.
5. Let A_i, $i < \omega$ be finite sets and let $f_k \in \prod_{i<k} A_i$ for $k = 0, 1, \ldots$. Then there is an $f \in \prod_{i<\omega} A_i$ such that on any finite set $S \subseteq \omega$ the function f agrees with one of the f_k's (i.e., $f|_S = f_k|_S$).
6. An infinite bounded set of reals has a limit point.
7. Given the natural numbers r, k, and s there is a natural number n such that if all r-tuples of $\{0, 1, \ldots, n-1\}$ are colored with k colors, then there is a homogeneous subset increasingly enumerated as $\{a_1, \ldots, a_p\}$ with $p \geq s$ and also with $p \geq a_1$.

8. A domino is a one-by-one square, where the four sides are colored. Given a collection D of dominoes with finitely many different color types, we want to tile the plane with them, i.e., to place a domino on each lattice point with its center on the lattice point, in a horizontal-vertical position such that the common sides of neighboring dominoes have the same color.
   (a) If for every $n < \omega$ an $n \times n$ square has a tiling from D, then so has the plane.
   (b) If the plane has a tiling from D, then it has from D', where D' is obtained from D by omitting those types that contain only finitely many pieces.
9. The vertex set of a locally finite graph can be partitioned into two sets, A and B such that if for v, a vertex, $d_A(v)$, $d_B(v)$ denote the number of vertices joined to v in A, B, respectively, then $d_A(v) \le d_B(v)$ if $v \in A$ and $d_A(v) \ge d_B(v)$ if $v \in B$.
10. (a) If $a_1 + \cdots + a_n$ is a sum of positive reals, then there are indices $0 = k(0) < k(1) < \cdots < k(r) = n$ such that $S_1 \ge \cdots \ge S_r$ holds for the subsums $S_i = a_{k(i-1)+1} + \cdots + a_{k(i)}$ and $S_1 < 2\sqrt{a_1^2 + \cdots + a_n^2}$.
    (b) If $\sum_1^\infty a_i$ is a divergent series of positive terms and $\sum a_i^2 < \infty$, then there are indices $0 = k(0) < k(1) < \cdots$ such that $S_1 \ge \cdots \ge S_r$ holds for the subsums $S_i = a_{k(i-1)+1} + \cdots + a_{k(i)}$.
11. There is an Aronszajn-tree.
12. There is a special Aronszajn-tree.
13. Every special ω_1-tree is Aronszajn.
14. If $\langle T, \prec\rangle$ is a tree, then $\langle T, <_{\text{lex}}\rangle$ is an ordered set.
15. If $\langle T, \prec\rangle$ is an Aronszajn-tree, then the order type of $\langle T, <_{\text{lex}}\rangle$ is a Specker type.
16. There exist functions $\{e_\alpha : \alpha < \omega_1\}$ such that each $e_\alpha : \alpha \to \omega$ is injective and for $\beta < \alpha$ the functions e_β and $e_\alpha|_\beta$ are identical at all but finitely many points.
17. The tree $T = \{e_\alpha|_\beta : \beta \le \alpha < \omega_1\}$ (with the functions of the previous problem) is an Aronszajn-tree, where $g \prec g'$ if and only if g' properly extends g.
18. Let e_α from Problem 16, and set $S = \{e_\alpha : \alpha < \omega_1\}$, where $\prec$ is the lexicographic ordering. Then the order type of S is a Countryman type.
19. Every Countryman type is a Specker type.
20. An ω_1-tree $\langle T, \prec\rangle$ is special if and only if there is an order preserving $f : \langle T, \prec\rangle \to \langle \mathbf{Q}, <\rangle$.
21. Assume that $\langle T, \prec\rangle$ is an ω_1-tree with a function $f : T \setminus T_0 \to T$ such that $f(t) \prec t$ and for every t and for every element $s \in T$ the set $f^{-1}(s)$ is the union of countably many antichains. Then $\langle T, \prec\rangle$ is special.

22. If a normal ω_1-tree $\langle T, \prec\rangle$ has no uncountable antichain, then it is a Suslin tree.
23. If $\langle T, \prec\rangle$ is a Suslin tree then for all but countably many $x \in T$, the set $T_{\succeq x}$ is uncountable.
24. If there is a Suslin tree, then there is a normal Suslin tree.
25. There is a Suslin tree if and only if there is a Suslin line.
26. If $\langle T, \prec\rangle$ is a Suslin tree, $D \subseteq T$ is dense, open then D is co-countable in T.
27. If $\langle T, \prec\rangle$ is a normal Suslin tree, $D_0, D_1, \ldots \subseteq T$ are dense, open sets, then $D_0 \cap D_1 \cap \cdots$ is also a dense, open set.
28. If $\langle T, \prec\rangle$ is a Suslin tree, $A \subseteq T$ is uncountable then A is somewhere dense, i.e., there is some $t \in T$ such that for every $x \succeq t$ there is $y \succeq x$, $y \in A$.
29. If $\langle T, \prec\rangle$ is a normal Suslin tree, $f : T \to \mathbf{R}$ preserves $\preceq$, then f has countable range. There is no such f that preserves $\prec$.

In Problems 30–31 we consider the topology of the tree $\langle T, \prec\rangle$ generated by the open intervals, i.e., of the sets of the form $(p, q) = \{t \in T : p \prec t \prec q\}$. This amounts to declaring $t \in T_\alpha$ isolated if $\alpha = 0$ or successor, and if α is limit then the sets of the form $(s, t]$ $(s \prec t)$ give a neighborhood base of t.

30. If $\langle T, \prec\rangle$ is a normal Suslin tree, $f : T \to \mathbf{R}$ is continuous, then f has countable range.
31. If $\langle T, \prec\rangle$ is a normal Suslin tree, then it is a normal topological space.
32. On a normal ω_1-tree $\langle T, \prec\rangle$ two players, I and II alternatively pick the successive elements of the sequence $t_0 \prec t_1 \prec \cdots$ with I choosing t_0. I wins if and only if there is an element of T above all of $t_0, t_1, \ldots$.
    (a) I has no winning strategy.
    (b) If $\langle T, \prec\rangle$ is special, II has winning strategy.
    (c) If $\langle T, \prec\rangle$ is Suslin, II has no winning strategy.
33. If κ is regular, $\lambda < \kappa$, $\langle T, \prec\rangle$ is a κ-tree with $|T_\alpha| < \lambda$ for $\alpha < \kappa$ then $\langle T, \prec\rangle$ has a κ-branch. This is not true if κ is singular.
34. If, for some regular $\kappa \geq \omega$, there is a κ-Aronszajn tree, then there is a normal one.
35. If $\langle T, \prec\rangle$ is a κ-tree for some regular cardinal κ, then the following are equivalent.
    (a) $\langle T, \prec\rangle$ has a κ-branch.
    (b) $\langle T, <_{\text{lex}}\rangle$ includes a subset of order type κ or κ^*.

36. There exists a κ^+-Aronszajn tree if $\Box_\kappa$ holds, that is, for every limit $\alpha < \kappa^+$ there is a closed, unbounded subset $C_\alpha \subseteq \alpha$ of order type $\leq \kappa$ such that if $\beta < \alpha$ is a limit point of C_α, then $C_\beta = C_\alpha \cap \beta$.
37. There exists a κ^+-Aronszajn tree if κ is regular and $2^\mu \leq \kappa$ holds for $\mu < \kappa$.
38. κ has the tree property if κ is real measurable (see Chapter 28).
39. Assume that κ is a singular cardinal such that for every $\lambda < \kappa$ there is an ultrafilter D_λ on the subsets of κ^+ such that if $A \in D_\lambda$ then $|A| = \kappa^+$ and if $A_\alpha \in D_\lambda$ $(\alpha < \lambda)$ then $\bigcap_{\alpha<\lambda} A_\alpha \in D_\lambda$. Then κ^+ has the tree property.
40. If $\kappa \to (\kappa)^2_2$ then every ordered set of cardinality κ includes either a well-ordered or a reversely well-ordered subset of cardinality κ.
41. If every ordered set of cardinality κ includes either a subset of order type κ or a subset of order type κ^*, then κ is strongly inaccessible.
42. If κ has the tree property, then κ is regular.
43. If κ is the smallest strong limit regular cardinal bigger than ω, then κ does not have the tree property.
44. For an infinite cardinal κ the following are equivalent.
    (a) $\kappa \to (\kappa)^2_2$,
    (b) $\kappa \to (\kappa)^n_\sigma$ for any $\sigma < \kappa$ and $n < \omega$,
    (c) κ is strongly inaccessible and has the tree property,
    (d) in any ordered set of cardinality κ there is either a well-ordered or a reversely well-ordered subset of cardinality κ.

28

The measure problem

It has always been an important problem to measure length, area, volume, etc. In the 19th and 20th centuries various measure and integral concepts (like Riemann and Lebesgue measures and integrals) were developed for these purposes and they have proved adequate in most situations. However, it is natural to ask what their limitations are, e.g., to what larger classes of sets can the notion of Lebesgue measure be extended by preserving its well-known properties. The standard proof for the existence of not Lebesgue measurable set in $\mathbf{R}$ (using the axiom of choice!) shows that there is no nontrivial translation invariant σ-additive measure on all subsets of $\mathbf{R}$. It was S. Banach who proved that in $\mathbf{R}$ and $\mathbf{R}^2$ there is a finitely additive nontrivial isometry invariant measure. If we go to $\mathbf{R}^3$, then the situation changes: by the Banach–Tarski paradox (Chapter 19) a ball can be decomposed into two balls of the same size; therefore, there is no nontrivial finitely additive isometry invariant measure on all subsets of $\mathbf{R}^n$ with $n \geq 3$.

In this chapter we discuss the problem when we do not care for translation invariance, but want to keep σ-additivity or some kind of higher-order additivity. Let X be an infinite set. By the phrase "μ is a measure on X" we mean a measure $\mu : \mathcal{P}(X) \to [0,1]$ on all subsets of X. Such a measure is called nontrivial if $\mu(X) = 1$ and $\mu(\{x\}) = 0$ for each $x \in X$. Since we shall only be interested in nontrivial measures, in what follows we shall always assume that the measures in question are nontrivial (hence we exclude discrete measures, which are completely additive). μ is called κ-additive if for any disjoint family Y_i, $i \in I$ of fewer than κ sets (i.e., $|I| < \kappa$) we have $\mu(\cup_{i\in I} Y_i) = \sum_{i \in I} \mu(Y_i)$. The right-hand side is defined as the supremum of its finite partial sums, and, as a consequence, on the right-hand side only countably many $\mu(Y_i)$ can be positive. Instead of ω-additivity we shall keep saying "finite additivity" and instead of ω_1-additivity we say "σ-additivity".

It turns out (see Problems 8, 9) that the first cardinal κ on which there is a σ-additive measure has also the stronger property that it carries a κ-additive measure as well. A cardinal $\kappa > \omega$ is called real measurable if there is a κ-additive $[0,1]$-valued measure on κ. It is called measurable if there is such a measure taking only the values 0 and 1. Real measurable but not measurable cardinals are at most as large as the continuum (Problem 7), but measurable cardinals are very large, their existence cannot be proven in ZFC (Zermelo–Fraenkel axiom system with the axiom of choice). On the other hand, R. Solovay proved in 1966 that if ZFC is consistent with the existence of a real measurable cardinal, then

- ZFC is consistent with the existence of a measurable cardinal,
- ZFC is consistent with $\mathbf{c}$ being real measurable,
- ZF is consistent with the statement that all subsets of $\mathbf{R}$ are Lebesgue-measurable.

In the present chapter we discuss a few properties of measurable cardinals. One of the main results in this subject is the existence of a normal ultrafilter on any measurable cardinal (Problem 14), which has the easy consequence that all measurable cardinals are weakly compact, that is, $\kappa \to (\kappa)^2_2$ holds for them. A stronger Ramsey property will be established in Problem 16.

In analogy with κ-additivity of measures let us call an ideal κ-complete if it is closed for $< \kappa$ unions and a filter κ-complete if it is closed for $< \kappa$ intersections. Recall that an ideal/filter on a ground set X is called a prime ideal/ultrafilter if for all $Y \subset X$ either Y or $X \setminus Y$ belongs to it (and this is equivalent to the maximality of the ideal/filter). A prime ideal $\mathcal{I} \subset \mathcal{P}(X)$ is called nontrivial if it contains all singletons $\{x\}$, $x \in X$, and an ultrafilter $\mathcal{F} \subset \mathcal{P}(X)$ is called nontrivial if it does not contain any of the $\{x\}$, $x \in X$.

In the problems below all measures, prime ideals, and ultrafilters will be assumed to be nontrivial.

1. On any infinite set there is a finitely additive nontrivial 0–1-valued measure.
2. Let X be an infinite set and $\kappa \geq \omega$ a cardinal. The following are equivalent:
   - there is a κ-additive 0–1-valued measure on X;
   - there is a κ-complete prime ideal on X;
   - there is a κ-complete ultrafilter on X.
3. There is no σ-additive $[0,1]$-valued measure on ω_1 (i.e., $\aleph_1$ is not real measurable).
4. If $\mathbf{R}$ is decomposed into a disjoint union of $\aleph_1$ sets of Lebesgue measure zero, then some of these sets have nonmeasurable union.
5. If κ is real measurable, then it is a regular limit cardinal.
6. If there is a $[0,1]$-valued σ-additive measure μ on $[0,1]$ then there is such a $\overline{\mu}$ extending the Lebesgue measure. Furthermore, if μ is κ-additive for some κ, then so is $\overline{\mu}$.

7. If $\kappa > \mathbf{c}$ is real measurable, then it is measurable.
8. If κ is the smallest cardinal on which there is a σ-additive $[0,1]$-valued measure, then κ is real measurable.
9. If κ is the smallest cardinal on which there is a σ-additive 0–1-valued measure, then κ is measurable.
10. There is no σ-additive 0–1-valued measure on $\mathbf{R}$.
11. If κ is measurable, then it is a strong limit regular cardinal.

If $\kappa > 0$ is a regular cardinal, then a filter $\mathcal{F}$ on κ is called a normal filter if for every $F \in \mathcal{F}$ and every $f : F \to \kappa$ regressive function f there is an $\alpha < \kappa$ such that $f^{-1}(\alpha) \in \mathcal{F}$.

12. Let κ be regular. An ultrafilter $\mathcal{F}$ on κ is normal if and only if it is closed for diagonal intersection (see Problem 21.5).
13. Let κ be regular and $\mathcal{F}$ a normal ultrafilter on κ. Then $\mathcal{F}$ is κ-complete if and only if no element of $\mathcal{F}$ is of cardinality smaller than κ.
14. If κ is measurable, then on κ there is a κ-complete normal ultrafilter. Prove this via the following outline.
    (a) Let μ be a κ-additive measure on κ, and for $f, g \in {}^{\kappa}\kappa$ set $f \equiv g$ if $f(\alpha) = g(\alpha)$ for a.e. α (i.e., the μ-measure of the set of the exceptional α is 0). Then this is an equivalence relation, and between the equivalence classes $\overline{f}$ and $\overline{g}$ of f and g set $\overline{f} \prec \overline{g}$ if $f(\alpha) < g(\alpha)$ a.e. This is a well-ordering on the set of equivalence classes ${}^{\kappa}\kappa/_{\equiv}$.
    (b) Let Y be the set of those functions $f \in {}^{\kappa}\kappa$ for which $f^{-1}(\alpha)$ is of measure 0 for all $\alpha \in \kappa$, and let $f_0 \in Y$ be such that its equivalence class is minimal in $Y/_{\equiv}$. Then $\mathcal{F} = \{F : f_0^{-1}[F] \text{ is of measure } 1\}$ is a κ-complete normal ultrafilter on κ.
15. If κ is measurable, then $\kappa \to (\kappa)^r_\sigma$ for any $r < \omega$ and $\sigma < \kappa$.
16. If κ is measurable, then $\kappa \to (\kappa)^{<\omega}_\sigma$ for any $\sigma < \kappa$, i.e., if we color the finite subsets of κ by $\sigma < \kappa$ colors then there is a set A of cardinality κ that is homogeneous in the sense that for every fixed $r < \omega$ all the r tuples of A have the same color (cardinals with the property $\kappa \to (\kappa)^{<\omega}_\sigma$ for $\sigma < \kappa$ are called Ramsey cardinals).

The following problems lead to the existence of finitely additive isometry invariant measures on all subsets of $\mathbf{R}$ and $\mathbf{R}^2$. First we deal with the case when the whole space has measure 1, and then with the case that extends Jordan measure (in this case the measure necessarily is extended-valued, i.e., it is infinite on the whole space). Such measures are called Banach measures. Note that by the Banach–Tarski paradox (see Chapter 19) in $\mathbf{R}^3$ (and in $\mathbf{R}^n$ with $n \geq 3$) there is no such measure.

The construction of finitely additive isometry invariant measures on all subsets runs parallel with the construction of additive positive linear functionals on the space of bounded functions, which is the analogue of integration. We shall also construct these so-called Banach integrals in $\mathbf{R}$ and $\mathbf{R}^2$ both in the normalized case (when the identically 1 function has integral 1) and also in the case which extends the Riemann integral. Actually, Banach measures are obtained by taking the Banach integral of characteristic functions.

Let $\mathcal{B}_A$ denote the set of all bounded real-valued functions on the set A equipped with the supremum norm $\|f\| = \sup_{a\in A}|f(a)|$. We call a function $I : \mathcal{B}_A \to \mathbf{R}$

- *linear* if for any $f_1, f_2 \in \mathcal{B}_A$, $c_1, c_2 \in \mathbf{R}$ we have $I(c_1f_1 + c_2f_2) = c_1I(f_1) + c_2I(f_2)$,
- *nontrivial* if $I(1) = 1$, where 1 denotes the identically 1 function,
- *normed* if it is nontrivial and $|I(f)| \le \|f\|$ for all $f \in \mathcal{B}_A$,
- *positive* if it is nonnegative for nonnegative functions: $I(f) \ge 0$ if $f \ge 0$.

Positivity is clearly equivalent to monotonicity: if $f \le g$, then $I(f) \le I(g)$. In what follows in statements **(a)**–**(k)** the adjective "normed" can be replaced everywhere by "positive", since a linear functional I for which $I(1) = 1$ is positive if and only if $|I(f)| \le \|f\|$.

If Φ is a family of automorphisms of A, then we say that I is *Φ-invariant* if $I(f) = I(f_\varphi)$ for all $f \in \mathcal{B}_A$ and $\varphi \in \Phi$, where $f_\varphi(x) = f(\varphi(x))$.

17. (a) There is a normed linear functional on $\mathcal{B}_{\mathbf{N}}$.

    (b) There is a translation invariant normed linear functional I on $\mathcal{B}_{\mathbf{N}}$, i.e., if $g(n) = f(n+1)$, $n \in \mathbf{N}$, then $I(f) = I(g)$ (such a functional is called a Banach limit).

    (c) There is a translation invariant normed linear functional on $\mathcal{B}_{\mathbf{Z}}$.

    (d) For any finite n there is a translation invariant normed linear functional on $\mathcal{B}_{\mathbf{Z}^n}$.

    (e) If A is an Abelian group and $s_1, \ldots, s_n \in A$ are finitely many elements, then there is a normed linear functional I on $\mathcal{B}_A$ that is invariant for translation with any s_j (i.e., if $f_j(x) = f(s_j + x)$, then $I(f_j) = I(f)$ for all $1 \le j \le n$).

    (f) If A is an Abelian group, then there is a translation invariant normed linear functional on $\mathcal{B}_A$.

    (g) If A is an Abelian group, then there is a finitely additive translation invariant measure μ on all subsets of A such that $\mu(A) = 1$. In particular, there is a finitely additive translation invariant measure μ on all subsets of $\mathbf{R}^n$ such that $\mu(\mathbf{R}^n) = 1$.

    (h) There is an isometry invariant normed linear functional on $\mathcal{B}_{\mathbf{R}}$.

(i) There is a finitely additive isometry invariant measure μ on all subsets of $\mathbf{R}$ such that $\mu(\mathbf{R}) = 1$.

(j) There is an isometry invariant normed linear functional on $\mathcal{B}_{\mathbf{R}^2}$.

(k) There is a finitely additive isometry invariant measure μ on all subsets of $\mathbf{R}^2$ such that $\mu(\mathbf{R}^2) = 1$.

In statements **(l)–(p)** we allow the measure to take infinite values, and in these statements $\mathcal{B}^b_{\mathbf{R}^n}$ denotes the set of bounded functions on $\mathbf{R}^n$ with bounded support.

(l) There is a translation invariant positive linear functional on $\mathcal{B}^b_{\mathbf{R}}$ that extends the Riemann integral.

(m) For every n there is a translation invariant positive linear functional on $\mathcal{B}^b_{\mathbf{R}^n}$ that extends the Riemann integral.

(n) There is a translation invariant finitely additive measure on all subsets of $\mathbf{R}^n$ that extends the Jordan measure.

(o) For $n = 1, 2$ there is an isometry invariant positive linear functional on $\mathcal{B}^b_{\mathbf{R}^n}$ that extends the Riemann integral (Banach integral).

(p) For $n = 1, 2$ there is a finitely additive isometry invariant measure on all subsets of $\mathbf{R}^n$ that extends the Jordan measure (Banach measure).

29

Stationary sets in $[\lambda]^{<\kappa}$

In this chapter we consider subsets of $[\lambda]^{<\kappa}$ where $\kappa > \omega$ is regular and $\lambda > \kappa$. $X \subseteq [\lambda]^{<\kappa}$ is called

- *unbounded* if for every $P \in [\lambda]^{<\kappa}$ there exists some $Q \in X$ with $P \subseteq Q$,
- *closed* if whenever $\alpha < \kappa$ and $\{P_\beta : \beta < \alpha\}$ is an increasing transfinite sequence of elements of X then $\bigcup\{P_\beta : \beta < \alpha\} \in X$,
- a *club set* when it is both closed and unbounded.

If something is true for the elements of a closed, unbounded set, then we say that it holds for *almost every* $P \in [\lambda]^{<\kappa}$ (a.e. P). Similarly, if $X \subseteq [\lambda]^{<\kappa}$, then some property holds for *almost every element of* X if there is a closed, unbounded set C such that it holds for the elements of $C \cap X$. $S \subseteq [\lambda]^{<\kappa}$ is *stationary* if it intersects every closed, unbounded set. Otherwise, it is *nonstationary.*

As we shall see these notions extend the classical notion of club sets and stationary sets. Most of the classical results from Chapters 20–21 have an analogue in this setting, and the present generalization opens space for some other questions as well.

We define $\kappa(P) = P \cap \kappa$ whenever it is $< \kappa$, i.e., when P intersects κ in an initial segment.

1. $[\lambda]^{<\kappa}$ is the union of κ bounded sets.
2. The union of $< \kappa$ bounded sets is bounded again.
3. For every $\alpha < \lambda$ the cone $\{P \in [\lambda]^{<\kappa} : \alpha \in P\}$ is a closed, unbounded set. In general, if $Q \in [\lambda]^{<\kappa}$, then $\{P \in [\lambda]^{<\kappa} : Q \subseteq P\}$ is a closed, unbounded set.
4. Every stationary set is unbounded.
5. As all ordinals, specifically all ordinals $< \kappa$, are identified with the initial segment determined by them, $\kappa \subseteq [\kappa]^{<\kappa}$ holds. A set $A \subseteq \kappa$ is stationary, (or closed, unbounded) in the sense of κ exactly if it is in the sense of $[\kappa]^{<\kappa}$.

6. $X \subseteq [\lambda]^{<\kappa}$ is closed if and only if for every directed set $Y \subseteq X$ of cardinality $< \kappa$, $\bigcup Y \in X$ holds (Y is called directed if for any $P_1, P_2 \in Y$ there is a $P \in Y$ such that $P_1 \cup P_2 \subseteq P$).
7. If $f : [\lambda]^{<\omega} \to [\lambda]^{<\kappa}$, then define $C(f) = \{P \in [\lambda]^{<\kappa} : P \text{ is closed under } f\}$.
   (a) $C(f)$ is a closed, unbounded set.
   (b) If C is a closed, unbounded set, then $C(f) \setminus \{\emptyset\} \subseteq C$ holds for an appropriate f.
8. The intersection of $< \kappa$ closed, unbounded sets is a closed, unbounded set again.
9. For a.e. P, $\kappa \cap P < \kappa$ holds (that is, P intersects the interval κ in an initial segment).
10. Given an algebraic structure with countably many operations (group, ring, etc.) on λ, a.e. $P \in [\lambda]^{<\kappa}$ is a substructure.
11. Almost every $P \in [\lambda]^{<\kappa}$ is the disjoint union of intervals of the type $[\kappa \cdot \alpha, \kappa \cdot \alpha + \beta)$ with $\beta = \kappa(P)$.
12. If $\{C_\alpha : \alpha < \lambda\}$ are closed, unbounded sets, then so is their diagonal intersection

$$\nabla_{\alpha<\lambda} C_\alpha = \{P \in [\lambda]^{<\kappa} : \alpha \in P \longrightarrow P \in C_\alpha\}.$$

13. Assume that $S \subseteq [\lambda]^{<\kappa}$ is stationary, $f(P) \in P$ holds for every $P \in S$, $P \neq \emptyset$. Then for some $\alpha < \lambda$, $f^{-1}(\alpha)$ is stationary.
14. Assume that $S \subseteq [\lambda]^{<\kappa}$ is stationary, $f(P) \in [P]^{<\omega}$ holds for every $P \in S$. Then for some s, $f^{-1}(s)$ is stationary.
15. If $X \subseteq [\lambda]^{<\kappa}$ is a nonstationary set, then there exists a function f with $f(P) \in [P]^{<\omega}$ for every $P \in X$ such that $f^{-1}(s)$ is bounded for every finite set s.
16. If $C \subseteq \kappa$ is a closed, unbounded set, then so is $\{P \in [\lambda]^{<\kappa} : \kappa(P) \in C\}$.
17. If λ is regular, $C \subseteq \lambda$ is a closed, unbounded set, then

$$A = \{P \in [\lambda]^{<\kappa} : \sup(P) \in C\}$$

    is again a closed, unbounded set.
18. If $S \subseteq \kappa$ is a stationary set, then so is $\{P \in [\lambda]^{<\kappa} : \kappa(P) \in S\}$.
19. There is a stationary set in $[\omega_2]^{<\aleph_1}$ of cardinality $\aleph_2$.
20. Every closed, unbounded set in $[\omega_2]^{<\aleph_1}$ is of maximal cardinality $\aleph_2^{\aleph_0}$.
21. Set $Z = \{P \in [\lambda]^{<\kappa} : \kappa(P) = |P|\}$. (Remember the identification of cardinals with ordinals!)
    (a) Z is stationary.
    (b) If $S \subseteq Z$ is a stationary set, then it is the disjoint union of λ stationary sets.

22. Every stationary set in $[\lambda]^{<\kappa}$ is the union of κ disjoint stationary sets. Prove this via the following steps. Let S be a counterexample.
    (a) Every stationary $S' \subseteq S$ is also a counterexample.
    (b) For almost every $P \in S$, $\kappa(P) < |P|$ holds.
    (c) Assume that $f(P) \in P$ holds for every $P \in S$, $P \neq \emptyset$. Then there is some $Q \in [\lambda]^{<\kappa}$ such that $f(P) \in Q$ holds for a. e. $P \in S$.
    (d) κ is weakly inaccessible (a regular limit cardinal).
    (e) If $S' \subseteq S$ is stationary, $f(P) \subseteq P$, $|f(P)| < \kappa(P)$ holds for $P \in S'$ then there is some $Q \in [\lambda]^{<\kappa}$ such that $f(P) \in Q$ holds for a. e. $P \in S'$.
    (f) For a. e. $P \in S$, $\kappa(P)$ is weakly inaccessible.
    (g) For a. e. $P \in S$, $S \cap [P]^{<\kappa(P)}$ is stationary in $[P]^{<\kappa(P)}$.
    (h) Get the desired contradiction.
23. (GCH) Set $\lambda = \aleph_\omega$, $\kappa = \aleph_2$. There is a stationary set $S \subseteq [\lambda]^{<\kappa}$ such that every unbounded subset of S is stationary.
24. For any nonempty set A call $S \subseteq \mathcal{P}(A)$ A-stationary if for every function $f : [A]^{<\omega} \to [A]^{\leq\aleph_0}$ there is some $B \in S$, $B \neq \emptyset$ which is closed under f.
    (a) $S = \{A\}$ is A-stationary on A.
    (b) If S is A-stationary on A, then $A = \bigcup S$.
    (c) If $A = \lambda \geq \omega_1$ is a cardinal, $S \subseteq [\lambda]^{<\aleph_1}$ then S is λ-stationary on λ if and only if it is stationary.
    (d) If S is A-stationary, $\emptyset \neq B \subseteq A$, then $T = \{P \cap B : P \in S\}$ is B-stationary.
    (e) If S is A-stationary, $B \supseteq A$, then $T = \{P \subseteq B : P \cap A \in S\}$ is B-stationary.
    (f) If S is A-stationary, $F(P) \in P$ holds for every $P \in S$, $P \neq \emptyset$, then for some x, the set $F^{-1}(x)$ is A-stationary.

30

The axiom of choice

In this chapter we do not assume the axiom of choice.

We now enter a strange and interesting world. Strange, as our everyday tools cannot be used; we no longer have the trivial rule for addition and multiplication of two cardinals, and as some sets may not be well orderable, we cannot always apply transfinite induction or recursion. Interesting, as we are still able to prove some statements similar to the corresponding statements under the axiom of choice, only it requires delicate arguments, and in some cases we discover phenomena that can only hold if AC fails.

We can use the notion of a cardinal, in the naive sense, that is, without the von Neumann identification of cardinals with ordinals. That is, we can speak of the equality, sum, etc., of two cardinals.

AC_ω is the axiom of choice for countably many nonempty sets.

1. For no cardinal κ does $2^\kappa = \aleph_0$ hold.
2. If φ is an ordinal, then there is a sequence $\langle f_\alpha : \omega \leq \alpha < \varphi \rangle$ such that $f_\alpha : \alpha \times \alpha \to \alpha$ is an injection.
3. If $0 < \alpha < \omega_2$, then there is a surjection $\mathbf{R} \to \alpha$.
4. There is a mapping from the set of reals onto a set of cardinality *greater* than continuum if either
   (a) every uncountable set of reals has a perfect subset, or
   (b) every set of reals is measurable, or else
   (c) (AC_ω) there are no two disjoint stationary subsets of ω_1.
5. Let C_n denote the axiom of choice for n-element sets. Then C_m implies C_n if m is a multiple of n.
6. C_2 implies C_4.
7. C_2 and C_3 imply C_6.
8. If every set carries an ordering then $\mathrm{C}_{<\omega}$ (the axiom of choice for families of finite sets) holds.

9. Let κ, λ be cardinals, n a natural number, and assume that $\kappa+n = \lambda+n$ holds. Then $\kappa = \lambda$.
10. If $\kappa \geq \aleph_0$, then $\kappa + \aleph_0 = \kappa$.
11. If $\kappa > 1$, then $\kappa + 1 < 2^\kappa$.
12. If $\kappa \geq \aleph_0$, then $\kappa + 2^\kappa = 2^\kappa$.
13. Set $\kappa \ll \lambda$ if and only if $\kappa + \lambda = \lambda$. This $\ll$ is transitive. Furthermore, $\kappa \ll \lambda$ holds if and only if $\aleph_0\kappa \leq \lambda$.
14. If κ is of the form either $\kappa = \aleph_0\lambda$ for some cardinal λ or $\kappa = 2^\lambda$ for some cardinal $\lambda \geq \aleph_0$, then $\kappa + \kappa = \kappa$.
15. If a, b are cardinals and $2a = 2b$, then $a = b$.
16. If κ is an infinite cardinal then $\aleph_0 \leq 2^{2^\kappa}$.
17. $\aleph_1 \leq 2^{2^{\aleph_0}}$.
18. $\kappa \cdot \kappa \leq 2^{2^\kappa}$ holds for every cardinal κ.
19. (Hartogs' lemma) If κ is a cardinal then there is an ordinal $H(\kappa)$ with $|H(\kappa)| \leq 2^{2^{2^\kappa}}$ such that $|H(\kappa)| \not\leq \kappa$.
20. If $\kappa^2 = \kappa$ holds for every infinite cardinal κ then the axiom of choice is true.
21. The generalized continuum hypothesis implies the axiom of choice. That is, if for no infinite κ exists a cardinal λ with $\kappa < \lambda < 2^\kappa$ then the AC holds.
22. AC is implied by the following statement: if $\{A_i : i \in I\}$ is a set of nonempty sets, then there is a function that selects a nonempty finite subset of each.
23. If every vector space has a basis, then the axiom of choice holds.

In the following problem, the *chromatic number* of graph $G = (V, E)$ is the minimal cardinality (if it exists) of the form $|A|$ for which there is a surjection $f: V \to A$ which is a good coloring, i.e., if x, $y \in V$ are joined, then $f(x) \neq f(y)$.

24. The axiom of choice is equivalent to the statement that every graph has a chromatic number.
25. Hajnal's set mapping theorem (Problem 26.8) implies the axiom of choice.
26. If $\mathbf{R}$ is the union of countably many countable sets, then so is ω_1 and $\mathrm{cf}\,(\omega_1) = \omega$.
27. ω_2 is not the union of countably many countable sets.

31

Well-founded sets and the axiom of foundation

In this chapter we investigate well-founded sets. These are partially ordered sets where every nonempty subset has a least element (one with no predecessor in the subset). These sets share many properties with the well-ordered sets. We can, therefore, use some techniques developed for well-ordered sets, as transfinite induction. In applications, e.g., in descriptive set theory, important facts can be transformed into the existence (or nonexistence) of an infinite decreasing chain in some specific partially ordered sets, which we call trees. That these two properties are equivalent for any given partially ordered set follows from the axiom of dependent choice (a weakening of the axiom of choice), which says that if A is a nonempty set, R is a binary relation on A with the property that for every element $x \in A$ there is some $y \in A$ such that $R(x, y)$ holds, then there is an infinite sequence $x_0, x_1, \ldots$ of elements of A such that $R(x_0, x_1), R(x_1, x_2), \ldots$ hold.

The axiom of foundation (or regularity) says that if A is a nonempty set, then there is some element x of it with $x \cap A = \emptyset$. This claims that the universe is well founded under $\in$ and that implies that it is possible to create every set from the empty set by iterating the power set operation (cumulative hierarchy).

In this chapter, we assume the axioms of choice and regularity, unless indicated otherwise.

A class is a defined part of the universe which is not necessarily a set. If a class is indeed not a set, then we call it a proper class. An operation is a well-defined mapping on some part of the universe which is possibly not a function, that is, it does not necessarily go between sets.

1. The following statements are equivalent:
   (a) DC, the axiom of dependent choice;
   (b) If the nonempty partially ordered set $\langle P, < \rangle$ has no minimal element, then there is an infinite descending chain in $\langle P, < \rangle$,

(c) A partially ordered set is well founded iff there is no infinite descending chain in it.

2. If $\langle P, <\rangle$ is a partially ordered set, then there is an order-preserving ordinal-valued function f on P, that is, $x < y$ implies $f(x) < f(y)$ if and only if $\langle P, <\rangle$ is well founded.
3. If $\langle P, <\rangle$ is a partially ordered set, then there exists a cofinal subset $Q \subseteq P$ such that $\langle Q, <\rangle$ is well founded.
4. Let $\langle P, <\rangle$ be a partially ordered set that does not include an infinite increasing or decreasing sequence. Is it true that P is the union of countably many antichains (an antichain is a set of pairwise incomparable elements)?
5. If $\langle P, <\rangle$ is a well-founded set, then there is a unique ordinal-valued function r (the rank function of $\langle P, <\rangle$) with the properties

   (a) if $x < y$, then $r(x) < r(y)$,

   (b) if $\alpha = r(x)$ and $\beta < \alpha$, then there exists some $y < x$ with $r(y) = \beta$.

For κ a cardinal let $\text{FS}(\kappa)$ be the set of all finite strings of ordinals less than κ. We think the elements of $\text{FS}(\kappa)$ as finite functions from n to κ for some $n < \omega$ and simply write $s = s(0)s(1)\cdots s(n-1)$ (rather than using e.g., the ordered sequence notation). If $s, t \in \text{FS}(\kappa)$ we set $s < t$ if t properly extends s, and $s \lhd t$ if t is a one-step extension of s. $s\hat{\ }t$ is the *juxtaposition* of s and t; that is, if $s = s(0)s(1)\cdots s(n-1)$ $t = t(0)t(1)\cdots t(m-1)$, then $s\hat{\ }t = s(0)s(1)\cdots s(n-1)t(0)t(1)\cdots t(m-1)$.

For Problems 6–10 we define a set $T \subseteq \text{FS}(\kappa)$ a *tree* if it is closed under restriction, i.e., $s < t \in T$ implies that $s \in T$. The nth level of T is formed by those elements of length n. T is well founded if it does not include an infinite branch, that is, if $(T, >)$ is well founded in the original sense. In this case, let $R(T)$ be the ordinal assigned to the root (the empty sequence) by Problem 5. (Notice that these trees are trees in the sense of Chapter 27, only turned upside down.)

6. If $T \subseteq \text{FS}(\kappa)$ is a well-founded tree, then $R(T) < \kappa^+$. For every ordinal $\alpha < \kappa^+$ there is a well-founded tree $T \subseteq \text{FS}(\kappa)$ with $R(T) = \alpha$.
7. If T, T' are well-founded trees and $R(T) \leq R(T')$ then $T \preceq T'$, i.e., there is a level and extension preserving (but not necessaily one–one) map from T into T'.
8. For any two trees, T and T' either $T \preceq T'$ or $T' \preceq T$ holds.
9. Define the Kleene–Brouwer ordering $<_{\text{KB}}$ on $\text{FS}(\kappa)$ as follows. If $s = s(0)s(1)\cdots s(n)$ and $t = t(0)\cdots t(m)$, then $s <_{\text{KB}} t$ if and only if either s properly extends t or $s(i) < t(i)$ holds for the least i where they differ. This is an ordering on $\text{FS}(\kappa)$. A tree $T \subseteq \text{FS}(\kappa)$ is well founded if and only if it is well ordered by $<_{\text{KB}}$.

10. (Galvin's tree game) Two players, W and B, play the following game. They play on the isomorphic well-founded trees, T_W and T_B. At the beginning both players have a pawn at the root of his/her own tree. At every round first W makes a move with either pawn, i.e., moves it to one of the immediate extensions of its current position, then B does the same with one of the pawns. B may pass but W may not. The winner is whose pawn first reaches a leaf (that is, queens).
    (a) One of the players has a winning strategy.
    (b) W has a winning strategy.
11. Exhibit two well-founded sets such that neither has an order-preserving (not necessarily injective) mapping into the other.
    A set (or possibly a class) A is *transitive* if $x \in A$, $y \in x$ imply that $y \in A$.
12. There is no set x with $x \in x$.
13. There are no sets x, y with $x \in y$ and $y \in x$.
14. For every natural number n, there is an n-element set A with the following properties: if $x, y \in A$, then either $x \in y$, or $x = y$, or $y \in x$, and if $x \in A$, $y \in x$, then $y \in A$. For a given n, can there be more than one such sets?
15. What are the transitive singletons?
16. The intersection and union of transitive sets are transitive.
17. Let A be a set. Define $A_0 = \{A\}$, $A_{n+1} = \bigcup A_n$ for $n = 0, 1, \ldots$, $\mathrm{TC}(A) = A_0 \cup A_1 \cup \cdots$ (the transitive closure of A). $\mathrm{TC}(A)$ is transitive and if $A \in B$, B is transitive, then $\mathrm{TC}(A) \subseteq B$.
18. (Cumulative hierarchy) Construct, by transfinite recursion, the following sets. $V_0 = \emptyset$. $V_{\alpha+1} = \mathcal{P}(V_\alpha)$. If α is a limit ordinal, then $V_\alpha = \bigcup\{V_\beta : \beta < \alpha\}$.
    If a set x is an element of some V_α then x is a *ranked set*, and $\mathrm{rk}(x)$ (the rank of x) is the least α with $x \in V_\alpha$.
    (a) Every V_α is a transitive set.
    (b) $V_\beta \subseteq V_\alpha$ holds for $\beta < \alpha$.
    (c) $\mathrm{rk}(x)$ is always a successor ordinal.
    (d) If x is ranked and $y \in x$, then y is also ranked and $\mathrm{rk}(y) < \mathrm{rk}(x)$.
    (e) If every element of x is ranked, then so is x.
    (f) The axiom of foundation holds if and only if every set is ranked.
19. Solve the equation $X \times Y = X$ in sets X, Y.
20. If C is a proper class, then there is a surjection from C onto the class of ordinals such that the inverse image of every ordinal is a
    (a) set,
    (b) proper class.

21. Assume that $\mathcal{C}$ is a class, $\sim$ is an equivalence relation on it. Then there is an operation $\mathcal{F}$ defined on $\mathcal{C}$ such that $\mathcal{F}(x) = \mathcal{F}(y)$ holds iff $x \sim y$ is true.

22. The axiom of choice is equivalent to the statement that every set can be embedded into every proper class.

23. The following are equivalent.

    (a) (The axiom of global choice) There is an operation $\mathcal{F}$ defined on all nonempty sets, such that $\mathcal{F}(X) \in X$ holds for every such set X.

    (b) The universe has a well-ordering, that is, a relation $<$ such that every nonempty class has a $<$-least minimal element.

    (c) Moreover, $<$ is set-like, that is, the predecessors of every set form a set.

    (d) If $\mathcal{A}$, $\mathcal{B}$ are proper classes, then there is an injection of $\mathcal{A}$ into $\mathcal{B}$.

    (e) If $\mathcal{A}$, $\mathcal{B}$ are proper classes, then there is a bijection between $\mathcal{A}$ and $\mathcal{B}$.

24. If κ is an infinite cardinal, then $H_\kappa = \{x : |\mathrm{TC}(x)| < \kappa\}$ is a set (here $\mathrm{TC}(x)$ is the transitive closure of x; see Problem 17).

25. (Mostowski's collapsing lemma) Assume that M is a class, E is a binary relation on M which is

    (a) irreflexive, that is, xEx holds for no $x \in M$;

    (b) extensional: if $\{z : zEx\} = \{z : zEy\}$, then $x = y$;

    (c) well founded: there is no infinite E-decreasing chain, i.e., a sequence $\{x_n : n < \omega\}$ with $x_{n+1}Ex_n$ for $n = 0, 1, \ldots$.

    (d) set-like: for every $x \in M$, $\{y : yEx\}$ is a set.

    Then there are a unique transitive class N, and a unique isomorphism $\pi : (M, E) \to (N, \in)$.

Part II

Solutions

1

Operations on sets

1. If an element a is contained in exactly $s \geq 1$ of the sets $A_1, \ldots, A_n$, then on the right-hand side this a is counted exactly

$$\binom{s}{1} - \binom{s}{2} + \binom{s}{3} + \cdots + (-1)^{s-1}\binom{s}{s}$$

times, and this is 1 because the binomial theorem gives that

$$0 = (1-1)^s = 1 - \binom{s}{1} + \binom{s}{2} - \binom{s}{3} + \cdots + (-1)^s\binom{s}{s}.$$

To prove the second identity, set $X = \cup A_i$, apply the first identity to the sets $A_i^* = X \setminus A_i$ and subtract the resulting equation from $N = |X|$, the number of elements of X:

$$|A_1 \cap \cdots \cap A_n| = N - |A_1^* \cup \cdots \cup A_n^*|$$

$$= N - \sum_i |A_i^*| + \sum_{i<j} |A_i^* \cap A_j^*| - \sum_{i<j<k} |A_i^* \cap A_j^* \cap A_k^*| - \cdots$$

$$= \sum_i (N - |A_i^*|) - \sum_{i<j} (N - |A_i^* \cap A_j^*|) + \sum_{i<j<k} (N - |A_i^* \cap A_j^* \cap A_k^*|) - \cdots,$$

and since

$$N - |A_i^* \cap A_j^* \cap \cdots \cap A_k^*| = |A_i \cup A_j \cup \cdots \cup A_k|,$$

we are done.

2. Both the commutativity and the associativity of Δ can be directly verified. It is also easy to see that $\cap$ is distributive with respect the Δ:

$$A \cap (B \Delta C) = (A \cap B) \Delta (A \cap C).$$

In fact,

- an element belongs to the left-hand side if and only if it belongs to A and to exactly one of B and C,
- an element belongs to the right-hand side if and only if it belongs to A and B or to A and C, but not to A, B, and C,

and it is clear that these two statements are the same.

Thus, $\mathcal{H}$ is a ring. Clearly, $A\Delta\emptyset = A$, so the empty set $\emptyset$ plays the role of zero for Δ. Furthermore, $A\Delta A = \emptyset$, hence every set is its own additive inverse.

3. The statement is clearly true for $n = 2$ and from here we can proceed by induction. Suppose we know its validity for some n. Writing $B = A_1\Delta A_2\Delta\cdots\Delta A_n\Delta A_{n+1}$ as $C\Delta A_{n+1}$ with $C = A_1\Delta A_2\Delta\cdots\Delta A_n$, we can see that an element a belongs to B if and only if either it belongs to A_{n+1} and not to C, or it belongs to C and not to A_{n+1}. In either case the induction hypothesis gives that $a \in B$ if and only if it belongs to an odd number of the A_i's.

4. We apply the characterization given in Problem 3. If a belongs to s of the A_i's, then it is counted on the right-hand side

$$\binom{s}{1} - 2\binom{s}{2} + 4\binom{s}{3} - \cdots = \frac{1}{2}(1 - (1-2)^s)$$

times, and this is 0 if s is even and 1 if s is odd.

5. Since $A^c = A \downarrow A$, we can see that $A \cup B = (A \downarrow B)^c = (A \downarrow B) \downarrow (A \downarrow B)$. Using that $A \cap B = (A^c \cup B^c)^c$, it follows that $\cap$ can also be expressed via $\downarrow$. Finally, $A \setminus B = A \cap B^c$.

One can proceed similarly with $|$.

6. Consider part **a**). If a belongs to the left-hand side then there is an $i_0 \in I$ such that a belongs to all the sets $A_{i_0,j}$, $j \in J_{i_0}$. But then a belongs to every $\bigcup_{i\in I} A_{i,f(i)}$, so it belongs to the right-hand side as well.

Conversely, if a does not belong to the left-hand side, then for every $i \in I$ there is $j \in J_i$, which we shall denote by $f_0(i)$, such that $a \notin A_{i,f_0(i)}$. But then this f_0 is in $\prod_{i\in I} J_i$, hence a does not belong to the right-hand side.

The other identities can be verified in the same manner.

7. Let $\mathcal{H}(X; A_1, \ldots, A_n)$ be the collection of those sets that can be obtained from $A_1, \ldots, A_n$ using the operations $\cap$, $\cup$, and $\cdot^c$ (complementation with respect to X). We have to show that

$$|\mathcal{H}(X; A_1, \ldots, A_n)| \leq 2^{2^n}. \tag{1.1}$$

This is clearly true for $n = 1$, and we can proceed by induction. Thus, suppose that (1.1) is true for an n. Note that $\mathcal{H}(X; A_1, \ldots, A_n)$ is nothing else than the smallest set containing $X; A_1, \dots, A_n$ that is closed under union, intersection, and complementation. Therefore, it immediately follows that

$$\mathcal{H}(X; A_1, \dots, A_n, A_{n+1}) = \{S \cup T\},$$

where on the right we take all possible unions with $S \in \mathcal{H}(A_{n+1}; A_1 \cap A_{n+1}, \dots, A_n \cap A_{n+1})$ and $T \in \mathcal{H}(A_{n+1}^c; A_1 \cap A_{n+1}^c, \dots, A_n \cap A_{n+1}^c)$. By the induction hypothesis these latter sets have at most 2^{2^n} elements, so there are that many choices for S and T. Thus, for $S \cup T$ we have at most $2^{2^n} \cdot 2^{2^n} = 2^{2^{n+1}}$ choices, and this proves (1.1) with n replaced by $(n+1)$.

8. The hyperplanes $x_i = 1/2$ divide the unit cube into 2^n pairwise disjoint subcubes $C_1, \dots, C_{2^n}$ of side length 1/2. Clearly, each of $C_1, \dots, C_{2^n}$ can be obtained from the sets A_k using the operations $\cap$ and $\cdot^c$, and so taking the union of any possible subcollection of $C_1, \dots, C_{2^n}$ (there are 2^{2^n} different such subcollections), one can construct 2^{2^n} different sets from $A_1, A_2, \dots, A_n$.

9. Let $\mathcal{H}$ be the collection of all sets that can be obtained from $A_1, A_2, \dots, A_n$ using the operations $\setminus$, $\cap$, and $\cup$. Note that each such set is a subset of $A_1 \cup \dots \cup A_n$. Let us also choose a set X that is strictly larger than $A_1 \cup \dots \cup A_n$, and consider the set $\mathcal{H}(X; A_1, \dots, A_n)$ from the solution of Problem 7. Note that since $A \setminus B = A \cap B^c$, we have $\mathcal{H} \subseteq \mathcal{H}(X; A_1, \dots, A_n)$. Thus, if $H \in \mathcal{H}$, then $H \in \mathcal{H}(X; A_1, \dots, A_n)$, and since this latter set is closed for complementation, we also get $H^c \in \mathcal{H}(X; A_1, \dots, A_n)$. Moreover, it is not possible that $H^c \in \mathcal{H}$, for then X would belong to $\mathcal{H}$. Thus, for every $H \in \mathcal{H}$ there are two different sets (H and H^c) in $\mathcal{H}(X; A_1, \dots, A_n)$, and so the first statement is a consequence of Problem 7.

To show that the bound $2^{2^n - 1}$ can be achieved, consider $A_1, \dots, A_n$ from Problem 8. It is easy to see that using $\cup$, $\cap$, and $\setminus$, all but one of the cubes $C_1, \dots, C_{2^n}$ from the solution (namely the one with one vertex at the point $(0, 0, \dots, 0)$) can be constructed, and we can form again the union of all possible subcollections of these $2^n - 1$ cubes to get $2^{2^n - 1}$ different sets.

10. If there is a solution to

$$\text{(a)} \quad A_i \cap X = B_i, \qquad i \in I,$$

then we must have $\cup_j B_j \subseteq X$, and then it is easy to see that $X' = \cup_j B_j$ is also a solution. But then substituting this into the equations we can see that we must have $\cup_j (A_i \cap B_j) = B_i$, which holds if and only if $B_i \subseteq A_i$ and $A_i \cap B_j \subseteq B_i$ for all i and $j \neq i$. Thus, the system is solvable if and only if these two conditions are satisfied, and then one solution is $X = \cup_j B_j$. One can always add elements from outside $\cup_j A_j$ to X, so the solution is never unique.

In a similar manner (or take the complement of all sides with respect to a large set and reduce the problem to Problem (a))

$$(\mathbf{b}) \quad A_i \cup X = B_i, \qquad i \in I,$$

is solvable if and only if $A_i \subseteq B_i$ and $B_i \subseteq A_i \cup B_j$ for all i and $j \neq i$. In this case one solution is $X = \cap_j B_j$.

In dealing with

$$(\mathbf{c}) \quad A_i \setminus X = B_i, \qquad i \in I,$$

let Z be the union of all the sets A_i and B_j, and let $Y = Z \setminus X$. Then the system takes the form

$$(\mathbf{c}') \quad A_i \cap Y = B_i, \qquad i \in I,$$

i.e., the one we have considered in (a).

In a similar manner, the system

$$(\mathbf{d}) \quad X \setminus A_i = B_i, \qquad i \in I.$$

can be reduced to the case (a) if we write $X \setminus A_i$ as $(Z \setminus A_i) \cap X$.

11. Let

$$B_i = A_i \setminus \left(\bigcup_{j<i} A_j \right).$$

It is immediate that these sets are pairwise disjoint and $\cup_i B_i \subseteq \cup_i A_i$. Furthermore, if for an $a \in \cup_i A_i$ the first index i with $a \in A_i$ is i_0, then clearly $a \in B_{i_0}$, so we actually we have $\cup B_i = \cup_i A_i$.

12. If the C and D with the prescribed properties exist, then clearly $A_i \cap B_j$ is finite for all i and j.

Conversely, suppose that $A_i \cap B_j$ is finite for all i, j. The sets

$$C = \bigcup_{i=0}^{\infty} \left(A_i \setminus \bigcup_{j \le i} B_j \right), \qquad D = \bigcup_{i=0}^{\infty} \left(B_i \setminus \bigcup_{j \le i} A_j \right)$$

are disjoint since $A_i \setminus \bigcup_{k \le i} B_k$ and $B_j \setminus \bigcup_{k \le j} A_k$ are disjoint for all i, j. That $A_i \setminus C$ is finite follows from the finiteness of $A_i \cap B_j$ for all j and hence for all $j \le i$. We get analogously that $B_i \setminus D$ are finite for all i.

13. Let $\mathcal{S} \subseteq \mathcal{P}(X)$ be the smallest family of sets including $\mathcal{A}$ and closed under countable intersection and countable disjoint union (this is the intersection of all such families). It is clear that S is also closed under finite intersection and finite disjoint union. Set

$$\mathcal{B} = \{A \in \mathcal{S} \ : \ X \setminus A \in \mathcal{S}\}.$$

By assumption $\mathcal{A} \subseteq \mathcal{B}$.

If $A, B \in \mathcal{B}$, then $B \setminus A = B \cap (X \setminus A) \in \mathcal{S}$, $A \cup B = A \cup (B \setminus A) \in \mathcal{S}$, and $X \setminus (A \cup B) = (X \setminus A) \cap (X \setminus B) \in \mathcal{S}$. These latter two show that $\mathcal{B}$ is closed under two-term union, and hence under finite union. Finally, since $A \setminus B = A \cap (X \setminus B) \in \mathcal{S}$, $X \setminus (B \setminus A) = (X \setminus B) \cup (A \cap B) \in \mathcal{S}$, $\mathcal{B}$ is also closed under difference ($\mathcal{B}$ is a so-called algebra of sets).

If $A_n \in \mathcal{B}$, $n = 0, 1, \ldots$, then, as in the solution of Problem 11, we have

$$\bigcup_{n=0}^{\infty} A_n = \bigcup_{n=0}^{\infty} \left(A_n \setminus \bigcup_{j<n} A_j \right),$$

and this latter one is a countable disjoint union of elements of $\mathcal{B}$, hence it belongs to $\mathcal{S}$. Furthermore,

$$X \setminus \left(\bigcup_{n=0}^{\infty} A_n \right) = \bigcap_{n=0}^{\infty} (X \setminus A_n) \in \mathcal{S}.$$

These show that $\mathcal{B}$ is closed under countable union, hence it is a σ-algebra including $\mathcal{A}$. Therefore, it includes the σ-algebra $\mathcal{A}^*$ generated by $\mathcal{A}$. On the other hand, $\mathcal{B} \subseteq \mathcal{S}$, and clearly $\mathcal{S}$ is a subset of the σ-algebra $\mathcal{A}^*$, and these show that $\mathcal{B} = \mathcal{S} = \mathcal{A}^*$.

14. All the statements are immediate consequences of the definitions.

15. Clearly, two subsets of X are the same if and only if their characteristic functions are the same. Furthermore, if $g \in {}^X\{0,1\}$ is arbitrary, then $g = \chi_A$, where A is the set of those $x \in X$ where $g(x) = 1$. Thus, $A \mapsto \chi_A$ is a 1-to-1 correspondence.

The statements concerning the lim inf and lim sup sets immediately follow from parts b) and c) of the preceding problem.

16. By the definition $\{A_n\}_{n=1}^{\infty}$ is convergent if and only if every element a that is contained in infinitely many of the A_i's is contained in all but finitely many of the them. This is the same as saying that there is no element a and two infinite subsequences $\{m_i\}$ and $\{n_i\}$ of the natural numbers with $a \in A_{m_i}$ and $a \notin A_{n_i}$, and this is the same as the condition in the problem.

17. See the solution of the preceding problem.

18. Of the infinitely many sets A_i either infinitely many contain 0, or infinitely many do not contain 0. In the first case let $A_0^{(0)}, A_1^{(0)}, \ldots$ be the sequence of those A_i's that contain 0, and in the second case let $A_0^{(0)}, A_1^{(0)}, \ldots$ be the sequence of those A_i's that do not contain 0. Now of $A_0^{(0)}, A_1^{(0)}, \ldots$, either infinitely many contain 1, or infinitely many do not contain 1. In the first case let $A_0^{(1)}, A_1^{(1)}, \ldots$ be the sequence of those $A_i^{(0)}$'s that contain 1, and in

the second case let $A_0^{(1)}, A_1^{(1)}, \ldots$ be the sequence of those $A_i^{(0)}$'s that do not contain 1. Proceeding similarly with the numbers $2, 3, \ldots$ we get infinitely many infinite subsequences $\{A_i^{(j)}\}_{i=0}^{\infty}$, $j = 0, 1, \ldots$ of the original sequence. It is immediate (see also Problem 14) that the diagonal sequence $\{A_i^{(i)}\}_{i=0}^{\infty}$ is convergent.

19. Let A_i be the set of those real numbers the ith decimal digit (after the decimal point) of which is 0 (warning: some rational numbers have two decimal expansions, one finite and one infinite, e.g., $0.1 = 0.09999\cdots$, but in this solution it does not matter which one we fix). We claim that there is no convergent subsequence of $\{A_i\}_{i=1}^{\infty}$. In fact, let $0 < n_1 < n_2 < \cdots$ be any subsequence of the natural numbers, and consider the number

$$x = \sum_{j=1}^{\infty} \frac{1}{10^{n_{2j}}}.$$

The n_{2j+1}th decimal digit of this is 0, so x belongs to all the sets $A_{n_{2j+1}}$. However, the n_{2j}th decimal digit of x is 1, so x does not belong to any of the sets $A_{n_{2j}}$. Thus, x belongs to $\limsup_j A_{n_j}$, but does not belong to $\liminf_j A_{n_j}$, i.e., the subsequence $\{A_{n_j}\}_{j=1}^{\infty}$ is not convergent.

20. It is clear that $\subset$ (proper subset) is irreflexive and transitive (but in general not trichotomous, i.e., in general for $A \neq B$ we do not have either $A \subset B$ or $B \subset A$), hence it is a partial ordering.

Conversely, let $\langle A, \prec \rangle$ be a partially ordered set, and consider the family $\mathcal{A}$ of those subsets H_a of A of the form $H_a = \{b \in A : b \preceq a\}$. It is clear that $a \prec b$ exactly if $H_a \subset H_b$, hence $\langle A, \prec \rangle$ is isomorphic with $\langle \mathcal{A}, \subset \rangle$.

21. Let (V, E) be a graph where V denotes the set of vertices and E denotes the set of edges. To every vertex $x \in V$ associate the subset E_x of E that consists of the edges that are adjacent to x. It is clear that E_x and E_y intersect if and only if there is an edge between x and y, so $x \mapsto E_x \cup \{x\}$ is an appropriate isomorphism.

22. Clearly, $A \Delta \emptyset = A$, so the empty set $\emptyset$ plays the role of zero for Δ. Furthermore, $A \Delta A = \emptyset$, so every set is its own additive inverse. All the other ring properties follow from Problem 2.

23. Let $(A, +, \cdot, 0)$ be a ring in which every element is idempotent ($a \cdot a = a$). Then

$$a + a \cdot b + b \cdot a + b = a \cdot a + a \cdot b + b \cdot a + b \cdot b = (a + b) \cdot (a + b) = a + b,$$

hence $a \cdot b + b \cdot a = 0$. Putting here $b = a$ we get $a + a = a \cdot a + a \cdot a = 0$ for every a. Using this in the preceding formula we obtain

$$a \cdot b = a \cdot b + (a \cdot b + b \cdot a) = (a \cdot b + a \cdot b) + b \cdot a = b \cdot a.$$

Thus, the ring is commutative, in which every element is its own additive inverse.

Call a subring $I \subset A$ a prime ideal if it is not the whole ring A and $a \in I$, $b \in A$ implies $a \cdot b \in I$ (that is it is an ideal) and if $a \cdot b \in I$ implies that one of a or b belongs to I. Let the set of prime ideals be X and to every element $a \in A$ associate the set

$$H_a = \{I \in X \ : \ a \notin I\},$$

the set of prime ideals not containing a. We claim that the set $\mathcal{H} = \{H_a\}_{a \in A}$ is closed for the operations $\cap$ and Δ, and that $a \mapsto H_a$ is a ring isomorphism.

First we show that $a \mapsto H_a$ is a 1-to-1 mapping. Let a and b be two different elements in A, and first assume that $b \cdot a = b$. There is an ideal containing b but not a, e.g., the set $\{c \in A \ : \ b \cdot c = c\}$ is such an ideal. Now it is easy to see that if M is a set of ideals ordered with respect to inclusion such that every member of M contains b but does not contain a, then their union also has this property. Thus, by Zorn's lemma (see Chapter 14) there is a maximal (with respect to inclusion) ideal I containing b but not containing a. We claim that this is a prime ideal. In fact, if that was not the case then we would have $c, d \notin I$ with $c \cdot d \in I$. The ideal generated by I and c consists of all elements $c \cdot p + q$ with $p \in A$ and $q \in I$ (check the ideal properties for the set of these elements). Thus, by the maximality of I, there are $p_1 \in A$ and $q_1 \in I$ such that $a = c \cdot p_1 + q_1$. In a similar fashion, there are $p_2 \in A$ and $q_2 \in I$ such that $a = d \cdot p_2 + q_2$. But then

$$a = a \cdot a = (c \cdot p_1 + q_1) \cdot (d \cdot p_2 + q_2) = c \cdot d \cdot (p_1 \cdot p_2) + q_1 \cdot (d \cdot p_2 + q_2) + q_2 \cdot (c \cdot p_1)$$

belongs to I, for all the products on the right-hand side are in I (they are the products of elements of I with some elements of A). This contradiction shows that, in fact, I is a prime ideal containing b but not a.

If $a \cdot b = a$, then by the same argument there is a prime ideal containing a but not b. Finally, if $a \cdot b \neq a, b$, then $(a \cdot b) \cdot a = a \cdot b$, and by what we have just proven, then there is a prime ideal containing $a \cdot b$ but not a. But then the prime property shows that I must contain b.

Thus, for different elements there are prime ideals containing exactly one of them, so the mapping $a \mapsto H_a$ is 1-to-1.

It is clear that $H_0 = \emptyset$, and

$$H_{a \cdot b} = \{I \in X \ : \ a \cdot b \notin I\} = \{I \in X \ : \ a \notin I \text{ and } b \notin I\} = H_a \cap H_b.$$

It is also clear that if I is a prime ideal and $a \in I$ and $b \notin I$ or $b \in I$ and $a \notin I$, then $a + b \notin I$. Furthermore, if $a \notin I$ and $b \notin I$, then $a \cdot b \notin I$, but $(a \cdot b) \cdot (a + b) = a \cdot b + a \cdot b = 0$ is in I, hence $a + b$ must be in I. Thus, $a + b \notin I$ if and only if exactly one of a and b is not in I. Hence $H_{a+b} = H_a \Delta H_b$, and this completes the proof that the mapping $a \mapsto H_a$ is an isomorphism.

24. The intersection of a finite set with any set and symmetric difference of two finite sets is finite, hence $\mathcal{I}$ is a subring which is also an ideal. If $H \subseteq X$ is infinite, then we can write H as a disjoint union of two infinite sets H_1 and H_1. Thus, if $\overline{H}$ denotes the image of H under the ring homomorphism $\mathcal{H} \to \mathcal{H}/\mathcal{I}$, then $\overline{H_1} \neq \overline{\emptyset}$ is different from $\overline{H}$, and $\overline{H_1} \cdot \overline{H} = \overline{H_1}$, and this proves that $\overline{H}$ is not an atom.

25. All the lattice properties are easy to check. The distributivity is also true, since $A \cap (B \cup C) = (A \cap B) \cup (A \cap C)$ and $A \cup (B \cap C) = (A \cup B) \cap (A \cup C)$ (see also the general distributivity laws in Problem 6).

26. Let L be a distributive lattice with the operations $\wedge$ and $\vee$, and for $b, a \in L$ set $a \leq b$ if $a \wedge b = a$. It is easy to see that this is a partial ordering on L.

We call a subset $I \neq \emptyset$ of L an ideal if it is closed for $\vee$, and it is also true that if $a \in I$ and $b \leq a$, then $b \in I$. Call an ideal I prime ideal if it is not the whole L and $a \wedge b \in I$ implies that either a or b belongs to I. We denote the set of prime ideals by X, and for $a \in L$ set

$$H_a = \{I \in X \ : \ a \notin I\}.$$

We claim that the family $\mathcal{H} = \{H_a\}_{a \in L}$ of sets is closed under two-term intersection and union, and that the mapping $a \mapsto H_a$ is an isomorphism from L onto $\{H_a\}_{a \in L}$ considered as a lattice with $\cap$ and $\cup$ for operations.

First we show that $a \mapsto H_a$ is 1-to-1, and to this end it is sufficient to show that for any two $a \neq b$ in L there is a prime ideal I which contains exactly one of a and b. First assume that $a < b$, and let $\mathcal{S}$ be the set of all ideals that contain a but do not contain b. $\mathcal{S}$ is not empty, for $\{c \in L \ : \ c \leq a\}$ is such an ideal. It is easy to show that if M is an ordered subset of $\mathcal{S}$ with respect to inclusion, then the union of the ideals in M is again in M, hence by Zorn's lemma (see Chapter 14) there is a maximal element I in $\mathcal{S}$. We claim that I is a prime ideal. In fact, suppose to the contrary that $c \wedge d \in I$ but $c, d \notin I$. The ideal generated by the set $I \cup \{c\}$ consists of those elements $p \in L$ for which there is a $q \in I$ with the property that $p \leq c \vee q$ (just check that the set of all these elements form an ideal). Thus, by the maximality of I there must be an $e \in I$ such that $b \leq c \vee e$. In a similar manner there is an $f \in I$ such that $b \leq d \vee f$. But then $b \leq c \vee (e \vee f)$ and $b \leq d \vee (e \vee f)$, hence

$$b \leq [c \vee (e \vee f)] \wedge [d \vee (e \vee f)] = (c \wedge d) \vee (e \vee f) \in I,$$

since both $c \wedge d \in I$ and $e \vee f \in I$. Thus, we must have $b \in I$, which is not the case, hence the claim that I is a prime ideal follows. This verifies that for $a < b$ there is a prime ideal containing a but not b.

If $b < a$, then the argument is similar. Finally, if neither a nor b is smaller than the other one, then $a \wedge b$ is strictly smaller than a, hence, according to what we have just proven, there is a prime ideal I that contains $a \wedge b$ but does not contain a. The primeness of I shows that we must then have $b \in I$, and the existence of I has been verified in this case as well.

The proof that $\mathcal{H}$ is closed for union and intersection and that $a \mapsto H_a$ is an isomorphism is easy:

$$\begin{aligned} H_{a\wedge b} &= \{I \in X \ : \ a \wedge b \notin I\} = \{I \in X \ : \ a, b \notin I\} \\ &= \{I \in X \ : \ a \notin I\} \cap \{I \in X \ : \ b \notin I\} = H_a \cap H_b, \end{aligned}$$

and similarly,

$$\begin{aligned} H_{a\vee b} &= \{I \in X \ : \ a \vee b \notin I\} = \{I \in X \ : \ a \notin I \text{ or } b \notin I\} \\ &= \{I \in X \ : \ a \notin I\} \cup \{I \in X \ : \ b \notin I\} = H_a \cup H_b. \end{aligned}$$

27. For every $H \in \mathcal{H}$ there is a $K \in \mathcal{H}$ with $H \cdot K = 0$ and $H + K = 1$, namely the complement $X \setminus H$ of H with respect to X has this property. All the other Boolean algebra properties are easy consequences of properties of set operations.

28. Let $\langle B, +, \cdot, ', 0, 1\rangle$ be a Boolean algebra. Then $\langle B, \wedge, \vee\rangle$ with $\vee = +$ and $\wedge = \cdot$ is a distributive lattice, hence it can be represented in the prime ideal space X as in Problem 26. Following the notation of the proof of Problem 26 it is clear that $H_0 = \emptyset$ and $H_1 = X$. Thus, all that is left is to show that $H_{a'} = X \setminus H_a$. But this follows from the other properties that we know of the mapping $a \mapsto H_a$:

$$X = H_1 = H_{a\vee a'} = H_a \cup H_{a'}$$

and

$$\emptyset = H_0 = H_{a\wedge a'} = H_a \cap H_{a'},$$

hence $H_{a'} = X \setminus H_a$ as was claimed.

29. $\mathcal{P}(X)$ is a Boolean algebra by Problem 27, and clearly the union $\cup_{i\in I} H_i$ of any set of subsets H_i, $i \in I$ of X is a subset of X, which is the smallest set U with $U \cap H_i = H_i$ for all i. In a similar fashion, $\cap_{i\in I} H_i$ is the infimum of the sets H_i, $i \in I$. Thus, the completeness of $\mathcal{P}(X)$ as a Boolean algebra follows. Complete distributivity was proved in Problem 6.

30. Let $(A, +, \cdot, ', 0, 1)$ be a complete and completely distributive Boolean algebra. Let us denote the smallest majorant and the greatest minorant of a subset $B \subseteq A$ by $\vee B$ and $\wedge B$, respectively. It is clear that $\vee\{a, b\} = a + b$ and $\wedge\{a, b\} = a \cdot b$, and for two elements we shall use $\vee$ and $+$ and $\wedge$ and $\cdot$ interchangeably.

We call an element $x \in A$ an atom if there is no $a \neq 0, x$ with $a \cdot x = a$. As in the solution to Problem 26, we set $a \preceq b$ if $a \cdot b = a$. With this partial ordering an element x is an atom if there is no element between 0 and x; i.e., if $0 \prec a \preceq x$ implies $a = x$.

Let $\mathcal{F} = {}^{A}\{0,1\}$ be the set of all functions from A to $\{0,1\}$, and for any element a of A set $a^0 = a$ and $a^1 = a'$. For any $f \in \mathcal{F}$ consider the greatest minorant x_f of the elements $a^{f(a)}$, i.e., we set

$$x_f = \bigwedge_{a \in A} a^{f(a)}.$$

This may be 0, but if it is not zero, then it is an atom. In fact, if $a_0 \neq 0$ and $a_0 \preceq x_f$, then $a_0 \preceq a_0^{f(a_0)}$, hence $f(a_0) = 0$, and then $a_0 \preceq x_f \preceq a_0$, so $a_0 = x_f$, which shows that x_f is, in fact, an atom. Let X be the set of all the atoms x_f.

Assign to any element $a \in A$ the set

$$H_a = \{x_f \in X \ : \ x_f \preceq a\}.$$

We claim that $a \mapsto H_a$ is an isomorphism from $(A, +, \cdot, ', 0, 1)$ onto $\mathcal{P}(X)$.

By complete distributivity we have

$$1 = \wedge\{a \vee a' \ : \ a \in A\} = \bigvee_{f \in \mathcal{F}} \bigwedge_{a \in A} a^{f(a)} = \bigvee_{f \in \mathcal{F}} x_f,$$

and so for every $b \in A$ we get (recall that $a \cdot b = \inf\{a, b\} = a \wedge b$)

$$b = b \cdot 1 = b \cdot (\bigvee_{f \in \mathcal{F}} x_f) = \bigvee_{f \in \mathcal{F}} b \cdot x_f,$$

and here on the right-hand side the nonzero elements $b \cdot x_f$ are exactly the atoms $x_f \preceq b$. Thus, every element in the algebra is the least upper bound of the atoms below it. This shows that $a \mapsto H_a$ is a 1-to-1 mapping. Conversely, if $C \subseteq X$ is a subset of the set of the atoms, and $c = \vee C$, then for an x_f we have

$$x_f \cdot c = x_f \cdot \left(\bigvee C\right) = \bigvee \{x_f \cdot x_g \ : \ x_g \in C\},$$

and this is 0 if $x_f \notin C$ and is x_f if $x_f \in C$. Thus, $a \mapsto H_a$ is a mapping onto $\mathcal{P}(X)$. It is also clear that $x_f \preceq a \cdot b$ if and only if $x_f \preceq a$ and $x_f \preceq b$, thus $H_{a \cdot b} = H_a \cap H_b$. Furthermore, $x_f \preceq a$ if and only if $x_f \not\preceq a'$, so $X \setminus H_a = H_{a'}$. Finally, $x_f \preceq a + b$ if and only if $x_f \preceq a$ or $x_f \preceq b$ (because if $x_f \not\preceq a, b$ then $x_f \preceq a', b'$, which implies $x_f \not\preceq (a' \cdot b')' = a + b$), and so $H_{a+b} = H_a \cup H_b$. Since $H_0 = \emptyset$ and $H_1 = X$, we are done.

Naturally it is also true that the mapping $a \mapsto H_a$ preserves the greatest minorant and the smallest majorant as well.

31. Let H_0 be the smallest element of $\mathcal{H}$ (there is such, just apply the condition to $\mathcal{H}^* = \mathcal{H}$). Then for this we have $H_0 \subseteq f(H_0)$. Let

$$\mathcal{B} = \{H \in \mathcal{H} \ : \ H \subseteq f(H)\}.$$

This set is not empty ($H_0 \in \mathcal{H}$), and let F be the smallest element in $\mathcal{H}$ that contains all elements of $\mathcal{B}$. We have $H \subseteq F$ for all $H \in \mathcal{B}$, hence $H \subseteq f(H) \subseteq f(F)$ for all $H \in \mathcal{B}$, and by taking union we can see that $F \subseteq f(F)$. On applying f to both sides we get $f(F) \subseteq f(f(F))$, so $f(F)$ is an element in $\mathcal{B}$, and hence $f(F) \subseteq F$. Thus, $f(F) = F$, and F is a fixed point.

* * *

32. Suppose to the contrary that, e.g., there is a subfamily $\mathcal{H}^*$ of sets in $\mathcal{H}$ such that there is no smallest element in $\mathcal{H}$ including all the sets in $\mathcal{H}^*$.

Let $A_0 \in \mathcal{H}^*$ be arbitrary, and by transfinite recursion we select sets $A_\xi \in \mathcal{H}^*$, $\xi < \alpha$ as follows. If A_ξ, $\xi < \eta$ have already been selected, and there is no smallest set in $\mathcal{H}$ that includes all A_ξ, $\xi < \eta$, then terminate the construction, and set $\alpha = \eta$. If, however, there is a smallest set $K_\eta \in \mathcal{H}$ including all the sets A_ξ, $\xi < \eta$, then K_η cannot include all the sets in $\mathcal{H}^*$, hence there is a set $K_\eta^* \in \mathcal{H}^*$ that is not included in K_η. Now let A_η be the set $K_\eta \cup K_\eta^*$. It is clear that this process terminates (in fewer than $|\mathcal{H}^*|^+$ steps), α is a limit ordinal (otherwise $\mathcal{H}^*$ would have a largest element), and $\{A_\xi\}_{\xi<\alpha}$ is a strictly increasing sequence of sets in $\mathcal{H}$. The way we defined α shows that if $\mathcal{B}$ is the set of all sets in $\mathcal{H}$ that include all A_ξ, $\xi < \alpha$ as a subset, then there is no smallest set in $\mathcal{B}$. If $\mathcal{B}$ is not empty, then we define a transfinite sequence $\{B_\xi\}_{\xi<\beta}$ of elements of $\mathcal{B}$. Let $B_0 \in \mathcal{B}$ be arbitrary, and if B_ξ, $\xi < \eta$ have already been defined for some ordinal η, then let B_η be an element of $\mathcal{B}$ that is strictly included in all sets B_ξ, $\xi < \eta$ if there is one, and if there is no such set then we put $\beta = \eta$, and the process terminates. It is clear that this process has to terminate in fewer than $|\mathcal{B}|^+$ steps, and by the assumption on $\mathcal{B}$, β is a limit ordinal.

It is also clear that there cannot be any set $H \in \mathcal{H}$ that includes all A_ξ, $\xi < \alpha$ and is included in all B_ξ, $\xi < \beta$, for such an H would belong to $\mathcal{B}$, and then it would be the smallest element of $\mathcal{B}$. Thus, for all sets H either there exists a smallest $\alpha_H < \alpha$ such that $A_{\alpha_H} \not\subseteq H$, or there is a smallest $\beta_H < \beta$ such that $H \not\subseteq B_{\beta_H}$.

Now we define a mapping $f : \mathcal{H} \to \mathcal{H}$ as follows. If α_H is defined, then let $f(H) = A_{\alpha_H}$, otherwise set $f(H) = B_{\beta_H}$. It is clear by the definition of the ordinals α_H and β_H that this f does not have a fixed point. Thus, if we can show that f preserves $\subseteq$, then the statement in the problem follows from the contradiction to the hypothesis in the problem.

Let $H \subseteq K$ be two elements of $\mathcal{H}$. If α_K is defined, then α_H is also defined, and $\alpha_H \le \alpha_K$, hence we have $f(H) = A_{\alpha_H} \subseteq A_{\alpha_K} = f(K)$. In a similar way, if α_H is not defined then α_K is not defined and $\beta_K \le \beta_H$, so in this case $f(H) = B_{\beta_H} \subseteq B_{\beta_K} = f(K)$. The only remaining case is when α_H is defined but β_K is not, in which case we have $f(H) = A_{\alpha_H} \subset B_{\beta_K} = f(B)$, because every A_ξ is a subset of every B_η. This proves that f preserves $\subseteq$.

33. Follow the solution of Problem 24, and let H be an infinite subset of X. It is easy to prove that there is a family $\mathcal{F}$ of cardinality continuum of subsets of H such that if $F_1, F_2 \in \mathcal{F}$, then both $F_1 \setminus F_2$ and $F_2 \setminus F_1$ are infinite; e.g., this follows from Problem 4.41. It is now clear that if we take the images of the sets in $\mathcal{F}$ under the ring homomorphism $\mathcal{H} \to \mathcal{H}/\mathcal{I}$ used in the solution of Problem 24, then these images are all different and satisfy the condition that for them $b \cdot a = b$ but $b \neq 0$.

34. Just follow the proofs of Problems 24 and 33, and use that if X is a set of cardinality κ, then there are 2^κ subsets of X any two differing in at least κ elements; see Problem 18.3.

2

Countability

1. Let the sets be $A_0, A_1, \ldots$. We can assume that neither of these is empty, and let $A_i = \{a_0^{(i)}, a_1^{(i)}, \ldots\}$ be an enumeration of the elements of A_i. Then

$$a_0^{(0)}, a_1^{(0)}, a_0^{(1)}, a_2^{(0)}, a_1^{(1)}, a_0^{(2)}, \ldots$$

is an enumeration of the union.

2. It is enough to prove that the product of two countable sets is countable. Let the sets be

$$A = \{a_0, a_1, \ldots\} \quad \text{and} \quad B = \{b_0, b_1, \ldots\}.$$

Then the elements of the product can be enumerated as

$$(a_0, b_0), (a_0, b_1), (a_1, b_0), (a_0, b_2), (a_1, b_1), (a_2, b_0), \ldots.$$

3. The set of k element sequences of a set A is nothing else than the k-fold product of A with itself. Apply Problem 2.

4. The set of finite sequences is the union of the sets of k-element sequences for all $k = 0, 1, \ldots$. Now the result follows from Problems 3 and 1.

5. Identify each polynomial with the sequence of its coefficients (starting with the nonzero highest coefficient), and then apply the preceding problem.

6. Recall that a complex number is called algebraic if it is the zero of a not identically zero polynomial with integer coefficients. Each nonzero polynomial has at most a finite number of zeros. Hence the set of all zeros of nonzero polynomials with integer coefficients is countable by Problems 5 and 1.

7. Suppose $\mathbf{R}$ is countable. Then $(0,1)$ is also countable. Let $x_0, x_1, \ldots$ be an enumeration of the elements of $(0,1)$, and let $x_i = 0.\alpha_1^{(i)}\alpha_2^{(i)}\ldots$ be the decimal representation of x_i (some reals have two decimal representations; in that case choose either one). Now let $b_i = 4$ if $a_i^{(i)} \neq 4$, and let $b_i = 6$ if $a_i^{(i)} = 4$. The number $x = 0.b_1b_2\ldots$ is in $(0,1)$ and is different from any of the numbers $x_0, x_1, \ldots$, which is a contradiction since in this last sequence we have listed all numbers in $(0,1)$. That $x \neq x_i$ follows from the fact that the ith digits of these numbers differ (which in itself does not prove that $x \neq x_i$ as is seen from $0.1000\ldots = 0.099999\ldots$), and x does not have 0 or 9 among its digits (if two different decimal expansions represent the same number, then one of them contains only 0's and the other one contains only 9's from a certain point on).

8. This follows from Problems 6 and 7.

9. **a)** Enumerate the rationals as $0,\ 1/1,\ -1/1,\ 1/2,\ 2/1,\ -2/1,\ -1/2,\ 1/3,\ 2/2,\ 3/1,\ -3/1,\ -2/2,\ -1/3,\ 1/4,\ 2/3,\ 3/2,\ \ldots$.

b) If $S \subset A$ is a finite set, then let H_S be the set of mappings of S into B. If S has k elements, then clearly H_S is equivalent to B^k, hence it is countable by Problem 2. Now the set in the problem is the union of all the H_S's for finite subsets S of A, and there are at most countably many such S's (see Problem 4). Hence the statement follows from Problem 1.

c) If $A = \{a_i\}_{i=0}^{\infty}$ is a convergent sequence consisting of natural numbers, then there is a j such that $a_k = a_j$ for all $k \geq j$. If j is the smallest index with this property, then associate with S the finite sequence $S^* = \{a_0, a_1, \ldots, a_j\}$. It is clear that S^* uniquely determines S, hence the statement follows from Problem 4.

10. For every $a \in \mathbf{N}$ if $a \in A_i$ for some i then select such an A_{i_a}, and if $a \notin A_i$ for some i then select such an $A_{i_a^*}$. It is clear that $\{A_{i_a}, A_{i_a^*} : a \in \mathbf{N}\}$ is an appropriate subfamily.

11. Let m be the supremum of all those real numbers r for which $A \cap (-\infty, r)$ is countable (if there is no such r, then let $m = -\infty$). We cannot have $m = \infty$, since then A, as the union of the countable sets $A \cap (-\infty, k)$, $k = 0, 1, 2, \ldots$, would itself be countable. It is also clear that $A \cap (-\infty, m) = \cup_n A \cap (-\infty, m - 1/n)$ is also countable.

In a similar fashion, let M be the infimum of all those real numbers r for which $A \cap (r, \infty)$ is countable. Then this M is bigger than $-\infty$, and $A \cap (M, \infty)$ is countable. These imply that we cannot have $m \geq M$. But then any number $a \in (m, M)$ has the desired property.

12. By Problem 4 the set $\mathbf{N}$ has at most countably many subsets consisting of less than $(K+1)$ elements, so it is enough to prove that the set $\mathcal{T}$ of those $H \in \mathcal{H}$ that have at least $(K+1)$ elements is also countable. Let $\mathcal{B}_{K+1}$ be

the set of $(K+1)$ element subsets of $\mathbf{N}$. As we have just mentioned, $\mathcal{B}_{K+1}$ is countable. But every element $H \in \mathcal{T}$ includes a set $B \in \mathcal{B}_{K+1}$ as a subset, and the condition on the family $H \in \mathcal{H}$ implies that no $B \in \mathcal{B}_{K+1}$ can be contained in more than k such H. All these imply that $\mathcal{T}$ is countable (see Problem 1), and we are done.

13. Every subinterval (a, b) in question can be identified with the pair $(a, b) \in \mathbf{Q} \times \mathbf{Q}$. Hence the statement follows from Problems 2 and 9, a).

14. Select a rational point from every interval. Thus, there are at most as many intervals as rational numbers.

The argument is the same for $\mathbf{R}^n$, since the points with rational coordinates are dense and form a countable set.

15. Let A be a discrete set. Write a ball B_{r_x} of radius r_x around every point $x \in A$ in such a way that B_{r_x} contains only the point x from A. Then the balls $B_{r_x/2}$ are disjoint. Apply Problem 14.

16. Let $G \subset \mathbf{R}$ be open. For $x, y \in G$ let $x \sim y$ if the interval $[x, y]$ lies in G. It is easy to see that this is an equivalence relation, and the equivalence classes are open intervals. Since the different equivalence classes are disjoint, and since by Problem 14 there are at most countably many of them, we are done.

17. Every open disk with rational center (a, b) and rational radius r can be identified with the triplet (a, b, r). Use now Problems 2 and 9, a). The argument is the same for $\mathbf{R}^n$.

18. Let $G \subset \mathbf{R}^2$ be an open set, and let $\mathcal{H}$ be the collection of all disks with rational center and rational radius that lie in G. We are going to show that these disks cover G (see also Problem 17). For $P \in G$ let ρ_P be the supremum of all radii $\rho \leq 1$ for which the disk $B_\rho(P)$ with center at P and of radius ρ is included in G, and select a rational number $\rho_P/3 < r_P < 2\rho_P/3$. If S is a point with rational coordinates that lies closer to P than $\rho_P/3$, then the ball $B_{r_P}(S)$ belongs to $\mathcal{H}$ (use the triangle inequality) and clearly covers P.

19. Let $\mathcal{H}_n$ be the set of those circles in $\mathcal{H}$ that have radius $\geq 1/n$. Since $\mathcal{H}$ is uncountable (see Problem 7) and $\mathcal{H} = \cup_n \mathcal{H}_n$, at least one of the sets $\mathcal{H}_n$, say $\mathcal{H}_{n_0}$, is uncountable. Let k be an integer, and let $\mathcal{H}_{n_0,k}$ be the set of those circles in $\mathcal{H}_{n_0}$ that touch the real line in a point of the interval $((k-1)/2n, k/2n]$. Since $\cup_{k \in \mathbf{Z}} \mathcal{H}_{n_0,k} = \mathcal{H}_{n_0}$, at least one of the sets, say $\mathcal{H}_{n_0,k_0}$ is uncountable, hence this set contains infinitely many circles that lie on the same side of the real axis. But it is easy to see that if two circles of $\mathcal{H}_{n_0,k_0}$ lie on the same side of the real axis, then they intersect.

An alternative way is to select for each $x \in \mathbf{R}$ a circle C_x from $\mathcal{H}$ touching $\mathbf{R}$ at x and for each C_x select a point with rational coordinates inside C_x. Then two of these selected points must be the same, and then the corresponding circles intersect.

20. The answer is no: consider the family of circles C_r, $1 \le r < \infty$, where C_r is the circle with center at the point $(0, r)$ and of radius $2r - 1$.

21. Let H_n be the set of touching points where two circles of radius bigger than $1/n$ touch each other. It is enough to prove that each H_n is countable. Let us divide the plane into the squares

$$Q_{j,k} = \{(x, y) \ : \ j/2n \le x < (j+1)/2n, \ k/2n \le y < (k+1)/2n\}$$

with $k, l = 0, \pm 1, \pm 2, \ldots$ of side length $1/2n$, and let $H_{n,j,k} = H_n \cap Q_{j,k}$. Simple geometry shows that each $H_{n,j,k}$ can contain at most one point where two circles of radius bigger than $1/n$ touch each other from the outside. Associate with every other point $P \in H_{n,j,k}$ the region between the two circles of radius bigger than $1/n$ that touch each from the inside at the point P. Then simple inspection shows that these regions are pairwise disjoint, so by Problem 14 their number is countable. Thus, each $H_{n,j,k}$ is countable, and we can conclude that $H_n = \cup_{j,k=-\infty}^{\infty} H_{n,j,k}$, as a countable union of countable sets, is countable.

22. A letter T is a Y-set in the sense of the next problem, hence the statement follows from the next problem.

23. Let $\mathcal{H}$ be a set of disjoint Y-sets on the plane, and let $\mathcal{H}_n$ be the set of those elements in $\mathcal{H}$ that consist of segments that are longer than $1/n$ and for which each angle formed by the segments is also bigger than $2\pi/n$. It is enough to show that each set $\mathcal{H}_n$ is countable. Let us divide the plane into the squares

$$Q_{j,k} = \{(x, y) \ : \ j/2n \le x < (j+1)/2n, \ k/2n \le y < (k+1)/2n\}$$

with $k, l = 0, \pm 1, \pm 2, \ldots$ of side length $1/2n$, and let $\mathcal{H}_{n,j,k}$ be the set of those Y-sets in $\mathcal{H}_n$ for which the common point (call it the vertex) of the segments lies in $Q_{j,k}$. Simple geometry shows that each $Q_{j,k}$ can contain at most finitely many vertices of Y-sets from $\mathcal{H}_n$ (actually at most $5n$), hence $\mathcal{H}_n = \cup_{j,k=-\infty}^{\infty} \mathcal{H}_{n,j,k}$, as a countable union of finite sets, is countable.

24. Let $X = \{x_i\}_{i=0}^{\infty}$ and $Y = \{y_i\}_{i=0}^{\infty}$ be a separate enumeration of all the x- and all y-coordinates of the points in A, and put a point $(x_i, y_k) \in A$ into B if $k \le i$, otherwise put it into C. Now if a vertical line cuts A then it must be of the form $x = x_{i_0}$ for some i_0, and on this line there are at most $i_0 + 1$ points from B (namely only those (x_{i_0}, y_k), $k \le i_0$ points that lie in A). In a similar manner, any horizontal line that intersects A is of the form $y = y_{k_0}$, and there are at most k_0 points of C on such a line.

25. First we verify the sufficiency of the condition, so let $A \times A = B \cup C$ be an appropriate decomposition. We have to show that then A must be countable. In fact, suppose to the contrary that A is uncountable. Take a countably infinite subset $K \subset A$. Then $(A \times K) \cap C$ is countable, since for each $y \in K$ the number of (x, y) with $(x, y) \in C$ is finite. But for every $x \in A$ there is a $y \in K$ such that $(x, y) \in C$, because the number of those y for which $(x, y) \in B$ is finite. Thus, $(A \times K) \cap C$ has to be uncountable. This contradiction shows that, indeed, A is countable.

The necessity of the condition is easily established, namely if $\{x_0, x_1, \ldots\}$ is an enumeration of the points of A, then $B = \{(x_i, x_j) \ : \ j \le i\}$ and $C = \{(x_i, x_j) \ : \ i < j\}$ is clearly an appropriate decomposition.

26. The set S of numbers of the form $b - c$ with $b, c \in A$ is countable (see Problem 2), hence there are real numbers outside S. If $a \notin S$, then $(a+A) \cap A = \emptyset$.

27. Fix two different points R, S of A, and let $\mathcal{C}_R$ resp. $\mathcal{C}_S$ be the family of all circles with rational radius and with center at R resp. S. The assumption implies that any point of A lies on one of the circles in $\mathcal{C}_R$ and also on one of the circles in $\mathcal{C}_S$, hence all points of A are among the points of intersection of the pairs of circles $C_R \in \mathcal{C}_{\mathcal{R}}$ and $C_S \in \mathcal{C}_S$. There are only countably many pairs (see Problems 2 and 9, a)) and each such pair has at most two common points, hence the number of points in A is countable.

The answer to the last question is 'YES': there is such a set lying on the circle, namely select an angle $\alpha \neq 0$, and let A be the set of those points that are obtained by counterclockwise rotating the point $(1, 0)$ about the origin by angles $n\alpha$, $n = 0, 1, \ldots$. Using trigonometric identities it is easy to show that if both $\sin(\alpha/2)$ and $\cos(\alpha/2)$ are rational numbers, then the distances between points of A are rationals. That there is an $0 < \alpha < \pi/2$ for which both $\sin(\alpha/2)$ and $\cos(\alpha/2)$ are rational numbers follows from the existence of Pythagorean triplets. The fact that by selecting α this way all the points of A are different (hence A is infinite) lies somewhat deeper, and it follows from the irrationality of α/π.

An alternative way of constructing an infinite set not lying on a straight line but having all distances rational is to choose infinitely many different Pythagorean triples (a_n, b_n, c_n), i.e., $a_n > b_n > c_n$ positive integers with $a_n^2 = b_n^2 + c_n^2$ and no common factors, and consider the points $(0, 1)$, $(0, 0)$, $(b_n/c_n, 0)$, $n = 0, 1, \ldots$. The only thing we have to check is the distance from $(0, 1)$ to $(b_n/c_n, 0)$, but it is $\sqrt{1 + (b_n/c_n)^2} = a_n/c_n$, a rational number.

28. The sequence $\{a_n\}$ with $a_n = n(\max_{i \le n} b_n^{(i)})$ does the job.

29. The sequence $\{s_n\}$ with $s_n = 1 + \max_{i \le n} s_n^{(i)}$ does the job. [W. Sierpiński, Cardinal and Ordinal Numbers, Polish Sci. Publ., Warszawa, 1965, III.6/1]

30. Take those sequences that contain only finitely many nonzero elements. Their number is countable (see Problem 9), and since we can match any initial segment of any sequence with such a sequence, the property required in the problem follows. [W. Sierpiński, Cardinal and Ordinal Numbers, Polish Sci. Publ., Warszawa, 1965, III.6/2]

31. The answer is no: if $\{s_n^{(i)}\}_{n=0}^{\infty}$, $i = 0, 1, \ldots$ are any sequences of natural numbers, then there is a sequence $\{s_n\}$ of natural numbers with the property that for some subsequence $\{s_n^{(i_k)}\}_{n=0}^{\infty}$, $0 \le i_1 < i_2 < i_3 < \cdots$ it is true that $s_n \ne s_n^{(i_k)}$ for all n and k. In fact, if there are infinitely many i's with $s_0^{(i)} \ne 0$, then let $s_0 = 0$, and let I_0 be the set of all i's for which $s_0^{(i)} \ne 0$. If, on the other hand, there are only finitely many i's with $s_0^{(i)} \ne 0$, then let $s_0 = 1$, and let I_0 be the set of all i's for which $s_0^{(i)} = 0$. In either case let i_0 be the smallest element of I_0.

Next we define s_1, I_1, and i_1. Choose a natural number a_1 bigger than $s_1^{(i_0)}$. If there are infinitely many $i \in I_0$ with $s_1^{(i)} \ne a_1$, then let $s_1 = a_1$, and let I_1 be the set of all $i \in I_0$ for which $s_1^{(i)} \ne a_1$. If, however, there are only finitely many i's with $s_1^{(i)} \ne a_1$, then let $s_1 = a_1 + 1$, and let I_1 be the set of all i's for which $s_1^{(i)} = a_1$. Now let i_1 be the smallest element of I_1 larger than i_0.

In defining s_2, I_2 and i_2, choose a natural number a_2 bigger than $s_2^{(i_0)}$ and $s_2^{(i_1)}$. If there are infinitely many $i \in I_1$ with $s_1^{(i)} \ne a_2$, then let $s_2 = a_2$, and let I_2 be the set of all $i \in I_1$ for which $s_2^{(i)} \ne a_2$. If, however, there are only finitely many i's with $s_2^{(i)} \ne a_2$, then let $s_2 = a_2 + 1$ and let I_2 be the set of all i's for which $s_2^{(i)} = a_2$, and let i_2 be the smallest element of I_2 that is larger than both i_0 and i_1. If we continue this process, then the construction shows that $s_n \ne s_n^{(i_k)}$ for all n and k.

32. We can inductively define the permutations π_1, π_2, and π_3. Let $\pi_1(0) = 0$ and $\pi_2(0)$ and $\pi_3(0)$ be arbitrary two values for which $r_{\pi_2(0)} + r_{\pi_3(0)} = x_0 - r_0$.

Now suppose that $\pi_1(k), \pi_2(k)$, and $\pi_3(k)$ have already been defined for $k < n$. If n is divisible by 3, then let $\pi_1(n)$ be the smallest natural number that is not of the form $\pi_1(k)$ for some $k < n$. Note that for any s there is a unique t such that $r_s + r_t = x_n - r_{\pi_1(n)}$, so we can select $\pi_2(n) = s$ and $\pi_3(n) = t$ where s, t is such a pair that s is different from every $\pi_2(k)$, $k < n$, and t is different from every $\pi_3(k)$, $k < n$.

If n is of the form $3l + 1$ then do the same, just select first $\pi_2(n)$ to be the smallest natural number different from every $\pi_2(k)$, $k < n$, and then select $\pi_1(n)$ and $\pi_3(n)$ according to the above process, and similarly if n is of the form $3l + 2$, then select first $\pi_3(n)$ to be the smallest natural number different from every $\pi_3(k)$, $k < n$, and then select $\pi_1(n)$ and $\pi_2(n)$ according to the above process. It is clear that this procedure produces three permutations of $\mathbf{N}$, and the equation $x_n = r_{\pi_1(n)} + r_{\pi_2(n)} + r_{\pi_3(n)}$ holds for all n.

33. Consider as $\{x_n\}$ the sequence $1, 0, 0, 0, \ldots$. Suppose that for two permutations π_1 and π_2 we had $x_n = r_{\pi_1(n)} + r_{\pi_2(n)}$ for all n. Let π be the permutation of the natural numbers for which $r_{\pi(n)} = -r_n$ for all n. Then since $x_n = 0$ for $n = 1, 2, \ldots$, we have $\pi_2(n) = \pi \circ \pi_1(n)$ for all $n = 1, 2, 3, \ldots$, and since both π_2 and $\pi \circ \pi_1$ are permutations of $\mathbf{N}$, it follows that we must also have $\pi_2(0) = \pi \circ \pi_1(0)$. But this means that $r_{\pi_1(0)} + r_{\pi_2(0)} = 0 \neq 1 = x_0$, which is a contradiction.

34. First of all we prove that the number of elements in a finite Boolean algebra is a power of 2, and two finite Boolean algebras having the same number of elements are isomorphic. In fact, if $\langle A, +, \cdot, ', 0, 1\rangle$ is a finite Boolean algebra, and S is the set of its atoms (i.e., the elements $a \in A$ with the property that there is no $b \in A$ such $a \cdot b \neq 0, a$), then it is easy to see that every element is obtained by taking the sum of the elements in some subset C of S, and for different C's we get different elements in the Boolean algebra. Thus in this case, A has 2^n elements. If $\langle A^*, +^*, \cdot^*, '', 0^*, 1^*\rangle$ is another Boolean algebra with 2^n elements, then the set S^* of its atoms is of cardinality n, and it is easy to see that any 1-to-1 correspondence $f : S \to S^*$ extends in a natural way to an isomorphism from $\langle A, +, \cdot, ', 0, 1\rangle$ to $\langle A^*, +^*, \cdot^*, '', 0^*, 1^*\rangle$.

Now let $\langle A, +, \cdot, ', 0, 1\rangle$ and $\langle A^*, +^*, \cdot^*, '', 0^*, 1^*\rangle$ be two countably infinite Boolean algebras, and let $A = \{a_0, a_1, \ldots\}$ and $A^* = \{a_0^*, a_1^*, \ldots\}$ be an enumeration of the different elements in them. We use a back-and-forth argument, and for simpler notation we shall write $+, \cdot, '$ instead of $+^*, \cdot^*, ''$. Let $A_0 = \{0, 1\}$ and $A_0^* = \{0^*, 1^*\}$, and by induction we define increasing subalgebras A_n and A_n^* of some 2^{k_n} elements as follows. Suppose that A_{n-1} and A_{n-1}^* have already been defined, and $f_{n-1} : A_{n-1} \to A_{n-1}^*$ is an isomorphism between them. If n is even, then let $a_j \in A \setminus A_{n-1}$ be the element with smallest index j, and let A_n be the subalgebra generated by a_j and A_{n-1}. We claim that there is an element $a_m^* \in A^* \setminus A_{n-1}^*$ such that the subalgebra A_n^* generated by a_m^* and A_{n-1}^* is isomorphic to A_n, and what is more, the isomorphism f_{n-1} can be extended to an isomorphism f_n of A_n onto A_n^*. This will prove the statement in the problem. In fact, if n is odd then first select $a_m^* \in A^* \setminus A_{n-1}^*$ to be the element with smallest index m, and let A_n^* be the subalgebra generated by a_m^* and A_{n-1}^*, and to this select in a similar fashion as above an $a_j \in A \setminus A_{n-1}$ so that the subalgebra generated by a_j and A_{n-1} is isomorphic to A_n^*, and an isomorphism f_n can be obtained from an appropriate extension of f_{n-1}. Repeating this process it is clear that $\cup_n A_n = A$, $\cup_n A_n^* = A^*$, and if we define $f(a) = f_n(a)$ with an n for which $a \in A_n$, then this is a correct definition, and f establishes an isomorphism from A to A^*.

To simplify notation let us denote a_j by a. Since $a \notin A_{n-1}$, if s is an atom of A_{n-1}, then there are three possibilities: $s \cdot a = 0$, $s \cdot a \neq 0, s$ and $s \cdot a = s$. Let $s_1, s_2, \ldots, s_{k_{n-1}}$ be the atoms of A_{n-1} arranged in such an order that for $1 \leq i \leq p$ we have $s_i \cdot a = 0$, for $p < i \leq q$ we have $s_i \cdot a \neq 0, s_i$, and for $q < i \leq k_{n-1}$ we have $s_i \cdot a = s_i$ (some of these index sets may be empty, but we shall just discuss the general case). It is easy to see the atoms in A_n, which is the Boolean algebra generated by a and A_{n-1}, are the elements

$$s_1, \dots, s_p, s_{p+1} \cdot a, \dots s_q \cdot a, s_{p+1} \cdot a', \dots s_q \cdot a', s_{q+1}, \dots, s_{k_{n-1}}.$$

In fact, if s is any of these elements, then consider the set B of all elements $b \in A_n$ for which $s \cdot b = 0, s$ and $s \cdot b' = 0, s$. These elements form a subalgebra that contains a and all of the s_i's, so $B = A_n$. Thus, all these s's are atoms in A_n, and clearly the subalgebra generated by them contains A_{n-1} as well as a, hence there cannot be any other atom in A_n.

Note that $q > p$, for otherwise we would have $a \in A_{n-1}$. Let $s_i^* = f_{n-1}(s_i)$ be the corresponding atoms of A_{n-1}^*. We claim that there is an element $a^* \notin A_{n-1}^*$ such that for $1 \le i \le p$ we have $s_i^* \cdot a^* = 0^*$, for $p < i \le q$ we have $s_i^* \cdot a^* \ne 0^*, s_i^*$, and for $q < i \le k_{n-1}$ we have $s_i^* \cdot a^* = s_i^*$. In fact, since we assumed that the algebras are non-atomic, for every $p < i \le q$ there is an element $b_i^* \in A^*$ such that $b_i^* \cdot s_i^* \ne 0^*, s_i^*$, and then

$$a^* = b_{p+1}^* \cdot s_{p+1}^* + \dots + b_q^* \cdot s_q^* + s_{q+1}^* + \dots + s_{k_{n-1}}^*$$

is appropriate. Thus, the atoms of the Boolean algebra generated by a^* and A_{n-1}^* are

$$s_1^*, \dots, s_p^*, s_{p+1}^* \cdot a^*, \dots s_q \cdot a^*, s_{p+1}^* \cdot a^{*\prime}, \dots s_q^* \cdot a^{*\prime}, s_{q+1}^*, \dots, s_{k_{n-1}}^*,$$

$f_n(s_i) = s_i^*$ for $1 \le i \le p$ and $q < i \le k_{n-1}$, and if we define $f_n(s_i \cdot a) = s_i^* \cdot a^*$, $f_n(s_i \cdot a') = s_i^* \cdot a^{*\prime}$, then it is easy to see that this defines an isomorphism of A_n onto A_n^*, which is an extension of f_{n-1}.

35. In proving that **a)** implies **b)**, let us assume that $\mathcal{A}$ has uncountably many automorphisms $\varphi \in \Phi$ and let $B \subset A$ be an arbitrary finite subset. Then the restrictions of the automorphisms $\varphi \in \Phi$ to B cannot all be different (recall that there are only countably many mappings from B into A; see Problem 9), hence there are two distinct automorphisms φ_1 and φ_2 that agree on B. But then the non-identity automorphism $\varphi_2^{-1} \circ \varphi_1$ leaves all elements of B fixed, and this proves property **b)**.

Now let us assume that **b)** holds. Without loss of generality, we can assume that the ground set A of the algebra is $\mathbf{N}$. We set $N_0 = 0$, $\varphi_0 =$ identity, and inductively define the numbers N_n and the automorphisms φ_n as follows. Suppose that these are known for all indices not bigger than n. By assumption there is a non-identity automorphism φ_{n+1} that is the identity on the set $[0, N_n]$. Let a_{n+1} be an element with $\varphi_{n+1}(a_{n+1}) \ne a_{n+1}$, and let C_{n+1} be the set of the inverse images of a_{n+1} under the finitely many mappings $\varphi_n^{\epsilon_n} \circ \dots \circ \varphi_1^{\epsilon_1}$, where $e_i = 0$ or 1 independently of each other, and φ^ϵ is φ if $\epsilon = 1$ and φ^ϵ is the identity automorphism if $\epsilon = 0$. We also set D_{n+1} to be the set of all the images of the elements $j \le N_n$ under the mappings $\varphi_n^{\epsilon_n} \circ \dots \circ \varphi_1^{\epsilon_1}$ where again $\epsilon_i = 0$ or 1 independently of each other.

Let N_{n+1} be a number bigger than $N_n + 1$, the elements of C_{n+1} and D_{n+1}, a_{n+1} and $\varphi_{n+1}(a_{n+1})$. We claim that if $\epsilon_1, \epsilon_2, \dots$ is any 0–1 sequence, then the automorphism

$$\varphi_{\epsilon_1,\epsilon_2,\ldots} = \cdots \circ \varphi_n^{\epsilon_n} \circ \cdots \circ \varphi_1^{\epsilon_1} \tag{2.1}$$

is well defined, and for different 0–1 sequences this defines a different automorphism of $\mathcal{A}$. This will prove a), for this way we get as many automorphisms as infinite 0–1 sequences, and the infinite 0–1 sequences form an uncountable set (see the solution of Problem 7 or apply Problems 7 and 3.11).

Note that if B is an arbitrary finite subset of A, say $B \subset [0, N_m]$, then for all $n \geq m$ the automorphisms

$$\varphi_n^{\epsilon_n} \circ \cdots \circ \varphi_1^{\epsilon_1}$$

agree on B. In fact, the image B' of B under

$$\varphi_m^{\epsilon_m} \circ \cdots \circ \varphi_1^{\epsilon_1}$$

is part of $D_{m+1} \subseteq D_{n+1}$, hence all the authomorphisms φ_{n+1}, $n \geq m$ are the identities on that image set B'. This proves that the right-hand side of (2.1) is well defined and is a 1-to-1 homomorphism of $\mathcal{A}$ into itself. But it is actually a mapping of A onto A, and hence it is an automorphism. Indeed, if $a \in A$ is given, then let n_a be so large that for $n > n_a$ we have $\varphi_n(a) = a$, and choose b in such a way that $\varphi_{n_a}^{\epsilon_n} \circ \cdots \circ \varphi_1^{\epsilon_1}(b) = a$ (such a b exists, for $\varphi_{n_a}^{\epsilon_n} \circ \cdots \circ \varphi_1^{\epsilon_1}$ is an automorphism). It is clear that the image of b under the mapping (2.1) is a.

Thus, we have found that each $\varphi_{\epsilon_1,\epsilon_2,\ldots}$ is an automorphism of $\mathcal{A}$, and it is left to show that for different 0–1 sequences we obtain different automorphisms this way. In fact, let $\epsilon_1, \epsilon_2, \ldots$ and $\epsilon'_1, \epsilon'_2, \ldots$ be two different 0–1 sequences, and let, say, $\epsilon_1 = \epsilon'_1$, …, $\epsilon_n = \epsilon'_n$ but $\epsilon_{n+1} = 1$ while $\epsilon'_{n+1} = 0$. If b_{n+1} is the element in A such that $\varphi_n^{\epsilon_n} \circ \cdots \circ \varphi_0^{\epsilon_0}(b_{n+1}) = a_{n+1}$, then, by the choice of the numbers a_{n+1}, N_{n+1} and of the automorphisms φ_j with $j > n+1$, we have $\varphi_{\epsilon_1,\epsilon_2,\ldots}(b_{n+1}) \neq a_{n+1}$, while $\varphi_{\epsilon'_1,\epsilon'_2,\ldots}(b_{n+1}) = a_{n+1}$, hence the two automorphisms $\varphi_{\epsilon_1,\epsilon_2,\ldots}$ and $\varphi_{\epsilon'_1,\epsilon'_2,\ldots}$ are different. [M. Makkai, see G.J. Székely (editor), Contests in Higher Mathematics, Problem Books in Mathematics, Springer-Verlag, 1996, pp. 74–75.]

36. The possible starting points of the rabbit are the lattice points (a, b), $a, b \in \mathbf{Z}$, and the possible jumps are the vectors (p, q), $p, q \in \mathbf{Z}$ (which means that if at a certain time the rabbit is in a position (n, m), then in the next minute it will be in $(n + p, m + q)$). Thus, the motion of the rabbit can be described by the quadruple (a, b, p, q), and the set of all such quadruples is a countable set (see Problem 2). Let us enumerate all these possible motions into a sequence $\{(a_i, b_i, p_i, q_i)\}_{i=1}^{\infty}$. If the motion of the rabbit is according to the quadruple (a_i, b_i, p_i, q_i), then after k hours from the start the rabbit will be in the position $(a_i + 60kp_i, b_i + 60kq_i)$. Thus, if we test with a trap at the ith hour the coordinate $(a_i + 60ip_i, b_i + 60iq_i)$, then we catch the rabbit. Since we can do that for every i, we will eventually catch it.

37. Let $A = \{a_i\}_{i=0}^{\infty}$, and let $a_i = 0.\alpha_1^{(i)}\alpha_2^{(i)}\alpha_2^{(i)}\dots$ be the decimal representation for a_i (either one if there are two such representations). Let II select $y_j = 4$ if $a_{2j}^{(j)} \neq 4$, otherwise it selects $y_j = 6$. Then whatever numbers $x_1, x_2, \dots$ the player I selects, the number $0.x_1y_1x_2y_2\dots$ does not coincide with any of the a_j's so it is not in A (see also the proof of Problem 7).

38. I can force winning only if he lists only one digit infinitely many times. In fact, suppose that he lists both the digits a and b infinitely many times. Let $A = \{a_i\}_{i=0}^{\infty}$, and let $a_i = 0.\alpha_1^{(i)}\alpha_2^{(i)}\alpha_2^{(i)}\dots$ be the decimal representation of a_i (either one if there are two such representations). Then II can play in the following way: he makes sure that $y_{2j} = a$ if $a_{2j}^{(j)} \neq a$, otherwise he puts $y_{2j} = b$. It is easy to see that II can form such a permutation, and then II wins, for the number $0.y_1y_2\dots$ does not coincide with any one of the a_i's, so it is not in A.

Thus, I can have a winning strategy only if he selects some finitely many digits $x_1, x_1, \dots, x_{m_0}$, and then on he always selects the same digit, say a (in other words, for $i > m_0$ he chooses $x_i = a$). In this case II can still form any permutations, and I wins only if all the (countably many) numbers

$$\sum_{i=1}^{m_0} \frac{x_i}{10^{l_i}} + \left(\sum_{j=1}^{\infty} \frac{a}{10^j} - \sum_{i=1}^{m_0} \frac{a}{10^{l_i}}\right) = \frac{a}{9} + \sum_{i=1}^{m_0} \frac{x_i - a}{10^{l_i}},$$

where $1 \le l_1, \dots, l_{m_0} < \infty$ are arbitrary different integers, lie in A. Thus, I can force winning only if there are a digit $0 \le a \le 9$ and finitely many digits $x_1, x_2, \dots, x_{m_0}$ such that A contains all numbers of the form

$$\frac{a}{9} + \sum_{i=1}^{m_0} \frac{x_i - a}{10^{l_i}}, \qquad 1 \le l_1, \dots l_{m_0},\ l_i \neq l_j \text{ if } i \neq j.$$

By letting here l_i tend to infinity for all $i = 1, 2, \dots, m_0$ we get that A must contain the number $a/9$ (recall that A is closed).

On the other hand, it is obvious that if A contains a number of the form $a/9$, $a = 0, 1, \dots, 9$, and I chooses the sequence

$$a, a, \dots,$$

then he wins.

Thus, the answer to the problem is that I has a winning strategy if and only if A contains one of the numbers $0, 1/9, 2/9, \dots, 8/9$.

39. This is a special case of Problem 8.48.

* * *

40. Suppose first that H has cardinality at most κ, and without loss of generality we may assume $H = \kappa$. It is clear that the representation $H \times H = B \cup C$ with $B = \{(\xi, \eta) \ : \ \xi, \eta < \kappa, \eta < \xi\}$ and $C = \{(\xi, \eta) \ : \ \xi, \eta < \kappa, \xi \leq \eta\}$ is such that B intersects every vertical line $\{(x, y) \ : \ x = \xi\}$ in the set $\{(\xi, \eta) \ : \ \eta < \xi\}$, which is of cardinality smaller than κ, and similarly C intersects every horizontal line in less than κ points.

Conversely, suppose that $H^2 = B \cup C$, where B resp. C intersect every vertical resp. horizontal lines in less than κ points, and suppose that to the contrary to what we have to prove, the cardinality of H is bigger than κ. Take a subset $K \subset H$ of cardinality κ. Then $(H \times K) \cap C$ is of cardinality at most κ, since for each $y \in K$ the number of $(x, y) \in C$ is of cardinality smaller than κ. But for every $x \in H$ there is a $y \in K$ such that $(x, y) \in C$, since the number of those y for which $(x, y) \in B$ is of cardinality smaller than κ. Thus, $(H \times K) \cap C$ has to be at least of the cardinality of H, i.e., it has to be of cardinality bigger than κ. This contradiction shows that, indeed, H is of cardinality at most κ.

3

Equivalence

1. By considering $A \times \{0\}$ and $B \times \{1\}$ instead of A and B, we may assume that A and B are disjoint. Let $x \sim y$ if x or y can be reached from the other one by alternatively applying f and g finitely many times. Then this $\sim$ is an equivalence relation on $A \cup B$. Every equivalence class is a finite, one-way infinite or two-way infinite path $\ldots x_j, x_{j+1}, \ldots$, where $x_{j+1} = f(x_j)$ if $x_j \in A$ and $x_{j+1} = g(x_j)$ if $x_j \in B$. Let us call the equivalence class C of type I if it is a finite path (actually, a cycle), of type II if it is a two-way infinite path, of type III if it is a one-way infinite path that starts in A, and of type IV if it is a one-way infinite path that starts in B. Note that if C is of class I, II, or III, then the restriction of f to $C \cap A$ maps $C \cap A$ onto $C \cap B$, and similarly, if C is of class IV, then the restriction of g to $C \cap B$ maps $C \cap B$ onto $C \cap A$. Thus, if U is the union of all equivalence classes of type I, II, and III, and $F : A \to B$ is defined as $F(x) = f(x)$ if $x \in U \cap A$ and $F(x) = g^{-1}(x)$ if $x \in A \setminus U$, then this F is a 1-to-1 mapping of A onto B. Thus, the selection $A_1 = U$, $A_2 = A \setminus U$, $B_1 = f[U]$, $B_2 = B \setminus B_1$ is a decomposition that satisfies the requirements. [G. Cantor, this proof is due to Gy. König]

2. See the preceding problem.

3. If $f : A \to B$ is 1-to-1, and the range of f in B is B^*, then let $g(x) = f^{-1}(x)$ if $x \in B^*$, and otherwise let $g(x) = a_0$ where a_0 is a fixed element of A. Then this g is a mapping from B onto A.

Conversely, let $g : B \to A$ be a mapping of B onto A. The relation "$x \sim y$ if $g(x) = g(y)$" is an equivalence relation on B. Let h be a choice function on the set of equivalence classes, i.e., if C is an equivalence class, then $h(C)$ is an element of C. It is clear that the map $f(x) = h(g^{-1}[x])$ is a 1-to-1 mapping of A into B.

4. A includes an infinite sequence $a_0, a_1, \ldots$ of different elements (just select the elements $a_0, a_1 \ldots$ from A one after another). Now $B \cup \{a_0, a_1 \ldots\}$

is countable, so it is equivalent to $\{a_0, a_1 \ldots\}$. Let $g : B \cup \{a_0, a_1 \ldots\} \to \{a_0, a_1 \ldots\}$ be a 1-to-1 correspondence. Clearly, the mapping $h(x) = g(x)$ if $x \in B \cup \{a_0, a_1 \ldots\}$ and $h(x) = x$ otherwise is a 1-to-1 mapping of $B \cup A$ onto A.

5. The set $A \setminus B$ cannot be countable, for then A would also be countable. Thus, it is uncountable, and the previous problem shows that $A = (A \setminus B) \cup B$ is equivalent to $A \setminus B$.

6. Use the previous problem and the facts that the set of real numbers is uncountable, while the set of rational numbers is countable.

7. Recall that the Cantor set is precisely the set of those $x \in [0, 1]$ that have a ternary expansion that does not contain the digit 1. Therefore, the correspondence

$$(\epsilon_0, \epsilon_1, \ldots) \mapsto 0.(2\epsilon_0)(2\epsilon_1)\ldots,$$

where the number on the right-hand side is given by its ternary expansion, establishes an equivalence between the set of infinite 0–1 sequences and the Cantor set.

8. **a)** $f(n, m) = 2^n 3^m$.

b) $f(x) = 1/2 + 2(\arctan x)/\pi$.

c) In view of b), it is enough to give a 1-to-1 mapping from $(0, 1)$ into the set of infinite 0–1 sequences. If $x \in (0, 1)$, and its binary expansion is $x = 0.\alpha_1\alpha_2\ldots$ (fix any one if x has two binary expansions), then the mapping $x \mapsto (\alpha_1, \alpha_2, \ldots)$ is clearly appropriate.

d) As in Problem 7, associate with an infinite 0–1 sequence $\epsilon_0, \epsilon_1, \ldots$ the number $0.(2\epsilon_0)(2\epsilon_1)\ldots$ in ternary form. (Warning: it would be wrong to associate with it the number $0.\epsilon_0\epsilon_1\ldots$ in binary form, for then the sequences $1, 0, 0, 0, \ldots$ and $0, 1, 1, 1, \ldots$ would have the same image.)

e) With a sequence $n_0, n_1, \ldots$ of natural numbers associate the 0–1 sequence, in which $n_0 + 1$ zeros are followed by a single 1, then $n_1 + 1$ zeros are followed by a 1, etc.

f) Let $S = \{x_0, x_1, \ldots\}$ be a sequence of real numbers, and let

$$x_j = \pm \cdots \alpha_{-2}^{(j)}\alpha_{-1}^{(j)}.\alpha_1^{(j)}\alpha_2^{(j)}\ldots$$

be the binary representation of x_j, where $\alpha_{-k}^{(j)} = 0$ except for a finite number of the k's (thus, we put infinitely many zeros in front of the standard binary representation). Let also $\alpha_0^{(j)} = 1$ if x_j is positive, and otherwise $\alpha_0^{(j)} = 0$. Now associate with S the sequence

$$\alpha_0^{(0)}, \alpha_{-1}^{(0)}, \alpha_0^{(1)}, \alpha_1^{(0)}, \alpha_{-2}^{(0)}, \alpha_{-1}^{(1)}, \alpha_0^{(2)}, \alpha_1^{(1)}, \alpha_2^{(0)}, \alpha_{-3}^{(0)}, \alpha_{-2}^{(1)}, \alpha_{-1}^{(2)}, \alpha_0^{(3)}, \alpha_1^{(2)} \ldots .$$

This is a 1-to-1 mapping.

The equivalence of the two sets in **a)–f)** immediately follows from the equivalence theorem.

9. **a)** $f(n) = (k, m)$, where the prime decomposition of n is of the form $n = 2^k \cdot 3^m \cdots$ (here we allow k and m to be equal to 0).

b) $f(n) = (-1)^k l/(m+1)$, where $n = 2^k \cdot 3^l \cdot 5^m \cdots$

c) If x is in the Cantor set, then it has a ternary representation $x = 0.\alpha_1\alpha_2\ldots$, where each α_j is 0 or 2. Let $f(x) = 0.(\alpha_1/2)(\alpha_2/2)\ldots$, where the number on the right is understood in binary form.

d) With a 0–1 sequence $\alpha_1, \alpha_2, \ldots$ associate $0.\alpha_1\alpha_2\ldots$ in binary form.

10. **a)** If (a, b) and (c, d) are bounded intervals, then let $f(x) = c + (d - c)(x - a)/(b - a)$. If, say, a is finite, $b = \infty$ and (c, d) is finite, then let $f(x) = c + 2(d-c)(\arctan(x-a))/\pi$. The other cases can be similarly handled.

b) Let $g(n, m) = (n+m)\cdot(n+m+1)/2+n$ (this g is called the Gödel pairing function). It is easy to see that g is a 1-to-1 mapping of $\mathbf{N}\times\mathbf{N}$ onto $\mathbf{N}$. In fact, we have $g(n, m) = k$ if and only if $n + m$ is the unique nonnegative integer a with $a(a+1) \le 2k < (a+1)(a+2)$, and then n is equal to $k - a(a+1)/2$ and m is $a - (k - a(a+1)/2)$.

c) Associate with any subset $A \subseteq X$ its characteristic function: $\chi_A(x) = 1$ if $x \in A$ and $\chi_A(x) = 0$ if $x \in X \setminus A$. The mapping $A \mapsto \chi_A$ is a 1-to-1 correspondence (bijection) between the elements of the power set $\mathcal{P}(X)$ and ${}^X\{0, 1\}$.

d) If $a_0, a_1, \ldots$ is an infinite sequence of the numbers $0, 1, 2$, then let us write in it instead of 1 the sequence $1, 0$, and instead of 2 the sequence $1, 1$. Then we get an infinite 0–1 sequence, and it is easy to see that every infinite 0–1 sequence is obtained from a unique 0–1–2 sequence $a_0, a_1, \ldots$.

e) Let $x \in [0, 1)$ and let $x = 0.a_1a_2\ldots$ be its decimal expansion, where infinitely many of the a_i's is different from 9. Let us group consecutive 9's in the expansion with the first digit after them that is different from 9, and all other digits form a single group, e.g., if $x = 0.12979996659999793\ldots$, then the grouping is (indicating the groups by brackets)

$$x = 0.(1)(2)(97)(9996)(6)(5)(99997)(93)\ldots,$$

and let us call the blocks in this grouping by $x_1, x_2, \ldots$, i.e.,

$$x = 0.(x_1)(x_2)(x_3)\ldots,$$

where the harmless brackets are added only to show the grouping. Now let $f(x) = (y, z)$, where $y = 0.(x_1)(x_3)(x_5)\ldots$ and $z = 0.(x_2)(x_4)(x_6)\ldots$. Note that this form of y and z is the one that we obtain after the aforedescribed grouping, and conversely, if $y = 0.(y_1)(y_2)(y_3)\ldots$ and $z = 0.(z_1)(z_2)(z_3)\ldots$ are given in grouped from, then so is $x = 0.(y_1)(z_1)(y_2)(z_2)\ldots$, this x belongs to $[0, 1)$ and it is the unique number with $f(x) = (y, z)$. [Gy. König]

11. **a)** Use the equivalence theorem and Problems 8, c) and d).

b) In view of parts c) and f) of Problem 8, **R** is equivalent to the set of infinite real sequences (recall that in Problem 8 the given pairs of sets are actually equivalent, as is stated in the last part of the problem). Hence the claim follows from the equivalence theorem, for $\mathbf{R}^n$ is the set of real sequences of length n.

c) As it has just been said, this follows from the equivalence theorem if we use parts c) and f) of Problem 8.

12. **a)** With a function $f : B \cup C \to A$ associate the pair $(f|_B, f|_C)$.

b) With a $g : C \times B \to A$ associate $f : C \to {}^B A$, where $f(c)(b) = g((c, b))$.

c) With a $(g, h) \in {}^C A \times {}^C B$ associate $f : C \to A \times B$ where $f(c) = (g(c), h(c))$.

13. For **a)** consider the imbedding $x \to \{x\}$ of X into $\mathcal{P}(X)$.

To verify **b)** we want to show that there is no mapping from X onto $\mathcal{P}(X)$ (see Problem 3). Let $f : X \to \mathcal{P}(X)$ be any mapping. We have to show that f is not onto $\mathcal{P}(X)$. Let $A = \{a \in X \ : \ a \notin f(a)\}$. We claim that A does not have a preimage under f. In fact, suppose that is not the case, and $f(a_0) = A$ with some $a_0 \in X$. Then there are two possibilities:

1. $a_0 \in A$, i.e., $a_0 \in f(a_0)$ which is not possible for then a_0 cannot be in A by the definition of A,

2. $a_0 \notin A$, which is again not possible, for then $a_0 \notin f(a_0)$, so a_0 should belong to A.

Thus, in either case we have arrived at a contradiction, which means that a_0 with the property $f(a_0) = A$ does not exist.

4

Continuum

1. Let $\mathcal{H}$ be a family of lines in the plane such that $\mathcal{H}$ has fewer elements than $\mathbf{R}$. Consider the vertical lines $x = r$, $r \in \mathbf{R}$. Not all of them can belong to $\mathcal{H}$, say the line $l_0 : x = r_0$ is not in $\mathcal{H}$. But then every element of $\mathcal{H}$ intersects the line l_0 in at most 1 point, so there are fewer than continuum many intersections on l_0, hence some points of l_0 are not covered by any line in $\mathcal{H}$.

2. See Problem 3.11, a).

3. This follows from Problems 2 and 3.8, f).

4. An $x \in [0, 1]$ is in the Cantor set if and only if it can be represented in base 3 as $x = 0.\alpha_1\alpha_2\dots$ with $\alpha_i = 0$ or $\alpha_i = 2$. Thus, the Cantor set is equivalent to the set of 0–2 sequences. Apply Problem 2.

5. Let $A = \{x_0, x_1, \dots\}$ be an enumeration of the elements in the set so that we list each element exactly once. Clearly every subset $X \subseteq A$ is uniquely determined by the function $f(j) = 1$ if $x_j \in X$ and $f(j) = 0$ if $x_j \notin X$. Such an f is nothing else than a 0–1 sequence, so we can apply Problem 2.

6. It is sufficient to show the claim for $\mathbf{R}$. But $\mathbf{R}$ has at most as many countable subsets as sequences, hence the claim follows from Problems 3 and 5.

7. Let $\mathcal{B}$ be the set of all balls in $\mathbf{R}^n$ with rational center and rational radius. Then $\mathcal{B}$ is countable (see Problem 2.17), and every open set is a union of a subset of $\mathcal{B}$ (Problem 2.18). Thus, by the preceding problem, there are at most continuum many of them. It is also clear that there are at least as many open sets as real numbers, so there are exactly continuum many open sets by the equivalence theorem.

The closed sets are the complements of the open ones, so their number is also continuum.

8. Let $\{B_i : i < \omega\}$ be a countable base for the Hausdorff space X. The mapping $x \mapsto \{i < \omega : x \in B_i\}$ is an injective mapping of X into $\mathcal{P}(\omega)$, a set of size $\mathbf{c}$, hence $|X| \le \mathbf{c}$.

9. Let $(X, \mathcal{T})$ be an infinite topological space with the Hausdorff separation property, i.e., any two points have disjoint neighborhoods. It is clear that then any finitely many points can be simultaneously separated by disjoint neighborhoods.

The solution is based on the following observation: let $x_0, \ldots, x_n$ be different points in X such that with some neighborhoods $G_0, \ldots, G_n$ of them there is an infinite set A_n that does not intersect any G_i. Then there is a point $x_{n+1} \in A_n$, a neighborhood G_{n+1} of it and an infinite subset $A_{n+1} \subset A_n$ such that $x_i \notin G_{n+1}$ for all $0 \le i \le n$ and $G_{n+1} \cap A_{n+1} = \emptyset$. In fact, select any two points $y_1, y_2 \in A_n$ and two disjoint neighborhoods U_1, U_2 for them. We can also achieve that $x_i \notin U_1, U_2$ for all $0 \le i \le n$. Then either $U_1 \cap A_n$ is finite, or $A_n \setminus U_2$ is infinite. In any case, one of $A_n \setminus U_1$ or $A_n \setminus U_2$ is infinite. Suppose, e.g., that $A_n \setminus U_1$ is infinite. Then the $x_{n+1} = y_1$, $G_{n+1} = U_1$ and $A_{n+1} = A_n \setminus U_1$ is an appropriate choice.

Now starting from the empty set, construct the above points and neighborhoods for all n. Then clearly $G_n \cap \{x_0, x_1, \ldots\} = \{x_n\}$, which shows that if $I, J \subseteq \mathbf{N}$ are two different subsets of $\mathbf{N}$, then $\cup_{n\in I} G_n \ne \cup_{n\in J} G_n$. Thus, all the open sets $\cup_{n\in I} G_n$, $I \subseteq \mathbf{N}$ are different, and so there are at least continuum many open sets in X by Problem 5.

10. Without loss of generality, we may assume $A = \mathbf{N}$ and $B = \mathbf{R}$. The set of functions $f : \mathbf{N} \to \mathbf{R}$ is the set of all sequences of real numbers. Now apply Problem 3.

11. Any continuous $f : \mathbf{R} \to \mathbf{R}$ is uniquely determined by its restriction to $\mathbf{Q}$. Apply the preceding problem.

12. It is enough to prove that $\mathbf{R} \times \mathbf{R} \times \cdots$ is of cardinality continuum. But this set is the same as ${}^{\mathbf{N}}\mathbf{R}$, the set of infinite real sequences. Now apply Problem 3.

13. It is enough to show the claim for disjoint sets. Let the sets be A_γ, $\gamma \in \Gamma$, and let $f_\gamma : A_\gamma \to \mathbf{R}$ be a 1-to-1 mapping. Then the union $\cup_{\gamma\in\Gamma} A_\gamma$ can be mapped into $\mathbf{R} \times \Gamma$ by the 1-to-1 mapping $F(a) = (f_\gamma(a), \gamma)$ if $a \in A_\gamma$. Now apply the preceding problem, according to which $\mathbf{R} \times \Gamma$ is of cardinality at most continuum.

14. **a)** Apply Problem 12.

b) See Problem 3.

c) A continuous curve γ is $\gamma(t) = (\gamma_1(t), \gamma_2(t))$, $t \in (0,1)$, where $\gamma_1, \gamma_2 : (0,1) \to \mathbf{R}$ are continuous functions. Apply Problems 11 and 12 (see also Problem 3.10, a)).

d) Let f be a monotone real function, and let S_f be the set of its discontinuity points. Then S_f is countable (see Problem 5.6). Now let $X \subset \mathbf{R}$ be countable, and let $\mathcal{M}_X$ be the collection of all monotone functions f with $S_f \subset X$. Every $f \in \mathcal{M}_X$ is uniquely determined by its restriction to the set $X \cup \mathbf{Q}$, and there are only continuum many functions $f : X \cup \mathbf{Q} \to \mathbf{R}$ (see Problem 10). Thus, $\mathcal{M}_X$ is of cardinality at most continuum.

By Problem 6 there are at most continuum many possibilities for X. Thus, by Problem 13 the union $\cup_X \mathcal{M}_X$, which is the set of monotone functions, is of cardinality at most continuum. Since clearly there are at least as many monotone functions as real numbers, the set of monotone functions is of cardinality continuum by the equivalence theorem.

e) See the preceding proof, but apply Problem 5.4 instead of 5.6 in the proof.

f) See the solution to Problem d).

g) This problem cannot be solved along the lines of the preceding three problems. In fact, a lower semi-continuous function can have more than countably many discontinuity points (consider, e.g., the characteristic function of the complement of the Cantor set).

The key to the solution is the observation that a function f is lower semi-continuous if and only if all its level sets of the form $\{x : f(x) > r\}$ are open. Furthermore, each f is determined by its level sets $\{x : f(x) > r\}$ with rational r. Thus, there are at most as many lower semi-continuous functions as sequences of open subsets of $\mathbf{R}$, and since there are continuum many open sets in $\mathbf{R}$ (see Problem 7), there are continuum many sequences of them (see Problem 3).

h) Every permutation is a mapping from $\mathbf{N}$ into $\mathbf{N}$, so there are at most continuum many of them in view of Problem 10. To show that there at least continuum many permutations, consider the transpositions $\pi_i = ((2i)(2i+1))$ that interchange $2i$ and $(2i+1)$, and leave everything else fixed. For any 0–1 sequence $\epsilon := (\epsilon_0, \epsilon_1, \ldots)$ consider the permutation π_ϵ that is the product of all those π_i's for which $\epsilon_i = 1$. For different ϵ's we get different π_ϵ's hence, by Problem 2, there are at least continuum many permutations of $\mathbf{N}$.

i) An ordering of $\mathbf{N}$ is a subset of $\mathbf{N} \times \mathbf{N}$, hence there are at most continuum many of them in view of Problems 2.2 and 5. Now every permutation π of $\mathbf{N}$ defines a well-ordering of $\mathbf{N}$ (set $x \prec y$ if $\pi(x) < \pi(y)$), so there are at least continuum many well-orderings by the previous problem.

j) A closed additive subgroup is a closed set. Apply Problem 7 to deduce that there are at most continuum many closed additive subgroups. But their number is exactly continuum by the equivalence theorem and by the fact that all the sets $\{nx\}_{n=-\infty}^{\infty}$, $x \in \mathbf{R}$ are closed additive subgroups of $\mathbf{R}$.

k) For $x \in (0,1)$ let $f_x \in C[0,1]$ be the piecewise linear function on $[0,1]$ that vanishes outside $(0,x)$ and for which $f(x/2) = 1$ (thus, the curve of f starts from the origin, goes straight to the point $(x/2,1)$, from then to the point $(x,0)$, and follows the real line from then on). Since each set $\{\lambda f_x\}_{\lambda \in \mathbf{R}}$, $x \in (0,1)$ is a closed subspace of $C[0,1]$ that are different for different $x \in (0,1)$, there are at least continuum many closed subspaces in $C[0,1]$. To show that their number is exactly continuum, it is enough to prove that there are only continuum many closed sets in $C[0,1]$, and by the proof of Problem 7 this will be accomplished if we show a countable set $\mathcal{B}$ of open balls such that every open set is a union of some balls in $\mathcal{B}$. Clearly as $\mathcal{B}$ we can choose the set of balls $B_r(P) = \{g \ : \ |g - P| < r\}$ with rational radius r and with center at P where P is a polynomial with rational coefficients (cf. Problem 2.5 and the fact that $\mathbf{Q} \sim \mathbf{Z}$). This construction works by the Weierstrass approximation theorem.

l) First of all we should make the clarification that functions in $L^2[0,1]$ are considered the same if they agree almost everywhere. This makes $L^2[0,1]$ into a set of power continuum. In fact, we know that $L^2[0,1]$ is isomorphic with l_2, the set of all real sequences $(x_0, x_1, \ldots)$ with $\sum_i x_i^2 < \infty$, and by Problem 3 there are at most continuum many such sequences.

Every bounded linear transformation is uniquely determined by its restriction to a dense subset, hence, in view of Problem 10, it is enough to show a countable dense subset in l_2. But that is easy, just take the set of all sequences $(x_0, x_1, \ldots)$ such that $x_i = 0$ for all $i \geq m$ with some m, and $x_0, \ldots, x_{m-1}$ are rational numbers (see Problem 2.4).

To show that there are at least continuum many bounded linear operators on l_2, just take the constant multiples of the identity operator.

15. Since $\mathbf{R}$ and $\mathbf{R}^\infty$ are equivalent, it is enough to show that $\mathbf{R}^\infty$ cannot be represented as the union of countably many sets none of which is equivalent to $\mathbf{R}$. Let $A_0, A_1, \ldots$ be subsets of $\mathbf{R}^\infty$ not equivalent to $\mathbf{R}$. Let A_j^* be the projection of A_j onto the jth coordinate axis, i.e., A_j^* consists of those numbers $a \in \mathbf{R}$ for which there is an $(x_0, x_1, \ldots) \in A_j$ with $x_j = a$. Since A_j is of power less than continuum, it follows that A_j^* cannot be equal to $\mathbf{R}$. Thus, for each j there is an $a_j \in \mathbf{R} \setminus A_j^*$. But then the sequence $(a_0, a_1, \ldots)$ does not belong to any of the sets $A_0, A_1, \ldots$, but it belongs to $\mathbf{R}^\infty$, which means that, as we have claimed, $\cup_n A_n$ cannot be the whole $\mathbf{R}^\infty$. [W. Sierpiński, Cardinal and Ordinal Numbers, Polish Sci. Publ., Warszawa, 1965, , VI.7. Theorem 15]

16. Consider the lines $l_n := \{(x,y) \ : \ x = n\}$, $n = 0, 1, \ldots$, and their union H. If none of them intersects $\mathbf{R}^2 \setminus A$ in continuum many points then $H \cap (\mathbf{R}^2 \setminus A)$ is of cardinality less than continuum by Problem 15. But this is not possible, for each horizontal line intersects A in at most finitely many points, so each

such line has to intersect $H \cap (\mathbf{R}^2 \setminus A)$. [P. Erdős, Proc. Amer. Math. Soc., **1**(1950), 127–141]

17. Since, according to Problem 15, countable union of subsets of $\mathbf{R}$ each of cardinality less than continuum is again of cardinality less than continuum, the proof of Problem 2.11 can be copied by replacing "uncountable" by "of power continuum" everywhere. [W. Sierpiński, Cardinal and Ordinal Numbers, Polish Sci. Publ., Warszawa, 1965, , VI.7/2]

18. The statement follows from the solution of Problem 2.35.

19. Let $\mathcal{A}$ be an infinite σ-algebra. We are going to show that there is an infinite family $\mathcal{S}$ of pairwise disjoint sets in $\mathcal{A}$. Since the union of any countable subset of $\mathcal{S}$ is in $\mathcal{A}$, and there are at least continuum many such unions/subsets (see Problem 5), it follows that the cardinality of $\mathcal{A}$ is at least continuum.

Call a nonempty $A \in \mathcal{A}$ an atom if it cannot be written as the union of two nonempty disjoint sets in $\mathcal{A}$. Two different atoms cannot have a nonempty intersection, thus if there are infinitely many atoms, then their collection can serve as $\mathcal{S}$. If there are only finitely many atoms, then let them be $A_0, A_1, \ldots, A_m$. Since $\mathcal{A}$ is infinite, there must be an element B in $\mathcal{A}$ which is not a union of some of these atoms, and then considering $B \setminus (A_0 \cup \cdots \cup A_m)$ instead of B, we can even assume that B is nonempty and does not include as a subset any atom. Thus, B can be decomposed into nonempty disjoint sets as $B = B_1 \cup C_1$. Here B_1 has the same property as B, hence it can be written as $B_1 = B_2 \cup C_2$ with disjoint and nonempty B_2 and C_2. Do the same thing with B_2, etc. The sets $C_1, C_2, \ldots$ will be nonempty and pairwise disjoint, so we can take as $\mathcal{S}$ their collection.

20. See Problem 12.24.

21. The set of Borel sets is the σ-algebra generated by the open intervals (open sets in $\mathbf{R}^n$). Thus, there are continuum many Borel sets by Problems 19, 20, and 7.

A real function f is a Borel function if and only if all of its level sets $\{x : f(x) > r\}$ are Borel sets. Furthermore, each f is determined by its level sets $\{x : f(x) > r\}$ with rational r. Thus, there are at most as many Borel functions as sequences of Borel sets, so there are at most continuum many of them (see the solution to Problem 14, g)).

22. Every Baire function is a Borel function. Use the preceding problem.

23. See Problem 3.13.

24. Let $a \neq b$ be two elements in A, and to a subset $Y \subset X$ assign the function f_Y which maps the elements in Y to a and the elements in $X \setminus Y$ to b. For

different Y's these f_Y's are different, so we have at least as many functions in ${}^XA$ as subsets of X. Apply now Problem 23.

25. **a)** This is ${}^{\mathbf{R}}\mathbf{R}$, apply Problem 24.

b) Let $f : [0,1] \to [0,1]$ be an arbitrary function. The mapping $F(x) = (x, f(x))$, $x \in [0,1]$ can be extended to a 1-to-1 correspondence between $\mathbf{R}$ and $\mathbf{R}^2$. Thus, there are at least as many 1-to-1 correspondences as functions $f : [0,1] \to [0,1]$, and we can apply Problem 24.

c) We use that if B is a basis of $\mathbf{R}$ considered as a linear space over $\mathbf{Q}$ (i.e., a Hamel basis), then B is of power continuum (see Problem 15.3). Now let $Y \subset B$ be arbitrary, and consider the set B_Y consisting of all numbers x in $B \setminus Y$ and all $2x$ with $x \in Y$. Clearly, this is again a basis, and we have as many such bases as possible choices of Y, i.e., more than continuum many (see Problem 23).

For more on Hamel bases, see Chapter 15. In particular, Problem 15.4 says that there are $2^{\mathfrak{c}}$ Hamel bases.

d) Let C be the Cantor set, and $X \subset C$. The characteristic function χ_X is Riemann integrable. Since C is of power continuum (Problem 4), we get more than continuum many such functions by taking all subsets of X (Problem 23).

e) Every subset of the Cantor set is Jordan measurable. Since C is of power continuum (Problem 4), we can apply Problem 23.

f) Let B be a basis of $\mathbf{R}$ considered as a linear space over $\mathbf{Q}$ (i.e., a Hamel basis). Then B is of power continuum (see Problem 15.3). Now every $X \subset B$ generates an additive subgroup of R, and these subgroups are different for different X's. Apply Problem 23.

g) Let $x \in (0,1)$ be a number and $f_x(t)$ be the piecewise linear function that vanishes outside $(0,x)$ and takes the value 1 at $t = x/2$ (see the solution to Problem 14, k)). It is easy to see that these functions are linearly independent, and any subset $Y \subset \{f_x : x \in (0,1)\}$ generates a linear subspace C_Y which are different for different Y's. Thus, there are at least as many such subspaces as subsets of $\{f_x : x \in (0,1)\}$, and since this set is of power continuum, we can apply Problem 23.

h) Consider the set $\mathcal{F} = \{f_x : x \in (0,1)\}$ from the preceding solution. This is a linearly independent subset of $L^2[0,1]$, and any mapping $F : \mathcal{F} \to \mathbf{R}$ can be uniquely extended to a linear functional on the (linear) subspace generated by $\mathcal{F}$, and then (non-uniquely) to a linear functional on $L^2[0,1]$. Since there are more than continuum many such F's (Problem 24), we are done.

26. **a)** This set is of bigger cardinality than continuum. To prove that it is enough to show that there is a closed set E of cardinality continuum which does not contain a rational point. In fact, then for any subset $X \subset E$ its

characteristic function χ_X is continuous at every rational point, and there are more than continuum many such characteristic functions by Problem 23.

To show the existence of E, let $\{r_j\}_{j=0}^{\infty}$ be an enumeration of the rational numbers. Do now the Cantor construction with the following modification: choose a closed interval I of length 1 that does not contain r_0, then choose two disjoint closed subintervals I_0 and I_1 of I of length $< 1/2$ such that neither of them contains r_1, then choose disjoint closed subintervals $I_{00}, I_{01} \subset I_0$, and $I_{10}, I_{11} \subset I_1$ of length $< 1/2^2$ such that neither of them contains x_2, etc. Let J_n be the union of all intervals at level n (e.g., $J_2 = I_{00} \cup I_{01} \cup I_{10} \cup I_{11}$), and set $E = \cap_n J_n$. This is a closed set and clearly $E \cap \mathbf{Q} = \emptyset$. Since every 0–1 sequence $\epsilon = (\epsilon_0, \epsilon_1, \ldots)$ defines a point $x_\epsilon = \cap_n I_{\epsilon_0\epsilon_1\ldots\epsilon_n}$, and these points are different for different 0–1 sequences, E has continuum many points by Problem 2.

b) This set is of power continuum. This follows from Problem 14, part f).

c) This set is of cardinality bigger than continuum. In fact, let B be a basis of $\mathbf{R}$ considered as a linear space over $\mathbf{Q}$ (i.e., a Hamel basis). Then B is of power continuum (see 15.3). But (see also the solution to Problem 15.13,(a))) every mapping $f : B \to \mathbf{R}$ can be extended to a linear functional (with scalar space $\mathbf{Q}$) on $\mathbf{R}$, and clearly every linear functional satisfies the Cauchy equation. Finally use Problem 24 to deduce that there are more than continuum many f's.

27. It is enough to prove the result for a particular A of cardinality continuum. Let $A = {}^{\mathbf{N}}\{0,1\}$ be the set of infinite 0–1 sequences, and let A_m be the set of 0–1 sequences of length m. Any mapping of $g : A_m \to \mathbf{N}$ generates a mapping $f_g : A \to \mathbf{N}$ defined as

$$f_g((\epsilon_0, \epsilon_1, \ldots)) = g((\epsilon_0, \epsilon_1, \ldots, \epsilon_{m-1})).$$

Now there are countably many ways to map A_m into $\mathbf{N}$ (see Problem 2.9, b)), so if $\mathcal{F}$ is the set of all f_g's with all possible $g : A_m \to \mathbf{N}$ and all possible $m = 1, 2, \ldots$, then $\mathcal{F}$ is countable. This set $\mathcal{F}$ of functions satisfies the requirements in the problem. Indeed, assume that we are given finitely many different 0–1 sequences $\mathbf{e}_i = \{\epsilon_{i,j}\}_{j=0}^{\infty}$, $0 \le i \le n$, and let $f(\mathbf{e}_i) = a_i \in \mathbf{N}$ be given. Let m be so large that all the initial sequences $\mathbf{e}_i^{(m)} = \{\epsilon_{i,j}\}_{j=0}^{m-1}$, $0 \le i \le n$ are different. Let $g : A_m \to \mathbf{N}$ be an arbitrary mapping for which $f(\mathbf{e}_i^{(m)}) = a_i$ is satisfied for all $i = 0, 1, \ldots, n$. Then for $f_g \in \mathcal{F}$ we have $f_g(\mathbf{e}_i) = a_i$, as required.

28. Let A be of power continuum, and for every $a \in A$ let $\mathcal{T}_a$ be a separable topological space. Let $\{x_j^{(a)}\}_{j=0}^{\infty}$ be a countable dense set in $\mathcal{T}_a$. Consider the functions f_k from the preceding problem and the corresponding elements F_k in the product space with $F_k(a) = x_{f_k(a)}^{(a)}$ for all a. This is a countable set in the product space, and using the definition of product topology and the

definition of the functions f_k it is easy to see that $\{F_k\}_{k=0}^{\infty}$ is dense in the product space.

29. **First solution.** For every $x \in (1/10, 1)$ let

$$A_x = \{[10x], [10^2 x], \ldots, [10^k x], \ldots\}$$

where $[\cdot]$ denotes the integral part. Note that if $x \in (1/10, 1)$ then $x = 0.\alpha\ldots$ with $\alpha \neq 0$, hence the sequence $[10x], [10^2 x], \ldots$ consists of positive integers and it contains for every $k = 0, 1, 2, \ldots$ exactly one number from the range $10^k \leq z < 10^{k+1}$ (i.e., its decimal form consists of exactly $k+1$ digits). If x and y are different elements of $(1/10, 1)$, then their decimal expansions differ, say the mth decimal digit in x and y are different. Then $[10^k x] \neq [10^k y]$ for $k \geq m$, hence the two sets A_x and A_y have only finitely common elements.

Second solution. Let $\mathcal{P}$ be the set of prime numbers, and for an infinite subset $\Sigma = \{p_0, p_1, \ldots\}$ of $\mathcal{P}$ arranged in increasing order assign

$$A_\Sigma = \{p_0, p_0 p_1, p_0 p_1 p_2, \ldots\}.$$

The prime factorization for integers is unique, hence if $\Sigma' \subseteq \mathcal{P}$ is another infinite subset of $\mathcal{P}$ different from Σ, then A_Σ and $A_{\Sigma'}$ have only finitely many common terms. Since the number of different Σ's is continuum (see Problems 5 and 2.4), we are done.

Third solution. It is sufficient to show the result for $\mathbf{Q}$ rather than for $\mathbf{N}$, i.e., that there are continuum many sets $A_\gamma \subset \mathbf{Q}$ such that if $\gamma_1 \neq \gamma_2$, then $A_{\gamma_1} \cap A_{\gamma_2}$ is a finite set. Now choose for every $\gamma \in \mathbf{R}$ a rational sequence $A_\gamma = \{r_k^{(\gamma)}\}_{k=0}^{\infty}$ converging to γ. These A_γ sets clearly satisfy the requirements, for two sequences converging to different limits can have only finitely many terms in common.

Fourth solution. Instead of $\mathbf{N}$ work with the set of lattice points $\mathbf{N} \times \mathbf{N}$ on the plane, and for $m \in \mathbf{R}$ let A_m be the set of points $(x, y) \in \mathbf{N} \times \mathbf{N}$ that are of distance ≤ 1 from the line $y = mx$. It is clear that A_m is infinite (it has a point on every vertical line $x = k$, $k = 0, 1, 2, \ldots$) and for any two lines $y = mx$ and $y = m'x$ there can be only a finite number of lattice points lying of distance ≤ 1 from both, i.e., $A_m \cap A_{m'}$ is finite.

[G. Fichtenholz and L. Kantorovich, Studia Math., **5**(1934), 69–98]

30. See Problem 2.12.

31. Since $\mathbf{R}$ can be mapped onto $(1, 2)$ by a monotone increasing function, it is enough to construct the sequences in question for $x \in (1, 2)$. But the sequences $\{s_n^{(x)}\}$ with $s_n^{(x)} = [10^n x]$ (where $[\cdot]$ denotes integral part) clearly

satisfy the requirements (cf. the first solution to Problem 29). [W. Sierpiński, Cardinal and Ordinal Numbers, Polish Sci. Publ., Warszawa, 1965, , IV.14/8]

32. Let $\mathcal{H}$ be an almost disjoint family of cardinality continuum of infinite subsets of $\mathbf{N}$. For any $H \in \mathcal{H}$ let $h_0^H < h_1^H < \dots$ be the listing of different elements of H, and let $s_n^H = 2^{h_n^H}$. It is clear that the family $\{\{s_n^H\}_{n=0}^\infty\}_{H\in\mathcal{H}}$ satisfies the requirements.

An alternative way is to consider the first solution to Problem 29.

33. Let $a_0, a_1, \dots, a_k$ be any sequence of length k of natural numbers. The assumption implies that there is at most one $\{s_n^\gamma\}_{n=0}^\infty$ with $s_n^\gamma = a_n$ for all $n = 0, 1, \dots k$. Thus, there are at most as many sequences $\{s_n^\gamma\}_{n=0}^\infty$, $\gamma \in \Gamma$ as $(k+1)$-element sequences of the natural numbers, so Γ is countable by Problem 2.3.

34. Let $\mathcal{H}$ be an almost disjoint family of cardinality continuum of infinite subsets of $\mathbf{N}$, and for each $H \in \mathcal{H}$ we set $H^* = \cup_{n\in H} A_n$, where A_n is the set $\{k : 2^{2^n} \le k < 2^{2^{n+1}}\}$. It is clear that the family $\{H^* : H \in \mathcal{H}\}$ is almost disjoint, and since each H^* includes as its subset infinitely many A_n's, the upper density of every H^* is 1 (note that $2^{2^n}/2^{2^{n+1}} \to 0$ as $n \to \infty$).

35. We shall show the $k = 3$ case; the general case can be verified along similar lines. Since $\mathbf{N}$ and $\mathbf{N} \times \mathbf{N}$ are equivalent, it is enough to show a family of cardinality continuum of subsets of $\mathbf{N} \times \mathbf{N}$ such that the intersection of any 2 members of the family is infinite, but the intersection of any 3 members is finite. For $x \in (1, 2)$ set

$$A_x = \cup_{n=1}^\infty \{([10^n x], k), (k, [10^n x]) : 10^n \le k < 10^{n+1}\}.$$

Note that if $(u, v) \in A_x$ then there is an n with $10^n \le u, v < 10^{n+1}$, and one of u or v must be equal to $[10^n x]$. It is clear that if $x, y \in (1, 2)$, then $A_x \cap A_y$ contains all pairs $([10^n x], [10^n y])$, $n = 1, 2, \dots$. On the other hand, if $x, y, z \in (1, 2)$ are all different, and n is sufficiently large, then the numbers $[10^n x]$, $[10^n y]$, and $[10^n z]$ are all different, so $A_x \cap A_y \cap A_z$ cannot contain any pair (u, v) with $10^n \le u, v < 10^{n+1}$. This proves that $A_x \cap A_y \cap A_z$ is finite.

36. One of $0, 1, 2, \dots$ must be contained in uncountably many members of $\mathcal{H}$, say a_0 is in every $H \in \mathcal{H}_0$ where $\mathcal{H}_0$ is an uncountable subfamily. Let $H_0 \in \mathcal{H}_0$ be any set in $\mathcal{H}_0$, and let $a_0^0, a_1^0, \dots$ be the listing of different elements of H_0 (one of them is a_0). Since every $H \in \mathcal{H}_0$ intersects H_0 in an infinite set, there must be an $a_1 \ne a_0$ among $a_0^0, a_1^0, \dots$ that is contained in uncountably many $H \in \mathcal{H}_0$, and the set of all such H be $\mathcal{H}_1$. Choose $H_1 \in \mathcal{H}_1$ arbitrarily. By the assumption the set $H_0 \cap H_1$ is infinite, and let $a_0^1, a_1^1, \dots$ be the listing of different elements of $H_0 \cap H_1$ (one–one of them is a_0 and a_1). Then there must be an $a_2 \ne a_0, a_1$ among $a_0^1, a_1^1, \dots$ that is contained in uncountably

many $H \in \mathcal{H}_1$, and the set of all such H be $\mathcal{H}_2$, etc. We can continue this process indefinitely, and it is clear that the intersection $H_1 \cap H_2 \cap \cdots$ contains all elements $a_0, a_1, \ldots$.

37. This immediately follows from Problem 43.

38. Let $f(H)$ be a countable subset of H for every $H \in \mathcal{H}$. By condition, the mapping $H \mapsto f(H)$ is an injection of $\mathcal{H}$ into the set of countable subsets of $\mathbf{R}$, which is a set of power continuum (see Problem 6). Thus, there are at most continuum many sets in $\mathcal{H}$.

39. See the solution to Problem 18.2.

40. It is again enough to use $\mathbf{Q}$ instead of $\mathbf{N}$ (see the second solution to Problem 29), and then we can set for $\gamma \in (0,1)$ $A_\gamma = (0,\gamma) \cap \mathbf{Q}$.

41. Consider the preceding solution, but set $A_\gamma = [(0,\gamma) \cup (1+\gamma, 2)] \cap \mathbf{Q}$.

42. Instead of $\mathbf{N}$ we work again with $\mathbf{Q}$ (see the second solution to Problem 29). For every $x \in \mathbf{R}$ let A_x be a rational sequence converging to x, and let $B_x = \mathbf{Q} \setminus A_x$. It is clear that these sets satisfy the requirements.

43. Since $(0,1)$ is equivalent with $\mathbf{R}$, it is enough to give A_x for $x \in (0,1)$. Let $[y]$ denote the integral part of y, and for $x \in (0,1)$ let $A_m(x)$ be the set of all those integers $2^{2^m} \le k < 2^{2^{m+1}}$ for which the $[mx]$th binary digit (counted from the right) is 1, and set

$$A_x = \cup_{m=1}^{\infty} A_m(x).$$

If $x_1, \ldots, x_n$ are different numbers, then there is an m_0 such that for $m \ge m_0$ the numbers $[mx_1], \ldots, [mx_n]$ are all different. For each such m a set of the form $A_m(x_1)^{\epsilon_1} \cap \cdots A_m(x_n)^{\epsilon_n}$ consists of those numbers $2^{2^m} \le k < 2^{2^{m+1}}$ for which n different binary digits are prescribed (the $[mx_i]$th binary digit is ϵ_i for $i = 1, 2, \ldots, n$), hence the number of elements in such a set is

$$\frac{2^{2^{m+1}} - 2^{2^m}}{2^n}.$$

Thus, if $m > m_0 + 1$ and $2^{2^{m+1}} < N \le 2^{2^{m+2}}$, then the number of elements of the set $A_{x_1}^{\epsilon_1} \cap \cdots A_{x_n}^{\epsilon_n}$ in the interval $[0, N]$ is

$$\frac{N - 2^{2^{m+1}} + O(2^m)}{2^n} + \frac{2^{2^{m+1}} - 2^{2^m}}{2^n} + O(2^{2^m})$$

and this divided by N tends to $1/2^n$ as $N \to \infty$.

44. Let $f(x,x) = 0$, and for $x \ne y$ let $f(x,y) = 1 + \min(A_x \cap B_y)$ where A_x, B_y are the sets from Problem 42. Since $B_y \cap A_y = \emptyset$, the equality $f(x,y) = f(y,z)$ can occur only for $x = y = z$.

5

Sets of reals and real functions

1. Let A_1 and A_2 be the set of those points $a \in A$ for which $(a, a+\delta_a) \cap A = \emptyset$ and $(a - \delta_a, a) \cap A = \emptyset$, respectively. Notice that if $a_1, a_2 \in A_1$, then the intervals $(a_1, a_1 + \delta_{a_1})$ and $(a_2, a_2 + \delta_{a_2})$ are disjoint. Hence A_1 is countable by Problem 2.14. In a similar manner, A_2 is also countable.

2. Let $A \subset \mathbf{R}$ be uncountable. By Problem 1 there is an $a \in A$ such that $(a, a + \delta) \cap A$ is nonempty for all $\delta > 0$. Now let $a_0 \in A$ be a point with $a < a_0 < a + 1$, then $a_1 \in A$ a point with $a < a_1 < \min(a_0, a + 1/2)$, etc.. Clearly the sequence $\{a_n\}$ selected this way converges to a.

3. For $\mathbf{R}$ this follows from the preceding problem. For $\mathbf{R}^n$ apply, e.g., Problem 12 to an open cover $\cup_{a\in A} G_a$ of the discrete set A, where each G_a contains only the point a from A. Since this includes a countable subcover the countability of A follows.

See also (the solution of) Problem 2.14.

4. Let f be right continuous, and let

$$\mathrm{osc}_x = \limsup_{y_1, y_2 \to x} |f(y_1) - f(y_2)|$$

be the oscillation of f at x. f is continuous at x if and only if $\mathrm{osc}_x = 0$. Thus, if $A_m = \{x \ : \ \mathrm{osc}_x > 1/m\}$, then $\cup_m A_m$ is the set of discontinuity points of f, and it is enough to prove that each set A_m is countable. Because of the right continuity of f, for every $x \in \mathbf{R}$ there is a $\delta_{x,m} > 0$ such that for $y \in (x, x+\delta_{x,m})$ we have $|f(y) - f(x)| < 1/2m$. It follows that in $(x, x+\delta_{x,m})$ there cannot be any point from A_m. Thus, the countability of A_m follows from Problem 1.

5. Follow the preceding solution, and let A_m^+ be the set of those points in A_m where f is continuous from the right, and in a similar manner let A_m^- be

the set of those points in A_m where f is continuous from the left. Now the preceding solution gives that both A_m^+ and A_m^- are countable, hence the result follows.

6. Let f be a monotone real function. Then f has a limit $f(x-0)$ from the left and a limit $f(x+0)$ from the right at every point $x \in \mathbf{R}$, and an x is a discontinuity point x if and only if $f(x+0) > f(x-0)$. Let us assign the interval $(f(x-0), f(x+0))$ to every discontinuity point x of f. These intervals are disjoint: if $x_1 < x_2$ are distinct points and $x_1 < x_3 < x_2$, then by monotonicity we have $f(x_1+0) \le f(x_3) \le f(x_2-0)$, so $(f(x_1-0), f(x_1+0)) \cap (f(x_2-0), f(x_2+0)) = \emptyset$. Now the result follows from Problem 2.14.

7. Let f be a real function that has right and left derivatives, which we denote by $f'_+(x)$ and $f'_-(x)$, at every point x. Let $r_0, r_1, \ldots$ be an enumeration of the rational numbers, and for $r_m < r_n$ let $A_{n,m} = \{x : f'_+(x) > r_n,\ f'_-(x) < r_m\}$. It is enough to show that each $A_{n,m}$ is countable. In fact, then $\cup_{n,m} A_{n,m}$ is also countable, and this is the set of those points in which the right derivative is bigger than the left derivative. In a similar manner it follows that the set where the left derivative is bigger than the right derivative is countable, and these two statements prove the claim.

Let $A_{n,m,k}$ be the set of those points $x \in A_{n,m}$ for which it is true that if $x < y < x + 1/k$ then

$$\left| \frac{f(y) - f(x)}{y - x} - f'_+(x) \right| < \frac{r_n - r_m}{2},$$

and if $x - 1/k < y < x$, then

$$\left| \frac{f(y) - f(x)}{y - x} - f'_-(x) \right| < \frac{r_n - r_m}{2}.$$

Since $\cup_k A_{n,m,k} = A_{n,m}$, it is enough to show that each $A_{n,m,k}$ is countable.

From the preceding inequalities and the definition of the set $A_{n,m}$ it is clear that if $x \in A_{n,m,k}$ and $0 < h < 1/k$, then the expression $(f(x+h) - f(x))/h$ is bigger than $r_n - (r_n - r_m)/2 = (r_n + r_m)/2$, while $(f(x) - f(x-h))/h$ is smaller than $r_m + (r_n - r_m)/2 = (r_n + r_m)/2$. On applying this to the point x and $x - h$ we can see (use that $f(x) - f(x-h) = f((x-h)+h) - f(x-h))$) that it is not possible to simultaneously have $x, (x-h) \in A_{n,m,k}$. But this means that for any $x \in A_{n,m,k}$ the interval $(x - 1/k, x)$ does not contain any point of $A_{n,m,k}$, and we can apply Problem 1 to deduce the countability of $A_{n,m,k}$.

8. Use the preceding problem and the fact that a convex function has left and right derivatives at every point.

9. Let f be a real function and let A be the set of maximum values of f. Thus, $a \in A$ if there is a point $x_a \in \mathbf{R}$ and a positive δ_a such that $f(x_a) = a$, and

there is no larger value of f in the interval $(x_a - \delta_a, x_a + \delta_a)$. Let A_n be the set of those $a \in A$ for which $\delta_a > 1/n$. It is obvious from the definitions that if $a, b \in A_n$ are different points, then the distance between x_a and x_b is at least $1/n$. Thus A_n is a discrete set, and hence it is countable (see Problem 3). Since $A = \cup A_n$, the set A is also countable.

10. If a is a strict maximum point of f, then there is a $\delta_a > 0$ such that for every $y \in (a - \delta_a, a + \delta_a)$, $y \neq a$ the inequality $f(y) < f(a)$ holds. If A_n is the set of all such a's for which $\delta_a > 1/n$, then clearly for $a, b \in A_n$ we must have $|a - b| > 1/n$. Hence A_n is countable by Problem 3, and so is $\cup_n A_n$, the set of strict maximum points of f.

11. If f is continuous and non-constant, then, by the intermediate value theorem, its image covers a whole interval. Thus, in this case not every point in the image can be a minimum or maximum value by Problem 9.

12. Let $\{B_j,\ j = 0, 1, \ldots\}$ be the collection of open balls in $\mathbf{R}^n$ of rational center and rational radii (cf. Problem 2.17). Represent each G_γ, $\gamma \in \Gamma$ as a union some of the B_j's as in Problem 2.18: $G_\gamma = \cup_{j \in \Delta_\gamma} B_j$, where Δ_γ is a subset of the natural numbers. Then

$$\bigcup_{\gamma \in \Gamma} G_\gamma = \bigcup_{j \in \cup_{\gamma \in \Gamma} \Delta_\gamma} B_j,$$

hence if for each $j \in \cup_{\gamma \in \Gamma} \Delta_\gamma$ we select a $\gamma_j \in \Gamma$ such that $j \in \Delta_{\gamma_j}$, then clearly

$$\bigcup_{\gamma \in \Gamma} G_\gamma = \bigcup_{j \in \cup_{\gamma \in \Gamma} \Delta_\gamma} G_{\gamma_j},$$

so the subfamily G_{γ_j}, $j \in \cup_{\gamma \in \Gamma} \Delta_\gamma$ covers whatever is covered by the family G_γ, $\gamma \in \Gamma$.

13. There are two kinds of semi-open intervals, namely those of the form $[a, b)$ and of the form $(a, b]$. Let G_γ, $\gamma \in \Gamma_1$ be the set of those intervals in $\{G_\gamma\}_{\gamma \in \Gamma}$ that are of the first kind, and let G_γ, $\gamma \in \Gamma_2$ be the set of those intervals in $\{G_\gamma\}_{\gamma \in \Gamma}$ that are of the second kind. It is clearly enough to prove the claim separately for the families G_γ, $\gamma \in \Gamma_1$ and G_γ, $\gamma \in \Gamma_2$ and for the sets $E_1 = \cup_{\gamma \in \Gamma_1} G_\gamma$ and $E_2 = \cup_{\gamma \in \Gamma_2} G_\gamma$, respectively. Thus, we may assume that all the intervals G_γ are of the first kind.

Let $\mathrm{Int}(\Gamma_\gamma)$ be the interior of G_γ. On applying the previous problem (to $\mathbf{R}$), we can see that the union of these interiors can be covered by countably many of them, thus we only have to show that the same is true of the set

$$F = \left(\bigcup_{\gamma \in \Gamma} G_\gamma\right) \setminus \left(\bigcup_{\gamma \in \Gamma} \mathrm{Int}(G_\gamma)\right).$$

It is clear that for every $x \in F$ there is a $\delta_x > 0$ such that $(x, x+\delta_x)$ is part of the interior of a G_γ, hence in $(x, x+\delta_x)$ there is no point from F. By Problem 1 F is countable, and so if we select for each of its points a G_γ that covers it, then we get a countable subcover of F.

14. This problem can be reduced to the preceding one. In fact, every nondegenerated interval can be written as a union of two semi-open intervals. Thus, we write $G_\gamma = G_{\gamma,1} \cup G_{\gamma,2}$ with sets $G_{\gamma,1}$ and $G_{\gamma,2}$ of semi-open intervals. Now apply Problem 13 to each family $G_{\gamma,1}$, $\gamma \in \Gamma$ and $G_{\gamma,2}$, $\gamma \in \Gamma$ with E as their union, and then unite the so obtained two countable subcovers.

15. Let Y be the set of those y for which $f^{-1}(y) \cap H$ is uncountable. We have to show that Y is of measure zero, and to this end it is enough to show that if for $M = 1, 2, \ldots$ we denote by Y_M is the set of those points y for which $f^{-1}(y) \cap (H \cap [-M, M])$ is uncountable, then Y_M is of measure zero.

Let us pick for each $y \in Y_M$ a point $t_y \in f^{-1}(y) \cap (H \cap [-M, M])$ such that t_y is a limit point of the set $f^{-1}(y) \cap (H \cap [-M, M])$. By Problem 2 such a t_y exists. Since f is constant on $f^{-1}(y) \cap (H \cap [-M, M])$, the differentiability of f at t_y implies that $f'(t_y) = 0$. If T_M denotes the set of all these t_y's, then $Y_M = f[T_M]$.

Let $\epsilon > 0$. For every $x \in T_M$ there is a $1 > \delta_x > 0$ such that if $0 < h < \delta_x$ then

$$\left| \frac{f(x) - f(x \pm h)}{h} \right| \le \epsilon.$$

The intervals $I_x = (x - \delta_x, x + \delta_x)$, $x \in T_M$ cover T_M, so, by Problem 12, we can select a countable subcover $U = \cup_{i=0}^{\infty} I_{x_i}$. Then U is open, $Y_M \subset f[U]$, hence, as $\epsilon > 0$ is arbitrary, it is enough to prove that the measure of $f[U]$ is at most $4(M+1)\epsilon$. Since U is the union of an increasing sequence of compact sets, it is sufficient to show that if $K \subset U$ is compact, then the measure of $f[K]$ is at most $4(M+1)\epsilon$. But for a compact K there is an N such that $K \subset \cup_{i=0}^{N} I_{x_i}$, and without loss of generality we may assume that in this union each point is covered at most twice (in fact, if three intervals intersect in a point then one of them is included in the union of the other two). By the definition of the numbers δ_{x_i}, every point of the set $f[I_{x_i}]$ is of distance at most $\epsilon\delta_{x_i}$ from $f(x_i)$, hence $f[I_{x_i}]$ is of measure at most $2\epsilon\delta_{x_i}$. But then $f[U]$ is of measure at most $2\epsilon \sum_{i=0}^{N} \delta_{x_i}$, and since every point of $\cup_{i=0}^{N} I_{x_i}$ is covered at most twice, the sum $\sum_{i=0}^{N} 2\delta_{x_i}$ is at most twice the measure of U, i.e., at most $2 \cdot 2(M+1)$. This shows that $f[K]$ has measure at most $4(M+1)\epsilon$ as claimed.

16. Let G_γ, $\gamma \in \Gamma$ be a family of almost closed rectangles, and let G_γ, $\gamma \in \Gamma_n$ be the subfamily that consists of those elements in G_γ, $\gamma \in \Gamma$ that have side lengths bigger than $1/n$. It is enough to verify the problem for each subfamily G_γ, $\gamma \in \Gamma_n$, for then we can unite for $n = 1, 2, \ldots$ the so obtained countable subcovers of $\cup_{\gamma \in \Gamma_n} G_\gamma$ to get a countable subcover of $\cup_{\gamma \in \Gamma} G_\gamma$.

Call a rectangle semi-closed if it is obtained from an open rectangle by adding (without the endpoints) one of the sides of that rectangle, and accordingly we can speak of left–, right–, down– and up semi-closed rectangles. Every almost closed rectangle G_γ is the union of four semi-closed rectangles. Thus, if we can prove the countable subcover property for semi-closed rectangles of the same type (e.g., for left semi-closed rectangles), then the claim follows by uniting these four countable subcovers.

Thus, in what follows we can assume that each G_γ, $\gamma \in \Gamma$ is a left semi-closed rectangle with sidelengths bigger than $1/n$.

The set which is covered by the interiors of the left semi-open rectangles $G_\gamma, \gamma \in \Gamma$ can be covered by countably many of them (see Problem 12), hence it is enough to show that the same is true of the set

$$F = \left(\bigcup_{\gamma \in \Gamma} G_\gamma \right) \setminus \left(\bigcup_{\gamma \in \Gamma} \mathrm{Int}(G_\gamma) \right).$$

This will follow if we can prove that the set F lies on countably many vertical lines. In fact, if l is a vertical line, then every G_γ intersects l in an open interval. Thus, we can apply the Lindelöf property (Problem 12 for $\mathbf{R}$) to l and the family $l \cap G_\gamma$, $\gamma \in \Gamma_n$ of open intervals to conclude that $l \cap F$ can be covered by countably many G_γ. Since this is true for every vertical line, eventually we get a countable subcover of F.

The points of F are covered by the left-hand sides of some rectangles G_γ, and let F_k, $k = 1, 2, \ldots$ be the set of those points x in F that are covered by the left-hand side of a rectangle G_γ with vertices of distance $> 1/k$ from x. Again it is enough to show that each set F_k lies on countably many vertical lines. Let L be the set of those vertical lines that intersect F_k, and for every $l \in L$ select from $l \cap F_k$ a point x_l. From the definition of F_k it follows that if we place a small disk D_l of radius $1/4kn$ and of center at x_l to every point x_l, $l \in L$, then these disks are disjoint. In fact, if, say, $D_l \cap D_s \neq \emptyset$ and l lies to the left of s, then x_s is covered by the rectangle G_γ that contains x_l on its left side and has vertices of distance $> 1/k$ from x_l, hence x_s could not be in F. Now by Problem 2.14 there are countably many D_l's, so there are countably many x_l's, and this is what we had to show.

The same result is not true for closed rectangles. In fact, if we cover each point on the line $y = x$ by the vertex of a closed rectangle with sides parallel with the coordinate axes, then from this cover one cannot omit a single rectangle to remain a cover of the whole line $y = x$.

17. If the claim was not true, then for every $a \in A$ there would be a ball B_a that intersects A only in a countable set. The set $\{B_a \ : \ a \in A\}$ is an open cover of A, hence by Problem 12 there is a countable subcover $A \subset \cup_{i=0}^{\infty} B_{a_i}$. But since $A \cap B_{a_i}$ is countable for all i, this would mean that A can have only countably many points.

18. Let $A^{**} \subseteq A$ be the set of accumulation points of A lying in A. The set $A \setminus A^{**}$ must be countable, for otherwise it would contain by Problem 17 an accumulation point of itself, and that would belong to A^{**}, a contradiction. The set B_1 of those $a \in A^{**}$ for which there is a $\delta > 0$ such that $A \cap (a-\delta, a)$ is countable clearly has the property that $B_1 \cap (a, a-\delta) = \emptyset$, hence, by Problem 1, it is countable. In a similar fashion countable is the set B_2 of those $a \in A^{**}$ for which there is a $\delta > 0$ such that $A \cap (a, a+\delta)$ is countable. These show that $A^* = A^{**} \setminus (B_1 \cup B_2)$ has the property that $A \setminus A^*$ is countable.

If $a, b \in A^*$, then $A \cap (a, b)$ is uncountable, therefore, by what we have just proven, it contains an element of $(A \cap (a, b))^* = A^* \cap (a, b)$, hence A^* is densely ordered.

19. It is clear that the set X of accumulation points of any set A is closed. We have to show that if it is not empty, then it is dense in itself, i.e., every neighborhood U of any point x in X contains a point in X different from x. This follows from the previous problem, for $(U \setminus \{x\}) \cap X$ is uncountable, hence one of its points is an accumulation point of this set by Problem 17.

20. Let E be closed, and let X be the set of its accumulation points. Then $X \subset E$ and X is perfect by Problem 19. Thus, it is enough to show that $E \setminus X$ is countable. If this was not the case, then, by Problem 17, the set $E \setminus X$ would have an accumulation point x in $E \setminus X$. But that is not possible, for then x would be an accumulation point of E, and so it would have to belong to X.

21. Let $E \subset \mathbf{R}^n$ be nonempty and perfect. Choose two disjoint nonempty closed subsets E_0 and E_1 of E in the following way: select two points P_0 and P_1 in E and two disjoint closed balls B_0 and B_1 around them of diameter $< 1/2$, and set $E_0 = E \cap B_0$ and $E_1 = E \cap B_1$. Then choose disjoint nonempty closed subsets $E_{00}, E_{01} \subset E_0$ and $E_{10}, E_{11} \subset E_1$ in the following way: select two points P_{00} and P_{01} in E_0 that lie *inside* B_0 and two disjoint closed balls B_{00} and B_{01} around them of diameter $< 1/2^2$ in such a way that both of them lie in B_0, and set $E_{00} = E \cap B_{00}$ and $E_{01} = E \cap B_{01}$ (and the choice of E_{10} and E_{11} is similarly done relative to E_1 and B_1). Continue this process. The perfectness of the set E guarantees that this process does not terminate. Let J_n be the union of all subsets at level n (e.g., $J_2 = E_{00} \cup E_{01} \cup E_{10} \cup E_{11}$), and set $E^* = \cap_n J_n$. This is a closed subset of E. Every 0–1 sequence $\epsilon = \{\epsilon_0, \epsilon_1, \ldots\}$ defines a point $\{x_\epsilon\} = \cap_n E_{\epsilon_0\epsilon_1\ldots\epsilon_n}$, and these points are different for different 0–1 sequences: if $\epsilon' = \{\epsilon_i'\}_{i=0}^\infty$ is another sequence and we select m in such a way that $\epsilon_0 = \epsilon_0'$, ..., $\epsilon_{m-1} = \epsilon_{m-1}'$ but $\epsilon_m \neq \epsilon_m'$, say $\epsilon_m = 0$ and $\epsilon_m' = 1$, then x_ϵ resp. $x_{\epsilon'}$ lie in the disjoint sets $E_{\epsilon_0\epsilon_1\cdots\epsilon_{n-1}0}$ resp. $E_{\epsilon_0\epsilon_1\cdots\epsilon_{n-1}1}$. Thus, E^* has continuum many points by Problem 4.3, and so E has at least that many points. But $\mathbf{R}^n$ is of cardinality continuum (Problem 4.14, a)), and we are done because of the equivalence theorem.

22. Apply Problems 21 and 20.

23. That d is a metric is easily established. It is also easy to see that if $\{a_j^{(n)}\}_{j=0}^{\infty}$, $n = 0, 1, \ldots$ is a sequence of elements of $\mathbf{R}^\infty$, then this sequence converges to an $\{a_j\}_{j=0}^{\infty} \in \mathbf{R}^\infty$ if and only if for each j we have

$$\lim_{n\to\infty} a_j^{(n)} = a_j,$$

i.e., the metric describes the topology of pointwise convergence (recall that $\mathbf{R}^\infty$ is the set of mappings $f : \mathbf{N} \to \mathbf{R}$, therefore the statement is that if $f, f_n \in R^\infty$, $n = 0, 1, \ldots$, then $d(f_n, f) \to 0$ as $n \to \infty$ if and only if for all j the limit $f_n(j) \to f(j)$ holds as $n \to \infty$, which is pointwise convergence). This, and the completeness of $\mathbf{R}$, easily imply the completeness of $\mathbf{R}^\infty$.

Finally, there is a countable dense subset of $\mathbf{R}^\infty$, namely the set of sequences of rational numbers that contain only finitely many nonzero terms (cf. Problem 2.4).

24. Let H be a countable dense subset of $\mathbf{R}^\infty$ and let $\mathcal{G}$ be the set of open balls of rational radius and with center in H. Exactly as in Problem 2.17, this set is countable, and exactly as in Problem 2.18, any open subset of $\mathbf{R}^\infty$ is a union of countably many open balls from $\mathcal{G}$. This is enough for the Lindelöf property (cf. Problem 12) to hold in $\mathbf{R}^\infty$, i.e., any open cover of any subset of $\mathbf{R}^\infty$ includes a countable subcover (see the solution to Problem 12). Now the notion of accumulation point (see Problem 17) can be carried over to $\mathbf{R}^\infty$, and using this exactly as in the solutions of Problems 17–20 we get that any closed set in $\mathbf{R}^\infty$ is the union of a perfect and a countable set.

25. This follows from the preceding problem, since a nonempty perfect set is of cardinality continuum (recall also Problem 4.14, b), according to which $\mathbf{R}^\infty$ is of power continuum).

26. First we prove the claim for open sets in $\mathbf{R}^n$. For $\mathbf{R}^n$ this is clear, and if $O \subseteq \mathbf{R}^n$, $O \neq \mathbf{R}^n$ is an open set, then for $\mathbf{x} = (x_0, x_1, \ldots, x_{n-1}) \in O$, consider the point $f_\mathbf{x} \in \mathbf{R}^\infty$ for which $f_\mathbf{x}(m) = x_m$ for $m < n$, $f_\mathbf{x}(m) = 0$ for $m > n$, and $f_\mathbf{x}(n) = 1/\mathrm{dist}(\mathbf{x}, \partial O)$ (the reciprocal of the distance from $\mathbf{x}$ to the boundary of O). It is easy to see that the set $F_O = \{f_\mathbf{x} : \mathbf{x} \in O\}$ is a closed subset of $\mathbf{R}^\infty$, and $f_\mathbf{x} \mapsto \mathbf{x}$ is a continuous and one-to-one mapping from F_O onto of O.

Next we use that the family of Borel sets in $\mathbf{R}^n$ is the smallest family of sets containing the open sets and closed under countable intersection and countable disjoint union (see Problem 1.13 or 12.25). Therefore, it is sufficient to show that the property of being the continuous and one-to-one image of a closed subset of $\mathbf{R}^\infty$ is preserved under countable intersection and disjoint countable union.

Let $A = \cap_j A_j$, and suppose that each $A_j \subseteq \mathbf{R}^n$ is a continuous and one-to-one image of a closed subset of $\mathbf{R}^\infty$. Let $N = \cup_{i=0}^{\infty} N_i$ be a disjoint decomposition of $\mathbf{N}$ into some infinite sets N_i. Then ${}^{N_j}\mathbf{R}$ is homeomorphic to $\mathbf{R}^\infty$, and let $f_j : F_j \to A_j$ be a continuous and one-to-one mapping of a closed subset F_j of ${}^{N_j}\mathbf{R}$ onto A_j. The set

$$F^* = \left\{ g \in \mathbf{R}^\infty \ : \ g_{|N_i} \in F_i, \ i = 0, 1, \dots \right\}$$

is a closed subset of $\mathbf{R}^\infty$, and

$$F = \left\{ g \in F^* \ : \ f_i(g_{|N_i}) = f_k(g_{|N_k}) \text{ for all } i, k = 0, 1, \dots \right\}$$

is a closed subset of F^*, and hence of $\mathbf{R}^\infty$. It is clear that $g \mapsto f_0\left(g_{|N_0}\right)$ is a continuous one-to-one mapping of F onto $\cap_j A_j$.

Next let $A = \cup_{j=0}^\infty A_j$, $A_j \cap A_k = \emptyset$ for $j \neq k$ be a countable disjoint union, and suppose that each $A_j \subseteq \mathbf{R}^n$ is a continuous one-to-one image of a closed subset F_j of $\mathbf{R}^\infty$. We may assume that $g(0) = j$ for all $g \in F_j$ (clearly, for fixed j the set of points $g \in \mathbf{R}^\infty$ with $g(0) = j$ is isomorphic and homeomorphic to $\mathbf{R}^n$). But then the set $F = \cup_j F_j$ is closed in $\mathbf{R}^\infty$ (note that the distance between different F_j's is at least $1/2$), and if we define $f : F \to \cup_j A_j$ by $f(u) = f_j(u)$ for $u \in F_j$, then we get a continuous one-to-one mapping of F onto $\cup_j A_j$.

27. This is an immediate consequence of Problems 25 and 26.

28. Suppose to the contrary that no A_i is dense in any interval. Then for every interval I and every i there is a closed subinterval $J \subset I$ such that $J \cap A_i = \emptyset$. Now starting with $J_0 = [a, b]$ inductively select nondegenerated closed intervals $J_1, J_2 \dots$ such that $J_{n+1} \subseteq J_n$ and $J_{n+1} \cap A_n = \emptyset$. Then $\cap_n J_n$ is nonempty, and if $x \in \cap_n J_n$, then $x \notin \cup_n A_n$, which contradicts the assumption. This contradiction proves the claim.

29. See the proof of the more general result in Problem 31.

30. The proof of Problem 28 shows that if $A = \cup_{i=0}^\infty A_i$, then there are a ball B and an i such A_i is dense in B. But then this A_i is not nowhere dense.

31. Let $B^* \subseteq A$ be a closed ball. Suppose to the contrary that for any ball $B \subseteq B^*$ and for any i there is a ball $B' \subset B$ such that $B' \cap A_i$ is of power less than continuum. Choose two disjoint closed balls $B_0 \subseteq B^*$ and $B_1 \subseteq B^*$ (say two smaller balls from the B' above for $i = 0$) of (positive) diameter $< 1/2$ such that $(B_0 \cup B_1) \cap A_0$ is of power smaller than continuum. Then choose disjoint closed balls $B_{00}, B_{01} \subset B_0$ and $B_{10}, B_{11} \subset B_1$ of radius $< 1/2^2$ so that the set $(B_{00} \cup B_{01} \cup B_{01} \cup B_{11}) \cap A_1$ is of power smaller than continuum. Continue this process. As in the solution of Problem 21, for every sequence

$$\epsilon = \{\epsilon_0, \epsilon_1, \dots\} \in {}^{\mathbf{N}}\{0, 1\}$$

of zeros and ones the intersection

$$\bigcap_{n=0}^{\infty} B_{\epsilon_0\epsilon_1\cdots\epsilon_n}$$

contains a single point x_ϵ, and different sequences generate different points. Thus, the set

$$X = \{x_\epsilon \,:\, \epsilon \in {}^{\mathbf{N}}\{0,1\}\}$$

is of power continuum and for any n we have

$$X \subseteq \bigcup_{\epsilon_j=0,1 \quad j=0,1,\ldots,n} B_{\epsilon_0\epsilon_1\cdots\epsilon_n}.$$

But by the construction this latter set intersects A_n in a set of power smaller than continuum, hence $A_n \cap X$ is of power smaller than continuum. Since $\cup_n A_n$ includes X (recall that $X \subset B^* \subset A$), it follows that $X = \cup_{n=0}^{\infty}(X \cap A)$, i.e., a set of power continuum is represented as countable union of sets each of power less than continuum. But this contradicts Problem 4.15, and this contradiction proves the claim.

32. Instead of $x \in \mathbf{R}$ we shall index our sets by infinite 0–1 sequences (their number is equally continuum; see Problem 4.3). For an infinite 0–1 sequence $\epsilon = \{\epsilon_i\}_{i=0}^{\infty}$, let A_ϵ be the set of real numbers that have decimal expansion of the form

$$\cdots \diamond . \diamond \cdots \diamond 0004\epsilon_0\beta_0 4\epsilon_1\beta_1 4\epsilon_2\beta_2 \cdots,$$

where $\diamond$ stand for any digits and $\beta_i = 2$ or 3. Since for each $\{\epsilon_i\}_{i=0}^{\infty}$ we can select the β_i's in continuum many ways (see Problem 4.3), it is clear that this A_ϵ is of cardinality continuum in every interval.

33. Consider the sets A_x, $x \in \mathbf{R}$ from the preceding problem. If we set $f(u) = x$ if $u \in A_x$, then this f takes any real value x continuum many times in every interval I (namely in the points of the set $I \cap A_x$).

We can also get a concrete f as follows. We define $f(x)$ using the decimal expansion $x = \cdots .x_1x_2\cdots$ of x (if x has two such representations, then fix the one that has infinitely many zero digits). We shall only consider the digits after the decimal point. Let $f(0) = 0$, and if in the expansion of x there are infinitely many blocks of length ≥ 2 consisting of the digit 5 or if there is no such block at all, then also let $f(x) = 0$. Otherwise let $l \geq 2$ be the length of the longest block of consecutive fives, m the number of 0's following the last one of the longest block of fives, $\beta_1, \beta_2, \ldots$ the digits in the expansion of x after these zeros, and set

$$f(x) = (-1)^l 10^m \cdot 0.\beta_3\beta_6\beta_9\ldots.$$

If I is any interval,

$$a = (-1)^s \ldots a_{-1}.a_1a_2\ldots$$

is its middle point, $k \geq 1$ is a number such that $10^{-k} < |I|/10$, and if $y = (-1)^p 10^q \cdot 0.y_1y_2\ldots$ ($p = 0, 1$, $q \geq 1$) is any nonzero real number, then let

$$x = (-1)^s \ldots a_{-1}.a_1a_2\ldots a_k 4\overbrace{555\ldots55}^{p+2k}\overbrace{000\ldots00}^{q}4\beta_1y_14\beta_2y_24\beta_3y_34\ldots,$$

where $\beta_i = 2$ or 3 independently of each other. It is clear that there are continuum many such numbers (the β_i's can be selected in continuum many ways by Problem 4.3), each of them lies in the interval I and each satisfies $f(x) = y$.

34. Let $f : [0,1] \to [0,1] \times [0,1]$ be the mapping from the next problem. Then f is of the form $f(t) = (g(t), h(t))$, with some continuous $g, h : [0,1] \to [0,1]$, and it is clear that, e.g., g takes every value $y \in [0,1]$ continuum many times (since all the points (y,u), $u \in [0,1]$ are in the range of f).

35. Recall that each point x in the Cantor set C has a triadic representation $x = 0.\alpha_1\alpha_2\ldots$ where each α_i is 0 or 2. It is also easy to see that if $y \in C$ is another point with similar representation $y = 0.\beta_1\beta_2\ldots$, and $|x - y| < 3^{-n}$, then the first n digits in the expansions of x and y are the same, i.e., $\alpha_i = \beta_i$ for all $1 \leq i \leq n$ (a warning is appropriate here: for $x, y \in \mathbf{R}$ two numbers can be close without having many common digits, e.g., if we use decimal expansion and $x = 0.1000\cdots00111\ldots$ and $y = 0.0999\cdots9900\ldots$ where $\cdots$ represent sufficiently many identical digits, then x and y can be arbitrary close without having a single common decimal digit). In fact, just take into account that $\alpha_1 = \beta_1$ exactly if x, y lie in the same subinterval of the Cantor construction at the first level, then $\alpha_2 = \beta_2$ exactly if x, y lie in the same subinterval of the Cantor construction at the second level, etc. Thus, if for $x \in C$ we set

$$g(x) = 0.(\alpha_1/2)(\alpha_3/2)(\alpha_5/2)\ldots, \qquad h(x) = 0.(\alpha_2/2)(\alpha_4/2)(\alpha_6/2)\ldots,$$

where the numbers represent binary expansions, then we get that both g and h are continuous functions on C. It is also clear that $f(x) = (g(x), h(x))$ maps C onto $[0,1] \times [0,1]$. In fact, if $P = (0.\gamma_1\gamma_2\ldots, 0.\delta_1\delta_2\ldots)$ is any point in $[0,1] \times [0,1]$ (with binary expansion for the coordinates), then with $x = 0.(2\gamma_1)(2\delta_1)(2\gamma_2)(2\delta_2)\ldots$ we have $f(x) = P$.

Extend now both g and h to the contiguous intervals of C linearly, i.e., if (a, b) is a subinterval of $[0,1]\setminus C$, then let $g(t) = (g(b)-g(a))(t-a)/(b-a)+g(a)$ for $t \in (a, b)$. It is easy to see that these extended functions are continuous on $[0,1]$. This way we get a continuous extension of f, and this f has all the desired properties.

36. Let $f(t) = (g(t), h(t))$ be the function from the preceding problem. We claim that the functions

$$f_0(t) = g(t),\ f_1(t) = g(h(t)),\ f_2(t) = g(h(h(t))), \ldots$$

are appropriate. To this end it is sufficient to verify that for every n and arbitrary real numbers $x_0, x_1, \ldots, x_n$ from $[0,1]$ there is a $t_n \in [0,1]$ such that $f_i(t_n) = x_i$ for all $0 \le i \le n$. In fact, then if $x_0, x_1, \ldots$ is an arbitrary infinite sequence, then we can select a convergent subsequence of the aforementioned sequence $t_0, t_1, \ldots$ converging to some number $t \in [0,1]$, and then it is clear that we have for all i the equality $f_i(t) = x_i$.

We show the existence of t_n by induction. For $i = 0$ it clearly exists, and let us suppose that we know the existence of t_{n-1} for all sequences $z_0, z_1, \ldots, z_{n-1}$. Then, by this induction hypothesis, there is a $t^*_{n-1} \in [0,1]$ with the property that

$$g(t^*_{n-1}) = x_1,\ g(h(t^*_{n-1})) = x_2, \ldots\ g(\overbrace{h(h\cdots(h(}^{n-1}t^*_{n-1}))\cdots)) = x_n,$$

where the last function is composed of g and $n-1$ copies of h. By the property of the function f, there is a t_n such that $g(t_n) = x_0$ and $h(t_n) = t^*_{n-1}$. Thus, for this t_n we get

$$g(t_n) = x_0,\ g(h(t_n)) = x_1, \ldots\ g(\overbrace{h(h\cdots h(}^{n}t_n)\cdots)) = x_n,$$

where the last function is composed of g and n copies of h, and so the induction step has been verified.

* * *

37. Set $\epsilon = 1/m$ with $m = 1, 2, \ldots$. If $\nu = \nu_m$ is the corresponding number in the definition of convergence, then there is a countable ordinal τ larger than any of the countably many countable ordinals ν_m. It follows that if A is the limit, then for $\xi > \tau$ we have $a_\xi = A$, and this proves the claim.

38. Assume the sequence to be increasing. The statement is a consequence of Problem 1, for there is no point of the set $A = \{a_\xi\}_{\xi<\alpha}$ in the interval $(a_\gamma, a_{\gamma+1})$ for any $a_\gamma \in A$.

39. Consider α as an ordered set. Since it is countable, it is similar to a subset of $\mathbf{Q} \cap [0,1]$ (see Problems 6.26 and 6.28), thus there is a mapping $f : \alpha \to [0,1]$ that is monotone. If for $\xi < \alpha$ we set $a_\xi = f(\xi)$, then $\{a_\xi\}_{\xi<\alpha}$ is a strictly increasing sequence, and it is easy to see that if $A = \sup_{\xi<\alpha} a_\xi$, then this sequence converges to A.

6

Ordered sets

1. Let $\langle A, \prec \rangle$ be an infinite ordered set, and let $B = \{a_0, a_1, \ldots\}$ be any sequence in A consisting of different elements. We are going to show that in B there is a monotone subsequence. Consider the set C of all elements $a_j \in B$ for which there is no a_k, $k > j$ with $a_j \prec a_k$. The elements in C form a decreasing sequence; therefore, if C is infinite, then we are done. If C is finite, then there is an N such that $a_j \notin C$ for $j \geq N$. This means that for every a_j with $j \geq N$ there is an index $k > j$ such that $a_j \prec a_k$. But then starting from a_N we can select larger and larger elements, and we obtain an infinite increasing sequence in B.

2. See the solution to Problem 3.10, a).

3. Consider the set $\{-1/n, 1/n, 1 - 1/n \ : \ n = 2, 3, \ldots\}$.

4. Let $[x]$ resp. $\{x\}$ denote the integral resp. fractional part of x, and set $x \prec y$ if $\{x\} < \{y\}$ or if $\{x\} = \{y\}$ and $[x] < [y]$. It is clear that in this ordering $x - 1$ is the predecessor, and $x + 1$ is the successor of x.

5. The necessity is obvious. Now suppose that $\langle A, \prec \rangle$ is such that for every $a \in A$ there are only finitely many elements $b \in A$ with $b \prec a$. To an $a \in A$ associate the number n_a of those $b \in A$ with $b \prec a$. By the assumption the mapping $a \mapsto n_a$ is a mapping from A into $\mathbf{N}$, and it is immediate that it is a monotone mapping. Let B be the set of all n_a's. Since A is infinite, B is also infinite (note that a monotone mapping is 1-to-1). If $n_a \in \mathbf{N}$, and $m < n_a$, then there is an element $c \in A$ with $m = n_c$. In fact, the set $\{b \ : \ b \prec a\}$ is finite and has n_a elements, so if we select as c the $(m + 1)$st element of $\{b \ : \ b \prec a\}$, then for this $n_c = m$ as we claimed. Thus, the mapping $a \mapsto n_a$ is a similarity mapping from A onto $\mathbf{N}$.

6. The answer is that $\langle A, \prec \rangle$ is similar to $\mathbf{N}$ or to the set of the negative integers, $\mathbf{Z} \setminus \mathbf{N}$. The sufficiency of this condition is clear, so now suppose that

$\langle A, \prec\rangle$ has the property that every infinite subset is similar to the whole set. By Problem 1 in $\langle A, \prec\rangle$ there is a monotone infinite sequence S. Thus, S is either similar to $\mathbf{N}$ or to the set of the negative integers $\mathbf{Z}\setminus\mathbf{N}$. The assumption is that S is similar to A, thus A must be similar to either $\mathbf{N}$ or to $\mathbf{Z}\setminus\mathbf{N}$, as we claimed.

7. The necessity of the condition is clear, so let us suppose that $\langle A, \prec\rangle$ has no smallest or largest element, and every interval $\{c : a \prec c \prec b\}$, $a, b \in A$ is finite. This implies that every element has a predecessor as well as a successor. Thus, starting from any element of A and successively taking predecessors and successors, we can define a two-way infinite sequence $\{a_j\}_{j=-\infty}^{\infty}$ in A with the property that $a_j \prec a_k$ if $j < k$. Now if A had any additional element a, then that would have to be either bigger than all a_j or smaller than all a_j, and in both cases we would have infinitely many elements between a_0 and a, which is not possible. Thus, $A = \{a_j\}_{j=-\infty}^{\infty}$, and the proof is over.

8. It is clear that every set similar to $\mathbf{Z}$, $\mathbf{N}$ or $\mathbf{Z}\setminus\mathbf{N}$ has this property, and we show that this condition is also necessary. If A does not have a smallest and largest element, then by the previous problem it is similar to $\mathbf{Z}$. If A has a smallest element, then it cannot have a largest element, for then there would be only finitely many elements of A between them. Now as in the previous proof, starting from the smallest element we can form an infinite increasing sequence $a_0 \prec a_1 \prec \cdots$ by taking successors one after the other, and it is also clear that there cannot be any additional element a of A, for then this a would have to be larger than any a_j, and there would be infinitely many elements between a_0 and a. Thus, in this case $\langle A, \prec\rangle$ is similar to $\mathbf{N}$. In an analogous manner if A has a largest element, then it is similar to $\mathbf{Z}\setminus\mathbf{N}$.

9. The set $\mathbf{Q}$ has the continuum many different initial segments $\{r \in \mathbf{Q} : r < x\}$, $x \in \mathbf{R}$.

10. See Problem 90.

11. Let B_m be the set $B_m = \{m - 1/k : k = 1, 2, 3, \ldots\}$, which is similar to $\mathbf{N}$, and for $n = 1, 2, \ldots$ consider the set

$$A_n = \left\{\bigcup_{m=-\infty}^{0} B_m\right\} \bigcup \{0, 1, \ldots, n-1\}.$$

It is clear that these are nonsimilar for different n's (in fact, in A_n there are exactly n elements a with the property that the set $\{b \in A_n : a < b\}$ is finite, and this property is preserved under similarity mapping). But it is also clear that A_n is similar to the initial segment $\{a \in A_m : a < -1/(n+1)\}$ of A_m. [W. Sierpiński, Cardinal and Ordinal Numbers, Polish Sci. Publ., Warszawa, 1965, XII.9/1]

12. This follows from Problem 3.1. [S. Banach, Fund. Math., **39**(1952), 236–239.]

13. Let $f : A \to B$ be a similarity mapping from A onto an initial segment of B and $g : B \to A$ be a similarity mapping from B onto an end segment of A. If f or g is an onto mapping, then we are done, so let us assume that they are not. Consider the set A' of all elements $a' \in A$ such that there is an $a' \leq a \in A$ for which it is true that $a < g(f(a))$. It is clear that then this a also belongs to A', A' is an initial segment of A, and it is not empty, for any element outside of the range of g (which is the same as preceding every element in the range) is in A'. Let B' be the image of A' under the mapping f. Then B' is an initial segment of B, and we claim that g maps $B \setminus B'$ onto $A \setminus A'$. With this the proof will be over, for then the mapping $h(x) = f(x)$ if $x \in A'$ and $h(x) = g^{-1}(x)$ if $x \in A \setminus A'$ is clearly a similarity mapping.

Let $b \in B \setminus B'$. Then for every $a' \in A'$ there is an $a \in A'$ such that $a' \leq a < g(f(a)) < g(b)$, thus $g(b) \notin A'$, and so g maps $B \setminus B'$ into $A \setminus A'$. If $d \in A \setminus A'$, then $g(f(d)) \leq d$, furthermore $f(d) \in B \setminus B'$. Since g is mapping $B \setminus B'$ onto an end segment of $A \setminus A'$, there is an element $b \in B \setminus B'$ with $g(b) = d$. Since d was an arbitrary element of $A \setminus A'$, this proves that g is mapping from $B \setminus B'$ onto $A \setminus A'$, and we are done. [A. Lindenbaum, see W. Sierpiński, Cardinal and Ordinal Numbers, Polish Sci. Publ., Warszawa, 1965, XII.9. Theorem 2]

14. Let $f : A \to B$ be a monotone mapping onto an initial segment of B, $g : A \to B$ a monotone mapping onto an end segment B, and $h : B \to A$ a monotone mapping of B onto an interval of A. We distinguish three cases.

Case I: there is a $b \in B$ such that $b \preceq f \circ h(b)$. Let

$$B_1 = \{c \in B \ : \ c \preceq b \text{ for some } b \in B \text{ with } b \preceq f \circ h(b)\}.$$

B_1 is an initial segment of B, and we claim that $f \circ h$ maps B_1 into B_1. In fact, if $c \in B_1$ and $b \in B$ is as in the definition of B_1, then $f \circ h(c) \preceq f \circ h(b)$, and here $f \circ h(b) \preceq (f \circ h)(f \circ h(b))$, so by the definition of B_1 we have $f \circ h(c) \in B_1$. Let

$$A_1 = \{a \in A \ : \ a \leq h(b) \text{ for some } b \in B_1\}.$$

Then A_1 is an initial segment of A, and f is mapping A_1 onto B_1. In fact, f maps A_1 into B_1, since if $a \leq h(b)$ for some $b \in B_1$, then $f(a) \preceq f \circ h(b) \in B_1$, so $f(a) \in B_1$. On the other hand, if $c \in B_1$ is arbitrary, and b is as in the definition of B_1, then $c \preceq b \preceq f \circ h(b) = f(h(b))$. But f maps initial segments into initial segments, so there is an $a \in A_1$ such that $c = f(a)$, which proves that f maps A_1 onto B_1, and incidentally, that A_1 and B_1 are similar. It is left to show that $A \setminus A_1$ and $B \setminus B_1$ are also similar. It is also clear that h maps B_1 onto an end segment of A_1. Thus, f maps $A \setminus A_1$ into an initial segment of $B \setminus B_1$, and h maps $B \setminus B_1$ into an initial segment of $A \setminus A_1$. Now if g maps $A \setminus A_1$ into $B \setminus B_1$, then it maps it into an end segment of it, so on applying the preceding problem to the ordered sets $A \setminus A_1$ and $B \setminus B_1$ and to (the restrictions of) the mappings g, h we can conclude that $A \setminus A_1$ and $B \setminus B_1$ are similar. If, however, g does not map $A \setminus A_1$ into $B \setminus B_1$, then $B \setminus B_1$ is in the range of the restriction of g onto $A \setminus A_1$, so g^{-1} is defined on $B \setminus B_1$ and maps it into an end segment of $A \setminus A_1$. Now the similarity of $A \setminus A_1$ and $B \setminus B_1$ follows again from the previous problem if we consider (the restrictions of) the mappings f and g^{-1}.

Case II: there is a $b \in B$ such that $g \circ h(b) \preceq b$. This case can be verified along the same lines as Case I, and actually follows from it if we consider the reverse orderings.

Case III: for every $b \in B$ we have $f \circ h(b) \prec b \prec g \circ h(b)$. Take any $b \in B$, and set

$$B_1 = \{c \in B \,:\, c \preceq b\}, \quad A_1 = \{a \in A \,:\, a \leq h(b) \text{ for some } b \in B_1\}.$$

By our assumption, A_1 is mapped by f into B_1, and $A \setminus A_1$ is mapped by g into $B \setminus B_1$. Thus, A_1 is similar to an initial segment of B_1 under f, and B_1 is similar to an end segment of A_1 under h, so A_1 and B_1 are similar (Problem 13). In a similar fashion, $B \setminus B_1$ is similar to an initial segment of $A \setminus A_1$ under h and $A \setminus A_1$ is similar to an end segment of $B \setminus B_1$ under g, so $A \setminus A_1$ and $B \setminus B_1$ are also similar. Since A_1 and B_1 are initial segments, this proves that A and B are similar.

15. For every n let A_n^1 be the set $(\mathbf{Q} \cap [n+1/3, n+2/3]) \cup \{n+1/4, n+1-1/4\}$, and let $A_n^0 = \{n+1/3, n+2/3\}$. For any 0–1 sequence $\epsilon = (\epsilon_0, \epsilon_1, \ldots)$ consider the set

$$A_\epsilon = \cup_{n=0}^{\infty} A_n^{\epsilon_n}.$$

Note that if $\epsilon_n = 1$ then the set $A_\epsilon \cap (n, n+1)$ contains a point followed by a countable densely ordered set with first and last elements which is followed by one more point. Since a similarity mapping maps successors into successors and densely ordered subset into densely ordered subsets, it is easy to see by induction that if there are two 0–1 sequences ϵ and ϵ' such that $\epsilon_0 = \epsilon'_0, \ldots, \epsilon_{m-1} = \epsilon'_{m-1}$, then a similarity mapping f between the sets A_ϵ onto $A_{\epsilon'}$ must map the set $A_\epsilon \cap [0, m]$ into the set $A_{\epsilon'} \cap [0, m]$. Thus, if in addition, say $\epsilon_m = 0$ but $\epsilon'_m = 1$, then f cannot exist, for it would have to map the three-point set $\{m + 1/3, m + 2/3, m + 1 + 1/4\}$, which is an initial segment of $A_\epsilon \cap (m, \infty)$, onto an initial segment of $A_{\epsilon'} \cap (m, \infty)$, which is not possible, for in this latter set the point $m + 1/4$ is followed by the dense set $\mathbf{Q} \cap [m + 1/3, m + 2/3]$.

Thus, the sets A_ϵ are not similar for different 0–1 sequences, and so we have found continuum many subsets of $\mathbf{Q}$ no two of which are similar. [W. Sierpiński, Cardinal and Ordinal Numbers, Polish Sci. Publ., Warszawa, 1965, XII.6/1]

16. See Problem 18.

17. By Problem 34 it is enough to show that if $\langle {}^{\mathbf{N}}\mathbf{N}, \prec \rangle$ is the set of all sequences of natural numbers with the lexicographic ordering, then there are continuum many disjoint subsets of ${}^{\mathbf{N}}\mathbf{N}$ similar to $\langle {}^{\mathbf{N}}\mathbf{N}, \prec \rangle$ (recall also that $\mathbf{R}$ and $(0,1)$ are similar). For a 0–1 sequence $\epsilon = (\epsilon_0, \epsilon_1, \ldots)$ let H_ϵ be the set of all those sequences $s = (n_0, n_1, \ldots)$ from ${}^{\mathbf{N}}\mathbf{N}$ for which $n_i \equiv \epsilon_i \pmod 2$. For different ϵ's these sets are disjoint, and it is obvious that each H_ϵ is similar to ${}^{\mathbf{N}}\mathbf{N}$. In fact, if $[x]$ denotes the integral part of x, then the mapping $f(s) = ([n_0/2], [n_1/2], \ldots)$ establishes a monotone correspondence between A_ϵ and ${}^{\mathbf{N}}\mathbf{N}$. [W. Sierpiński, Cardinal and Ordinal Numbers, Polish Sci. Publ., Warszawa, 1965, XI.10. Remark]

18. The mapping $x \mapsto \arctan x$ is monotone and maps A into a similar subset A' of the interval $(-\pi/2, \pi/2)$. Hence for every $a \in \mathbf{R}$ the set $A' + a$ is a set similar to A, and all these sets are different. This shows that there are at least continuum many subsets of $\mathbf{R}$ similar to A.

Next let $f : A \to \mathbf{R}$ be any similarity mapping from A onto a subset of $\mathbf{R}$. f can be extended to a nondecreasing real function F: select any point $a_0 \in A$, and set $F(x) = \inf_{x \le a,\ a \in A} f(a)$ if $x \le a_0$ and $F(x) = \sup_{x \ge a,\ a \in A} f(a)$ if $x > a_0$. Clearly, from different f's we get different F's, so there is at most as many subsets of $\mathbf{R}$ similar to A as nondecreasing functions F on $\mathbf{R}$, and by Problem 4.14, d) there are at most continuum many such functions. [W. Sierpiński, Cardinal and Ordinal Numbers, Polish Sci. Publ., Warszawa, 1965, XII.6/2]

19. Consider the Cantor set C, list the bounded intervals in $\mathbf{R} \setminus C$ as $I_1, I_2, \ldots$, and let $P_n \subset I_n$ be a set consisting of n points. Let P be the set that consists of the points of the sets P_n, of the endpoints of the intervals I_n and of the points $0, 1$. For an arbitrary subset $X \subseteq C \setminus P$ of cardinality continuum consider the set $X \cup P$. Since C is of cardinality continuum and P is countable, the set $C \setminus P$ is of cardinality continuum, therefore there are $2^{\mathbf{c}}$ such subsets. It is enough to show that if $X, Y \subseteq C \setminus P$ are different subsets of $C \setminus P$ (of cardinality $\mathbf{c}$), then $X \cup P$ and $Y \cup P$ are not similar. Let us assume that f is a similarity mapping from $X \cup P$ onto $Y \cup P$. Note that for any n there is exactly one pair $a, b \in X \cup P$ such that there are exactly n points in $X \cup P$ in between a and b but for any $a', b' \in X \cup P$ with $a' < a$ and $b < b'$ both sets $(a', b) \cap (X \cup P)$ and $(a, b') \cap (X \cup P)$ are infinite. In fact, this pair must be the one for which $a, b \in P$ and $(a, b) = I_n$, and then the portion of P in between a and b is exactly P_n. Since the same is true of $Y \cup B$, it follows that every point of P_n is a fixed point of f and it also follows that the same is true of the endpoints of the intervals I_n, i.e., every point of P is a fixed point of f. But the set of the endpoints of the intervals I_n is dense in C, hence P is dense in $X \cup P$ and $Y \cup P$. Now if a monotone mapping fixes a dense set then it must be the identity mapping (see, e.g., the proof of Problem 21), hence $X = Y$, and the proof is complete. [W. Sierpiński, Cardinal and Ordinal Numbers, Polish Sci. Publ., Warszawa, 1965, XI.10/5]

20. See the second part of Problem 22.

21. First observe, that the assumption implies that there can be only countably many pairs a_1, a_2 in A such that a_2 is a successor of a_1. In fact, any element from b can belong to at most two such sets $\{a_1, a_2\}$, and B is countable. Thus, if we add all such pairs to B, then B remains countable, and with this we achieve that for any two elements $a_1, a_2 \in A$ with $a_1 \prec a_2$ there are $b_1, b_2 \in B$ with $a_1 \preceq b_1 \prec b_2 \preceq a_2$.

By Problem 26 there is a similarity mapping f from B onto a subset C of $\mathbf{Q} \cap (0,1)$. Now for $a \in A$ define $F(a) = \sup_{b \preceq a,\ b \in B} f(b)$. This is well defined since $C \subset \mathbf{R}$ is bounded, and we claim that it is monotone. In fact, if $a_1 \prec a_2$, then there are $b_1, b_2 \in B$ with $a_1 \preceq b_1 \prec b_2 \preceq a_2$ and with them $F(a_1) \le f(b_1) < f(b_2) \le F(a_2)$. Thus, F maps $\langle A, \prec \rangle$ onto a subset of $\mathbf{R}$ in a monotone fashion, and we are done. [W. Sierpiński, Cardinal and Ordinal Numbers, Polish Sci. Publ., Warszawa, 1965, XI.10/5]

22. The set $A = \mathbf{R}$ is similar to $B = (0,1)$, but their complements are not similar.

If, however, A and B are two countable dense subsets of $\mathbf{R}$, then their complements are similar. In fact, by Problem 27 A and B are similar, and let $f : A \to B$ be a similarity mapping between them. If we set $F(x) = \sup_{a \le x,\ a \in A} f(a)$, then F is a strictly increasing monotone real function the range of which contains all points of the dense set B. Hence F cannot have any point of discontinuity (jump), so F is a monotone mapping from $\mathbf{R}$ onto $\mathbf{R}$. The restriction of F to $\mathbf{R} \setminus A$ is mapping from $\mathbf{R} \setminus A$ onto $\mathbf{R} \setminus B$, and so these sets are similar.

23. Let us enumerate the open subintervals of $\mathbf{R}$ with rational endpoints into a sequence $I_0, I_1, \ldots$ (cf. Problem 2.13), and for an arbitrary member G of $\mathcal{M}$ set $f(G) = \sum_{I_j \subseteq G} 10^{-j}$. Since every open subset of $\mathbf{R}$ is the union of some I_j's, it follows that if $G_1, G_2 \subset \mathcal{M}$ and $G_1 \subset G_2$, then there is an I_j with $I_j \subseteq G_2$ but $I_j \not\subseteq G_1$. This shows that $f(G_1) < f(G_2)$, and so f is a similarity mapping from $\mathcal{M}$ into $\mathbf{R}$.

24. Let $\{r_n\}_{n=1}^{\infty}$ be an enumeration of the rational numbers, and for $x \in \mathbf{R}$ let $F_x = \{0\} \cup \{1/n \ :\ r_n < x\}$. This is a closed set of measure zero, and it is clear that if $x < y$ then $F_x \subset F_y$.

25. This is a special case of Problem 90, since out of two initial segments one of them includes the other one.

26. Let $\langle A, \prec \rangle$ be any countable ordered set. We may assume A to be infinite, and let $A = \{a_0, a_1, \ldots\}$ be an enumeration of the elements in A, and also select an enumeration $\mathbf{Q} \cap (0,1) = \{r_0, r_1, \ldots\}$ of the rationals in $(0,1)$. Now let $f(a_0) = r_0$, and if $f(a_i)$ have already been selected for $i < n$, then let

$f(a_n) = r_m$, where m is the smallest index for which it is true that we have $f(a_j) < r_m$ for exactly those $0 \le j < n$ for which $a_j \prec a_n$ holds. Since $\mathbf{Q} \cap (0,1)$ is densely ordered, there is such an m, so this definition is sound. It is clear from the definition that f is a monotone mapping from A onto a subset of $\mathbf{Q} \cap (0,1)$.

27. Follow the preceding proof with the following modification (this is the so-called back-and-forth argument). For each n we select subsets $A_n \subset A$ and $Q_n \subset \mathbf{Q}$ and an $f_n : A_n \to Q_n$ monotone mapping in such a way that A_{n+1} and Q_{n+1} are obtained by adding one element $a'_{n+1} \in A$ and $r'_{n+1} \in Q$ to A_n and Q_n, respectively, and f_{n+1} is the extension of f_n by setting $f_{n+1}(a'_{n+1}) = r'_{n+1}$. Start from $A_0 = \{a_0\}$, $Q_0 = \{r_0\}$, $f_0(a_0) = r_0$ as before, and if A_i, Q_i, f_i have already been defined for $i \le n$, then for even n let a'_{n+1} be the element a_k in $A \setminus A_n$ with smallest index k, and let $f_{n+1}(a'_{n+1}) = r_m$, where m is the smallest index for which it is true that we have $f_n(a) < r_m$ for exactly those $a \in A_n$ for which $a \prec a'_{n+1}$ holds, and set $r'_{n+1} = r_m$. However, for odd n let r'_{n+1} be the element r_k in $\mathbf{Q} \setminus Q_n$ with smallest index k, and let m be the smallest index for which it is true that we have $a'_j \prec a_m$ for exactly those $a'_j \in A_n$, $0 \le j \le n$ for which $f(a'_j) < r'_{n+1}$ holds, and set $a'_{n+1} = a_m$. By the density of the sets A and $\mathbf{Q}$ and by the fact that neither of them has a smallest or largest element, the selection of a'_{n+1}, r'_{n+1} above is possible, and by the construction $a_k \in A_n$ and $r_k \in Q_k$ for $n > 2k$. Thus, $\cup_n A_n = A$, $\cup_n Q_n = \mathbf{Q}$. Now if we set $f(a_n) = f_{2n+1}(a_n)$, then, in view of the fact that the functions $f_0, f_1, \ldots$ extend each other, it follows that f is a monotone mapping from A onto $\mathbf{Q}$.[G. Cantor]

28. If the set $\langle A, \prec \rangle$ is countable and densely ordered and does not have a smallest and largest element, then by the preceding problem it is similar to $\mathbf{Q}$. The same is true of $\mathbf{Q} \cap (0,1)$, hence $\langle A, \prec \rangle$ is similar to $\mathbf{Q} \cap (0,1)$. If in $\langle A, \prec \rangle$ there is a smallest element a_0 but there is no largest element, then the set $\langle A \setminus \{a_0\}, \prec \rangle$ is densely ordered and is without a smallest and largest element, hence, as we have just seen, it is similar to $\mathbf{Q} \cap (0,1)$. But then clearly $\langle A, \prec \rangle$ is similar to $\mathbf{Q} \cap [0,1)$. In a similar fashion, if $\langle A, \prec \rangle$ has a largest element but no smallest one, then it is similar to $\mathbf{Q} \cap (0,1]$, and if it has both, then it is similar to $\mathbf{Q} \cap [0,1]$.

29. Let B be the set of countable ordinals with the usual ordering on the ordinals, and let $A = \mathbf{Q} \times B$ with the antilexicographic ordering. Any nonempty proper initial segment of $\langle A, \prec \rangle$ is an initial segment of $\mathbf{Q} \times C$, where $C \subset B$ is a countable set. Now apply Problem 28.

30. Let $\langle A, \prec \rangle$ be the set of the countable ordinals with the usual ordering on the ordinals. Any uncountable subset B of $\langle A, \prec \rangle$ is well ordered. By Problem 42 one of A or B is similar to an initial segment of the other one. But both A and B have countable proper initial segments, hence the initial segment in

question must be the whole set. Therefore, in either case we get that A and B are similar.

31. It is clear that the elements follow one another in the order:

$$\begin{aligned}&(0,0,\ldots),\\&(1,0,0\ldots),\\&(0,1,0,0\ldots),(1,1,0,0\ldots),\\&(0,0,1,0,0\ldots),(1,0,1,0,0\ldots),(0,1,1,0,0\ldots),(1,1,1,0,0\ldots),\\&\quad\vdots\end{aligned}$$

and this is the same how the numbers $0,1,2,\ldots$ follow one another. [W. Sierpiński, Cardinal and Ordinal Numbers, Polish Sci. Publ., Warszawa, 1965, XII.2/7]

32. Let $\langle A,\prec\rangle$ be the lexicographically ordered set of infinite 0–1 sequences that contain only a finite number of 1's. By Problem 2.9, c), $\langle A,\prec\rangle$ is countable, and it has a smallest element, namely the identically zero sequence. It is also clear that $\langle A,\prec\rangle$ has no largest element. Thus, by Problem 28 it is enough to show that $\langle A,\prec\rangle$ is densely ordered. Let $x=(x_0,x_1,\ldots)$ and $y=(y_0,y_1,\ldots)$ be two elements in A with $x\prec y$. Then there is an n such that $x_i=y_i$ for $i=0,1,\ldots,n-1$, but $x_n=0$ and $y_n=1$. Set $z=(x_0,x_1,\ldots,x_{n-1},0,1,1,1,\ldots,1,1,1,0,0,\ldots)$, where the last 1 appears so far out that there the numbers in the sequence x are already all zero. For this z we clearly have $x\prec z\prec y$, which proves that $\langle A,\prec\rangle$ is densely ordered. [W. Sierpiński, Cardinal and Ordinal Numbers, Polish Sci. Publ., Warszawa, 1965, XII.2/7]

33. To a 0–1 sequence $\epsilon_0,\epsilon_1,\ldots$ associate the number $0.(2\epsilon_0)(2\epsilon_1)\ldots$ in base 3 in the Cantor set. It is easy to see that this establishes a monotone correspondence between the lexicographically ordered set of infinite 0–1 sequences and the Cantor set.

34. To a sequence $s=(n_0,n_1,\ldots)$ of natural numbers associate the number

$$f(s)=1-2^{-n_0-1}-2^{-n_0-n_1-2}-2^{-n_0-n_1-n_2-3}-\cdots.$$

This is a monotone mapping: if $s\prec s'$ in the lexicographic ordering, then $f(s)<f(s')$. It is also clear that $f(s)\in[0,1)$. Furthermore, if $y\in[0,1)$, then $1-y$ can be uniquely written in the form

$$1-y=2^{-m_0}+2^{-m_1}\cdots,\qquad\text{with }\ 1\le m_0<m_1<m_2<\cdots.\tag{6.1}$$

In fact, select the integer m_0 according to $2^{-m_0}<y\le 2^{-m_0+1}$, then $0<y-2^{-m_0}\le 2^{-m_0}$, hence if m_1 is chosen according to $2^{-m_1}<y-2^{-m_0}\le 2^{m_1}$,

then $m_1 > m_0$. Continuing this process we get the representation (6.1). But then it is clear that for $s = (m_0 - 1, m_1 - m_0 - 1, m_2 - m_1 - 1, \ldots)$ we have $f(s) = y$, thus f is a mapping onto $[0, 1)$.

35. Associate with a sequence $s = (n_0, -n_1, n_2, -n_3, \ldots)$ the value of the continued fraction

$$f(s) = 1 - \cfrac{1}{n_0 + 1 + \frac{1}{n_1+1+\frac{1}{\cdots}}}.$$

On the right we have an infinite continued fraction, hence $f(s)$ is irrational and lies in $(0, 1)$. Conversely, the continued fraction expansion of every irrational number is of the preceding form, hence f is a mapping from A onto $(0, 1)\setminus\mathbf{Q}$. It is also clear that if in two sequences s and s' we have $n_0 < n_0'$, or $n_0 = n_0'$ and $n_1 > n_1'$ (sic!), or $n_0 = n_0'$, $n_1 = n_1'$ and $n_2 < n_2'$ or $n_0 = n_0'$, $n_1 = n_1'$, $n_2 = n_2'$ and $n_3 > n_3'$ (sic!) etc., then $f(s) < f(s')$ (note that by increasing the bottom denominator at a k-level continued fraction built up from positive numbers increases the fraction if k is even and decreases it if k is odd, simply because by increasing the denominator in a fraction of positive numbers we decrease the fraction). Thus, f is a similarity mapping from $\langle A, \prec\rangle$ onto $(0, 1) \setminus \mathbf{Q}$. But this latter set is similar to the set of irrational numbers. Indeed, $(0, 1)$ is similar to $\mathbf{R}$, say under a mapping f, hence $(0, 1) \setminus \mathbf{Q}$ is similar to $\mathbf{R} \setminus f[\mathbf{Q}]$ where $f[\mathbf{Q}]$ is a countable dense subset of $\mathbf{R}$. Now just apply the second part of Problem 22 to deduce that $\mathbf{R} \setminus \mathbf{Q}$ and $\mathbf{R} \setminus f[\mathbf{Q}]$ are similar. [F. Hausdorff, Grundzüge der Mengenlehren, Leipzig, 1914; Set Theory, Second edition, Chelsea, New York, 1962]

36. In a well-ordered set there cannot be a decreasing infinite sequence, for then in the subset formed from the elements of the sequence there is no smallest element.

Conversely, if the set $\langle A, \prec\rangle$ is not well ordered, then there is a nonempty subset $B \subset A$ which does not have a smallest element, i.e., for any $b \in B$ there is a smaller element in B. But then we can select elements $b_0, b_1, \ldots$ from B such that each one is smaller than the previous one, and this $b_0, b_1, \ldots$ is then an infinite monotone decreasing subsequence of $\langle A, \prec\rangle$.

37. Apply the previous problem along with Problem 5.2.

38. Apply the preceding problem and Problem 23. If $\mathcal{U}$ consists of closed sets, then consider complements with respect to $\mathbf{R}$.

39. If we had $f(a) \prec a$ for some a, then by monotonicity $f(f(a)) \prec f(a)$, $f(f(f(a))) \prec f(f(a))$, etc., i.e., $a, f(a), f(f(a)), \ldots$ would be a monotone decreasing sequence, which is not possible in view of Problem 36.

40. If $\langle A, \prec\rangle$ and $\langle B, <\rangle$ are two well-ordered sets and f_1 and f_2 are similarity mappings from $\langle A, \prec\rangle$ onto $\langle B, <\rangle$, then $f_2^{-1} \circ f_1$ and $f_1^{-1} \circ f_2$ are mappings

of $\langle A, \prec\rangle$ into itself. Hence by the preceding problem for every a we have $a \preceq f_2^{-1} \circ f_1(a)$ and $a \preceq f_1^{-1} \circ f_2(a)$, which, when applying f_2 resp. f_1 to both sides, yields $f_2(a) \leq f_1(a)$ and $f_1(a) \leq f_2(a)$, i.e., $f_1(a) = f_2(a)$, and this shows that f_1 and f_2 are identical.

41. This is a consequence of Problem 39, for if the well-ordered set $\langle A, \prec\rangle$ was similar via a mapping f to a subset of a proper initial segment S of it, and $a \notin S$ is a point outside S, then $f(a)$ belongs to S, hence it is smaller than a, and by Problem 39 this is not possible.

42. Let $\langle A, \prec\rangle$ and $\langle B, <\rangle$ be two well-ordered sets. Let A' be the set of all $a \in A$ such that the initial segment $A_a := \{\alpha \in A \;:\; \alpha \prec a\}$ is similar to an initial segment C_a of B. By the previous problem, this C_a is uniquely defined by a. If for some $a \in A$ we have $C_a = B$, then we are done, so let us assume that this is not the case. Then $B \setminus C_a$ has a smallest element that we denote by $f(a)$. It is easy to see that $C_a = B_{f(a)} := \{b \in B \;:\; b < f(a)\}$.

In a similar manner let B' be the set of all $b \in B$ for which the initial segment B_b is similar to an initial segment D_b of A. This D_b is again uniquely determined by b, and we can assume again that $D_b \neq A$ for any $b \in B'$. It is clear that for $b \in B'$ we must have $D_b = A_a$ for some $a \in A'$, and $f(a) = b$. Thus, f maps A' onto B'. Since a similarity mapping maps an initial segment into an initial segment, it follows that both A' and B' are initial segments of A and B, respectively.

Let f_a be the similarity mapping from A_a onto $B_{f(a)}$ (cf. Problem 40). Then for $c \prec a$ the restriction of f_a maps A_c into $B_{f_a(c)}$, hence, by the unicity of the B_c, we must have $f_a(c) = f(c)$. But this means that $f(c) = f_a(c) < f(a)$, i.e., f is monotone.

Thus, we have obtained so far that $f : A' \to B'$ is a similarity mapping and so the claim follows if we can show that $A' = A$. If this is not the case, then the set $A \setminus A'$ is not empty, and so it has a smallest element, say a'. As above, we get again that $A' = A_{a'}$, hence $A_{a'}$ is similar (via f) to an initial segment (B') of B. But then we would have $a' \in A'$, which is not the case, and so this contradiction proves that, in fact, $A' = A$.

43. If $\langle A, \prec\rangle$ and $\langle B, <\rangle$ are the two well-ordered sets, then by the previous problem, one of them is similar to an initial segment of the other one. Suppose, for example, that $\langle B, <\rangle$ is similar to an initial segment S of $\langle A, \prec\rangle$. But we must have $S = A$, for otherwise $\langle A, \prec\rangle$, being similar to a subset of $\langle B, <\rangle$, would be similar to a subset of its proper initial segment S, which is not possible by Problem 41. Thus $S = A$, which means that the two sets are similar.

44. Let $\langle A, \prec\rangle$ be an ordered set, and let $<$ be a well-ordering of A. Let B be the set of those elements $b \in A$ which satisfy the property that $b \preceq \alpha$ implies

$b \leq \alpha$ for every $\alpha \in A$. Since on B the two ordering $\prec$ and $<$ coincide, and $<$ is a well-ordering, it follows that $\langle B, \prec \rangle$ is well ordered.

In order to show that B is cofinal, let $a \in A$ be arbitrary, and let b be the smallest element with respect to $<$ of the nonempty set $\{\alpha : a \preceq \alpha\}$. Then $a \preceq b$, so if $b \preceq \alpha$ we also have $a \preceq \alpha$, hence, by the choice of b, we get that $b \leq \alpha$. Since this is true for any $\alpha \in A$, this b belongs to B, and since we also have $a \preceq b$, the proof is over.

To see that the order type of $\langle B, \prec \rangle$ can be made to be at most $|B|$, just take the well-ordering of A above so that the order type of $\langle A, < \rangle$ is $|A|$. The order type of $\langle B, \prec \rangle$ is the same as the order type of $\langle B, < \rangle$, and it is at most the order type of $\langle A, < \rangle$, i.e., at most $|A|$.

45. Assume that $\langle A, \prec \rangle$ is an ordered set with the property in the problem. If A has a largest element a and $A \setminus \{a\}$ is the union of countably many well ordered sets, then obviously so is A as well. Assume, therefore, that A has no largest element, and let B be a well-ordered, cofinal subset of A (see Problem 44). For every $b \in B$, let A^b consist of those elements x of A for which this b is the least element $y \in B$ with $x \prec y$. Then $A = \cup\{A^b : b \in B\}$ is a partition and if $b \prec b'$ then $x \prec x'$ holds whenever $x \in A^b$, $x' \in A^{b'}$ (i.e., $\langle A, \prec \rangle$ is the ordered union of the ordered set $\{\langle A^b, \prec \rangle : b \in B\}$). As A^b is a subset of the initial segment determined by b, it is the union of countably many well-ordered sets: $A^b = A_0^b \cup A_1^b \cup \cdots$. If we set $A_i = \cup\{A_i^b : b \in B\}$, then on the one hand, $A = A_0 \cup A_1 \cup \cdots$, and on the other hand, A_i, as the well-ordered union of well-ordered sets, is well ordered for every $i = 0, 1, \ldots$.

46. Suppose first that A does not have a largest element. Let a_0 be the smallest element of A and for $a \in A$ let a^+ be the successor of a in A. By Problem 26 there is a monotone mapping $h : A \to \mathbf{Q} \cap (0, 1)$. The mapping $f(a) = \sup_{b<a} h(b^+)$ $(a \neq a_0)$, $f(a_0) = 0$ is monotone from A into $\mathbf{Q} \cap [0, 1)$ with the property that if $a \in A$, $a \neq a_0$ does not have a predecessor, then $\sup_{b<a} f(b) = f(a)$ (i.e., f is continuous in the order topologies). In particular, the union of the intervals $[f(a), f(a^+))$, $a \in A$ is $[0, b)$, where $b = \sup_{a \in A} f(a)$. Choose for each $a \in A$ a monotone mapping g_a of $\{a\} \times [0, 1)$ onto $[f(a)/b, f(a^+)/b)$, and for $(a, x) \in A \times [0, 1)$ define $g(a, x) = g_a(x)$. This is clearly a monotone mapping of $A \times [0, 1)$ onto $[0, 1)$.

If A has a largest element $a_{\max}$, then the only change in the above argument we have to make is to *define* both b and $f(a_{\max}^+)$ to be 1.

47. Let B be the set of countable ordinals with the usual ordering among ordinals, and let A be the ordered union of $(0, 1)$ with the set $[0, 1) \times B$ (the latter with antilexicographic ordering). Every initial segment of $\langle A, \prec \rangle$ is an initial segment of $(0, 1) \cup [0, 1) \times C$ with some countable set $C \subset B$, and this latter set is similar to $(-1, 0) \cup [0, 1)$ by Problem 46. Thus, the nonempty proper initial segments of $\langle A, \prec \rangle$ are similar to nonempty initial segments of $(-1, 1)$, and hence they are similar either to $(0, 1)$ or to $(0, 1]$. However, $\langle A, \prec \rangle$

is not similar to a subset of $\mathbf{R}$, for it includes an uncountable well-ordered subset: $\{0\} \times B$ (see Problem 37).

In the second part of the proof we show the unicity. Let $\langle A, \prec \rangle$ be an ordered set not similar to a subset of $\mathbf{R}$, but for which all proper initial segments are similar to $(0,1)$ or $(0,1]$. If $\langle A, \prec \rangle$ had a largest element, then the proper initial segment determined by that element would be similar to either $(0,1)$ or $(0,1]$; therefore, the set $\langle A, \prec \rangle$ would be similar to either $(0,1]$ or $(0,1] \cup \{2\}$, which is not possible. Thus, there is no largest element. Select a cofinal well-ordered set $B = \{b_\xi\}_{\xi<\alpha}$, $b_\xi \prec b_\eta$ for $\xi < \eta < \alpha$ in $\langle A, \prec \rangle$ as in Problem 44. Then B does not have a largest element, and for each $\xi < \alpha$ the interval $I_\xi = \{a \in A \ : \ b_\xi \preceq a \prec b_{\xi+1}\}$ is similar to either $[0,1)$ or to $[0,1]$. But actually, the latter is not possible, for then the proper initial segment $\{a \ : \ a \preceq b_{\xi+1}\}$ would have a largest element which has a predecessor, therefore it would not be similar to either $(0,1)$ or to $(0,1]$. In a similar fashion, the interval $J = \{a \ : \ a \prec b_0\}$ is similar to $(0,1)$. Now $\langle A, \prec \rangle$ is the ordered union of the intervals J, I_ξ, $\xi < \alpha$, just the same type that we gave in the beginning of the proof. The proof will be completed by showing that $\alpha = \omega_1$, i.e., $\langle B, \prec \rangle$ is similar to the set of countable ordinals. α cannot be countable, for then the just constructed ordered union would be similar to $(0,1)$ (see Problem 46). But $\alpha \geq \omega_1 + 1$ is not possible, either, for then the proper initial segment $\{a \in A \ : \ a \prec b_{\omega_1}\}$ would include an uncountable well-ordered set ($\{b_\xi\}_{\xi<\omega_1}$), and hence it could not be similar to either $(0,1)$ or $(0,1]$ (see Problem 37). Thus, $\alpha = \omega_1$, and the proof is over.

48. One direction is clear. Now suppose that there is a monotone mapping $f : \langle A, \prec \rangle \to \langle A, \prec \rangle$ with $f(x) \neq x$, say $x \prec f(x)$. Define $g : A \to A$ as

$$g(y) = \begin{cases} y & \text{if } y \prec x, \\ f(y) & \text{if } x \preceq y. \end{cases}$$

It is clear that g is monotone, and its range omits x.

49. Assume that $x \prec y$. x and y divide $A \setminus \{x, y\}$ into three parts; let them be in the order of $\prec$ the sets X, Y, Z. Thus, $A = X \cup \{x\} \cup Y \cup \{y\} \cup Z$. Suppose to the contrary, that y is not a fixed point of $A \setminus \{x\}$. Then, by the previous problem, there is an order-preserving mapping $f : A \setminus \{x\} \to A \setminus \{x, y\}$.

We now consider three cases, and in each case we construct an order-preserving mapping g from A into either $A \setminus \{x\}$ or $A \setminus \{y\}$, so x or y is not a fixed point for the set $\langle A, \prec \rangle$, and this contradiction proves the claim.

If $f^n(y) \in Z$ for some $n \geq 1$, then let

$$g(z) = \begin{cases} z & \text{if } z \prec y, \\ f^n(z) & \text{if } y \preceq z. \end{cases}$$

This g maps A into $A \setminus \{y\}$.

If for some n we have $f^n(y) \in X$, then set

$$g(z) = \begin{cases} f^n(z) \text{ if } z \prec x, \\ f^n(y) \text{ if } z = x, \\ \quad z \text{ if } x \prec z. \end{cases}$$

This g maps A into $A \setminus \{x\}$.

If, finally, for every $n = 1, 2, \ldots$, $f^n(y) \in Y$ is true, then let

$$g(z) = \begin{cases} \quad z \ \text{ if } z \prec f^n(y) \text{ for every } n, \\ f(z) \text{ if } f^n(y) \preceq z \text{ for some } n. \end{cases}$$

This g maps A into $A \setminus \{y\}$.

50. Assume to the contrary that $x_0, x_1, \ldots$ are different fixed points in $\langle A, \prec\rangle$. By the previous problem x_n is a fixed point of $A_n = A \setminus \{x_0, \ldots, x_{n-1}\}$, so (see Problem 48) for $i < j$ the set $\langle A_i, \prec\rangle$ cannot be mapped by a monotone mapping into $\langle A_j, \prec\rangle$. But this contradicts Problem 60. and this contradiction proves the claim.

51. Let $A = \{1 - 1/j\}_{j=1}^{\infty} \cup \{1, 2, \ldots, n\}$. A monotone mapping $f : A \to A$ cannot map any of $\{1, 2, \ldots, n\}$ into the set $\{1 - 1/j\}_{j=1}^{\infty}$. Thus, f maps $\{1, 2, \ldots, n\}$ into itself, and hence onto itself, and so the points $1, 2, \ldots, n$ are all fixed points.

52. If $\langle A, \prec\rangle$ does not have a subset similar to $\mathbf{Q}$, then we can apply the reasoning from the Problem 50 referring to Laver's theorem rather than to Problem 60.

53. See Problem 1.20. [W. Sierpiński, Cardinal and Ordinal Numbers, Polish Sci. Publ., Warszawa, 1965, XI.2]

54. It is clear that $\prec$ is irreflexive. It is transitive, since if $x \prec y$ and $y \prec z$, then there are sets $E, F \in \mathcal{M}$ with $x \in E$, $y \notin E$, $y \in F$, $z \notin F$. Since one of E and F includes the other one, we must have $E \subset F$, hence $x \in F$ but $z \notin F$, and so really we have $x \prec z$. Finally, we show that $\prec$ is trichotomous. Suppose to the contrary that $x \neq y$ are not comparable with respect to $\prec$. Then for every $E \in \mathcal{M}$ either both of them belong to E, or both of them belong to its complement. Now let M be the union of all those sets $E \in \mathcal{M}$ that omit x. If $F \in \mathcal{M}$ contains x, then it contains all sets in $\mathcal{M}$ that omit x, so $M \subset F$. On the other hand, clearly all sets in $\mathcal{M}$ that omit x are subsets of M. Hence M is comparable with respect to inclusion with every set in $\mathcal{M}$, hence, by the maximality of $\mathcal{M}$, we have $M \in \mathcal{M}$. Since x is not in M, we get that, $y \notin M$. But then $M \cup \{x\}$ is also comparable with every member of $\mathcal{M}$, and as such, it would have to belong to $\mathcal{M}$, which is not possible, for then it would have to contain y, which is not the case. This contradiction proves that any two elements are comparable.

Thus, $\prec$ is an ordering on X. It easily follows from what we have done above that the initial segments in $\langle X, \prec \rangle$ are the sets $E \in \mathcal{M}$. In fact, if $y \in E$ and $x \prec y$, then we must have $x \in E$, so E is an initial segment. Conversely, if S is an initial segment of $\langle X, \prec \rangle$, then, since any two initial segments are comparable via inclusion, S is comparable with any member of $\mathcal{M}$. Hence by the maximality of $\mathcal{M}$ it has to belong to $\mathcal{M}$.

55. Let $\langle A, < \rangle$ be an ordered set, and let $\mathcal{M}$ be the set of its initials segments. It is clear that the relation $\prec$ defined in the preceding problem and $<$ coincide on A, thus it is left to show that $\mathcal{M}$ is a maximal family with respect to inclusion. Let $M \subset A$ be any set that is comparable with every member of $\mathcal{M}$, and let S be the union of all initial segments of $\langle A, < \rangle$ that are inluded in M. Then S is an initial segment such that $S \subset M$, and we claim that it is actually equal to M, and this will prove that $M \in \mathcal{M}$. If we had $S \neq M$, then we could select an element $m \in M \setminus S$. Consider now the initial segment $\{a : a \leq m\}$. Since this cannot be equal to M (otherwise m would belong to S), there is a $b < m$ such that $b \notin M$. But then the initial segment $\{a : a < m\}$ is incomparable with M, since it contains b but omits m. This contradiction shows that actually we have $S = M$.

56. Take as $\langle A^*, \prec^* \rangle$ the product of $\langle A, \prec \rangle$ with itself with the lexicographic ordering. Let $A^* = B \cup C$ be an arbitrary decomposition. If there is a $b_0 \in A$ such that all (b_0, a) with $a \in A$ belong to B, then these elements form a subset of A^* similar to $\langle A, \prec \rangle$. If, however, no such b_0 exists, then for every $b \in A$ there is an $a = a_b$ such that $(b, a_b) \in C$. But then the elements (b, a_b), $b \in A$ form a subset similar to $\langle A, \prec \rangle$.

57. Let $\langle A, \prec \rangle$ be an infinite ordered set. By Problem 1 it includes an infinite monotone sequence. If $\langle A, \prec \rangle$ includes an infinite decreasing sequence, then let us choose a well-ordered set $\langle B, < \rangle$ of cardinality bigger than the cardinality of A. By Problem 36 the set $\langle A, \prec \rangle$ cannot be similar to a subset of $\langle B, < \rangle$, and it is also clear that $\langle B, < \rangle$ cannot be similar to a subset of $\langle A, \prec \rangle$ because it is of bigger cardinality than the latter.

If $\langle A, \prec \rangle$ includes an infinite decreasing sequence, then just reverse the order on B.

58. According to the previous proof, one of $\mathbf{N}$ and $\mathbf{Z} \setminus \mathbf{N}$ is suitable.

59. For $i = 1, 2, \ldots, n$ consider the sets

$$A_i = \{-i, -i+1, \ldots -1\} \cup \{-1 + 1/n, 1 - 1/n\}_{n=1}^{\infty} \cup \{1, 2, \ldots n + 1 - i\}.$$

No $\langle A_i, < \rangle$ is similar to a subset of any other $\langle A_j, < \rangle$. Indeed, if $1 \leq j < i \leq n$, then a monotone mapping $f : A_i \to A_j$ should map the first i elements of A_i into the first j elements of A_j, which is not possible. For $j < i$ work similarly with the $n + 1 - i$ largest elements.

60. If $\langle A_i, \prec_i \rangle$, for some $i = 1, 2, \ldots$ includes a densely ordered subset, then $\langle A_0, \prec_0 \rangle$ is similar to a subset of $\langle A_i, \prec_i \rangle$ by Problems 26–28. If, however, neither of $\langle A_i, \prec_i \rangle$, $i = 1, 2, \ldots$, includes a densely ordered subset, then we can apply Laver's theorem to them.

61. Let $\langle A, \prec \rangle$ be a countable set, and for $a, b \in A$ set $a \sim b$ if there are only finitely many elements between a and b. This is clearly an equivalence relation. By Problem 8 an infinite equivalence class C is similar to either $\mathbf{Z}$, $\mathbf{N}$, or $\mathbf{Z} \setminus \mathbf{N}$. In each case we can omit an element c from C and the remaining set will be still similar to C, and this similarity relation can be extended (defining as the identity elsewhere) to a similarity from $\langle A \setminus \{c\}, \prec \rangle$ to $\langle A, \prec \rangle$.

Thus, if there is an infinite equivalence class, then we are done. If all equivalence classes are finite, then between any two equivalence classes there must be at least one other equivalence class; in other words if a, b belong to different equivalence classes, then there is an $a \prec c \prec b$ that is not equivalent to either a or b (otherwise a and b were equivalent since their classes are finite). Thus, if we select one–one element from the equivalence classes, then the set S so obtained is densely ordered. Omit now any element $s \in S$ from S. The remaining set is still densely ordered, and so by Problems 26 and 28 $\langle A, \prec \rangle$ is similar to a subset of $\langle S \setminus \{s\}, \prec \rangle$, and the proof is over. [B. Dushnik and E. W. Miller, Bull. Amer. Math. Soc., **46**(1940), 322]

62. See Problem 88.

63. If the set is $\langle A, \prec \rangle$, and it is well ordered, then we can just move the first element of the set to be last element. Then the set $\langle A, \prec' \rangle$ so obtained is not similar to the original one. In fact, this is clear if $\langle A, \prec \rangle$ does not have a largest element. If, however, it has a largest element, then the number of elements that are followed by finitely many elements is finite, say n (recall Problem 36, according to which if we start from the largest element in a well-ordered set and repeatedly take predecessors, then we get stuck in finitely many steps). But then in $\langle A, \prec' \rangle$ there are $(n + 1)$ elements with this property, and this proves that these two sets are not similar.

Now let us suppose that $\langle A, \prec \rangle$ is not well ordered. The union of well-ordered initial segments is clearly a well-ordered initial segment, so $\langle A, \prec \rangle$ has a largest well-ordered initial segment S (which may be empty). Then in $A \setminus S$ there is no smallest element (otherwise we could add that smallest element to S). Now move any element s in $A \setminus S$ to lie between S and $A \setminus (S \cup \{s\})$. We claim that the set $\langle A, \prec' \rangle$ so obtained is not similar to $\langle A, \prec \rangle$. In fact, in a similarity mapping well-ordered initial segments are mapped into well-ordered initial segments, hence S would have to be mapped into $S \cup \{s\}$, which is not possible as we have just seen it. [Z. Chajot, Fund. Math. **16**(1930), 132–136; W. Sierpiński, Cardinal and Ordinal Numbers, Polish Sci. Publ., Warszawa, 1965, XIV.4/14]

64. Just follow the preceding proof.

For removing one element the claim is not true: if we remove any element from $\mathbf{N}$, the remaining set is still similar to $\mathbf{N}$.

65. If $\langle A, \prec\rangle$ is an ordered set, then $\langle A, \prec\rangle \times \langle \mathbf{Q}, <\rangle$ with the lexicographic ordering is densely ordered, and it clearly includes a subset (say the set of elements $(a, 0)$ with $a \in A$) similar to $\langle A, \prec\rangle$.

66. Let the ordered set be $\langle A, \prec\rangle$. Consider the set $\mathcal{S}$ of all the initial segments of $\langle A, \prec\rangle$ that do not have a largest element, and consider the inclusion ordering on $\mathcal{S}$. To every $a \in A$ we can associate the initial segment $S_a := \{x : x < a\}$, and this mapping $a \mapsto S_a$ is clearly monotone. Thus, it is sufficient to show that $\langle \mathcal{S}, \subset\rangle$ is continuously ordered and that $\{S_a : a \in A\}$ is a dense subset of it. Let $\mathcal{S} = \mathcal{S}_1 \cup \mathcal{S}_2$ be a disjoint decomposition of $\mathcal{S}$ in such a way that each initial segment in $\mathcal{S}_1$ is a subset of any initial segment in $\mathcal{S}_2$, and let S be the union of all initial segments in $\mathcal{S}_1$. S is again an initial segment of $\langle A, \prec\rangle$ without largest element, hence it belongs either to $\mathcal{S}_1$ or to $\mathcal{S}_2$. In the first case it is clear that S is the largest element in $\mathcal{S}_1$, and in the second case it is the smallest element in $\mathcal{S}_2$. It is not possible that simultaneously $\mathcal{S}_1$ has a largest element S_1 and $\mathcal{S}_2$ has a smallest element S_2. In fact, then there would be a point $a \in S_2 \setminus S_1$, and since S_2 does not have a largest element, there would also be another point $b \in S_2$ with $a \prec b$. But then the initial segment S_b would lie strictly between S_1 and S_2, which is not possible. This proves that $\langle \mathcal{S}, \subset\rangle$ is continuously ordered.

Now let $S_1 \subset S_2$ be two initial segments in $\mathcal{S}$, and let $b \in S_2 \setminus S_1$. Since S_2 does not have a largest element, there is a point $a \in S_2$ with $b \prec a$. Now the initial segment S_a lies strictly between S_1 and S_2 ($b \in S_a \setminus S_1$ and $a \in S_2 \setminus S_a$), which proves that $\{S_a : a \in A\}$ forms a dense subset of $\langle \mathcal{S}, \subset\rangle$. [W. Sierpiński, Cardinal and Ordinal Numbers, Polish Sci. Publ., Warszawa, 1965, XI.9. Theorem 1]

67. Let $\langle A, <\rangle$ and $\langle B, \prec\rangle$ be two continuously ordered sets such that some dense subsets $A' \subset A$ and $B' \subset B$ are similar. Let $f : A' \to B'$ be a similarity transformation between these two sets, and for any $a \in A$ let

$$B_1 = \{y \in B : y \prec f(x) \text{ for some } x \in A',\, x < a\}.$$

Then B_1 is a proper initial segment in B, hence either in B_1 there is a largest element, or $B \setminus B_1$ contains a smallest element. For any $x < a$ there are elements from A' between x and a, hence no y can be a largest element in B_1 (for if $y \in B_1$ and $y \prec f(x)$ with $x \in A'$, $x < a$, then $f(x)$ also belongs to B_1 by what we have just said). Thus, $B \setminus B_1$ must contain a smallest element, which we denote by $F(a)$. It is easy to prove that F is a monotone mapping from A into B and extends f. It is left to prove that it is a mapping onto B.

Now do the same reversing the role of A and B and with the mapping $f^{-1} : B' \to A'$. We get that f^{-1} has a monotone extension $G : B \to A$. Now $G \circ F$ is a monotone mapping of A into itself that extends $f^{-1} \circ f$, i.e., it fixes the dense set A'. Hence $G \circ F$ is the identity on A. In a similar fashion, $F \circ G$ is the identity mapping on B, hence G is the inverse of F. As a consequence, $\langle A, \prec \rangle$ and $\langle B, < \rangle$ are similar. [W. Sierpiński, Cardinal and Ordinal Numbers, Polish Sci. Publ., Warszawa, 1965, XI.9. Theorem 2]

68. Let $\langle B, \prec \rangle$ be continuously ordered such that it contains at least two points. Selecting the terms in the sequence $B' = \{a_0, a_1, \ldots\}$ one by one, we can easily construct a countable densely ordered subset B' of $\langle A, \prec \rangle$. Now repeat the procedure in the solution of Problem 67 with the role $A = \mathbf{R}$ and $A' = \mathbf{Q}$. The mapping F constructed there will be a monotone mapping from $\mathbf{R}$ into $\langle B, < \rangle$ (now we cannot claim that it is onto, since we do not repeat the process starting from B). [W. Sierpiński, Cardinal and Ordinal Numbers, Polish Sci. Publ., Warszawa, 1965, XI.10/6]

69. The nested property of the intervals implies that $a_0 \preceq a_1 \preceq \cdots$ and $b_0 \succeq b_1 \succeq \cdots$ and each b_j is bigger than any a_k. Now let $S = \{c \in A \ : \ c \preceq a_n \text{ for some } n \in N\}$. Then either in S there is largest element a or in $A \setminus S$ there is a smallest element a. In either case a is a common points of all the closed intervals A_n. [W. Sierpiński, Cardinal and Ordinal Numbers, Polish Sci. Publ., Warszawa, 1965, XI.5/3]

70. Let $\langle B, < \rangle$ be the set of the countable ordinals with the standard ordering on the ordinals and let $\langle B^*, <^* \rangle$ be an ordered set that is similar to the ordered set that we obtain when we reverse the ordering on B, and also for which $B^* \cap B = \emptyset$. We choose $\langle A, \prec \rangle$ as the ordered union of the sets (in this order) $B \times [0,1)$ and $B^* \times [0,1)$ (which are equipped with the lexicographic ordering). It is clear that $\langle A, \prec \rangle$ is not continuously ordered, since $A = (B \times [0,1)) \cup (B^* \times [0,1))$, every element of $B \times [0,1)$ precedes every element of $B^* \times [0,1)$, but there is no largest element in $B \times [0,1)$, nor a smallest element in $B^* \times [0,1)$.

It is easy to see that every subset $C \neq \emptyset$ of $B \times [0,1)$ has a greatest lower bound. In fact, if ζ is the smallest ordinal with the property that there is an $y \in [0,1)$ with $(\zeta, y) \in C$, and if $x = \inf\{y \ : \ (\zeta, y) \in C\}$, then clearly (ζ, x) is the greatest lower bound of C. Furthermore, every sequence

$$\{(\zeta_n, x_n) \ : \ \zeta_n \in B, \ x_n \in [0,1), \ n = 0, 1, \ldots\}$$

has a smallest upper bound. In fact, since there is a countable ordinal bigger than every ζ_n, the sequence $\{(\zeta_n, x_n)\}_{n=0}^{\infty}$ is bounded from above in $B \times [0,1)$. But then the smallest upper bound is just the greatest lower bound of all the upper bounds.

In a similar manner, every sequence in $B^* \times [0,1)$ has a smallest upper bound and a greatest lower bound, and it immediately follows that $\langle A, \prec \rangle$ also has this property. Thus, if $A_n = \{c \ : \ a_n \preceq c \preceq b_n\}$ is a sequence of closed intervals in A, and a is the smallest upper bound of the sequence $\{a_n\}_{n=0}^{\infty}$, then

a belongs to all the sets A_n. [W. Sierpiński, Cardinal and Ordinal Numbers, Polish Sci. Publ., Warszawa, 1965, XV.1]

71. Suppose that B and C are subsets of an ordered set, $B \cup C$ is not scattered, but B is. Then $B \cup C$ includes a densely ordered subset D. But $D \cap B$ is not densely ordered, so there are two elements b, d in it such that there are no further elements from $D \cap B$ between b and d. However, in a densely ordered set the elements lying in between two given elements form a densely ordered set, hence the elements from C that lie between b and d form a densely ordered set, and so C is not scattered.

72. First suppose that the closure $\overline{A}$ of $A \subseteq \mathbf{R}$ is not countable. Then by Problem 2.11 there is a point $a \in \overline{A}$ such that each of the sets $\overline{A} \cap (-\infty, a)$ and $\overline{A} \cap (a, \infty)$ is uncountable. Since $a \in \overline{A}$ is the limit of points in A, the solution to Problem 2.11 also gives that we can actually select a from A.

Thus, if A is such that its closure $\overline{A}$ is uncountable, then there is a point $a_0 \in A$ such that the closure of both $A \cap (-\infty, a_0)$ and of $A \cap (a_0, \infty)$ are uncountable. Apply this separately to the sets $A \cap (-\infty, a_0)$ and $A \cap (a_0, \infty)$; we obtain points $a_1, a_2 \in A$ such that $a_1 < a_0 < a_2$ and the closure of each of the sets $A \cap (-\infty, a_1)$, $A \cap (a_1, a_0)$, $A \cap (a_0, a_2)$, and $A \cap (a_2, \infty)$ is uncountable. Now apply the same reasoning separately to these sets, then we get $a_3, a_4, a_5, a_6 \in A$ such that $a_3 < a_1 < a_4 < a_0 < a_5 < a_2 < a_6$, etc. It is clear that this way we get a densely ordered subset of A, hence A is not scattered.

Conversely, suppose that A is not scattered, i.e., it has a densely ordered subset, which we can continue to denote by A. Let $a_0 < b_0 < b_1 < a_1$ be points from A, then select points $a_0 < a_{00} < b_{00} < b_{01} < a_{01} < b_0$ and points $b_1 < a_{10} < b_{10} < b_{11} < a_{11} < a_1$ from A, then for each $i, j = 0, 1$ with $a_{ij} < b_{ij}$ select four points $a_{ij} < a_{ij0} < b_{ij0} < b_{ij1} < a_{ij1} < b_{ij}$ from A, and for each $i, j = 0, 1$ with $b_{ij} < a_{ij}$ select four points $b_{ij} < a_{ij0} < b_{ij0} < b_{ij1} < a_{ij1} < a_{ij}$ from A, etc. This process can be continued indefinitely due to the dense ordering on A. Now if $\epsilon = (\epsilon_0, \epsilon_1, \ldots)$ is an arbitrary infinite 0–1 sequence, then consider the number

$$x_\epsilon = \liminf_{n \to \infty} a_{\epsilon_1 \epsilon_2 \ldots \epsilon_n}.$$

If $\epsilon' = (\epsilon'_0 \epsilon'_1 \ldots)$ is a different sequence, for example, $\epsilon_0 = \epsilon'_0, \ldots, \epsilon_n = \epsilon'_n$ but $\epsilon_{n+1} = 0$ and $\epsilon'_{n+1} = 1$, then

$$x_\epsilon \le b_{\epsilon_0 \ldots \epsilon_n 0} < b_{\epsilon'_0 \ldots \epsilon'_n 1} \le x_{\epsilon'}.$$

Thus, the numbers x_ϵ are all different. Since they all belong to the closure of A, we obtain that this closure is of cardinality continuum (see Problem 4.2), and this proves the sufficiency of the condition.

73. The necessity of the condition is clear from the preceding problem: if $\epsilon_0, \epsilon_1, \ldots$ are given, then, since the closure $\overline{A}$ of A is countable, we can enumerate its points in a sequence, and by covering the ith point with an open

interval I_i of length ϵ_i we get a cover of $\overline{A}$. But this set is compact, so we can select a finite subcover $\cup_{j=0}^{N} I_j$, and this proves the necessity.

Conversely, suppose that A is not scattered, and consider points $a_0 < b_0 < b_1 < a_1$, $a_0 < a_{00} < b_{00} < b_{01} < a_{01} < b_0$, $b_1 < a_{10} < b_{10} < b_{11} < a_{11} < a_1$, etc. selected in the preceding proof. Let ϵ_j be smaller than all the distances between all the points $a_{\alpha_0 \dots \alpha_{jm}}, b_{\alpha_0 \dots \alpha_{jm}}$, where $\alpha_0, \dots, \alpha_{jm}$ run through all possible choices of 0's and 1's (in other words, ϵ_j is smaller than the shortest distance between points at the $(j+1)$th level). We claim that then there is no natural number N such that A can be covered with some intervals $I_0, I_1, \dots, I_N$ of length $|I_i| = \epsilon_i$. In fact, suppose that $I_0, I_1, \dots$ are intervals with $|I_j| = \epsilon_j$. In what follows $[a, b]$ denotes the interval $[a, b]$ if $a \le b$ and the interval $[b, a]$ if $b < a$. By the choice of ϵ_0, if $I_0 \cap [a_0, b_0] \ne \emptyset$, then $I_0 \cap [a_1, b_1] = \emptyset$, and conversely, if $I_0 \cap [a_1, b_1] \ne \emptyset$, then $I_0 \cap [a_0, b_0] = \emptyset$. In other words, I_0 does not intersect one of the intervals $[a_0, b_0]$ and $[a_1, b_1]$, say I_0 does not intersect $[a_{\alpha_0}, b_{\alpha_0}]$. In a similar fashion, I_1 does not intersect one of the intervals $[a_{\alpha_0,0}, b_{\alpha_0,0}]$ and $[a_{\alpha_0,1}, b_{\alpha_0,1}]$, say I_1 does not intersect $[a_{\alpha_0\alpha_1}, b_{\alpha_0\alpha_1}]$. Note that $[a_{\alpha_0\alpha_1}, b_{\alpha_0\alpha_1}]$ is part of $[a_{\alpha_0} b_{\alpha_0}]$, so I_0 does not intersect this interval, either. We can continue this process and find that for each n there is a subinterval $[a_{\alpha_0 \dots \alpha_n}, b_{\alpha_0 \dots \alpha_n}]$ such that neither of $I_0, \dots I_n$ intersects this interval. Since this process can be carried out indefinitely, there cannot be an N such that the intervals $I_0, I_1, \dots, I_N$ cover A.

74. Assume to the contrary that $q \mapsto f_q$ is an order-preserving injection of $\mathbf{Q}$ into $\langle H(\alpha), \prec \rangle$. Let $\beta_0 < \alpha$ be the least ordinal that occurs as the largest ordinal where some $f_{q_0}, f_{q_0'}$ with $q_0 < q_0'$ differ. Now choose rational numbers $q_0 < q_1 < q_1' < q_0'$. Then all the functions $f_{q_0}, f_{q_0'}, f_{q_1}, f_{q_1'}$ agree above β_0 and some two at β_0, too. Hence for these two functions the largest difference would have to occur before β_0, but this contradicts the choice of β_0. [P. Komjáth and S. Shelah]

75. The product of $\langle A, \prec \rangle$ and $\langle B, < \rangle$ is similar to the ordered union with respect to $\langle B, < \rangle$ of disjoint copies of $\langle A, \prec \rangle$, hence this statement follows from the next problem.

76. Suppose $\langle A, \prec \rangle$ is the ordered union of the scattered sets $\langle A_b, \prec_b \rangle$ with respect to the scattered set $\langle B, < \rangle$, and suppose that there is a densely ordered subset $C \subset A$ of A. Consider the set

$$B_C = \{b \ : \ a_b \in C \text{ for some } a_b \in A_b\}.$$

This cannot have a densely ordered subset, so there are two elements $b_1, b_2 \in B_C$ such that there are no further elements from B_C between them. The elements $a \in C$ with $a_{b_1} \prec a \prec a_{b_2}$ form a densely ordered set, but all such elements are from the sets A_{b_1} and A_{b_2}, and by Problem 71, $A_{b_1} \cup A_{b_2}$ does not have a densely ordered subset. This contradiction proves that a densely ordered subset C cannot exist.

77. Let $\langle A, \prec\rangle$ be any ordered set, and for $x, y \in A$ let $x \sim y$ if the interval determined by x and y is scattered (i.e., for example, for $y \prec x$ the interval $\{a \in A : y \prec a \prec x\}$ is scattered). It easily follows from Problem 71 that this is an equivalence relation. It is also clear that every equivalence class is scattered, and if C and D are two different equivalence classes, then either all elements in C precede all elements in D or vice versa. Thus, there is a natural ordering $\prec^*$ on the set A^* of equivalence classes coming from the ordering $\prec$. Furthermore, if C and D are two equivalence classes then there must be an equivalence class between them, for otherwise $C \cup D$ would be scattered by Problem 71, and so it would be part of a single equivalence class. This means that the set of equivalence classes is a densely ordered set. But it is clear that $\langle A, \prec\rangle$ is the ordered union of the equivalence classes (with the ordering $\prec$ restricted to them) with respect to the densely ordered set $\langle A^*, \prec^*\rangle$, and this proves the claim.

78. Let $\langle A, \prec\rangle$ be an ordered set. Follow the preceding proof, just replace "scattered" everywhere by "belongs to $\mathcal{F}$". The proof remains valid if we show that every equivalence class belongs to $\mathcal{F}$. Let E be an equivalence class, $a \in E$, $E_+ = \{b \in E : a \prec b\}$ and $E_- = \{b \in E : b \prec a\}$. It is sufficient to show that $E_\pm$ belong to $\mathcal{F}$, for then E, as the ordered union of $E_-, \{a\}, E_+$, also belongs to $\mathcal{F}$. If E_+ has a largest element b, then $b \sim a$, so the definition of $\sim$ shows that $E_+ \in \mathcal{F}$. Suppose now that E_+ has no largest element. Let $\{a_\xi\}_{\xi<\alpha}$ be a well-ordered cofinal subset of E_+ (see Problem 44). Since any two a_ξ, a_ζ ($\xi, \zeta < \alpha$) are equivalent, the interval $[a_\xi, a_{\xi+1}) = \{b \in A : a_\xi \preceq b \prec a_{\xi+1}\}$ belongs to $\mathcal{F}$. But E_+ is a well-ordered union of the sets (a, a_0),$[a_0, a_1)$, ..., $[a_\xi, a_{\xi+1})$, ..., $\xi < \alpha$, hence it belongs to $\mathcal{F}$.

The proof that E_- belongs to $\mathcal{F}$ is similar if we use a reversely well-ordered coinitial subset of it.

79. First of all, by Problem 76 every set in $\mathcal{O}$ is scattered, hence it is enough to prove that every scattered set is in $\mathcal{O}$.

It is clear that the family $\mathcal{F} = \mathcal{O}$ satisfies the hypothesis in the preceding problem (prove by induction that $\mathcal{O}$ is closed for forming subsets, as well). Thus, if $\langle A, \prec\rangle$ is scattered, then either it belongs to $\mathcal{O}$, or it is similar to an ordered union of nonempty sets in $\mathcal{O}$ with respect to a densely ordered set. But the latter would mean that $\langle A, \prec\rangle$ includes a densely ordered set (just select one–one point from each summand), which is impossible. Hence $\langle A, \prec\rangle \in \mathcal{O}$, as was claimed. [F. Hausdorff, Grundzüge der Theorie der Geordnete Mengen, Math. Ann., 65(1908), 435–505]

80. Let $\mathcal{F}$ be the family of ordered sets that can be embedded into one of the $\langle H(\alpha), \prec\rangle$. By Problem 74 all these sets are scattered, hence it is left to show that every scattered set is in $\mathcal{F}$. Note that $\mathcal{F}$ is closed for well-ordered and reversely well-ordered union. In fact, suppose that each $\langle A_\xi, <_\xi\rangle$, $\xi < \gamma$, can be

embedded into some $\langle H(\alpha_\xi), \prec\rangle$. Selecting an α bigger than all α_ξ, $\xi < \gamma$, we may assume $\alpha_\xi = \alpha$ for all ξ, and then the ordered sum of the sets $\langle H(\alpha), <_\xi\rangle$, $\xi < \gamma$, is similar to the product $H(\alpha) \times \gamma$ with the antilexicographic ordering. Thus, it is enough to prove that this product can be embedded into $H(\alpha+\gamma)$. But that is easy: map a pair (f, β) with $f \in H(\alpha)$ and $\beta < \gamma$ into the function $g \in H(\alpha + \gamma)$ that agrees with f on the set α, and in the interval $[\alpha, \alpha + \gamma)$ it is everywhere zero except at $\alpha + \beta$ where it is 1 (and this one is the last nonzero element of g). It is easy to see that this is an embedding of $H(\alpha) \times \gamma$ into $H(\alpha + \gamma)$.

The proof that $\mathcal{F}$ is closed for reversely well-ordered unions is the same, just use the value -1 instead of 1 as a last nonzero element in the embedding.

Thus, the family $\mathcal{F}$ satisfies the hypothesis in Problem 78, and hence if $\langle A, <\rangle$ is a scattered set, then either it belongs to $\mathcal{F}$, or it is similar to an ordered union of nonempty sets in $\mathcal{F}$ with respect to a densely ordered set. This latter one is impossible (cf. the end of the solution to the preceding problem), hence $\langle A, <\rangle \in \mathcal{F}$ as was claimed. [P. Komjáth and S. Shelah]

81. We are going to show that there is an ordered set $\langle A, \prec\rangle$ with countable intervals and smallest element such that every ordered set with countable intervals and smallest element is similar to a subset of $\langle A, \prec\rangle$. This will already solve the problem. In fact, let $\langle A^*, \prec^*\rangle$ be the ordered set that we obtain by replacing every element a in A by an element a^* from a disjoint set A^* and let $a^* \prec^* b^*$ be precisely if $b \prec a$ (in other words, we take the reverse ordering of $\langle A, \prec\rangle$). It is clear that every ordered set with countable intervals and a largest element is similar to a subset of $\langle A^*, \prec^*\rangle$. Now let $\langle \mathcal{A}, \ll\rangle$ be the ordered union of $\langle A^*, \prec^*\rangle$ and $\langle A, \prec\rangle$, in which every element of A^* precedes every element of A. We claim that every ordered set $\langle B, <\rangle$ with countable intervals is similar to a subset of $\langle \mathcal{A}, \ll\rangle$. Choose an element $b_0 \in B$, and consider the sets $B_1 = \{b \in B \ : \ b \le b_0\}$ and $B_2 = \{b \in B \ : \ b_0 \le b\}$. Then B_2 has a smallest element and countable intervals, so it is similar to a subset of $\langle A, \prec\rangle$. In a similar fashion, B_1 is similar to a subset of $\langle A^*, \prec^*\rangle$, and since the elements in $B_1 \setminus \{b_0\}$ precede the elements in B_2, these two facts show that $\langle B, <\rangle$ is similar to a subset of $\langle \mathcal{A}, \ll\rangle$.

Thus, it is enough to construct $\langle A, \prec\rangle$. Let ω_1 be the set of countable ordinals with the standard ordering on the ordinals, and let $\langle A, \prec\rangle$ be the product $\omega_1 \times (\mathbf{Q} \cap [0, 1))$ with the lexicographic ordering. Clearly $(0, 0)$ is the smallest element in $\langle A, \prec\rangle$. If $(\xi, r) \in A$ is any element, then, since there are only countably many smaller ordinals than ξ, we have that the set of those elements in A that are smaller than (ξ, r) is countable (cf. Problem 2.2). This shows that $\langle A, \prec\rangle$ has countable intervals.

Now let $\langle C, <\rangle$ be any set with smallest element c_0 and countable intervals. If C has a largest element, then it is countable by the countable interval property, and on applying Problem 26 we can immediately see that then $\langle C, <\rangle$ is similar to a subset of $\{0\} \times (\mathbf{Q} \cap [0, 1))$. Thus, in what follows we are going to assume that C does not have a largest element.

By Hausdorff's theorem (Problem 44) there is a well-ordered cofinal subset B of C. Let $B = \{b_\alpha : \alpha < \gamma\}$ be the increasing enumeration of B. Clearly we must have $\gamma \le \omega_1$. Decompose C as $C = \cup\{C_\alpha : \alpha < \gamma\}$, where $x \in C_\alpha$ if and only if b_α is the least element of B that is grater than x. Then each C_α is a countable set and C is the ordered union of them. By Problem 26 $\langle C_\alpha, < \rangle$ can be monotonically embedded into $\{\alpha\} \times (\mathbf{Q} \cap [0,1))$, and these together give a monotone embedding of $\langle C, < \rangle$ into $\gamma \times (\mathbf{Q} \cap [0,1))$, which is a subset of $\omega_1 \times (Q \cap [0,1))$.

82. Let k be the largest exponent of 2 in the expansion of n_1 in base 2. Then in the expansion of any n_i in the appropriate base b_i (it is $[i/2]+2$) the highest exponent of b_i is at most k. Thus, if the coefficient of $(b_i)^j$ in this expansion is $c_j^{(i)}$, then

$$\mathbf{c}^{(i)} = \left(c_k^{(i)}, c_{k-1}^{(i)}, \dots, c_1^{(i)}, c_0^{(i)}\right)$$

is an element of the set $\mathbf{N}^{k+1}$, which we order by the lexicographic ordering $\prec$. It is clear that if $n_{2i-1} > 0$, then $\mathbf{c}^{(2i)} = \mathbf{c}^{(2i-1)}$, and it is easy to see that $\mathbf{c}^{(2i+1)} \prec \mathbf{c}^{(2i)}$. Thus, $\{\mathbf{c}^{(2i)}\}_i$ is a decreasing sequence in $\mathbf{N}^{k+1}$, and since the latter set is well ordered, it cannot be infinite, i.e., there must be an i with $n_i = 0$.

The proof of part **(b)** is identical.

* * *

83. Let $\langle A, \prec \rangle$ be densely ordered. It is sufficient to show a coloring of A by red and blue in such a way that elements of either colors form a dense subset of A. Let $\{a_\alpha\}_{\alpha<\kappa}$ be an enumeration of the elements of A into a transfinite sequence of type $\kappa = |A|$. The coloring is done by transfinite recursion on this enumeration: if $\{\alpha_\beta : \beta < \alpha\}$ is already colored, then let a_α be red if in the ordered set $\langle \{\alpha_\beta : \beta \le \alpha\}, \prec \rangle$ the element a_α has both a successor and a predecessor and both of them are blue, otherwise let the color of a_α be blue. We claim that this is an appropriate coloring. Let $a \prec b$ be two elements in A and let us show that there is a red element in between them. Let a_γ, a_δ, $\gamma < \delta$ be the two elements with smallest index lying in between a and b, and then let a_α be the element with smallest index α lying in between a_γ and a_δ. If either a_γ or a_δ is red, then we are ready. Otherwise in $\langle \{\alpha_\beta : \beta \le \alpha\}, \prec \rangle$ the element a_α has a predecessor and a successor (these are a_γ and a_δ) and both are blue, hence a_α is red.

In a similar fashion, if either a_γ or a_δ is red, then necessarily a_α is blue, hence there is a blue element in between a and b. [I. Juhász]

84. Let $\langle A, < \rangle$ be the ordered set, and set $a \sim b$ if there are only finitely many elements in between a and b. This is clearly an equivalence relation, and every equivalence class C is either finite or similar to either $\mathbf{Z}$, $\mathbf{N}$, or $\mathbf{Z} \setminus \mathbf{N}$. Hence we can color alternately the elements of any equivalence class

C consisting of at least two elements by red and blue so that in between any two elements of the same color there is an element with a different color. Between equivalence classes let $\prec'$ be the natural ordering inherited from $<$ (i.e., $C_1 \prec' C_2$ if for some—and then for all—$c_1 \in C_1$ and $c_2 \in C_2$ we have $c_1 < c_2$), and let B be the union of all the equivalence classes consisting of more than one point. This set is already colored. Color the set $A \setminus B$ by the method of the preceding problem: if $\{a_\alpha\}_{\alpha<\kappa}$ is an enumeration of $A \setminus B$, and if we have already colored the subset $\{a_\beta : \beta < \alpha\}$, then let the color of a_α be red if it has a blue predecessor and successor in $\{a_\beta : \beta \leq \alpha\}$, otherwise color it blue. We claim that this is a good coloring.

Let a, b be two elements belonging to the equivalence classes C_a and C_b. If C_a does not have a largest element or C_b does not have a smallest element, then there are points of both colors in between a and b. The same is true if there is an equivalence class of size ≥ 2 in between C_a and C_b. In the remaining case there are only one-element equivalence classes in between C_a and C_b, and the elements of these form a densely ordered set. Hence, by the proof of the preceding problem, both colors occur among these elements, and we are done. [I. Juhász]

85. Let $\langle A, \prec\rangle$ be an ordered set of the following structure: there is a largest element a_0 and a decreasing sequence $\ldots \prec a_1 \prec a_0$ such that the interval $\{a \in A : a_{n+1} \preceq a \prec a_n)$ has order type ω^n, and these intervals together with $\{a_0\}$ cover the whole set. Assume that $B, C \subseteq A$ are different nonempty initial segments and $f : B \to C$ an isomorphism. There is an $n < \omega$ with $a_n \in B \cap C$. In A, and therefore also in B and C, a_n is the largest element, which is the supremum of a subset of type ω^n and this implies that $f(a_n) = a_n$. But then f is an isomorphism between the parts of B and C consisting of the elements that are larger than a_n, which is impossible, as they are distinct initial segments of the same well-ordered set.

86. Let A be the set of all limit ordinals smaller than ω_1, and for all $\alpha \in A$ fix a strictly increasing sequence $\alpha_0 < \alpha_1 < \cdots$ of ordinals with supremum α. Let us also agree that in this proof α_n, β_n, etc., mean the corresponding terms in the sequence associated with α, β, etc. For $\alpha \neq \beta$ let $\alpha \prec \beta$ if $\alpha_n < \beta_n$ for the smallest natural number n for which $\alpha_n \neq \beta_n$. This is clearly an ordering on A. We claim that this ordered set cannot be represented as a countable union of its well-ordered subsets, but every uncountable subset includes an uncountable well-ordered subset.

Since A is stationary in ω_1, and a countable union of nonstationary subsets of ω_1 is nonstationary, the fact that A cannot be represented as a countable union of its well ordered subsets follows if we show that no stationary subset X of A is well ordered under $\prec$. Let $Y_{-1} = X$, and suppose that we have already defined Y_{n-1}, and it is stationary in ω_1. The mapping $f_n(\alpha) = \alpha_n$ is regressive on Y_{n-1}, hence by Problem 20.16 there is a $\delta < \omega_1$, such that the set of those $\alpha \in Y_{n-1}$ for which $f_n(\alpha) = \delta$ is a stationary set in ω_1. Let δ_n be the smallest such ordinal, and set $Y_n = \{\alpha \in Y_{n-1} \ : \ \alpha_n = \delta_n\}$. This completes the definition of the sequences $Y_0 \supseteq Y_1 \supseteq \cdots$ and the ordinals $\delta_0 < \delta_1 < \cdots$. By definition, the set of those $\alpha \in Y_{n-1}$ for which $\alpha_n < \delta_n$ is nonstationary, so if we omit all these elements from X for all n, then the remaining set X' is still stationary. Let δ be the supremum of the δ_n's. Then δ is a countable ordinal, so there is a $\gamma \in X'$ bigger than δ. Note that for $\alpha, \beta \in Y_n$ we have $\alpha_0 = \beta_0 = \delta_0$, $\alpha_1 = \beta_1 = \delta_1$, ...,$\alpha_n = \beta_n = \delta_n$, and since $\sup_n \delta_n < \gamma$, there is a smallest $n = n_0$, such that $\delta_{n_0} \neq \gamma_{n_0}$. Thus, $\gamma \in Y_{n_0-1}$, and since in forming X' we have omitted all elements α from Y_{n_0-1} for which $\alpha_{n_0} < \delta_{n_0}$, we must have $\delta_{n_0} < \gamma_{n_0}$. All these imply that γ is bigger than all elements of Y_{n_0} with respect to $\prec$.

What we have proved is that if $X \subseteq A$ is stationary then there is an element $\gamma^0 \in X$ such that there is a stationary subset $X_1 = Y_{n_0}$ of X the elements of which are smaller than γ^0 with respect to $\prec$. Now repeat this process with $X_1 = Y_{n_0}$. Then we get a stationary set $X_2 \subseteq X_1$ and an ordinal $\gamma^1 \in X_1$ such that γ^1 is strictly larger (with respect to $\prec$) than any element in X_1. Clearly $\gamma^1 \prec \gamma^0$, and if we continue this process indefinitely, then we obtain an infinite monotone decreasing sequence $\ldots \prec \gamma^1 \prec \gamma^0$ in X, so X is not well ordered.

Next we show that every uncountable subset X of A has an uncountable well-ordered subset. Consider the sets

$$H_m = \{(\alpha_0, \alpha_1, \ldots, \alpha_m) \ : \ \alpha \in X\}.$$

All these sets cannot be countable, for then there was a countable ordinal ν with the property that $\alpha_n < \nu$ for all $\alpha \in X$ and all n, but then this would imply by the definition of the sequences $\{\alpha_n\}$ that all ordinals in X would be at most ν, and this is not the case. Thus, there is an $m = m_0$ such that the set H_{m_0} is uncountable. For every $s = (s_0, s_1 \ldots, s_{m_0}) \in H_{m_0}$ choose an $\alpha^s \in X$ with $(\alpha^s)_0 = s_0$, ..., $(\alpha^s)_{m_0} = s_{m_0}$. The H_{m_0} is part of $\overbrace{\omega_1 \times \omega_1 \times \cdots \times \omega_1}^{m_0+1}$ and this latter set is well ordered with respect to lexicographic ordering, hence H_{m_0} is also well ordered with respect to lexicographic ordering. But this then means that the elements $\{\alpha^s\}_{s \in H_{m_0}}$ are also well ordered with respect to the ordering $\prec$, so we have found an uncountable well-ordered subset of X.

87. We shall construct the sets A and B so that they will be dense in $\mathbf{R}$. The key to the construction of A and B is the observation that any monotone mapping of A or B into $\mathbf{R}$ can be extended to a (strictly) monotone real function (see the solution to Problem 18 and observe that if the domain of f is a dense set and f is strictly increasing on its domain, then the extension of f will be strictly increasing) and that the number of increasing real functions is continuum (see Problem 4.14, d)). Thus, let f_α, $\alpha < \mathbf{c}$ be an enumeration of the strictly increasing real functions. By transfinite recursion we define increasing sets A_α, B_α of cardinality at most $(|\alpha| + \aleph_0)$ as follows. Set $A_0 = \mathbf{Q}$, $B_0 = \mathbf{Q}$. Suppose we already know A_γ and B_γ for $\gamma < \alpha$. If α is a limit ordinal,

then let A_α and B_α be the union of these A_γ and B_γ sets, respectively. If α is a successor ordinal, say $\alpha = \beta + 1$, then consider the functions f_ξ, $\xi < \alpha$ and the set

$$H_\alpha = A_\beta \bigcup \left(\bigcup_{\xi<\alpha} f_\xi^{-1}[B_\beta] \right).$$

Since A_β and B_β are of cardinality at most $|\alpha| + \aleph_0$, this H_α is also of cardinality at most $|\alpha| + \aleph_0$. Thus, there is a number $a_\beta \in \mathbf{R} \setminus H_\alpha$, and let $A_\alpha = A_\beta \cup \{a_\beta\}$ and $B_\alpha = B_\beta \cup \{b_\beta\}$, where b_β is any number outside the set $B_\beta \cup \left(\cup_{\xi<\alpha} f_\xi[A_\alpha]\right)$.

Finally, set $A = \cup_{\alpha<\mathbf{c}} A_\alpha$ and $B = \cup_{\alpha<\mathbf{c}} B_\alpha$. It is clear that if $f : A \to \mathbf{R}$ is monotone, then f is the restriction of some f_ξ, $\xi < \mathbf{c}$ to A, and then for $\alpha > \xi$ we have

$$f(a_\alpha) = f_\xi(a_\alpha) \notin B;$$

thus, f can map only a subset of A of cardinality smaller than continuum into B.

88. We shall construct a set X of cardinality continuum with the desired property. Similarly as in the preceding proof we use that any monotone mapping of X into itself can be extended to a nondecreasing real function (see Problem 18), and the number of nondecreasing real functions is of power continuum (see Problem 4.14, d)). We can actually discard all those nondecreasing functions that have a range of cardinality smaller than continuum, since they cannot establish a monotone mapping of X into itself. Thus, let f_α, $\alpha < \mathbf{c}$ be an enumeration of those nondecreasing real functions that assume continuum many different values, and that are not the identity. By transfinite recursion we define disjoint sets X_α, Y_α of cardinality at most $|\alpha|$ as follows. Set $X_0 = \emptyset$, $Y_0 = \emptyset$. Suppose we already know X_γ and Y_γ for $\gamma < \alpha$. If α is a limit ordinal, then let X_α and Y_α be the union of these X_γ and Y_γ sets, respectively. If α is a successor ordinal, say $\alpha = \beta + 1$, then consider the function f_β. The range of it is of power continuum, thus there is an $x_\beta \notin X_\beta \cup Y_\beta$ such that $f_\beta(x_\beta) \notin X_\beta$ (note that there are $\mathbf{c}$ values satisfying the second property, and all but $< \mathbf{c}$ of them satisfy the first one, as well). Set $X_\alpha = X_\beta \cup \{x_\beta\}$ and $Y_\alpha = Y_\beta \cup \{f_\beta(x_\beta)\}$.

Finally, set $X = \cup_{\alpha<\mathbf{c}} X_\alpha$. It is clear that if $f : X \to \mathbf{R}$ is monotone and not the identity function, then f is the restriction to X of some f_α, $\alpha < \mathbf{c}$, and then $f(x_\alpha) = f_\alpha(x_\alpha) \in Y_{\alpha+1} \subset \mathbf{R} \setminus X$, thus f is not mapping X into X.

89. Exactly as in the proof of Problems 57 and 58, either κ with the usual ordering on the ordinals, or κ with the reverse ordering is suitable (according to whether the ordered set includes an infinite decreasing sequence or not).

90. It is enough to show an ordered set $\langle B, \prec \rangle$ of cardinality bigger than κ and a dense subset $A \subset B$ in it of cardinality κ. In fact, then every element $b \in B$

determines the initial segment $S_b = \{a \in A : a \prec b\}$ of $\langle A, \prec\rangle$, hence $\langle A, \prec\rangle$ has more than κ initial segments (note that for $b_1 \neq b_2$ the initial segments S_{b_1} and S_{b_2} are different by the density of A in B).

Let ρ be the smallest cardinal for which $\kappa^\rho > \kappa$. We have $\rho \leq \kappa$ (see Problem 10.16). Let $\langle B, \prec\rangle$ be the set ${}^\rho\kappa$ of mappings $f : \rho \to \kappa$ with lexicographic ordering and let $A \subset B$ be the set of mappings $f : \rho \to \kappa$ for which only less than ρ elements are mapped into a nonzero element. The cardinality of B is $\kappa^\rho > \kappa$, while the cardinality of A is at most $\sum_{\alpha<\rho} \kappa^{|\alpha|} = \sum_{\alpha<\rho} \kappa = \kappa\rho = \kappa$, and it is clearly at least κ, so $|A| = \kappa$. Finally, it is easy to prove that A is dense in B, and we are done.

91. For finite κ the statement is clear, and for an infinite one consider the ordered set from the previous problem and the initial segments discussed there.

92. Let $\{H_\alpha\}_{\alpha<\nu}$ be a family of subsets of a set X of cardinality κ well ordered with respect to inclusion: $H_\alpha \subset H_\beta$ if $\alpha < \beta < \nu$. Then for every $\alpha < \nu$, except perhaps for the ordinal immediately preceding ν, there is an element $x_\alpha \in H_{\alpha+1} \setminus H_\alpha$, and it is clear that for $\alpha < \beta$ the elements x_α and x_β are different ($x_\alpha \in H_\beta$, but $x_\beta \notin H_\beta$). Thus, the mapping $\alpha \mapsto x_\alpha$ is a 1-to-1 mapping of ν into X (if ν has a largest element μ then a 1-to-1 mapping of $\nu \setminus \{\mu\}$ into X), hence ν is of cardinality at most κ.

93. Suppose to the contrary that a family $\{f_\alpha : \alpha < \kappa^+\}$ of more than κ elements of ${}^\kappa\kappa$ are well ordered: $f_\alpha \prec f_\gamma$ for $\alpha < \gamma < \kappa^+$, where $\prec$ denotes the lexicographic ordering. The sequence $\{f_\alpha(0) : \alpha < \kappa^+\}$ is a weakly (i.e., not strictly) increasing sequence of ordinals smaller than κ, so it stabilizes from some point onward: $f_\xi(0) = g(0)$ for all $\kappa^+ > \xi \geq \xi_0$ with some $g(0) < \kappa$ and $\xi_0 < \kappa^+$ (recall that $\kappa^2 = \kappa$, therefore there must be a value $\gamma < \kappa$ and a set $A \subset \kappa^+$ of cardinality κ^+ such that $f_\xi(0) = \gamma$ for all $\xi \in A$, but then the weak monotonicity gives for the smallest element ξ_0 of A that $f_\xi(0) = \gamma$ for all $\xi \geq \xi_0$). Restricting to these functions, the sequence $\{f_\alpha(1) : \xi_0 \leq \alpha < \kappa^+\}$ is a weakly increasing sequence of ordinals smaller than κ, so again $f_\xi(1) = g(1)$ for $\kappa^+ > \xi \geq \xi_1$ with some $g(1) < \kappa$ and $\xi_1 < \kappa^+$. Proceeding by induction, we get the values $g(\alpha) < \kappa$ and $\xi_\alpha < \kappa^+$ for all $\alpha < \kappa$ such that $f_\xi(\alpha) = g(\alpha)$ for $\xi \geq \xi_\alpha$ (note also that the supremum of at most κ ordinals each smaller than κ^+ is smaller than κ^+, so we never get stuck). But then $\xi^* = \sup\{\xi_\alpha : \alpha < \kappa\}$ is an ordinal smaller than κ^+ and the functions $\{f_\xi : \xi^* < \xi < \kappa^+\}$ are all equal to g, which is absurd. This contradiction proves the claim.

94. Let $A \subset T$. By transfinite recursion we define an element $g : \kappa \to \{0, 1\}$ of T which will be the least upper bound of the elements in A. Let $g(0) = 1$ if there is an element $f \in A$ with $f(0) = 1$, otherwise let $g(0) = 0$. If $g(\xi)$ has already been defined for $\xi < \eta$, and there is an element $f \in A$ such that $f(\xi) = g(\xi)$ for all $\xi < \eta$ and $f(\eta) = 1$, then let $g(\eta) = 1$, otherwise set

$g(\eta) = 0$. It is easy to see that the g we obtain this way is an upper bound for the set A, and it is smaller than any other upper bound.

The proof of the existence of a greatest lower bound follows the same lines or apply that the greatest lower bound is the smallest upper bound of all the lower bounds.

By Problem 44, part **b)** follows from part **c)**, which in turn is a direct consequence of the preceding problem.

95. Let the ordered set be $\langle A, \prec \rangle$. Without loss of generality, we may assume that $A = \kappa$, and then for $a, b \in A$ let $f_a(b) = 1$ if $b \prec a$ and let $f_a(b) = 0$ otherwise. This f_a belongs to ${}^{\kappa}\{0,1\}$, and all we have to show is that the mapping $a \mapsto f_a$ is monotone. But that is clear: if $a \prec a'$, and if for a b we have $f_{a'}(b) = 0$, then we also have $f_a(b) = 0$, hence f_a must precede $f_{a'}$ in the lexicographic ordering. [W. Sierpiński, Pontificia Acad. Sc., **4**(1940), 20[illegible]–208, N. Cuesta, Revista Mat. Hisp.-Amer., **4**(1947), 130–131]

96. Let $\langle A, \prec \rangle$ be an ordered set of cardinality κ, $A = \{a_\alpha\}_{\alpha<\kappa}$ an enumeration of the elements of A into a transfinite sequence of length κ, and let $<_\kappa$ be the lexicographic ordering on $\mathcal{F}_\kappa$. For $\alpha < \kappa$ set

$$f_\alpha(\gamma) = \begin{cases} 1 \text{ if } \gamma \le \alpha \text{ and } a_\gamma \preceq a_\alpha, \\ 0 \text{ otherwise.} \end{cases}$$

It is clear that $f_\alpha \in \mathcal{F}_\kappa$. We claim that $a_\alpha \mapsto f_\alpha$ is an embedding of $\langle A, \prec \rangle$ into $\mathcal{F}_\kappa$. It is clear that this mapping is one-to-one.

Let $\beta < \alpha$, and set $K = \{\gamma \le \beta \; : \; a_\gamma \preceq a_\beta\}$, $L = \{\gamma \le \beta \; : \; a_\gamma \preceq a_\alpha\}$. If $K = L$ then $\beta \in K$ implies $a_\beta \prec a_\alpha$, and it is clear that $f_\beta \le_\kappa f_\alpha$, hence $f_\beta <_\kappa f_\alpha$.

If $K \ne L$, then let γ be the first difference between K and L. If $\gamma \in K \setminus L$, then $a_\alpha \prec a_\gamma \preceq a_\beta$ and because of $f_\alpha(\gamma) = 0$, $f_\beta(\gamma) = 1$, we have $f_\alpha <_\kappa f_\beta$. On the other hand, if $\gamma \in L \setminus K$, then $a_\beta \prec a_\gamma \preceq a_\alpha$, and because of $f_\beta(\gamma) = 0$, $f_\alpha(\gamma) = 1$, we have $f_\beta <_\kappa f_\alpha$.

Thus, in all cases $a_\alpha \prec a_\beta \Leftrightarrow f_\alpha <_\kappa f_\beta$, hence $\alpha \mapsto f_\alpha$ is a monotone embedding.

97. For every $\xi < \kappa$ let $\langle A_\xi, \prec_\xi \rangle$ be an ordered set similar to $\langle A, \prec \rangle$, and let $\langle B, < \rangle$ be the lexicographically ordered product of them (i.e., $B = \prod_{\xi<\kappa} A_\xi$, and the ordering $<$ on B is the lexicographic one). Let $B = \cup_{\xi<\kappa} B_\xi$ be any decomposition. If for each $a \in A_0$ there is an $f_{0,a} \in B_0$ such that $f_{0,a}(0) = a$, then these $f_{0,a}$'s form a subset of B_0 similar to $\langle A, \prec \rangle$, and we are done. Suppose therefore that this is not the case, and let a_0 be an element of A_0 such that for no $f \in B_0$ is it true that $f(0) = a_0$. Thus, all the elements $f \in B$ with $f(0) = a_0$ belong to $\cup_{\xi>0} B_\xi$. We continue this process. Suppose that the elements a_γ have already been selected for all $\gamma < \alpha$ where $\alpha < \kappa$ is an ordinal, and they have the property that there is no $f \in B_\gamma$ with $f(\xi) = a_\xi$ for all $\xi \le \gamma$. Consider the set C_α of elements $f \in B$ such that $f(\gamma) = a_\gamma$ for all

$\gamma < \alpha$. Then $C_\alpha \subseteq \cup_{\xi \geq \alpha} B_\xi$, and if for each $a \in A_\alpha$ there is an $f_{\alpha,a} \in C_\alpha \cap B_\alpha$ such that $f_{\alpha,a}(\alpha) = a$, then these $f_{\alpha,a}$'s form a subset of B_α similar to $\langle A, \prec \rangle$, and we are done. If this is not the case, then let a_α be an element of A_α such that for no $f \in C_\alpha \cap B_\alpha$ is it true that $f(\alpha) = a_\alpha$. Thus, then the elements $f \in C_\alpha$ with $f(\alpha) = a_\alpha$ all belong to $\cup_{\xi > \alpha} B_\xi$.

To finish the proof all we have to mention is that for some $\alpha < \kappa$ the first possibility will happen. In fact, in the opposite case the elements a_α would be defined for all $\alpha < \kappa$. Consider now the function f for which $f(\alpha) = a_\alpha$ for all $\alpha < \kappa$, and the smallest $\alpha < \kappa$ for which $f \in B_\alpha$. Then, by the definition of the set C_α, we have $f \in C_\alpha$, and therefore $f(\alpha) = a_\alpha$ is not possible, since $f \in C_\alpha \cap B_\alpha$.

7

Order types

1. The nonempty initial segments of the set of rational numbers are densely ordered sets with or without largest element, so their order is η or $\eta+1$ (see Problem 6.28 and its proof).

2. The order types in a)–c) are the order types of a densely ordered countable sets without smallest and largest element, so they are η (see Problem 6.27). d) is the order type of $\mathbf{Q}\cap((0,1)\cup\{2,3\}\cup(4,5))$, and this set is not densely ordered. e) is the order type of $(0,1)\cup\{1\}\cup(1,2)=(0,2)$ so it is λ by Problem 6.2. f) is the order type of $(-\infty,0)\cup(0,\infty)$, and here there is no largest element in $(-\infty,0)$ and there is no smallest element in $(0,\infty)$, i.e., this set is not continuously ordered, thus the type in f) is not λ. g) is the order type of the lexicographically ordered set $\mathbf{R}\times\mathbf{R}$. But $\mathbf{R}\times\mathbf{R}=\mathbf{R}\times(-\infty,0)\cup\mathbf{R}\times[0,\infty)$, and here there is no largest element in $\mathbf{R}\times(-\infty,0)$ and there is no smallest element in $\mathbf{R}\times[0,\infty)$, i.e., this set is not continuously ordered, thus the type in g) is not λ. Finally, in h) in an ordered set of type $\eta\cdot\lambda$ there are points such that there are only countably many points lying between them, but in a set of type $\lambda\cdot\eta$ there are continuum many points between any two points.

3. We shall just consider the nontrivial solutions. By Problem 1 if $\theta_1+\theta_2=\eta$, then θ_1 is either η or $\eta+1$, and similarly θ_2 is either η or $1+\eta$. Thus, the solution is that θ_1 is either η or $\eta+1$ and θ_2 is either η or $1+\eta$, except that $\theta_1=\eta+1$ and $\theta_2=1+\eta$ cannot hold simultaneously.

In a similar way, the equation $\theta_1+\theta_2=\lambda$ holds if and only if $\theta_1=\lambda+1$, $\theta_2=\lambda$ or if $\theta_1=\lambda$ and $\theta_2=1+\lambda$.

4. Since

$$(1+\eta)\cdot(\eta+1)=(1+\eta)\cdot\eta+(1+\eta)=\eta+1+\eta=\eta$$

(see Problem 2), this is an appropriate representation, for $1+\eta\neq\eta$ and $\eta+1\neq\eta$.

5. Since $\mathbf{R} = \cup_{-\infty}^{\infty}[i, i+1)$, and this is an ordered union, one possibility is $\lambda = (1+\lambda)\cdot(\omega^* + \omega)$.

6. Let $\langle A, \prec\rangle$ have order type θ. If $\langle A, \prec\rangle$ has smallest element a, and τ is the type of $\langle A \setminus \{a\}, \prec\rangle$, then $\theta = 1 + \tau$ is an appropriate representation. If, however, $\langle A, \prec\rangle$ does not have a smallest element, then let $a \in A$ be any element, and τ_1 resp. τ_2 be the types of the sets $\{x \in A : x \prec a\}$ and $\{x \in A : a \preceq x\}$. Then $\theta = \tau_1 + \tau_2$ is an appropriate representation (note that $\tau_1 \neq \tau_2$, for in the second set there is a smallest element ($= a$), but in the first one there is no smallest element). [W. Sierpiński, Cardinal and Ordinal Numbers, Polish Sci. Publ., Warszawa, 1965, XII.9/5]

7. $\omega + 1$ is an example. In fact, if S_1 and S_2 are ordered sets and the product $S_1 \times S_2$ has order type $\omega + 1$, then S_2 must have a largest element s, and one of S_1 or S_2 must be infinite. If S_1 is infinite, then S_2 can have only one element, since in a set of type $\omega + 1$ there is just one element that is preceded by infinitely many elements. For the same reason, if S_2 is infinite, then S_1 can consist of at most one element (for otherwise there would be at least two elements that follow the infinite set $S_1 \times (S_2 \setminus \{s\})$.

8. The statement for η follows from the solution of Problem 3. As for ω, one can check easily that the only nontrivial solutions of the equation $\theta_1 + \theta_2 = \omega$ are $\theta_1 = n$, $\theta_2 = \omega$ where n is a natural number.

9. If a product $S_1 \times S_2$ of ordered sets is of type ω, then one of the sets is infinite. If S_1 is infinite, then S_2 can have only one element, for in a set of type ω no element is preceded by infinitely many elements. If, however, S_2 is infinite, then for the same reason every element in it is preceded by at most finitely many elements, hence we can apply Problem 6.5 to deduce that the order type of S_2 is ω.

If a product $S_1 \times S_2$ of ordered sets is of type $\eta + 1$, then S_1 and S_2 are countable, and if S_1 has at least two elements, then it is densely ordered. If its type is η, then $S_1 \times S_2$ has type η. If its type is $1 + \eta$, then depending on if S_2 has a smallest element or not, the type of $S_1 \times S_2$ is $1 + \eta$ or η. Finally, if S_1 is of order type $1 + \eta + 1$ and S_2 is not densely ordered, then $S_1 \times S_2$ is not densely ordered. However, if S_2 is also densely ordered but not of type $\eta + 1$, then the type of $S_1 \times S_2$ is $1 + \eta$, η, or $1 + \eta + 1$ depending on if S_2 is of type $1 + \eta$, η, or $1 + \eta + 1$. Thus, the only remaining possibility is that either S_1 or S_2 is of type $\eta + 1$.

The order types $1+\eta$ and $1+\eta+1$ can be similarly handled. [W. Sierpiński, Cardinal and Ordinal Numbers, Polish Sci. Publ., Warszawa, 1965, XII.10/5]

10. All infinite cardinals have this property (see Problem 10.3).

11. All infinite cardinals have this property (see Problem 10.3).

12. By Problems 6.31 and 6.32 the answer is $1+\eta$ if the ordering is lexicographic and it is ω if the ordering is antilexicographic.

13. If the ordering is lexicographic, then the answer is still $1+\eta$, for we are speaking of the order type of a densely ordered countable set with smallest element. If the ordering is antilexicographic, then the answer is $\omega+\omega^2+\omega^3+\cdots$. In fact, first come the elements

$$(0,0,\ldots),\ (1,0,0,\ldots),\ (2,0,0,\ldots),\ldots,$$

then come the elements

$$(0,1,0,0\ldots),\ (1,1,0,0,\ldots),\ (2,1,0,0,\ldots),\ldots,$$

$$(0,2,0,0\ldots),\ (1,2,0,0,\ldots),\ (2,2,0,0,\ldots),\ldots,$$

$$\vdots$$

etc..

14. See Problem 6.34.

15. We show by induction on n that the order type is $(\omega^n)^*$. It is easier to work with the set

$$-A_n = \left\{-\left(\frac{1}{k_1}+\cdots+\frac{1}{k_n}\right) : 1 \le k_1,\ldots,k_n < \omega\right\}.$$

We shall show that this is well ordered, and the order type is ω^n. The case $n=1$ is obvious. Also, if we have the result for n then there is, in the interval $(-\frac{1}{i},-\frac{1}{i+1})$, a subset of $-A_{n+1}$ of type ω^n (choose $k_1=i+1$, $k_2,\ldots,k_{n+1}=i(i+2)nj$, $j=1,2,\ldots$) so (pending that $-A_{n+1}$ is well ordered) the order type of $-A_{n+1}$ is at least ω^{n+1}.

To get an upper bound for the type of $-A_{n+1}$ we investigate the initial segments. For any $-\frac{1}{i}<0$ if $1\le k_1\le\cdots\le k_{n+1}<\omega$ and

$$-\left(\frac{1}{k_1}+\cdots+\frac{1}{k_{n+1}}\right) < -\frac{1}{i},$$

then $k_1 < ni$, so by the induction hypothesis the initial segment of $-A_{n+1}$ determined by $-\frac{1}{i}$ is the union of finitely many well-ordered sets of order type $\le\omega^n$, therefore itself is a well-ordered set of order type $<\omega^{n+1}$. Then, $-A_{n+1}$ is well ordered of order type at most ω^{n+1}.

16. The order type in question is clearly a product, where the second factor is the order type of a densely ordered set without smallest or largest element, and the first factor is ω. Thus, the answer is $\omega\cdot\eta$. [W. Sierpiński, Cardinal and Ordinal Numbers, Polish Sci. Publ., Warszawa, 1965, XII.3/8]

17. Let τ be an order type and let it be the type of $\langle A, \prec\rangle$. Consider in A the set B of those elements b for which the initial segment $\{a : a \preceq b\}$ is well ordered. It is easy to see that B is an initial segment, it is well ordered, and $A \setminus B$ cannot have a smallest element, for then it could be added to B. Thus, $\tau = \alpha + \theta$, where α is the order type of B and θ is the order type of $A \setminus B$. [W. Sierpieński, Fund. Math., **35**(1948), 1–12]

18. This is the same as Problem 6.13.

19. This is the same as Problem 6.14.

20. Let $\langle A_1, <_1\rangle$ and $\langle A_2, <_2\rangle$ be ordered sets of type θ_1 and θ_2, and let $A_i^* = \{0, 1, \ldots, n-1\} \times A_i$ the cross product of the ground sets with $\{0, 1, \ldots, n-1\}$. Then A_i^* with the antilexicographic ordering has order type $n \cdot \theta_i$, so if $n \cdot \theta_1 = n \cdot \theta_2$, then there is a similarity mapping $f : A_1^* \to A_2^*$. For every $a \in A_1$ there is a unique $c \in A_2$ such that $f((0, a)) \in \{0, 1, \ldots, n-1\} \times \{c\}$, and let us denote this c by $F(a)$. If $a <_1 b$ are two elements of A_1, then $(0, a)$ is smaller in the antilexicographic ordering on A_1^* than $(0, b)$, so $F(a) \le_2 F(b)$. But $F(a) = F(b)$ is not possible, since there are at least $n-1$ elements (namely $(1, a), \ldots, (n-1, a)$) in A_1^* lying between $(0, a)$ and $(0, b)$, and in any set $\{0, 1, \ldots, n-1\} \times \{c\}$ there are at most $n-2$ elements between any two elements. Thus, F is a monotone mapping from A_1 into A_2. We show that it actually maps A_1 onto A_2 by which $\theta_1 = \theta_2$ follows.

Let $c \in A_2$ be any element of A_2. There is an $a \in A_1$ such that for some $0 \le j < n$ we have $f((j, a)) = (0, c)$. If here $j = 0$, then $F(a) = c$, and we are done. If, however, $j > 0$, then the image of $\{0, 1, \ldots, n-1\} \times \{a\}$ under f does not contain $(n-1, c)$, and so there is an element $a^* \in A_1$ such that $a <_1 a^*$, and with some $0 \le i < n$ we have $f(i, a^*) = (n-1, c)$. Then clearly $f((0, a^*))$ must belong to $\{0, 1, \ldots, n-1\} \times \{c\}$, i.e., $F(a^*) = c$. [W. Sierpiński, Fund. Math., **35**(1948), 1–12]

21. For $i = 1, \ldots, n$ let $\langle A_i, \prec_i\rangle$ resp. $\langle B_i, <_i\rangle$ be pairwise disjoint ordered sets of type θ_1 resp. θ_2, and let $\langle A, \prec\rangle$ resp. $\langle B, <\rangle$ be their ordered unions. Then $\langle A, \prec\rangle$ has type $\theta_1 \cdot n$ and $\langle B, <\rangle$ has type $\theta_2 \cdot n$, so if these order types are the same, then there is a similarity mapping $f : A \to B$. We have to show that $\theta_1 = \theta_2$, i.e., one of the sets $\langle A_i, \prec_i\rangle$ is similar to $\langle B_i, <_i\rangle$.

If $f[A_1] = B_1$, then we are done, so we may assume that $f[A_1] \subset B_1$ (if $B_1 \subset f[A_1]$ then we consider f^{-1} and reverse the role of A_i and B_i). Thus, $\langle A_1, \prec_1\rangle$ is similar to an initial segment of $\langle B_1, <_1\rangle$, i.e., $\theta_2 = \theta_1 + \rho$ with some order type ρ. If $B_n \subseteq f[A_n]$, then $\langle B_n, <_n\rangle$ is similar to an end segment of $\langle A_n, \prec_n\rangle$ (under f^{-1}), hence $\theta_1 = \tau + \theta_2$ with some order type τ. Thus, in this case $\theta_1 = \theta_2$ by Problem 18.

If, however, $B_n \not\subseteq f[A_n]$, then $f[A_n] \subset B_n$, and so $\langle A_n, \prec_n\rangle$ is similar to an end segment of $\langle B_n, <_n\rangle$, i.e., $\theta_2 = \tau + \theta_1$ with some order type τ. Since f maps an interval of $\langle A, \prec\rangle$ into an interval of $\langle B, <\rangle$, and $f[A_0] \subset B_0$ and

$f[A_n] \subset B_n$, in this case there must be a $0 < j < n$ with $B_j \subseteq f[A_j]$. Thus, f^{-1} maps B_j into an interval of $\langle A_j, \prec_j \rangle$, which means that with some order types σ_1 and σ_2 we have $\theta_1 = \sigma_1 + \theta_2 + \sigma_2$. Now in this case $\theta_1 = \theta_2$ follows from Problem 19.

22. Clearly $\omega^* = 2 \cdot \omega^*$ and $\omega^* = 2 \cdot \omega^* + 1$. On the other hand, η cannot be written in either form $2 \cdot \tau$ or $2 \cdot \tau + 1$. [W. Sierpiński, Cardinal and Ordinal Numbers, Polish Sci. Publ., Warszawa, 1965, XII.3/18]

23. Clearly

$$\omega \cdot 2 + 1 = (\omega + 1) + (\omega + 1) = (\omega + 1) \cdot 2$$

so $\theta = \omega \cdot 2 + 1$ is suitable. On the other hand, ω cannot be written in either form $\tau_1 \cdot 2 + 1$ or $\tau_2 \cdot 2$.

24. Since $\eta \cdot 2 = \eta$, any order type $\tau \cdot \eta$ where τ is an arbitrary order type satisfies $\theta \cdot 2 = \theta$. [W. Sierpiński, Cardinal and Ordinal Numbers, Polish Sci. Publ., Warszawa, 1965, XII.3/13]

25. Since $2 \cdot \omega = \omega$, any order type $\omega \cdot \tau$ where τ is an arbitrary order type satisfies $2 \cdot \theta = \theta$.

26. Notice that if θ and τ are order types such that $2 \cdot \theta = \theta$ and $\tau \cdot 2 = \tau$, then $2 \cdot (\theta \cdot \tau) = (\theta \cdot \tau) \cdot 2 = \theta \cdot \tau$, hence products of order types from the preceding two problems satisfy the requirements. E.g., $\omega \cdot n \cdot \eta$ are all different and they are of the required property.

27. The types $n \cdot \eta$ where $n = 0, 1, \dots$ are all different (in an ordered set of this type n is the largest number of consecutive elements) and they satisfy $(n \cdot \eta) \cdot (n \cdot \eta) = n \cdot (\eta \cdot n) \cdot \eta = n \cdot (\eta \cdot \eta) = n \cdot \eta$.

28. $\tau = \theta \cdot (\omega + \omega^*)$ clearly satisfies $\tau + \theta = \theta + \tau = \tau$.

29. Let $\langle A, \prec \rangle$ have order type θ, and order antilexicographically the product $\cdots \times A \times A$. If the order type of this set is τ_1, which we can write as $\cdots \theta \cdot \theta$, then clearly $\tau_1 \cdot \theta = \tau_1$.

Similarly, let τ_2 be the order type of the lexicographically ordered $A \times A \times \cdots$. Then $\theta \cdot \tau_2 = \tau_2$.

Now if we set $\tau = \tau_2 \cdot \tau_1$, then $\theta \cdot \tau_2 = \tau_2$ and $\tau_1 \cdot \theta = \tau_1$ imply $\theta \cdot \tau = \tau \cdot \theta = \tau$. [W. Sierpiński, Cardinal and Ordinal Numbers, Polish Sci. Publ., Warszawa, 1965, XII.9/3]

30. The types $\omega^2 \cdot k$, $k = 1, \dots$ are all different and form an arithmetic progression. Furthermore

$$(\omega \cdot k) \cdot (\omega \cdot k) = \omega \cdot (k \cdot \omega) \cdot k = \omega \cdot \omega \cdot k = \omega^2 \cdot k,$$

so they are all squares. [W. Sierpiński, Cardinal and Ordinal Numbers, Polish Sci. Publ., Warszawa, 1965, XII.3/9]

31. The types $\theta_i = \eta+i+1$, $1 \le i \le n$ are appropriate, since if $\theta = \sum_{1\le i\le n} \theta_{\pi(i)}$ is their sum in any order and $\langle A, \prec\rangle$ is an ordered set of type θ, then the order can be recognized from $\langle A, \prec\rangle$: $\pi(1)+1$ is the length of the first (i.e., leftmost) maximal chain of consecutive elements in $\langle A, \prec\rangle$, $\pi(2)$ is the length of the next maximal chain of consecutive elements, etc. [W. Sierpiński, Cardinal and Ordinal Numbers, Polish Sci. Publ., Warszawa, 1965, XII.3/22]

32. The types $\theta_i = \eta + i + 1$, $1 \le i \le n$, are appropriate, since if $\theta = \prod_{1\le i\le n} \theta_{\pi(i)}$ is their product in any order and $\langle A, \prec\rangle$ is an ordered set of type θ, then the order can be recognized from $\langle A, \prec\rangle$. In fact, take any other order with the same product $\theta = \prod_{1\le i\le n} \theta_{\sigma(i)}$. Since $\pi(1)+1$ and $\sigma(1)+1$ are both the length of the longest chain of consecutive elements in $\langle A, \prec\rangle$, we have $\pi(1) = \sigma(1)$, say $\pi(1) = \sigma(1) = k_1$.

Now let $B = (\mathbf{Q} \cap (0,1)) \cup \{1, 2, \dots, k_1+1\}$ and $\langle C_1, <_1\rangle$ and $\langle C_2, <_2\rangle$ be two ordered sets of type $\prod_{2\le i\le n} \theta_{\pi(i)}$ and $\prod_{2\le i\le n} \theta_{\sigma(i)}$, respectively. Then $B \times C_1$ and $B \times C_2$ (with antilexicographic ordering) both have order types θ, so there is a similarity mapping $f : B \times C_1 \to B \times C_2$ between them. For any $c_1 \in C_1$ there is a unique $c_2 \in C_2$, that we are going to denote by $F(c_1)$, such that

$$f((2, c_1)) \in \{1, 2, \dots, k_1+1\} \times \{c_2\}.$$

Exactly as in the proof of Problem 20 it follows that this F is a monotone map from C_1 onto C_2, thus besides $\pi(1) = \sigma(1)$ we also have $\prod_{2\le i\le n} \theta_{\pi(i)} = \prod_{2\le i\le n} \theta_{\sigma(i)}$.

Now repeating this argument (or using induction) we can conclude that $\pi(i) = \sigma(i)$ for all $1 \le i \le n$, and the proof is over. [W. Sierpiński, Cardinal and Ordinal Numbers, Polish Sci. Publ., Warszawa, 1965, XII.3/35]

33. See the next problem.

34. First of all, the order types θ_k, $k = 1, 2, \dots$ are all different. In fact, if $\langle A, \prec\rangle$ is an ordered set of type θ_k, then for $a, b \in A$ let $a \sim b$ if there are only finitely many elements between a and b. This is an equivalence relation, and k is the number of equivalence classes with the property that there are only finitely many equivalence classes that follow them (in the ordering given in $\langle A, \prec\rangle$).

We claim that

$$\theta_k \cdot \theta_k = (\omega^* + \omega) \cdot \omega \cdot \eta$$

and

$$((\omega^* + \omega) \cdot \omega \cdot \eta) \cdot \theta_k = (\omega^* + \omega) \cdot \omega \cdot \eta,$$

and these imply that for all k and all $n \ge 2$ we have $\theta_k^n = (\omega^* + \omega) \cdot \omega \cdot \eta$.

Since $k + \omega = \omega$ and

$$(1+\eta)\cdot(\omega^*+\omega)=\cdots+1+\eta+1+\eta+\cdots=\eta,$$

we obtain

$$\begin{aligned}(\omega+\omega\cdot\eta+k)\ \cdot\ (\omega^*+\omega)&\\ &=\cdots+\omega+\omega\cdot\eta+k+\omega+\omega\cdot\eta+k+\cdots\\ &=\cdots+\omega\cdot(1+\eta)+\omega\cdot(1+\eta)+\cdots\\ &=\omega\cdot(1+\eta)(\omega^*+\omega)=\omega\cdot\eta,\end{aligned}$$

and hence

$$\begin{aligned}\theta_k\cdot\theta_k&=(\omega^*+\omega)\cdot(\omega+\omega\cdot\eta+k)\cdot(\omega^*+\omega)\cdot(\omega+\omega\cdot\eta+k)\\ &=(\omega^*+\omega)\cdot\omega\cdot\eta\cdot(\omega+\omega\cdot\eta+k)\\ &=(\omega^*+\omega)\cdot\omega\cdot\eta.\end{aligned}$$

In a similar fashion, since $\eta\cdot(\omega^*+\omega)=\eta$, we obtain

$$\begin{aligned}(\omega^*+\omega)\cdot\omega\cdot\eta\cdot\theta_k&=(\omega^*+\omega)\cdot\omega\cdot\eta\cdot(\omega^*+\omega)\cdot(\omega+\omega\cdot\eta+k)\\ &=(\omega^*+\omega)\cdot\omega\cdot\eta\cdot(\omega+\omega\cdot\eta+k)\\ &=(\omega^*+\omega)\cdot\omega\cdot\eta.\end{aligned}$$

[A. C. Davis, cf. W. Sierpiński, Cardinal and Ordinal Numbers, Polish Sci. Publ., Warszawa, 1965, XII.3/10]

35. Since $\eta^n=\eta$ and $(1+\eta)^n=1+\eta$, we have $1^n+\eta^n=(1+\eta)^n$.

36. Both the irreflexivity and the transitivity are clear. Since ω and ω^* are not comparable with respect to $\prec$, the trichotomy is not true.

37. If $\langle A_1,<_1\rangle$ is similar to a proper initial segment of $\langle A_2,<_2\rangle$, then, in view of Problem 6.41, $\langle A_2,<_2\rangle$ cannot be similar to a subset of $\langle A_1,<_1\rangle$, hence $\theta_1\prec\theta_2$. Conversely, suppose that $\theta_1\prec\theta_2$. Since either $\langle A_1,<_1\rangle$ or $\langle A_2,<_2\rangle$ is similar to an initial segment of the other one (Problem 6.42), the only possibility is that $\langle A_1,<_1\rangle$ is similar to a proper initial segment of $\langle A_2,<_2\rangle$.

The trichotomy of $\prec$ among ordinals is an immediate consequence of Problems 6.42 and 6.43.

38. See Problem 6.26.

39. For different 0–1 sequences $\{\epsilon_i\}$ the countable types

$$(\omega^*+\omega)+\epsilon_0+(\omega^*+\omega)+\epsilon_1+\cdots$$

are all different, so there are at least continuum many order types θ with $\theta\prec\eta$. But $\mathbf{Q}$ has only continuum many subsets, so their number is then exactly continuum.

40. See Problem 6.57. [cf. W. Sierpiński, Cardinal and Ordinal Numbers, Polish Sci. Publ., Warszawa, 1965, XII.11/4]

41. See Problem 6.58.

42. The order types $i + \omega^* + \omega + (n + 1 - i)$, $1 \le i \le n$ are appropriate; see Problem 6.59. [W. Sierpiński, Cardinal and Ordinal Numbers, Polish Sci. Publ., Warszawa, 1965, XII.11.4]

43. See Problem 6.60.

44. Since $1+\omega = \omega$, the sufficiency of the condition is clear. Now suppose that $1 + \theta = \theta$, and let $\langle A, \prec\rangle$ be an ordered set of type θ. Then A has a smallest element a_0 (since it is of type $1 + \theta$), and A is similar to $A \setminus \{a_0\}$. Let f be a similarity mapping from A onto $A \setminus \{a_0\}$. It is clear, that since $f(a_0)$ is the smallest element of $A \setminus \{a_0\}$, it is the successor of a_0 in A. In a similar fashion, $f(f(a_0))$ is a successor of $f(a_0)$, $f(f(f(a_0)))$ is a successor of $f(f(a_0))$, etc. All these mean that $\{a_0, f(a_0), f(f(a_0)), \ldots\}$ is an initial segment of $\langle A, \prec\rangle$, and if τ is the order type of $A \setminus \{a_0, f(a_0), f(f(a_0)), \ldots\}$, then it follows that $\theta = \omega + \tau$.

45. The proof is similar to the preceding one.

46. The sufficiency of the condition is clear since $\eta + \eta = \eta$.

Let $\langle A, \prec\rangle$ be an ordered set of type θ. Then since $\eta + \theta = \theta + \eta$, $\langle A, \prec\rangle$ has an initial segment A_1 of type η and has an end segment A_2 of type η. If $A_1 \cap A_2 \neq \emptyset$, then $\langle A, \prec\rangle$ is of type η, and in this case $\eta = \eta + 0 + \eta$. If, however, $A_1 \cap A_2 = \emptyset$ and if τ is the order type of $A \setminus (A_1 \cup A_2)$, then with this τ we clearly have $\theta = \eta + \tau + \eta$. [W. Sierpiński, Cardinal and Ordinal Numbers, Polish Sci. Publ., Warszawa, 1965, XII.3/23]

47. The sufficiency is clear, since $1 + \omega = \omega$ and $\omega^* + 1 = \omega^*$; thus, we only have to verify the necessity of the condition.

Suppose $\theta \neq 0$, and $\theta + \lambda = \lambda + \theta$. Let $\langle A, \prec\rangle$, $\langle A_1, \prec_1\rangle$ be of type θ, and $\langle B, <\rangle$, $\langle B_1, <_1\rangle$ of type λ such that these sets are pairwise disjoint. Since $\lambda+\theta = \theta+\lambda$, it follows that there is a similarity mapping $f : B \cup A \to A_1 \cup B_1$ between these ordered unions. If we have $A_1 \subseteq f[B]$, then A_1 is similar to an initial segment of B, which implies that $\theta = \lambda$ or $\theta = \lambda + 1$. At the same time, since then f maps A into B_1, we get that A is similar to an end segment of B_1, and hence $\theta = \lambda$ or $\theta = 1 + \lambda$. Thus, in this case $\theta = \lambda$.

Now let $f[B] \subset A_1$. If θ_1 is the order type of $A_1 \setminus f[B]$, then we have $\theta = \lambda+\theta_1$ and also $\theta = \theta_1+\lambda$ (consider that $(A_1 \cup B_1) \setminus f[B] = (A_1 \setminus f[B]) \cup B_1$ is similar to A).

Thus, we have proved that if $\theta \neq 0$, then either $\theta = \lambda$, or there is a $\theta_1 \neq \emptyset$ such that $\theta = \lambda + \theta_1$ and $\theta = \theta_1 + \lambda$. Thus, the same argument can be applied to θ_1 and we get that either $\theta_1 = \lambda$, or there is a $\theta_2 \neq \emptyset$ such that $\theta_1 = \lambda + \theta_2$ and $\theta_1 = \theta_2 + \lambda$. Repeat the same process as long as it is possible. It follows that either there is an n such that $\theta = \lambda \cdot n$, or for all n the set $\langle A, \prec \rangle$ has an initial and an end segment of type $\lambda \cdot n$. Since $\mathbf{R}$ is continuously ordered, it follows that an initial segment S_n of type $\lambda \cdot n$ has to be an initial segment of S_{n+1}, so the segments $S_1, S_2, \ldots, S_n$ are strictly increasing, and their union is an initial segment S of $\langle A, \prec \rangle$ of type $\lambda \cdot \omega$. In a similar manner, there are end segments E_n of $\langle A, \prec \rangle$ of type $\lambda \cdot n$ for all n, and their union E is an end segment of $\langle A, \prec \rangle$ of type $\lambda \cdot \omega^*$. Note also that it is not possible that $S_n \cap E_n \neq \emptyset$, for then $E_m \setminus E_n \subseteq S_n$ for all $m > n$ and S_n would have intervals of type $\lambda \cdot (m - n)$ for all $m > n$, and this is not the case. Thus, $S \cap E = \emptyset$, and if τ is the order type of $A \setminus (S \cup E)$, then we have $\theta = (\lambda \cdot \omega) + \tau + (\lambda \cdot \omega^*)$, as was claimed. [W. Sierpiński, Cardinal and Ordinal Numbers, Polish Sci. Publ., Warszawa, 1965, XII.3/24]

48. Since $(\tau^*)^* = \tau$ and $(\tau + \sigma)^* = \sigma^* + \theta^*$, the sufficiency of the condition follows.

Suppose now that $\theta = \theta^*$. Let $\langle A, \prec \rangle$ be of type θ, $\prec^*$ the reverse ordering on A, and let $f : A \to A$ be a similarity mapping between $\langle A, \prec \rangle$ and $\langle A, \prec^* \rangle$.

If there is an element $a \in A$ with $f(a) = a$, and $A_1 = \{b \in A : b \prec a\}$, then A_1 is mapped by f onto $A \setminus (A_1 \cup \{a\})$, so with τ equal to the order type of $\langle A_1, \prec \rangle$ we have $\theta = \tau + 1 + \tau^*$.

Suppose now that there is no element $a \in A$ with $a = f(a)$, and consider the set $A_1 = \{a \in A : a \prec f(a)\}$. This is an initial segment of A_1, for if $a \in A_1$ and $b \prec a$, then $b \prec a \prec f(a) \prec f(b)$, so $b \in A_1$ is also true. Now f maps A_1 into $A \setminus A_1$. In fact, if $a \prec f(a)$, then $f(a) \prec^* f(f(a))$, i.e., $f(f(a)) \prec f(a)$, and so $f(a) \in A \setminus A_1$. The same reasoning gives that f maps A_1 onto $A \setminus A_1$. Thus, if τ is the order type of A_1, then we have $\theta = \tau + \tau^*$.

* * *

49. Let X be a set of cardinality κ. Then every order type of cardinality κ is the order type of X with some ordering $\prec \subseteq X \times X$. Thus, there are at most as many order types of cardinality κ as subsets of $X \times X$, which is of cardinality $2^{|X|^2} = 2^{|X|} = 2^\kappa$.

On the other hand, let $\tau_0 = 2$ and $\tau_1 = \eta$, and for a transfinite sequence $\epsilon = \{\epsilon_\xi\}_{\xi<\kappa}$ of type κ of the numbers 0 and 1 consider the order type $\theta_\epsilon = \sum_{\xi<\kappa} \tau_{\epsilon_\xi}$. Let ϵ and ϵ' be two transfinite sequences and for each ξ let $\langle A_\xi, \prec_\xi \rangle$ and $\langle A'_\xi, \prec'_\xi \rangle$ be ordered sets of type τ_{ϵ_ξ} and $\tau_{\epsilon'_\xi}$, respectively, and let $\langle A, \prec \rangle$, resp. $\langle A', \prec' \rangle$, be their ordered union for $\xi < \kappa$. Then $\langle A, \prec \rangle$ has order type θ_ϵ and $\langle A', \prec' \rangle$ has order type $\theta_{\epsilon'}$, respectively. Now if for some $\alpha < \kappa$ we have $\epsilon_\xi = \epsilon'_\xi$ for all $\xi < \alpha$ and $f : A \to A'$ is a similarity mapping between these sets, then f maps each A_ξ into A'_ξ. The proof of this is the same as the analogous statement in the proof of Problem 6.15 and can be easily carried out by transfinite induction on α.

Thus, if in addition we have, say, $\epsilon_\alpha = 0$ and $\epsilon'_\alpha = 1$, then f cannot exist, for f cannot map the two-element set A_α onto an initial segment of the densely ordered set A'_α.

This proves that for different 0–1 transfinite sequences ϵ we get different order types θ_ϵ, and so there are exactly 2^κ different order types of cardinality κ. [W. Sierpiński, Cardinal and Ordinal Numbers, Polish Sci. Publ., Warszawa, 1965, XVII.5. Theorem 1]

8

Ordinals

1. **(a)** Let x be an N-set and $y \in x$. Then $y \subset x$. If $z \in y$, then $z \in x$ and $z <_\in y$. Now every element of z is smaller with respect to $<_\in$ than z, and hence than y, which gives that every such element must belong to y. This proves that y is transitive. That it is well ordered by $\in$ is a consequence of the fact that it is a subset of a well-ordered set.

(b) It is clear that y is transitive and well ordered by $\in$ (x is its largest element). If $x \in z$ where z is an N-set, then $x \subset z$, and hence $y \subset z$.

(c) This is clear from the definitions.

(d) If $y \in Y$ and $z \in y$, then $z <_\in y$, hence by the initial segment property $z \in Y$. This shows that Y is transitive. That it is well ordered by $\in$ is clear, since $Y \subset x$. Thus, Y is an N-set. If $Y \neq x$, then Y is a proper initial segment of x, hence it is the initial segment determined by an element p. Thus, $y \in Y \Leftrightarrow y <_\in p \Leftrightarrow y \in p$, i.e., $Y = p$, which shows $Y \in x$.

(e) Consider $z = x \cap y$. This is an initial segment of both x and y (cf. (c)), hence by (d) it is an N-set and either $z = x$ or $z \in x$ and either $z = y$ or $z \in y$. Thus, for the conclusion we only have to show that it is impossible to have simultaneously $z \in x$ and $z \in y$. Indeed, if that was true, then we had $z \in x \cap y = z$, i.e., z was an element of the N-set x such that $z \in z$, which is impossible by the irreflexivity of $\in$ on x.

(f) Irreflexivity of $<$ follows exactly as in part (e); transitivity of $<$ is due to the transitivity of N-sets, and trichotomy was proved in (e). Let $B \neq \emptyset$ be a set of N-sets. Pick any $x \in B$. Then either $x \cap B = \emptyset$, which means that B does not have a smaller element than x, i.e., x is its smallest element, or $x \cap B \neq \emptyset$, and then the smallest element (with respect to $\in$) of $x \cap B$ is the smallest element of B (note that there is a smallest element in $x \cap B$ because x is well ordered).

(g) If, say, $y \in x$, then y is a proper initial segment of x, hence it cannot be similar to x (see Problem 6.41).

(h) We follow the ideas from the solution of Problem 6.42. Let $\langle A, \prec\rangle$ be a well-ordered set, and let B be the collection of all N-sets that are similar to a proper initial segment of $\langle A, \prec\rangle$. For $x \in B$ let a_x be the unique element of A such that the initial segment determined by a_x is similar to x, and let g_x be the appropriate similarity mapping. Then any $y \in x$ is similar to the initial segment of $\langle A, \prec\rangle$ determined by $g_x(y)$, hence $y \in B$, i.e., B is transitive. The mapping $x \to a_x$ is monotone, hence B is similar to a subset of A, in particular B is a set. $\in$ is a well order on B by (f), hence B is an N-set. The mapping $f(x) = a_x$ maps B monotonically onto an initial segment of A, hence either $A = f[B]$, in which case $\langle A, \prec\rangle \sim \langle B, \in\rangle$, or B is similar to a proper initial segment of A. But the latter would imply $B \in B$, which is impossible (see (f)).

Unicity is a consequence of (g).

2. Each ordinal is well ordered. Thus, $\alpha_0 > \alpha_1 > \cdots$ is not possible, for otherwise we would have an infinite decreasing sequence in α_0 (see Problem 6.36).

3. This follows from the previous problem and from Problem 6.1.

4. $1 + \omega = \omega$ is clear, $\omega + 1 \neq \omega$ because in ω there is no largest element, but in $\omega + 1$ there is a largest element.

$2 \cdot \omega = \omega$ is true because $2 \cdot \omega$ is the order type of the product $2 \times \omega$ ordered antilexicographically, so the order of the elements is

$$(0,0) < (1,0) < (0,1) < (1,1) < (0,2) < \ldots,$$

Finally, $\omega \cdot 2 \neq \omega$ because in a set of type $\omega \cdot 2$ there are elements preceded by infinitely many elements, but in a set of type ω this is not possible.

5. Since

$$(\omega + a) \cdot (\omega + b) = (\omega + a) \cdot \omega + (\omega + a) \cdot b,$$

and $a + \omega = \omega$, it easily follows that $(\omega + a) \cdot (\omega + b) = \omega^2 + \omega \cdot b + a$ if $b \geq 1$ and $(\omega + a) \cdot \omega = \omega^2$ if $b = 0$.

6. One can easily see that

(a) $\omega + \xi = \omega \Longleftrightarrow \xi = 0$,

(b) $\xi + \omega = \omega \Longleftrightarrow \xi$ is finite,

(c) $\xi \cdot \omega = \omega \Longleftrightarrow \xi$ is finite and not 0,

(d) $\omega \cdot \xi = \omega \Longleftrightarrow \xi = 1$,

(e) $\xi + \zeta = \omega \Longleftrightarrow \xi < \omega,\ \zeta = \omega$ or $\xi = \omega,\ \zeta = 0$,

(f) $\xi \cdot \zeta = \omega \Longleftrightarrow 1 \leq \xi < \omega,\ \zeta = \omega$ or $\xi = \omega,\ \zeta = 1$.

7. The proper initial segments of ω^2+1 are ω^2 and the proper initial segments of ω^2, which are of the form $\omega\cdot n+m$ with some n,m natural numbers. Thus, either $\xi=\omega^2+1$, $\zeta=0$, or $\xi=\omega^2$, in which case ζ is 1, or else $\xi=\omega\cdot n+m$ with some natural numbers n,m, in which case ζ has to be ω^2+1.

8. **a)** $\omega+k>k+\omega$ since $k+\omega=\omega$, and this is a proper initial segment of $\omega+k$.

b) $k\cdot\omega<\omega\cdot k$ since $k\cdot\omega=\omega$, and this is a proper initial segment of $\omega\cdot k$.

c) $\omega+\omega_1<\omega_1+\omega$ since $\omega+\omega_1=\omega_1$, and this is a proper initial segment of $\omega_1+\omega$.

d) If $P(\omega)=\omega^n\cdot a_n+\omega^{n-1}\cdot a_{n-1}+\cdots+\omega\cdot a_1+a_0$, then on applying that $a_0<\omega$ implies

$$\omega\cdot a_1+a_0<\omega\cdot a_1+\omega=\omega\cdot(a_1+1)<\omega^2,$$

which implies

$$\omega^2\cdot a_2+\omega a_1+a_0<\omega^2\cdot a_2+\omega^2=\omega^2\cdot(a_2+1)<\omega^3,$$

etc., we can see that $P(\omega)<\omega^{n+1}$.

e) Similarly as in part d), $P(\omega)$ is larger than $Q(\omega)$ if and only if $n>m$, or $n=m$, and $a_i>a_i'$ for the smallest index i with $a_i\neq a_i'$.

9. If $\alpha_1\le\alpha_2$, then a set of type $\alpha_1+\beta$ is similar to a subset of type $\alpha_2+\beta$, so by Problem 7.37 we have $\alpha_1+\beta\le\alpha_2+\beta$.

The proof that $\beta+\alpha_1\le\beta+\alpha_2$ is similar. Finally, if $\alpha_1<\alpha_2$, then α_1 is an initial segment of α_2, hence a set of type $\beta+\alpha_1$ is similar to a proper initial segment of $\beta+\alpha_2$, and so we have $\beta+\alpha_1<\beta+\alpha_2$.

The proofs of the claims for multiplication are the same.

10. If $\gamma+\alpha=\gamma+\beta$, then $\alpha=\beta$ by the preceding problem. If $\alpha+\gamma=\beta+\gamma$ we do not need to have $\alpha=\beta$, an example is $0+\omega=1+\omega$. In a similar fashion, if $\gamma\cdot\alpha=\gamma\cdot\beta$ and $\gamma\neq0$, then $\alpha=\beta$ by the preceding problem. However, $\alpha\cdot\gamma=\beta\cdot\gamma$ does not imply $\alpha=\beta$, an example is $1\cdot\omega=2\cdot\omega$.

If $\gamma>0$ finite, then $\alpha+\gamma=\beta+\gamma$ clearly implies $\alpha=\beta$, and $\alpha\cdot\gamma=\beta\cdot\gamma$ also implies $\alpha=\beta$ by Problem 7.21.

11. Suppose that $\gamma=\delta+1$ and, say, $\alpha<\beta$. Then $\alpha\cdot\delta\le\beta\cdot\delta$, hence by Problem 9 we have $\alpha\cdot\delta+\alpha<\beta\cdot\delta+\beta$, so in this case $\alpha\cdot\gamma=\beta\cdot\gamma$ cannot hold.

12. If $\alpha<\beta$, then from Problem 9 we get by induction on k that $\alpha^k<\beta^k$. Thus, $\alpha^k=\beta^k$ implies $\alpha=\beta$.

13. **a)** It is clear that $\sup_{\eta<\xi}(\alpha+\eta)\le\alpha+\xi$. However, if $\gamma<\alpha+\xi$, then either $\gamma\le\alpha$ or there is a $\delta<\xi$ such that $\gamma<\alpha+\delta$, hence this γ cannot be an upper bound for the ordinals $\alpha+\eta$, $\eta<\xi$. This proves part a). Since

$$\sup_{n<\omega}(n+\omega) = \omega \neq \omega + \omega = (\sup_{n<\omega} n) + \omega,$$

the analogous statement for the reversed order is not true.

b) It is again clear that $\sup_{\eta<\xi}(\alpha \cdot \eta) \leq \alpha \cdot \xi$. If $\gamma < \alpha \cdot \xi$, then there is a $\delta < \xi$ such that $\gamma < \alpha \cdot \delta$, so then γ cannot be an upper bound of the ordinals $\alpha \cdot \eta$, $\eta < \xi$. Since

$$\sup_{n<\omega}(n \cdot \omega) = \omega \neq \omega \cdot \omega = (\sup_{n<\omega} n) \cdot \omega,$$

the analogous statement for the reversed order is not true.

14. Since α is an initial segment of β, we can write β as an ordered union $\alpha \cup C$, and so if ξ is the order type of C, then we have $\beta = \alpha + \xi$. The unicity of ξ follows from the strict monotonicity of addition in the second argument (Problem 9).

For the equation $\xi + \alpha = \beta$ neither the solvability nor the unicity can be guaranteed. In fact, $\xi + 1 = \omega$ is not solvable, and $\xi + \omega = \omega$ has infinitely many solutions, namely $\xi < \omega$.

15. Let ζ be the supremum of all ordinals τ with the property $\alpha \cdot \tau \leq \beta$. We claim that $\alpha \cdot \zeta \leq \beta$. For the case when ζ is a successor ordinal this is clear, for then ζ has to agree with one of the τ's, and for limit ζ the statement follows from Problem 13, a). Thus, by Problem 14 the equation $\beta = \alpha \cdot \zeta + \xi$ is uniquely solvable for ξ. Here we cannot have $\alpha \leq \xi$, for then we could write $\xi = \alpha + \sigma$ with some σ, and then $\alpha \cdot (\zeta + 1) = \alpha \cdot \zeta + \alpha \leq \alpha \cdot \zeta + \xi = \beta$ would hold, which is not possible by the choice of ζ. Thus, $\xi < \alpha$, and the existence of the representation has been proved.

To show unicity, suppose that $\zeta_1 < \zeta_2$ and $\xi_1, \xi_2 < \alpha$. Then

$$\alpha \cdot \zeta_1 + \xi_1 < \alpha \cdot \zeta_1 + \alpha = \alpha \cdot (\zeta_1 + 1) \leq \alpha \cdot \zeta_2 \leq \alpha \cdot \zeta_2 + \xi_2.$$

Thus, if $\alpha \cdot \zeta_1 + \xi_1 = \alpha \cdot \zeta_2 + \xi_2$, then we must have $\zeta_1 = \zeta_2$, and then $\xi_1 = \xi_2$ follows from Problem 14.

16. If $\alpha \cdot \omega \leq \beta$, then $\beta = \alpha \cdot \omega + \gamma$ with some ordinal γ (see Problem 14), and hence

$$\alpha + \beta = (\alpha + \alpha \cdot \omega) + \gamma = \alpha(1 + \omega) + \gamma = \alpha \cdot \omega + \gamma = \beta.$$

17. Choose a large ordinal β with $\alpha + \beta = \beta$ (see the preceding problem). Then the assumption implies that $\beta = \beta + \alpha$, so $\alpha = 0$ because of Problem 9.

18. Let α be an ordinal and let β be the supremum of all limit ordinals not bigger than α. Then β is zero or a limit ordinal, and by Problem 14 we can write $\alpha = \beta + \gamma$. Here we cannot have $\omega \leq \gamma$, for then $\gamma = \omega + \delta$, and as

$\alpha = (\beta + \omega) + \delta$, the ordinal $\beta + \omega > \beta$ would be a larger limit ordinal $\leq \alpha$. Thus, $\gamma < \omega$, and we are done.

19. It is clear that if $\gamma < \omega \cdot \beta$, then $\gamma + 1 < \omega \cdot \beta$, so $\omega \cdot \beta$ is a limit ordinal for $\beta \geq 1$. Conversely, let α be a limit ordinal. By Problem 15 we can write $\alpha = \omega \cdot \beta + n$ with some natural number n. Now here we must have $n = 0$, for otherwise α would be a successor ordinal (the successor of $\omega \cdot \beta + (n - 1)$).

20. If α is a limit ordinal, then using the representation in the preceding problem and $n \cdot \omega = \omega$ we get that $n \cdot \alpha = n \cdot \omega \cdot \beta = \omega \cdot \beta = \alpha$. Conversely, suppose that $n \cdot \alpha = \alpha$ for all n. Exactly as in the preceding solution we can write $\alpha = \omega \cdot \beta + m$ with some ordinal β and some natural number n (see Problem 14). But then $2 \cdot \alpha = 2 \cdot \omega \cdot \beta + 2m = \omega \cdot \beta + 2m$, and this can be at most α only if $m = 0$ (see Problem 9).

21. Write β in the form $\beta = \omega \cdot \gamma + m$ with some ordinal γ and natural number m (see Problem 15). If $m \neq 0$, then using that $(\alpha + n) \cdot \omega = \alpha \cdot \omega$, it follows that

$$\begin{aligned}(\alpha + n) \cdot \beta &= (\alpha + n) \cdot \omega \cdot \gamma + \overbrace{(\alpha + n) + \cdots + (\alpha + n)}^{m} \\ &= \alpha \cdot \omega \cdot \gamma + \alpha \cdot m + n = \alpha \cdot \beta + n.\end{aligned}$$

If, however, $m = 0$, then the same computation shows that

$$(\alpha + n) \cdot \beta = (\alpha + n) \cdot \omega \cdot \gamma = \alpha \cdot \omega \cdot \gamma = \alpha \cdot \beta.$$

22. We write α in the form $\omega \cdot \beta$ and use that $n \cdot \omega = \omega$ if $n \geq 1$ to conclude

$$(\alpha \cdot n)^k = \alpha \cdot \overbrace{(n \cdot \alpha) \cdots (n \cdot \alpha)}^{k-1} \cdot n = \alpha^k \cdot n.$$

23. Write $\alpha = \delta + k$, where k is a natural number and δ is 0 or a limit ordinal (see Problem 18). We also write β as $\omega \cdot \gamma + m$ with some ordinal γ and some natural number m (see Problem 15). Then $n \cdot \beta = n \cdot (\omega \cdot \gamma) + n \cdot m = \omega \cdot \gamma + nm$, so if this is α, then $k = mn$. Conversely, if $k = mn$, $n > 0$, then $\alpha = n \cdot (\delta + m/n)$. Thus, the answer to the problem is that n is a divisor of k.

24. **a)** If α is infinite then $1 + \alpha = \alpha$, but $\alpha + 1 > \alpha$, so α has to be finite.

b) If $\alpha > 0$ and $\alpha + \omega = \omega + \alpha$, then α is a limit ordinal, so it is of the form $\omega \cdot \beta$. But then $\alpha + \omega = \omega \cdot (\beta + 1)$ and $\omega + \alpha = \omega \cdot (1 + \beta)$, and exactly as in case a) here $1 + \beta < \beta + 1$ if β is infinite, so by Problem 9 in this case $\omega + \alpha < \alpha + \omega$. Thus, the answer is that $\alpha = \omega \cdot n$ with some finite n (which is clearly sufficient).

c) $\alpha \cdot \omega = \omega \cdot \alpha$ if and only if α is a power of ω (see Problem 9.11). The sufficiency is obvious, the necessity immediately follows from the normal form of α (see Problem 9.16).

d) If $\alpha + (\omega + 1) = (\omega + 1) + \alpha$ and α is finite, then it must clearly be zero. If α is infinite, then α is a successor ordinal, actually the successor of a limit ordinal. Thus, $\alpha = \omega \cdot \beta + 1$. But then $\alpha + (\omega + 1) = \omega \cdot (\beta + 1) + 1$ while $(\omega + 1) + \alpha = \omega \cdot (1 + \beta) + 1$, and from these we get $\omega \cdot (\beta + 1) = \omega \cdot (1 + \beta)$, which implies as in part b) that β is finite. Thus, the answer is that α is 0 or it is of the form $\omega \cdot n + 1$ with some natural number $n \geq 1$.

e) Clearly, $\alpha = 0$ and $\alpha = (\omega + 1)^n = \omega^n + \omega^{n-1} + \cdots + \omega + 1$ with $n < \omega$ are solutions, and we show that there are no other solutions.

Suppose the contrary, and let α be the smallest solution not listed above. This α can be written as $\alpha = (\omega + 1) \cdot \beta + \gamma$ with $\gamma \leq \omega$. If $\beta = 0$ and $0 < \gamma < \omega$, then the equation becomes $\omega + \gamma = \omega \cdot \gamma + 1$, so in this case $\gamma = 1$. If $\beta = 0$ and $\gamma = \omega$, then we have $\omega^2 + \omega = \omega^2$, an impossibility. Finally, if $\beta > 0$, then we get

$$(\omega + 1) \cdot (\beta \cdot \omega + \beta) + \gamma = (\omega + 1) \cdot [(\omega + 1) \cdot \beta + \gamma]. \tag{8.1}$$

Here we must have $\beta \cdot \omega + \beta \leq (\omega + 1) \cdot \beta + \gamma$, hence (see Problem 14) $(\omega+1) \cdot \beta + \gamma = \beta \cdot \omega + \beta + \zeta$ with some ζ. Writing this back into (8.1), Problem 14 gives that we must have $\gamma = (\omega+1) \cdot \zeta$, which, in view of $\gamma \leq \omega$, is possible only if $\gamma = \zeta = 0$. Thus, in this case $\beta \cdot (\omega+1) = (\omega+1) \cdot \beta$. Here we must have $\beta < \alpha$, for in the case $\beta = \alpha$ we would have $\alpha \cdot (\omega + 1) > \alpha = (\omega + 1) \cdot \beta = (\omega + 1) \cdot \alpha$, i.e., α would not be a solution. Therefore, this $\beta < \alpha$ is again a solution, and by the minimality of α it must of the form $\beta = (\omega + 1)^n$ for some $n < \omega$. But then $\alpha = (\omega + 1)^{n+1}$, which is a contradiction, and this contradiction proves that the only solutions are the ones listed above.

25. The statement is true for $n = 1$: $\sum_{\xi<\omega} \xi = \omega = \omega^{2\cdot 1-1}$, and we proceed by induction. Thus, suppose the validity of $\sum_{\xi<\omega^n} \xi = \omega^{2n-1}$ has been verified for some n. An ordinal $\omega^n \leq \xi < \omega^{n+1}$ can be written in the form $\xi = \omega^n \cdot m + \eta$ with some natural number m and with $\eta < \omega^n$. The latter implies that η is less than a number $\omega^{n-1} \cdot k$, $k = 1, 2, \ldots$ from which we obtain $\eta + \omega^n \leq \omega^{n-1} \cdot k + \omega^n = \omega^n$, hence

$$S_m := \sum_{\eta<\omega^n} (\omega^n \cdot m + \eta) = \sum_{\eta<\omega^n} \omega^n = \omega^{2n}.$$

Now these sums follow each other in $\sum_{\xi<\omega^{n+1}}$ in the order of m; thus,

$$\sum_{\omega^n \leq \xi < \omega^{n+1}} \xi = \sum_{m=1}^{\infty} S_m = \sum_{n=1}^{\infty} \omega^{2n} = \omega^{2n+1}.$$

This and the induction hypothesis gives

$$\sum_{\xi<\omega^{n+1}} \xi = \sum_{\xi<\omega^n} \xi + \sum_{\omega^n\le\xi<\omega^{n+1}} \xi = \omega^{2n-1} + \omega^{2n+1} = \omega^{2n+1},$$

and this verifies the induction step.

26. Suppose $\alpha = \xi_n + \gamma_n$, $n = 1, 2, \ldots$ where the γ_n's are different. We can assume that the numbers γ_n are increasing (see Problem 3). But then Problem 9 shows that $\{\xi_n\}_{n=1}^{\infty}$ has to be a strictly decreasing sequence, and this is not possible (Problem 2). So there are only finitely many different γ's in question.

Since $n + \omega = \omega$, the same is not true for the representation $\alpha = \gamma + \xi$.

27. The proof is identical with the preceding one, and again $n \cdot \omega = \omega$, $1 \le n < \omega$, furnishes a counterexample for the representation $\alpha = \gamma \cdot \xi$.

28. The proof goes by induction on m, and suppose that the claim has already been verified for $m-1$ factors. It is clear that if one of the factors in a finite product is a limit ordinal, then the product itself is a limit ordinal. Thus, in the representation in question all the factors must be successor ordinals. Now in a representation into m factors there can only be finitely many last factors by the preceding problem, and if in two representations the last factors are the same, then the product of the first $m-1$ factors must also be the same by Problem 11. Now, by the induction hypothesis, the number of representations is finite.

29. Suppose that $\xi^2 + \omega = \zeta^2$. Then $\zeta^2 > \xi^2$, so $\zeta > \xi$. Clearly, ξ must be infinite (otherwise ω would be a square), and so $\xi^2 + \omega = \zeta^2 \ge (\xi+1)^2 \ge \xi \cdot (\xi+1) = \xi^2 + \xi$, which implies $\xi \le \omega$. Thus, $\xi = \omega$ and $\zeta \ge \omega + 1$, but then

$$\zeta^2 \ge (\omega+1)^2 = \omega^2 + \omega + 1 > \omega^2 + \omega = \xi^2 + \omega,$$

and hence the equation cannot hold. [W. Sierpiński, Cardinal and Ordinal Numbers, Polish Sci. Publ., Warszawa, 1965, XIV.8/7]

30. Note that

$$(\omega \cdot n)^2 + \omega^2 = \omega^2 \cdot n + \omega^2 = \omega^2 \cdot (n+1) = (\omega \cdot (n+1))^2,$$

and here the ordinals $\omega \cdot n$ are all different. [W. Sierpiński, Cardinal and Ordinal Numbers, Polish Sci. Publ., Warszawa, 1965, XIV.8/8]

31. It is well known that the only finite solution is $\alpha = \beta = 0$, so from now on we assume that both α and β are infinite.

Write $\alpha = \omega \cdot \gamma + n$ and $\beta = \omega \cdot \delta + m$. Then

$$\alpha^2 \cdot 2 = (\omega \cdot \gamma \cdot \omega \cdot \gamma + \omega \cdot \gamma \cdot n + n) \cdot 2 = \omega \cdot \gamma \cdot \omega \cdot \gamma \cdot 2 + \omega \cdot \gamma \cdot n + n$$

and

$$\beta^2 = \omega \cdot \delta \cdot \omega \cdot \delta + \omega \cdot \delta \cdot m + m;$$

thus, we must have $m = n$, for the ordinals before them on the right-hand sides are limit ordinals. Therefore,

$$\omega \cdot \gamma \cdot \omega \cdot \gamma \cdot 2 + \omega \cdot \gamma \cdot n = \omega \cdot \delta \cdot \omega \cdot \delta + \omega \cdot \delta \cdot n,$$

which implies (see Problem 10)

$$\gamma \cdot \omega \cdot \gamma \cdot 2 + \gamma \cdot n = \delta \cdot \omega \cdot \delta + \delta \cdot n. \tag{8.2}$$

It is clear that we must have $\delta \geq \gamma$. If $\delta < \gamma \cdot 2$, then $\delta = \gamma + \tau$ with some $\tau < \gamma$, and

$$\begin{aligned}\delta \cdot \omega \cdot \delta + \delta \cdot n &\leq \gamma \cdot 2 \cdot \omega \cdot (\gamma + \tau) + (\gamma + \tau) \cdot n \\ &\leq \gamma \cdot \omega \cdot \gamma + \gamma \cdot \omega \cdot \tau + \gamma \cdot (2n) \\ &< \gamma \cdot \omega \cdot \gamma + \gamma \cdot \omega \cdot \tau + \gamma \cdot \omega \\ &\leq \gamma \cdot \omega \cdot \gamma + \gamma \cdot \omega \cdot \gamma = \gamma \cdot \omega \cdot \gamma \cdot 2,\end{aligned}$$

so in this case (8.2) cannot hold. Thus, we must have $\delta \geq \gamma \cdot 2$, and then

$$\delta \cdot \omega \cdot \delta + \delta \cdot n \geq \gamma \cdot 2 \cdot \omega \cdot \gamma \cdot 2 + \gamma \cdot 2 \cdot n,$$

which, compared with (8.2), yields $n = 0$. The same computation shows that $\delta > \gamma \cdot 2$ is not possible, either, so we must have $\delta = \gamma \cdot 2$.

So far we have shown that if $\alpha^2 \cdot 2 = \beta^2$, then both α and β are limit ordinals and $\beta = \alpha \cdot 2$. It is easy to see that conversely, if α and β are limit ordinals and $\beta = \alpha \cdot 2$, then $\beta^2 = \alpha \cdot (2 \cdot \alpha) \cdot 2 = \alpha^2 \cdot 2$, so these pairs are all solutions.

32. Since $n \cdot \omega = \omega$ for all positive integer n, we can set $\omega^k \cdot n = (\omega \cdot n)^k$ for $n = 1, 2, \ldots$.

33. Consider $\alpha = 2$ and $\beta = \omega + 1$. Since

$$(\omega + 1)^n = \omega^n + \omega^{n-1} + \cdots + \omega + 1,$$

we have

$$\alpha^n \cdot \beta^n = \omega^n + \omega^{n-1} + \cdots + \omega + 2^n, \tag{8.3}$$

and this cannot be the nth power of a limit ordinal. If, however, $\gamma = \omega \cdot \delta + m$ is a successor ordinal, then

$$\gamma^n = (\omega \cdot \delta)^n + (\omega \cdot \delta)^{n-1} \cdot m + (\omega \cdot \delta)^{n-2} \cdot m + \cdots + \cdot(\omega \cdot \delta) \cdot m + m, \tag{8.4}$$

so we would have to have $m = 2^n$, but then the ordinal in (8.4) is clearly bigger than the ordinal in (8.3).

In a similar manner,

$$\beta^n \cdot \alpha^n = \omega^n \cdot 2^n + \omega^{n-1} + \omega^{n-2} + \cdots + \omega + 1,$$

and if this is equal to the ordinal in (8.4), then we must have $m = 1$. If $\delta > 2^n$, then

$$(\omega \cdot \delta)^n + (\omega \cdot \delta)^{n-1} + (\omega \cdot \delta)^{n-2} + \cdots + (\omega \cdot \delta) + 1$$
$$> \omega^n \cdot 2^n + \omega^{n-1} + \omega^{n-2} + \cdots + \omega + 1,$$

and actually this inequality is also true for $\delta = 2^n$. In a similar fashion, if $\delta < 2^n$ then we have the reverse inequality. Thus, $\beta^n \cdot \alpha^n$ is not the nth power of $\omega \cdot \delta + m$. [W. Sierpiński, Cardinal and Ordinal Numbers, Polish Sci. Publ., Warszawa, 1965, XIV.8/12]

34. The sum in question is $\omega + \omega$, and since $n + \omega = \omega$, the sum clearly does not change if we change the order of finitely many terms in it.

35. Clearly, all the sums $1+2+3+\cdots(n-1)+(n+1)+\cdots+\omega+n = \omega+\omega+n$ are different.

36. Consider the sum $\omega^2 + \overbrace{\omega + \omega + \cdots + \omega}^{n-1} + 1 + 1 + \cdots$. If we move exactly k of the ω's in front of ω^2, then the value of the sum is $\omega^2 + (n-k)\omega$, and these are n different ordinals for $k = 0, 1, \ldots, n-1$ (moving any ω after some of the 1's makes no effect). [W. Sierpiński, Cardinal and Ordinal Numbers, Polish Sci. Publ., Warszawa, 1965, XIV.8/4]

37. It is clear that if the terms that follow ω in the sum are $2^{i_1}, \ldots, 2^{i_k}$, then the value of the sum is $\omega + (2^{i_1} + \cdots + 2^{i_k})$, and all numbers from 0 to $2^n - 1$ have one and only one form of the type $2^{i_1} + \cdots + 2^{i_k}$ with $i_k \le n-1$. [W. Sierpiński, Cardinal and Ordinal Numbers, Polish Sci. Publ., Warszawa, 1965, XIV.3/8]

38. This is an immediate consequence of Problem 9.55(d), which implies $g(n) \le C(\sqrt[5]{81})^n$ with some constant C.

A direct proof can run as follows. Let $\alpha_1 \ge \alpha_2 \ge \ldots \ge \alpha_n$ be arbitrary ordinals, and let $\omega^\beta \cdot a_1$ be the largest ordinal in this form that is $\le \alpha_1$ (in other words, $\omega^\beta \cdot a_1$ is the leading term in the normal form of α_1 (see Problem 9.16). Take a permutation of the α_i's and take the sum $\alpha_{\pi(1)} + \cdots + \alpha_{\pi(n)}$. If in this sum α_1 is the kth term (i.e., if $\pi(k) = 1$), then the sum does not change if we permute the preceding first $k-1$ terms. In fact, if $\omega^\beta \cdot a_i$ is the largest multiple of ω^β that $\le \alpha_i$ (i.e., a_i is the coefficient of ω^β in the normal expansion of α_i), then $0 \le a_i < \omega$, and due to the fact that $\omega^\gamma + \omega^\beta = \omega^\beta$ for $\gamma < \beta$, we have

$$\alpha_{\pi(1)} + \cdots + \alpha_{\pi(k)} = \omega^\beta \cdot (\sum_{\pi(i)<k} a_i) + \alpha_k$$

(cf. Problem 9.18). Therefore, out of the $(n-1)!$ permutations with $\pi(k) = 1$ at most $(n-1)!/(k-1)!$ will give different sums, hence the number of different sums is at most

$$g(n) = (n-1)! \sum_{k=1}^{n} \frac{1}{(k-1)!} < e(n-1)!,$$

from which $g(n)/n! < e/n \to 0$ as $n \to \infty$, follows.

39. Set $\alpha_i = \omega + i$, $i = 1, \ldots, n$. Easy computation shows that if $i_1, \ldots, i_n$ is any permutation of the numbers $1, 2, \ldots, n$, then

$$\alpha_{i_1} \cdots \alpha_{i_n} = \omega^n + \omega^{n-1} \cdot i_n + \cdots + \omega \cdot i_2 + i_1,$$

and all these ordinals are different. [E. Spanier, see P. Erdős, Some remarks on set theory, Proc. Amer. Math. Soc., **23**(1950), 127–141]

40. Suppose that the least upper bound of any increasing transfinite subsequence of A is in A or is equal to α, and let $\beta \in \alpha \setminus A$ be an element outside A. If $\beta = \gamma + 1$, then the interval $\{\xi : \gamma < \xi < \beta + 1\} = \{\beta\}$ is a neighborhood of β disjoint from A, and a small modification gives the same in case $\beta = 0$. If, however, β is a limit ordinal, then there is a $\gamma < \beta$ for which there is no element of A between γ and β (otherwise we could construct an infinite transfinite sequence the supremum of which would be β, and hence β would have to belong to A). But then the interval $\{\xi : \gamma < \xi < \beta + 1\}$ is a neighborhood of β that is disjoint from A. Thus, the complement of A is open in the interval topology, so A is closed in that topology.

Conversely, suppose that $A \subset \alpha$ is a closed subset of α in the interval topology, and let $\{\alpha_\xi\}_{\xi<\delta}$ be an increasing sequence from A, with supremum $\beta < \alpha$. If β is a successor ordinal, then the sequence has a largest element that equals β, and so $\beta \in A$. If, however, β is a limit ordinal, then no matter how we choose $\gamma < \beta$, there is an a_ξ, $\xi < \delta$, such that $\gamma < a_\xi \leq \gamma$. Thus, any interval $\{\xi : \gamma < \xi < \sigma\}$ that contains β contains an a_ξ, so β is in the closure of A. But then $\beta \in A$ since A was assumed to be closed, and this proves the equivalence of the two statements.

The proof that A is closed in α in the interval topology if and only if the supremum of every subset $B \subset A$ is in A, or is equal to α, is the same.

41. The statement is an immediate consequence of the preceding problem and of the definition of continuity (namely that the inverse image of any open set is open).

42. Let $\overline{A}$ be the closure of A. The statement is clear if $\overline{A} \setminus A$ is a finite set. So let $\overline{A} \setminus A$ be infinite, and enumerate $\overline{A} \setminus A$ into the increasing transfinite sequence $\{\alpha_\xi\}_{\xi<\gamma}$ with $\gamma \geq \omega$. For each $\xi < \gamma$ with $\xi + 1 < \gamma$ there must be an

element a_ξ of A lying in the interval $(\alpha_\xi, \alpha_{\xi+1})$, and these a_ξ's are different. Hence, $|\gamma| \le |A|$, which shows that $|\overline{A}| = |\gamma| + |A| = |A|$.

43. The statement is clear if σ is a successor ordinal, since in that case $\{\xi\}$ is an open neighborhood of σ. If, σ is a limit ordinal, then the cofinality of σ is ω, so there is a sequence $\{\beta_n\}_{n=0}^\infty$ the supremum of which is σ. For each β_n there is a γ_n such that for $\gamma_n < \xi < \omega_1$ the ordinals a_ξ lie in the neighborhood $\{\eta : \beta_n < \eta < \sigma + 1\}$ of σ. Thus, if ν is the supremum of the ordinals γ_n, $n = 0, 1, \ldots$, then $\nu < \omega_1$, and for $\nu < \xi < \omega_1$ we have $\sigma = \sup_n \beta_n \le a_\xi < \sigma + 1$, i.e., $a_\xi = \sigma$, as we claimed.

44. As

$$Z_f(\alpha, n) = \bigcup \{Z_f(\beta, n) : f(\beta, \alpha) \le n\},$$

induction on α proves the claim.

45. If $\alpha < \omega_1$ is enumerated as $\alpha = \{\gamma_n(\alpha) : n < \omega\}$, then let $g(\gamma_n(\alpha), \alpha) = n$. Clearly, this has the property mentioned in Problem 44 for f. We know that $Z_g(\alpha, m)$ is always finite. Now for $\beta = \gamma_m(\alpha) < \alpha$ let

$$f(\beta, \alpha) = \max\Big\{m, g(\beta, \alpha), \big|Z_g(\alpha, m)\big|\Big\}.$$

We claim that this satisfies the requirements. It is clear that every $\{\beta : \beta < \alpha,\ f(\beta, \alpha) \le n\}$ is finite. To show the second property, assume to the contrary that $\alpha_0 < \alpha_1 < \cdots$ and for some n it is always the case that $f(\alpha_k, \alpha_{k+1}) \le n$. Then $g(\alpha_k, \alpha_{k+1}) \le n$, hence $\alpha_i \in Z_g(\alpha_j, n)$ for $i < j < \omega$, and also this latter set has at most n elements, which is a contradiction if $j > n$.

46. **(a)** For every $\alpha < \omega_1$ fix an enumeration $\alpha = \{\gamma_n(\alpha) : n < \omega\}$. If $\alpha_0 \le \alpha_1 \le \cdots$ are the numbers selected by I then for the ith one, let II respond by the set

$$S_i = \{\gamma_j(\alpha_k) : j, k \le i\}.$$

This is clearly a winning strategy for player II.

(b) Let $f : \omega_1 \times \omega_1 \to \omega$ be a function as in Problem 45. Our strategy σ is the following. If $\alpha_{i-1} = \alpha_i$ then let

$$\sigma(i, \alpha_i, \alpha_i) = \{\beta < \alpha_i : f(\beta, \alpha_i) \le i\}.$$

If, however, $\alpha_{i-1} < \alpha_i$, say $f(\alpha_{i-1}, \alpha_i) = m$, then set $\sigma(i, \alpha_{i-1}, \alpha_i) = Z_f(\alpha_i, m)$. We show that $\cup_i S_i = \sup_i \alpha_i$, so this strategy is a win for II. This is clearly the case if $\alpha_i = \alpha_{i+1} = \cdots$ for some i. Assume now the contrary. Then there are $\alpha_{i_0} < \alpha_{i_1} < \cdots$ such that $\alpha_j = \alpha_{i_k}$ for $i_k \le j < i_{k+1}$. Let $\xi < \alpha_{i_r}$, say $n = f(\xi, \alpha_{i_r})$. Let $k \ge r$ be least number with the property $n \le f(\alpha_{i_{k-1}}, \alpha_{i_k})$ (such a k exists by the selection of f), and set

$m = f(\alpha_{i_{k-1}}, \alpha_{i_k}) = f(\alpha_{i_k - 1}, \alpha_{i_k})$. Then $\xi \in Z_f(\alpha_{i_k}, m) \subseteq \sigma(k, \alpha_{i_k-1}, \alpha_{i_k})$, and we are done.

47. **(a)** For every $\alpha < \omega_1$ fix an enumeration $\alpha + 1 = \{\gamma_n(\alpha) \ : \ n < \omega\}$. Let $\{\alpha_0, \alpha_1, \ldots\}$ be the sequence of the ordinals selected by I. In her $2^i(2n+1)$th step, let II choose $\gamma_n(\alpha_i)$. Then II selects exactly the numbers below $\sup_n(\alpha_n + 1)$ and since I's selections are also there, II wins.

(b) Now a strategy is a function $f : [\omega_1]^{<\omega} \to \omega_1$ from the finite subsets of ω_1 into ω_1. Let I select first an $\alpha < \omega_1$, and then always 0. Set $A_0 = \omega_1$. There are two possibilities: either there are a $\tau_0 < \omega_1$ and uncountably many $\alpha \in A_0$ such that with $\gamma_\alpha^0 := f(\{\alpha\})$ we have $\gamma_\alpha^0 = \tau_0$, or else $\gamma_\alpha^0 \to \omega_1$ as $\alpha \to \omega_1$ (which means that for every $\gamma < \omega_1$ there is a $\theta < \omega_1$ such that we have $\gamma_\alpha^0 > \gamma$ if $\alpha > \theta$). In the first case let A_1 be the set of those α with $\gamma_\alpha^0 = \tau_0$, while in the second case set $A_1 = A_0$ and $\tau_0 = -1$. Consider now the values $\gamma_\alpha^1 = f(\{\alpha, \gamma_\alpha^0, 0\})$ for $\alpha \in A_1$, for which there are again two possibilities: either there are a $\tau_1 < \omega_1$ and uncountably many $\alpha \in A_1$ such that $\gamma_\alpha^1 = \tau_1$, or else $\gamma_\alpha^1 \to \omega_1$ as $\alpha \to \omega_1$, $\alpha \in A_1$. In the first case let A_2 be the set of those α with $\gamma_\alpha^1 = \tau_1$, while in the second case set $A_2 = A_1$ and $\tau_1 = -1$. We proceed the same way with the values $\gamma_\alpha^2 = f(\{\alpha, \gamma_\alpha^0, \gamma_\alpha^1, 0\})$, $\alpha \in A_2$, etc., indefinitely.

Let $\gamma < \omega_1$ be bigger than all the values τ_n, $n < \omega$. For this γ for every n there is a θ_n such that if $\tau_n = -1$ (i.e., when $\gamma_\alpha^n \to \omega_1$ as $\alpha \to \omega_1$, $\alpha \in A_n$) and $\alpha > \theta_n$, then $\gamma_\alpha^n > \gamma$. Now if $\alpha > \gamma$ is bigger than all the θ_n, then the selected set $\{\alpha, 0, \gamma_\alpha^0, \gamma_\alpha^1, \gamma_\alpha^2, \ldots\}$ is not an initial segment since α is, but $\gamma < \alpha$ is not there (each γ_α^j is either $\tau_j < \gamma$ or bigger than γ).

(c) If II can select finitely many ordinals at any step, then she can do the following. At some step she sees a set H consisting of, say, n ordinals. Then she pretends that she plays the game in part (a) with the slight modification that she never selects already selected ordinals. Then she is at step at least $n/2$ and at most $n+1$, and there are only finitely many ways/orders how the set H could have been created in that many steps by the two players in game (a). For each such order let II select her choice from game (a), and her response for H be the set of all these finitely many elements. Since the strategy in part (a) produces an initial segment, eventually the set of the selected ordinals will be the union of initial segments, hence itself is an initial segment.

An alternative formalized strategy is as follows. For every $\alpha < \omega_1$ fix an enumeration $\alpha + 1 = \{\gamma_n(\alpha) \ : \ n < \omega\}$, and if II sees $H = \{\alpha_0, \ldots, \alpha_{n-1}\}$ then let her response be $(n+1) \cup \{\gamma_m(\alpha_i) \ : \ i, m \le n\}$. Now if I or II chooses α in the kth step and $\beta = \gamma_m(\alpha) < \alpha$ then II will choose β in her $\max(k, m)$th step (at the latest) so eventually α will be filled up.

* * *

48. Without loss of generality, let $K_0 = \kappa$, and for each n let $\{\xi_\alpha^{(n)}\}_{\alpha<\kappa}$ be an increasing enumeration of the elements of K_n. Now the strategy of the second player be that given K_n (n even) he keeps only the elements $\xi_\alpha^{(n)}$ with a successor ordinal α, i.e., he selects

$$K_{n+1} = \{\xi_\alpha^{(n)} \ : \ \alpha < \kappa \text{ is a successor ordinal}\}.$$

Note that then the index of an element in K_{n+1} is decremented by at least one, i.e., if $\xi \in K_n \cap K_{n+1}$ is a common element, $\xi = \xi_\alpha^{(n)}$ and $\xi = \xi_\beta^{(n+1)}$, then $\alpha > \beta$. Furthermore, no matter how the first player selects K_{n+2} in the next step, the index of an element is never incremented (see Problem 6.39). Now it is clear that $\cap_n K_n$ is empty, for if $\xi \in \cap K_n$ was for all n, then we would have $\xi = \xi_{\alpha_{2n}}^{(2n)}$ for some ordinals $\alpha_{2n} < \kappa$, and then these ordinals would form a strictly decreasing sequence, which is not possible by Problem 2.

9

Ordinal arithmetic

1. First we show the claim for two ordinals α and β. We shall repeatedly use the fact that if $\alpha = \gamma + \delta$, and α and γ are divisible from the left by τ, then δ is also divisible from the left by τ (the fact that if γ and δ are divisible from the left by τ, then α is also divisible from the left by τ is clear). If fact, write $\alpha = \tau \cdot \alpha_1$, $\gamma = \tau \cdot \gamma_1$, and $\delta = \tau \cdot \delta_1 + \delta_2$ with some $\delta_2 < \tau$. Then

$$\tau \cdot \alpha_1 = \alpha = \gamma + \delta = \tau \cdot \gamma_1 + \tau \cdot \delta_1 + \delta_2 = \tau \cdot (\gamma_1 + \delta_1) + \delta_2,$$

which, in view of the unicity of the representation in Problem 8.15, yields $\delta_2 = 0$ as we claimed.

Now let $\beta < \alpha$ and let δ be a common left divisor of these two ordinals. Based on Problem 8.15 we can carry out the Euclidean algorithm: we write $\alpha = \beta \cdot \gamma_1 + \beta_1$, $\beta_1 < \beta$. By what we have proven above, here β_1 is divisible from the left by δ. Now write $\beta = \beta_1 \cdot \gamma_2 + \beta_2$, $\beta_2 < \beta_1$, and again here β_2 is divisible from the left by δ. Continuing this process, we have to arrive to a β_{n+1} which is zero (recall that there is no infinite decreasing sequence of ordinals), and then the process terminates. Then δ is a left divisor of β_n. Conversely, since $\beta_{n-1} = \beta_n \cdot \gamma_n$, we get that β_n is a left divisor of β_{n-1}. Then, since $\beta_{n-2} = \beta_{n-1} \cdot \gamma_n + \beta_n$, we get that β_n is a left divisor of β_{n-2}, etc. Eventually we obtain that β_n is a common left divisor of α and β. All these mean that β_n is the greatest common left divisors of α and β, and since any common left divisor δ of α and β divides β_n, the claim has been verified for two ordinals.

After this, let A be an arbitrary set of nonzero ordinals. Let α_0, α_1 be two ordinals from A, and let δ_1 be their greatest common left divisor. If δ_1 divides every element of A, then we are done, δ_1 is the greatest common left divisor of the elements of A. If this is not the case, then there is an element in A, which we denote by α_2, which is not divisible from the left by δ_1. Thus, if δ_2 is the greatest common left divisor of δ_1 and α_2, then $\delta_2 < \delta_1$, and clearly it is the greatest common left divisor of the ordinals $\alpha_0, \alpha_1, \alpha_2$. If this δ_2 divides every element of A from the left, then we are done, otherwise let α_3 be an element of A not divisible from the left by δ_2, etc. Continuing this, the process has to terminate since $\{\delta_k\}$ is a decreasing sequence of ordinals, and if it terminates with δ_n, then δ_n is the greatest common left divisor of the elements of A.

2. See Problem 8.19.

3. It is clear that $(\omega+2)\cdot\omega = (\omega+3)\cdot\omega = \omega^2$, so the condition is sufficient. Conversely, suppose that α is not divisible by ω^2 from the left. Then it is of the form $\alpha = \omega^2\cdot\alpha_1 + \omega\cdot k_1 + k_2$ with $k_1 \neq 0$ or $k_2 \neq 0$ (see Problem 8.15). Thus, $\omega\cdot k_1 + k_2$ is divisible from the left by $\omega+2$ and by $\omega+3$, therefore $k_1 \geq 1$. Since $\omega\cdot k_1 + k_2 = (\omega+2)\cdot(k_1-1)+\omega+k_2$, it follows that $\omega+k_2$ is divisible by $\omega+2$, which is the case only if $k_2 = 2$. In a similar fashion from the divisibility by $(\omega+3)$ it would follow that $k_2 = 3$, and this is a contradiction. [W. Sierpiński, Cardinal and Ordinal Numbers, Polish Sci. Publ., Warszawa, 1965, XIV.12/4]

4. See Problem 7. [W. Sierpiński, Cardinal and Ordinal Numbers, Polish Sci. Publ., Warszawa, 1965, XIV.12/6]

5. See Problem 7. [W. Sierpiński, Cardinal and Ordinal Numbers, Polish Sci. Publ., Warszawa, 1965, XIV.12/2]

6. See Problem 8.27. If α is a successor ordinal then it has only finitely many left divisors by 8.28. [cf. W. Sierpiński, Cardinal and Ordinal Numbers, Polish Sci. Publ., Warszawa, 1965, XIV.11. Theorem 2]

7. We shall repeatedly use Problem 8.21.

Suppose that α and β are right divisors of $\gamma \geq 1$, say $\xi_0\cdot\alpha = \eta_0\cdot\beta$. In this equation we can divide through with the greatest common left divisors of ξ_0 and η_0, so we may assume that they do not have a common left divisor bigger than 1. Hence, if we write $\xi_0 = \omega\cdot\xi_1 + m_0$ and $\eta_0 = \omega\cdot\eta_1 + n_0$, then one of m_0 or n_0 is not zero.

First we show that *if* $\xi_1 \neq 0$ *and* $\eta_1 \neq 0$, *then* $\xi_1\cdot\alpha = \eta_1\cdot\beta$ *also holds, and here either* $\xi_1 < \xi_0$ *or* $\eta_1 < \eta_0$. In fact, if $m_0 \neq 0$ and α is a successor ordinal, then we must have $n_0 \neq 0$ and β must also be a successor ordinal. Hence by Problem 8.21 we have

$$\omega\cdot\xi_1\cdot\alpha + m_0 = \xi_0\cdot\alpha = \eta_0\cdot\beta = \omega\cdot\eta_1\cdot\beta + n_0,$$

and this implies first $m_0 = n_0$, then $\omega\cdot\xi_1\cdot\alpha = \omega\cdot\eta_1\cdot\beta$, and then $\xi_1\cdot\alpha = \eta_1\cdot\beta$. If, however, $m_0 \neq 0$ and α is a limit ordinal, then either $n_0 = 0$ or β must be a limit ordinal, and in each case

$$\omega\cdot\xi_1\cdot\alpha = \xi_0\cdot\alpha = \eta_0\cdot\beta = \omega\cdot\eta_1\cdot\beta,$$

which gives again $\xi_1 \cdot \alpha = \eta_1 \cdot \beta$. The same argument works if $n_0 \neq 0$. Since $\xi_0 = \omega \cdot \xi_1 + m_0$ and $\eta_0 = \omega \cdot \eta_1 + n_0$, we get $\xi_1 < \xi_0$ if $m_0 \neq 0$ and $\eta_1 < \eta_0$ if $n_0 \neq 0$.

Continuing this process we get ordinals ξ_k, η_k such that $\xi_k \cdot \alpha = \eta_k \cdot \beta$, and either $\xi_k < \xi_{k-1}$ or $\eta_k < \eta_{k-1}$, and this process terminates only if one of ξ_k or η_k is finite (in the italicized assertion above the assumption was that $\xi_1 \neq 0$ and $\eta_1 \neq 0$). But there is no infinite decreasing sequence of ordinals, so the process must terminate, and we get to a first stage when one of ξ_k or η_k is finite; suppose, for example, that $\xi_k = m > 0$. Thus, we have $m \cdot \alpha = \rho \cdot \beta$ with some ordinal ρ. We write $\alpha = \omega \cdot \alpha_1 + p$ and $\rho = \omega \cdot \rho_1 + k$.

Thus, we have $\omega \cdot \alpha_1 + pm = (\omega \cdot \rho_1 + k) \cdot \beta$. First we consider the case when $\rho_1 \neq 0$. If $p = 0$, then $\alpha = m \cdot \alpha = \rho \cdot \beta$, so β is a right divisor of α. If $p \neq 0$, then β must be a successor ordinal, and so we have $\omega \cdot \alpha_1 + pm = \omega \cdot \rho_1 \cdot \beta + k$, which implies $k = pm$, $\rho = \omega \cdot \rho_1 + k = \omega \cdot \rho_1 + pm = m \cdot (\omega \cdot \rho_1 + p)$, and hence $m \cdot \alpha = m \cdot (\omega \cdot \rho_1 + p) \cdot \beta$, which, upon dividing by m from the left, yields again that β is a right divisor of α.

It has only left to consider the case when $\rho_1 = 0$. In this case $m \cdot \alpha = k \cdot \beta$, and we can divide again from the left by the greatest common divisor of m and k, so we may assume that m and k are relative primes. If we write $\beta = \omega \cdot \beta_1 + q$, then the equation $m \cdot \alpha = k \cdot \beta$ takes the form $\omega \cdot \alpha_1 + pm = \omega \cdot \beta_1 + qk$, so $pm = qk$, and $\alpha_1 = \beta_1$. Thus, in this case $\alpha = \xi + p$, $\beta = \xi + q$, where $\xi = \omega \cdot \alpha_1$ is a limit ordinal or 0. If $p = 0$, then we must have $q = 0$, i.e., $\beta = \alpha$. If, however, $p \neq 0$, then $q \neq 0$, and $pm = qk$ is a common multiple of p and q. Thus, if $[p, q]$ denotes their least common multiple, then $\xi + [p, q]$ also divides $m \cdot \alpha = \xi + pm$ from the right: $m \cdot \alpha = (pm/[p, q]) \cdot (\xi + [p, q])$. Thus, $\xi_k \cdot \alpha$ is divisible from the right by $\xi + [p, q]$, say $\xi_k \cdot \alpha = \theta_k \cdot (\xi + [p, q])$. Now

$$\xi_{k-1} \cdot \alpha = \omega \cdot \xi_k \cdot \alpha + m_{k-1} = \omega \cdot \theta_k \cdot (\xi + [p, q]) + m_{k-1} = (\omega \cdot \theta_k + m_{k-1}) \cdot (\xi + [p, q]),$$

i.e., $\xi_{k-1} \cdot \alpha$ is also divisible from the right by $\xi + [p, q]$. Now going back in a similar fashion on the sequence ξ_s, $s = k, k-1, \ldots, 0$ we can see that each $\xi_s \cdot \alpha$ is divisible from the right by $\xi + [p, q]$, and for $s = 0$ this gives that γ is divisible from the right by $\xi + [p, q]$.

8. Let A be a set of positive ordinals and let $\alpha \in A$ be any element of A. α has finitely many right divisors (Problem 6), so there is a largest one δ among them that divides all ordinals in A. By Problem 7 any common right divisor of the ordinals in A divides this δ from the right.

9. If $\alpha > 1$ is any ordinal and κ is an infinite cardinal bigger than the cardinality of α, then $\alpha \cdot \kappa = \kappa$. Thus, if A is any set of ordinals and κ is an infinite cardinal bigger than all the elements in A, then this κ is a common right multiple of the ordinals in A. Thus, the ordinals in A have a smallest common right multiple σ. Suppose that $\gamma > \sigma$ is any common right multiple, and let us write γ in the form $\gamma = \sigma \cdot \xi + \eta$ with $\eta < \sigma$. Then any element of A divides both γ and σ from the left, hence, by the beginning of the proof of

Problem 1, it also divides η. Since $\eta < \sigma$, this can only happen if $\eta = 0$, so σ divides γ from the left.

10. By Problem 7 the ordinals 2 and $\omega+1$ do not have a common left multiple (note that 2 does not divide $\omega + 1$ from the right).

11. **(i)** $\gamma^\alpha \cdot \gamma^\beta = \gamma^{\alpha+\beta}$ is true for $\beta = 0$, and from here one can proceed by transfinite induction on β. Thus, suppose that $\gamma^\alpha \cdot \gamma^\delta = \gamma^{\alpha+\delta}$ is true for all $\delta < \beta$. If β is a successor ordinal, say $\beta = \delta + 1$, then

$$\gamma^\alpha \cdot \gamma^\beta = \gamma^\alpha \cdot \gamma^{\delta+1} = \gamma^\alpha \cdot \gamma^\delta \cdot \gamma = \gamma^{\alpha+\delta} \cdot \gamma = \gamma^{\alpha+\delta+1} = \gamma^{\alpha+\beta}.$$

If, however, β is a limit ordinal, then we can apply Problem 8.13 to write

$$\gamma^\alpha \cdot \gamma^\beta = \gamma^\alpha \cdot \sup_{\delta<\beta} \gamma^\delta = \sup_{\delta<\beta}(\gamma^\alpha \cdot \gamma^\delta) = \sup_{\delta<\beta} \gamma^{\alpha+\delta} = \sup_{\theta<\alpha+\beta} \gamma^\theta = \gamma^{\alpha+\beta},$$

where at the last but one equality we used the monotonicity of ordinal exponentiation to be proven in part (iii) below.

(ii) $(\gamma^\alpha)^\beta = \gamma^{\alpha\cdot\beta}$ is true for $\beta = 0$, and for general β one can use transfinite induction just as in case (i), during which one uses part (i), as well.

(iii) The definition shows that if $\alpha \le \beta$ then $\gamma^\alpha \le \gamma^\beta$. Thus, if $\alpha < \beta$, then (cf. Problem 8.9) $\gamma^\alpha < \gamma^{\alpha+1} \le \gamma^\beta$.

(iv) Using (iii), the inequality $\alpha \le \gamma^\alpha$ can again be easily proven by transfinite induction.

12. We prove by transfinite induction on α that $\Phi_{\alpha,\gamma}$ is well ordered and is of order type γ^α . Since $\Phi_{0,\gamma} = \{\emptyset\}$, the statement is true for $\alpha = 0$. Now suppose we know that $\Phi_{\beta,\gamma}$ is well ordered and its order type is γ^β for all $\beta < \alpha$, and first let us consider the case when $\alpha = \beta+1$ is a successor ordinal. For $\xi < \gamma$ let $H_\xi = \{f \in \Phi_{\alpha,\gamma} \ : \ f(\beta) = \xi\}$. This H_ξ is clearly similar to $\Phi_{\beta,\gamma}$, and $\Phi_{\alpha,\gamma}$ is the ordered union of the H_ξ's with respect to $\xi < \gamma$, thus in this case we get that the type of $\Phi_{\alpha,\gamma}$ equals the type of $\Phi_{\beta,\gamma}$ times γ, i.e., by the induction hypothesis the type is $\gamma^\beta \cdot \gamma = \gamma^{\beta+1} = \gamma^\alpha$.

For $\beta < \alpha$ we can think an $f : \beta \to \gamma$ to be extended to an $f : \alpha \to \gamma$ by setting $f(\xi) = 0$ for $\xi \in \alpha \setminus \beta$. In this sense if α is a limit ordinal, then $\Phi_{\alpha,\gamma}$ is the increasing union of the family $\{\Phi_{\beta,\gamma}\}_{\beta<\alpha}$, each $\Phi_{\beta,\gamma}$ being an initial segment of $\Phi_{\alpha,\gamma}$, which also implies that each proper initial segment of $\Phi_{\alpha,\gamma}$ is a subset of one of the $\Phi_{\beta,\gamma}$, $\beta < \gamma$ (this is where we use the finiteness of the supports of the functions in $\Phi_{\alpha,\beta}$). Thus, by the induction hypothesis $\Phi_{\alpha,\beta}$ is well ordered, and its order type is the supremum of the order types of $\Phi_{\beta,\gamma}$, $\beta < \alpha$, i.e., it is $\sup_{\beta<\alpha} \gamma^\beta = \gamma^\alpha$, and this is what we had to prove. [W. Sierpiński, Cardinal and Ordinal Numbers, Polish Sci. Publ., Warszawa, 1965, XIV.15]

13. **a)** Since $n^\omega = \omega$, we have $n^{\omega^k} = (n^\omega)^{\omega^{k-1}} = \omega^{\omega^{k-1}}$, the equality $n^{\omega^\omega} = \omega^{\omega^\omega}$ follows by taking the supremum of both sides for $k = 1, 2, \ldots$.

b) Since

$$(\omega+n)^m = \omega^m + \omega^{m-1}\cdot n + \cdots + \omega \cdot n + n,$$

we have $\omega^m \le (\omega+n)^m < \omega^{m+1}$. Now taking here supremum for $m = 1, 2, \ldots$ we obtain $(\omega+n)^\omega = \omega^\omega$.

14. $1^\alpha = 1$, and if $\alpha = \omega\cdot\beta$, then $2^\alpha = (2^\omega)^\beta = \omega^\beta$ and similarly $3^\alpha = \omega^\beta$. Whence $1 + \omega^\beta = \omega^\beta$.

15. **a)** $2^\omega = \sup_{n<o} 2^n = \omega$.

b) if α is countable, then so is 2^α by the definition of ordinal exponent.ation and by the fact that the supremum of countably many countable ordinals is countable.

c) By part d) and part (iv) of Problem 11 we have $\kappa \le 2^\kappa \le \kappa$, so $2^\kappa = \kappa$.

d) Using that $2^\omega = \omega$, it can be easily verified by transfinite induction on the infinite ordinal α that 2^α has cardinality at most $|\alpha|$. Now this together with part (iv) of Problem 11 shows that, in fact, 2^α has cardinality equal to $|\alpha|$.

e) Let α be an arbitrary ordinal, and let 2^{ζ_0} be the largest power of 2 that is not bigger than α. Then $\alpha < 2^{\zeta_0+1} = 2^{\zeta_0}\cdot 2$, hence if we write $\alpha = 2^{\zeta_0} + \alpha_1$, then $\alpha_1 < 2^{\zeta_0} \le \alpha$. Now repeat this process with α_1 to get a ζ_1 and an α_2 such that $\alpha_1 = 2^{\zeta_1} + \alpha_2$, and $\alpha_2 < 2^{\zeta_1} \le \alpha_1$, then repeat again and again. Since the α_n's are decreasing, this process has to terminate in finitely many steps, in which case we must have arrived at 0. Thus, $\alpha = 2^{\zeta_0} + \cdots + 2^{\zeta_k}$ with some k, and here $\zeta_0 > \zeta_1 \ldots > \zeta_k$, and this is just the form as in part e).

To establish the unicity, downward induction on $l = k, k-1, \ldots, 1$ shows that if $\zeta_0 > \zeta_1 > \cdots > \zeta_k$, then

$$2^{\zeta_l} + \cdots + 2^{\zeta_k} < 2^{\zeta_{l-1}},$$

the induction step being

$$2^{\zeta_l} + \cdots + 2^{\zeta_k} < 2^{\zeta_l} + 2^{\zeta_l} = 2^{\zeta_l+1} \le 2^{\zeta_{l-1}}.$$

The case $l = k$ gives $2^{\zeta_0} \le \alpha < 2^{\zeta_0} + 2^{\zeta_0} = 2^{\zeta_0+1}$, and so 2^{ζ_0} is the largest power of 2 that is not bigger than α. So if we have two representations, this largest power has to be the same in both. Now cancel this highest power, and repeat the same process to prove that actually, all powers have to coincide in the two representations.

The form (9.1) of the ordinal $\omega^4\cdot 6 + \omega^2\cdot 7 + \omega + 9$ is

$$2^{\omega\cdot 4+2} + 2^{\omega\cdot 4+1} + 2^{\omega\cdot 2+2} + 2^{\omega\cdot 2+1} + 2^{\omega\cdot 2} + 2^\omega + 2^3 + 2^0.$$

16. We proceed as in the preceding solution. Let α be any ordinal, and let γ^{ζ_0} be the largest power of γ that is not bigger than α. Then $\alpha < \gamma^{\zeta_0+1} = \gamma^{\zeta_0}\cdot\gamma$,

hence if we write $\alpha = \gamma^{\zeta_0} \cdot \theta_0 + \alpha_1$ with $\alpha_1 < \gamma^{\zeta_0}$, then we must have $\theta_0 < \gamma$. Now repeat this process with α_1 to get a ζ_1, θ_1 and an α_2 such that $\alpha_1 = \gamma^{\zeta_1} \cdot \theta_1 + \alpha_2$, and $\alpha_2 < \gamma^{\zeta_1} \le \alpha_1$, $\theta_1 < \gamma$, etc. Since the α_n's are decreasing, this process has to terminate in finitely many steps, in which case we must have arrived at 0. Thus, $\alpha = \gamma^{\zeta_0} \cdot \theta_0 + \cdots + \gamma^{\zeta_k} \cdot \theta_k$ with some k, and here $\zeta_0 > \zeta_1 > \cdots > \zeta_k$, so the existence of the representation in base γ has been established.

To verify the unicity, let $\alpha = \gamma^{\zeta_0} \cdot \theta_0 + \cdots + \gamma^{\zeta_k} \cdot \theta_k$ with $\zeta_0 > \zeta_1 \ldots > \zeta_k$ and $\theta_i < \gamma$. Then $\gamma^{\zeta_k} \cdot \theta_k < \gamma^{\zeta_k} \cdot \gamma \le \gamma^{\zeta_{k-1}}$, and so

$$\gamma^{\zeta_{k-1}} \cdot \theta_{k-1} + \gamma^{\zeta_k} \cdot \theta_k < \gamma^{\zeta_{k-1}} \cdot (\theta_{k-1} + 1) \le \gamma^{\zeta_{k-2}},$$

and continuing this process we can see that $\alpha < \gamma^{\zeta_0} \cdot (\theta_0 + 1) \le \gamma^{\zeta_0 + 1}$. On the other hand, $\alpha \ge \gamma^{\zeta_0}$, thus γ^{ζ_0} is the largest power of γ that is not bigger than α, and then θ_0 is the largest ordinal such that $\gamma^{\zeta_0} \cdot \theta_0 \le \alpha$. Thus, in any two representations in base γ the main terms are the same, and then we can cancel these main terms from both representations. Continuing the same process with the next-highest term, we get eventually that all terms in the two representations are the same.

17. Let (9.2) be the normal form of α. The inequality $\alpha < \omega^{\xi_n + 1}$ has been proven in the preceding proof. Now if $\omega^{\xi_n + 1} \le \beta$, then we can write $\beta = \omega^{\xi_n + 1} + \eta$, and since we have

$$\omega^{\xi_k} \cdot a_k + \omega^{\xi_n + 1} \le \omega^{\xi_n} \cdot a_k + \omega^{\xi_n} \cdot \omega = \omega^{\xi_n}(a_k + \omega) = \omega^{\xi_n + 1},$$

for all k, we obtain

$$\alpha + \beta = \alpha + \omega^{\xi_n + 1} + \eta = \omega^{\xi_n + 1} + \eta = \beta,$$

as was claimed.

18. We shall repeatedly use Problem 17.

Let α have normal form (9.2) and let β have normal form

$$\beta = \omega^{\zeta_m} \cdot b_m + \cdots + \omega^{\zeta_0} \cdot b_0. \tag{9.1}$$

If $\zeta_m > \xi_n$, then $\alpha + \beta = \beta$. If this is not the case, and there is a k such that $\xi_k = \zeta_m$, then

$$\alpha + \beta = \omega^{\xi_n} \cdot a_n + \cdots + \omega^{\xi_k} \cdot (a_k + b_m) + \omega^{\zeta_{m-1}} b_{m-1} + \cdots + \omega^{\zeta_0} \cdot b_0. \tag{9.2}$$

The representation is similar if there is no ξ_k that equals ζ_m, namely just add then $\omega^{\zeta_m} \cdot 0$ to the representation of α (i.e., consider as if the term ω^{ζ_m} was there with 0 coefficient).

Since $\alpha \cdot \omega = \omega^{\xi_n + 1}$, it follows that if $\zeta_0 > 0$, then

$$\alpha \cdot \beta = \omega^{\xi_n} \cdot \beta = \omega^{\xi_n + \zeta_m} \cdot b_m + \cdots + \omega^{\xi_n + \zeta_0} \cdot b_0. \tag{9.3}$$

If, however, $\zeta_0 = 0$, and we write $\beta = \beta' + b_0$, then $\alpha \cdot \beta = \alpha \cdot \beta' + \alpha \cdot b_0$, and hence according to what we have just said

$$\alpha \cdot \beta = \omega^{\xi_n} \cdot \beta + \alpha \cdot b_0 = \omega^{\xi_n + \zeta_m} \cdot b_m + \cdots + \omega^{\xi_n + \zeta_1} \cdot b_1 \\ + \omega^{\xi_n} \cdot (a_n b_0) + \omega^{\xi_{n-1}} \cdot a_{n-1} + \omega^{\xi_{n-2}} \cdot a_{n-2} \cdots + \omega^{\xi_0} \cdot a_0. \quad (9.4)$$

[W. Sierpiński, Cardinal and Ordinal Numbers, Polish Sci. Publ., Warszawa, 1965, XIV.19/4]

19. The limit case is an immediate consequence of the representation (9.3). In a similar manner, one can obtain the claim for successor ordinals from repeated application of (9.4).

20. By considering the terms in the normal form of α separately, it suffices to show that if $\gamma < \omega^\xi$, then γ is a left divisor of ω^ξ. If the highest power in the normal form of γ is ω^δ, then $\delta < \xi$, i.e., $\xi = \delta + 1 + \zeta$ with some ordinal ζ. But then $\gamma \cdot \omega \cdot \omega^\zeta = \omega^{\delta+1} \cdot \omega^\zeta = \omega^\xi$, i.e., γ is, indeed, a left divisor of ω^ξ.

Conversely, if $\alpha = \gamma \cdot \beta$, β has normal form (9.1) and β is a limit ordinal, then, by (9.3), $\gamma < \omega^{\xi_0}$. On the other hand, if β is a successor ordinal, then $\beta = \beta^* + 1$ with some ordinal β^*. Then $\alpha = \gamma \cdot \beta^* + \gamma$, and we know that for each α there are only finitely many ordinals σ such that with some ρ we have $\alpha = \rho + \sigma$ (Problem 8.26). Thus, there are only finitely many possibilities for γ.

21. It easily follows from the normal form for the sums of ordinals that if $\alpha = \beta \cdot k$ and the highest power of ω in the normal form of β is ω^{ζ_m} with coefficient b_m, then α has the same normal form as β, except that the coefficient of ω^{ζ_m} in its normal form is $b_m k$. Thus, the answer to the problem is that k is a divisor of the coefficient of the highest power in the normal form of α (with the notation (9.2) this amounts the same as k is a divisor of a_n).

22. The finite case has been considered in Problem 8.25, i.e., for finite α the sum in question is $\omega^{2\alpha-1}$. Since the sum $\sum_{\xi<\omega^\alpha} \xi$ is the order type of a set $\langle A, \prec \rangle$ that is the ordered union of ordered sets of type ξ, $\xi < \omega^\alpha$, it follows that for limit ordinal α this sum is the same as the supremum of the sums $\sum_{\xi<\omega^\beta} \xi$ for all $\beta < \alpha$ (since $\langle A, \prec \rangle$ is the union for all $\beta < \alpha$ of its initial segments that are the ordered unions of sets of type ξ, $\xi < \omega^\beta$). Thus, we have $\sum_{\xi<\omega^\omega} \xi = \omega^\omega$.

Next let $\alpha > \omega$ be a successor ordinal. Then it can be written in the form $\lambda + (k+1)$ with some limit ordinal λ and with some natural number k. It is clear that

$$\sum_{\xi<\omega^\alpha} \xi \le \sum_{\xi<\omega^\alpha} \omega^\alpha = \omega^\alpha \cdot \omega^\alpha = \omega^{\alpha \cdot 2}.$$

On the other hand, the set $\{\xi \ : \ \omega^\lambda < \xi < \omega^\alpha\}$ has order type ω^α, hence

$$\sum_{\xi<\omega^\alpha} \xi \geq \sum_{\omega^\lambda<\xi<\omega^\alpha} \omega^\lambda = \omega^\lambda \cdot \omega^\alpha = \omega^{\lambda+\alpha} = \omega^{\alpha\cdot 2}.$$

Thus, if α is a successor ordinal, then the sum in question is $\omega^{\alpha\cdot 2}$.

Next, if $\alpha > \omega$ is a limit ordinal, then according to what we have said before, $\sum_{\xi<\omega^\alpha} \xi = \sup_{\beta<\alpha} \sum_{\xi<\omega^\beta} \xi$, and here in the supremum we can take the supremum for successor ordinals β smaller than α. Thus, according to what we have just proved, in this case

$$\sum_{\xi<\omega^\alpha} \xi = \sup_{\beta<\alpha} \omega^{\beta\cdot 2} = \omega^\sigma,$$

where $\sigma = \sup_{\beta<\alpha} \beta \cdot 2$. Here if α equals one of the powers of ω, say $\alpha = \omega^\tau$, then

$$\omega^\tau = \sup_{\beta<\alpha} \beta \leq \sup_{\beta<\alpha} \beta \cdot 2 \leq \sup_{\beta<\alpha} \beta \cdot \omega = \omega^\tau,$$

i.e., then $\sigma = \alpha$. If, however, α is not a power of ω, then there are at least two-terms in its normal representation (9.2), and since α is a limit ordinal, we have $\xi_0 > 0$. Thus, then α is the supremum of the ordinals

$$\beta = \omega^{\xi_n} \cdot a_n + \cdots + \omega^{\xi_0} \cdot (a_0 - 1) + \delta,$$

where $\delta < \omega^{\xi_0}$, and here

$$\beta \cdot 2 = \omega^{\xi_n} \cdot (2a_n) + \omega^{\xi_{n-1}} \cdot a_{n-1} + \cdots + \omega^{\xi_0} \cdot (a_0 - 1) + \delta,$$

thus

$$\sigma = \sup_{\beta<\alpha} \beta \cdot 2 = \omega^{\xi_n} \cdot (2a_n) + \cdots + \omega^{\xi_0} \cdot a_0 = \alpha \cdot 2.$$

In summary, the sum in question is equal to $\omega^{2\alpha-1}$ if α is finite, it equals ω^α if α is a power of ω, and in all other cases it equals $\omega^{\alpha\cdot 2}$.

23. We prove the statement by transfinite induction on α, the case $\alpha = 0$ being trivial. Thus, suppose that the claim is true for all ordinals less than α; and we have to show that it is also true for α.

If α is a successor ordinal, $\alpha = \beta + 1$, then ω^α is the order type of the antilexicographically ordered set $\omega^\beta \times \omega$, and suppose that we have decomposed $\omega^\beta \times \omega$ as $A \cup B$. For each $n \in \omega$ let A_n be the set $\{\xi \in \omega^\beta : (\xi, n) \in A\}$, and similarly define B_n. Then $A_n \cup B_n = \omega^\beta$, so by the induction hypothesis either A_n or B_n has type ω^β. If for infinitely many n the set A_n has type ω^β, then $\cup_n A_n$ has type $\omega^\beta \cdot \omega = \omega^\alpha$ and then so does $\cup_n A_n \subseteq A \subseteq \omega^\alpha$. If this is not the case, then for infinitely many n the set B_n has type ω^β, and then $\cup_n B_n$ has type $\omega^\beta \cdot \omega = \omega^\alpha$, and together with it the same is true for $\cup_n B_n \subseteq B \subseteq \omega^\alpha$.

Now suppose that α is a limit ordinal, and for each $\beta < \alpha$ let $A_\beta = A \cap \omega^\beta$, $B_\beta = B \cap \omega^\beta$. By the induction assumption either A_β or B_β has order type

ω^β, and let C be the set of those $\beta < \alpha$ for which A_β is of type ω^β. If C is cofinal with α, then clearly the type of A is $\sup_{\beta\in C}\omega^\beta = \omega^\alpha$. If, however, C is not cofinal with α, then $\alpha \setminus C$ is cofinal with α, and then exactly as before, B has order type ω^α.

24. For $\alpha = \omega^\delta$ the statement follows from the previous problem. For other α we prove the result by induction on α, so suppose that it has been verified for all ordinals smaller than α. Let the normal form of α be (9.2), and consider the ordinal

$$\beta = \omega^{\xi_n} \cdot (a_n - 1) + \cdots + \omega^{\xi_0} \cdot a_0.$$

This is smaller than α, and $\alpha = \omega^{\xi_n} \cup B$, where in this ordered union B has type $\beta < \alpha$. Thus, by the induction hypothesis, there is an N such that if we decompose B into N sets, then one of the sets can be omitted, and the union of the remaining ones still has order type β. We claim that then the same is true if we decompose α into $2N$ parts. Thus, let $\alpha = A_1 \cup \cdots \cup A_{2N}$. We have sets $B = \cup_{i=1}^N B \cap (A_{2i-1} \cup A_{2i})$, hence there is an i_0 such that the order type of $B^* = \cup_{1\le i\le N,\ i\ne i_0} B \cap (A_{2i-1} \cup A_{2i})$ is β. If the order type of $\omega^{\xi_n} \cap (A_{2i_0-1} \cup A_{2i_0})$ is smaller than ω^{ξ_n}, then, by the preceding problem, the order type of $\cup_{1\le i\le N,\ i\ne i_0}\omega^{\xi_n} \cap (A_{2i-1} \cup A_{2i})$ is ω^{ξ_n}, and we are done. If, however, the order type of $\omega^{\xi_n} \cap (A_{2i_0-1} \cup A_{2i_0})$ is ω^{ξ_n}, then, again by the preceding problem, either the order type of $\omega^{\xi_n} \cap A_{2i_0-1}$ is ω^{ξ_n}, or the order type of $\omega^{\xi_n} \cap A_{2i_0}$ is ω^{ξ_n}. In the first case the order type of $\cup_{1\le i\le 2N,\ i\ne 2i_0} A_i$, while in the second case the order type of $\cup_{1\le i\le 2N,\ i\ne 2i_0-1} A_i$ is $\omega^{\xi_n} + \beta = \alpha$, and the induction step has been verified.

25. We show by transfinite induction on α the stronger claim that every infinite ordinal α of cardinality at most κ can be decomposed as $\alpha = A_0 \cup A_1 \cup \cdots$ such that the order type of A_n is at most κ^n. For $\alpha = \omega = 1 + 2 + \cdots$ this is clear, and suppose now that this claim has been verified for all infinite ordinals $\beta < \alpha$. If $\alpha = \beta + 1$ is a successor ordinal and the assumed decomposition of β is $B_0 \cup B_1 \cup \cdots$, then $\alpha = A_0 \cup A_1 \cup \cdots$ with $A_0 = \{\beta\}$, $A_{i+1} = B_i$, $i = 0, 1, \ldots$ is clearly an appropriate decomposition of α.

Assume now that $\alpha > 0$ is a limit ordinal, and let $\{\beta_\xi\}_{\xi<\mathrm{cf}\,(\alpha)}$ be an increasing sequence of type $\mathrm{cf}\,(\alpha)$ of ordinals smaller than α converging to α. Then α splits into the disjoint union of the sets $B_\xi = [\beta_\xi, \beta_{\xi+1})$, $\xi < \mathrm{cf}\,(\alpha)$. By the induction hypothesis for each $\xi < \mathrm{cf}\,(\alpha)$ there is a decomposition $B_\xi = B_0^\xi \cup B_1^\xi \cup \cdots$, where the order type of B_n^ξ is at most κ^n for each n. Now set $A_0 = \emptyset$ and $A_n = \cup_{\xi<\mathrm{cf}\,(\alpha)} B_{n-1}^\xi$ for $n = 1, 2, \ldots$. Then $\alpha = A_0 \cup A_1 \cup \cdots$ is a partition, and since $\mathrm{cf}\,(\alpha) \le \kappa$ the order type of A_n is $\le \kappa^{n-1} \cdot \kappa = \kappa^n$, which proves the induction step.

26. ω, ω^2, and ω^3 are the first three infinite indecomposable ordinals (cf. Problem 23).

27. See Problem 34.

28. This is clear from the definition of indecomposability and from the definition of γ. [W. Sierpiński, Cardinal and Ordinal Numbers, Polish Sci. Publ., Warszawa, 1965, XIV.5. Theorem 2]

29. If $\alpha = \gamma \cdot (\beta + 1)$ and here $\beta > 0$, then $\gamma \cdot \beta < \alpha$ and $\gamma < \alpha$, hence $\alpha = \gamma \cdot \beta + \gamma$ is decomposable. Conversely, if α is decomposable, then it is not a power of ω (see Problem 23), hence in its normal form representation (9.2) there are at least two-terms. Thus, if ω^{ξ_0} is the largest power of ω that divides α, then $\alpha = \omega^{\xi_0} \cdot \beta$, and here β is a successor ordinal bigger than 1.

30. For a $\xi < \alpha$ the equality $\xi + \alpha = \alpha$ holds if and only if for all $\eta < \alpha$ we have $\xi + \eta < \alpha$ (see, e.g., Problems 8.9 and 8.13). Thus α is indecomposable if and only if for all $\xi < \alpha$ the equality $\xi + \alpha = \alpha$ is true.

31. This is a consequence of Problems 30 and 8.13. [W. Sierpiński, Cardinal and Ordinal Numbers, Polish Sci. Publ., Warszawa, 1965, XIV.6. Theorem 3]

32. If $\xi < \beta \cdot \alpha$, then $\xi < \beta \cdot \eta$ for some $\eta < \alpha$, hence by Problem 30 $\xi + \beta \cdot \alpha \le \beta \cdot \eta + \beta \cdot \alpha = \beta \cdot (\eta + \alpha) = \beta \cdot \alpha$. Therefore, again by Problem 30, $\beta \cdot \alpha$ is indecomposable.

33. If we write $\alpha = \beta \cdot \gamma + \delta$ with some $\delta < \beta$ (see Problem 8.15), then by the indecomposability of α we must have $\delta = 0$.

34. By Problem 32 the ordinal $\alpha \cdot \omega$ is indecomposable because ω is. But if $\alpha < \beta < \alpha \cdot \omega$ and we write $\beta = \alpha \cdot m + \delta$ with some $\delta < \alpha$ and $m = 1, 2, \ldots$, then $\beta = \alpha + (\alpha \cdot (m-1) + \delta)$ is a decomposition of β into a sum of smaller ordinals, hence it is not indecomposable.

35. This is an immediate consequence of Problem 37 below and of the normal form of (9.2) of α. [W. Sierpiński, Cardinal and Ordinal Numbers, Polish Sci. Publ., Warszawa, 1965, XIV.6. Theorem 2]

36. This is a consequence of Problem 37 below and of the normal form of the sum of two ordinals found in (9.2).

37. This is immediate from the normal form representation (9.2) and from Problem 23. [W. Sierpiński, Cardinal and Ordinal Numbers, Polish Sci. Publ., Warszawa, 1965, XIV.19. Theorem 1]

38. The first three infinite primes are ω, $\omega + 1$, and $\omega^2 + 1$. This follows from the fact that any ordinal $\omega \cdot m + n$ with $m > 1$ can be written as $(\omega + n) \cdot m$, and every $\omega + n$ with $n > 1$ can be written as $n \cdot (\omega + 1)$.

39. If $\alpha > 1$ is prime and $\alpha = \beta \cdot \gamma$ with some $\gamma > 1$, then $\beta < \alpha$, hence we must have $\gamma = \alpha$. The converse is trivial.

40. Suppose that α is indecomposable, and $\alpha+1$ is the product of two positive ordinals. Then both of them have to be a successor ordinals, say $\alpha + 1 = (\beta+1)\cdot(\gamma+1) = (\beta+1)\cdot\gamma+\beta+1$. Hence $\alpha = (\beta+1)\cdot\gamma+\beta$, and by the indecomposability of α here either $\beta = \alpha$, in which case $\beta+1 = \alpha+1$, or $(\beta+1)\cdot\gamma = \alpha$, when $\alpha = (\beta+1)\cdot\gamma+\beta$ implies $\beta = 0$.

41. Let α be an infinite successor ordinal and consider its normal form (9.2). Then $\xi_0 = 0$, $a_0 > 0$, and if we write $\xi_k = \xi_1 + \zeta_k$ for $k = 1, \ldots, n$, then

$$\alpha = (\omega^{\xi_1} + a_0)\cdot(\omega^{\zeta_n}\cdot a_n + \cdots + \omega^{\zeta_2}\cdot a_2 + a_1),$$

and the last factor is smaller than α. If α is prime, then this can only happen if $\alpha = \omega^{\zeta_1} + a_0$, and then since $\alpha = \omega^{\zeta_1} + a_0 = a_0\cdot(\omega^{\zeta_1}+1)$, only if $a_0 = 1$.

That each of $\omega^{\xi}+1$ is a prime ordinal follows from Problems 37 and 40.

42. If α is a limit ordinal, then in its normal form (9.2) $\xi_0 > 0$, and α is divisible from the left by ω^{ξ_0} and from the right by $\omega^{\zeta_n}\cdot a_n + \cdots + \omega^{\zeta_1}\cdot a_1 + a_0$, where the ζ_k are the ordinals, for which $\xi_k = \xi_0 + \zeta_k$. If this last sum consists of more than one term, then

$$\begin{aligned}\omega^{\zeta_n}\cdot a_n + \cdots + \omega^{\zeta_2}\cdot a_1 + a_0 &\le \omega^{\xi_n}\cdot a_n + \cdots + \omega^{\xi_1}\cdot a_1 + a_0\\ &< \omega^{\xi_n}\cdot a_n + \cdots + \omega^{\xi_1}\cdot a_1 + \omega^{\xi_0}\cdot a_0 = \alpha,\end{aligned}$$

and clearly in this case $\omega^{\xi_0} < \omega^{\xi_n} \le \alpha$ also holds, so α cannot be a prime. Thus, α can have only one term in its normal form, and then obviously it has to be of the form $\alpha = \omega^{\beta}$. Now here β must be indecomposable, for if $\beta = \gamma+\delta$ with $\gamma, \delta < \beta$, then we would have $\alpha = \omega^{\gamma}\cdot\omega^{\delta}$ with $\omega^{\gamma}, \omega^{\delta} < \alpha$, so α could not be a prime. Thus, β is indecomposable, and hence by Problem 37 we have $\alpha = \omega^{\omega^{\xi}}$ with some ξ.

43. Since for limit ordinal ξ the ordinals $\xi + k$, $k = 2, 3, \ldots$ are non-primes $(\xi + k = k\cdot(\xi+1))$, the statement follows from Problem 7. [W. Sierpiński, Cardinal and Ordinal Numbers, Polish Sci. Publ., Warszawa, 1965, XIV.22. Theorem 4]

44. Suppose that $\alpha + 1 = \beta\cdot\gamma = \delta\cdot\eta$, where β and δ are infinite primes. Then $\beta, \gamma, \delta, \eta$ have to be successor ordinals, hence by Problem 41 we have $\beta = \omega^{\xi}+1$, $\delta = \omega^{\zeta}+1$, and if we also write $\gamma = \omega\cdot\gamma_1 + k$ and $\eta = \omega\cdot\eta_1 + l$ with some positive natural numbers l and k, then $\beta\cdot\gamma = \omega^{\xi+1}\cdot\gamma_1 + \omega^{\xi}\cdot k + 1$, $\delta\cdot\eta = \omega^{\zeta+1}\cdot\eta_1 + \omega^{\zeta}\cdot l + 1$, which shows that in the normal representation of $\beta\cdot\gamma$ the last two-terms are $\omega^{\xi}\cdot k + 1$, while in the normal representation of $\delta\cdot\eta$ the last two-terms are $\omega^{\zeta}\cdot l + 1$. Since the normal representation is unique, we must have $\xi = \zeta$, hence $\beta = \delta$.

To show that the statement is not necessarily true for limit ordinals, consider ω^{ω}, which has the infinitely many primes $\omega^n + 1$, $n = 1, 2, \ldots$ as its left divisors: $(\omega^n + 1)\cdot\omega^{\omega} = \omega^{\omega}$.

45. The proof is by transfinite induction on α. If α is prime, then there is nothing to do. If it is not, then it is the product of two smaller ordinals for which we can apply the induction hypothesis to conclude that α is the product of finitely many prime ordinals.

Since $\omega^2 = \omega \cdot \omega = (\omega + 1) \cdot \omega$, the representation is not unique. [W. Sierpiński, Cardinal and Ordinal Numbers, Polish Sci. Publ., Warszawa, 1965, XIV.23. Theorem 1]

46. By the normal form representation every ordinal α can be written in a unique way as the product $\alpha = \omega^\beta \cdot \gamma$ of a power of ω and of a successor ordinal γ. Thus, if we write α as a product of primes in such a way that limit prime factors precede successor prime factors, then the product of the limit prime factors (which are powers of ω) must be ω^β, and the product of the successor prime factors must be γ.

Thus, it is enough to prove the existence and unicity of the prime representation in question in the two cases when α is a power of ω and when α is a successor ordinal.

Suppose first that $\alpha = \omega^\beta$. If α is the product of prime factors $\omega^{\omega^{\zeta_m}} \geq \omega^{\omega^{\zeta_{m-1}}} \geq \cdots \geq \omega^{\omega^{\zeta_0}}$, then $\beta = \omega^{\zeta_m} + \omega^{\zeta_{m-1}} + \cdots + \omega^{\zeta_0}$ must be the normal form of β (some terms may be repeated), and both the existence and the unicity of the representation follow from the existence and unicity of normal form representation.

Next we prove the unicity of the representation when α is a successor ordinal. Let $\alpha = \gamma_m \cdot \cdots \cdot \gamma_0$ with $\gamma_m \geq \gamma_{m-1} \geq \ldots \geq \gamma_0$, where γ_i are prime ordinals, so they are either prime natural numbers or ordinals of the form $\omega^\xi + 1$. Let $\gamma_s, \gamma_{s-1}, \ldots, \gamma_0$ be all finite, but γ_{s+1} infinite. Since the largest term in the normal form of the product $\gamma_n \cdots \gamma_{s+1}$ has coefficient 1, it follows that $\gamma_s \cdots \gamma_0$ must be equal to a_n, the largest coefficient (the coefficient of the highest power) in the normal expansion of α. Thus, the finite prime factors on the right are uniquely determined by α, therefore we can cancel them (Problem 8.10), and we may assume that α does not have a right prime divisor, which is finite. But then by Problem 43 in two representations of α of the kind we are discussing the last (rightmost) prime factor is uniquely determined. Thus, we can factor out this common rightmost prime factor from both representations (see Problem 8.11), and we get the unicity by induction.

Finally, we prove the existence of the prime representation in question for successor ordinals. Since for $\xi < \zeta_0$ we have

$$\left(\omega^{\zeta_m} \cdot b_m + \cdots + \omega^{\zeta_0} \cdot b_0\right) \cdot \left(\omega^\xi + 1\right) = \omega^{\zeta_m + \xi} + \omega^{\zeta_m} \cdot b_m + \cdots + \omega^{\zeta_0} \cdot b_0,$$

we can successively change the normal form of α into such a representation: first note that

$$\begin{aligned} \alpha &= \omega^{\xi_n} \cdot a_n + \cdots + \omega^{\xi_0} \cdot a_0 \\ &= \left(\omega^{\xi_{n-1}} \cdot a_{n-1} + \omega^{\xi_{n-2}} a_{n-2} + \cdot + \omega^{\xi_1} \cdot a_1 + a_0\right) \cdot \left(\omega^{\delta_n} + 1\right) \cdot a_n, \end{aligned}$$

where δ_n is the ordinal for which $\xi_n = \xi_{n-1} + \delta_n$. Here a_n can be uniquely written as a nonincreasing sequence of finite prime factors. Now repeat this with the factor

$$\omega^{\xi_{n-1}} \cdot a_{n-1} + \omega^{\xi_{n-2}} a_{n-2} + \cdot + \omega^{\xi_1} \cdot a_1 + a_0,$$

etc., to obtain the required form.

Actually, the method we used for the existence can be easily extended to yield both the existence and unicity of the representation. In fact, if the normal form of α is $\alpha = \omega^{\xi_n} \cdot a_n + \cdots + \omega^{\xi_0} \cdot a_0$ and set $\delta_0 = \xi_0$ and choose δ_i, $1 \le i \le n$, so that $\xi_i = \xi_{i-1} + \delta_i$, then $\delta_0 \ge 0$ and $\delta_i > 0$ for $1 \le i \le n$. Now

$$\alpha = \omega^{\delta_0} \cdot a_0 \cdot (\omega^{\delta_1} + 1) \cdot a_1 \cdots a_{n-1} \cdot (\omega^{\delta_n} + 1) \cdot a_n,$$

and if $\delta_0 = \omega^{\gamma_m} + \cdots + \omega^{\gamma_0}$ with $\gamma_m \ge \cdots \ge \gamma_0$, then

$$\alpha = \omega^{\omega^{\gamma_m}} \cdots \omega^{\omega^{\gamma_0}} \cdot a_0 \cdot (\omega^{\delta_1} + 1) \cdot a_1 \cdots a_{n-1} \cdot (\omega^{\delta_n} + 1) \cdot a_n$$

is the required decomposition. Unicity is also clear since in order that this formula should hold, the choice of δ_i must be what was given above. [W. Sierpiński, Cardinal and Ordinal Numbers, Polish Sci. Publ., Warszawa, 1965, XIV.23]

47. This follows from Problem 53.

48. See Problem 53. [W. Sierpiński, Cardinal and Ordinal Numbers, Polish Sci. Publ., Warszawa, 1965, XIV.25. Corollary 1]

49. See Problem 53.

50. See Problem 53. [N. Aronszajn, Fund. Math., **39**(1952), 65–96]

51. See Problem 53. [W. Sierpiński, Cardinal and Ordinal Numbers, Polish Sci. Publ., Warszawa, 1965, XIV.25, Theorem 1]

52. See Problem 53. [W. Sierpiński, Cardinal and Ordinal Numbers, Polish Sci. Publ., Warszawa, 1965, XIV.25. Theorem 2]

53. This is an immediate consequence of the normal form of the sum of two ordinals given in the solution to Problem 18. In fact, if α has normal form (9.2) and β has normal form (9.1), then, e.g., for $\beta < \alpha$ we cannot have $\zeta_m < \xi_n$, for then $\beta + \alpha = \alpha$, and $\alpha = \beta + \alpha = \alpha + \beta$ gives $\beta = 0$. Thus, $\zeta_m = \xi_n$, and then

$$\alpha + \beta = \omega^{\xi_n} \cdot (a_n + b_m) + \omega^{\zeta_{m-1}} \cdot b_{n-1} + \cdots + \omega^{\zeta_0} \cdot b_0$$

while

$$\beta + \alpha = \omega^{\xi_n} \cdot (a_n + b_m) + \omega^{\xi_{n-1}} \cdot a_{n-1} + \cdots + \omega^{\xi_0} \cdot a_0$$

and these are the same exactly when $m = n$, $\xi_i = \zeta_i$ for all $i \le n$ and $a_i = b_i$ for $i < n$.

54. The sum of n nonzero ordinals $\alpha_1, \ldots, \alpha_n$ is independent of their order if and only if each two are additively commutative. Now apply Problems 53 and 18.

55. **(a)** Let α_i, $i = 1, \ldots, n$ be n ordinals, and let us write each α_i as $\alpha_i = \omega^{\gamma_i} \cdot a_i + \beta_i$, where $a_i = 1, 2, \ldots$ and $\beta_i < \omega^{\gamma_i}$. Such a decomposition follows from the normal representation of α_i. Let $\gamma = \min_i \gamma_i$, and let k be the number of those α_i for which $\gamma_i = \gamma$. Without loss of generality, we may assume $\gamma_1 = \ldots = \gamma_k = \gamma$ and $\gamma_i > \gamma$ for $i > k$. If $i_1, \ldots, i_n$ is any permutation of the numbers $1, \ldots, n$ and $i_n > k$, then each of $\alpha_1, \ldots, \alpha_k$ gets absorbed in the following summands in $\alpha_{i_1} + \cdots + \alpha_{i_n}$, hence those sums are as if the numbers $\alpha_1, \ldots, \alpha_k$ were all missing. Hence there are exactly $g(n-k)$ such sums. If, however, $i_n \le k$ and r is chosen so that $i_{n-1} \le k, \ldots, i_{n-r+1} \le k$ but $i_{n-r} > k$, then again all $\alpha_1, \ldots, \alpha_n$ but the r ones at the end (i.e., $\alpha_{i_{n-r+1}}, \ldots, \alpha_{i_n}$) get absorbed in the following summands in $\alpha_{i_1} + \cdots + \alpha_{i_n}$. Thus, in this case we obtain that

$$\alpha_{i_1} + \cdots + \alpha_{i_n} = \delta + \omega^{\gamma} \cdot (a_{i_{n-r+1}} + \cdots + a_n) + \beta_{i_n},$$

where $\delta \ge \omega^{\gamma+1}$. Here there are at most $\binom{k}{r}$ possibilities for the selection of the indices $i_{n-r+1}, \ldots, i_n$ and hence for the sum $a_{i_{n-r+1}} + \cdots + a_{i_n}$, and for a given selection of these indices there are at most r possibilities for i_n. As we have just seen δ can be obtained in at most $g(n-k)$ ways, hence the number of possibilities for the sum $\alpha_{i_1} + \cdots + \alpha_{i_n}$ is at most

$$\left[1 + \sum_{r=1}^{k} r \binom{k}{r}\right] g(n-k) = (k2^{k-1} + 1) g(n-k).$$

This gives the upper bound

$$g(n) \le \max_k (k2^{k-1} + 1) g(n-k).$$

It is clear from the given consideration that if for a particular k we set

$$\alpha_i = \omega \cdot 2^i + i, \qquad i = 1, \ldots, k,$$

and

$$\alpha_{k+j} = \omega^2 \cdot \alpha'_j, \qquad j = 1, \ldots, n-k,$$

where α'_j, $j = 1, 2, \ldots, n-k$, is a system of ordinals for which we get $g(n-k)$ possible sums, then the γ above is 1 and all possible choices listed above are different, so in this case the bound $(k2^{k-1} + 1) g(n-k)$ is achieved. This gives

$$g(n) \ge \max_k (k2^{k-1} + 1) g(n-k),$$

and part (a) is proved.

(b) can be obtained by direct computations from the formula in part (a).

(c) is a consequence of (a), (b), and part (d).

(d) Consider the ratios $g(n)/81^{n/5}$. Part (a) implies that

$$\frac{g(n)}{81^{n/5}} = \max_{1\le k\le n-1} \frac{k2^{k-1}+1}{81^{k/5}} \cdot \frac{g(n-k)}{81^{(n-k)/5}}.$$

Since the fraction

$$\frac{k2^{k-1}+1}{81^{k/5}}$$

increases as k increases for $k \le 4$ and decreases as k increases for $k \ge 5$ and for $k = 5$ it takes its maximum value 1, it follows that if for some m we have

$$\frac{g(m-i)}{81^{(m-i)/5}} = \frac{g(m-i-5)}{81^{(m-i-5)/5}} \qquad \text{for } i = 0, 1, \ldots, 8, \tag{9.5}$$

then

$$\begin{aligned}
\frac{g(m+1)}{81^{(m+1)/5}} &= \max_{1\le k\le m+1} \frac{k2^{k-1}+1}{81^{k/5}} \cdot \frac{g(m+1-k)}{81^{(m+1-k)/5}} \\
&\le \max\left\{ \max_{1\le k\le 9} \frac{k2^{k-1}+1}{81^{k/5}} \cdot \frac{g(m+1-k-5)}{81^{(m+1-k)/5}}, \right. \\
&\qquad \left. \max_{10\le k\le m} \frac{(k-5)2^{(k-5)-1}+1}{81^{(k-5)/5}} \cdot \frac{g(m+1-k)}{81^{(m+1-k)/5}} \right\} \\
&= \frac{g(m+1-5)}{81^{(m+1-5)/5}}.
\end{aligned}$$

Since here the term on the right-hand side appears as the 5th term in the first maximum, we obtain

$$\frac{g(m+1)}{81^{(m+1)/5}} = \frac{g(m+1-5)}{81^{(m+1-5)/5}}.$$

Thus, the property (9.5) is inherited from m to $m+1$ and we obtain that for all $n \ge m-8$

$$\frac{g(n)}{81^{n/5}} = \frac{g(n-5)}{81^{(n-5)/5}}.$$

But based on the values in part (b) and on similar computations (resulting from part (a)) for the values $g(16)$—$g(27)$) it is easy to check that (9.5) is true for $m = 27$, hence the preceding formula proves part (d). [P. Erdős, Some remarks on set theory, Proc. Amer. Math. Soc., **23**(1950), 127–141]

56. This follows from Problem 60.

57. Let $\alpha > 1$ be a successor ordinal. If α is finite, say $\alpha = a$, but β is infinite, say β is of the form $\beta = \omega\cdot\gamma+k$, then $\alpha\cdot\beta = \omega\cdot\gamma+ak$, while $\beta\cdot\alpha = \omega\cdot\gamma\cdot a+k$, and this latter ordinal is clearly bigger than the former one.

Thus, let α and β be infinite, α a successor and β a limit ordinal. By Problem 18 if the normal expansion of α has k terms and the normal expansion of β has l terms, then the normal expansion of $\alpha \cdot \beta$ has l terms, while the normal expansion of $\beta \cdot \alpha$ has $l + (k - 1)$ terms. Thus, if $\alpha \cdot \beta = \beta \cdot \alpha$, then we must have $k = 1$, which means that α is finite (α was assumed to be a successor ordinal), but this is not the case. [W. Sierpiński, Cardinal and Ordinal Numbers, Polish Sci. Publ., Warszawa, 1965, XIV.26]

58. See Problems 60 and 8.12. [W. Sierpiński, Cardinal and Ordinal Numbers, Polish Sci. Publ., Warszawa, 1965, XIV.26. Corollary 1]

59. This follows from the next problem, Problem 60. [W. Sierpiński, Cardinal and Ordinal Numbers, Polish Sci. Publ., Warszawa, 1965, XIV.10/3]

60. First let us consider the case when α and β are successor ordinals. If α and β are multiplicatively commutative, then by Problem 63 there is a ξ such that $\alpha = \xi^m$ and $\beta = \xi^n$ with some natural numbers n, m. Thus, in this case $\alpha^n = \beta^m$.

Conversely, suppose that for some n, m we have $\alpha^n = \beta^m$, and let ξ be the smallest ordinal bigger than 1 that is multiplicatively commutative with $\alpha^n = \beta^m$. Since α is multiplicatively commutative with α^n, by Problem 62 we must have $\alpha = \xi^k$ for some k, and for similar reasons $\beta = \xi^l$ for some l, and this shows that α and β are multiplicatively commutative.

Now let $\alpha < \beta$ be limit ordinals. If they are multiplicatively commutative, then, according to Problem 61, there is a θ and positive integers p, r such that $\beta = \omega^{\theta\cdot r}\alpha$, and the highest power of ω in the normal representation of α is $\omega^{\theta\cdot p}$. The latter property implies by the solution of Problem 18 that $\alpha^s = \omega^{\theta\cdot(p(s-1))}\alpha$ for $s = 1, 2, \ldots$, hence $\alpha^{p+r} = \omega^{\theta\cdot(p+r-1)p}\alpha = \omega^{\theta\cdot(p+r)(p-1))}\omega^{\theta\cdot r}\alpha = \beta^p$.

Conversely, suppose that for some positive natural numbers n, m we have $\alpha^n = \beta^m$, and let ω^τ and ω^σ be the highest powers of ω in the normal representation of α and β, respectively. Then the highest power of ω in the normal form of α^n is $\omega^{\tau\cdot n}$ and in the normal form of β^m it is $\omega^{\sigma\cdot m}$. Thus, $\tau \cdot n = \sigma \cdot m$, and hence, by Problems 51, 50 there is a θ and some positive integers k, l such that $\tau = \theta \cdot k$ and $\sigma = \theta \cdot l$. We also have $\alpha^n = \omega^{\tau\cdot(n-1)} \cdot \alpha$ and $\beta^m = \omega^{\sigma\cdot(m-1)} \cdot \beta$, thus $\omega^{\theta\cdot k(n-1)} \cdot \alpha = \omega^{\theta\cdot l(m-1)} \cdot \beta$. Here $\alpha < \beta$ implies $\theta \cdot l(m-1) \le \theta \cdot k(n-1)$, so by Problem 8.10 we can cancel with the common factor $\omega^{\theta\cdot l(m-1)}$ from the left to obtain $\beta = \omega^{\theta\cdot(k(n-1)-l(m-1))} \cdot \alpha$. This and Problem 61 show that α and β are multiplicatively commutative. [W. Sierpiński, Cardinal and Ordinal Numbers, Polish Sci. Publ., Warszawa, 1965, XIV.26. Theorem 1]

61. Let $\alpha < \beta$ be two limit ordinals with the respective normal forms

$$\alpha = \omega^{\xi_n} \cdot a_n + \cdots + \omega^{\xi_0} \cdot a_0$$

and

$$\beta = \omega^{\zeta_m} \cdot b_m + \cdots + \omega^{\zeta_0} \cdot b_0.$$

Then (see Problem 18)

$$\beta \cdot \alpha = \omega^{\zeta_m + \xi_n} \cdot a_n + \cdots + \omega^{\zeta_m + \xi_0} \cdot a_0$$

and

$$\alpha \cdot \beta = \omega^{\xi_n + \zeta_m} \cdot b_m + \cdots + \omega^{\xi_n + \zeta_0} \cdot b_0,$$

and these are normal forms. Thus, these two numbers are the same if and only if $n = m$, $a_i = b_i$ for all i, and $\zeta_m + \xi_i = \xi_n + \zeta_i$. For $i = n$ this means that ζ_n and ξ_n are additively commutative, and hence (see Problem 51) there is a θ and natural numbers k, l such that $\xi_n = \theta \cdot k$ and $\zeta_m = \theta \cdot l$. Since $\alpha < \beta$, we must have $l > k$, and so $\zeta_m = \xi_n + \theta \cdot (l-k)$. But then $\xi_n + \theta \cdot (l-k) + \xi_i = \xi_n + \zeta_i$, which means that $\theta \cdot (l-k) + \xi_i = \zeta_i$ for all i, i.e., $\omega^{\theta \cdot (l-k)} \alpha = \beta$, and this proves the necessity of the condition.

It is also clear that if $\xi_n = \theta \cdot k$, $\zeta_m = \theta \cdot l$ and $\omega^{\theta \cdot (l-k)} \alpha = \beta$, then $\zeta_n + \xi_i = \xi_n + \zeta_i$, hence, as we have mentioned above, α and β are multiplicatively commutative, i.e., the condition is also sufficient.

62. According to Problem 57, ξ is a successor ordinal.

First we prove that α is a finite power of ξ.

We have seen in Problem 46 that α can be uniquely written in the form $\alpha = a_{m+1} \cdot \beta_m \cdot a_m \cdot \beta_{m-1} \cdots a_1 \cdot \beta_0 \cdot a_0$ where $a_i \geq 1$ are natural numbers and $\beta_m \geq \beta_{m-1} \geq \ldots \geq \beta_0$ are infinite prime ordinals of the form $\omega^\tau + 1$. Now let $\xi = c_{n+1} \cdot \gamma_n \cdot c_n \cdot \gamma_{n-1} \cdots c_1 \cdot \gamma_0 \cdot c_0$ be the corresponding representation of ξ. Then $\xi \cdot \alpha = \alpha \cdot \xi$ implies that both a_0 and c_0 are coefficients of the largest power of ω in the normal form representation of $\xi \cdot \alpha = \alpha \cdot \xi$, therefore $a_0 = c_0$. We can cancel this common right factor from the equation (see Problem 8.11) to obtain

$$\begin{aligned} &a_{n+1} \cdot \beta_m \cdot a_{n-1} \cdot \beta_{n-1} \cdots a_1 \cdot \beta_0 \cdot (a_0 c_{n+1}) \cdot \gamma_n \cdot c_n \cdot \gamma_{n-1} \cdots c_1 \cdot \gamma_0 \\ &= c_{n+1} \cdot \gamma_n \cdot c_n \cdot \gamma_{n-1} \cdots c_1 \gamma_0 \cdot (c_0 a_{m+1}) \cdot \beta_m \cdot a_m \cdot \beta_{m-1} \cdots a_1 \cdot \beta_0. \end{aligned}$$

Now γ_0 and β_0 are infinite prime right divisors of the same ordinal, so they must be the same by Problem 43. Cancelling them (see Problem 8.11) and continuing this process we can see that the numbers a_i and c_i are equal and so are the prime ordinals β_i and γ_i for $i = 0, 1, \ldots$. This process terminates only when i reaches m or n. If $n \neq m$, then this implies that either ξ or α is a right divisor of the other one. If, however, $m = n$, then we obtain from this procedure that $\alpha = \xi$ (recall that $a_0 = c_0$ has been verified above). Thus, in any case one of α and ξ is a right divisor of the other one, and since $\xi \leq \alpha$, it must be ξ: $\alpha = \alpha_1 \cdot \xi$ with some ordinal α_1. Here $\alpha_1 \cdot \xi \cdot \xi = \alpha \cdot \xi = \xi \cdot \alpha = \xi \cdot \alpha_1 \cdot \xi$, and so $\alpha_1 \cdot \xi = \xi \cdot \alpha_1$, i.e., ξ and α_1 are also multiplicatively commutative. Now continue this process with α_1 and ξ. It follows that either $\alpha_1 < \xi$ or $\alpha_1 = \alpha_2 \cdot \xi$, and α_2 and ξ are multiplicatively commutative. Repeat this process again and again. Since the sequence $\{\alpha_i\}$ is strictly decreasing, there will be a smallest index i_0 such that $\alpha_{i_0} < \xi$. Since α_{i_0} is multiplicatively commutative with ξ, and $\alpha_{i_0-1} = \alpha_{i_0} \cdot \xi$, it follows that α_{i_0} and α_{i_0-1} are also multiplicatively commutative. Going back this way, we get that α_{i_0} and α are multiplicatively commutative. But in view of the choice of ξ, this can only happen for $\alpha_{i_0} < \xi$ if $\alpha_{i_0} = 1$. Hence $\alpha_{i_0-1} = \xi$, $\alpha_{i_0-2} = \xi^2$, ..., $\alpha = \xi^{i_0}$, and this proves the claim.

Now we prove by induction on β that if β is multiplicatively commutative with α then it is a finite power of ξ. ξ and β are right divisors of the the same ordinal (namely $\alpha \cdot \beta = \beta \cdot \alpha$), and we apply Problem 7. By the minimality of ξ if β is a right divisor of ξ then $\beta = \xi$, and we are done. If ξ is a right divisor of β, say $\beta = \gamma \cdot \xi$, then $\gamma < \beta$ and

$$\alpha \cdot \gamma \cdot \xi = \alpha \cdot \beta = \beta \cdot \alpha = \gamma \cdot \xi \cdot \alpha = \gamma \cdot \alpha \cdot \xi,$$

and since ξ is a successor ordinal we can cancel the ξ on the right of the two extreme sides, and we get that γ is multiplicatively commutative with α. So in this case the induction hypothesis gives that γ is a finite power of ξ, and hence so is $\beta = \gamma \cdot \xi$. The only remaining possibility in Problem 7 is that $\beta = \zeta + p$, $\xi = \zeta + q$ with some limit ordinal ζ and $0 < q < p$ integers. Since $\alpha = (\zeta + q)^m$ for some m, an application of Problem 8.21 yields

$$\alpha \cdot \beta = (\zeta + q)^m \cdot (\zeta + p) = \zeta \cdot (\zeta + q)^{m-1} \cdot (\zeta + p) + q = \text{a limit ordinal} + q$$

while

$$\beta \cdot \alpha = (\zeta + p) \cdot (\zeta + q)^m = \zeta \cdot (\zeta + q)^m + p = \text{a limit ordinal} + p$$

and these are different when $q < p$. Thus, this possibility cannot occur, and the proof is complete.

63. This follows from Problem 62.

64. Since

$$(\omega^2 + \omega) \cdot (\omega^3 + \omega^2) = \omega^5 + \omega^4 = (\omega^3 + \omega^2) \cdot (\omega^2 + \omega),$$

these ordinals are multiplicatively commutative. But there is no ordinal ξ for which $\omega^2 + \omega = \xi^n$ was true with some $n \geq 2$. In fact, then it would have to be of the form $\omega \cdot k + l$ and n would have to be 2. Furthermore $\omega^2 + \omega$ is a limit ordinal, which means $l = 0$, but then $(\omega \cdot k)^2 = \omega^2 \cdot k \neq \omega^2 + \omega$. Thus, if $\omega^2 + \omega = \xi^n$, then $n = 1$, $\xi = \omega^2 + \omega$, in which case $\xi^m = \omega^3 + \omega^2$ is an impossibility. [W. Sierpiński, Cardinal and Ordinal Numbers, Polish Sci. Publ., Warszawa, 1965, XIV.26. (26.4)]

65. The product of n ordinals $\alpha_1, \dots, \alpha_n$, $\alpha_i \geq 2$ is independent of their order if and only if each two are multiplicatively commutative. Now the statement follows from Problem 60.

66. $\alpha_i = \omega + i$, $i = 1, \dots, n$ will do; see Problem 8.39.

67. Suppose that α and β are different infinite ordinals that are additively commutative. Then (see Problem 51) there is a ξ and $n \neq m$ such that $\alpha = \xi \cdot n$ and $\beta = \xi \cdot m$. If ξ is a limit ordinal, then $\alpha \cdot \beta = \xi^2 \cdot m$, while $\beta \cdot \alpha = \xi^2 \cdot n$, and these are different. If, however, ξ is a successor ordinal, say $\xi = \omega \cdot \gamma + k$ with $k > 0$, then

$$\beta \cdot \alpha = (\omega \cdot \gamma + k) \cdot m \cdot (\omega \cdot \gamma + k) \cdot n = \omega \cdot \gamma \cdot \omega \cdot \gamma \cdot n + \omega \cdot \gamma \cdot mk + k,$$

while

$$\alpha \cdot \beta = (\omega \cdot \gamma + k) \cdot n \cdot (\omega \cdot \gamma + k) \cdot m = \omega \cdot \gamma \cdot \omega \cdot \gamma \cdot m + \omega \cdot \gamma \cdot nk + k,$$

and these are different: e.g., if $n > m$, then

$$\begin{aligned} \omega \cdot \gamma \cdot \omega \cdot \gamma \cdot m + \omega \cdot \gamma \cdot nk &< \omega \cdot \gamma \cdot \omega \cdot \gamma \cdot m + \omega \cdot \gamma \cdot \omega \\ &= \omega \cdot \gamma \cdot \omega \cdot (\gamma \cdot m + 1) \leq \omega \cdot \gamma \cdot \omega \cdot \gamma \cdot n \\ &< \omega \cdot \gamma \cdot \omega \cdot \gamma \cdot n + \omega \cdot \gamma \cdot mk. \end{aligned}$$

Thus, α and β are not multiplicatively commutative. [W. Sierpiński, Cardinal and Ordinal Numbers, Polish Sci. Publ., Warszawa, 1965, XIV.26/12]

68. b) clearly implies a) (cf. Problem 8.13). Furthermore, if $\alpha = \omega^{\omega^\beta}$, $\beta > 0$, and $1 \leq \xi < \alpha$, then in the normal expansion of ξ the highest power is ω^γ with some $\gamma < \omega^\beta$. Thus, then $\xi \cdot \alpha \leq \omega^{\gamma+1} \cdot \omega^{\omega^\beta} = \omega^{\gamma+1+\omega^\beta} = \omega^{\omega^\beta} = \alpha \leq \xi \cdot \alpha$, and this is b). Thus, c) implies b) (the case $\beta = 0$ is trivial).

Finally, suppose that a) holds, and let $\alpha = \omega^{\xi_n} \cdot a_n + \cdots + \omega^{\xi_0} \cdot a_0$ be the normal form of α. If this has more than one term, then selecting $\xi = \theta = \omega^{\xi_n}$, we get two ordinals that are smaller than α such that their product $\omega^{\xi_n \cdot 2}$ is bigger than α, thus α must be of the form $\omega^\gamma \cdot a$, and for the same reason as before, here we must have $a = 1$, $\gamma > 0$. Finally, selecting $\xi = \omega^\rho$ and $\zeta = \omega^\sigma$ with $\rho, \sigma < \gamma$ the condition in part a) implies that $\rho + \sigma$ has to be smaller than γ, i.e., γ has to be an indecomposable ordinal. Thus, by Problem 37 γ is of the form ω^β for some β, and this proves that a) implies c). [G. Hessenberg, Grundbegriffe der Mengenlehre, Göttingen 1906, W. Sierpiński, Cardinal and Ordinal Numbers, Polish Sci. Publ., Warszawa, 1965, XIV.20]

69. Let μ be any infinite ordinal. Inductively define $\mu_0 = \mu$, $\mu_{n+1} = \omega^{\mu_n}$, $n = 0, 1, \dots$, and let ν be supremum of all the ordinals μ_n. If ζ is any epsilon-ordinal such that $\mu \leq \zeta$, then by induction we find that $\mu_{n+1} = \omega^{\mu_n} \leq \omega^\zeta = \zeta$,

thus all μ_n are at most ζ, and so $\nu \le \zeta$. But it is clear that ζ is an epsilon-ordinal, since $\omega^\zeta = \sup_n \omega^{\mu_n} = \sup_n \mu_{n+1} = \zeta$. This proves that ν is the smallest epsilon-ordinal that is at least as large as μ.

If we start from $\mu = \omega$, then we get that the smallest epsilon-ordinal is the limit of the sequence $\omega, \omega^\omega, \omega^{\omega^\omega}, \ldots$, which can also be written as the sum

$$\omega + \omega^\omega + \omega^{\omega^\omega} + \ldots.$$

70. See the preceding proof, and notice that if μ is countable, then so are $\mu_1, \mu_2, \ldots$, and also $\sup_n \mu_n$.

71. **(i)** If $\xi < \alpha = \omega^\alpha$, then there is an ordinal $\beta < \alpha$ such that $\xi < \omega^\beta$, and hence (see Problem 17) $\xi + \alpha \le \omega^\beta + \omega^\alpha = \omega^\alpha = \alpha$.

(ii) In the same fashion as before, $\xi \cdot \alpha \le \omega^\beta \cdot \omega^\alpha = \omega^{\beta+\alpha} = \omega^\alpha = \alpha$, where we used part (i).

(iii) With the notation before, $\alpha \le \xi^\alpha \le \left(\omega^\beta\right)^\alpha = \omega^{\beta\cdot\alpha} = \omega^\alpha = \alpha$, where we used part (ii). [W. Sierpiński, Cardinal and Ordinal Numbers, Polish Sci. Publ., Warszawa, 1965, XIV.21. Theorem 2]

72. This is clear, for $\omega^\alpha \le \beta^\alpha = \alpha \le \omega^\alpha$ (see Problem 11(iv)), hence $\omega^\alpha = \alpha$.

73. If α is an epsilon-ordinal and $\beta, \gamma < \alpha$, then $\beta^\gamma < \beta^\alpha = \alpha$ where we used Problem 71(iii).

Conversely, if $\alpha > \omega$ is a limit ordinal but not an epsilon-ordinal, then $\alpha < \omega^\alpha$ (cf. Problem 11(iv)) and so there is a $\beta < \alpha$ such that $\alpha < \omega^\beta$, and here both ω and β are smaller than α. If, however, $\alpha > \omega$ is a successor ordinal, $\alpha = \beta + 1$, then clearly $\alpha < \beta \cdot 2 \le \beta^2 \le \beta^\omega$, and here again both ω and β are smaller than α.

74. If α is a limit ordinal and $\beta = \gamma \cdot \alpha$, where $\gamma > \alpha$ is an epsilon-ordinal, then (use Problem 71(iii)) $\alpha^\beta = \alpha^{\gamma\cdot\alpha} = (\alpha^\gamma)^\alpha = \gamma^\alpha$, while $\beta^\alpha = (\gamma\cdot\alpha)^\alpha$. Now $\gamma^\alpha \le (\gamma\cdot\alpha)^\alpha \le (\gamma^2)^\alpha = \gamma^{2\cdot\alpha} = \gamma^\alpha$ gives that $\alpha^\beta = \beta^\alpha$ holds.

Now suppose that $\alpha^\beta = \beta^\alpha$. First we prove that α and β cannot be simultaneously successor ordinals. Suppose, to the contrary, that they are successor ordinals. If $\alpha = \omega^{\xi_n} \cdot a_n + \cdots + \omega^{\xi_0} \cdot a_0$ and $\beta = \omega^{\zeta_m} \cdot b_m + \cdots + \omega^{\zeta_0} \cdot b_0$ is their normal form, then $\xi_0 = \zeta_0 = 0$, $a_0, b_0 > 0$. If $\alpha = \alpha' + a_0$ and $\beta = \beta' + b_0$, then it easily follows that $\alpha^\beta = \omega^{\xi_n\cdot\beta'} \cdot \alpha^{b_0}$ and $\beta^\alpha = \omega^{\zeta_m\cdot\alpha'} \cdot \beta^{a_0}$, and since the last factors are successor ordinals, the smallest power of ω in the normal form of α^β resp. β^α is $\omega^{\xi_n\cdot\beta'}$, resp. $\omega^{\zeta_m\cdot\alpha'}$. Hence we must have $\xi_n \cdot \beta' = \zeta_m \cdot \alpha'$. Together with this it also follows that $\alpha^{b_0} = \beta^{a_0}$, hence, by Problem 60, α and β are multiplicatively commutative. But then (see Problem 63) there is a ξ such that $\alpha = \xi^p$ and $\beta = \xi^q$ with some natural numbers p and q. Therefore, $\xi^{p\cdot\xi^q} = \xi^{q\cdot\xi^p}$, which implies $p \cdot \xi^q = q \cdot \xi^p$. Now ξ must be a successor ordinal since α and β are, so it is of the form $\xi = \gamma + k$, where γ is a limit ordinal and $k \geq 1$ is a natural number. Since then $\xi^s = \gamma^s + \gamma^{s-1} \cdot k + \cdots + \gamma \cdot k + k$, the equation $p \cdot \xi^q = q \cdot \xi^p$ means that

$$\gamma^q + \gamma^{q-1} \cdot k + \cdots + \gamma \cdot k + kp = \gamma^p + \gamma^{p-1} \cdot k + \cdots + \gamma \cdot k + kq,$$

which immediately implies $p = q$, and hence $\alpha = \beta$.

Next we show that it is not possible that, say, α is a limit ordinal and β is a successor ordinal (in this part of the proof we will not use the assumption $\alpha < \beta$, so this proof also proves that it is equally impossible that β is a limit ordinal and α is a successor ordinal). In fact, then using the preceding notation we have $\xi_0 > 0$ but $\zeta_0 = 0$. α^β is still $\omega^{\xi_n \cdot \beta'} \cdot \alpha^{b_0}$, and β^α is $\omega^{\zeta_m \cdot \alpha}$, and these imply first of all that $\xi_n \cdot \beta' < \zeta_m \cdot \alpha$, and then, since the equation $\xi_n \cdot \beta' + \sigma = \zeta_m \cdot \alpha$ is solvable for σ, that $\alpha^{\beta_0} = \omega^\sigma$ is a power of ω, i.e., its normal form has only one component. Now apply Problem 19 to conclude that the normal form of α also has only one component, say $\alpha = \omega^\xi \cdot a$. Thus, then $\alpha^\beta = \omega^{\xi \cdot \beta} \cdot a$ and $\beta^\alpha = \omega^{\zeta_m \cdot \alpha}$, from which we obtain first that $a = 1$, and then that $\xi \cdot \beta = \zeta_m \cdot \alpha$. Thus, β and α are right divisors of the same ordinal, and it follows from Problem 7 that one of them divides the other one from the right (the third possibility from Problem 7 cannot hold here, since α is a limit ordinal). If $\alpha = \gamma \cdot \beta$, then, since the normal form of α consists of a single term, the same must be true of β. But then β cannot be a successor ordinal. Thus, we must have $\beta = \gamma \cdot \alpha$, which is not possible either, since then β again would be a limit ordinal.

Thus, the only possibility that is left is that both α and β are limit ordinals. In this case $\xi_0 > 0$ and $\zeta_0 > 0$, and $\omega^{\xi_n \cdot \beta} = \alpha^\beta = \beta^\alpha = \omega^{\zeta_m \cdot \alpha}$, so we get again $\xi_n \cdot \beta = \zeta_m \cdot \alpha$, i.e., again α and β are the right divisors of the same ordinal. Thus, exactly as before we can conclude that α is a right divisor of β (recall that we have assumed $\alpha < \beta$), say $\beta = \gamma \cdot \alpha$. With this we also have $\xi_n \cdot \gamma \cdot \alpha = \zeta_m \cdot \alpha$.

Let ω^δ be the highest power in the normal expansion of γ. Then $\zeta_m = \delta + \xi_n$, and the preceding equation takes the form

$$(\delta + \xi_n) \cdot \alpha = \xi_n \cdot \gamma \cdot \alpha.$$

Here we cannot have $\delta \leq \xi_n$, for then

$$(\delta + \xi_n) \cdot \alpha \leq \xi_n \cdot 2 \cdot \alpha < \xi_n \cdot \gamma \cdot \alpha$$

because $2 \cdot \alpha = \alpha < \beta = \gamma \cdot \alpha$. Thus, $\delta \geq \xi_n$, and

$$\xi_n \cdot \omega^\delta \cdot \alpha \leq \xi_n \cdot \gamma \cdot \alpha = (\delta + \xi_n) \cdot \alpha \leq \delta \cdot 2 \cdot \alpha = \delta \cdot \alpha \leq \omega^\delta \cdot \alpha \leq \xi_n \cdot \omega^\delta \cdot \alpha,$$

which shows that we must have equality everywhere. In particular, $\xi_n \cdot \omega^\delta \cdot \alpha = \xi_n \cdot \gamma \cdot \alpha$, which implies that $\omega^\delta \cdot \alpha = \gamma \cdot \alpha = \beta$, i.e., we may assume without loss of generality that $\gamma = \omega^\delta$.

Our aim is to show that δ is an epsilon-ordinal (which, in view of $\gamma = \omega^\delta$ amounts the same as γ being an epsilon-ordinal), and at the end of the proof we shall also verify that it is bigger than α. If $\delta = \gamma = \omega^\delta$, then we are done. In the opposite case $\gamma \geq \delta + 1$, and hence

$$\xi_n \cdot \gamma \cdot \alpha = (\delta + \xi_n) \cdot \alpha \leq \xi_n \cdot (\delta + 1) \cdot \alpha \leq \xi_n \cdot \gamma \cdot \alpha,$$

which shows that $\gamma \cdot \alpha = (\delta + 1) \cdot \alpha = \delta \cdot \alpha$ (recall that α is a limit ordinal). Thus, we have arrived at the equation $\delta \cdot \alpha = \omega^\delta \cdot \alpha$. If ω^σ is the largest power of ω in the normal form of δ, then the preceding equation yields

$$\omega^\sigma \cdot \alpha \leq \omega^{\omega^\sigma} \cdot \alpha \leq \omega^\delta \cdot \alpha = \delta \cdot \alpha = \omega^\sigma \cdot \alpha,$$

giving $\omega^\delta \cdot \alpha = \omega^{\omega^\sigma} \cdot \alpha$, i.e., in $\beta = \gamma \cdot \alpha = \omega^\delta \cdot \alpha = \omega^{\omega^\sigma} \cdot \alpha$ we may assume $\delta = \omega^\sigma$ (and $\gamma = \omega^{\omega^\sigma}$). Now looking at the largest exponent in the normal form of $\delta \cdot \alpha = \omega^\delta \cdot \alpha$ we obtain

$$\sigma + \xi_n = \delta + \xi_n. \tag{9.6}$$

Let us go back to the equation $(\gamma \cdot \alpha)^\alpha = \alpha^{\gamma \cdot \alpha}$. It is not possible that here $\gamma \leq \alpha^m$ for some natural number m, for then

$$\beta^\alpha = (\gamma \cdot \alpha)^\alpha \leq (\alpha^{m+1})^\alpha = \alpha^{(m+1) \cdot \alpha} = \alpha^\alpha < \alpha^\beta.$$

Therefore, $\gamma \geq \alpha^m$ for all $m = 1, 2, \ldots$, and so $\gamma \geq \alpha^\omega = \omega^{\xi_n \cdot \omega}$ is also satisfied. This gives for δ that $\delta \geq \xi_n \cdot \omega$. Now $\sigma \leq \xi_n \cdot m$ is not possible for some $m = 1, 2, \ldots$, because then $\sigma + \xi_n \leq \xi_n \cdot (m+1) < \xi_n \cdot \omega \leq \delta + \xi_n$ holds contradicting (9.6). Thus, $\sigma \geq \xi_n \cdot \omega$. Therefore, if ω^τ is the largest power of ω in the normal form of σ, then on the left-hand side of (9.6) the highest exponent is τ, while on the right-hand side it is $\sigma \geq \omega^\tau$, which gives $\omega^\tau \leq \tau$. Since $\tau \leq \omega^\tau$ always holds, we obtain $\tau = \omega^\tau$, i.e., τ is an epsilon-ordinal. Using the inequality $\xi_n \leq \sigma$ we can see that if $\sigma > \tau$ then

$$\sigma + \xi_n \leq \sigma \cdot 2 < \sigma \cdot \omega = \omega^\tau \cdot \omega = \omega^{\tau+1} \leq \omega^\sigma = \delta < \delta + \xi_n,$$

which contradicts (9.6). Thus, we must have $\sigma = \tau$, and so $\delta = \omega^\sigma = \omega^\tau = \tau$ and $\gamma = \omega^\delta = \omega^\tau = \tau$.

Thus, so far we have verified that $\beta = \gamma \cdot \alpha$, where γ is an epsilon-ordinal. We have also seen that $\gamma \geq \alpha^m$ for all finite m, which yields $\gamma \geq \alpha^2 > \alpha$. This proves the claim. [W. Sierpiński, Cardinal and Ordinal Numbers, Polish Sci. Publ., Warszawa, 1965, XIV.27]

75. The product $\prod_{\xi<\theta} \alpha_\xi$ can be defined by transfinite induction in the same way as ordinal exponentiation was defined in Problem 11. Transfinite multiplication is associative but not commutative or distributive. Ordinal exponentiation is just repeated multiplication, i.e., $\gamma^\theta = \prod_{\xi<\theta} \gamma$. More generally, if $\beta = \sum_{\xi<\theta} \alpha_\xi$, then $\gamma^\beta = \prod_{\xi<\theta} \gamma^{\alpha_\xi}$.

76. The case when there are only finitely many nonzero terms is obvious, so we may assume the opposite. With no harm we may also discard all the zero

terms, as well, i.e., we may assume $\alpha_i > 0$ for all i. We replace every α_i with a finite sum of powers of ω. Then we get a sum $\beta_0+\beta_1+\cdots$ of powers of ω such that every permuted sum of $\alpha_0+\alpha_1+\cdots$ is a permuted sum of $\beta_0+\beta_1+\cdots$ (but not vice versa). If $\beta_i = \omega^{\gamma_i}$, then let δ be the minimal ordinal for which the set $\{i : \gamma_i \geq \delta\}$ is finite. Call β_i of the first (second) type if $\gamma_i \geq \delta$ ($\gamma_i < \delta$). The finitely many β_i of the first type can produce only finitely many permuted sums. Every permuted sum of $\beta_0+\beta_1+\cdots$ can be written as $x+y$ where x is a finite sum ending with a β_i of the first type, and all terms in y are of the second type. If some $\beta_i = \omega^{\gamma_i}$ of the second type is a term in x, then there is a later $\beta_j = \omega^{\gamma_j}$ with $\gamma_j > \gamma_i$, so we can discard this term, as well. Therefore, there are only finitely many possibilities for x. We show that $y = \omega^\delta$, and this will conclude the proof. Indeed, on the one hand every term in y is smaller than ω^δ, so $y \leq \omega^\delta$. On the other hand, if $\tau < \delta$, then there are infinitely many β_i with $\gamma_i \geq \tau$, so y is at least $\omega^\tau + \omega^\tau + \cdots = \omega^{\tau+1}$. As this holds for every $\tau < \delta$, $y \geq \sup_{\tau<\delta} \omega^{\tau+1} = \omega^\delta$. [W. Sierpiński, Sur les séries infinies de nombres ordinaux. (French) Fund. Math. **36**(1949), 248–253]

77. We use the notations from the preceding proof. Deleting finitely many α_i means deleting finitely many β_i, so we may work with the series $\beta_0+\beta_1+\cdots$. Let us delete finitely many terms, and let the remaining terms be $\beta'_0, \beta'_1, \ldots$. If we do not delete all β_i of the first type, then every permuted sum of $\beta'_0+\beta'_1+\cdots$ can be written as $x'+y'$, where x' is a finite sum ending with a β_i of the first type, and all terms in y' are of the second type, and just as before, there are only finitely many possibilities for x'. The preceding proof also gives $y' = \omega^\delta$, and this concludes the proof in the case when there are non-deleted terms β_i of the first type.

If, however, all terms of the first kind are deleted, then x' is the empty sum, and $\beta'_0 + \beta'_1 + \cdots$ in any order is ω^δ.

78. Consider the sum $\omega^4 + \overbrace{\omega^3 + \omega^3 + \cdots + \omega^3}^{n-1} + \omega + \omega + \cdots$. If we move exactly k of the ω^3's in front of ω^4, then the value of the sum is $\omega^4 + \omega^3 \cdot (n-1-k) + \omega^2$, and these are n different ordinals for $k = 0, 1, \ldots, n-1$. (See also Problem 8.36.)

79. See the next proof.

80. We may assume $\alpha_i > 0$ for all i, otherwise the product is 0, unless all zero terms are deleted.

Let ω^{ξ_i} be the highest power of ω in the normal form of α_i, and let $\eta = \sup_i \xi_i$. We shall prove the statement by transfinite induction on η. The statement is clearly true if $\eta = 0$ (in which case all α_i are positive natural numbers, and their product is ω unless only finitely many α_i's are different from 1). Thus, suppose that the claim has been verified for all ordinals (in place of η) that are smaller than η.

Now we distinguish three cases. First suppose that no ξ_i equals η. Then no matter in what order we take the product and which finitely many terms we delete, we always get ω^η. Next, if infinitely many of the ξ_i's agree with η, then, just as before, no matter in what order we take the product and which finitely many terms we delete, the product is always $\omega^{\eta+1}$. Thus, we only have to consider the case when there are only finitely many terms in the product, say $\alpha_0, \alpha_1, \dots, \alpha_k$, for which the corresponding ξ's are equal to η. Then $\eta_0 = \sup_{i>k} \xi_i$ is less than η, so we can apply the induction hypothesis for the ordinals $\alpha_{k+1}, \alpha_{k+2}, \dots$. Let us now take any permutation π of the natural numbers, consider the product $\alpha_{\pi(0)} \cdot \alpha_{\pi(1)} \cdots$, and let us delete the elements $\alpha_{\pi(i)}$, $i \in I$, from this product, where $I \subset \mathbf{N}$ is an arbitrary finite set. If all the α_i, $i \le k$, are deleted, then, by the induction hypothesis, we can only get finitely many different values for such a product. If l of the α_i's, $i \le k$, are still in the product, and the one with the largest index $\pi(\sigma)$ of them is $\alpha_{i_{\pi,I}}$ ($i_{\pi,I} \in \{1, 2, \dots, k\}$), then $\prod_{\pi(j) \le \pi(i_{\pi,I}),\ i \notin I} \alpha_{\pi(j)} = \omega^{\eta \cdot (l-1)} \alpha_{i_{\pi,I}}$, and the rest of the product, namely $\prod_{\pi(j) > \pi(i_{\pi,I}),\ i \notin I} \alpha_{\pi(j)}$, can take only finitely many different values by the induction hypothesis. Since there are only finitely many choices for $l \le k$ and $i_{\pi,I} \in 0, 1, \dots, k$, we can conclude that there are only finitely many different values for the product.

81. Let $\sum_{i=0}^{\infty} \beta_i$, $\beta_i > 0$, be a sum from which one can get exactly n different sums by taking permutations of the terms (see Problem 78). Then clearly $\omega^{\beta_0} \cdot \omega^{\beta_1} \cdots = \omega^{\sum_i \beta_i}$ is a product, from which one can get exactly n different values by permuting the terms in the product.

82. Consider the sum $\omega + \omega^2 + \omega^3 + \dots + 0$. If we switch the position of ω^k and 0, then the sum becomes $\omega^\omega + \omega^k$, and these are different for different k's.

83. We show by transfinite induction on $\gamma < \omega_1$ that if A is a countable set of ordinals, then the set $S_\gamma(A)$ of sums $\sum_{\beta<\gamma} y_\beta$ of type γ with $y_\beta \in A$, is countable. This is obvious for finite γ. Let $S_{<\gamma}(A) = \cup_{\delta<\gamma} S_\delta(A)$. The finite sums in A form the countable set $S_{<\omega}(A)$, and the infinite sums of type ω are limits of finite sums, hence the claim for $\gamma = \omega$ is true, since then $S_\omega(A)$ is in the closure of $S_{<\omega}(A)$, which is a countable set by Problem 8.42.

Assume now that the claim is known for all ordinals smaller than $\gamma < \omega_1$. Then $S_{<\gamma}(A)$ is countable. γ can be written as a finite or ω type sum of smaller ordinals, so $S_\gamma(A) \subseteq S_\omega(S_{<\gamma}(A))$, and the last set is a countable set by the induction hypothesis. [J. L. Hickman, J. London Math. Soc. (2), 9(1974), 239–244]

84. Consider the product $\omega \cdot \omega^2 \cdot \omega^3 \cdot \dots \cdot 1$. If we switch the position of ω^k and 1, then the product becomes $\omega^{\omega+k}$, and these are different for different k's.

85. The proof is identical with the proof of Problem 83, just say "product" instead of "sum" everywhere.

86. We have $\Gamma(\omega) = 1 \cdot 2 \cdot 3 \cdot = \omega$ and $\Gamma(\omega+1) = \Gamma(\omega) \cdot \omega = \omega^2$. To calculate $\Gamma(\omega \cdot 2)$ consider that $\omega^k \le (\omega+1) \cdots (\omega+k) \le \omega^{k+1}$ for each $k = 1, 2, \ldots$ and so $\Gamma(\omega \cdot 2) = \Gamma(\omega+1) \cdot (\omega+1) \cdot (\omega+2) \cdots = \omega^2 \cdot \lim_k \omega^k = \omega^2 \cdot \omega^\omega = \omega^\omega$. In a similar fashion as before one can see that $\prod_{l=0}^{\infty} (\omega \cdot k + l) = \omega^\omega$ for all $k = 1, 2, \ldots$, and hence $\Gamma(\omega^2) = \Gamma(\omega) \cdot \omega^\omega \cdot \omega^\omega \cdot = \omega \cdot (\omega^\omega)^\omega = \omega^{\omega^2}$. [W. Sierpiński, Cardinal and Ordinal Numbers, Polish Sci. Publ., Warszawa, 1965, XIV.17]

87. Since $\mathcal{F}(0) = \mathcal{F}(0) + \mathcal{F}(0)$, we have $\mathcal{F}(0) = 0$. If we set $\mathcal{F}(1) = \gamma$, then for all α the equality $\mathcal{F}(\alpha+1) = \mathcal{F}(\alpha) + \gamma$ is true, and for limit α we have by continuity $\mathcal{F}(\alpha) = \sup_{\beta<\alpha} \mathcal{F}(\beta)$. Thus, we get by transfinite induction that $\mathcal{F}(\alpha) = \gamma \cdot \alpha$ for all α. Conversely, all these operations satisfy the functional equation $\mathcal{F}(\alpha + \beta) = \mathcal{F}(\alpha) + \mathcal{F}(\beta)$ and they are continuous in the interval topology (see Problems 8.13 and 8.41). [cf. W. Sierpiński, Cardinal and Ordinal Numbers, Polish Sci. Publ., Warszawa, 1965, XIV.18. Theorem 1]

88. For such an operation the equation $\omega^{\mathcal{F}(\alpha+\beta)} = \omega^{\mathcal{F}(\beta)} \cdot \omega^{\mathcal{F}(\alpha)}$ would be true, and by Problem 90 there is no such operation.

89. We have $\mathcal{F}(0) = \mathcal{F}(0) \cdot \mathcal{F}(0)$, so either $\mathcal{F}(0) = 0$ or $\mathcal{F}(0) = 1$. In the former case $\mathcal{F}(\alpha) = \mathcal{F}(\alpha+0) = \mathcal{F}(\alpha) \cdot \mathcal{F}(0) = 0$, i.e., $\mathcal{F}$ is identically 0. Thus, suppose that $\mathcal{F}(0) = 1$, and set $\mathcal{F}(1) = \gamma$. Then $\mathcal{F}(\alpha+1) = \mathcal{F}(\alpha) \cdot \gamma$, so one can easily get by transfinite induction that $\mathcal{F}(\alpha) = \gamma^\alpha$ for all α. Thus, $\mathcal{F}$ must be such an exponential operation, and clearly all these satisfy the equation $\mathcal{F}(\alpha + \beta) = \mathcal{F}(\alpha) \cdot \mathcal{F}(\beta)$ and are continuous in the interval topology.

90. We show that there is no such operation. In fact, from the functional equation $\mathcal{F}(\alpha+\beta) = \mathcal{F}(\beta) \cdot \mathcal{F}(\alpha)$ it follows just as in the preceding proof that if $\mathcal{F}$ is not identically zero, then $\mathcal{F}(0) = 1$, and if we set $\gamma = \mathcal{F}(1)$, then for all finite ordinals k we have $\mathcal{F}(k) = \gamma^k$. Thus, by continuity we get $\mathcal{F}(\omega) = \gamma^\omega$. Then $\mathcal{F}(\omega+1) = \mathcal{F}(1) \cdot \mathcal{F}(\omega) = \gamma \cdot \gamma^\omega = \gamma^\omega$, and proceeding this way we obtain that $\mathcal{F}(\omega + k) = \gamma^\omega$ for all $k < \omega$, and so $\mathcal{F}(\omega + \omega) = \sup_k \mathcal{F}(\omega + k) = \gamma^\omega$. But then $\gamma^\omega = \mathcal{F}(\omega + \omega) = \mathcal{F}(\omega) \cdot \mathcal{F}(\omega) = \gamma^\omega \cdot \gamma^\omega = \gamma^{\omega+\omega}$, which is possible only if $\gamma = 1$. In this case transfinite induction shows that $\mathcal{F}$ is the identically one operation.

91. **(a)** is straightforward from the definition ot $\oplus$.

(b) If α and β are as in (9.3) and

$$\gamma = \omega^{\delta_n} \cdot c_n + \cdots + \omega^{\delta_0} \cdot c_0,$$

then $\beta < \gamma$ implies $b_m < c_m$ for the largest m with $b_m \neq c_m$. But then the coefficients of $\omega^{\delta_n}, \cdots, \omega^{\delta_{m+1}}$ in $\alpha \oplus \beta$ and in $\alpha \oplus \gamma$ are the same, while $a_m + b_m < a_m + c_m$, hence $\alpha \oplus \beta < \alpha \oplus \gamma$.

(c) If $\alpha = 0$, then $x = y = 0$. If, however,

$$\alpha = \omega^{\delta_n} \cdot a_n + \cdots + \omega^{\delta_0} \cdot a_0$$

with nonzero a_i, then the solutions are

$$x = \omega^{\delta_n} \cdot p_n + \cdots + \omega^{\delta_0} \cdot p_0, \qquad y = \omega^{\delta_n} \cdot (a_n - p_n) + \cdots + \omega^{\delta_0} \cdot (a_0 - p_0),$$

where $0 \leq p_i \leq a_i$ are integers, and the number of solutions is $(a_0+1)\cdots(a_n+1)$.

(d) The answer is NO: set $\alpha = 1$ and $x_n = n$ for $n < \omega$. Then $\lim_n x_n = \sup_n x_n = \omega$, while

$$\lim_n (1 \oplus x_n) = \lim_n (n+1) = \omega \neq \omega + 1 = 1 \oplus \lim_n x_n.$$

(e) $\alpha_1 + \cdots \alpha_n$ is some sum of some powers of ω, while $\alpha_1 \oplus \cdots \oplus \alpha_n$ is the sum of the same powers of ω but in nonincreasing order. As $\omega^\gamma + \omega^\delta = \omega^\delta$ for $\gamma < \delta$, it easily follows that $\alpha_1 + \cdots + \alpha_n$ can only be smaller than $\alpha_1 \oplus \cdots \oplus \alpha_n$ if they differ.

Equality holds if and only if for all $1 \leq i < n$ the smallest exponent in the normal form of α_i is at least as large as the largest exponent in the normal form of α_{i+1}.

(f) Assume that

$$\alpha = \omega^{\delta_n} \cdot a_n + \cdots + \omega^{\delta_0} \cdot a_0$$

is the largest of $\alpha_1, \ldots, \alpha_n$. Using (a) and (b) we get

$$\alpha_1 \oplus \cdots \oplus \alpha_n \leq \alpha_1 \oplus \cdots \oplus \alpha_1 = \omega^{\delta_n} \cdot (a_n n) + \cdots + \omega^{\delta_0}(a_0 n).$$

The last ordinal is at most

$$\omega^{\delta_n} \cdot (a_n(n+1)) + \omega^{\delta_{n-1}} \cdot a_{n-1} + \cdots + \omega^{\delta_0}(a_0) = (\omega^{\delta_n} \cdot a_n + \cdots + \omega^{\delta_0} \cdot a_0) \cdot (n+1).$$

[G. Hessenberg, Grundbegriffe der Mengenlehre, Abh. der Friesschen Schule, N. S. **1**(1906), 220]

92. To get $\alpha_1 \oplus \cdots \oplus \alpha_n$ we split every α_i into the ordered union of subsets of order type of the form ω^γ and then consider the nonincreasing sum of all these components. This shows that $\alpha_1 \oplus \cdots \oplus \alpha_n$ does occur as the order type of some set described in the problem.

For the other direction suppose to the contrary that $\alpha = \alpha_1 \oplus \cdots \oplus \alpha_n$, and the order type of some $S = A_1 \cup \cdots \cup A_n$ is bigger than α (where A_i is of order type α_i), and assume that α is minimal with this property. Let $x \in S$ be the element such that the initial segment T of S determined by x has order type α. We have $T = B_1 \cup \cdots \cup B_n$ with $B_i = A_i \cap T$. If β_i is the order type of B_i, then $\beta_i \leq \alpha_i$ for every i, and $\beta_i < \alpha_i$ for at least one i (namely for the one with $x \in A_i$). But then this decomposition witnesses a decomposition of a set of order type α into parts of order types $\beta_1, \ldots, \beta_n$, and here, by Problem 91(a),(b) we have $\beta := \beta_1 \oplus \cdots \oplus \beta_n < \alpha_1 \oplus \cdots \oplus \alpha_n = \alpha$. However, this contradicts the minimality of α, and this contradiction proves the claim. [P. W. Carruth: Arithmetic of ordinals with applications to the theory of ordered Abelian groups, Bull. Amer. Math. Soc., **48**(1942), 262–271]

93. Monotonicity gives by transfinite induction on ξ the inequality $\mathcal{F}(\alpha, \beta + \xi) \geq \mathcal{F}(\alpha, \beta) + \xi$ for any α, β, ξ, and hence commutativity shows that $\mathcal{F}(\alpha + \xi, \beta) \geq \mathcal{F}(\alpha, \beta) + \xi$ is also true. If

$$\alpha = \omega^{\delta_n} \cdot a_n + \cdots + \omega^{\delta_0} \cdot a_0, \qquad \beta = \omega^{\delta_n} \cdot b_n + \cdots + \omega^{\delta_0} \cdot b_0,$$

we prove $\mathcal{F}(\alpha, \beta) \geq \alpha \oplus \beta$ by induction on n. For $n = 0$ we can write

$$\begin{aligned} \mathcal{F}(\omega^\delta \cdot a, \omega^\delta \cdot b) &\geq \mathcal{F}(\omega^\delta \cdot a, 0) + \omega^\delta \cdot b \geq \mathcal{F}(0,0) + \omega^\delta \cdot b + \omega^\delta \cdot a \\ &\geq \omega^\delta \cdot (a + b) = \omega^\delta \cdot a \oplus \omega^\delta \cdot b. \end{aligned}$$

If we have the statement for n terms, then set

$$\overline{\alpha} = \omega^{\delta_n} \cdot a_n + \cdots + \omega^{\delta_1} \cdot a_1, \qquad \overline{\beta} = \omega^{\delta_n} \cdot b_n + \cdots + \omega^{\delta_1} \cdot b_1.$$

Now using the induction hypothesis for $\overline{\alpha}$ and $\overline{\beta}$, we can write

$$\begin{aligned} \mathcal{F}(\alpha, \beta) \geq \mathcal{F}(\alpha, \overline{\beta}) + \omega^{\delta_0} \cdot b_0 &\geq \mathcal{F}(\overline{\alpha}, \overline{\beta}) + \omega^{\delta_0} \cdot (a_0 + b_0) \\ &\geq (\overline{\alpha} \oplus \overline{\beta}) + \omega^{\delta_0} \cdot (a_0 + b_0) = \alpha \oplus \beta. \end{aligned}$$

94. **(a)** In each step exchange in the actual "superbase" form of n_i written in base b_i the base b_i by ω. This gives an ordinal ζ_i. For example, if $n_1 = 23 = 2^{2^2} + 2^2 + 2 + 1$ then $\zeta_1 = \omega^{\omega^\omega} + \omega^\omega + \omega + 1$. If $n_{2i-1} > 0$, then clearly $\zeta_{2i} = \zeta_{2i-1}$, and it is easy to see that because $n_{2i+1} = n_{2i} - 1$, we have $\zeta_{2i+1} < \zeta_{2i}$ (see Problem 8.8(e)). Thus, $\{\zeta_{2i}\}_i$ is a decreasing sequence of ordinals, so it cannot be infinite, i.e., there must be an i with $n_i = 0$.

The proof of part **(b)** is identical. [Goodstein, R. L., J. Symbolic Logic 9, (1944). 33–41]

10

Cardinals

1. If only finitely many a_i's are different from 1, then the product is equal to their product. If, however, there are infinitely many a_i's with $a_2 \geq 2$, then the product is at least $2^{\aleph_0} = \mathbf{c}$. On the other hand, it is clearly not bigger than

$$\mathbf{c}^{\aleph_0} = (2^{\aleph_0})^{\aleph_0} = 2^{\aleph_0^2} = 2^{\aleph_0} = \mathbf{c},$$

and we find that then the product in question is $\mathbf{c}$.

2. Since $\kappa \leq \kappa \cdot \kappa$ is clear, it is enough to show that for all infinite cardinal κ we have $\kappa \cdot \kappa \leq \kappa$. This is true for $\kappa = \aleph_0$, and from here we shall prove the claim by transfinite induction. Thus, suppose that we already know for all infinite cardinals $\sigma < \kappa$ that $\sigma^2 = \sigma$. It is enough to give a well-ordering $\prec$ on $\kappa \times \kappa$ such that every proper initial segment has order type smaller than κ. In fact, then the order type of $\langle \kappa \times \kappa, \prec \rangle$ is at most κ, and so $\kappa \times \kappa$ is similar, and hence equivalent to a subset of κ.

Let $\prec$ be defined as follows: $(\tau_1, \eta_1) \prec (\tau_2, \eta_2)$ if and only if with $\zeta_1 = \max\{\tau_1, \eta_1\}$, $\zeta_2 = \max\{\tau_2, \eta_2\}$ we have $\zeta_1 < \zeta_2$ or $\zeta_1 = \zeta_2$ and $\eta_1 < \eta_2$, or $\zeta_1 = \zeta_2$ and $\eta_1 = \eta_2$ and $\tau_1 < \tau_2$. For $\xi < \kappa$ let $A_\xi = \{(\tau, \eta) : \max\{\tau, \eta\} = \xi\}$. On A_ξ the ordering given by $\prec$ is the following:

$$(\xi, 0) \prec (\xi, 1) \prec \cdots \prec (0, \xi) \prec (1, \xi) \prec \cdots \prec (\xi, \xi),$$

and this is well ordered and of order type $\xi + \xi + 1$. It is clear that $\langle \kappa \times \kappa, \prec \rangle$ is the ordered union of the sets $\langle A_\xi, \prec \rangle$, $\xi < \kappa$, and hence it is well ordered. Any proper initial segment is included in a union $\cup_{\xi < \alpha} A_\xi$ for some $\omega \leq \alpha < \kappa$, and hence it is of cardinality at most

$$\sum_{\xi \leq \alpha} (|\xi| + |\xi| + 1) \leq \sum_{\xi \leq \alpha} (|\alpha| + |\alpha| + 1) = (|\alpha| + |\alpha| + 1)|\alpha| \leq (3|\alpha|)|\alpha|$$
$$= 3|\alpha|^2 = 3|\alpha| \leq |\alpha|^2 = |\alpha| < \kappa,$$

where we have also used that $|\alpha| \le \alpha < \kappa$, and hence by the induction hypothesis we have $|\alpha|^2 = |\alpha|$.

This verifies the induction step, and together with it the claim, as well.

3. Let, e.g., $\kappa \ge \lambda$. Then using the fundamental theorem of cardinal arithmetic, we can write

$$\kappa \le \kappa + \lambda \le 2\kappa \le 2\kappa\lambda \le 2\kappa^2 = 2\kappa \le \kappa^2 = \kappa.$$

4. **a)** The set of sequences of elements of X of length $k = 1, 2, \ldots$ is of cardinality $|X|^k = \kappa^k = \kappa$, and so the set of finite sequences is of cardinality $\omega \cdot \kappa = \kappa$.

b) The set of those functions that map a given finite subset S of X into X is of cardinality $|X|^{|S|} = |X| = \kappa$, and since by part a) there are κ possible choices for S, the cardinality in question is $\kappa\kappa = \kappa$.

5. Let $X = A \cup B$ be a decomposition of X into two disjoint subsets of cardinality κ, and further let us decompose both A and B into κ disjoint subsets of cardinality κ: $A = \cup_{\xi<\kappa} A_\xi$, $B = \cup_{\xi<\kappa} B_\xi$. If $f : \kappa \to A$ and $g : B \to \kappa$ are 1-to-1 mappings, then the decompositions $X = \cup_{\xi<\kappa}(B_\xi \cup \{f(\xi)\})$ and $X = \cup_{\xi<\kappa}(A_\xi \cup \{g(\xi)\})$ show that both A and B are "small". (cf. [W. Sierpiński, Cardinal and Ordinal Numbers, Polish Sci. Publ., Warszawa, 1965, IV.8. Exercise])

6. An ordinal α is a cardinal if and only if no $\xi < \alpha$ is equivalent to α. Now if A is any set of cardinals and α is its supremum, then for $\xi < \alpha$ there is a $\beta \in A$ such that $\xi < \beta$. Therefore, the cardinality of ξ is smaller than that of β, and hence it is smaller than that of α.

7. The statement is clear for finite ρ_1 and ρ_2, so suppose that $\rho_1 \le \rho_2$ and ρ_2 is infinite. We can also suppose that each λ_ξ is at least 1.

Let $I \subset \alpha$ be the set of those ξ for which λ_ξ is infinite. If $\rho_2 = \sum_{\xi\in I}\lambda_\xi$, then using the fact that $\lambda_\xi = \lambda_\xi + \lambda_\xi$, we get $\rho_2 = \sum_{\xi<\alpha}\lambda_\xi$, and $\rho_1 \le \sum_{\xi\in I}\lambda_\xi$. This last inequality means that there is a set A of cardinality ρ_1 which is a subset of a union $\cup_{\xi<\alpha} B_\xi$ where the B_ξ's are disjoint and the cardinality of B_ξ is λ_ξ. But then by looking at the sets $A \cap B_\xi$ it is immediate that $\rho_1 = \sum_{\xi\in I}\lambda'_\xi$ with some $\lambda'_\xi \le \lambda_\xi$. Thus, in this way we can set $\lambda_\xi^{(2)} = \lambda_\xi$ for all ξ and $\lambda_\xi^{(1)} = \lambda'_\xi$ if $\xi \in I$ and $\lambda_\xi^{(1)} = 0$ if $\xi \notin I$.

If, however, $\rho_2 > \sum_{\xi\in I}\lambda_\xi$ (the third possibility $\sum_{\xi\in I}\lambda_\xi > \rho_2$ is not possible, for it would imply $\rho_1 + \rho_2 = \rho_2 < \sum_{\xi\in I}\lambda_\xi$), then $\rho_2 = \sum_{\xi\notin I}\lambda_\xi = \sum_{\xi\notin I} 1 = |\alpha \setminus I|$. Now select an $I_1 \subset \alpha \setminus I$ of cardinality ρ_1 such that the set $(\alpha \setminus I) \setminus I_1$ is still of cardinality ρ_2. If we set $\lambda_\xi^{(2)} = \lambda_\xi$, $\lambda_\xi^{(1)} = 0$ for all $\xi \notin I_1$ and $\lambda_\xi^{(2)} = \lambda_\xi - 1$, $\lambda_\xi^{(1)} = 1$ for $\xi \in I_1$, then these cardinals are appropriate.

8. Let $\langle A, \prec\rangle$ be an ordered set with cofinality α and let $B \subset A$ be a cofinal subset of order type α. It is clear that if $C \subset B$ is cofinal with B, then it is cofinal with A, hence the order type of C is not smaller than α. This gives $\mathrm{cf}(\alpha) \geq \alpha$, and since $\alpha \geq \mathrm{cf}(\alpha)$ is always true, we get $\mathrm{cf}(\alpha) = \alpha$. Since $\mathrm{cf}(\alpha) \leq |\alpha| \leq \alpha$ always holds, it follows that $\mathrm{cf}(\alpha) = |\alpha| = \alpha$, i.e., α is a regular cardinal.

9. Let $\alpha = \mathrm{cf}(\kappa)$, and let γ be the smallest ordinal α for which there is a transfinite sequence $\{\kappa_\xi\}_{\xi<\gamma}$ of cardinals smaller than κ with the property $\kappa = \sum_{\xi<\gamma} \kappa_\xi$.

Choose in κ a sequence $\{\beta_\xi\}_{\xi<\alpha}$ that is cofinal with κ. Then $\kappa = \cup_{\xi<\alpha}\beta_\xi$, and so $\kappa \leq \sum_{\xi<\alpha} |\beta_\xi|$, and all cardinals $|\beta_\xi|$ are smaller than κ. This shows that $\gamma \leq \alpha$. Thus, if $\gamma = \kappa$, then $\kappa = \gamma \leq \alpha \leq \kappa$, and so we must have equality.

If, however, $\gamma < \kappa$, then the sequence $\{\kappa_\xi\}_{\xi<\gamma}$ for which $\sum_{\xi<\gamma} \kappa_\xi = \kappa$, $\kappa_\xi < \kappa$, is cofinal with κ. In fact, in the opposite case there was an ordinal $\zeta < \gamma$ such that $\kappa_\xi < \zeta$ for all $\xi < \gamma$. But then $\sum_{\xi<\gamma} \kappa_\xi \leq \sum_{\xi<\gamma} |\zeta| = |\gamma||\zeta| = \max\{|\gamma|, |\zeta|\} < \kappa$, which contradicts the choice of γ and of the cardinals κ_ξ. Thus, $\{\kappa_\xi\}_{\xi<\gamma}$ is cofinal with κ, and so $\alpha \leq \gamma$.

All these prove that, indeed, $\alpha = \gamma$.

10. See the preceding problem, and note that if $\kappa = \lambda^+$, then $\lambda\lambda < \kappa$.

11. If $\kappa = \lambda^+$, then the sum of fewer than κ (i.e., at most λ) cardinals all of which are smaller than κ (i.e., at most λ) is at most $\lambda\lambda = \lambda$.

12. $\omega = \aleph_0$ is regular, and so are each $\aleph_n$, $n = 1, 2, \ldots$, since they are successor cardinals. But $\aleph_\omega$ is not regular, since its cofinality is ω, hence this is the smallest infinite singular cardinal. In a similar fashion, the next two ones are $\aleph_{\omega+\omega}$ and $\aleph_{\omega+\omega+\omega}$.

13. If $\alpha = \beta+1$ is a successor ordinal, then $\aleph_\alpha = (\aleph_\beta)^+$ is a successor cardinal, and so by Problem 11 it is regular, hence $\mathrm{cf}(\aleph_\alpha) = \aleph_\alpha$. If α is a limit ordinal, then $\{\aleph_\beta\}_{\beta<\alpha}$ is a cofinal sequence in $\aleph_\alpha$, hence it has the same cofinality as $\aleph_\alpha$ (see the proof of Problem 8), and this gives $\mathrm{cf}(\aleph_\alpha) = \mathrm{cf}(\alpha)$.

14. First we prove the necessity of the condition, i.e., assume that the cardinality of H is at most $\aleph_n$. Without loss of generality, we may assume $H = \aleph_n$. The $n = 0$ case was the content of Problem 2.25, and from here we proceed by induction. Thus, suppose that if the cardinality of a set K is at most $\aleph_{n-1}$, then K^{n+1} can be represented in the form $A_1 \cup \cdots \cup A_{n+1}$, where A_k is finite in the direction of the kth coordinate. Consider H^{n+2}. It can be written as the union of the sets S_j and $R_{i,j}$, $i, j = 1, \ldots n+2$, where S_j is the set of those $(n+2)$-tuples $(\xi_1, \ldots \xi_{n+2})$, $\xi_i < \aleph_n$ for which all ξ_i, $i \neq j$ are smaller than ξ_j, and $R_{i,j}$ is the set of those $(n+2)$-tuples in which $\xi_j = \xi_i$ and all other $\xi_k \leq \xi_i$. It is enough to represent each S_j and each $R_{i,j}$ in the

form $A_1 \cup \cdots \cup A_{n+2}$, where A_k is finite in the direction of the kth coordinate. For $R_{i,j}$ this follows from the induction hypothesis, for $R_{i,j}$ can be identified with a subset of H^{n+1}, since the ith and jth coordinates are the same. The set S_j is the disjoint union of the sets $S_{j,\xi}$, $\xi < \aleph_n$, where $S_{j,\xi}$ is the set of those $(n+2)$-tuples in S_j for which $\xi_j = \xi$ (and recall that this was the largest component). Since the jth component is fixed in the elements of $S_{j,\xi}$, each $S_{j,\xi}$ can be identified with a subset of H^{n+1}, and so by the induction hypotheses $S_{j,\xi} = \cup_{1\le k\le n+2,\ k\ne j} A_{k,\xi}$, where $A_{k,\xi}$ is finite in the direction of the kth coordinate. Now set $A_k = \cup_{\xi<\aleph_n} A_{k,\xi}$ (for $k = j$ this gives $A_j = \emptyset$). It is clear that $S_j = \cup_{k=1}^{n+2} A_k$. Now A_j is actually the empty set, and if $k \ne j$ and $\xi_1, \ldots, \xi_{k-1}, \xi_{k+1}, \ldots \xi_{n+2} < \aleph_n$ are fixed, then $(\xi_1, \ldots, \xi_{k-1}, \xi, \xi_{k+1}, \ldots \xi_{n+2}) \in A_k$ exactly when this $(n+2)$-tuple belongs to A_{k,ξ_j}. Hence by the selection of the sets A_{k,ξ_j}, there are only finitely many such ξ's. This proves the induction step, and the necessity is proved.

The sufficiency is also proved by induction. The $n = 0$ case was done in Problem 2.25. Now we verify the induction step, so suppose the sufficiency has already been verified for $(n-1)$ instead of n. Let H^{n+2} be represented in the form $A_1 \cup \cdots \cup A_{n+2}$, where A_k is finite in the direction of the kth coordinate, and suppose to the contrary that H has cardinality at least $\aleph_{n+1}$. Select a subset K of H of cardinality $\aleph_n$. Then $\left(H \times (\overbrace{K \times \cdots \times K}^{n+1})\right) \cap A_1$ is of cardinality at most $\aleph_n$, since for each $x_1, \ldots, x_{n+1} \in K$ the number of y's with $(y, x_1, \ldots x_{n+1}) \in A_1$ is finite. Thus, there is a $y_0 \in H$ for which there are no $x_1, \ldots x_{n+1} \in K$ such that $(y_0, x_1, \ldots x_{n+1}) \in A_1$. But then $\{y\} \times K^{n+1} = \cup_{k=2}^{n+2} A_k$, and here each A_k is finite in the direction of the kth coordinate. Thus, by the induction hypothesis, K must be of cardinality at most $\aleph_{n-1}$, and this contradiction proves the claim.

15. This problem can be solved along the same lines as the previous one. Since the induction steps are the same, we only have to verify the $n = 0$ case, from where the induction starts, but actually that was the content of Problem 2.40.

16. Since 2^κ is the cardinality of the set of subsets of κ (see Problem 3.10, c)), the statement is equivalent to what was proved in Problem 3.13.

17. Let A_i be disjoint sets of cardinality ρ_i and B_i be sets of cardinality κ_i. We have to show that if $F : \cup_{i\in I} A_i \to \prod_{i\in I} B_i$ is any mapping, then F is not onto. For each $i \in I$ the set $\{F(a)(i) \ :\ a \in A_i\}$ is of cardinality at most ρ_i; thus, there is a point $a_i \in B_i \setminus \{F(a)(i) \ :\ a \in A_i\}$. Now for the choice function $f(i) = a_i$ belonging to $\prod_{i\in I} B_i$ there is no $b \in \cup_{i\in I} A_i$ such that $F(b) = f$. In fact, if $b \in \cup_{i\in I} A_i$, then $b \in A_{i_0}$ for some $i_0 \in I$, and then $f(i_0) = a_{i_0} \ne F(b)(i_0)$ by the choice of a_{i_0}. This proves the claim.

18. Clearly θ is a limit ordinal. We may assume that $\{\kappa_\xi\}_{\xi<\theta}$, is an increasing sequence. Then $\kappa_\xi < \kappa_{\xi+1}$, and so by the preceding problem $\kappa = \sup_{\xi<\rho} \kappa_\xi \le \sum_{\xi<\theta} \kappa_\xi < \prod_{\xi<\theta} \kappa_{\xi+1} = \prod_{\xi<\theta} \kappa_\xi$.

19. If κ is regular, then

$$\prod_{\xi<\kappa} \kappa_\xi \le \kappa^\kappa \le (2^\kappa)^\kappa = 2^{\kappa^2} = 2^\kappa \le \prod_{\xi<\kappa} \kappa_\xi.$$

Let us now assume κ to be singular. Then the supremum of the cardinals κ_ξ must be κ, therefore we may assume that $\{\kappa_\xi\}_{\xi<\kappa}$ is an increasing transfinite sequence. Using the fundamental theorem of cardinal arithmetic we can write $\mathrm{cf}\,(\kappa) = \cup_{\alpha<\mathrm{cf}\,(\kappa)} I_\alpha$, where the I_α's are disjoint and each of them is of cardinality $\mathrm{cf}\,(\kappa)$. Thus,

$$\begin{aligned}\kappa^{\mathrm{cf}\,(\kappa)} &= \prod_{\alpha<\mathrm{cf}\,(\kappa)} \kappa = \prod_{\alpha<\mathrm{cf}\,(\kappa)} \left(\sup_{\xi\in I_\alpha} \kappa_\xi\right) \le \prod_{\alpha<\mathrm{cf}\,(\kappa)} \left(\prod_{\xi\in I_\alpha} \kappa_\xi\right) \\ &= \prod_{\xi<\kappa} \kappa_\xi \le \prod_{\xi<\kappa} \kappa = \kappa^{\mathrm{cf}\,(\kappa)}.\end{aligned}$$

20. If κ is regular, then $\kappa^{\mathrm{cf}\,(\kappa)} = \kappa^\kappa \ge 2^\kappa > \kappa$ (see Problem 16). If, however, $\kappa > \omega$ is singular, then κ is the supremum of all the infinite cardinals smaller than κ, hence by Problems 18 and 19 we have $\kappa^{\mathrm{cf}\,(\kappa)} > \kappa$.

21. If we had $\mathrm{cf}(\lambda^\kappa) \le \kappa$, then we would have $(\lambda^\kappa)^{\mathrm{cf}\,(\lambda^\kappa)} \le (\lambda^\kappa)^\kappa = \lambda^{\kappa^2} = \lambda^\kappa$, and this would contradict the preceding problem.

22. κ^λ is the cardinality of ${}^\lambda\kappa$; the set of functions $f : \lambda \to \kappa$. But since $\lambda < \mathrm{cf}\,(\kappa)$, the range of such a function cannot be cofinal with κ, so such an f actually maps λ into some ordinal $\xi < \kappa$. Thus,

$$\begin{aligned}\kappa^\lambda &= |{}^\lambda\kappa| = |\cup_{\xi<\kappa} {}^\lambda\xi| \le \sum_{\rho<\kappa} \sum_{|\xi|=\rho} |{}^\lambda\xi| \\ &= \sum_{\rho<\kappa} \sum_{|\xi|=\rho} \rho^\lambda \le \left(\sum_{\rho<\kappa} \rho^\lambda\right) \kappa \le \left(\sum_{\rho<\kappa} \kappa^\lambda\right) \kappa \le \kappa^\lambda \kappa\kappa = \kappa^\lambda.\end{aligned}$$

23. It easily follows that $\mathrm{cf}\,(\kappa) = \mathrm{cf}\,(\alpha)$; therefore, we can apply the preceding problem. Now

$$\sum_{\xi<\alpha} \kappa_\xi^\lambda \le \sum_{\rho<\kappa} \rho^\lambda \le \kappa \sum_{\xi<\alpha} \kappa_\xi^\lambda = \sum_{\xi<\alpha} \kappa_\xi^\lambda,$$

hence by the formula in Problem 22

$$\kappa^\lambda = \kappa \sum_{\xi<\alpha} \kappa_\xi^\lambda = \sum_{\xi<\alpha} \kappa_\xi^\lambda.$$

24. Clearly, $2^\lambda \geq \kappa$ (by monotonicity of exponentation). As λ is singular, it can be written as $\lambda = \sum\{\lambda_\alpha : \alpha < \mathrm{cf}(\lambda)\}$ with $\lambda_\alpha < \lambda$. Then $2^\lambda = \prod_{\alpha<\mathrm{cf}(\lambda)} 2^{\lambda_\alpha} \leq \kappa^{\mathrm{cf}(\lambda)} = \kappa$ since the assumption gives that $\kappa^\tau = \kappa$ for every $\tau < \lambda$: if $\mu < \tau < \lambda$ then $\kappa^\tau = (2^\tau)^\tau = 2^\tau = \kappa$.

25. Assume that $\gamma \geq \omega$. Let δ be the minimal ordinal such that $\delta + \gamma > \gamma$. Clearly, $\omega \leq \delta \leq \gamma$ and δ is a limit ordinal. Pick a cardinal $\aleph_\alpha > \delta$ and let $\lambda = \aleph_{\alpha+\delta}$, a singular cardinal (note that $\mathrm{cf}\,(\aleph_{\alpha+\delta}) = \mathrm{cf}\,(\delta) \leq |\delta| < \aleph_\alpha < \aleph_{\alpha+\delta}$). Then for any $\tau < \delta$

$$2^{\aleph_{\alpha+\tau}} = \aleph_{\alpha+\tau+\gamma} = \aleph_{\alpha+\gamma}.$$

As $\lambda = \aleph_{\alpha+\delta}$ is singular, we get from Problem 24 that $2^\lambda = \aleph_{\alpha+\gamma}$, and at the same time

$$2^\lambda = \aleph_{\alpha+\delta+\gamma} > \aleph_{\alpha+\gamma},$$

because $\delta + \gamma > \gamma$, hence $\alpha + \delta + \gamma > \alpha + \gamma$. This contradiction shows that δ has to be finite.

26. **(a)** We show how to calculate, by transfinite recursion, 2^κ. If κ is regular, then $2^\kappa = \kappa^\kappa = \kappa^{\mathrm{cf}(\kappa)}$. Assume that κ is a singular cardinal and we know 2^τ for $\tau < \kappa$. Set $\kappa = \sum_{\xi<\mathrm{cf}(\kappa)} \kappa_\xi$ with $\kappa_\xi < \kappa$. If the sequence $\{2^{\kappa_\xi}\}$ is eventually constant then this eventual constant value will be 2^κ by Problem 24. Otherwise, if $\lambda = \sum_{\xi<\mathrm{cf}(\kappa)} 2^{\kappa_\xi}$, then λ is a singular cardinal with cofinality $\mathrm{cf}(\kappa)$ and

$$2^\kappa = \prod_{\xi<\mathrm{cf}(\kappa)} 2^{\kappa_\xi} \leq \lambda^{\mathrm{cf}(\lambda)} \leq (2^\kappa)^{\mathrm{cf}(\lambda)} \leq (2^\kappa)^\kappa = 2^{\kappa\kappa} = 2^\kappa$$

so $2^\kappa = \lambda^{\mathrm{cf}(\lambda)}$.

(b) We determine κ^λ by transfinite recursion on κ, and inside that, by transfinite recursion on λ. $\kappa^n = \kappa$ for $1 \leq n < \aleph_0$. If $\lambda < \mathrm{cf}(\kappa)$, we use Problem 22. If $\lambda = \mathrm{cf}(\kappa)$, then $\kappa^{\mathrm{cf}(\kappa)}$ is given. If $\lambda \geq \kappa$, then $\kappa^\lambda = 2^\lambda$ an already calculated value (see part (a)). If $\mathrm{cf}(\kappa) < \lambda < \kappa$, then κ is singular. In this case if there is some $\tau < \kappa$ with $\tau^\lambda > \kappa$, then $\kappa^\lambda = \tau^\lambda$. If, on the other hand, $\tau^\lambda < \kappa$ holds for every $\tau < \kappa$ then let $\{\kappa_\xi : \xi < \mathrm{cf}(\kappa)\}$ be cofinal in κ, and then

$$\kappa^\lambda \leq \left(\prod_{\xi<\mathrm{cf}(\kappa)} \kappa_\xi\right)^\lambda = \prod_{\xi<\mathrm{cf}(\kappa)} \kappa_\xi^\lambda \leq \prod_{\xi<\mathrm{cf}(\kappa)} \kappa = \kappa^{\mathrm{cf}(\kappa)} \leq \kappa^\lambda.$$

27. (a) For $n = 0$ the statement is clear, and we use induction on n. Thus, suppose that (a) has been proven for some n. If $\lambda \geq \aleph_{\alpha+n+1}$, then

$$2^\lambda \leq \aleph_{\alpha+n+1}^\lambda \leq \left(2^{\aleph_{\alpha+n+1}}\right)^\lambda = 2^{\lambda\aleph_{\alpha+n+1}} = 2^\lambda,$$

and a similar computation show that the right-hand side in (a) is also 2^λ. Thus, in this case the claim is true.

If, however, $\lambda < \aleph_{\alpha+n+1}$, then λ is smaller than the cofinality of $\aleph_{\alpha+n+1}$ (see Problem 13), and so we obtain from Problem 22 and the induction hypothesis that

$$\aleph^\lambda_{\alpha+n+1} = \left(\sum_{\rho\le\aleph_{\alpha+n}} \rho^\lambda\right)\aleph_{\alpha+n+1} \le \aleph^\lambda_{\alpha+n}\aleph_{\alpha+n}\aleph_{\alpha+n+1}$$
$$= \aleph^\lambda_\alpha\aleph^2_{\alpha+n}\aleph_{\alpha+n+1} = \aleph^\lambda_\alpha\aleph_{\alpha+n+1}.$$

(b) If λ is finite, then (b) is clear. On the other hand, for infinite λ it follows from the $\alpha = 0$ special case of part (a), since then, as we have seen in the solution of part (a) above, $\aleph_0^\lambda = 2^\lambda$. [F. Hausdorff, Jahresb. deutschen Math. Ver., **13**(1940), 569–571, F. Bernstein, Math. Annalen, **61**(1905), 117–155]

28. If $\prod_{n<\omega}\aleph_n = 2^{\aleph_0}$, then $2^{\aleph_0}$ must be larger than every $\aleph_n$, i.e., $2^{\aleph_0} \ge \aleph_\omega$. But by Problem 21 we must have $\mathrm{cf}(2^{\aleph_0}) > \omega_0 = \mathrm{cf}(\aleph_\omega)$, hence $2^{\aleph_0} > \aleph_\omega$. On the other hand, if $2^{\aleph_0} > \aleph_\omega$ holds, then $2^{\aleph_0} \le \prod_n \aleph_n \le \left(2^{\aleph_0}\right)^{\aleph_0} = 2^{\aleph_0}$. Thus, we have $\prod_{n<\omega}\aleph_n = 2^{\aleph_0}$ if and only if $\mathbf{c} > \aleph_\omega$.

29. Monotonicity gives $\prod_n \aleph_n \le \aleph_\omega^{\aleph_0}$. For the other inequality notice that $\aleph_\omega^{\aleph_0} = (\sum_{n<\omega}\aleph_n)^{\aleph_0}$, when multiplied out, is the sum of c many terms, each the product of infinitely many of the $\aleph_n$'s. We have, therefore,

$$\aleph_\omega^{\aleph_0} \le 2^{\aleph_0}\prod_{n<\omega}\aleph_n.$$

As the first term of the right-hand side is clearly less than or equal to the second, we have the desired inequality $\aleph_\omega^{\aleph_0} \le \prod_n \aleph_n$.

30. If $2^{\aleph_k} = \aleph_{n(k)}$ then $2^{\aleph_\omega} = \prod_{k<\omega} 2^{\aleph_k} = \prod_{k<\omega}\aleph_{n(k)}$, and this is equal to $\prod_{k<\omega}\aleph_k$ as it is the product of infinitely many $\aleph_k$'s (we can cut out repetitions). Now apply Problem 29.

31. Clearly, for all cardinal κ of the form λ^ρ we have $\kappa^\rho = (\lambda^\rho)^\rho = \lambda^{\rho^2} = \lambda^\rho = \lambda$. On the other hand, if $\rho \ge \mathrm{cf}\,(\kappa)$, then by Problem 20 we have $\kappa^\rho > \kappa$.

32. Let κ be an arbitrary cardinal. We find some $\lambda > \kappa$ with $\lambda^{\aleph_0} < \lambda^{\aleph_1}$. For $\alpha \le \omega_1$ construct the following sequence: $\lambda_0 = \kappa$, $\lambda_{\alpha+1} = 2^{\lambda_\alpha}$, and if α is a limit ordinal, then let $\lambda_\alpha = \sup\{\lambda_\beta : \beta < \alpha\}$. If $\lambda = \lambda_{\omega_1}$ then $\mu^{\aleph_0} < \lambda$ holds for $\mu < \lambda$, therefore $\lambda^{\aleph_0} = \lambda$ by Problem 22. On the other hand, $\mathrm{cf}(\lambda) = \aleph_1$, so $\lambda^{\aleph_1} = \lambda^{\mathrm{cf}(\lambda)} > \lambda$ by Problem 20.

33. Assume that κ is the least counterexample to the statement: there are $\tau_0 < \tau_1 < \cdots$ that $2^{\tau_n} < \kappa$ and $\kappa^{\tau_0} < \kappa^{\tau_1} < \cdots$. Assume first that for some n we have $\rho^{\tau_n} < \kappa$ for every $\rho < \kappa$. If $\tau_n < \mathrm{cf}(\kappa)$, then

$$\kappa^{\tau_n} = \Big(\sum_{\rho<\kappa} \rho^{\tau_n}\Big)\kappa = \kappa$$

by Problem 22. If, however, $\tau_n \geq \mathrm{cf}(\kappa)$, then κ is singular, $\kappa = \sum_{\xi<\mathrm{cf}\,(\kappa)} \kappa_\xi$, $\kappa_\xi < \kappa$, and then by Problem 19

$$\kappa^{\mathrm{cf}(\kappa)} \leq \kappa^{\tau_n} = \kappa^{\tau_n \mathrm{cf}(\kappa)} = \prod \kappa_\xi^{\tau_n} \leq \kappa^{\mathrm{cf}(\kappa)}.$$

Taking into account that $\{\kappa^{\tau_n}\}_{n<\omega}$ is an increasing sequence, it follows that for all but 2 of the τ_n's there is some $\rho < \kappa$ that $\rho^{\tau_n} \geq \kappa$, so for $n \geq N$ we have a minimal ρ_n with $\rho_n^{\tau_n} \geq \kappa$. As $\rho_N \geq \rho_{N+1} \geq \cdots$, this sequence is constant from some point, so there is some $\rho < \kappa$ such that $\rho^{\tau_n} \geq \kappa$ for $n \geq M$. Now $2^{\tau_n} \geq \rho$ is impossible, for then we would have $2^{\tau_n} = 2^{\tau_n \tau_n} \geq \rho^{\tau_n} \geq \kappa$. Hence $2^{\tau_n} < \rho$ and clearly $\rho^{\tau_n} = \kappa^{\tau_n}$ also hold, so ρ is a smaller counterexample, a contradiction. [A. Hajnal]

34. By Problems 2 and 20 we have $\aleph_0 \leq \rho \leq \mathrm{cf}\,(\kappa)$. If ρ was singular, then we could write $\rho = \sum_{\xi<\lambda} \rho_\xi$, where $\lambda := \mathrm{cf}\,(\rho) < \rho$ and each ρ_ξ is smaller than ρ. Thus, by the definition of ρ we would have

$$\kappa^\rho = \prod_{\xi<\lambda} \kappa^{\rho_\xi} = \prod_{\xi<\lambda} \kappa = \kappa^\lambda = \kappa,$$

which contradicts the definition of ρ. Thus, ρ must be regular.

Clearly, $\rho_\omega = \rho_{\omega_\omega} = \aleph_0$ because the cofinality of these cardinals is ω.

35. Assume that κ is singular with cofinality μ. Then $\kappa = \sum_{\xi<\mu} \kappa_\xi$ for some infinite cardinals $\kappa_\xi < \kappa$. By assumption, $2^{\kappa_\xi} = \mathbf{c}$ for every $\xi < \mu$, and $2^\mu = \mathbf{c}$. This gives $2^\kappa = \prod_{\xi<\mu} 2^{\kappa_\xi} = \prod_{\xi<\mu} \mathbf{c} = \mathbf{c}^\mu = (2^\mu)^\mu = 2^\mu = \mathbf{c}$, a contradiction.

36. Clearly, $\aleph_\kappa = \sup_n \aleph_{\kappa_n} = \sup_n \aleph_{n+1} = \kappa$. On the other hand, if $\lambda = \aleph_\lambda$, then $\lambda \geq \omega = \kappa_0$, and by induction $\lambda = \aleph_\lambda \geq \aleph_{\kappa_n} = \kappa_{n+1}$, i.e., $\lambda \geq \sup_n \kappa_n$. Thus, this supremum is indeed the smallest cardinal with the stated property.

37. Given any cardinal λ construct as in the preceding problem $\kappa_0 = \lambda$, $\kappa_{n+1} = \aleph_{\kappa_n}$ and $\kappa = \sup_n \kappa_n$. Then exactly as there, $\kappa = \aleph_\kappa$, so for any cardinal λ there is a "large" cardinal $\kappa \geq \lambda$. Now if we enumerate the large cardinals as $\lambda_0 < \lambda_1 < \cdots < \lambda_\alpha < \cdots$, and starting with $\kappa = \lambda_0$ we define $\kappa_{n+1} = \lambda_{\kappa_n}$, then $\lambda = \sup_n \kappa_n$ will be a cardinal with index $\geq \sup_n \kappa_n = \lambda$, so it is a "large" cardinal, and there are $\sup_n \kappa_n = \lambda$ "large" cardinals that are smaller than λ.

38. Assume GCH. If $2 \leq \kappa \leq \aleph_1$ then $\kappa^{\aleph_0} = \aleph_1$, $\kappa^{\aleph_1} = \aleph_2$, $\kappa^{\aleph_2} = \aleph_3$. If, however, $\kappa \geq \aleph_2$, then $\kappa \leq \kappa^{\aleph_0} \leq \kappa^{\aleph_1} \leq \kappa^{\aleph_2} \leq \kappa^\kappa \leq (2^\kappa)^\kappa = 2^\kappa = \kappa^+$, so each one of the values of $\kappa^{\aleph_0}, \kappa^{\aleph_1}, \kappa^{\aleph_2}$ is either κ or κ^+; therefore, there cannot be three different values of them. Thus, the answer is $2 \leq \kappa \leq \aleph_1$.

39. Assume first that α is a limit ordinal. Then, as

$$\aleph_\alpha = \sum_{\beta<\alpha} \aleph_\beta < \prod_{\beta<\alpha} \aleph_\beta$$

(see Problem 17), we find that the product is at least $\aleph_{\alpha+1}$. On the other hand, it is at most $\aleph_{\alpha+1}^{\aleph_\alpha} = \left(2^{\aleph_\alpha}\right)^{\aleph_\alpha} = 2^{\aleph_\alpha} = \aleph_{\alpha+1}$ by GCH; therefore, in this case $\prod_{\beta<\alpha} \aleph_\beta = \aleph_{\alpha+1}$.

Assume next that $\alpha = \gamma + 1$, γ is a limit ordinal. Then, using the already proved limit case, we get

$$\prod_{\beta<\gamma+1} \aleph_\beta = \Big(\prod_{\beta<\gamma} \aleph_\beta\Big) \cdot \aleph_\gamma = \aleph_{\gamma+1}\aleph_\gamma = \aleph_{\gamma+1} = \aleph_\alpha.$$

Assume now that $\alpha = \gamma + 2$, γ is arbitrary. Then the product is at least $\aleph_{\gamma+1}$ as this is a factor. On the other hand,

$$\prod_{\beta<\gamma+2} \aleph_\beta \le \aleph_{\gamma+1}^{|\gamma+2|} \le \aleph_{\gamma+1}^{\aleph_\gamma} = \left(2^{\aleph_\gamma}\right)^{\aleph_\gamma} = 2^{\aleph_\gamma} = \aleph_{\gamma+1}.$$

The uncovered cases are $\alpha = 0, 1$ when the result is immediate.

40. Assume κ is infinite. If $\lambda = 0$, then $\kappa^\lambda = 1$.

If $1 \le \lambda < \mathrm{cf}(\kappa)$, then $\kappa^\lambda = \kappa$ as (see Problem 22)

$$\kappa^\lambda = \kappa\Big(\sum_{\tau<\kappa} \tau^\lambda\Big) \le \kappa\Big(\sum_{\tau<\kappa} 2^{\max(\tau,\lambda)}\Big) = \kappa\Big(\sum_{\tau<\kappa} \max(\tau^+, \lambda^+)\Big) \le \kappa^2 = \kappa$$

(the inequality $\kappa^\lambda \ge \kappa$ is trivial).

If $\mathrm{cf}(\kappa) \le \lambda \le \kappa$, then $\kappa^\lambda = \kappa^+$ as (see Problem 20)

$$\kappa < \kappa^{\mathrm{cf}(\kappa)} \le \kappa^\lambda \le \kappa^\kappa \le (2^\kappa)^\kappa = 2^\kappa = \kappa^+.$$

If $\kappa < \lambda$, then $\kappa^\lambda = \lambda^+$ as

$$\lambda^+ = 2^\lambda \le \kappa^\lambda \le \lambda^\lambda \le (2^\lambda)^\lambda = 2^{\lambda\lambda} = 2^\lambda = \lambda^+.$$

11

Partially ordered sets

1. Suppose that $\langle A, \prec \rangle$ is partially ordered and it does not include an infinite antichain. For $a \in A$ let $B_a^A = \{x \in A \ : \ a \prec x\}$ and $S_a^A = \{x \in A \ : \ x \prec a\}$ be the set of elements that are bigger or smaller than a, respectively. Let C_0 be a maximal antichain (see Zorn's lemma in Chapter 14). Then every element of $A \setminus C_0$ is either bigger or smaller than an element of C_0; thus, there is an element $x_0 \in C_0$ such that either $B_{x_0}^A$ or $S_{x_0}^A$ is infinite. In the former case set $A_0 = B_{x_0}^A$, in the latter case set $A_0 = S_{x_0}^A$. Now repeat this process with A_0 to get an element $x_1 \in A_0$ and an infinite set $A_1 \subset A_0$ such that $A_1 = B_{x_1}^{A_0}$ or $A_1 = S_{x_1}^{A_0}$. Repeat again the same thing with A_1 instead of A_0, etc. It is clear that the process never terminates, and we get a set $\{x_0, x_1, \ldots\}$ that is ordered. [W. Sierpiński, Cardinal and Ordinal Numbers, Polish Sci. Publ., Warszawa, 1965, XI.2/2]

For an alternative proof, see Problem 24.3.

2. The statement immediately follows from the next problem.

3. Let $\langle A, \prec \rangle$ be the partially ordered set in question. For $a \in A$ let $\rho(a)$ be the length of the longest increasing chain having a as its smallest element. By the assumption $1 \le \rho(a) \le k$ for all $a \in A$, and the statement follows if we show that the elements that have the same ρ value are incomparable. But that is clear: if $a \prec b$, then the definition of ρ shows that $\rho(a) \ge \rho(b) + 1$.

4. First we prove the claim for finite sets.

Let $\langle A, \prec \rangle$ be a partially ordered set of n elements with at most k pairwise incomparable elements. The statement is clear if $n = 1$, and from here the proof goes on by induction of n. Thus, suppose we know the statement for all sets with at most $n-1$ elements, and let C be a maximal chain in $\langle A, \prec \rangle$. If $A \setminus C$ has at most $k-1$ pairwise incomparable elements, then by the induction hypothesis $A \setminus C = \cup_{j=1}^{k-1} C_j$ with some chains C_j, and we are done. In the opposite case let $X = \{x_1, \ldots, x_k\}$ be a maximal antichain of k elements in $A \setminus C$, and set

$$A_1 = \{x \in A \, : \, x \preceq x_j \text{ for some } j\}, \quad A_2 = \{x \in A \, : \, x_j \preceq x \text{ for some } j\}.$$

By the maximality of X we have $A = A_1 \cup A_2$, and by the maximality of C the largest element of C is not in A_1, and the smallest element of C is not in A_2. Thus, $|A_1|, |A_2| < n$, and we can use the induction hypothesis to write $A_1 = \cup_{j=1}^k B_j^{(1)}$ and $A_2 = \cup_{j=1}^k B_j^{(2)}$ with some chains $B_j^{(1)}$ and $B_j^{(2)}$. Note that each x_l belongs to exactly one of the $B_j^{(1)}$'s and to exactly one of the $B_j^{(2)}$'s, so we may assume that $x_j \in B_j^{(1)}$ and $x_j \in B_j^{(2)}$. But then $C_j = B_j^{(1)} \cup B_j^{(2)}$ is a chain, and since $A = \cup_{j=1}^k C_j$, the induction step is complete.

Now we turn to the general case where we allow $\langle A, \prec \rangle$ to be infinite. We use induction on k, the case $k = 1$ being trivial. Thus, suppose that the claim has already been verified for $k - 1$, and we are going to prove it for k. Let $\mathcal{M}$ be the set of all subsets H of A for which it is true that if $S \subset A$ is any finite subset, then there is a decomposition of S into k chains such that $H \cap S$ is included in one of them. By the finite case of the problem that we have already verified above, each one element subset is in $\mathcal{M}$, and clearly every element of $\mathcal{M}$ is a chain. It is easy to see that the union of any subset of $\mathcal{M}$ that is ordered by inclusion is again in $\mathcal{M}$, hence by Zorn's lemma (Chapter 14) there is a maximal (with respect to inclusion) set H^* in $\mathcal{M}$. We claim that every antichain in $A \setminus H^*$ has at most $k - 1$ elements. Since then the induction hypothesis says that then $A \setminus H^*$ can be represented as the union of at most $k - 1$ chains, and these chains with H^* form a family of at most k chains that cover A, the proof will be over.

Let us suppose to the contrary that $A \setminus H^*$ has a k-element antichain $K = \{a_1, \ldots, a_k\}$. By the maximality of H^*, for each $a_j \in K$ there is a finite subset S_j of A such that S_j does not have a representation as the union of k chains such that one of them includes $S_j \cap (H^* \cup \{a_j\}$. Since on the other hand, H^* does have this property, it follows that necessarily we have $a_j \in S_j$. Apply again the same property of H^* for the finite set $S = S_1 \cup \cdots \cup S_k$, to conclude that there is a representation $S = C_1 \cup \cdots \cup C_k$ of S as a union of k chains such that for some j_0 we have $H^* \cap S \subseteq C_{j_0}$. Note that each C_j contains exactly one of the points $a_1, a_2, \ldots, a_k$ (the C_j's are chains and $C_1, \ldots, C_k$ cover S), and we may number them in such a way that $a_j \in C_j$ for all $j = 1, 2, \ldots, k$.

But then

$$S_{j_0} = (C_1 \cap S_{j_0}) \cup \cdots \cup (C_k \cap S_{j_0})$$

is a representation of S_{j_0} into the union of k chains such that $H^* \cap S_{j_0} \subseteq C_{j_0} \cap S_{j_0}$, which implies

$$(H^* \cup \{a_{j_0}\}) \cap S_{j_0} \subseteq C_{j_0} \cap S_{j_0},$$

which is not possible in view of the definition of a_{j_0}. This contradiction proves that there are at most $k-1$ pairwise incomparable elements in $A \setminus H^*$, as was claimed.

The reduction of the general case to the finite one can be also done via the de Bruijn–Erdős theorem (Problem 23.8). In fact, consider the graph with vertex set A where two points are connected if they are incomparable. In a coloring a set of points with the same color forms a chain, hence a subgraph is colorable with k colors if and only if it is the union of k chains. Now the de Bruijn–Erdős theorem asserts that if every subset of A is the union of k chains then so is the set itself. [R. P. Dilworth, A decomposition theorem for partially ordered sets, Ann. Math. **51**(1950), 161–165]

5. The counterexample will be built on the Cartesian product $\omega_1 \times \omega_1$. We make $(\alpha, \beta) \prec (\alpha', \beta')$ if and only if $\alpha < \alpha'$ and $\beta > \beta'$. In a supposed infinite decreasing/increasing sequence the first/second coordinates would give an infinite decreasing sequence of ordinals, which is impossible. For a contradiction assume that $\omega_1 \times \omega_1 = A_0 \cup A_1 \cup A_2 \cup \cdots$ is a decomposition into countable many antichains. For every $\alpha < \omega_1$ there is some natural number $i(\alpha)$ such that for uncountably many β we have $\langle \alpha, \beta \rangle \in A_{i(\alpha)}$. By the pigeon hole principle there are ordinals $\alpha < \alpha'$ and some number i such that $i = i(\alpha) = i(\alpha')$ holds. Pick an $\langle \alpha', \beta' \rangle \in A_i$. As there are arbitrarily large β with $\langle \alpha, \beta \rangle \in A_i$ we can select with $\beta > \beta'$ and then we get $\langle \alpha, \beta \rangle, \langle \alpha', \beta' \rangle \in A_i$ that is $\langle \alpha, \beta \rangle < \langle \alpha', \beta' \rangle$, a contradiction.

6. Consider the partially ordered set $\langle \omega_1 \times \omega_1, \prec \rangle$ from Problem 5. Set $(\alpha, \beta) \ll (\alpha', \beta')$ if and only if $\alpha \le \alpha'$ and $\beta \le \beta'$ with equality at most in one place. This is a partially ordered set, and two different pairs (α, β) and (α', β') are comparable with respect to $\ll$ if and only if they are incomparable with respect to $\prec$. Thus, the chains of $\langle \omega_1 \times \omega_1, \ll \rangle$ are precisely the antichains of $\langle \omega_1 \times \omega_1, \prec \rangle$, and the antichains of $\langle \omega_1 \times \omega_1, \ll \rangle$ are precisely the chains of $\langle \omega_1 \times \omega_1, \prec \rangle$. Now the statement follows from Problem 5.

7. Suppose that in the partially ordered set $\langle A, \prec \rangle$ all antichains are countable, but the set is not countable. We are going to show that $\langle A, \prec \rangle$ includes an infinite chain. We use the notation of the solution to Problem 1, and follow the argument given there. Let C_0 be a maximal antichain (see Zorn's lemma in Chapter 14). Then every element of $A \setminus C_0$ is either bigger or smaller than an element of C_0, thus there is an element $x_0 \in C_0$ such that either $B^A_{x_0}$ is uncountable, or $S^A_{x_0}$ is uncountable. In the first case set $A_0 = B^A_{x_0}$ and in the second case set $A_0 = S^A_{x_0}$. Now repeat this process with A_0 to get an element $x_1 \in A_0$ and an uncountable set $A_1 \subset A_0$ such that $A_1 = B^{A_0}_{x_1}$ or $A_1 = S^{A_0}_{x_1}$. Repeat again the same thing with A_1 instead of A_0, etc. It is clear that the process never terminates, and we get an infinite set $\{x_0, x_1, \ldots\} \subset A$, which is ordered.

8. The proof is similar to what we have done in the preceding problem. Suppose that in the partially ordered set $\langle A, \prec\rangle$ all chains are countable, but the set is not countable. We are going to show that $\langle A, \prec\rangle$ includes an infinite antichain. Let C_0 be a maximal chain (see Zorn's lemma in Chapter 14). Then every element of $A \setminus C_0$ is incomparable to an element of C_0, thus there is an element $x_0 \in C_0$ such that the set B_{x_0} of elements that are incomparable with x_0 is uncountable. Now repeat this process with B_0 (select a maximal chain, etc.) to get an element $x_1 \in B_0$ and an uncountable set $B_1 \subset B_0$ such that all elements of B_1 are incomparable to x_1. Repeat the same thing with B_1 instead of B_0, etc. It is clear that the process never terminates, and we get an infinite set $\{x_0, x_1, \ldots\}$ of pairwise incomparable elements.

9. Consider $\mathbf{R}$ with its usual ordering $<$ and also let $\prec$ be a well-ordering on $\mathbf{R}$ (see Problems 14.1 and 14.3). For $x, y \in \mathbf{R}$ put $x \ll y$ if $x < y$ and $x \prec y$. In the partially ordered set $\langle R, \ll\rangle$ every chain is a well-ordered subset of $\langle \mathbf{R}, <\rangle$ (since on a chain $<$ and $\prec$ are the same), and every antichain is a well-ordered subset of $\langle \mathbf{R}, <^*\rangle$ with the reverse ordering on $\mathbf{R}$. But $\mathbf{R}$ has only countable well-ordered subsets (see Problem 6.37), so in $\langle \mathbf{R}, \ll\rangle$ every chain and every antichain is countable.

10. The case when κ is finite follows from Problem 2, thus we may suppose that κ is infinite. We show that if $\langle A, \prec\rangle$ is a partially ordered set of cardinality bigger than 2^κ, then it includes either a chain or an antichain of cardinality bigger than κ. Consider the graph G with vertex set A where two points are connected if they are comparable in $\langle A, \prec\rangle$. By Problem 24.19 either G or its complement includes a complete subgraphs of cardinality bigger than κ. But it is clear that a complete subgraph of G is a chain, and a complete subgraph of its complement is an antichain in $\langle A, \prec\rangle$.

11. We can copy the argument that was used in the proof of Problem 9 (which is the $\kappa = \aleph_0$ case) if we can construct an ordered set of cardinality 2^κ such that all of its well-ordered subsets as well as its reversely well-ordered subsets are of cardinality at most κ. But such a set was given in Problem 6.94.

12. Without loss of generality, κ is infinite. We will use the notation $L(x) = \{y : y \preceq x\}$. We call a subset $A \subseteq P$ good if $|A| \le \kappa$, and $L(x) \cap L(y) = \emptyset$ holds for distinct $x, y \in A$. We construct $\langle P, \prec\rangle$ as follows. $P = \bigcup\{P_\alpha : \alpha < \kappa^+\}$, an increasing, continuous union. P_0 is a set of 2^κ incomparable elements. $P_{\alpha+1}$ is obtained from P_α by adding for every good $A \subseteq P_\alpha$ an element $u_\alpha(A)$ with $u_\alpha(A) \succ x$ for $x \in A$, incomparable with the other elements of $P_{\alpha+1} \setminus P_\alpha$ and comparable with only those elements of P_α with which it must be, i.e., with the elements in $\bigcup\{L(x) : x \in A\}$. Notice that for $x, y \in P_\alpha$, $L(x) \cap L(y) = \emptyset$ holds in P if and only if it holds in P_α.

We claim that $\langle P, \prec\rangle$ is as required.

For the first property assume that $x \prec y$. There is a unique $\alpha < \kappa^+$ such that $x \in P_\alpha$ and $y \in P_{\alpha+1} \setminus P_\alpha$. We show that $[x, y]$ is finite by transfinite induction on α. Indeed, if x, y are as above, then $y = u(A)$ for some good $A \subseteq P_\alpha$, so there is a unique $z \in A$ with $x \in L(z)$. Now obviously $[x, y] = [x, z] \cup \{y\}$ and $[x, z]$ is finite by the induction hypothesis since $z \in P_{\beta+1} \setminus P_\beta$ for some $\beta < \alpha$, and if $x \neq z$, then $x \in P_\beta$.

For the other property assume on the contrary that $f : P \to \kappa$ is a coloring in which every color class is an antichain. If $A \subseteq P$ is a good set, call it ξ-good for some $\xi < \kappa$ if for every $x \notin A$, if $A \cup \{x\}$ is good, then $f(x) \neq \xi$. Notice that if $A \subseteq B$ are good sets and A is ξ-good, then so is B.

We claim that there is a good set $A \subseteq P$ such that for every $\xi < \kappa$ if there is a ξ-good $B \supseteq A$, then A is already ξ-good. For this, construct the increasing, continuous union $A = \bigcup\{A_\xi : \xi < \kappa\}$ with $A_0 = \emptyset$, and if A_ξ is given we let $A_{\xi+1} \supseteq A_\xi$ be ξ-good, if there is a ξ-good set extending A_ξ, and $A_{\xi+1} = A_\xi$, otherwise. Now A clearly has the required property.

Let $U \subseteq \kappa$ be the set of those ξ's for which A is ξ-good, and let $V = \kappa \setminus U = \{v_\xi : \xi < \kappa\}$. By transfinite recursion on $\xi < \kappa$ we choose x_ξ such that $A \cup \{x_\eta : \eta \leq \xi\}$ is good and $f(x_\xi) = v_\xi$. This is possible, as $A \cup \{x_\eta : \eta < \xi\}$ is good (by induction) and by our conditions it is not v_ξ-good. There is some $\alpha < \kappa^+$ such that $A \cup \{x_\xi : \xi < \kappa\} \subseteq P_\alpha$. Finally, set $B = \{x_\xi : \xi < \kappa\}$, a good set, and $y = u_\alpha(B)$. Note that $y \in P_{\alpha+1} \setminus P_\alpha$, hence $y \notin A$, and actually $A \cup \{y\}$ is good, because $L(y) = \cup_{x \in B} L(x)$, and $A \cup B$ was good. The color $f(y)$ cannot be in U, for if $u \in U$, then, as A is u-good, $f(y) \neq u$. And $f(y)$ cannot be in V, either, for if $f(y) = v_\xi \in V$, then $f(x_\xi) = v_\xi = f(y)$, and so the comparable $x_\xi \in B$ and $y = v_\alpha(B)$ get the same color, a contradiction.

13. **(a)** The statement is obvious if $c(P, \prec)$ is a successor cardinal. Assume that $\kappa = c(P, \prec)$ is a singular cardinal and there is no strong antichain with size κ. For $x \in P$ let $c(x)$ be the supremum of the size of those strong antichains that consist of elements that are smaller than x. First we claim that for every x there is some $y \prec x$ with $c(y) < \kappa$. If this was not the case, then we could choose below x a strong antichain of size at least $\mathrm{cf}(\kappa)$, and below its elements larger and larger strong antichains with sizes converging to κ. Their union would then be a strong antichain of cardinality κ (note that if x, y are strongly incompatible the so are any x', y' with $x' \preceq x$ and $y' \preceq y$), which is not possible.

Choose a nonextendable set A of incompatible elements with $c(x) < \kappa$ (possible by Zorn's lemma, see Chapter 14). Notice that A is a maximal strong antichain. Indeed, if $x \notin A$, then if $y \prec x$ is such that $c(y) < \kappa$, then y is strongly compatible with some element z of A, but then x and z are also strongly compatible, so we cannot add x to A. By our indirect hypothesis $|A| < \kappa$. If $\kappa = \sum\{c(x) : x \in A\}$, then the argument from the preceding paragraph shows that there is a strong antichain of size κ (below the elements of A). Therefore, $\sum\{c(x) : x \in A\} < \kappa$, and there is a strong antichain B with $|B| > \sum\{c(x) : x \in A\}$. For every $y \in B \setminus A$, as A is a maximal strong antichain, there is some $x \in A$ such that for some element, denote it by $f(x, y)$, we have $f(x, y) \preceq x, y$. For some $x \in A$ there is a $B' \subseteq B$ such that x is selected for $y \in B'$ and $|B'| > c(x)$. But this is a contradiction as $\{f(x, y) : y \in B'\}$ is a strong antichain (for B' is a strong antichain) below x of cardinality $> c(x)$. This contradiction proves part (a).

(b) Let P consist of those regressive functions which are defined on a finite subset of κ. Set $f \prec g$ if f properly extends g. Notice that f, g are incompatible in $\langle P, \prec\rangle$ exactly if they are incompatible as functions, that is, they assume distinct values at a certain point. For every cardinal $\lambda < \kappa$ there is a strong antichain of cardinality λ: $\{f_\alpha : \alpha < \lambda\}$ where the domain of f_α is $\{\lambda\}$ $(\alpha < \lambda)$ and $f_\alpha(\lambda) = \alpha$.

It is left to show that there is no strong antichain of cardinality κ. Assume, in order to get a contradiction, that $\{f_\alpha : \alpha < \kappa\}$ is a strong antichain. Applying Problem 25.3 to the finite sets formed by the domains of these functions, we get a set $Z \subseteq \kappa$ of cardinality κ such that for $\alpha \in Z$ the domain is of the form $s \cup s_\alpha$ where the sets $\{s\} \cup \{s_\alpha : \alpha \in Z\}$ are disjoint. As the functions are required to be regressive, the number of possibilities for $f_\alpha|_S$ is less than κ (namely, the product of the cardinalities of the elements of s). But then, if $\alpha, \beta \in Z$ and $f_\alpha|_S = f_\beta|_S$, the functions f_α, f_β are compatible, which is a contradiction. [P. Erdős, A. Tarski]

14. By the well-ordering theorem we can enumerate A as $A = \{p_\alpha : \alpha < \varphi\}$ for some ordinal φ. Put p_α into B if and only if there is no $\beta < \alpha$ with $p_\alpha \prec p_\beta$.

We show that $B \subseteq A$ is as required. $\langle B, \prec\rangle$ is well founded: if there is a decreasing chain $\cdots \prec q_1 \prec q_0$ in B, that is, $\cdots \prec p_{\alpha_1} \prec p_{\alpha_0}$ then, by the well-ordering property of ordinals we have $\alpha_n < \alpha_{n+1}$ for some n, that is, p_{α_n} is greater than the later $p_{\alpha_{n+1}}$, a contradiction.

B is cofinal: assume that $p \in A$. Choose $p_\alpha \geq p$ with α minimal. Then $p_\alpha \in B$, indeed, otherwise, there is some $p \preceq p_\alpha \prec p_\beta$ with $\beta < \alpha$, but that contradicts the minimal choice of α.

15. For $x \in P$ set $U(x) = \{y : y \succ x\}$. Notice that $U(x)$ is infinite for every $x \in P$. Call $x \in P$ good, if $|U(y)| = |U(x)|$ holds for every $y \succ x$. By Zorn's lemma (see Chapter 14) we can find a set A of good elements such that $U(x) \cap U(y) = \emptyset$ holds and A cannot be extended with this property preserved. We claim that $B = \bigcup\{U(y) : y \in A\}$ is cofinal. Indeed, if $x \in P$ is arbitrary, choose $y \succeq x$ with $|U(y)|$ least possible. Clearly, y is good. As A is nonextendable, there is some $z \in A$ with $U(z) \cap U(y) \neq \emptyset$, say, $t \in U(z) \cap U(y)$. Then $t \in B$ and $t \succeq x$, so we showed that B is cofinal.

As cofinal subsets of cofinal subsets are cofinal, it suffices to find disjoint cofinal subsets in B. This again reduces to finding two disjoint cofinal subsets Y_x, Z_x in $U(x)$, then $Y = \bigcup\{Y_x : x \in A\}$, $Z = \bigcup\{Z_x : x \in A\}$ will be two disjoint cofinal subsets in P. Given $x \in A$, enumerate $U(x)$ as $U(x) = \{x_\alpha : \alpha < \kappa\}$ with some cardinal κ. By transfinite recursion on $\alpha < \kappa$ choose $y_\alpha, z_\alpha \succ x_\alpha$ such that they differ from each other and from all earlier y_β, z_β. This is possible, as $2|\alpha| < \kappa$ elements have been selected so far, and $|U(x_\alpha)| =$

κ as x is good. Finally, both $Y_x = \{y_\alpha : \alpha < \kappa\}$ and $Z_x = \{z_\alpha : \alpha < \kappa\}$ are cofinal in $U(x)$, and they are disjoint, so we are done.

16. Let $\mathcal{G}$ be the set those open sets on $\mathbf{R}$ that can be written as the union of finitely many intervals with rational endpoints. Clearly, $|\mathcal{G}| = \aleph_0$. We define $\langle P, \prec\rangle$ as follows. $\langle s, G\rangle \in P$ if s is a finite set of reals, $G \in \mathcal{G}$ with Lebesgue-measure $\lambda(G)$ less than $1/|s|$, and $s \subseteq G$. We make $\langle s', G'\rangle \preceq \langle s, G\rangle$ if and only if $s' \supseteq s$, $G' \subseteq G$.

We show that $\langle P, \prec\rangle$ is the union of countably many centered sets. For this, we put the elements with identical second coordinate into one class. We have to show that every $\langle s_1, G\rangle, \ldots, \langle s_n, G\rangle$ has a common lower bound. Set $s = s_1 \cup \cdots \cup s_n$ and choose a $G' \in \mathcal{G}$ with $s \subseteq G' \subseteq G$ and measure less than $1/|s|$. This $\langle s, G'\rangle$ is clearly a common lower bound.

Assume that $F \subseteq P$ is a filter. Set $X = \bigcup\{s : \langle s, G\rangle \in F\}$. We claim that X has Lebesgue-measure zero. This is obvious if X is finite. If not, let $x_1, x_2, \ldots$ be distinct elements of X. There is $\langle s_n, G_n\rangle \in F$ with $s_n \supseteq \{x_1, \ldots x_n\}$ and $\lambda(G_n) < 1/n$. If now $\langle s, G\rangle \in F$ is arbitrary, there is in F an element $\langle s', G'\rangle \preceq \langle s, G\rangle, \langle s_n, G_n\rangle$, $s \subseteq G' \subseteq G_n$; therefore, $X \subseteq \bigcap_{n=1}^{\infty} G_n$, which is of measure zero.

This concludes the proof: indeed, if $F_0, F_1, \ldots \subseteq P$ are filters, then their first coordinates, the union of countably many sets of measure zero, cannot cover the reals, therefore $\bigcup F_n \neq P$. [I. Juhász, K. Kunen: On σ-centred posets, in: *A Tribute to Paul Erdős*, (eds. A. Baker, B. Bollobás, A. Hajnal), Cambridge University Press, 1990, 307–311]

17. Consider the sets from Problem 6.25. Their characteristic functions form an ordered subset (with respect to $\prec$) of cardinality bigger than continuum of the partially ordered set of real functions.

Now suppose that $\mathcal{F}$ is a well-ordered set of real functions with respect to $\prec$. For every $f \in \mathcal{F}$, except for the largest element in $\mathcal{F}$ if there is one, there is a successor $\tilde{f}$. But then there is a real number x_f such that $f(x_f) < \tilde{f}(x_f)$. If $x_f = x_g = x$, then the intervals $(f(x), \tilde{f}(x))$ and $(g(x), \tilde{g}(x))$ are disjoint because, e.g., $g \prec f$, and then $\tilde{g}(x) \leq f(x)$. Thus, by Problem 2.14, the set of those $f \in \mathcal{F}$ for which $x_f = x$ is countable. Hence there are at most continuum times countably infinite many elements in $\mathcal{F}$, so it is of cardinality at most continuum (see Problem 4.12).

18. The identity mapping is an order-preserving mapping from $\langle {}^{\omega}\omega, \prec\rangle$ into $\langle {}^{\omega}\omega, \ll\rangle$, and if for an $f \in {}^{\omega}\omega$ we set $F(n) = (f(n) + n)^2$, then $f \to F$ is an order-preserving mapping from $\langle {}^{\omega}\omega, \ll\rangle$ into $\langle {}^{\omega}\omega, \prec\rangle$.

It is clear that there is no element in $\langle {}^{\omega}\omega, \ll\rangle$ is lying in between $f_0(n) \equiv 0$ and $f_1(n) \equiv 1$, while it is easy to see that if $f \prec g$, then there are infinitely many elements in between them in $\langle {}^{\omega}\omega, \prec\rangle$ (in fact, uncountably many elements as is shown by $h_x(n) = f(n) + [x|g(n) - f(n)|]$, $x \in (0, 1)$). Thus, these two sets are not isomorphic.

19. See Problem 2.28.

20. Based on the preceding problem, by transfinite induction build a family $f_\alpha \in {}^\omega\omega$, $\alpha < \omega_1$ such that each f_α is bigger than any f_β with $\beta < \alpha$. Then this family has order type ω_1 in $\langle {}^\omega\omega, \prec\rangle$.

21. For $k = 1, 2, \ldots$ and $x \in (0,1)$ we set $f_{k,x}(n) = [xn^{1/k}]$, where $[\cdot]$ denotes the integral part. It is clear that for $x < y$ we have $f_{k,x} \prec f_{k,y}$, and for $k < l$ we have $f_{k,x} \prec f_{l,y}$ for all $x, y \in (0,1)$. Consider now for $\mathbf{x} = (x_1, \ldots, x_m) \in (0,1)^m$ the function $F_{\mathbf{x}} = \sum_{k=1}^m f_{k,x_k}$. Based on what was said it is easy to see that they form a subset of $\langle {}^\omega\omega, \prec\rangle$ of order type λ^m.

22. Consider a representation of θ in $\langle {}^\omega\omega, \prec\rangle$, i.e., an $\langle A, \prec\rangle$ with order type θ. For $f \in A$ replacing $f(n)$ by $F(n) = \max_{k\le n} f(k) + n$ we get a representation of θ consisting of strictly monotone functions, and then replacing $f(n)$ by $2^{f(n)}$ we get a representation by functions with $f(n) \ge 2^n$. We define $\mathrm{Inv}_f(k) = \min\{n \,:\, f(n) \ge k\}$. Then this $\le [\log_2 k] + 1$, and if $f \prec g$, then for k large $\mathrm{Inv}_g(k) \le \mathrm{Inv}_f(k)$ with strict inequality if $k = g(s)$ and s is large. Therefore, $H_f(m) = \sum_{k=0}^{[\sqrt{m}]} \mathrm{Inv}_f(k)$ is less than m (for large m it is less than $C\sqrt{m}\log m$), and for $f \prec g$ we have $H_g \prec H_f$. Thus, the functions $G_f(m) = m - H_f(m)$, $f \in A$ are smaller than the identity and they form a subset of $\langle {}^\omega\omega, \prec\rangle$ similar to $\langle A, \prec\rangle$.

This proof actually shows that there is an order-preserving mapping of $\langle {}^\omega\omega, \prec\rangle$ into its subset lying below the identity function.

23. For θ^* see the functions H_f, $f \in A$ in the preceding proof.

Let the type of $\langle A_i, \prec\rangle$ be θ_i for $i = 1, 2$. A_1 can be chosen to lie below the identity function, and similarly A_2 can be chosen to lie above it (Problem 22). Then clearly $\langle A_1 \cup A_2, \prec\rangle$ has order type $\theta_1 + \theta_2$. Next consider the functions of the form $H_{f,g}(n) = 2^{g(n)} + f(n)$, where $f \in A_1$ and $g \in A_2$. Clearly, if $g_1 \prec g_2$, $g_1, g_2 \in A_2$, then irrespective of $f_1, f_2 \in A_1$ we have $H_{f_1,g_1} \prec H_{f_2,g_2}$, and if $f_1 \prec f_2$, then $H_{f_1,g} \prec H_{f_2,g}$. Thus, with respect to $\prec$ these functions $H_{f,g}$ form a subset of ${}^\omega\omega$ of order type $\theta_1 \cdot \theta_2$.

24. By now it is clear what we have to do (see the previous solution). Represent θ_i by $\langle A_i, \prec\rangle$ where A_i lies below the identity function, and let $\{f_i \,:\, i \in I\}$ be a set lying above the identity function such that the ordered set $\langle I, <,\rangle$ is similar to $\langle \{f_i \,:\, i \in I\}, \prec\rangle$ under the mapping $i \to f_i$. Consider the set of functions of the form $h_{i,g}(n) = 2^{f_i(n)} + g(n)$, where $g \in A_i$. Exactly as in the preceding solution it follows that if $i < j$, $g_1 \in A_i$ and $g_2 \in A_j$, then $h_{i,g_1} \prec h_{i,g_2}$, and for $g_1, g_2 \in A_i$ we have $h_{i,g_1} \prec h_{i,g_2}$. Thus, these functions form a set of order type $\sum_{i\in I(<)} \theta_i$.

The last statement is proved by transfinite induction on $\alpha < \omega_2$. Each such α can be written as a sum $\alpha = \sum_{\beta<\varphi} \alpha_i$ of ordinals α_i smaller than α

with some $\varphi \leq \omega_1$. Now the induction is easy to carry out based on the first part of the problem (recall also Problem 20 for the representability of ω_1).

25. By selecting a cofinal sequence $\{\alpha_k\}_{k=0}^{\infty}$ in φ and considering $\{f_{\alpha_k}\}_{k=0}^{\infty}$ and $\{g_{\alpha_k}\}_{k=0}^{\infty}$, it is sufficient to work with the case $\varphi = \omega$. Let N_k be an increasing sequence such that $f_{i+1}(n) - f_i(n) > k$, $g_i(n) - g_{i+1}(n) > k$ and $g_i(n) - f_i(n) > k$ for all $0 \leq i \leq k$ and $n \geq N_k$, and set $F(n) = f_k(n)$ if $N_k \leq n < N_{k+1}$. This function F clearly lies strictly in between the f_α's and g_α's.

26. We construct the functions with the following extra properties:

1. For $\alpha < \beta < \omega_1$ we have $f_\beta \prec f_\alpha \prec g_\alpha \prec g_\beta$;
2. for $\alpha < \omega_1$, $n < \omega$ the set

$$A(n, \alpha) = \{\beta < \alpha : x \geq n \longrightarrow f_\beta(x) < g_\alpha(x)\}$$

is finite.

To show that this suffices, assume that $f_\alpha \prec f \prec g_\alpha$ holds for every $\alpha < \omega_1$. Then there are some $n < \omega$ and $\aleph_1$ ordinals α such that for $x \geq n$ we have $f_\alpha(x) < f(x) < g_\alpha(x)$. Select, among those, one that is preceded by at least ω many. Then $A(n, \alpha)$ is infinite, which is a contradiction.

We construct the functions by transfinite recursion on α. Set $f_0(n) = 0$, $g_0(n) = n$. The successor case is easy, given f_α and g_α, for x large enough we have $f_\alpha(x) < g_\alpha(x)$ and the interval $\big(f_\alpha(x), g_\alpha(x)\big)$ gets wider and wider. Therefore, we can split this intervals into larger and larger intervals with values $f_{\alpha+1}(x)$ and $g_{\alpha+1}(x)$ in such a way that $g_{\alpha+1}(x) \leq g_\alpha(x)$ always holds. A moment's reflection shows that for every n the set $A(n, \alpha+1)$ can contain at most one more element (namely: α) than $A(n, \alpha)$.

Assume that $\alpha < \omega_1$ is a limit ordinal, and assume that we have already constructed f_β, g_β for $\beta < \alpha$. We first determine g_α. Let the increasing sequence $\alpha_0 < \alpha_1 < \cdots$ converge to α. We select inductively the natural numbers $k_0 < k_1 < \cdots$ in such a way that

$$f_{\alpha_i}(x) + i < f_{\alpha_{i+1}}(x) < f_{\alpha_{i+1}}(x) + i < g_{\alpha_{i+1}}(x) < g_{\alpha_i}(x),$$

holds for $x \geq k_i$ moreover if $\beta \in A(k_i, \alpha_{i+1})$, $\alpha_i \leq \beta < \alpha_{i+1}$, then there is an x with $k_i \leq x < k_{i+1}$, for which $f_{\alpha_i}(x) \leq f_\beta(x)$ holds. (This can obviously be done.) Define g_α as follows: below k_0 it is arbitrary, and for $k_i \leq x < k_{i+1}$ we set $g_\alpha(x) = f_{\alpha_i}(x)$.

For the g_α so defined we quickly get the first property required as for $x \geq k_i$ the values $g_\alpha(x) - f_{\alpha_i}(x)$ and $g_{\alpha_i}(x) - g_\alpha(x)$ are at least i.

To check the other property assume that $\beta \in A(k_i, \alpha)$. If now $\beta < \alpha_i$ then, as for $x \geq k_i$ we have $g_\alpha(x) < g_{\alpha_i}(x)$, we will also have $\beta \in A(k_i, \alpha_i)$ which is satisfied by only finitely many β. On the other hand, we show that $\beta \geq \alpha_i$

is not possible. Assume, toward a contradiction, that β is such an ordinal. Then there is a $j \geq i$, such that if $\alpha_j \leq \beta < \alpha_{j+1}$, then for $x \geq k_j$ we have $f_\beta(x) < g_{\alpha_{j+1}}(x)$, that is, $\beta \in A(k_j, \alpha_{j+1})$. But we selected k_{j+1} in such a way that there is a $k_j \leq x < k_{j+1}$ for which $f_{\alpha_j}(x) \leq f_\beta(x)$ and the first value is $g_\alpha(x)$. We get, therefore, that no β of the above type exists.

And finally with a diagonal process (or use Problem 25) construct f_α in such a way that that for every i the inequalities $f_{\alpha_i} \prec f_\alpha \prec g_\alpha$ hold.

12

Transfinite enumeration

1. Let $I = \{i_\xi\}_{\xi<\alpha}$ be an enumeration of the elements of I, and for $\xi < \alpha$ recursively set

$$B_{i_\xi} = A_{i_\xi} \setminus \left(\cup_{\zeta<\xi} B_{i_\zeta}\right).$$

It is clear that these B_i's are pairwise disjoint and $\cup_{i\in I} B_i \subseteq \cup_{i\in I} A_i$. But if $a \in \cup_{i\in I} A_i$ and $\xi < \alpha$ is the first index for which $a \in A_{i_\xi}$, then clearly $a \in B_{i_\xi}$, so the union of the B_i's is equal to the union of the A_i's.

2. First consider the case $\kappa > \aleph_0$. Let the sets be X_ξ, $\xi < \kappa$. Choose, by recursion on $\alpha < \kappa$, and within this recursion by recursion on $\xi < \alpha$ distinct elements $a_{\xi,\alpha} \in X_\xi$. This is possible, since $a_{\xi,\alpha}$ can be any element of

$$X_\xi \setminus (\{a_{\eta,\beta} \ : \ \eta < \beta < \alpha\} \cup \{a_{\eta,\alpha} \ : \ \eta < \xi\}),$$

which is not empty, since it is the difference of a set of cardinality κ and of a set of cardinality

$$\leq |\alpha|^2 + |\alpha| \leq |\alpha| + \aleph_0 < \kappa. \tag{12.1}$$

If we set $Z_\xi = \{a_{\xi,\alpha} \ : \ \xi < \alpha < \kappa\}$, then $|Z_\xi| = \kappa$ and the sets Z_ξ are pairwise disjoint. As $2\kappa = \kappa$, we can split Z_ξ into two disjoint parts Y_ξ and Y'_ξ each of cardinality κ. Now the system $\{Y_\xi \ : \ \xi < \kappa\}$ is as required: $|X_\xi \setminus Y_\xi| \geq |Y'_\xi| = \kappa$.

For $\kappa = \aleph_0$ the above argument works, only the calculation in (12.1) reads as

$$\leq |\alpha|^2 + |\alpha| < \aleph_0 = \kappa,$$

since in this case every α is finite.

3. Select pairwise disjoint sets $Z_\xi \subseteq X_\xi$, $\xi < \kappa$ of cardinality κ as in the preceding problem, and let $Z_\xi = \cup_{\alpha<\kappa} Z_{\xi,\alpha}$ be a decomposition of Z_ξ into κ pairwise disjoint sets (since $\kappa^2 = \kappa$, this is possible). Now it is clear that the

sets $Y_\alpha = \cup_{\xi<\kappa} Z_{\xi,\alpha}$, $\alpha < \kappa$ are pairwise disjoint and they intersect all X_ξ in a set of power continuum.

4. Let $\{x_\xi\}_{\xi<\kappa}$ be an enumeration of the elements of X in a sequence of type κ, and let $\{f_\xi\}_{\xi<\kappa}$ be a similar enumeration of those functions g in $\mathcal{F}$ which have the property that the intersection of the range of g with X is of cardinality κ.

First of all we remark that if $Y \subset X$ is of cardinality smaller than κ, and the function g with domain X is such that its range intersects X in κ points, then there is an element $z \in X \setminus Y$ with $g(z) \in X \setminus Y$. This immediately implies that if $Y \subset X$ is of cardinality smaller than κ, and $\mathcal{G}$ is a family of cardinality smaller than κ of functions g such that each $g \in \mathcal{G}$ has domain X and a range intersecting X in κ points, then there is a set $Z \subseteq X \setminus Y$ of cardinality at most $2|\mathcal{G}|$ such that for every $g \in \mathcal{G}$ there is a $z \in Z$ with $g(z) \in Z$. In fact, all we have to do is to select for each $g \in \mathcal{G}$ an element $z_g \in X \setminus Y$ with $g(z_g) \in X \setminus Y$, and then take the union of all pairs $\{z_g, g(z_g)\}$ for all $g \in \mathcal{G}$. It is obvious that this Z has cardinality at most $2|\mathcal{G}|$.

After this we define by transfinite recursion pairwise disjoint sets A^0_ξ, A^1_ξ, $\xi < \kappa$ of cardinality at most $\max(|\xi|, \aleph_0)$ as follows. Let $A^0_0 = A^1_0 = \emptyset$, and suppose that for all $\alpha < \xi$ the sets A^0_α, A^1_α have already been defined for $\alpha < \xi$, where $\xi < \kappa$, and let $Y_\xi = \cup_{\alpha<\xi}(A^0_\alpha \cup A^1_\alpha)$. Select a set $A^0_\xi \subset X \setminus Y_\xi$ of cardinality at most $2|\xi|$ such that for all $\alpha < \xi$ there is a $z \in A^0_\xi$ such that $f_\alpha(z) \in A^0_\xi$, and then select a set $A^1_\xi \subset X \setminus (Y_\xi \cup A^0_\xi)$ of cardinality at most $2|\xi|$ such that for all $\alpha < \xi$ there is a $z \in A^1_\xi$ such that $f_\alpha(z) \in A^1_\xi$. Since the cardinality of A^0_ξ, A^1_ξ is at most $2|\xi|$, the set Y_ξ has cardinality smaller then κ, and the induction can be carried out.

Let $\mathcal{E}$ be a family of cardinality 2^κ of transfinite 0–1 sequences $\epsilon = \{\epsilon_\xi\}_{\xi<\kappa}$ such that for $\epsilon, \epsilon' \in \mathcal{E}$ there are κ indices ξ with $\epsilon_\xi \neq \epsilon'_\xi$. By Problem 18.3 there is such a family (apply Problem 18.2 to the set κ and identify a subset by a 0–1 sequence in the standard manner). For each $\epsilon \in \mathcal{E}$ consider the set $H_\epsilon = \cup_{\xi<\kappa} A^{\epsilon_\xi}_\xi$. Let $\epsilon, \epsilon' \in \mathcal{E}$ be two different sequences in $\mathcal{E}$, and let $f \in \mathcal{F}$ be arbitrary. If the range of f intersects X in a set of cardinality smaller than κ, then clearly $f[H_\epsilon] = H_{\epsilon'}$ cannot hold, for each $H_{\epsilon'}$ is of cardinality κ. On the other hand, if the range of f intersects X in a set of cardinality κ, then $f = f_\alpha$ for some $\alpha < \kappa$. There are κ indices ξ with $\epsilon_\xi \neq \epsilon'_\xi$; therefore, there is such an index with $\xi > \alpha$. Let, e.g., $\epsilon_\xi = 0$, $\epsilon'_\xi = 1$. By the construction of the sets H_ϵ and A_ξ, there is an element $z \in A^0_\xi \subset H_\epsilon$ such that $f_\alpha(z) \in A^0_\xi \subset X \setminus H_{\epsilon'}$, and this shows that $f[H_\epsilon] \neq H_{\epsilon'}$. Thus, the 2^κ sets H_ϵ, $\epsilon \in \mathcal{E}$ satisfy the requirements. [W. Sierpiński, Cardinal and Ordinal Numbers, Polish Sci. Publ., Warszawa, 1965, XII.4]

5. Since X is equivalent to $X \times X$, it is sufficient to construct a similar family on $X \times X$. We obtain $\mathcal{H}$ by taking the graphs of some functions $f_\xi : X \to X$, $\xi < \kappa^+$, i.e., taking the sets $G(f_\xi) := \{(x, f_\xi(x)) \ : \ x \in X\}$, $\xi < \kappa^+$ (technically speaking, this $G(f_\xi)$ is f_ξ itself, but it is instructive to consider

it as a graph). Suppose f_η is known for all $\eta < \xi$ where $\xi < \kappa^+$. Then there is an f_ξ such that its graph G_ξ is almost disjoint from all $G(f_\eta)$, $\eta < \xi$. In fact, to this end it is enough to show that if g_τ, $\tau < \kappa$ are κ functions from X into X, then there is a function $f : X \to X$ such that its graph $G(f)$ is almost disjoint from all the graphs $G(g_\tau)$. Let x_α, $\alpha < \kappa$ be an enumeration of the points in X. We can define by transfinite recursion the values $f(x_\alpha)$ so that $f(x_\alpha) \neq g_\tau(x_\alpha)$ for $\tau < \alpha$. This is possible, for we can always select a value for $f(x_\alpha)$ from the nonempty set $X \setminus \{g_\tau(x_\alpha) \ : \ \tau < \alpha\}$. Then clearly, $G(f) \cap G(g_\tau) \subseteq \{(x_\alpha, g_\tau(x_\alpha)) \ : \ \alpha \leq \tau\}$, so the graphs $G(f)$ and $G(g_\tau)$ are almost disjoint. [P. Erdős]

6. We define N_ξ with the additional property that the intersection of finitely many N_ξ's is infinite. Then N_ξ can be easily defined by transfinite recursion if we can show that if $M_0, M_1, \ldots$ are subsets of $\mathbf{N}$ such that for all m the intersection $M_0 \cap \cdots \cap M_m$ is infinite, then there is a subset $M \subset \mathbf{N}$ such that $M \cap M_0 \cap \cdots \cap M_m$ is infinite for all m, and $M \setminus M_k$ is finite but $M_k \setminus M$ is infinite for all k. But that is easy, namely if we select numbers $x_m \in M_0 \cap \cdots \cap M_m$ such that $x_m \neq x_0, x_1, \ldots x_{m-1}$, then clearly the set $M = \{x_0, x_2, x_4, \ldots\}$ satisfies all the requirements.

7. Let ℓ_ξ, $\xi < \mathbf{c}$ be an enumeration of the lines of the plane into a transfinite sequence of type $\mathbf{c}$. We shall set $A = \cup_{\alpha<\mathbf{c}} A_\alpha$, where the A_α's are increasing sets (by this we mean that $A_\alpha \subseteq A_\beta$ for $\alpha < \beta$) in such a way that A_α has at most two points on every line of the plane, and for $\xi \leq \alpha$ the set A_α has exactly two points on ℓ_ξ. Then clearly A will have exactly two points on every line.

The construction of the A_α's is given by transfinite recursion on α, and it will be done in such a way that $|A_\alpha| \leq 2(|\alpha|+1)$ is satisfied for all α. Suppose that A_β, $\beta < \alpha$, have already been constructed with the above properties. Then the set $B_\alpha = \cup_{\beta<\alpha} A_\beta$ also has at most two points on every line and it has cardinality at most $2|\alpha|$. Let L_α be the set of lines connecting any two points of B_α. Then L_α has cardinality at most $4|\alpha|^2$. If B_α has exactly two points on ℓ_α, then we set $A_\alpha = B_\alpha$. If B_α has one point on ℓ_α, then let $A_\alpha = B_\alpha \cup \{x_\alpha\}$, where $x_\alpha \in \ell_\alpha \setminus (\cup_{\ell \in L_\alpha} \ell)$. Since two different lines in the plane intersect in at most one point the set $\ell_\alpha \cap (\cup_{\ell \in L_\alpha} \ell)$ has cardinality at most $4|\alpha|^2 < \mathbf{c}$, thus $\ell_\alpha \setminus (\cup_{\ell \in L_\alpha} \ell)$ is not empty, and the selection of x_α is possible. In a similar way, if B_α has no point on ℓ_α, then let $A_\alpha = B_\alpha \cup \{x_\alpha, y_\alpha\}$, where $x_\alpha, y_\alpha \in \ell_\alpha \setminus (\cup_{\ell \in L_\alpha} \ell)$ are two different points. This completes the definition, and by the choice of the set A_α we can see that $|A_\alpha \cap \ell| \leq 2$ for every line ℓ, and $|A_\alpha \cap \ell_\alpha| = 2$. Since the sets A_α are increasing, these also prove that $|A_\alpha \cap \ell_\xi| = 2$ for all $\xi \leq \alpha$. It is also clear that $|A_\alpha| \leq |B_\alpha| + 2 \leq 2|\alpha| + 2$, so the induction runs through. [S. Mazurkiewicz, C. R. Soc. Sc. et Lettres de Varsovie **7**(1914), 382–383]

8. We extend the argument of the previous proof. Let ℓ_ξ, $\xi < \mathbf{c}$ be a listing of the lines on the plane so that each line is repeated continuum many times,

i.e., $C_\xi = \{\zeta < \mathbf{c} \ : \ \ell_\zeta = \ell_\xi\}$ is of cardinality $\mathbf{c}$. A will be the increasing union of sets A_α, $\alpha < \mathbf{c}$, where the A_α's are of cardinality at most $|\alpha + 1| < \mathbf{c}$. The inductive conditions are: $|\ell_\xi \cap A_\alpha| \le m_{\ell_\xi}$ for all $\xi < \mathbf{c}$, $|\ell_\alpha \cap A_\alpha| \ge \min(|C_\alpha \cap \alpha|, m_{\ell_\alpha})$, and A_α has at most 2 points on every line ℓ_ξ with $C_\xi \cap \alpha = \emptyset$.

As in Problem 7 we let $B_\alpha = \cup_{\beta<\alpha} A_\beta$ but now let L_α be the set of lines ℓ except ℓ_α such that $|\ell \cap B_\alpha| \ge 2$. At step α, if $|B_\alpha \cap \ell_\alpha| = m_{\ell_\alpha}$, then we set $A_\alpha = B_\alpha$. If, however, $|B_\alpha \cap \ell_\alpha| < m_{\ell_\alpha}$, then we add one more point to $B_\alpha \cap \ell_\alpha$ doing as little harm as possible. As $|L_\alpha| \le |\alpha + 1|^2 < \mathbf{c}$, the lines in L_α hit the line ℓ_α in less than $\mathbf{c}$ points, and we can select a point $P \in \ell_\alpha$ different from them and not lying in $B_\alpha \cap \ell_\alpha$. This point P can be added to B_α to form A_α. Clearly, with this step all the induction properties are preserved, and the construction runs through all $\alpha < \mathbf{c}$.

Since each line is listed continuum many times, eventually we will have $|\ell \cap A| = m_\ell$ for all lines ℓ.

9. Let $L_1 = \{a_\xi\}_{\xi<\alpha_1}$, $\alpha_1 \le \mathbf{c}$ and $L_2 = \{b_\eta\}_{\eta<\alpha_2}$, $\alpha_2 \le \mathbf{c}$ be the enumeration of the sets L_1 and L_2 into two transfinite sequences of length at most $\mathbf{c}$. For a point P on the plane let $\xi(P)$ resp. $\eta(P)$ be the smallest $\xi < \alpha$ resp. $\eta < \beta$ such that $P \in a_\xi$ resp. $P \in b_\eta$, and if there is no such ξ or η then let $\xi(P)$ resp. $\eta(P)$ be equal to $\mathbf{c}$. Finally, let A_1 be the set of points P for which $\eta(P) \le \xi(P)$, and A_2 its complement in $\mathbf{R}^2$. If $a_\xi \in L_1$ and $P \in A_1 \cap a_\xi$, then by the definition of A_1, there is an $\eta \le \xi$ such that $P \in b_\eta$, i.e., P is the common point of a_ξ and b_η. But this means that there can be at most $|\xi| + 1 < \mathbf{c}$ points on a_ξ from A_1. In the same fashion, there can be at most $|\eta|$ points on any b_η from A_2. [P. Erdős, cf. W. Sierpiński, Cardinal and Ordinal Numbers, Polish Sci. Publ., Warszawa, 1965, XVII.4/5]

10. A nonempty perfect set is of cardinality continuum and there are continuum many perfect subsets of $\mathbf{R}$. Therefore, this problem is a consequence of Problem 3.

A direct construction runs as follows. The number of nonempty perfect subsets of $\mathbf{R}$ is of power continuum; therefore, we can list them into a transfinite sequence of type $\mathbf{c}$:

$$A_0, A_1, \ldots, A_\xi, \ldots, \qquad \xi < \mathbf{c},$$

in such a manner that each nonempty perfect subset of $\mathbf{R}$ is repeated continuum many times in this sequence. We also know that each A_ξ is of power continuum (see Problem 5.21). Now for fixed α and for $\beta < \alpha < \mathbf{c}$ select different points $P_{\alpha,\beta} \in A_\alpha$ by transfinite recursion in such a manner that the $P_{\alpha,\beta}$ are different from all $P_{\alpha',\beta'}$ with $\beta' < \alpha' < \alpha$. Since the number of points $P_{\alpha',\beta'}$ with $\beta' < \alpha' < \alpha$ is at most $|\alpha|^2 < \mathbf{c}$, such a selection is possible. It is clear that if for $\beta < \mathbf{c}$ we set $H_\beta = \{P_{\alpha,\beta}\}_{\beta<\alpha<\mathbf{c}}$, then each H_β is of cardinality continuum and each H_β intersects every A_ξ. To get a decomposition of $\mathbf{R}$ just add the points outside $\cup_{\beta<\mathbf{c}} H_\beta$ to, say, H_0.

11. The sets in the decomposition $\mathbf{R} = \cup_{\beta<\mathbf{c}} H_\beta$ in the preceding solution are nonmeasurable. In fact, a measurable set of positive measure includes a compact set K of positive measure. Now by Problem 5.20 we have $K = A \cup B$, where A is perfect and B is countable. Thus, A must have positive measure, and this shows that a measurable set of positive measure includes a nonempty perfect set. The nonmeasurability of H_β follows, since neither H_β nor its complement includes a nonempty perfect set.

12. A set of the first category is included in the countable union of closed sets with empty interior. Now a construction similar to the one in the solution of Problem 5.21 shows that if $F_0, F_1, \ldots$ are closed sets with empty interior, then $\mathbf{R} \setminus \left(\cup_{i=0}^{\infty} F_i\right)$ includes a nonempty perfect set. In fact, all we have to make sure is that when selecting the nth-level intervals (in the notation of the solution of Problem 5.21 the sets $E_{i_1 \ldots i_n}$) we select them from the complement $\mathbf{R} \setminus F_n$ of F_n. Thus, the complement of a set of first category includes a nonempty perfect set. Therefore, the sets H_β from the solution of Problem 10 must be of second category.

13. First of all we note that if a set on the plane intersects every compact set of positive (planar) measure, then it cannot be of measure zero. In fact, in a set of positive measure (in particular in the complement of a set of zero measure) there is a compact set of positive measure.

Next we note that a compact set of positive (planar) measure cannot be covered by less than continuum many lines. In fact, let $L = \{\ell\}$ be a set of less than continuum many lines, and let us choose a line $\mathbf{l}$ that is not parallel with any line in L. If K is a compact set of positive measure, then, by Fubini's theorem, there is a line $\mathbf{l_0}$ parallel with $\mathbf{l}$ that intersects K in a set of positive measure, and hence in a set of power continuum (cf. the solution of Problem 11). But since each $\ell \in L$ intersects $\mathbf{l_0}$ in at most one point, the set $K \cap \mathbf{l_0}$ is not covered by the lines in L.

After these let K_ξ, $\xi < \mathbf{c}$ be an enumeration of the compact sets of positive measure on $\mathbf{R}^2$ into a transfinite sequence of type $\mathbf{c}$. We shall construct by transfinite recursion increasing sets A_ξ, $\xi < \mathbf{c}$ in such a way that each A_ξ is of cardinality at most $|\xi| + 1$, each A_ξ intersects every line in at most two points and $A_\xi \cap K_\xi \neq \emptyset$. Then clearly $A = \cup_{\xi<\mathbf{c}} A_\xi$ will be a set that has at most two points on every line and that intersects every K_ξ, hence it is not of measure zero.

Let A_0 be a one-point set containing a point from K_0, and suppose that all A_α are already known with the above property for all $\alpha < \xi < \mathbf{c}$. Then the set $\cup_{\alpha<\xi} A_\alpha$ is of cardinality at most $|\xi|$, hence if L is the set of lines determined by the points in $\cup_{\alpha<\xi} A_\alpha$, then L is of cardinality smaller than continuum. Therefore, according to what we have said before, L cannot cover the set K_ξ, so there is a point $P_\xi \in K_\xi$ that is not on any of the lines in L. But then the set $A_\xi = (\cup_{\alpha<\xi} A_\alpha) \cup \{P_\xi\}$ intersects K_ξ and has at most two points on every line, and this completes the construction.

14. Let $\mathcal{H}$ be the collection of those subsets of $\mathbf{R}^2$ that are a countable union of closed sets without interior points. Every set of first category is included in a set in $\mathcal{H}$, and $\mathcal{H}$ is of cardinality continuum (see Problems 4.6 and 4.7). Thus, if a set intersects the complement of every set in $\mathcal{H}$ then it is not of the first category.

We can copy the preceding proof provided we can show that the complement of any $H \in \mathcal{H}$ cannot be covered by less than continuum many lines, and this follows exactly as before if we can show that with any line $\mathbf{l}$ there is a parallel line $\mathbf{l_0}$ that intersects the complement of H in continuum many points. Let $H = \cup_{n=0}^{\infty} F_n$, where each F_n is a closed set with empty interior. Without loss of generality, we may assume that $\mathbf{l}$ is horizontal, and for each interval I with rational endpoints let $Y_{I,n}$ be the set of those $y \in \mathbf{R}$ for which $I \times \{y\} \subset F_n$. Since F_n is closed, we can infer that $Y_{I,n}$ must be nowhere dense (for otherwise F_n would include a rectangle $I \times J$); therefore, $Y = \cup_{I,n} Y_{I,n}$ is of the first category. Thus, there is a horizontal line $\mathbf{l_0}$ such that its intersection with every F_n is a closed set that does not include a segment. But then $\mathbf{l_0} \setminus \cup_{n=0}^{\infty} F_n$ includes a nonempty perfect set (cf. the solution of Problem 12), hence it is of power continuum.

15. Let x_ξ, $\xi < \mathbf{c}$ be an enumeration of the real numbers. By transfinite recursion we define increasing sets A_α, $\alpha < \mathbf{c}$ in such a way that $|A_\alpha| \le 2(|\alpha|+1)$ and every real number can be represented in at most one way in the form $a+b$, $a, b \in A_\alpha$, and x_α has such a representation. Then clearly the set $A = \cup_{\alpha<\mathbf{c}} A_\alpha$ will satisfy the requirements. Let $A_0 = \{a, b\}$ where $a + b = x_0$, and suppose that the sets A_β, $\beta < \alpha < \mathbf{c}$ have already been defined and satisfy the above properties. Then $\cup_{\beta<\alpha} A_\beta$ is of cardinality smaller than continuum, hence the set of numbers of the form $a+b-c$, $(a+b)/2$, $(x_\alpha + a - c)/2$, $(a + x_\alpha)/3$, $(2x_\alpha - c)/3$ with $a, b, c \in \cup_{\beta<\alpha} A_\beta$ is also of cardinality smaller than continuum. Therefore, there is a real number y_α such that neither y_α, nor $x_\alpha - y_\alpha$ is of the aforementioned form. Now if x_α can be represented as $a+b$ with $a, b \in \cup_{\beta<\alpha} A_\beta$, then let $A_\alpha = \cup_{\beta<\alpha} A_\beta$, and otherwise let $A_\alpha = (\cup_{\beta<\alpha} A_\beta) \cup \{y_\alpha, x_\alpha - y_\alpha\}$. Since each A_β had cardinality at most $2(|\beta|+1)$ and these sets are increasing, A_α has cardinality at most $2(|\alpha|+1)$. Furthermore, it is impossible to have two different representations $a+b$ and $c+d$ with $a, b, c, d \in A_\alpha$ for any number x. In fact, since $\cup_{\beta<\alpha} A_\beta$ possessed this property, we may have $a+b = c+d$ only if at least one of these numbers, say d, is y_α or $x_\alpha - y_\alpha$. But then by the choice of y_α either a or b also has to be y_α, resp. $x_\alpha - y_\alpha$, i.e., the two representations are the same. In fact, the excluded numbers were exactly those for which we would have two different representations; e.g., $y_\alpha = a+b-c$ was excluded to avoid $a+b = c+y_\alpha$, or $(2x_\alpha - c)/3$ was excluded to avoid $(x_\alpha - y_\alpha) + (x_\alpha - y_\alpha) = c + y_\alpha$, etc.

16. First of all we mention that if $A \cap (a, x)$ has the same cardinality, say $\kappa \ge \aleph_0$, for all $x \in (a, b)$, then there is a 1-to-1 map $g : (a, b] \cap A \to (a, b] \cap A$ such that $g(x) < x$ for all $x \in (a, b] \cap A$. In fact, enumerate the elements of

$(a,b]\cap A$ into a transfinite sequence x_α, $\alpha<\kappa$. Then it is easy to define $g(x_\alpha)$ by transfinite recursion in such a way that $g(x_\alpha)<x_\alpha$ and $g(x_\alpha)\neq g(x_\beta)$ for $\beta<\alpha$.

Now for $x\in A$ let $\kappa(x)=|\{y\in A\ :\ y<x\}|$. This is a mapping from A with cardinal values such that $\kappa(x_1)<\kappa(x_2)$ implies $x_1<x_2$, hence by Problem 6.37 its range is countable. Let Y be the subset of the range that consists of infinite cardinals. For $\kappa\in Y$ let $A_\kappa=\{x\in A\ :\ \kappa(x)=\kappa\}$. This set is of the form $I\cap A$ with an interval I (I can be closed, open, or semi-open). It is clear that if I has endpoints a_κ,b_κ, then $(a_\kappa,x)\cap A$ has cardinality κ for all $x\in(a_\kappa,b_\kappa)$, hence there is a 1-to-1 mapping g_κ from $(a_\kappa,b_\kappa]\cap A_\kappa$ into itself that maps every element into a smaller one. Now let A' be the one-point set consisting of the smallest element of A if A has a smallest element, and otherwise let A' be a countable subset of A that is coinitial with A. Let f agree with g_κ on each $(a_\kappa,b_\kappa]\cap A_\kappa$, and all other elements of A be mapped by f into A' in such a way that $f(x)<x$ for all x, except perhaps for the smallest element of A. Since $A\setminus\cup_{\kappa\in Y}\big((a_\kappa,b_\kappa]\cap A_\kappa\big)$ consists of the smallest elements of A_κ (if there are such) and of those elements x of A for which $\{y\in A\ :\ y<x\}$ is finite, this set is countable, and hence the claim follows with this f.

17. Let x_α, $\alpha<\mathbf{c}$, be an enumeration of the reals. If f is a real function, then define by transfinite recursion two functions g,h in such a way that $g(x_\alpha)+h(x_\alpha)=f(x_\alpha)$ for all α, and the values $g(x_\alpha)$ resp. $h(x_\alpha)$ are different from every $g(x_\beta)$ resp. $h(x_\beta)$, $\beta<\alpha$. Then g,h will be 1-to-1 functions and $f=g+h$. [A. Lindenbaum, Ann. Soc. Pol. Math., **15**(1936), 185]

18. There are continuum many monotone real functions (Problem 4.14, d)), therefore we can enumerate them into a transfinite sequence f_ξ, $\xi<\mathbf{c}$. Let us also enumerate the reals into a transfinite sequence x_ξ, $\xi<\mathbf{c}$, and by transfinite recursion define the real function f in such a way that $f(x_\xi)$ is different from every $f_\alpha(x_\xi)$ with $\alpha<\xi$. Then f agrees with any f_α only on a set of cardinality smaller then continuum, hence it can be monotone only on a set of cardinality smaller than continuum, for any function that is defined and monotone on a subset of the reals can be extended to a monotone real function (see the solution to Problem 6.18).

19. There are continuum many triplets (I,f,y) consisting of a nondegenerate interval $I\subseteq\mathbf{R}$, a continuous real function f, and a real number y (cf. Problems 4.11 and 4.12), hence we can enumerate them into a transfinite sequence (I_ξ,f_ξ,y_ξ), $\xi<\mathbf{c}$. Now define by transfinite recursion a sequence x_ξ, $\xi<\mathbf{c}$, in such a way that the x_ξ's are different, and $x_\xi\in I_\xi$. Now set $F(x_\xi)=y_\xi-f_\xi(x_\xi)$, $\xi<\mathbf{c}$, and define F arbitrarily for other values. Clearly, if f is a real continuous function, $I\subseteq\mathbf{R}$ is an interval and $y\in\mathbf{R}$ is a number, then there is an $x\in I$ with $(F+f)(x)=y$, namely $x=x_\xi$ for the index ξ for which $(I,f,y)=(I_\xi,f_\xi,y_\xi)$.

20. Consider the pairs of real sequences $(\{x_n\}_{n=0}^{\infty}, \{y_n\}_{n=0}^{\infty})$ where the numbers x_n are different and the numbers y_n are arbitrary. Their number is continuum (see Problems 4.3 and 4.12), therefore we can enumerate them into a transfinite sequence $\left(\{x_n^{\xi}\}_{n=0}^{\infty}, \{y_n^{\xi}\}_{n=0}^{\infty}\right)$, $\xi < \mathbf{c}$. By transfinite recursion we define increasing sets $X_\xi \subset \mathbf{R}$ of cardinality at most $|\xi| + \aleph_0$ and real numbers x_ξ in the following way: let $X_0 = \{x_n^0\}_{n=0}^{\infty}$, $x_0 = 0$ and if X_η, x_η, $\eta < \xi$ have already been defined, then let x_ξ be a real number for which $x_\xi + x_n^{\xi}$, $n = 0, 1, \ldots$ all lie outside the set $\cup_{\eta<\xi} X_\eta$, and set

$$X_\xi = \{x_\xi + x_n^{\xi}\}_{n=0}^{\infty} \bigcup (\cup_{\eta<\xi} X_\eta) .$$

It is clear that the property $|X_\xi| \le |\xi| + \aleph_0$ is preserved, hence the induction can be carried out and the numbers $x_\xi + x_n^{\xi}$, $\xi < \mathbf{c}$, $n \in \mathbf{N}$ are all different. Now all we have to do is to define $f(x_\xi + x_n^{\xi}) = y_n^{\xi}$ for all ξ and n, and set $f(x) = 0$ otherwise. The definition of the numbers x_ξ guarantee that f is uniquely defined, and if $\{x_n\}_{n=0}^{\infty}$, $\{y_n\}_{n=0}^{\infty}$ are arbitrary with $x_n \ne x_m$ for $n \ne m$, then there is an x with $f(x + x_n) = y_n$, namely $x = x_\xi$ is appropriate where ξ is the index for which $\left(\{x_n^{\xi}\}_{n=0}^{\infty}, \{y_n^{\xi}\}_{n=0}^{\infty}\right) = (\{x_n\}_{n=0}^{\infty}, \{y_n\}_{n=0}^{\infty})$.

21. The sets $X_1, X_2, \ldots, X_\xi, \ldots$ $\xi < \omega_1$ are closed and form a nonincreasing transfinite sequence, hence by Problem 6.38 there must be a $0 < \theta < \omega_1$ with $X_{\theta+1} = X_\theta$. But then $X_{\theta+2} = X_{\theta+1}^L = X_\theta^L = X_{\theta+1} = X_\theta$, and in a similar fashion we obtain by transfinite induction $X_{\theta+\alpha} = X_\theta$ for all α. It is also clear that X_θ is either empty or perfect, for it is closed and coincides with the set of its limit points. Since $X_{\beta+1} \setminus X_\beta$ consists of the isolated points of X_β, it is a discrete set, hence it is countable (Problem 5.3). Furthermore, $X \setminus X_\theta \subseteq \cup_{\beta<\theta}(X_\beta \setminus X_{\beta+1})$. In fact, if $x \in X \setminus X_\theta$, and α is the smallest index with $x \notin X_\alpha$, then by the definition of the set X_α, the ordinal $\alpha \le \theta$ is not a limit ordinal, i.e., $\alpha = \beta + 1$, and then $x \in X_\beta \setminus X_{\beta+1}$. All these prove that $X \setminus X_\theta$ is countable. [G. Cantor]

22. If $X \subseteq \mathbf{R}^n$ is closed, then all the sets X_α in the preceding problem are included in X, in particular $X_\theta \subseteq X$. Now the claim follows from the representation $X = X_\theta \cup (X \setminus X_\theta)$.

23. It is clear that the $\mathcal{H}_\alpha$'s form an increasing family.

Let $\mathcal{S}$ be the σ-algebra generated by $\mathcal{H}$. Then each $\mathcal{H}_\alpha$ is included in $\mathcal{S}$, hence it is enough to show that $\mathcal{H}_{\omega_1}$ is a σ-algebra, i.e., it is closed for countable union and complementation. If $A \in \mathcal{H}_{\omega_1}$, then $A \in \mathcal{H}_\alpha$ for some $\alpha < \omega_1$, and then its complement $X \setminus A$ is contained in $\mathcal{H}_{\alpha+1} \subseteq \mathcal{H}_{\omega_1}$. In a similar manner, if $A_i \in \mathcal{H}_{\omega_1}$ for $i = 0, 1, \ldots$, say $A_i \in H_{\alpha_i}$, $\alpha_i < \omega_1$, and $\alpha = \sup_i \alpha_i$, then $\alpha < \omega_1$ and $\cup_i A_i \in H_{\alpha+1} \subseteq \mathcal{H}_{\omega_1}$. These prove that $\mathcal{H}_{\omega_1}$ is indeed a σ-algebra.

24. Let $\mathcal{H}$ be a family of sets of cardinality at most continuum. Consider the families $\mathcal{H}_\alpha$ from the preceding solution. Since a set of power continuum

includes at most continuum many countable subsets (Problem 4.6) and since the union of continuum many sets of cardinality at most continuum is of cardinality at most continuum, we obtain by transfinite induction that each $\mathcal{H}_\alpha$, $\alpha \le \omega_1$, is of cardinality at most continuum. For $\alpha = \omega_1$ this is the statement of the problem.

25. Let $\mathcal{H}$ be the set of open subsets of $\mathbf{R}^n$, and consider the following hierarchy $\mathcal{H}_\alpha$, $\alpha < \omega_1$: let $\mathcal{H}_0 = \mathcal{H}$ and for every ordinal $0 < \alpha < \omega_1$ let $\mathcal{H}_\alpha$ be the family of sets that can be obtained as a countable intersection or a countable disjoint union of sets in $\cup_{\beta<\alpha}\mathcal{H}_\beta$. Exactly as in Problem 23 one can easily show that $\mathcal{H} = \cup_{\alpha<\omega_1}\mathcal{H}_\alpha$ is the smallest family of sets containing the open sets and closed under countable intersection and countable disjoint union. All we have to show is that $\mathcal{H}$ is closed under taking complement with respect to $\mathbf{R}^n$, for then it is closed under countable union (recall that $\cup_j A_j = \mathbf{R}^n \setminus (\cap_j(\mathbf{R}^n \setminus A_j)))$.

We prove by transfinite induction that if $A \in \mathcal{H}_\alpha$, then $(\mathbf{R}^n \setminus A) \in \mathcal{H}$, and this will complete the proof. For $\alpha = 0$ this is clear, for the complement of an $A \in \mathcal{H}_0$ is a closed set, and it can be represented as a countable intersection of open sets. Suppose now that we know this property for all $\beta < \alpha$. Let $A \in \mathcal{H}_\alpha$. If A is obtained from $A_j \in \cup_{\beta<\alpha}\mathcal{H}_\beta$, $j = 0, 1, \ldots$ by disjoint union, then

$$\mathbf{R}^n \setminus A = \bigcap_j (\mathbf{R}^n \setminus A_j) \in \mathcal{H}$$

by the induction hypothesis. If, however, A is obtained from $A_j \in \cup_{\beta<\alpha}\mathcal{H}_\beta$, $j = 0, 1, \ldots$ by intersection, then

$$\mathbf{R}^n \setminus A = \bigcup_j (\mathbf{R}^n \setminus A_j) = (\mathbf{R}^n \setminus A_0) \cup \Big(A_0 \cap (\mathbf{R}^n \setminus A_1)\Big) \cup \Big(A_0 \cap A_1 \cap (\mathbf{R}^n \setminus A_2)\Big) \cup \cdots,$$

and here on the right-hand side we have a disjoint union. Therefore, we get from the induction hypothesis that $(\mathbf{R}^n \setminus A) \in \mathcal{H}$, and the proof is over.

For an alternative proof see the solution to Problem 1.13.

26. It is clear that the $\mathcal{B}_\alpha$'s form an increasing family of functions, and that if $\mathcal{B}$ is the smallest set of functions that is closed for pointwise limits and that includes $C[0,1]$, then $\mathcal{B}_\alpha \subseteq \mathcal{B}$ for all α. Thus, it is enough to show that $\mathcal{B}_{\omega_1}$ is closed for pointwise limit. Let $f_i \in \mathcal{B}_{\omega_1}$ for $i = 0, 1, \ldots$, and let f be the pointwise limit of the functions f_i. We have, say, $f_i \in B_{\alpha_i}$, $\alpha_i < \omega_1$. But then, if $\alpha = \sup_i \alpha_i$, then $\alpha < \omega_1$ and $f \in B_{\alpha+1} \subseteq \mathcal{B}_{\omega_1}$, and this proves that $\mathcal{B}_{\omega_1}$ is closed for pointwise limit.

27. Let $\mathcal{F}$ be the set of all the operations in the algebra $\langle \mathcal{A}, \cdots \rangle$, and let $B \subset \mathcal{A}$ be a subset of cardinality at most $\kappa \ge \aleph_0$. Then the set of finite subsets of B is again of cardinality at most κ, hence if B^* is the set that we obtain by adding to B all elements of the form $g(b_1, \ldots, b_m)$ with $b_i \in B$ and $g \in \mathcal{F}$, then B^* is of cardinality at most $\kappa \cdot \rho \le \max(\kappa, \rho, \aleph_0)$. Now starting from $B_0 = B$

construct the sets B_k as $B_{k+1} = (B_k)^*$, $k = 0, 1, 2, \ldots$, and let $B_\infty = \cup_{k=0}^{\infty} B_k$. As we have just mentioned, each B_k is of cardinality at most $\max(\kappa, \rho, \aleph_0)$, hence B_∞ is of cardinality at most $\max(\kappa, \rho, \aleph_0) \cdot \aleph_0 = \max(\kappa, \rho, \aleph_0)$. It is clear that each B_k is contained in the subalgebra generated by B, hence the same is true of B_∞, and what is left to prove is that B_∞ is a subalgebra, since then it must be the subalgebra generated by B. Let $g \in \mathcal{F}$ be an operation of arity m, and let $b_1, \ldots, b_m \in B_\infty$ be arbitrary elements. Then $b_i \in B_{k_i}$ for some $k_i \in \mathbf{N}$, hence with $k = \max_i k_i$ we have $b_i \in B_k$ for all $i = 1, \ldots, m$, and then $g(b_1, \ldots, b_n) \in B_{k+1} \subset B_\infty$, verifying that B_∞ is closed for all the operations.

28. First of all we note that if $\mathcal{F}$ is any field of cardinality at most $\kappa \geq \aleph_0$ and $p(x) = a_n \cdot x^n + \cdots + a_1 \cdot x + a_0$ is any polynomial with coefficients in $\mathcal{F}$, then there is a field $\mathcal{F} \subseteq \mathcal{F}_1$ of cardinality at most κ such that p has a zero in $\mathcal{F}_1$. This is well known, but a sketch of the construction runs as follows. We may assume that p is irreducible over $\mathcal{F}$ (if not, just work with an irreducible factor of it). Let ξ be a symbol, and consider the set $\mathcal{F}_1$ of all formal expressions $b_0 + b_1 \cdot \xi + \cdots + b_{n-1} \cdot \xi^{n-1}$, $b_i \in \mathcal{F}$, with termwise addition and multiplications except that in multiplication we simplify with $a_n \cdot \xi^n + \cdots a_1 \cdot \xi + a_0 = 0$. It is easy to see that with these operations $\mathcal{F}_1$ is a field. For example, the existence of the multiplicative inverse of an element $b_0 + b_1 \cdot \xi + \cdots + b_{n-1} \cdot \xi^{n-1}$ with not all $b_i = 0$ runs as follows. Since $p(x)$ is irreducible, $p(x)$ and $b_0 + b_1 \cdot x + \cdots + b_{n-1} \cdot x^{n-1}$ have only constant (elements of $\mathcal{F}$) common divisors. Hence by carrying out the Euclidean algorithm, we get that there are polynomials $r(x)$ and $s(x)$ such that

$$r(x)p(x) + s(x)(b_0 + b_1 \cdot x + \cdots + b_{n-1} \cdot x^{n-1}) = 1.$$

Substituting here $x = \xi$ we obtain

$$r(\xi)(b_0 + b_1 \cdot \xi + \cdots + b_{n-1} \cdot \xi^{n-1}) = 1,$$

i.e., $r(\xi)$ is the multiplicative inverse of $b_0 + b_1 \cdot \xi + \cdots + b_{n-1} \cdot \xi^{n-1}$. It is also clear that $\mathcal{F}$ can be considered to be part of $\mathcal{F}_1$, and that $\mathcal{F}_1$ has cardinality at most κ.

The set of polynomials with coefficients in $\mathcal{F}$ is of cardinality at most κ (see Problem 10.4(a)), and let us enumerate them into a sequence p_ξ, $\xi < \kappa$ (with possible repetition). Starting from $\mathcal{F}_0 = \mathcal{F}$ we recursively define increasing fields $\mathcal{F}_\xi$, $\xi < \kappa$, where $\mathcal{F}_\xi$ is a field of cardinality at most κ that is an extension of the field $\cup_{\alpha<\xi} \mathcal{F}_\alpha$ in such a way that p_ξ has a zero in $\mathcal{F}_\xi$. Based on what we have said in the beginning of this solution, this F_ξ can be easily defined by transfinite recursion for all $\xi < \kappa$, and it has cardinality at most κ. Now let $\mathcal{F}_1^* = \cup_{\xi<\kappa} \mathcal{F}_\xi$. Then $\mathcal{F}_1^*$ is a field of cardinality at most κ, and every polynomial with coefficients in $\mathcal{F}$ has a zero in $\mathcal{F}_1^*$. Now repeat the same process starting from $\mathcal{F}_1^*$ rather than $\mathcal{F}$, to obtain a field $\mathcal{F}_1^* \subseteq \mathcal{F}_2^*$ such that every polynomial with coefficients in $\mathcal{F}_1^*$ has a zero in $\mathcal{F}_2^*$. In a similar manner we get fields $\mathcal{F}_k^* \subseteq \mathcal{F}_{k+1}^*$ for all $k = 1, 2, \dots$ such that every polynomial with coefficients in $\mathcal{F}_k^*$ has a zero in $\mathcal{F}_{k+1}^*$. Now it is clear that if $\mathcal{F}^* = \cup_{k=0}^{\infty} \mathcal{F}_k^*$, then $\mathcal{F}^*$ is a field of cardinality at most κ that includes $\mathcal{F}$. Furthermore it is algebraically closed. In fact, every polynomial with coefficients in $\mathcal{F}^*$ has coefficients in $\mathcal{F}_k^*$ for some $k \in \mathbf{N}$, and so it has a zero in $\mathcal{F}_{k+1}^* \subseteq \mathcal{F}^*$.

29. Let $\langle A, \prec \rangle$ be an infinite ordered set, $\kappa = |A|$, and let $T = {}^{\kappa}\{0, 1\}$ be the set of transfinite 0–1 sequences of type κ with lexicographic ordering $\prec^*$. Let $A = \{a_\xi\}_{\xi < \kappa}$ be an enumeration of the elements of A in type κ. By transfinite recursion we define a monotone mapping F from A into T.

Actually we shall show more, namely let T^* be the set of those elements f of T that contain a largest 1, i.e., for which there is a $\nu < \kappa$ such that $f(\nu) = 1$ but $f(\xi) = 0$ for all $\nu < \xi < \kappa$. We are going to construct F so that it maps A monotonically into T^*.

First we consider the case when κ is regular. Before the actual construction we establish a few facts about T^* for regular κ.

a) If $B \subset T^*$ is of cardinality smaller than κ, then there is an $h \in T^*$ that is smaller than any element of B. In fact, let us select $\nu < \kappa$ in such a way that $f(\xi) = 0$ for all $f \in B$ and $\xi \geq \nu$, and set $h(\nu) = 1$ and $h(\xi) = 0$ for $\xi \neq \nu$. This h is clearly smaller than any element of B.

b) If $B \subset T^*$ is of cardinality smaller than κ, then there is an $h \in T^*$ that is bigger than any element of B. In fact, again let $\nu < \kappa$ be such that $f(\xi) = 0$ for all $f \in B$ and $\xi \geq \nu$, and let $h(\xi) = 1$ for $\xi \leq \nu$ and $h(\xi) = 0$ for $\xi < \nu$. This h is bigger than any element of B.

c) If $B, C \subset T^*$ are of cardinality smaller than κ such that every element of B is smaller than any element of C, then there is an $h^* \in T^*$ that is bigger than any element of B and smaller than any element of C. First we construct a function $h \in T$. We define $h(\alpha)$, $\alpha < \kappa$, by transfinite recursion. Let $h(0) = 1$ if there is an $f \in B$ with $f(0) = 1$, and otherwise let $h(0) = 0$. Suppose now that $\alpha < \kappa$ and $h(\beta)$ is already defined for all $\beta < \alpha$. Then let $h(\alpha) = 1$ if there is an $f \in B$ such that $f|_\alpha = h|_\alpha$ and $f(\alpha) = 1$, and otherwise let $h(\alpha) = 0$. By transfinite induction we prove that for all $f \in B$ and $\alpha < \kappa$ the inequality $f|_\alpha \preceq^* h|_\alpha$ holds, where $\preceq^*$ denotes again lexicographic ordering (ordering with respect to first difference). This is clear for $\alpha = 0$, and if it is true that $f|_\beta \preceq^* h|_\beta$ for all $\beta < \alpha$ and α is a limit ordinal, then clearly $f|_\alpha \preceq^* h|_\alpha$. If, however, $\alpha = \beta + 1$, then $h|_\alpha \prec^* f|_\alpha$ together with $f|_\beta \preceq^* h|_\beta$ would imply $f|_\beta = h|_\beta$ and $h(\beta) = 0$, $f(\beta) = 1$, but this contradicts the choice of $h(\beta)$. Thus, $f|_\alpha \preceq^* h|_\alpha$ in all cases. In a similar manner, by transfinite induction we verify that $h|_\alpha \preceq^* g|_\alpha$ for all $\alpha < \kappa$ and $g \in C$. This is clear for $\alpha = 0$, and if it is true that $h|_\beta \preceq^* g|_\beta$ for all $\beta < \alpha$ and α is a limit ordinal, then clearly $h|_\alpha \preceq^* g|_\alpha$. If, however, $\alpha = \beta + 1$, then $g|_\alpha \prec^* h|_\alpha$ together with $h|_\beta \preceq^* g|_\beta$ implies that $h|_\beta = g|_\beta$ and $g(\beta) = 0$, $h(\beta) = 1$. This latter one means that there is an $f \in B$ such that $f|_\beta = h|_\beta$ and $f(\beta) = 1$. But this is impossible, for then we would have $g \prec^* f$, contradicting the assumption on the sets B and C. Thus, $h|_\alpha \preceq^* g|_\alpha$ in all cases.

What we have verified so far implies that $f \preceq^* h \preceq^* g$ for all $f \in B$ and $g \in C$. Next note that $h = g$ is impossible for $g \in C$. In fact, in the opposite case if ν is such that $g(\nu) = 1$ but $g(\xi) = 0$ for $\xi > \nu$, then $h(\nu) = 1$ would imply an $f \in B$ with $f|_\nu = h|_\nu$ and $f(\nu) = 1$, but then we would have $g \preceq^* f$ contradicting the assumption on the sets B and C. Thus, $h \prec^* g$ for all $g \in C$. Now let $\nu < \kappa$ be an ordinal such that $f(\xi) = 0$ and $g(\xi) = 0$ for all $\xi \geq \nu$ and $f \in B$, $g \in C$. Then clearly $h(\xi) = 0$ for $\xi \geq \nu$; therefore, if we set $h^*(\xi) = h(\xi)$ if $\xi \neq \nu$ and $h^*(\nu) = 1$, then this h^* will be strictly bigger than any element in B and smaller than any element in C.

After these preparations let us return to the construction of the mapping F for regular κ. As in the beginning of the proof, let $A = \{a_\xi\}_{\xi<\kappa}$ be an enumeration of the different elements of A in type κ. We are going to define $F(a_\alpha)$ by transfinite recursion on α. Let $F(a_0)$ be any element in T^*, and suppose that for some $\alpha < \kappa$ all the values $F(a_\xi)$, $\xi < \alpha$ have already been defined, and F is monotone on its domain. This domain is divided into two parts by a_α: $H_0 = \{a_\xi \ : \ \xi < \alpha,\ a_\xi \prec a_\alpha\}$ and $H_1 = \{a_\xi \ : \ \xi < \alpha,\ a_\alpha \prec a_\xi\}$. We set $B = \{F(a_\xi) \ : \ \xi \in H_0\}$ and $C = \{F(a_\xi) \ : \ \xi \in H_1\}$. Then B and C are subsets of T^* of cardinality smaller than κ, and every element of B is smaller than any element of C. Now let $F(a_\alpha) = h^*$, where $h^* \in T^*$ is the element constructed in part c) above for this B and C. If one of the sets, say B, happens to be empty, then just select an element $h \in T^*$ as $F(a_\alpha)$ that is smaller than any element of C (see property a) above), and in a similar manner if $C = \emptyset$, then let $F(a_\alpha)$ be an element of T^* that is bigger than any element of B (see property b)).

This recursion runs through $\alpha < \kappa$, and this proves the existence of F for regular κ. Note that for regular κ we have also shown the following: if $|A| = \kappa$ and $A' \subset A$ is of cardinality smaller than κ, and $G : A' \to T^*$ is a monotone mapping, then this can be extended to a monotone mapping of A into T^*. Now let κ be singular, and let $\kappa_0 < \kappa_1 < \cdots < \kappa_\alpha \cdots < \kappa$, $\alpha < \mathrm{cf}\,(\kappa)$ be infinite cardinals with sum equal to κ. By considering κ_α^+ instead of κ_α if necessary, we may assume that each κ_α is a regular cardinal, and by similar method one can achieve that for each α we have $\kappa_\alpha > \sum_{\beta<\alpha} \kappa_\beta$. Also let $A = \cup_{\alpha<\mathrm{cf}\,(\kappa)} A_\alpha$ be an appropriate representation of A as a disjoint union of some sets A_α of cardinality κ_α. We shall define by transfinite recursion on α a monotone mapping F_α from $\cup_{\beta\leq\alpha} A_\alpha$ into T_α^*, where T_α^* is the set of those elements of $f \in T^*$ for which $f(\xi) = 0$ for all $\xi \geq \kappa_\alpha$ (note that T_α^* is isomorphic with T^* constructed for the cardinal κ_α). We shall define F_α in such a way, that for $\beta < \alpha$ the mapping F_α is an extension of F_β. In fact, the mapping F_0 has just been constructed above. Suppose we know F_β for all $\beta < \alpha$. If α is a limit ordinal, then the mappings F_β, $\beta < \alpha$, have a common extension G_α defined on $\cup_{\beta<\alpha} A_\beta$: just set $G_\alpha(\xi) = F_\beta(\xi)$ for $\xi \in A_\beta$, $\beta < \alpha$. Now $|\cup_{\beta<\alpha} A_\beta| = \sum_{\beta<\alpha} \kappa_\beta < \kappa_\alpha$, hence, as we have seen above, this G_α

can be extended to a monotone mapping of $\cup_{\beta\le\alpha}A_\beta = (\cup_{\beta<\alpha}A_\beta)\cup A_\alpha$ into T^*_α. If, however, α is a successor ordinal, $\alpha=\beta+1$, then again, the mapping $F_\beta : \cup_{\gamma\le\beta}A_\gamma \to T^*_\beta$ can be extended to a monotone mapping F_α from $\cup_{\gamma\le\alpha}A_\gamma$ to T^*_α.

Finally, we set $F(\xi)=F_\alpha(\xi)$ if $\xi\in A_\alpha$. This is clearly a monotone mapping from $\langle A,\prec\rangle$ into T^*.

30. Let $\langle A,\prec\rangle$ be an ordered set, and let $\{x_\xi\}_{\xi<\kappa}$ be an enumeration of the elements of A. Since every ordered set is a subset of a densely ordered set (Problem 6.65), without loss of generality we may assume $\langle A,\prec\rangle$ to be densely ordered. Let $\langle A_\xi,<_\xi\rangle$ be an ordered set of type ω_ξ, and let $\langle B,<\rangle$ be the ordered union of the ordered sets $\langle A_\xi,<_\xi\rangle$, $\xi<\kappa$, with respect to $\langle A,\prec\rangle$, i.e., the element x_ξ in $\langle A,\prec\rangle$ is replaced by A_ξ with the order $<_\xi$ on it, and these well-ordered sets $\langle A_\xi,<_\xi\rangle$ follow each other exactly as the elements x_ξ follow one another in $\langle A,\prec\rangle$. It is clear that $\langle A,\prec\rangle$ can be considered as part of $\langle B,<\rangle$ (in $\langle B,<\rangle$ the smallest elements of the sets A_ξ form a subset similar to $\langle A,\prec\rangle$). It is also clear that for each $\xi<\kappa$ there is a unique maximal well-ordered subinterval of $\langle B,<\rangle$ with order type ω_ξ, namely A_ξ (use that $\langle A,\prec\rangle$ is densely ordered).

We claim that no two different initial segments of $\langle B,<\rangle$ are similar. In fact, let S_1 and S_2 be two similar initial segments of $\langle B,<\rangle$, and let $f : S_1\to S_2$ be a similarity mapping. The initial segment S_1 has the following structure: it is the ordered union of two sets S_1^1 and S_1^2, where S_1^1 is the ordered union of some of the sets $\langle A_\xi,<_\xi\rangle$ with respect to ξ's lying in an initial segment A_1 of $\langle A,\prec\rangle$, and S_1^2 is an initial segment of one of the sets $\langle A_{\xi_1},<_{\xi_1}\rangle$. Also, since $\langle A,\prec\rangle$ is densely ordered, there is no end segment of S_1^1 that is well ordered. Thus, S_1^2 can be recognized as the (possibly empty) largest end segment of S_1 that is well ordered. Now let $S_2=S_2^1\cup S_2^2$ be the analogous representation of S_2. Since a similarity mapping maps a well-ordered interval/end segment into a well-ordered interval/end segment, it follows that f maps S_1^2 into S_2^2, and hence it also maps S_1^1 into S_2^1. If $A_\xi\subseteq S_1^1$, i.e., $\xi\in A_1$, then A_ξ is a maximal interval in S_1^1 that is well ordered (recall that $\langle A,\prec\rangle$ was assumed to be densely ordered). Thus, its image is also a maximal interval in S_2^1 of type ω_ξ, which is possible only if $\xi\in A_2$. The argument can be reversed with $\xi\in A_2$, and it follows that the two initial segments A_1 and A_2 of $\langle A,\prec\rangle$ are the same. Thus, $S_1^1=S_2^1$, and then S_1^2 and S_2^2 are similar initial segments of the well-ordered set $\langle A_{\xi_1},<_{\xi_1}\rangle$, which is possible only if they are the same: $S_1^2=S_2^2$. Thus, $S_1=S_2$, which means that different initial segments of $\langle B,<\rangle$ are nonsimilar.

13

Euclidean spaces

1. Let $f : \mathbf{R} \to \mathbf{R}^n$ be a continuous mapping from $\mathbf{R}$ onto $\mathbf{R}^n$. Based on Problems 5.35 and 5.36 it is easy to see that there is such a mapping. Now the family of sets $\{f^{-1}[U] : U \in \mathcal{U}\}$ is a family of open subsets of $\mathbf{R}$ that is well ordered with respect to inclusion. Thus, the claim follows from Problem 6.38.

2. In the proof we need the following simple fact: for different $t_1, \ldots, t_M > 0$ and different $\alpha_1, \ldots, \alpha_M$ a determinant of the form $|t_i^{\alpha_j}|_{i,j=1}^M$ is nonzero. One can prove this by induction, the case $M = 1$ being trivial. Now suppose that the claim is true for $M - 1$. Replace t_M by a free variable t. Then the determinant becomes a generalized polynomial $S(t) = \sum_{i=1}^M a_i t^{\alpha_i}$, which, by the induction hypothesis, has nonzero coefficients. This $S(t)$ vanishes for $t = t_1, \ldots, t_{M-1}$, and so it is sufficient to show that a nontrivial S of the above form cannot have M positive zeros. This is proved again by induction on M. If S had M positive zeros, then so would $S(t)/t^{\alpha_M} = \sum_{i=1}^M a_i t^{\alpha_i - \alpha_M}$, and hence by Rolle's theorem its derivative

$$S_1(t) = \sum_{i=1}^{M-1} a_i(\alpha_i - \alpha_M)t^{\alpha_i - \alpha_M - 1}$$

would have $M - 1$ positive zeros. This S_1 is of the same form as S just M is replaced by $M - 1$, so we can apply induction to conclude that S can have only at most $M - 1$ zeros, by which we have proved the claim.

After this let us choose rationally independent positive real numbers $\alpha_1, \alpha_2, \ldots, \alpha_n$, and consider the set

$$B = \{(t^{\alpha_1}, \ldots, t^{\alpha_n}) : t \in [0, 1]\}.$$

We claim that every algebraic variety A intersects B in at most finitely many points. In fact, let

$$P(x_1,\dots,x_n)=\sum_{i_1,\dots,i_n=0}^{N} a_{i_1,\dots,i_n}x_1^{i_1}\cdots x_n^{i_n}$$

be a nontrivial polynomial with zero set A. For $t\in[0,1]$ we have

$$P(t^{\alpha_1},\dots,t^{\alpha_n})=\sum_{i_1,\dots,i_n=0}^{N} a_{i_1,\dots,i_n}t^{i_1\alpha_1+\cdots+i_n\alpha_n},$$

and since the numbers α_i are rationally independent, all the exponents on the right are different. Thus, if there are $(N+1)^n$ different points $t_1,\dots,t_{(N+1)^n}\in[0,1]$ that lie in A, then at these points we have

$$\sum_{i_1,\dots,i_n=0}^{N} a_{i_1,\dots,i_n}t_j^{i_1\alpha_1+\cdots+i_n\alpha_n}=0,\quad j=1,2,\dots,(N+1)^n.$$

But, according to what we proved above, the determinant of this $(N+1)^n\times(N+1)^n$ linear system of equations is nonzero, which implies that all $a_{i_1,\dots,i_n}$ are zero, i.e., the polynomial P is identically zero. This contradiction proves that $B\cap A$ can have at most $(N+1)^n-1$ points.

Now it is clear that $\mathbf{R}^n$ cannot be covered by less than continuum many algebraic varieties, for less than continuum many algebraic varieties can cover less than continuum many points of the set B, and B is of power continuum.

3. Consider the set $\mathcal{H}$ of all subsets A of $\mathbf{R}^3$ which have the property that if we connect the different points of A by a segment then all these segments are disjoint. It is easy to see that if $\mathcal{F}$ is a subset of $\mathcal{H}$ ordered with respect to inclusion, then the union of the sets in $\mathcal{F}$ also belong to $\mathcal{H}$. Thus, by Zorn's lemma (see Chapter 14) there is a maximal (with respect to inclusion) set A in $\mathcal{H}$. All we have to show is that A is of cardinality continuum.

Suppose that to the contrary that A is of cardinality less than continuum. If we consider all three points of A and the planes that they span, then we get less than continuum many planes (more precisely, $|A|^3=|A|$ many planes). By Problem 2 the space $\mathbf{R}^3$ cannot be covered by less than continuum many planes. Thus, there is a point P in $\mathbf{R}^3$ that does not lie on any plane spanned by any three points of A. But then it is easy to see that P can be added to A, because the lines through P and through points of A do not intersect lines that connect two other points of A. This contradicts the maximality of A, and this contradiction verifies the claim.

There are also easy constructions for the set in question. For example $A=\{(t,t^2,t^3)\ :\ t\in[0,1]\}$ is appropriate, for no plane intersects A in more than 3 points. [W. Sierpiński, Cardinal and Ordinal Numbers, Polish Sci. Publ., Warszawa, 1965, IV.7/6]

4. Let H be an uncountable set in $\mathbf{R}^n$, and first suppose that each sphere in $\mathbf{R}^n$ contains only countably many points from H. We select points P_ξ, $\xi<\omega_1$

by transfinite induction in such a way that all the distances between them are different. Suppose P_ξ, $\xi < \eta$, have already been selected for some $\eta < \omega_1$. Let D_η be the set of all distances between the points in the set $\{P_\xi\}_{\xi<\eta}$, and let S_η be the union of all the spheres with center at some P_ξ, $\xi < \eta$ and with radius $d \in D_\eta$. Then S_η is the union of countably many spheres, so our assumption implies that in S_η there are only countably many points from the set H, hence we can select a point $\xi_\eta \in H \setminus S_\eta$. This procedure can be carried out for all $\eta < \omega_1$, and it is clear that all the distances between the selected points are different.

Now suppose that there is a sphere S such that $S \cap H$ is uncountable. Then work on S with the set $S \cap H$ instead of H in the same fashion as we have done above. It is still possible that for this set there is a sphere S' different from S such that on S' the set $S \cap H$ has uncountably many points, but then on the lower-dimensional sphere $S \cap S'$ the set H has uncountably many points, Thus, if we choose a sphere S with the smallest possible dimension on which H has uncountably points, then the previous procedure can be carried out on S with the set $S \cap H$ instead of H.

5. Let M be the set of all finite 0–1 sequences, and let $f : M \to \mathbf{N}$ be a 1-to-1 and onto mapping; furthermore, let $\mathbf{b}_j = (0, \dots, 0, 1, 0, \dots)$ be the element of ℓ_2 which has zero coordinates except for the jth coordinate, which is 1. With $a_n = \sqrt{3/2} \cdot 2^{-n}$ for an infinite 0–1 sequence $\epsilon = (\epsilon_1, \epsilon_2, \dots)$ define the element $h_\epsilon \in \ell_2$ as

$$h_\epsilon = \sum_{n=1}^{\infty} a_n \mathbf{b}_{f((\epsilon_1,\epsilon_2,\dots,\epsilon_n))}.$$

This way we define continuum many elements of ℓ_2, and we claim that if $\epsilon' = (\epsilon'_1, \epsilon'_2, \dots)$ is another 0–1 sequence, then the distance between h_ϵ and $h_{\epsilon'}$ is rational. In fact, if m is the smallest index with $\epsilon_m \neq \epsilon'_m$, then the distance between h_ϵ and $h_{\epsilon'}$ is

$$\left(\sum_{n=m}^{\infty} 2a_n^2\right)^{1/2} = \left(\sum_{n=m}^{\infty} 2\frac{3}{2}2^{-2n}\right)^{1/2} = 2^{-m+1}.$$

6. This is an immediate consequence of the separability of ℓ_2, i.e., that there is a countable dense set (e.g., the set of those elements that have rational coefficients of which only finitely many are nonzero). In fact, if all the distances between points of a set H are the same, say ρ, then the balls about points of H of radius $\rho/3$ are disjoint, and each such ball contains at least one point from our countable dense subset.

7. Assume that $\ell_2 = A_0 \cup A_1 \cup \cdots$ is a decomposition. Let $\{\mathbf{b}_s : s\}$ be an orthonormal basis where we index with all finite strings of natural numbers. If $f : \omega \to \omega$ is an infinite sequence of natural numbers we let

$$\mathbf{a}_f = \sum_{n=0}^{\infty} \frac{1}{2^n} \mathbf{b}_{f|n},$$

where $f|n$ denotes the string of the first n elements of f. It is easy to see that if f, g are infinite sequences of natural numbers, then the square distance between $\mathbf{a}_f$ and $\mathbf{a}_g$ is $2(4^{-n} + 4^{-(n+1)} + \cdots) = 2/(3 \cdot 4^{n-1})$, where f and g first differ at the nth place.

We are done if we can find some $i < \omega$ and some finite string s such that for every extension $s' = sk$ $(k = 0, 1, \ldots)$ of s there is some $f_k : \omega \to \omega$ such that $f_k|(n+1) = sk$ and $\mathbf{a}_{f_k} \in A_i$. Indeed, then these $\mathbf{a}_{f_k}$'s will all have the same distance from one another. Assume therefore that on the contrary, this latter statement fails. Then the choice $i = 0$, $s = \emptyset$ is not good, i.e., there is some k_0 such that $\mathbf{a}_f \notin A_0$ for any f with $f(0) = k_0$. Next, the choice $i = 1$, $s = k_0$ is not good either, hence there is some k_1 such that $\mathbf{a}_f \notin A_1$ for any f with $f(0)f(1) = k_0k_1$. Continuing this way, we get a sequence $k_0, k_1, \cdots$ such that if $f(i) = k_i$, $i = 0, 1, \ldots$, then $\mathbf{a}_f$ is not in any of the sets A_i, which is a contradiction, and this contradiction proves the claim.

8. There is a family $\mathcal{H}$ of cardinality continuum of subsets of $\mathbf{N}$ such that the intersection of any two members of $\mathcal{H}$ is infinite, but the intersection of any three members is finite (see Problem 4.35). For $H \in \mathcal{H}$ consider the point $\mathbf{b}_H = (\pm 1, \pm 1/2, \pm 1/4, \ldots) \in \ell_2$, where the nth coordinate in $\mathbf{b}_H$ is $1/2^n$ if $n \in H$, and otherwise it is $-1/2^n$. If $\mathbf{b}_H$ and $\mathbf{b}_K$ are two such points, then in $\mathbf{b}_H - \mathbf{b}_K$ the nth coordinate is $0, \pm 2/2^n$, and it can be $2/2^n$ only if the nth coordinate in $\mathbf{b}_K$ is $-1/2^n$, and it can be $-2/2^n$ only if the nth coordinate in $\mathbf{b}_K$ is $1/2^n$. It follows that if $\mathbf{b}_H, \mathbf{b}_K, \mathbf{b}_S$ are three different points of the above type, then it is not possible to have simultaneously $2/2^n$ for the nth coordinate in $\mathbf{b}_H - \mathbf{b}_K$ and at the same time to have $-2/2^n$ for the nth coordinate in $\mathbf{b}_S - \mathbf{b}_K$. But this means that the inner product of $\mathbf{b}_H - \mathbf{b}_K$ and $\mathbf{b}_S - \mathbf{b}_K$ is nonnegative. It is actually positive, since $H \cap S$ is infinite but $H \cap S \cap K$ is finite, so there is an $n \in (H \cap S) \setminus K$, and for this n the nth coordinate both in $\mathbf{b}_H - \mathbf{b}_K$ and $\mathbf{b}_S - \mathbf{b}_K$ is $2/2^n$.

Thus, the inner product $(\mathbf{b}_H - \mathbf{b}_K, \mathbf{b}_S - \mathbf{b}_K)$ of $\mathbf{b}_H - \mathbf{b}_K$ and of $\mathbf{b}_S - \mathbf{b}_K$ is positive, which means that the angle at $\mathbf{b}_K$ in the triangle $(\mathbf{b}_H, \mathbf{b}_K, \mathbf{b}_S)$ is acute (recall that if φ is this angle, then

$$\cos\varphi = (\mathbf{b}_H - \mathbf{b}_K, \mathbf{b}_S - \mathbf{b}_K)/\|\mathbf{b}_H - \mathbf{b}_K\|\|\mathbf{b}_S - \mathbf{b}_K\|).$$

9. We show that there is a well ordering $\prec$ of $\mathbf{R}^2$ such that for every point $x \in \mathbf{R}^2$ the set

$$\{y \prec x : d(y, x) \in \mathbf{Q}\}$$

is finite. (Here $d(x, y)$ is the Euclidean distance.) This suffices, as then we can color $\mathbf{R}^2$ by a simple transfinite recursion along $\prec$ with countably many colors, since at every point x we can extend the previously defined coloring on

$\{y : y \prec x\}$ to the point x by omitting that finitely many colors that appear at rational distances from x.

In order to prove the existence of $\prec$ we show that for every $X \subseteq \mathbf{R}^2$ there is such a well-ordering, and this we do by transfinite induction on $\kappa = |X|$.

For $\kappa \leq \omega$ any well order into type $\leq \omega$ will do, and now assume that $\kappa > \omega$ and that the claim has already been verified for sets of cardinality smaller than κ. Call a set $S \subset \mathbf{R}^2$ 'closed' if $x_1, x_2 \in S$ and $d(x_1, y) \in \mathbf{Q}, d(x_2, y) \in \mathbf{Q}$ imply $y \in S$. If S is any subset of $\mathbf{R}^2$, its 'closure' is $\cup_{i=0}^{\infty} S_i$, where $S_0 = S$, and each S_i is obtained from S_{i-1} by adding all points that are of rational distance from some two points of S_{i-1}. As for any pair (x_1, x_2) of points the set

$$\{y \in \mathbf{R}^2 : d(x_1, y) \in \mathbf{Q},\ d(x_2, y) \in \mathbf{Q}\}$$

is countable, it follows from Problem 10.4 that each S_i is of cardinality $\max(|S|, \aleph_0)$, and hence the cardinality of the 'closure' of S is also at most $\max(|S|, \aleph_0)$.

Now we can decompose X as the union $X = \cup_{\alpha<\kappa} X_\alpha$ of increasing sets X_α, $\alpha < \kappa$, of cardinality less than κ such that each X_α is 'closed' and for limit ordinal α we have $X_\alpha = \cup_{\beta<\alpha} X_\beta$. This is easily achieved by transfinite induction from any enumeration of X into a transfinite sequence of type κ if we apply the 'closure' procedure and for limit ordinals α we set $X_\alpha = \cup_{\beta<\alpha} X_\beta$ (note that the union of increasing 'closed' sets is again 'closed'). By the inductive hypothesis, each $X_{\alpha+1} \setminus X_\alpha$ possesses a well-ordering $\prec_\alpha$ as needed. We can now define $\prec$ on X as follows: let $x \prec y$ if either $x \in X_\alpha$, $y \notin X_\alpha$ for some $\alpha < \kappa$ or if $x, y \in X_{\alpha+1} - X_\alpha$ and $x \prec_\alpha y$ for some $\alpha < \kappa$. This is a well-ordering (in fact, $\langle X, \prec \rangle$ is the ordered union of well-ordered sets with respect to $\alpha < \kappa$). Furthermore if $x \in X_{\alpha+1} \setminus X_\alpha$, then, since X_α is 'closed', there is at most one point $y \in X_\alpha$ of rational distance from x, and, by the choice of $\prec_\alpha$, there are only finitely many points $y \in X_{\alpha+1} \setminus X_\alpha$, $y \prec x$ of rational distance from x.

10. Similarly as in the preceding solution, we show that there is a well-ordering $\prec$ of $\mathbf{R}^n$ such that for every $x \in \mathbf{R}^n$ the value

$$\delta(x) = \inf\{d(y, x) : y \prec x, d(y, x) \in \mathbf{Q}\} \tag{13.1}$$

is positive. This done, we can color the points as follows. The color of $x \in \mathbf{R}^n$ be an ordered pair $(\epsilon, \mathbf{q})$ where $0 < \epsilon < \delta(x)$ is a rational number and $\mathbf{q} \in \mathbf{Q}^n$ is a rational point with $d(x, \mathbf{q}) < \epsilon/2$. This is indeed a good coloring. In fact, if $d(x, y)$ is rational and x and y would have the same color $(\epsilon, \mathbf{q})$, then we would have $d(x, \mathbf{q})$, $d(y, \mathbf{q}) < \epsilon/2$, and (say) $x \prec y$, which give $d(x, y) < \epsilon < \delta(y)$, which is a contradiction.

We prove the existence of the well-ordering $\prec$ for every $X \subseteq \mathbf{R}^n$ by transfinite induction on $\kappa = |X|$. For $\kappa \leq \omega$ any well-ordering into type $\leq \omega$ will do.

Given X of cardinality $\kappa = |X| > \omega$ we first decompose X into the union $X = \cup_{\alpha<\kappa} X_\alpha$ of increasing subsets X_α, $\alpha < \kappa$ of cardinality less then κ such that for limit α we also have $X_\alpha = \cup_{\beta<\alpha} X_\beta$. We shall also need one additional property.

First of all we mention that if $a_0, \ldots, a_d$ ($d \le n$) are points in $\mathbf{R}^n$ in general position (which means that there is no $(d-1)$-dimensional hyperplane containing them), H is the d-dimensional hyperplane spanned by these points and $r_0, \ldots, r_d$ are any rational numbers, then there can be at most one $x \in H$ with $d(x, a_0) = r_0, \ldots, d(x, a_d) = r_d$. Indeed, if $x, y \in H$ are both good then $y - x$ is orthogonal to $a_1 - a_0, \ldots, a_d - a_0$, therefore to every vector in $H - a_0$. In particular, it is orthogonal to itself, hence $x = y$. We can therefore require (see the preceding proof), that each X_α is 'closed' in the following sense: if $a_0, \ldots, a_d \in X_\alpha$ ($d \le n$) are points in X_α in general position, H is the d dimensional hyperplane spanned by them and $r_0, \ldots, r_d$ are any rational numbers, and if there is an $x \in H$ with $d(x, a_0) = r_0, \ldots, d(x, a_d) = r_d$, then this x belongs to X_α.

By the inductive hypothesis, each $X_{\alpha+1} \setminus X_\alpha$ has a well ordering $\prec_\alpha$ as required. We show that we can take $\prec$ as follows. If $x \in X_\alpha$, $y \notin X_\alpha$ for some α, then set $x \prec y$. If, however, $x, y \in X_{\alpha+1} \setminus X_\alpha$ for some α, then set $x \prec y$ if and only if $x \prec_\alpha y$. This is clearly a well-ordering, and we have to show that this $\prec$ satisfies the property that $\delta(x)$ from (13.1) is positive for all x, and this boils down to proving that if $x \notin X_\alpha$, then there cannot be points in X_α in rational distance from x and arbitrarily close to x. Assume to the contrary, that this is not true, and that $a_i \to x$, as $i \to \infty$, where $a_i \in X_\alpha$ and $d(a_i, x) \in \mathbf{Q}$. We can assume that $a_0, \ldots, a_d$ is a maximal subsystem of the points a_i in general position. Let H be the hyperplane spanned by $a_0, \ldots, a_d$. For $i > d$ we have $a_i \in H$, and, as $a_i \to x$, we get $x \in H$. But then we would have $x \in X_\alpha$ by the construction, and the proof is over.

11. Identify the plane with $\mathbf{C}$, and note that $x, y, z \in \mathbf{C}$ are the nodes of an equilateral triangle if and only if $z = \omega x + \overline{\omega} y$ or $z = \omega y + \overline{\omega} x$, where $\omega = (1 + \sqrt{3}i)/2$ and $\overline{\omega} = (1 - \sqrt{3}i)/2$. Thus, our task is to decompose $\mathbf{C}$ into countably many classes in such a way that the equation $z = \omega x + \overline{\omega} y$ has no solution in any of the classes.

Let $\mathbf{Q}(\sqrt{3})$ be the set of numbers of the form $a + b\sqrt{3}$ where $a, b \in \mathbf{Q}$. This is easily seen to be a subfield of $\mathbf{C}$, and $\mathbf{C}$ is a vector space over $\mathbf{Q}(\sqrt{3})$. Let $\mathcal{B}$ be a basis of this vector space, and let $\prec$ be an ordering on $\mathcal{B}$. Then every nonzero $x \in \mathbf{C}$ has a unique representation

$$x = \lambda_{i_0} b_{i_0} + \cdots + \lambda_{i_n} b_{i_n}, \tag{13.2}$$

where the coefficients $\lambda_{i_0}, \ldots, \lambda_{i_n} \in \mathbf{Q}(\sqrt{3})$ are nonzero numbers and $b_{i_0} \prec \cdots \prec b_{i_n}$ are from the basis $\mathcal{B}$. Notice that there are countably many possible ordered $\langle \lambda_{i_0}, \ldots, \lambda_{i_n} \rangle$, $n = 1, 2, \ldots$ sequences from $\mathbf{Q}(\sqrt{3})$, so we can decompose $\mathbf{C}$ into countably many classes in such a way that numbers in the same class have the same ordered coefficient sequence, and let 0 alone form a class. We show that this decomposition of $\mathbf{C}$ is as required.

Assume that the elements x, y, z of some class, say the one associated with $\langle \lambda_{i_0}, \ldots, \lambda_{i_n} \rangle$, satisfy $z = \omega x + \overline{\omega} y$. Although the sequence $\langle \lambda_{i_0}, \ldots, \lambda_{i_n} \rangle$ is the same for x, y, z, the associated sequences of basis vectors $\langle b^x_{i_0}, \ldots, b^x_{i_n} \rangle$, $\langle b^y_{i_0}, \ldots, b^y_{i_n} \rangle$, $\langle b^z_{i_0}, \ldots, b^z_{i_n} \rangle$, can be different. Let b be the smallest (with respect to the ordering $\prec$ on $\mathcal{B}$) of all the occurring basis elements, that is the minimal element of

$$\{b^x_{i_0}, \ldots, b^x_{i_n}, b^y_{i_0}, \ldots, b^y_{i_n}, b^z_{i_0}, \ldots, b^z_{i_n}\}.$$

Let the coefficient of b in x, y, and z be respectively α, β, and γ. Each of α, β, γ is either 0 or one of the numbers $\lambda_{i_0}, \ldots, \lambda_{i_n}$, and since b is the smallest of the basis vectors appearing in the representation of x, y, z we can conclude that of α, β, γ is either 0 or λ_{i_0} (recall that the b_{i_j}'s in the representation (13.2) are in increasing order). Also, since the representation (13.2) is unique, we must have $\gamma = \omega\alpha + \overline{\omega}\beta$. But these imply that either $\alpha = \beta = \gamma = 0$ (which is impossible) or $\alpha = \beta = \gamma = \lambda_{i_0}$. We have, therefore, $b^x_{i_0} = b^y_{i_0} = b^z_{i_0}$, and this common term can be cancelled from x, y, and z. We can continue in the same fashion, and get $b^x_{i_1} = b^y_{i_1} = b^z_{i_1}$, etc., finally all the components of x, y, z are equal, that is, $x = y = z$. This proves the claim.

12. We call a set $C \subset \mathbf{R}^2$ a partial circle if there is a point P, called the center of that partial "circle", such that every half-line emanating from P intersects C in at most one point. It is enough to cover the plane by countably many partial "circles".

Since the real line is part of a partial "circle", it is enough to cover the complement $\mathbf{R}^2 \setminus \mathbf{R}$. We shall prove that for any countably infinite set $K = \{P_1, P_2, \ldots\}$ on the real line and for any set $H \subset \mathbf{R}^2 \setminus \mathbf{R}$ there are partial "circles" with different centers in K that cover H, and we shall do that by induction on the cardinality of H. The case $|H| \leq \aleph_0$ being trivial, let us assume that $|H| = \kappa > \aleph_0$ and that the claim has been verified for all sets of cardinality smaller than κ.

Let us call H 'closed' if it contains every point that is the intersection of two lines determined by one–one points of H and K (i.e., the lines go through at least one points of H and K). Exactly as in the proof of Problem 9 one can easily show that H is included in a 'closed' set of cardinality κ, hence without loss of generality we may assume H to be 'closed'. Represent H as $H = \cup_{\alpha<\kappa} H_\alpha$, where the sets H_α are of cardinality smaller than κ, they are 'closed' and increasing, and for limit α we have $H_\alpha = \cup_{\beta<\alpha} H_\beta$ (see the proof of Problem 9). We shall define by transfinite recursion on α an allocation of the points of H_α into partial "circles" $C_{i,\alpha}$ with center in $P_i \in K$ in such a way that we keep previously defined allocations (i.e., $C_{i,\beta} \subseteq C_{i,\alpha}$ for $\beta < \alpha$), and the partial "circles" $C_{i,\alpha}$ themselves are also defined during the process. There is nothing to prove if α is a limit ordinal; therefore, suppose that $\alpha = \gamma + 1$, and let $H_\gamma = C_{1,\gamma} \cup \cdots \cup C_{j,\gamma} \cup \cdots$, where $C_{i,\gamma}$ is a partial "circle" with center at $P_i \in K$. The induction hypothesis gives that the set $H_{\gamma+1} \setminus H_\gamma$, which has cardinality smaller than κ, can be covered by "circles" $D_1, D_3, D_5, \ldots$ with center at $P_1, P_3, P_5, \ldots$, and also by "circles" $E_2, E_4, E_6, \ldots$ with center at $P_2, P_4, P_6, \ldots$. Thus, an arbitrary point $P \in H_{\gamma+1} \setminus H_\gamma$ is contained in a "circle" D_{2j+1} and also in a "circle" E_{2k}. Let the corresponding half-lines emanating from P_{2j+1} resp. from P_{2k} and containing P be l_1 and l_2. It is not possible that both l_1 and l_2 intersect H_γ, for then we would have $P \in H_\gamma$ because H_γ is 'closed'. But if, say, $l_1 \cap H_\gamma = \emptyset$, then P can be added to the partial "circle" $C_{2j+1,\gamma}$, i.e., we can put $P \in C_{2j+1,\alpha}$. This gives the allocation of the points in $H_{\gamma+1} \setminus H_\gamma$ into the partial circles C_j, and the proof is complete.

13. Let P_α, $\alpha < \mathbf{c}$ be an enumeration of the points of $\mathbf{R}^3$ into a sequence of type $\mathbf{c}$. By transfinite recursion we define sets C_α, $\alpha < \mathbf{c}$ where either $C_\alpha = \emptyset$ or C_α is a circle of radius 1 disjoint from every C_β, $\beta < \alpha$, and in any case $P_\alpha \in \cup_{\beta \leq \alpha} C_\beta$. Clearly, then $\cup_{\alpha < \mathbf{c}} C_\alpha$ is an appropriate decomposition of $\mathbf{R}^3$. The induction step is clear: for $\alpha < \mathbf{c}$ if $P_\alpha \in \cup_{\beta < \alpha} C_\beta$, then set $C_\alpha = \emptyset$, otherwise select as C_α a circle of radius 1 through P_α that is disjoint from $\cup_{\beta < \alpha} C_\beta$. That this is possible can be seen as follows. The nonempty circles C_β, $\beta < \alpha$ lie in less than continuum many planes, therefore there is a plane S through P_α different from all of them. Thus, S intersects every C_β, $\beta < \alpha$, in at most 2 points, so $S \cap \left(\cup_{\beta < \alpha} C_\beta\right)$ is of cardinality smaller than continuum. Therefore, there is a circle C of radius 1 that lies in S, goes through P_α but does not go through any of the points belonging to the set $S \cap \left(\cup_{\beta < \alpha} C_\beta\right)$. Then, clearly, $C \cap C_\beta = \emptyset$ for all $\beta < \alpha$, and so we can select $C_\alpha = C$.

14. The proof is along the same lines as in the previous problem. Let P_α, $\alpha < \mathbf{c}$, be an enumeration of the points of $\mathbf{R}^3$ into a sequence of type $\mathbf{c}$ and by transfinite recursion we define sets l_α, $\alpha < \mathbf{c}$, where either $l_\alpha = \emptyset$ or l_α is a line not parallel with any line l_β, $\beta < \alpha$, and in any case $P_\alpha \in \cup_{\beta \leq \alpha} l_\beta$. Again, for $\alpha < \mathbf{c}$ if $P_\alpha \in \cup_{\beta < \alpha} l_\beta$, then set $l_\alpha = \emptyset$, otherwise select as l_α a line through P_α that is not parallel with any of the lines l_β, $\beta < \alpha$. This selection is possible, just select l_α so that $P_\alpha \in l_\alpha$ and l_α is different from the fewer than continuum many lines l'_β that go through P_α and are parallel with the corresponding l_β's.

15. The proof below shows that it is indifferent if the intervals in question are open, closed, or semiclosed, hence without loss of generality we may assume $A = [0, 1]$, $B = [0, b]$, $b > 1$. It is enough to prove that there are a disjoint decomposition $B = \cup_{i=0}^\infty B_i$ and a 1-to-1 mapping $F : B \to A$ such that the restriction $F|_{B_i}$ of F to any B_i is a translation. In fact, let $G : A \to B$ be the identity mapping. By Problem 3.1 there are disjoint decompositions $A = A' \cup A''$ and $B = B' \cup B''$ such that F maps B' onto A' and G maps A'' onto B''. Let $B'_i = B_i \cap B'$, $B''_i = B_i \cap B''$, $A'_i = F[B'_i]$, and $A''_i = G^{-1}[B''_i]$. Since F is translation on B_i and G is the identity, we obtain that A'_i is a translated copy of B'_i and A''_i is a translated copy of B''_i, which, together with the disjoint representations

$$A = \left(\cup_{i=0}^\infty A'_i\right) \bigcup \left(\cup_{i=0}^\infty A''_i\right), \qquad B = \left(\cup_{i=0}^\infty B'_i\right) \bigcup \left(\cup_{i=0}^\infty B''_i\right),$$

verifies the claim in the problem.

To get the representation $B = \cup_{i=0}^{\infty} B_i$ and the 1-to-1 mapping $F : B \to A$, consider on $\mathbf{R}$ the equivalence relation $x \sim y \leftrightarrow x - y \in \mathbf{Q}$. Clearly, each equivalence class intersects the interval $[0, 1/2]$; therefore, we get from the axiom of choice that there is a set $H \subset [0, 1/2]$ such that H intersects every equivalence class in exactly one point. Since the sets $H + r$ with different $r \in \mathbf{Q}$ are disjoint and every real number belongs to exactly one of these sets, it follows that

$$\bigcup_{r\in\mathbf{Q}\cap[0,1/2]} (H + r) \subseteq [0, 1],$$

while

$$\bigcup_{r\in\mathbf{Q}\cap[-1/2,b]} (H + r) \supseteq [0, b],$$

and this latter shows that

$$B = \bigcup_{r\in\mathbf{Q}\cap[-1/2,b]} \big((H + r) \cap [0, b]\big)$$

is a disjoint representation. Now let $g : \mathbf{Q} \cap [-1/2, b] \to \mathbf{Q} \cap [0, 1/2]$ be a 1-to-1 mapping, and define F as $F(x) = x + (g(r) - r)$ if $x \in (H + r) \cap [0, b]$, $r \in \mathbf{Q} \cap [-1/2, b]$. This F is a translation on $(H + r) \cap [0, b]$, and maps this set into $H + g(r) \subset [0, 1]$. Thus, F maps $[0, b]$ into $[0, 1]$ and it is left to show that it is 1-to-1. In fact, g is an injection, hence if $F(x) = F(y)$, then x and y both belong to the same $(H + r) \cap [0, b]$, and since F is a translation on $(H + r) \cap [0, b]$, it follows that $x = y$.

14

Zorn's lemma

1. Let $(\mathcal{P}, <)$ be a partially ordered set satisfying the condition on chains. By the well-ordering theorem it can be well ordered as $\mathcal{P} = \{p_\alpha : \alpha < \varphi\}$ for some ordinal φ. We construct a chain L by determining with transfinite recursion if $p_\alpha \in L$ holds. First, put p_0 into L. For $\alpha > 0$, add p_α to L if and only if p_α is greater than any p_β selected into L, with $\beta < \alpha$. This obviously gives a chain L. By condition, there is an upper bound p_γ to L. We claim that p_γ is a maximal element. Assume not. Then some $p_\delta > p_\gamma$. When we considered p_δ, we observed that it was bigger than every p_β selected into L with $\beta < \delta$, so we must have chosen it into L, that is, $p_\delta \leq p_\gamma$, which is a contradiction.

2. Let $\{A_i : i \in I\}$ be a system of nonempty sets. Define the partially ordered set $(\mathcal{P}, <)$ as follows. $f \in \mathcal{P}$ if and only if f is a function with $\mathrm{Dom}(f) \subseteq I$ and $f(i) \in A_i$ holds for every $i \in \mathrm{Dom}(f)$. Set $f < f'$ if f' is a proper extension of f, i.e., $\mathrm{Dom}(f')$ is a proper superset of $\mathrm{Dom}(f)$ and $f(i) = f'(i)$ holds for every $i \in \mathrm{Dom}(f)$. Notice that $\mathcal{P}$ is nonempty as the empty function is in it.

Let $L \subseteq \mathcal{P}$ be a chain in $(\mathcal{P}, <)$. That is, if f, f' are two elements of L, then either $f < f'$ or $f' < f$ holds. In either case, if $f(i)$, $f'(i)$ are both defined, then they are equal. With this in mind, we can define a function F as follows. $\mathrm{Dom}(F) = \bigcup\{\mathrm{Dom}(f) : f \in L\}$, and for an i in this set we let $F(i)$ be the unique value $f(i)$ assume by all $f \in L$ which are defined at i. Clearly, $F \in \mathcal{P}$ and $f \leq F$ holds for every $f \in L$.

We can now apply Zorn's lemma and get a maximal element F of $(\mathcal{P}, <)$. We claim that $\mathrm{Dom}(F) = I$ (and that finishes the argument). If not, then there is some $i \in I \setminus \mathrm{Dom}(F)$. Pick an element $x \in A_i$ and extend F to F' as follows. $\mathrm{Dom}(F') = \mathrm{Dom}(F) \cup \{i\}$ and $F'(i) = x$. Then $F' > F$ and that contradicts the maximality of F.

3. Let A be a set for which we show, with the help of Zorn's lemma, that it has a well-ordering. We first define a partially ordered set $\langle \mathcal{P}, < \rangle$. The elements of $\mathcal{P}$ will be the ordered sets of the form $\langle B, <_B \rangle$, where $B \subseteq A$, $<_B$ is a

well order on B. $\mathcal{P}$ is a set as it is a subset of $\mathcal{P}(A) \times \mathcal{P}(A \times A)$. Partially order $\mathcal{P}$ the following way: $\langle B_1, <_{B_1} \rangle \le \langle B_2, <_{B_2} \rangle$ if and only if $B_1 \subseteq B_2$ and $\langle B_2, <_{B_2} \rangle$ *end-extends* $\langle B_1, <_{B_1} \rangle$, that is, the ordering $<_{B_2}$ extends $<_{B_1}$ and for $x \in B_1$, $y \in B_2 \setminus B_1$, we have $x <_{B_2} y$. Let $L \subseteq \mathcal{P}$ be a chain, we show that it has an upper bound. Indeed, set $C = \bigcup \{B : \langle B, <_B \rangle \in L\}$, and for $x, y \in C$, set $x \prec y$ if and only if $x < y$ holds for some/all $\langle B, <_B \rangle \in L$ with $x, y \in B$.

We show that $\langle C, \prec \rangle$ is a well-ordered set. Pick some $x \in C$ (if the set is empty, it is obviously well ordered). There is some $\langle B, <_B \rangle \in L$ with $x \in B$. As every $\langle B', <_{B'} \rangle \in L$ with $\langle B, <_B \rangle \le \langle B', <_{B'} \rangle$ end extends $\langle B, < \rangle$, we have $\langle C|x, \prec \rangle = \langle B|x, < \rangle$, so every initial segment of $\langle C, \prec \rangle$ determined by an element is well ordered. Hence by Problem 6.36 $\langle C, \prec \rangle$ is well ordered.

Next we show that $\langle B, <_B \rangle \le \langle C, \prec \rangle$ holds for every $\langle B, <_B \rangle \in L$. For every $\langle B', <_{B'} \rangle \in L$ with $\langle B, <_B \rangle \le \langle B', <_{B'} \rangle$ we have that $\langle B', <_{B'} \rangle$ end extends $\langle B, <_B \rangle$, so $\langle C, \prec \rangle$ end extends $\langle B, <_B \rangle$, as well.

We now apply Zorn's lemma and get some maximal $\langle B, <_B \rangle \in \mathcal{P}$. We claim that $B = A$, and so we are done. Assume otherwise, so $B \ne A$. Pick an element $a \in A \setminus B$. Define $\langle B', <_{B'} \rangle$ as follows. $B' = B \cup \{a\}$ and let $\langle B', <_{B'} \rangle$ extend $\langle B, <_B \rangle$ with making a greater than every element of B. Then clearly $\langle B, <_B \rangle < \langle B', <_{B'} \rangle$, so $\langle B, <_B \rangle$ is not maximal, a contradiction.

4. Assume that $(\mathcal{P}, \le)$ is a counterexample. By the axiom of choice there are functions F and G such that if $L \subseteq \mathcal{P}$ is a chain, then $F(L)$ is an upper bound for L and if $p \in \mathcal{P}$, then $G(p) > p$ (we use the axiom of choice to choose an element from the nonempty set of elements bigger than p and similarly for the chains). Using transfinite recursion, define for every ordinal α the element $p_\alpha \in \mathcal{P}$ as follows. Let $p_0 \in \mathcal{P}$ be arbitrary. For α limit then $\{p_\beta : \beta < \alpha\}$ is a chain and then let $p_\alpha = F(\{p_\beta : \beta < \alpha\})$. Further, let $p_{\alpha+1} = G(p_\alpha)$. It is easy to see that $\alpha \mapsto p_\alpha$ is a strictly increasing operation, and so it is defined for every α. But it is impossible to inject a proper class into a set; see the argument in Problem 3.

5. From the previous problems.

6. (a) Consider the partially ordered set $(\mathcal{P}, <)$ of the disjoint pairs (A, B) where $A, B \subseteq \mathbf{R}^+$, neither A nor B is empty, both are closed under addition and multiplication by a positive rational number. There are such pairs, for example, we can take $A = \mathbf{Q} \cap \mathbf{R}^+$ and $B = \mathbf{Q}\sqrt{2} \cap \mathbf{R}^+$. Order $\mathcal{P}$ as follows. $(A, B) \le (A', B')$ if and only if $A \subseteq A'$ and $B \subseteq B'$. It is easy to see that the condition on chains holds, so Zorn's lemma applies, and there is a maximal $(A, B) \in \mathcal{P}$. We claim that $A \cup B = \mathbf{R}^+$. Assume not, say, $a \notin A \cup B$, $a > 0$. We cannot extend A by a, so there is a rational number $0 < q \in \mathbf{Q}$ and $x \in A$ such that $qa + x \in B$. Similarly, we get a $0 < q' \in \mathbf{Q}$, and a $y \in B$ such that $q'a + y \in A$. But then $qq'a + q'x \in B$, and $qq'a + qy \in A$. As A, B are closed under addition and multiplication by positive rational numbers we get

that $qq'a + q'x + qy$ is both in A and B, so (A, B) is not an element of $\mathcal{P}$, a contradiction. [Zsigmond Nagy]

(b) Assume that R is a ring with a unity and I_0 is a proper ideal. Consider the following partially ordered set $(\mathcal{P}, \leq)$. $I \in \mathcal{P}$ if I is an ideal of R, $I_0 \subseteq I$ and $1 \notin I$. Clearly, $I_0 \in \mathcal{P}$, so $\mathcal{P}$ is nonempty. We show that the condition on chains holds. Assume that $L = \{I_a : a \in A\}$ is a chain. Then $I = \bigcup\{I_a : a \in A\}$ is an upper bound. Indeed, it is a proper ideal, as $1 \notin I$ holds. Let I be a maximal element of $(\mathcal{P}, \leq)$. Then clearly I is a maximal ideal in R.

(c) Let A be some set, $F \subseteq \mathcal{P}(A)$ a filter on A. Then F has the *finite intersection property* (f.i.p.), that is, if $X_1, \dots, X_n \in F$, then $X_1 \cap \cdots \cap X_n \neq \emptyset$. Set $p \in \mathcal{Q}$ if and only if $F \subseteq p \subseteq \mathcal{P}(A)$ and p has the f.i.p. Partially order $\mathcal{Q}$ by putting $p \leq q$ if and only if $p \subseteq q$. Then clearly, $\langle \mathcal{Q}, \leq \rangle$ is a nonempty partially ordered set.

We show the condition on chains. Let $L \subseteq \mathcal{Q}$ be a nonempty chain. We show that $q = \bigcup L$ is an element of $\mathcal{Q}$ (then obviously it will be an upper bound for L). Assume that $X_1, \dots, X_n \in q$. Then for appropriate $p_1, \dots, p_n \in L$ we have $X_i \in p_i$ $(1 \leq i \leq n)$. We can as well assume that $p_1 \leq \cdots \leq p_n$. Then $X_1, \dots, X_n \in p_n$ so $X_1 \cap \cdots \cap X_n \neq \emptyset$, and we are done.

We can, therefore, apply Zorn's lemma, and get a maximal $p \in \mathcal{Q}$. We show that it is an ultrafilter. First, assume that $X \in p$, $X \subseteq Y \subseteq A$ but $Y \notin p$. Then, $p \cup \{Y\}$ has the f.i.p., as if $X_1, \dots, X_n \in p$, then $X_1 \cap \cdots \cap X_n \cap Y \supseteq X_1 \cap \cdots \cap X_n \cap X \neq \emptyset$, so $p \cup \{Y\}$ was a proper extension of p, contradicting maximality. Next, assume that $X, Y \in p$ but $X \cap Y \notin p$. Again, $p \cup \{X \cap Y\}$ has the f.i.p., so it was a proper extension of p. Finally, assume that $X \subseteq A$ yet neither X nor $A \setminus X$ is an element of p. Then, both $p \cup \{X\}$ and $p \cup \{A \setminus X\}$ fail to have the f.i.p., so there are $Y_1, \dots, Y_n \in p$ and $Z_1, \dots, Z_m \in p$ such that $Y_1 \cap \cdots \cap Y_n \cap X = \emptyset$ and $Z_1 \cap \cdots \cap Z_m \cap (A \setminus X) = \emptyset$. But then

$$Y_1 \cap \cdots \cap Y_n \cap Z_1 \cap \cdots \cap Z_m = \emptyset$$

and so p fails to have the f.i.p.

(d) Assume that V is a vector space, I_0 a set of linearly independent vectors. Let $\mathcal{P}$ be the partially ordered set of all linearly independent sets $I_0 \subseteq I$. We show the condition on chains. Indeed, if $L \subseteq \mathcal{P}$ is a chain, then the union of the elements of L is also a set of linearly independent vectors as any finite subset is in some $I \in L$, therefore a supposed counterexample to independence would appear in some $I \in L$. Applying Zorn's lemma, we get a nonextendable $I \in \mathcal{P}$. It is a basis, as should it not generate some $x \in V$ then $I \cup \{x\}$ would extend I.

(e) Let G be a generating system of the vector space V. We cannot work with the reversely ordered generating subsets of G and seek for a minimal element (the intersection of decreasing sequence $G_0 \supseteq G_1 \supseteq \cdots$ of generating sets may be empty). Instead we let $\mathcal{P}$ be the partially ordered set of linearly independent subsets $I \subseteq G$. We can now repeat the previous argument. If a maximal element $I \in \mathcal{P}$ is not a basis then it does not generate some $x \in G$ (recall that if every element of G is generated then so is the whole space) so we can extend it to $I \cup \{x\}$.

(f) Let $\mathcal{P}$ be the set of those isomorphisms φ that map some B_1 onto some B_2 where $A \leq B_1 \leq D_1$ and $A \leq B_2 \leq D_2$ and φ is the identity on A. Set $\varphi \leq \psi$ if and only if ψ extends φ. $\mathcal{P}$ is nonempty as it contains, for example, the identity on A. The condition on chains holds: if L is a chain, then $\bigcup L$ (the union of all elements of L) is an isomorphism extending every element of L. Let φ be a maximal element of $\langle \mathcal{P}, \leq \rangle$. We show, and that suffices, that φ is defined on D_1. Assume not, and $a \notin B_1$ for some element $a \in D_1$. Let $n > 0$ be the least natural number with $na \in B_1$. Then $b = \varphi(na) \in B_2$; therefore, there is some $a' \in D_2$, such that $na' = b$. We can extend φ to the generated subgroup $\langle B_1, a \rangle$ as follows: $\psi(x + ka) = \varphi(x) + ka'$ for $x \in B_1$, $0 \leq k < n$. We claim that this is sum preserving, i.e., $\psi\left((x + ka) + (y + k'a)\right) = \psi(x+ka)+\psi(y+k'a)$. This is immediate if $k + k' < n$. However, if $k + k' = n + \ell$ for some $0 \leq \ell < n$ then $ka+k'a = na+\ell a$ and this indeed is mapped to $b+\ell a' = na'+\ell a' = ka'+k'a'$.

(g) Assume that $(F, 0, +, \cdot)$ is a field. To get some elbow space let $S \supseteq F$ be a set of cardinality greater than that of F if F is infinite, and of cardinality $\aleph_1$ if F is finite. Let $\mathcal{P}$ be the set of those fields $(K, 0, +, \cdot)$ where $F \subseteq K \subseteq S$ and $(K, 0, +, \cdot)$ is an algebraic extension of $(F, 0, +, \cdot)$. Set $(K, 0, +, \cdot) \leq (K', 0, +, \cdot)$ if $(K', 0, +, \cdot)$ is indeed an extension of $(K, 0, +, \cdot)$. Notice that by the condition on algebraicity over $(F, 0, +, \cdot)$, the inequality $|K| < |S|$ always holds. It is easy to see that the condition on chain holds (i.e., if $\{(K_i, 0, +, \cdot) : i \in I\}$ are algebraic extensions of $(F, 0, +, \cdot)$ then so is their union). Now let $(K, 0, +, \cdot)$ be a maximal element in $(\mathcal{P}, \leq)$. If it is not algebraically closed, then there is some irreducible $p(x)$ such that $p(x) = 0$ is not solvable in K. As $|K| < |S|$ we can extend K in the usual way to some $(K', 0, +, \cdot)$ in which there is a solution to $p(x) = 0$. So this would be a proper algebraic extension of $(K, 0, +, \cdot)$, a contradiction.

(h) Let F be an algebraically closed field. Let $\mathcal{P}$ be the set of all subsets $X \subseteq F$ that are algebraically independent, i.e., if $a_1, \ldots, a_n$ are distinct elements of X and $p(x_1, \ldots, x_n)$ is a nonzero polynomial over the prime field, then $p(a_1, \ldots, a_n) \neq 0$. Partially order $\mathcal{P}$ by making $X_0 \leq X_1$ if and only if $X_0 \subseteq X_1$. $\mathcal{P}$ is nonempty as $\emptyset \in \mathcal{P}$.

We show that $\langle \mathcal{P}, \leq \rangle$ satisfies the condition on chains. Let $L \subseteq \mathcal{P}$ be a chain. Set $Y = \bigcup L$. If we show that $Y \in \mathcal{P}$, then it will be obvious that Y is an upper bound for L, and so we have our claim. Indeed, if $a_1, \ldots, a_n \in Y$, then $a_1 \in X_1, \ldots, a_n \in X_n$ for some elements $X_1, \ldots, X_n$ of L, and as L is ordered, one of them, say X_n is the largest among them. So we have $a_1, \ldots, a_n \in X_n$, and therefore they are algebraically independent.

Applying Zorn's lemma we get a nonextendable $B \in \mathcal{P}$. B is a transcendence basis. Indeed, if $a \notin B$, then $B \cup \{a\}$ cannot be in $\mathcal{P}$, so $p(a_1, \ldots, a_n, a) = 0$ holds for some nonzero polynomial $p(x_1, \ldots, x_n, y)$ and elements $a_1, \ldots, a_n \in B$. Therefore, a is the root of a nonzero polynomial over B.

(i) We first argue that it suffices to find a subset $P \subseteq F$ which is closed under addition, multiplication, and division, $0 \notin P$, and for every $0 \neq a \in F$ either $a \in P$ or $-a \in P$. Indeed, given such a $P \subseteq F$ we can define $x < y$ exactly when $y - x \in P$. Easy arguments show that $<$ gives an ordered field on F.

In order to find such a set $P \subseteq F$, let $\mathcal{P}$ be the collection of those sets $P_0 \subseteq P \subseteq F$ for which $0 \notin P$, $x, y \in P$ implies $x + y, xy, x/y \in P$, and P_0 is the set of nonzero elements that can be written as the sum of finitely many squares. Observe that P_0 has the above closure properties: addition is trivial, multiplication follows from the identity $(\sum a_i^2)(\sum b_j^2) = \sum(a_i b_j)^2$, and for x/y we argue that $x/y = (xy)/y^2$ so if $xy = \sum a_i^2$ then $x/y = \sum(a_i/y)^2$. $\mathcal{P}$ is, therefore, nonempty, and if we order it by $P \leq P'$ if $P \subseteq P'$, then it obviously satisfies the condition on chains. We must show that if P is maximal, then $F = P \cup \{0\} \cup (-P)$. Assume indirectly that $a \neq 0$ is such that $a \notin P$ and $-a \notin P$. We show that $-a \notin P$ implies that P can be extended with a, and that, with the maximality of P, implies $a \in P$.

Let $P' = \{x + ya : x, y \in P\}$. We have to show that $P' \in \mathcal{P}$. Indeed, if $0 = x + ya$, then $-a = x/y \in P$, a contradiction to our assumption. P' is manifestly closed under addition. If $x + ya, x' + y'a \in P'$, then $(x + ya)(x' + y'a) = (xx' + yy'a^2) + (x'y + xy')a \in P'$, and finally for division we argue that $(x + ya)^{-1} = (x + ya)(x + ya)^{-2} \in P'$.

(k) Let B be a subgroup of G maximal with respect to the property that $A \cap B = 0$ (the trivial subgroup). Such a B exists by Zorn's lemma. We claim that $A + B = G$ and therefore G is the direct sum of A and B. Assume that this is not the case, and $x \notin A + B$ for some element x. Then (B, x), the subgroup generated by $B \cup \{x\}$, properly extends B, therefore by the maximality of the latter group we have $A \cap (B, x) \neq 0$. That is, for some $a \in A$, $a \neq 0$, we have $a = b + nx$ with $b \in B$ and n a nonzero integer. We found that there is some element $x \notin A + B$ such that $nx \in A + B$ holds for some positive integer.

Let p be the least positive number that occurs as such an n. Necessarily p is prime. Let x be such that $x \notin A + B$ yet $px = a + b$ for some $a \in A$, $b \in B$. As A is divisible, there is some $a' \in A$ such that $pa' = a$. Then $py = b$ holds for $y = x - a'$. Notice that $y \notin A + B$ as otherwise we had that $x = a' + y \in A + B$. Once again, $A \cap (B, y) \neq 0$, so $a'' = b'' + ky$ for some $0 \neq a'' \in A$, $b'' \in B$. k is not divisible by p, as otherwise $b'' + ky$ and therefore a'' would be in B, which is not the case. As p is prime, $mk + p\ell = 1$ holds for some integers m, ℓ. But then $y = (mk + p\ell)y = m(a'' - b'') + \ell \cdot b \in A + B$, a contradiction.

(l) Let X be (the edge set of) a connected graph. Consider the partially ordered set $\mathcal{P}$ of circuitless subgraphs Y of X with $Y_0 \leq Y_1$ if and only if Y_0 is a subgraph of Y_1. The condition on chains holds for this partially ordered set. Indeed, if $\{Y_i : i \in I\}$ is a chain, then $Y = \bigcup\{Y_i : i \in I\}$ is in $\mathcal{P}$ (every purported circuit of Y would be in some Y_i). Let Y be a maximal element of $(\mathcal{P}, \leq)$. Y has no circuits. If it is not a spanning tree, then, as X is connected, there is some edge e such that $Y \cup \{e\}$ is still curcuitless, so properly extends Y, a contradiction.

(m) Let $\mathcal{P}$ be the partially ordered set of partitions of V that are good colorings (that is, vertices in the same class are not joined). Define $P \leq Q$ in $\mathcal{P}$ if P is finer than Q, i.e., every class of P is a subset of some class of Q. Redefine $\mathcal{P}$ as those good colorings that are above a certain P which is a κ-coloring. (This will ensure that every element of $\mathcal{P}$ is a κ-coloring.) We show that $(\mathcal{P}, \leq)$ satisfies the condition on chains. Indeed, assume, that $L = \{P_i : i \in I\}$ is a chain. Define $x \sim y$ in the graph if there is some P_i in which they are in the same class. Clearly, $\sim$ is an equivalence relation. It is equally clear that if $x \sim y$, then x and y are not joined in X. $\sim$ therefore defines a partition in $\mathcal{P}$ which is an upper bound for every element of L. By Zorn's lemma, there is a maximal P in $\mathcal{P}$. Clearly, P is a partition, as required.

(n) Let $(\mathcal{P}, \leq)$ be the partially ordered set of all closed, nonempty subsets $F \subseteq X$ with $F + F \subseteq F$ with the reverse inclusion as partial ordering. $\mathcal{P}$ is nonempty, as $X \in \mathcal{P}$. We show that the condition on chains holds for $(\mathcal{P}, \leq)$. Indeed, if $\{F_i : i \in I\}$ is a chain of nonempty closed subsets of X with the above property, then, by compactness, $F = \bigcap\{F_i : i \in I\}$ is closed and nonempty, and for every $i \in I$ we have $F + F \subseteq F_i + F_i \subseteq F_i$, so $F + F \subseteq F$ indeed holds. Applying Zorn's lemma, there is some minimal, nonempty F with $F + F \subseteq F$. Pick $p \in F$. Clearly, $p + F \neq \emptyset$ and by right continuity $p + F$ is closed. Furthermore, $(p+F)+(p+F) \subseteq p+F+F+F \subseteq p + F$, and $p + F \subseteq F + F \subseteq F$ and so by minimality $p + F = F$. Hence there is some $q \in F$ with $p + q = p$. Set $F' = \{q \in F : p + q = p\}$. F' is nonempty, by the right continuity of $+$, it is a closed set in X, and obviously $F' + F' \subseteq F'$, so again by the minimality of F we have $F' = F$, therefore $p + p = p$. [S. Glazer, see: W. W. Comfort: Ultrafilters—some old and new results, Bull. Amer. Math. Soc., **83**(1977), 417–455]

7. Define the partially ordered set $(\mathcal{P}, \leq)$ as follows. $\mathcal{G} \in \mathcal{P}$ if and only if $\mathcal{G} \subseteq \mathcal{F}$ and for every finite $X \subseteq S$ there is a subfamily of $\mathcal{G}$ which is an exact cover of X. $\mathcal{P}$ is nonempty, as $\mathcal{F} \in \mathcal{P}$. Set $\mathcal{G} \leq \mathcal{G}'$ if and only if $\mathcal{G}' \subseteq \mathcal{G}$, that is, we consider the reverse of the natural order.

We show that the condition for chains holds. Assume that some $\{\mathcal{G}_i : i \in I\}$ is a chain in $(\mathcal{P}, \leq)$. We have to find a $\mathcal{G} \in \mathcal{P}$ such that $\mathcal{G}_i \leq \mathcal{G}$ holds for every $i \in I$, that is, $\mathcal{G} \subseteq \mathcal{G}_i$ holds for every $i \in I$. Therefore, we have to show that $\mathcal{G} = \bigcap\{\mathcal{G}_i : i \in I\}$ is an element of $\mathcal{P}$. Let X be a finite subset of S. Consider some $x \in X$. Let $i = i(x) \in I$ be such that $\{F : x \in F \in \mathcal{G}_i\}$ has the least possible number of elements. Then, if $i \geq i(x)$ and $x \in F \in \mathcal{G}_{i(x)}$, then necessarily $F \in \mathcal{G}_i$ holds as well. Set $i^* = \max\{i_x : x \in X\}$ (exists, as we consider the maximum of finitely many elements of an ordered set). By condition, $\mathcal{G}_{i^*}$ includes a subfamily $F_1, \ldots, F_t$ which is an exact cover of X. By the above arguments each F_j is in every $\mathcal{G}_i$, so each F_j is in $\mathcal{G} = \bigcap\{\mathcal{G}_i : i \in I\}$, so $\mathcal{G}$ itself includes a subfamily which is an exact cover for X, therefore we proved that $\mathcal{G} \in \mathcal{P}$.

We can therefore apply Zorn's lemma and let $\mathcal{G} \in \mathcal{P}$ be a maximal element. We argue that $\mathcal{G}$ is an exact cover of S. It is clearly a cover (that is, $S = \bigcup \mathcal{G}$).

Assume that some $x \in S$ is covered twice: $x \in F_1 \in \mathcal{G}$, $x \in F_2 \in \mathcal{G}$. Then by the maximality of $\mathcal{G}$, neither $\mathcal{G} \setminus \{F_1\}$ nor $\mathcal{G} \setminus \{F_2\}$ is an element of $\mathcal{P}$, that is, there are finite $X_1, X_2 \subseteq S$ that $\mathcal{G} \setminus \{F_1\}$, resp. $\mathcal{G} \setminus \{F_2\}$ does not include an exact cover of. But, by condition, some $\mathcal{G}' \subseteq \mathcal{G}$ is an exact cover of $X_1 \cup X_2$ and $\mathcal{G}'$ surely misses either F_1 or F_2, and we reached a contradiction.

8. (a) Let $\mathcal{P}$ be the set of partial orders on P that extend $<$. That is, $R \in \mathcal{P}$ if $R \subseteq P \times P$, $\langle x, x\rangle \notin R$ $(x \in P)$, $\langle x, y\rangle \in R$, $\langle y, z\rangle \in R$ imply $\langle x, z\rangle \in R$, and if $x < y$, then $\langle x, y\rangle \in R$. Set $R_1 \leq R_2$ if and only if $R_1 \subseteq R_2$. The condition on chains holds for $(\mathcal{P}, \leq)$: indeed, if $L \subseteq \mathcal{P}$ is a chain, then $\bigcup L$ (the union of the elements of L) is an upper bound for L. By Zorn's lemma, there is a maximal element $R \in \mathcal{P}$. In order to show that R is an order on P, assume that $x \neq y$ are elements of P with $\langle x, y\rangle, \langle y, x\rangle \notin R$. Let R' be the partial order "generated" by $\langle x, y\rangle$, that is,

$$R' = R \cup \{\langle u, v\rangle : \langle u, x\rangle \in R \text{ or } u = x, \langle y, v\rangle \in R \text{ or } v = y\}.$$

Inspection shows that $R' \in \mathcal{P}$ and it is strictly larger than R. As this is impossible, R is indeed an order of P.

(b) If $x, y \in P$ are incomparable, by the closing argument in part (a) there is a partial ordering $<'$ on P extending $<$ and with $x <' y$, and another one $<''$ for which $y <'' x$. As $<'$ and $<''$ can both be extended to an order, we are done.

(c) Let $(P, <)$ be a well-founded partially ordered set. Let r be a rank function on P, i.e., an order-preserving map from P to the ordinals (see Problem 31.5). Let $<_w$ be any well-ordering of the set P. Define $x <' y$ if and only if either $r(x) < r(y)$ or else $r(x) = r(y)$ and $x <_w y$. As the well-ordered union of well-ordered sets is well ordered, this will give a well-ordering of P. Also, if $x < y$, then $r(x) < r(y)$ and so certainly $x <' y$.

(d) It doesn't. If $(P, <)$ consists of incomparable elements and P happens to be an unorderable set, then (a) is false for $(P, <)$ yet (b) holds vacuously.

9. Assume that X is not compact. There is a base $\mathcal{B}$ such that every element of $\mathcal{B}$ is the intersection of finitely many members of $\mathcal{S}$ and by our indirect assumption there is some $\mathcal{U}_0 \subseteq \mathcal{B}$ that covers X but includes no finite subcover. Let $\mathcal{P}$ be the partially ordered set of those covers $\mathcal{U}_0 \subseteq \mathcal{U} \subseteq \mathcal{B}$, which do not include finite subcovers. Set $\mathcal{U} \leq \mathcal{U}'$ if $\mathcal{U} \subseteq \mathcal{U}'$. The partially ordered set $(\mathcal{P}, \leq)$ satisfies the condition on chains: indeed, if $\{\mathcal{U}_i : i \in I\}$ is a chain in $(\mathcal{P}, \leq)$ and we set $\mathcal{U} = \bigcup\{\mathcal{U}_i : i \in I\}$, then $\mathcal{U} \in \mathcal{P}$ as any possible finite subcover would be included in some $\mathcal{U}_i$. By Zorn's lemma there is a maximal element $\mathcal{U}$ in $\mathcal{P}$. Pick some $G \in \mathcal{U}$. G can be written as $G = S_1 \cap \cdots \cap S_n$ with $S_1, \ldots, S_n \in \mathcal{S}$. We claim that one of $S_1, \ldots, S_n$ must be in $\mathcal{U}$. Otherwise, by maximality of $\mathcal{U}$, for every $1 \leq i \leq n$ there would be a finite subfamily $\mathcal{U}_i$ of $\mathcal{U}$ such that $\mathcal{U}_i \cup \{S_i\}$ is a cover of X. But then $\mathcal{U}_1 \cup \cdots \cup \mathcal{U}_n$ is a finite cover of $X \setminus G$,

so $\mathcal{U}_1 \cup \cdots \cup \mathcal{U}_n \cup \{G\}$, a finite subfamily of $\mathcal{U}$ covering X, a contradiction. We have that for every $G \in \mathcal{U}$ there is some $G \subseteq S \in \mathcal{S} \cap \mathcal{U}$, so $\mathcal{S} \cap \mathcal{U}$ covers X, but no finite subfamily covers X, which is in contradiction with the assumption. [J. W. Alexander: Ordered sets, complexes, and the problem of compactification, *Proc. Nat. Acad. Sci. USA*, **25**(1939), 296–298]

10. Assume that X is the topological product of the spaces $\{X_i : i \in I\}$ so the elements of X are the choice functions $f(i) \in X_i$ (for $i \in I$). By the previous problem it suffices to find a subbase which has the property that every cover includes a finite subcover. We show that $\mathcal{S} = \bigcup\{\mathcal{S}_i : i \in I\}$ is such a subbase where $G \in \mathcal{S}_i$ if there is a nonempty open set U in X_i such that $G = G_i(U) = \{f \in X : f(i) \in U\}$. Notice that if $\{U_j : j \in J\}$ cover X_i, then $\{G_i(U_j) : j \in J\}$ cover X. Assume that some $\mathcal{S}' \subseteq \mathcal{S}$ covers X. Clearly, $\mathcal{S}' = \bigcup\{\mathcal{S}'_i : i \in I\}$, where $\mathcal{S}'_i \subseteq \mathcal{S}_i$. Let $\mathcal{S}'_i = \{G_i(U) : U \in \mathcal{C}_i\}$.

If for some $i \in I$ $\mathcal{C}_i$ is a cover of X_i, then we can choose a finite subcover $\mathcal{C}'_i$ of $\mathcal{C}_i$, (as X_i is compact) and then $\{G_i(U) : U \in \mathcal{C}'_i\}$ is a finite subcover of X. Otherwise, for every $i \in I$ there is some $f(i) \in X_i$ uncovered by $\mathcal{C}_i$, so $f \in X$ is not covered by $\mathcal{S}'$, a contradiction. [A. N. Tychonoff: Über einen Funktionenraum, *Math. Ann.*, **111**(1935), 5]

15

Hamel bases

1 See Problem 14.6(d).

2 See Problem 14.6(e)

3. Let the cardinality of some Hamel basis be κ. We can easily calculate the cardinality of the generated vector space: it is $\aleph_0(\kappa+\kappa^2+\cdots)=\aleph_0\kappa=\kappa$ and since this must be equal to $\mathbf{c}$, we obtain $\kappa=\mathbf{c}$

4. Assume $\{b_i : i\in I\}$ is a Hamel basis. From the previous problem we know that I is of cardinality $\mathbf{c}$. Observe that if $X\subseteq I$ then

$$\{2b_i : i\in X\}\cup\{b_i : i\in I\setminus X\}$$

is also a Hamel basis. As we have produced one Hamel basis per every subset of I, there are at least $2^{\mathbf{c}}$ Hamel bases. On the other hand, the total number of subsets of $\mathbf{R}$ is $2^{\mathbf{c}}$, so there cannot be more than $2^{\mathbf{c}}$ Hamel bases.

5. Let B be a Hamel basis and separate some infinitely many elements $\{b_i\}$ so that as $B=\{b_0,b_1,\ldots\}\cup B'$ be a Hamel basis. Enumerate the intervals with rational endpoints as $I_0,I_1,\ldots$. Choose the rational numbers $\lambda_0,\lambda_1,\ldots$ in such a way that $\lambda_i b_i\in I_i$ holds for $i=0,\ldots$. Then $\{\lambda_0 b_0,\lambda_1 b_1,\ldots\}\cup B'$ is an everywhere-dense Hamel basis.

6. Let C be the Cantor middle-third set. It is well known that C is nowhere dense and of measure zero. It is also known that $C+C$ contains every real in $[0,1]$ so C is a generating set in $\mathbf{R}$. By Problem 2 it includes a Hamel basis, which then must be of measure zero.

7. B is a Hamel basis with full outer measure if B intersects every perfect set of positive measure. As the number of perfect sets of positive measure

is continuum, we can enumerate them in a well-ordered sequence of length continuum: $\{P_\alpha : \alpha < c\}$. In a transfinite recursion of length c we select the elements b_α the following way. If $Y = \{b_\beta : \beta < \alpha\}$ have already been selected, let X be the linear hull of Y. As $|Y| < c$, we have $|X| < c$, as well, so we can pick $b_\alpha \in P_\alpha \setminus X$. This will give a linearly independent set $\{b_\alpha : \alpha < c\}$ intersecting every perfect set of positive measure. Extend it to a Hamel basis (see Problem 1). [W. Sierpiński]

8. Assume that B is a measurable Hamel basis with positive measure. Pick $b_0 \in B$. $B' = B \setminus \{b_0\}$ is still measurable with the same measure. By Steinhaus's theorem, if $h > 0$ is small enough, then h is the difference of two elements of B'. But then this is true for some qb_0 with $q \neq 0$ rational, so B is not linearly independent.

9. Assume that $B \subseteq \mathbf{R}$ is a Hamel basis that is an analytic set. Let $b_0 \in B$ be an arbitrary element, and A the set linearly generated by $B' = B \setminus \{b_0\}$ over $\mathbf{Q}$. We claim that A is also analytic. In fact, B' is analytic. Now if H, K are analytic sets, then $H + K$, being the projection of the plane analytic set $H \times K$ onto the line $y = x$, is also analytic. By induction, if $H_1, H_2, \ldots$ are analytic, then so is $H_1 + \cdots + H_n$ for finite n. Finally, $A = \bigcup\{\lambda_1 B' + \cdots + \lambda_n B' : \lambda_1, \ldots, \lambda_n \in \mathbf{Q}\}$, hence it is analytic. Every analytic set is measurable, in particular A is measurable. Since $A - A = A \neq \mathbf{R}$, A must be of measure zero (recall Steinhaus' theorem that the difference set of any set of positive measure includes an interval). But then $\mathbf{R} = \bigcup\{qb_0 + A : q \in \mathbf{Q}\}$ would be the union of countably many sets of measure zero, a contradiction. [The results of the last three problems are from W. Sierpiński: Sur la question de la the mesurabilité de la base de M. Hamel, *Fund. Math.*, **1**(1920), 105–111; see also F. B. Jones: Measure and other properties of a Hamel basis, *Bull. Amer. Math. Soc.* **48**(1942), 472–481. A. Miller proved that if the axiom of constructibility is assumed, then there is a coanalytic Hamel basis. A. Miller: Infinite combinatorics and definability, *Annals of Pure and Appl. Logic* **41** (1989), 179- 203]

10. By CH, there is a Hamel basis of the form $\{b_\alpha : \alpha < \omega_1\}$. Every nonzero real x can be written as

$$x = \lambda_1(x) b_{\alpha_1(x)} + \cdots + \lambda_n(x) b_{\alpha_n(x)}$$

with nonzero rational numbers $\lambda_1(x), \ldots, \lambda_n(x)$ and ordinals $\alpha_1(x) < \cdots < \alpha_n(x)$ for some natural number n. Denote $\alpha_n(x)$ by $\beta(x)$. We define the decomposition $\mathbf{R} \setminus \{0\} = A_0 \cup A_1 \cup \cdots$ as follows. For every ordinal $\alpha < \omega_1$ there are exactly $\aleph_0$ reals with $\beta(x) = \alpha$. We distribute them such that every A_i gets one and only one of them.

We claim that each A_i is a Hamel basis. Indeed, let $\mu_1, \ldots, \mu_n$ be nonzero rationals and $x_1, \ldots, x_n \in A_i$ different elements. The ordinals $\beta(x_1), \ldots, \beta(x_n)$ are different, and if $\beta(x_n)$ is the largest of them, then the coefficient of $b_{\beta(x_n)}$

is $\mu_n\lambda \neq 0$ with some $\lambda \neq 0$ in the linear combination $\mu_1 x_1 + \cdots + \mu_n x_n$, hence this linear combination cannot be zero. Thus, A_i is a rationally independent set.

To show that A_i is a generating set, it suffices to verify that it generates every b_α. This we prove by induction on α. Assume we have reached b_α, and we have already proved the statement for all earlier basis elements. There is one $x \in A_i$ with $\beta(x) = \alpha$. x can be written as $x = y + \lambda b_\alpha$ with y generated by earlier elements, hence, in view of the induction hypothesis, by A_i. Therefore, $b_\alpha = (1/\lambda)(x - y)$ is also generated by A_i. [P. Erdős, S. Kakutani: On non-denumerable graphs, *Bull. Amer. Math. Soc.*, **49**(1943), 457–461]

11. Assume indirectly that $c \geq \aleph_2$ yet $\mathbf{R} \setminus \{0\} = B_0 \cup B_1 \cup \cdots$ is the union of countably many Hamel bases. As $\aleph_1 + \aleph_2 = \aleph_2 \leq c$ we can find sets $X, Y \subseteq \mathbf{R}$ such that $|X| = \aleph_1$, $|Y| = \aleph_2$, and even $X \cup Y$ is independent. Color the complete bipartite graph on classes X, Y as follows. For $x \in X$, $y \in Y$ let the color of $\{x, y\}$ be that $n < \omega$ for which $x + y \in B_n$. By Problem 24.27 there are $x_1, x_2 \in X$, $y_1, y_2 \in Y$, and some $n < \omega$ such that $x_i + y_j \in B_n$ holds for $i, j = 1, 2$. But then, as $(x_1 + y_1) + (x_2 + y_2) - (x_1 + y_2) - (x_2 + y_1) = 0$, B_n is not independent, a contradiction.

12. We are going to construct a Hamel basis B such that B^+ is a Lusin set, i.e., it is of power continuum but intersects every set of first category in a countable set. This suffices, as every Lusin set is of measure zero (see Problem 16.20(b)). Let $D = \{d_\alpha : \alpha < \omega_1\}$ be an arbitrary Hamel basis and enumerate the first-category F_σ sets as $\{H_\alpha : \alpha < \omega_1\}$. Recall that every first-category set is included in a first-category F_σ set; therefore, it is sufficient to consider the sets H_α. Suppose that at step $\alpha < \omega_1$ we have already constructed a countable part B_0 of B, and we have countably many sets $\{H_\beta : \beta < \alpha\}$ to worry about in the sense that in the continuation of the construction we should not select any point from these sets. That is, we have to ensure that no element of B^+ with positive coefficients in $B \setminus B_0$ will be in $H = \bigcup\{H_\beta : \beta < \alpha\}$. Let d be the first element of D, not generated by B_0. The idea is that we add two elements to B_0, namely x and $x + d$ for some $x \in \mathbf{R}$. This way, B will generate every element of D, so it will be a Hamel basis. Of course, we need to make sure that $x, x + d$ are not linear combinations of elements of B_0, moreover, no element of the form $u + px + q(x + d)$ is in H where $u \in B_0^+$, $p, q \in \mathbf{Q}$, $p, q \geq 0$, $p + q > 0$. The first condition excludes countably many real numbers x. The second can be rewritten as $x \notin K$, where K is a first-category set. Hence x can be chosen to satisfy all conditions above, and then we can set $B_{\alpha+1} = B_\alpha \cup \{x, x + d\}$. This completes the construction (for limit α's let B_α be the union of all B_β with $\beta < \alpha$). [P. Erdős and S. Kakutani, Bull. Amer. Math. Soc., **49**(1943), 457–461, P. Erdős, *Coll. Math.* **X**(1963), 267–269]

13. (a) Let $B = \{b_i : i \in I\}$ be a Hamel basis, $\{c_i : i \in I\}$ arbitrary reals, indexed with the same index set I. We claim that there is one and only

one additive function f with $f(b_i) = c_i$. As the mapping $\sum \lambda_i b_i \mapsto \sum \lambda_i c_i$ is additive, one direction is clear. For the other direction we have to show that if f is additive and the coefficients λ_i are rational, then $f\left(\sum \lambda_i b_i\right) = \sum \lambda_i f(b_i)$, which boils down to showing that $f(\lambda x) = \lambda f(x)$ if λ is rational. From additivity, we get $f(nx) = nf(x)$ for $n = 1, 2, \ldots$, and $f(0) = 0$ is also clear. As $f(-x) = -x$, the equality $f(nx) = nf(x)$ also holds for negative integers. Finally, if $\frac{p}{q}$ is a rational number, $x \in \mathbf{R}$, then $f(\frac{1}{q}x) = \frac{1}{q}f(x)$ and $f(\frac{p}{q}x) = \frac{p}{q}f(x)$ by the previous remarks.

(b) If there is some x with $f(x) = 0$, then f is identically 0. Otherwise, as $f(x) = f\left(\frac{x}{2}\right)^2$, f is everywhere positive. Then $f(x) = e^{g(x)}$, where $g : \mathbf{R} \to \mathbf{R}$ is additive, and so is described in (a).

(c) $f(0)$ is either 0 or 1 and its value is independent of the other values of f. If for some $x \neq 0$ we have $f(x) = 0$, then f is identically 0 on all nonzero reals. As for $x > 0$ we have $f(x) = f\left(\sqrt{x}\right)^2$, we may assume that $f(x) > 0$ for $x > 0$. $f(-1) = \pm 1$ so either $f(-x) = f(x)$ or $f(-x) = -f(x)$ holds. Therefore, we can restrict to the calculation of f on positive reals. There, if we set $f(x) = e^{g(\log x)}$ then $g(\log x + \log y) = g(\log x) + g(\log y)$, that is, g is additive, and is described in (a).

(d) If $g(x) = f(x) - f(0)$ then we find that $g(\frac{x+y}{2}) = \frac{g(x)+g(y)}{2}$ holds, and $g(0) = 0$. Substituting $y = 0$ we obtain $g(x/2) = g(x)/2$ and that transforms the identity into $g(x + y) = g(x) + g(y)$. So the general solution is $f(x) = g(x) + c$, where g is an additive function, described in part (a).

(e) For $F(x) = f(x) + c$ the functional equation takes the form $F(x + y) = F(x) + F(y)$, hence part (a) can be applied. Thus, the solutions are the ones from part (a) with some constant c added to them.

(f) $f(0) = g(0) + h(0)$, hence

$$f(x) - f(0) = f(x+0) - f(0) = g(x) + h(0) - (g(0) + h(0)) = g(x) - g(0),$$

and similar computation gives $f(x) - f(0) = h(x) - h(0)$. Thus, for the function $F(x) = f(x) - f(0)$ we have $F(x + y) = F(x) + F(y)$, hence part (a) can be applied. Thus, the solutions are as follows: take any solution F from part (a) and let $f(x) = F(x) + a$, $g(x) = F(x) + b$ and $h(x) = F(x) + c$, where $a = b + c$ are constants.

(g) We have $f(0) = (a + b)f(0)$; thus, if $a + b \neq 1$, then $f(0) = 0$, and we get from the equation (by setting $y = 0$) $f(x) = af(x)$, and similarly $f(x) = bf(x)$. Thus either $f(x) \equiv 0$ or $a = b = 1$ and f is an arbitrary solution from part (a). On the other hand, if $a + b = 1$, then for $F(x) = f(x) - f(0)$ we get the equation $F(x + y) = aF(x) + bF(y)$, and as here already $F(0) = 0$ we obtain $F(x) \equiv 0$ as before (in this case $a = b = 1$ is not possible). In summary: if $a + b = 1$, then f is constant; if $a = b = 1$, then f is a solution from part (a); and for all other a, b, the function f is identically zero.

14. α, β are not commeasurable exactly when they are rationally independent, hence by Problem 1 they can be embedded into a Hamel basis, and by Problem 13(a), we can arbitrarily prescribe f on that basis.

15. Let a, b be two noncommensurable reals. By Problem 1 there is a Hamel basis B with $a, b \in B$. Every real x can uniquely be written as $x = \lambda_0 b_{i_0} + \cdots + \lambda_n b_{i_n}$ where $B = \{b_i : i \in I\}$, $\lambda_i \in \mathbf{Q}$. Separate the term containing a; $x = \lambda a +$ (the remaining terms) $= f(x) + g(x)$. As the first term of $x + b$ is $\lambda a = f(x)$, the first term of $x + a$ is $(\lambda + 1)a$, and the remaining terms are unchanged, we get that $f(x)$ is periodic with period b and $g(x)$ is periodic with period a

16. Let a, b, c be 3 reals, linearly independent over $\mathbf{Q}$. By Problem 1 there is a Hamel basis containing them. Every real can be written in this Hamel basis as

$$x = \lambda_1 a + \lambda_2 b + (\text{some other terms}) = f(x) + g(x) + h(x)$$

and here (see the preceding proof) $f(x)$ is periodic with period b and c, $g(x)$ is periodic with period a and c and $h(x)$ is periodic with period a and b. So x^2 can be written as the sum of nine terms (like $f(x)g(x)$, $g(x)h(x)$, etc.), each periodic by either a, or b, or c (e.g., $f(x)h(x)$ is periodic with period b). Grouping this representation of x^2 so that the functions with the same period get into a single group, we get the desired representation as the sum of three periodic functions.

To prove that $F(x) = x^2$ is not the sum of two periodic functions, assume that $F(x) = f(x) + g(x)$, where $f(x)$ is periodic with period $a > 0$ and $g(x)$ is periodic with period $b > 0$. We claim that for every real x,

$$F(x + a + b) - F(x + a) - F(x + b) + F(x) = 0$$

holds. Indeed, by rearranging, we get

$$\begin{aligned}
&F(x + a + b) - F(x + a) - F(x + b) + F(x) \\
&= (f(x + a + b) - f(x + b) - f(x + a) + f(x)) \\
&+ (g(x + a + b) - g(x + a) - g(x + b) + g(x)) = 0.
\end{aligned}$$

But the left-hand side is

$$(x + a + b)^2 - (x + a)^2 - (x + b)^2 + x^2 = 2ab \neq 0,$$

which is a contradiction, and this contradiction proves the claim.

17. The proof is similar to the previous one. As in the preceding proof let $a_1, a_2, \ldots, a_{k+1}$ be $k + 1$ reals linearly independent over $\mathbf{Q}$. By Problem 1 there is a Hamel basis containing them. Every real can be written in this Hamel basis as

$$x = \lambda_1 a_1 + \lambda_2 a_2 + \cdots + \lambda_k a_k + (\text{some other terms})$$
$$= f_1(x) + f_2(x) + \cdots + f_k(x) + (f_{k+1}(x))$$

and here (see the preceding proof) $f_i(x)$ is periodic with period a_j, $j = 1, 2, \ldots, k+1$ for every $j \neq i$. Now raising this expression to the kth power we find that x^k can be written as the sum of $(k+1)^k$ terms each periodic by either $a_1, a_2, \ldots a_k$ or a_{k+1} (the point is that when we multiply out $x \cdot \cdots \cdot x$, no product can contain all of the f_i's, and if f_i is missing from a particular product then this product is periodic with period a_i). As we have seen in the preceding proof, that is enough if we collect the terms with the same period.

To prove that $F(x) = x^k$ is not the sum of k periodic functions, let $\Delta_a f(x) = f(x+a) - f(x)$. Note that for $a \neq 0$ if f is a polynomial of degree m with leading term cx^m then $\Delta_a f(x)$ is a polynomial of degree $m-1$ with leading coefficient cma. It is also clear that if f is periodic with period b, then $\Delta_a f$ is also periodic with period b, while if f is periodic with period a then $\Delta_a f(x) \equiv 0$. These imply that if $x^k = f_1(x) + \cdots + f_k(x)$, where $f_i(x)$ is periodic with period $a_i \neq 0$, then on the one hand

$$\Delta_{a_1} \Delta_{a_2} \cdots \Delta_{a_k} x^k = k! a_1 a_2 \cdots a_k \neq 0,$$

and on the other hand,

$$\Delta_{a_1} \Delta_{a_2} \cdots \Delta_{a_k} x^k = \Delta_{a_1} \Delta_{a_2} \cdots \Delta_{a_k} (f_1(x) + \cdots + f_k(x)) \equiv 0.$$

This contradiction completes the proof.

18. Let B be a Hamel basis, $b \in B$, and let A be the linear span of $B \setminus \{b\}$ over $\mathbf{Q}$. Then $\mathbf{R} = \cup_{\lambda \in \mathbf{Q}}(A + \lambda b)$ is a disjoint decomposition and $A + \lambda b$, $\lambda \in \mathbf{Q}$ are the only subsets of $\mathbf{R}$ that are congruent to A (note that $-A = A$). [W. Sierpiński, Fund. Math., **35**(1948), 159–164]

19. Let B be a Hamel basis, and let A be the linear span of B over $\mathbf{Z}$, i.e., A consists of those elements $y \in \mathbf{R}$ such that if $y = \gamma_1 b_1 + \cdots + \gamma_m b_m$ is the representation of x in terms of the basis B with rational coefficients, then all γ_i are integers. Clearly, if $b \in B$, then $b/2 \notin A$, so $A \neq \mathbf{R}$. Now let $x \in \mathbf{R}$ be arbitrary, and let $x = \lambda_1 b_1' + \cdots + \lambda_n b_n'$ be a representation of x in terms of elements from B with nonzero rational coefficients λ_i. If N denotes the common denominator of $\lambda_1, \ldots, \lambda_n$, then $Nx \in A$, and since A is closed for addition it follows that $A + Nx = A$, hence $A + (k+N)x = A + kx$ for all $k \in \mathbf{Z}$. Thus, only A, $A + x$, $A + 2x$, $\ldots$, $A + (n-1)x$ can be different in the sequence A, $A + x$, $A + 2x$, $A + 3x, \ldots$. [E. Cech, see W. Sierpiński, Cardinal and Ordinal Numbers, Polish Sci. Publ., Warszawa, 1965, XVII.1]

20. Let $B = \{b_i : i \in I\}$ be a Hamel basis and assume that $\prec$ is an ordering of I. If $x \in \mathbf{R}$, $x \neq 0$, write it as $x = \lambda_0 b_{i_0} + \cdots + \lambda_n b_{i_n}$, where $i_0 \prec \cdots \prec i_n$ and none of the rational coefficients $\lambda_0, \ldots, \lambda_n$ is zero. Set $x \in A$ if and only if

$\lambda_0 > 0$. Assume that $a \in \mathbf{R} \setminus \{0\}$. If the least (by $\prec$) coefficient of a is positive then $A + a \subseteq A$, if it is negative, then $A + (-x) \subseteq A$, so $A \subseteq A + a$, and it is easy to see that A and $\mathbf{R} \setminus A$ are both everywhere dense.

21. Let $B = \{b_i : i \in I\}$ be a Hamel basis with $b_j = 1$. Let $\prec$ be an ordering of I in such a way that j is the maximal element (but otherwise $\prec$ is arbitrary). If $x \in \mathbf{R} \setminus \mathbf{Q}$, write it as $x = \lambda_0 b_{i_0} + \cdots + \lambda_n b_{i_n}$ where $i_0 \prec \cdots \prec i_n$ and none of the rational coefficients $\lambda_0, \ldots, \lambda_n$ is zero. Set $x \in A$ if and only if $x \notin \mathbf{Q}$ and $\lambda_0 > 0$, and let $B = (\mathbf{R} \setminus \mathbf{Q}) \setminus A$ (that is, when $\lambda_0 < 0$). A and B are both closed under addition, as if λ_0 and λ_0' are the leftmost coefficients of x and y, respectively, and they are both positive/negative, then the leftmost coefficient of $x + y$ is either λ_0, λ_0', or $\lambda_0 + \lambda_0'$.

22. Let $B = \{b_i : i \in I\}$ be a Hamel basis that contains positive as well as negative elements. If $x \in \mathbf{R}^+$, write it as $x = \lambda_0 b_{i_0} + \cdots + \lambda_n b_{i_n}$ where $b_{i_0} < b_{i_1} < \cdots < b_{i_n}$ and none of the rational coefficients $\lambda_0, \ldots, \lambda_n$ is zero. Set $x \in A$ if and only if $x > 0$ and $\lambda_0 > 0$, and let $C = \mathbf{R}^+ \setminus A$. Both A and C are closed under addition. Indeed, if λ_0, λ_0' are the leftmost coefficients of x, y, respectively, and they are both positive/negative then the leftmost coefficient of $x+y$ is either λ_0, λ_0', or $\lambda_0 + \lambda_0'$. If a is a positive element of B then $a \in A$, while if c is a negative element of B then $-c \in C$. Thus, A and C are not empty.

23. We remark first of all that if $\{a_1, \ldots, a_{17}\}$ satisfy the property in the problem, then so do the systems $\{a_1 - b, \ldots, a_{17} - b\}$ and $\{ca_1, \ldots, ca_{17}\}$, where b, c are real numbers. Assume first that the numbers $\{a_1, \ldots, a_{17}\}$ are integers. By adding the same integer to them, we can achieve that they are natural numbers and one of them is zero. The decomposition property implies that upon removal any of them the remaining 16 numbers have an even sum, so all numbers have the same parity, in this case, they are even. Dividing by 2, we get a family of 17 numbers with exactly the same properties, i.e., they are natural numbers, one of them is zero, and they have the decomposition property. Again, they are even, we can divide by 2, etc. Division by 2 unboundedly many times is only possible if all the initial numbers are equal to zero, so we have the result for integers.

Assume now that the numbers $a_1, \ldots, a_{17}$ are rational. By multiplying them with an appropriate natural number we get a system of 17 integers that must be equal by the preceding argument so our original system also consists of equal numbers.

Assume finally that we have a system $a_1, \ldots, a_{17}$ of real numbers. If $B = \{b_i : i \in I\}$ is a Hamel basis, then our numbers can be written as

$$a_j = \sum_{i \in I} \lambda_i^j b_i$$

and now for each $i \in I$ the system $\{\lambda_i^j : 1 \le j \le 17\}$ is a system of 17 rational numbers with the original property. We get, therefore, that $\lambda_i^j = \lambda_i$, that is, our original numbers are equal.

24. Let $B = \{b_i : i \in I\}$ be a Hamel basis. Every nonzero real number x can be uniquely written as

$$x = \lambda_{i_0} b_{i_0} + \cdots + \lambda_{i_n} b_{i_n},$$

where $\lambda_{i_0}, \ldots, \lambda_{i_n}$ are nonzero rational numbers and $b_{i_0} < \cdots < b_{i_n}$. Notice that there are countably many possible ordered $\langle \lambda_{i_0}, \ldots, \lambda_{i_n} \rangle$ sequences, so we can decompose $\mathbf{R}$ into countably many classes in such a way that reals in the same class have the same ordered sequence of rational numbers (and let 0 alone form a class). We show that this decomposition of $\mathbf{R}$ is as required.

Assume that the distinct elements x, y, z of some class, say the one associated with $\langle \lambda_{i_0}, \ldots, \lambda_{i_n} \rangle$ form a 3-element arithmetic progression, i.e., $2y = x + z$. Although the sequence $\langle \lambda_{i_0}, \ldots, \lambda_{i_n} \rangle$ is the same for x, y, z, the associated sequences of reals $\langle b^x_{i_0}, \ldots, b^x_{i_n} \rangle$, $\langle b^y_{i_0}, \ldots, b^y_{i_n} \rangle$, $\langle b^z_{i_0}, \ldots, b^z_{i_n} \rangle$, can be different.

Let b be the *least* of all the occurring elements, that is the minimal element of $\{b^x_{i_0}, \ldots, b^x_{i_n}, b^y_{i_0}, \ldots, b^y_{i_n}, b^z_{i_0}, \ldots, b^z_{i_n}\}$. Let the coordinate of b in x, y, and z be α, β, and γ. Each of α, β, γ is either 0 or λ_{i_0}. Also, $2\beta = \alpha + \gamma$. But these two latter properties imply that either $\alpha = \beta = \gamma = 0$ (which is impossible) or $\alpha = \beta = \gamma = \lambda_{i_0}$. We have, therefore, that $b^x_{i_0} = b^y_{i_0} = b^z_{i_0}$. We can continue, and get $b^x_{i_1} = b^y_{i_1} = b^z_{i_1}$, etc.; finally, all the coordinates of x, y, z are equal, that is, $x = y = z$. [R. Rado]

25. In this solution we only consider rectangles with sides parallel to the xy-axes. First we remark that every rectangle with commensurable sides can be decomposed into the union of squares so what the problem states is to show that if a rectangle can be split into squares then it has commensurable sides. Let f and g be two additive functions on the reals, cf. Problem 13(a).

We associate with a rectangle R the value $t(R) = f(a)g(b)$, where a, b are the lengths of the sides of R parallel to the x-, resp. y-axis. We claim that this function is additive on the rectangles, that is, if some rectangle R is split into $R_1, \ldots, R_n$, then $t(R) = t(R_1) + \cdots + t(R_n)$. In fact, draw all the lines that include one of the sides of one of the rectangles R_i. These lines divide the rectangle R into smaller rectangles, say $Q_1, \ldots, Q_m$, and each R_j is the union of some of the Q_i's. Actually, these representations in terms of the Q_i's are regular in the sense that if $R_i = [a, b] \times [c, d]$ then R_i is the union of rectangles of the form $[p, q] \times [c, d]$ (i.e., they have their $[c, d]$ side equal to the $[c, d]$ side of R_i), and each such rectangle $[p, q] \times [c, d]$ is the union of rectangles Q_i of the form $[p, q] \times [r, s]$ (i.e., the $[p, q]$ side of Q_i equals the $[p, q]$ side of $[p, q] \times [c, d]$). Since the same is true of R, the additivity can be reduced to the case when a rectangle R is split with side-to-side cuts to smaller rectangles (i.e., with m

horizontal and n vertical cuts into the union of mn rectangles) and that can further be reduced to the case when a rectangle is split with either horizontal or vertical cuts. Then the statement follows from the additivity of f and g.

If an $a \times b$ rectangle is divided into the union of squares $S_1, \ldots, S_n$, then these squares can be rearranged to form a $b \times a$ rectangle (just make a 90-degree rotation of the whole picture). With our previous statement this implies that $f(a)g(b) = f(b)g(a)$. In particular, with the choice $g(x) = x$ it follows that

$$f(a) = \frac{a}{b} f(b)$$

must be true. At this point f is still an arbitrary additive function. By Problem 14 if a and b are not commensurable, then we can choose the additive function f so that this relation does not hold, and this proves that, indeed, a and b are commensurable. [Max Dehn: Über die Zerlegung von Rechtecken in Rechtecke *Math. Ann.*, **57**(1903), 314–332]

26. We first show that for every natural number $n \geq 1$ the set $\{1, 2, \ldots, n\}$ carries such an ordering. This we do by induction. It is clear for $n = 1, 2$. We assume that $\{1, 2, \ldots, n\}$ has such an ordering $\prec$ and define one $\prec'$ for $\{1, 2, \ldots, 2n\}$. The idea is to put first the even numbers and then the odd numbers, that is, if $i \prec j$, set $2i \prec' 2j$ and $2i - 1 \prec' 2j - 1$, and for any i, j, make $2i \prec' 2j - 1$. If x, y, z form a 3-element arithmetic progression, and all three of them have the same parity, then $x \prec' y \prec' z$ is not possible because of the induction hypothesis (on the even and odd numbers $\prec'$ is a transform of $\prec$). If not, then $x + z = 2y$ shows that only y can have a different parity from the other two, and in this case $x \prec' y \prec' z$ is again not possible, for any number lying (with respect to $\prec'$) in between two numbers of the same parity has the same parity.

From this case we immediately get the statement for every finite subset of **Q**, and from that, using König's lemma on infinity (Problem 27.1), for **Q**.

Assume now that $B = \{b_i : i \in I\}$ is a Hamel basis. Fix $\prec$, an ordering with the required property for **Q**. If x, y are real numbers, write them in the form $x = \lambda_1 b_1 + \cdots + \lambda_n b_n$, $x = \lambda_1' b_1 + \cdots + \lambda_n' b_n$ with $b_1 < \cdots < b_n$ and rational λ_i, λ_i' (notice that for each i one of λ_i, λ_i' may be zero). Put $x \prec' y$ if $\lambda_i \prec \lambda_i'$ for the first i with $\lambda_i \neq \lambda_i'$. It is easy to see that this is an ordering. Assume that x, y, z form a 3-element arithmetic progression, and $x \prec' y \prec' z$. Write them as $x = \lambda_1 b_1 + \cdots + \lambda_n b_n$, $y = \lambda_1' b_1 + \cdots + \lambda_n' b_n$, $z = \lambda_1'' b_1 + \cdots + \lambda_n'' b_n$ and let i be the first coordinate with some two of $\lambda_i, \lambda_i', \lambda_i''$ different. Then, as x, y, z form a 3-element arithmetic progression, the values $\lambda_i, \lambda_i', \lambda_i''$ are different, only one of them can be zero, and this is the coordinate which is decisive in the comparison of x, y, and z, i.e., we must have $\lambda_i \prec \lambda_i' \prec \lambda_i''$. Furthermore, $2\lambda_i' = \lambda_i + \lambda_i''$ and such a $\lambda_i, \lambda_i', \lambda_i''$ triplet is impossible by the choice of $\prec$. This contradiction proves the claim. [Géza Kós, Gyula Károlyi]

27. Let B be a Hamel basis in **R** and C a similar basis in **C**, i.e., C is a basis of the vector space **C** over the field **Q**. Both B and C are of cardinality

continuum (see Problem 3); therefore, there is a one-to-one correspondence $f : B \to C$ between them. Now it is clear that the mapping

$$F(\lambda_1 b_1 + \cdots + \lambda_n b_n) = \lambda_1 f(b_1) + \cdots + \lambda_n f(b_n)$$

$(\lambda_1, \ldots, \lambda_n \in \mathbf{Q})$ is an addition-preserving bijection between $\mathbf{R}$ and $\mathbf{C}$.

16

The continuum hypothesis

1. If CH holds we can enumerate $\mathbf{R}$ as $\mathbf{R} = \{r_\alpha : \alpha < \omega_1\}$. If we are given $(x,y) \in \mathbf{R} \times \mathbf{R}$, then $x = r_\alpha$, $y = r_\beta$ for some countable ordinals α and β. Set $(x,y) \in A$ if and only if $\alpha < \beta$. Assume that L is a horizontal line, $L = \{(x,c) : x \in \mathbf{R}\}$ for some $c \in \mathbf{R}$. If $\beta < \omega_1$ is the ordinal such that $r_\beta = c$, then $(x,c) \in A$ if and only if $x = r_\alpha$ for some $\alpha < \beta$ and there are countably many ordinals like that. Assume now that L is a vertical line, $L = \{(c,x) : x \in \mathbf{R}\}$ for some $c \in \mathbf{R}$. If $\alpha < \omega_1$ is that ordinal for which $r_\alpha = c$, then $(c,x) \in B$ if and only if $x = r_\beta$ for some $\beta \leq \alpha$ and there are countably many ordinals like that.

For the other direction assume that $c \geq \aleph_2$ and there is a decomposition $\mathbf{R}^2 = A \cup B$ as above. Pick a subset $U \subseteq \mathbf{R}$ of cardinality $\aleph_1$. By condition on A, for every $y \in \mathbf{R}$ there is some $u = u(y) \in U$ such that $(u,y) \notin A$, so $(u,y) \in B$. As $|\mathbf{R}| > |U|$, there is some $u \in U$ that occurs uncountably many times as $u(y)$, so in this case the vertical line $L = \{(u,y) : y \in \mathbf{R}\}$ has uncountably many points in B, a contradiction. [W. Sierpiński]

2. If CH holds, there is a Sierpiński decomposition, $\mathbf{R}^2 = A \cup B$ (see the previous problem). By adding points to the sets A and B we may assume that A intersects every horizontal line and B intersects every vertical line in $\aleph_0$ points. For every $y \in \mathbf{R}$ the countably infinite set $\{x : (x,y) \in A\}$ can be counted as $\{g_0(y), g_1(y), \ldots\}$ and similarly, for every $x \in \mathbf{R}$ the countably infinite set $\{y : (x,y) \in B\}$ can be counted as $\{f_0(x), f_1(x), \ldots\}$. Now $\mathbf{R}^2 = A \cup B$ is the union of the graphs of the partial functions $x \mapsto f_n(x)$ and $y \mapsto g_n(y)$.

For the other direction, if $\mathbf{R}^2$ is the union of countably many $x \mapsto y$ and $y \mapsto x$ functions, then letting A be the union of the graphs in the second class, B that of in the first class, we get a Sierpiński decomposition, and we conclude with the second part of the previous problem.

3. Assume first that CH holds and $\mathbf{R} = \{r_\alpha : \alpha < \omega_1\}$. Fix, for every $\alpha < \omega_1$, an injection $\varphi_\alpha : \alpha + 1 \to \omega$. Assume we are given $(x_1, x_2, x_3) \in \mathbf{R}^3$ we

determine where to put it. Assume x_1, x_2, x_3 are $r_\alpha, r_\beta, r_\gamma$ in some order with $\alpha, \beta \le \gamma$. Compare $\varphi_\gamma(\alpha)$ and $\varphi_\gamma(\beta)$. Assume that $\varphi_\gamma(\alpha) \le \varphi_\gamma(\beta)$ (say). If now $r_\alpha = x_i$ then put (x_1, x_2, x_3) into A_i. We show that $A_i \cap L$ is finite if L is a line in the direction of the x_i-axis. For definiteness' sake assume that $i = 1$. The elements of L are triples of the form (x, b, c) with some fixed $b, c \in \mathbf{R}$. If $(x, b, c) = (r_\alpha, r_\beta, r_\gamma)$, then it is added to A_1 if either $\alpha, \beta \le \gamma$ and $\varphi_\gamma(\alpha) \le \varphi_\gamma(\beta)$ or else $\alpha, \gamma \le \beta$ and $\varphi_\beta(\alpha) \le \varphi_\beta(\gamma)$. Given β, γ there are only finitely many α that satisfy either one of the requirements.

For the other direction assume that $c \ge \aleph_2$ and $\mathbf{R}^3 = A_1 \cup A_2 \cup A_3$ is a decomposition as claimed. Pick $U, V, W \subseteq \mathbf{R}$ of cardinality $\aleph_0, \aleph_1, \aleph_2$, respectively. For any given $(u, v) \in U \times V$ there are finitely many $z \in W$ with $(u, v, z) \in A_3$ so, as $|U \times V| = \aleph_0\aleph_1 < \aleph_2 = |W|$, we can find some $c \in W$ that $(u, v, c) \notin A_3$ for $u \in A_1$, $v \in A_2$. For any given $u \in U$ there are only finitely many $y \in V$ that $(u, y, c) \in A_2$, so, as $|V| = \aleph_1 > \aleph_0 = |U|$, we can choose some $b \in V$ that $(u, b, c) \notin A_2$ holds for every $u \in U$. Finally, the set $\{u \in U : (u, b, c) \in A_1\}$ is finite, so we can choose some $a \in U$ not in it, and then, (a, b, c) is not in any of A_1, A_2, A_3. But this contradicts the choice of the sets A_i and this contradiction shows that we must have $\mathbf{c} = \aleph_1$. [W. Sierpiński, Cardinal and Ordinal Numbers, Polish Sci. Publ., Warszawa, 1965, XIV.9. Theorem 1]

4. Assume that there is such a decomposition $\mathbf{R}^3 = A_1 \cup A_2 \cup A_3$. Pick $U \subseteq \mathbf{R}$ with $|U| = 3m + 1$. Then $|U \times U \times U| = (3m + 1)^3$ but for $i = 1, 2, 3$ we have that $|A_i \cap (U \times U \times U)| \le m(3m + 1)^2$, i.e., $(3m + 1)^3 \le 3m(3m + 1)^2$, a contradiction.

5. The proof is similar to that of Problem 3. For an alternative proof utilizing induction, see the solution to Problem 10.15.

Let us assume first that $\mathbf{c} \le \omega_n$. For each $U \subseteq \mathbf{R}$ let $<_U$ be a well-ordering of U in order type $|U|$. For $(x_0, \dots, x_{n+1}) \in \mathbf{R}^{n+2}$ let i_0 be such that x_{i_0} is the maximal element of $\{x_0, \dots, x_{n+1}\}$ in the ordering $<_{\mathbf{R}}$, and set $U_0 = \{x \in \mathbf{R} : x \le x_{i_0}\}$. Note that $|U_0| < \omega_n$ and $\{x_j : j \ne i_0\}$ is a subset of this set. Let us suppose that for some $0 \le k \le n - 1$ the numbers $i_0, \dots, i_k$ and the sets $U_0, \dots, U_k$ have already been selected, $|U_k| < \omega_{n-k}$, and the set $\{x_j : j \ne i_0, \dots, i_k\}$ is part of U_k. Let i_{k+1} be such that $x_{i_{k+1}}$ is the maximal element of $\{x_j : j \ne i_0, \dots, i_k\}$ in the ordering $<_{U_k}$, and set $U_{k+1} = \{x \in \mathbf{R} : x \le_{U_k} x_{i_{k+1}}\}$. Since the index of $x_{i_{k+1}}$ with respect to $<_{U_k}$ is necessarily smaller than $|U_k| < \omega_{n-k}$, we get $|U_{k+1}| < \omega_{n-k-1}$, and the induction runs through. It follows that U_n is finite, and if $0 \le i_{n+1} \le n + 1$ is the index that differs from every i_j, $j \le n$, then $x_{j_{n+1}}$ is an element of U_n. Note that everything (i_j, U_j, $j = 0, \dots, n$) depends on the point $X = (x_0, \dots, x_n)$, and to show this dependence we write i_j^X, U_j^X.

This way we get an ordering $i_0^X, \dots, i_{n+1}^X$ of the set $0, 1, \dots, n + 1$, and let us put the point $X = (x_0, \dots, x_{n+1})$ into the class $A_{i_{n+1}}$. We show that each A_i is finite in the x_i-direction. For simpler notation let $i = 0$, and l be a line in the direction of the x_0- axis. The points on l are of the form $X = (x, c_1, \ldots, c_n)$ where $c_1, \cdots, c_n$ are fixed reals. Such a point belongs to A_0 if and only if $i_j^X \geq 1$ for all $j \leq n$ and $i_{n+1}^X = 0$, which implies $x \in U_n^X$. There are only finitely many permutations of the form $i_0, i_1, \ldots, i_n, 0$ of the numbers $0, 1, \ldots, n+1$, and if for another point $X' = (x', c_1, \ldots, c_n)$ on l we have the same permutation, i.e., $i_0^X = i_0^{X'}$, ..., $i_n^X = i_n^{X'}$, then for these two points the sets U_j are the same for all $j \leq n$, in particular $U_n^X = U_n^{X'}$. Then we have $x' \in U_n^{X'} = U_n^X$, and since U_n^X is finite, there are only finitely many such points X' in A_0. Since this is true for all permutations $i_0, i_1, \ldots, i_n, 0$, altogether there are only finitely many points of A_0 on the line l. This completes the existence of the decomposition.

Suppose now that $R^{n+2} = A_0 \cup \cdots \cup A_{n+1}$ and each A_i is finite in the x_i-direction. On the contrary to the claim let us suppose that $\mathbf{c} > \aleph_n$, and for $i \leq n$ let $X_i \subset \mathbf{R}$ be a set of cardinality $\aleph_i$. Every line of the form $(x_0, \ldots, x_n, y)$, $x_i \in X_i$, $y \in \mathbf{R}$ intersects A_{n+1} in finitely many points, and since there are only $\aleph_0 \aleph_1 \cdots \aleph_n = \aleph_n < \mathbf{c}$ such lines, there is a $c_{n+1} \in \mathbf{R}$ such that all points $(x_0, \ldots, x_n, c_{n+1})$, $x_i \in X_i$, $i \leq n$, lie outside A_{n+1}. In a completely analogous manner there is a point $c_n \in X_n$ such that all points $(x_0, \ldots, x_{n-1}, c_n, c_{n+1})$, $x_i \in X_i$, $i \leq n-1$, lie outside A_n, etc.. This way we get numbers $c_1, \ldots, c_{n+1}$ such that all points $(x_0, c_1, \ldots, c_{n+1})$, $x_0 \in X_0$, lie outside $A_1, \ldots, A_{n+1}$. Hence all these points should lie in A_0, which is impossible since A_0 intersects the line $(x, c_1, \ldots, c_{n+1})$, $x \in \mathbf{R}$ in only finitely many points. This contradiction proves that we must have $\mathbf{c} \leq \aleph_n$. [W. Sierpiński, Cardinal and Ordinal Numbers, Polish Sci. Publ., Warszawa, 1965, XV.9]

6. Let us assume first CH, and let A, B be the sets from Problem 1. We may assume that A has countably infinitely many points $x_0^r, x_1^r, \ldots$ on every horizontal line $y = r$ and B has countably infinitely many points $y_0^s, y_1^s, \ldots$ on every vertical line $x = s$. We set $f_1(t) = t \sin t$ for $t \in (-\infty, 1)$ and $f_2(t) = t \sin t$ for $t \in (-1, \infty)$. Then whatever the definition of these functions are on the rest of the real line, one of them is always differentiable. The idea of the proof is to choose $f_1(t)$ on $[1, \infty)$ in such a way that $(f_1(t), f_2(t))$, $t \in [1, \infty)$, cover the points of the set A, while to choose $f_2(t)$ on $(-\infty, -1]$ in such a way that $(f_1(t). f_2(t))$, $t \in (-\infty, -1]$, cover the points of the set B. For example, the first one can be done as follows: the function $f_2(t) = t \sin t$ takes every value r infinitely many times on the interval $[1, \infty)$, let us list them as $t_{r,0}, t_{r,1}, \ldots$. Now let $f_1(t_{r,j}) = x_j^r$, i.e., if $f_2(t)$ takes a particular value r jth time, then we choose $f_1(t)$ in such a way that $(f_1(t), f_2(t))$ be the jth point (x_j^r, r) from A on the line $y = r$. With this choice of f_1 we clearly cover the set A by the points $(f_1(t), f_2(t))$, $t \in [1, \infty)$. The selection of f_2 for $t \in (-\infty, -1]$ is similar: if the points $t \in (-\infty, -1]$ with $f_1(t) = t \sin t = s$ are listed as $t_{s,0}^*, t_{s,1}^*, \ldots$, then let $f_2(t_{j,s}^*) = y_j^s$. With this choice of f_2 we cover the set B

by the points $(f_1(t), f_2(t))$, $t \in (-\infty, -1]$, and the first part of the problem has been verified.

Now let us assume that there is a surjection $t \to (f_1(t), f_2(t))$ of $\mathbf{R}$ onto the plane in such a way that for all t one of the functions f_1 or f_2 is differentiable at t. Let H_i, $i = 1, 2$ be the set of points where f_i is differentiable. Then $\mathbf{R} = H_1 \cup H_2$. By Problem 5.15 the set Y_i of those y for which the intersection $f_i^{-1}(y) \cap H_i$ is uncountable is of measure zero. Let $\mathbf{R}^* = \mathbf{R} \setminus (Y_1 \cup Y_2)$. Then $\mathbf{R}^*$, as the complement of a set of measure zero, is of cardinality continuum, and if

$$A^* = \{(f_1(t), f_2(t)) \ : \ t \in H_2\}, \qquad B^* = \{(f_1(t), f_2(t)) \ : \ t \in H_1\},$$

then for every horizontal line ℓ the set $\mathbf{R}^* \times \mathbf{R}^* \cap A^* \cap \ell$ is countable: if ℓ has the form $y = r$, $r \in \mathbf{R}^*$, then there are only countably many $t \in H_2$ with $f_2(t) = r$ by the choice of the set $\mathbf{R}^* \subseteq \mathbf{R} \setminus Y_2$. In an analogous manner, for every vertical line ℓ the set $\mathbf{R}^* \times \mathbf{R}^* \cap B^* \cap \ell$ is countable. Thus, $\mathbf{R}^* \times \mathbf{R}^* = (\mathbf{R}^* \times \mathbf{R}^* \cap A^*) \cup (\mathbf{R}^* \times \mathbf{R}^* \cap B^*)$ is a decomposition of the "plane" $\mathbf{R}^* \times \mathbf{R}^*$ as in Problem 1, hence CH must hold (if we want to apply 1 directly to $\mathbf{R}^2$ then let $g : \mathbf{R}^* \to \mathbf{R}$ be a bijection between $\mathbf{R}^*$ and $\mathbf{R}$ and consider the sets $A = \{(g(x), g(y)), \ : \ (x, y) \in \mathbf{R}^* \times \mathbf{R}^* \cap A^*\}$ and $B = \{(g(x), g(y)), : \ (x, y) \in \mathbf{R}^* \times \mathbf{R}^* \cap B^*\})$ [M. Morayne: On differentiability of Peano type functions I, Colloq. Math., **53** (1988), 129–132]

7. If CH holds, then $\mathbf{R} = \{r_\alpha : \alpha < \omega_1\}$. If we set $A_\alpha = \{r_\beta : \beta < \alpha\}$, then $\{A_\alpha : \alpha < \omega_1\}$ is an increasing chain of countable sets with $\mathbf{R}$ as the union.

Assume now that $\{A_i : i \in I\}$ is an increasing chain (i.e., there is an ordering $\prec$ on I and if $i \prec j$ then $A_i \subseteq A_j$) of countable sets, and $\bigcup\{A_i : i \in I\} = \mathbf{R}$. Let $B \subseteq \mathbf{R}$ be a set of cardinality $\aleph_1$. For $x \in B$ there is some $i(x) \in I$ such that $x \in A_{i(x)}$. Should there be an index $j \in I$ such that $i(x) \preceq j$ held for every $x \in B$ we would get $B \subseteq A_j$, a contradiction. We have, therefore, that for every $j \in I$ there is some $x \in B$ that $j \prec i(x)$, so $\mathbf{R} = \bigcup\{A_i : i \in I\} \subseteq \bigcup\{A_{i(x)} : x \in B\}$, a set of cardinality at most $\aleph_1\aleph_0 = \aleph_1$.

8. For the forward direction if CH holds and $\mathbf{R} = \{r_\alpha : \alpha < \omega_1\}$, then we can set $f(r_\alpha) = \{r_\beta : \beta < \alpha\}$. If $X \subseteq \mathbf{R}$ is uncountable, then for every $r_\alpha \in \mathbf{R}$ there is some $\beta > \alpha$, $r_\beta \in X$, and so $r_\alpha \in f(r_\beta) \subseteq f[X]$.

For the other direction, if f is as required, choose some $X \subseteq \mathbf{R}$ of cardinality $\aleph_1$ then, as $f[X] = \mathbf{R}$, we get $c = |\mathbf{R}| \leq |X|\aleph_0 = \aleph_1\aleph_0 = \aleph_1$.

9. Suppose CH, and let $\{x_\alpha\}_{\alpha<\omega_1}$ be an enumeration of the reals, $\{y_\alpha\}_{\alpha<\omega_1}$ another enumeration of them in which each number is listed infinitely often and for each $\alpha < \omega_1$ let $\{\xi_k^\alpha\}_{k=0}^\infty$ be an enumeration of the set $\{\beta : \beta \leq \alpha\}$. Define $f_k(x_\alpha)$ as $f_k(x_\alpha) = y_{\xi_k^\alpha}$. If a is a real number then there are $\beta_0, \beta_1, \ldots$ such that $y_{\beta_i} = a$ for all $i = 0, 1, 2, \ldots$, and for every $\alpha > \sup_i \beta_i$ there are

$k_0^\alpha, k_1^\alpha, \ldots$ such that $\xi_{k_i^\alpha}^\alpha = \beta_i$. For each such $k = k_i^\alpha$ we have $f_k(x_\alpha) = y_{\xi_k^\alpha} = y_{\beta_i} = a$, hence for all such α the set $A_{x_\alpha,a} = \{n < \omega : f_n(x) = a\}$ is infinite.

Conversely, let us suppose that the sequence $f_0, f_1, \ldots$ with the stated properties exists, but $c > \aleph_1$. Let $K \subset \mathbf{R}$ be a set of cardinality $\aleph_1$, and for each $a \in K$ let H_a be the set of those $x \in \mathbf{R}$ for which $A_{x,a}$ is finite. By the properties of the functions f_i then each H_a is countable, so $\cup_{a\in K} H_a$ is of cardinality at most $\aleph_1$, hence, by the assumption $c > \aleph_1$, there is an $x^* \in \mathbf{R} \setminus (\cup_{a\in K} H_a)$. Now each $A_{x^*,a}$, $a \in K$ is infinite, which is impossible since these are disjoint subsets of ω. This contradiction proves the claim that $c \le \aleph_1$.

10. Suppose CH, let $\{x_\alpha\}_{\alpha<\omega_1}$ be an enumeration of the reals, and let $\{y_k^\alpha\}_{k=0}^\infty$, $\alpha < \omega_1$ be an enumeration of all real sequences in such a way that every real sequence is listed infinitely often. For each $\alpha < \omega_1$ let $\{\xi_k^\alpha\}_{k=0}^\infty$ be an enumeration of the set $\{\beta \ : \ \beta \le \alpha\}$. Define $f_k(x_\alpha)$ as $f_k(x_\alpha) = y_k^{\xi_k^\alpha}$. If $\{a_k\}_{k=0}^\infty$ is a real sequence then there are $\beta_0, \beta_1, \ldots$ such that $\{y_k^{\beta_i}\}_{k=0}^\infty = \{a_k\}_{k=0}^\infty$ for all $i = 0, 1, 2, \ldots$, and for every $\alpha > \sup_i \beta_i$ there are $k_0^\alpha, k_1^\alpha, \ldots$ such that $\xi_{k_i^\alpha}^\alpha = \beta_i$. For each such $k = k_i^\alpha$ we have

$$f_k(x_\alpha) = y_k^{\xi_k^\alpha} = y_k^{\beta_i} = a_k,$$

hence for all such α the set $A_{x_\alpha,\underline{a}} = \{k < \omega : f_k(x) = a_k\}$ is infinite.

The converse follows from the preceding problem if we just consider constant sequences.

11. If CH holds then with the functions f_k from the preceding problem the family $\mathcal{F} = \{\{f_k(x)\}_{k=0}^\infty \ : \ x \in \mathbf{R}\}$ is clearly appropriate. Conversely, suppose that $\mathcal{F}$ with the stated properties exists. Then $\mathcal{F}$ must be of cardinality continuum (otherwise we can define a sequence $\{a_n\}$ such that a_n is different from the nth element in the sequences in $\mathcal{F}$). Thus, we can index the sequences in $\mathcal{F}$ by the elements of $\mathbf{R}$, say $\mathcal{F} = \{\{a_n^x\}_{n=0}^\infty \ : \ x \in \mathbf{R}\}$. Now if we set $f_n(x) = a_n^x$ for $x \in \mathbf{R}$ and $n = 0, 1, \ldots$, then this sequence $\{f_n\}$ of functions satisfies the properties set forth in Problem 10. Now we can conclude CH from Problem 10.

12. Suppose CH, and let $\{x_\alpha\}_{\alpha<\omega_1}$ be an enumeration of the reals, and $(\{y_k^\alpha\}_{k=0}^\infty, \{n_k^\alpha\}_{k=0}^\infty)$, $\alpha < \omega_1$, enumeration of all the pairs consisting of a real sequence and a subsequence of ω. For each $\alpha < \omega_1$ let $\{\xi_k^\alpha\}_{k=0}^\infty$ be an enumeration of the set $\{\beta \ : \ \beta \le \alpha\}$ in such a way that $\xi_0^\alpha = 0$. We define the values $f_i(x_\alpha)$ as follows. For each $k = 0, 1, 2, \ldots$ we define a natural number m_k^α and together with it the function value $f_{m_k^\alpha}(x_\alpha)$: let $m_0^\alpha = 0$, $f_0(x_\alpha) = 0$, and suppose that $m_0^\alpha, m_1^\alpha, \ldots, m_{k-1}^\alpha$ are already defined. Let m_k^α be any element in the sequence $\{n_j^{\xi_k^\alpha}\}_{j=0}^\infty$ different from every m_i^α, $i < k$, say $m_k^\alpha = n_j^{\xi_k^\alpha}$, in which case we define $f_{m_k^\alpha}(x_\alpha) = y_j^{\xi_k^\alpha}$ (note that by the choice of m_k^α the

value of $f_{m_k^\alpha}(x_\alpha)$ has not been defined before). If m is not of the form m_k, $k = 0, 1, \ldots$, then let $f_m(x_\alpha)$ be arbitrary.

We claim that this system of functions satisfies the requirements. Suppose to the contrary that $X \subset \mathbf{R}$ is an uncountable set such that $f_{n_k}[X] \neq \mathbf{R}$ for $k = 0, 1, \ldots$, and for each j let $y_j \in \mathbf{R} \setminus f_{n_j}[X]$. The pair $(\{y_k\}_{k=0}^\infty, \{n_k\}_{k=0}^\infty)$ is listed above, say it is $\left(\{y_k^\beta\}_{k=0}^\infty, \{n_k^\beta\}_{k=0}^\infty\right)$. Let $x_\alpha \in X$ be a number with $\alpha > \beta$. Then β is one of the numbers ξ_k^α, say $\beta = \xi_{k_0}^\alpha$. Now $m_{k_0}^\alpha$ is one of the numbers $n_j^\beta = n_j$, say $m_{k_0}^\alpha = n_{j_0}$. But then $f_{n_{j_0}}(x_\alpha) = y_{j_0}^\beta = y_{j_0}$, which is impossible since $y_{j_0} \notin f_{n_{j_0}}[X]$. This contradiction proves the necessity direction in the problem.

Conversely, suppose that the f_n's with the stated property exist. If X is any subset of $\mathbf{R}$ of cardinality $\aleph_1$, then there is an n (actually all but finitely many n are such) with $f_n[X] = \mathbf{R}$, hence $\mathbf{c} = |\mathbf{R}| \leq |X| = \aleph_1$.

13. One direction is clear, for if CH holds then there are only ω_1 infinite subsets of ω, so we can list all of them in $\{A_\alpha : \alpha < \omega_1\}$. Conversely, suppose that there is a family $\{A_\alpha : \alpha < \omega_1\}$ of infinite subsets of ω such that if $X \subseteq \omega$ is infinite then there is some $\alpha_X < \omega_1$ with $A_{\alpha_X} \setminus X$ finite. If X and Y are infinite subsets such that $X \cap Y$ is finite, then we must have $\alpha_X \neq \alpha_Y$, for X and Y contain all but finitely many points of A_{α_X} and A_{α_Y}, respectively. But there is a family $\mathcal{F}$ of cardinality $\mathbf{c}$ of almost disjoint subsets of ω (see Problem 4.29), and since then the mapping $X \to \alpha_X$, $X \in \mathcal{F}$ is an injection of $\mathcal{F}$ into ω_1, we must have $\mathbf{c} \leq \omega_1$. [F. Rothberger, Fund. Math., **35**(1948), 29–46]

14. If CH holds and x_α, $\alpha < \omega_1$ is an enumeration of the reals, then $A_\alpha = \{x_\beta : \beta < \alpha\}$, $\alpha < \omega_1 = \mathbf{c}$ is clearly appropriate. Conversely, if $\mathbf{c} > \omega_1$ and $J \subset I$ is a subset of cardinality $\aleph_1$ of the index set I, then $\cup_{i \in J} A_i$ is of cardinality at most $\aleph_1$, hence $B = \mathbf{R} \setminus (\cup_{i \in J} A_i)$ is infinite but it does not intersect any of the A_i, $i \in J$, and the number of these latter sets is not countable.

15. Assume CH. Let $\{b_\alpha : \alpha < \omega_1\}$ be a Hamel basis (see Problem 15.3). If $x \neq 0$, it can be written as $x = \lambda_1 b_{\alpha_1} + \cdots + \lambda_n b_{\alpha_n}$ where the coefficients $\lambda_1, \ldots, \lambda_n$ are nonzero rational numbers and $\alpha_1 < \cdots < \alpha_n$. Denote by $\mu(x)$ the largest index α_n. Put $x \in A$ if and only if $\mu(x)$ is an even ordinal (i.e., of the form $\alpha + 2k$ where $k < \omega$ and α is a limit ordinal). To show the property required one has to notice that if $a \in \mathbf{R}$ is given, then $\mu(x + a) = \mu(x)$ holds if $\mu(x) > \mu(a)$, which in turn holds for all but countably many $x \in \mathbf{R}$. Thus, if $x \in A$ ($x \in B$), then $x + a \in A$ ($x + a \in B$) for all but countably many x. It is clear that both A and B are of cardinality continuum, so these sets satisfy the requirements.

For the other direction assume that $c \geq \aleph_2$ and $\mathbf{R} = A \cup B$ is a decomposition as claimed. Select $A' \subseteq A$ and $B' \subseteq B$ of cardinality $\aleph_1$. Let $\{r_\alpha : \alpha < \omega_2\}$

be distinct reals. By the assumption on the sets A and B for every $\alpha < \omega_2$ there are $a_\alpha \in A'$ and $b_\alpha \in B'$ such that $a_\alpha + r_\alpha \in A$, $b_\alpha + r_\alpha \in B$. There are $a \in A'$, $b \in B'$ such that for $\aleph_2$ many α we have $a_\alpha = a$, $b_\alpha = b$. Then, for these α, $b + r_\alpha \in (A + (b - a)) \cap B$ so the latter set is uncountable. But this contradicts the hypotheses on A and B, and this contradiction proves the claim. [St. Banach: Sur les transformations biunivoques, *Fund. Math.* **19**(1932), 10–16. L. Trzeciakiewicz: Remarque sur les translations des ensembles linéaires, *Comptes Rendus de la Société des Sciences et des Lettres de Varsovie Cl. III.*, **25**(1932), 63–65]

16. Assume CH, and let x_α, $\alpha < \omega_1$ be an enumeration of the reals. For a set $A \subseteq \mathbf{R}^2$ let $D(A)$ denote the set of distances between points of A. Let us call a set $C \subset \mathbf{R}^2$ "closed" if the following is true: if y is a point such that there are two points $u, v \in A$ with $\mathrm{dist}(y, u)$, $\mathrm{dist}(y, v) \in D(A)$, then $y \in A$. First of all let us remark that for any countable set B there is a "closed" countable set B^* including B (the smallest of which may be called the "closed" hull of B). In fact, starting from $B_0 = B$, for each $k = 0, 1, 2, \ldots$ let B_{k+1} be obtained by adding to B_k all points y for which there are two points $u, v \in B_k$ with $\mathrm{dist}(y, u)$, $\mathrm{dist}(y, v) \in D(B_k)$. Clearly each B_k is countable, and it is easy to see that $B^* = \cup_{k=0}^{\infty} B_k$ is the smallest "closed" set including B, and clearly B^* is countable.

Now we define an increasing sequence of "closed" and countable subsets C_α, $\alpha < \omega_1$ of $\mathbf{R}^2$: let $C_0 = \emptyset$, for limit ordinal α let $C_\alpha = \cup_{\beta<\alpha} C_\beta$ and otherwise for $\alpha = \beta + 1$ let C_α be the "closed" hull generated by C_β and the point x_α. Induction shows that each C_α is countable. Using these "closed" sets we can define the decomposition $\mathbf{R}^2 = A_0 \cup A_1 \cup \cdots$ by defining a decomposition $C_\alpha = A_0^\alpha \cup A_1^\alpha \cup \cdots$ in such a way that each A_i^α, $\alpha < \omega_1$ is increasing in α, and neither of these sets contains 4 distinct points a, b, c, and d such that $\mathrm{dist}(a, b) = \mathrm{dist}(c, d)$. Clearly, if we can do that, then $A_i = \cup_{\alpha<\omega_1} A_i^\alpha$ will be an appropriate decomposition of $\mathbf{R}^2$.

Suppose that A_i^β have already been defined for all $i = 0, 1, \ldots$ and all $\beta < \alpha$ with the property above. If α is a limit ordinal, then set $A_i^\alpha = \cup_{\beta<\alpha} A_i^\beta$. Since $C_a = \cup_{\beta<\alpha} C_\beta$, these give an appropriate decomposition of C_α. Now consider $\alpha = \beta + 1$. The set $C_\alpha \setminus C_\beta$ is countable. Furthermore, for each $y \in C_\alpha \setminus C_\beta$ there can be only one $j = j_y$ such that in A_j^β there is a point u such that $\mathrm{dist}(y, u) = d$ for some $d \in D(A_j^\beta) \subseteq D(C_\beta)$ (a second $j^* \neq j$ and $v \in A_{j^*}^\beta$ would imply $y \in C_\beta$ since C_β is "closed"). So y cannot be put to the set $A_{j_y}^\beta$, but it can be put to any other set A_i^β since $A_i^\beta \cup \{y\}$ will not have 4 distinct points a, b, c, and d such that $\mathrm{dist}(a, b) = \mathrm{dist}(c, d)$. Thus, we can put the points $y \in C_\alpha \setminus C_\beta$ into different sets $A_{k_y}^\beta$ with $k_y \neq j_y$, $k_y \neq k_z$ if $y, z \in A_\alpha \setminus A_\beta$, $y \neq z$, and setting $A_{k_y}^\alpha = A_{k_y}^\beta \cup \{y\}$ completes the definition of the sets A_i^α.

To prove the other direction let us assume that $\mathbf{c} \geq \aleph_2$ and let $\mathbf{R}^2 = \cup_{n=0}^{\infty} A_n$ be a decomposition of $\mathbf{R}^2$ into countably many classes. Consider the complete bipartite graph G with vertex sets $\{x : x \in (0,1)\}$ and $\{y : y \in (1,2)\}$, and let us color an edge (x, y) by the color i if the point (x, y) belongs to A_i. By Problem 24.27 there are $x_1, x_2 \in (0,1)$ and $y_1, y_2 \in (1,2)$ and an i such that all the edges (x_j, y_k), $j, k = 1, 2$, are of color i. But the points $a = (x_1, y_1)$, $b = (x_1, y_2)$, $c = (x_2, y_1)$, and $d = (x_2, y_2)$ form a rectangle and all belong to A_i. This shows that if CH is not true then there is no partition of $\mathbf{R}^2$ with the properties in the problem.

17. Assume CH. Then $\mathbf{R}$ is the union of an increasing family of countable sets A_α, $\alpha < \omega_1$ (Problem 16). By enlarging each A_α if necessary, we may assume that each A_α is closed for addition and subtraction. Furthermore, we may assume that if α is a limit ordinal, then $A_\alpha = \cup_{\beta<\alpha} A_\beta$ (if this is not the case, then rename each A_α as $A_{\alpha+1}$, and for limit α set $A_\alpha = \cup_{\beta<\alpha} A_\beta$). We define by transfinite recursion a coloring $f_\alpha : A_\alpha \to \omega$ in such a way that f_α extends f_β if $\beta < \alpha$, and there is no monochromatic solution of $x + y = u + v$ in A_α. For $\alpha = 0$ color the elements of A_0 with different colors. If α is a limit ordinal, just take $f_\alpha = \cup_{\beta<\alpha} f_\beta$. Finally, if $\alpha = \beta + 1$, then under f_α color the elements of $A_\alpha \setminus A_\beta$ by different colors arbitrarily (and on A_β keep the coloring f_β). This satisfies the requirement, for if $x + y = u + v$ with different $x, y, u, v \in A_\alpha$, then three of these numbers cannot belong to A_β for then the fourth would also belong to A_β. Hence at least two of them belong to $A_\alpha \setminus A_\beta$ and then these get different colors. Now $\cup_{\alpha<\omega_1} f_\alpha$ is a coloring of $\mathbf{R}$ without monochromatic solutions to the equation $x + y = u + v$.

If CH is not true, then under any coloring of $\mathbf{R}$ there is a monochromatic solution by Problem 24.37.

18. Call a subset $H \subseteq \mathbf{R}$ "closed" if $x, y \in H$ implies $(x + y)/2 \in H$ and $2y - x \in H$, i.e., if two of the points $x, x \pm \delta$ are in H, then the third one is also in H. It is clear that any countable set is included in a countable "closed" set; therefore, if we assume CH then, starting from $H_0 = \{0\}$, we can represent $\mathbf{R}$ as a strictly increasing union of countable "closed" sets: $\mathbf{R} = \cup_{\alpha<\omega_1} H_\alpha$. We may also assume that for limit α we have $H_\alpha = \cup_{\beta<\alpha} H_\beta$ (otherwise redefine H_α to this union). Let $f_0(0) = 1$, and by induction we define functions f_α on H_α in such a way that for $\beta < \alpha$ the function f_α is an extension of f_β, and for each α and $x \in H_\alpha$

$$\lim_{h_n \to 0,\ x \pm h_n \in H_\alpha} \max\left(f_\alpha(x - h_n), f_\alpha(x + h_n)\right) = \infty. \tag{16.1}$$

First let $\alpha = \beta + 1$, and let us assume that f_β with this property has already been defined. Let us enumerate H_α as $x_0, x_1, \ldots$, where the x_{2i}'s are the elements of H_β and the x_{2i+1}'s are the elements of $H_\alpha \setminus H_\beta$ (for which we have to define the value $f_\alpha(x_{2i+1})$, since $f_\alpha(x_{2i}) = f_\beta(x_{2i})$ are given). Define

$$f_\alpha(x_{2i+1}) = \left(\min_{j<2i+1} |x_j - x_{2i+1}|\right)^{-1}$$

Note that if $x = x_m$, and either $x + h_n = x_s \in H_\alpha \setminus H_\beta$, $s > m$, or $x - h_n = x_s \in H_\alpha \setminus H_\beta$, $s > m$, then

$$\max\{f_\alpha(x + h_n), f_\alpha(x - h_n)\} \geq 1/2h_n. \tag{16.2}$$

Furthermore, this is also true (regardless if $s > m$ or not) provided both $x+h_n$ and $x - h_n$ belong to $H_\alpha \setminus H_\beta$ (consider the maximum of the indices s, l for which $x+h_n = x_s$ and $x-h_n = x_l$). Now let $x \in H_\alpha$ and let $h_n \to 0$ in such a way that $x \pm h_n \in H_\alpha$. By selecting a subsequence we may assume that either $x \pm h_n \in H_\beta$ for all n, or for all n one of the points $x + h_n$ and $x - h_n$ belongs to $H_\alpha \setminus H_\beta$. In the former case $x \in H_\beta$ (recall that H_β is "closed"), so by the induction hypothesis (16.1) is true. In the latter case (16.2) is true for all but finitely many n, hence (16.1) holds again.

Next let α be a limit ordinal. To verify (16.1) it is enough to show that from any $h'_m \to 0$ with $x \pm h'_m \in H_\alpha$ we can select a subsequence $\{h_n\}$ for which (16.1) is true. Now for each h'_m let β_m be the smallest index with $x \pm h'_m \in H_{\beta_m}$. Then $\beta_m < \alpha$, and there are two possibilities: $\sup_m \beta_m < \alpha$ or $\sup_m \beta_m = \alpha$. In the former case for $\beta = \sup_m \beta_m$ the point x as well as all the points $x \pm h'_m$ lie in H_β, hence (16.1) is true for the whole sequence $h_n = h'_n$ by the induction assumption. In the second case there is an increasing sequence $m_0 < m_1 < m_2 < \ldots$ such that $\beta_{m_{n+1}} > \beta_{m_n}$ and $\sup_n \beta_{m_n} = \alpha$, and then we set $h_n = h'_{m_n}$. Since in this case both $x+h_n$ and $x-h_n$ belong to $H_{\beta_{m_n}} \setminus \cup_{\gamma<\beta_{m_n}} H_\gamma$, the inequality (16.2) is true for all $n \geq 1$, and this proves (16.1).

This completes the definition of the functions f_α. Set $f(x) = f_\alpha(x)$ for some α for which $f_\alpha(x)$ is defined. The proof that this satisfies

$$\lim_{h_n\to 0,\ x\pm h_n\in H_\alpha} \max\left(f(x - h_n), f(x + h_n)\right) = \infty$$

is completely analogous to what we have just done.

19. Assume first that $\mathbf{c} > \aleph_1$ and $\mathcal{F}$ is an uncountable family of entire functions. Select $\mathcal{F}' \subseteq \mathcal{F}$ of cardinality $\aleph_1$. If f, g are distinct entire functions then the set $\{x \in \mathbf{C} : f(x) = g(x)\}$ is countable so there are at most $\aleph_1 < \mathbf{c}$ points in which two members of $\mathcal{F}'$ may agree. If a is outside this set, then $\{f(a) : f \in \mathcal{F}'\}$ is uncountable (since all the values $f(a)$, $f \in \mathcal{F}'$ are different).

For the other direction assume the continuum hypothesis and enumerate $\mathbf{C}$ as $\{c_\alpha : \alpha < \omega_1\}$. Let $\mathbf{Q}^* = \mathbf{Q} + \mathbf{Q}i$ be the set of complex numbers with rational real and imaginary parts. Our goal is to define the distinct entire functions $\{f_\alpha : \alpha < \omega_1\}$ such that for any $\beta < \omega_1$

$$\{f_\alpha(c_\beta) : \alpha < \omega_1\} \subseteq \mathbf{Q}^* \cup \{f_\gamma(c_\beta) : \gamma \leq \beta\}.$$

As this set is countable, we will be finished. Assume we have arrived at the αth step. Reorder $\{c_\beta : \beta < \alpha\}$ as $\{d_n : n < \omega\}$ and $\{f_\beta : \beta < \alpha\}$ as $\{g_n : n < \omega\}$. Our function f_α will have the form

$$f_\alpha(z) = \varepsilon_0(z - d_0) + \varepsilon_1(z - d_0)(z - d_1) + \cdots$$

for some numbers $\varepsilon_0, \varepsilon_1, \ldots$ selected inductively. If $\varepsilon_0, \ldots, \varepsilon_{n-1}$ are selected, we choose ε_n in such a way that for

$$f_\alpha(d_{n+1}) = \epsilon_0(d_{n+1} - d_0) + \cdots + \epsilon_n(d_{n+1} - d_0)(d_{n+1} - d_1) \cdots (d_{n+1} - d_n)$$

we have $g_n(d_{n+1}) \neq f_\alpha(d_{n+1})$ and $f_\alpha(d_{n+1}) \in \mathbf{Q}^*$, and besides these also let $|\varepsilon_n|$ small enough to ensure that the series for $f_\alpha(z)$ converges for all z. For example, if we have

$$|\varepsilon_n|(2n)^n\big(1 + |d_0|\big) \cdots \big(1 + |d_n|\big) < 1,$$

then we have convergence: if $n > |z|$, then

$$\begin{aligned} \big|\varepsilon_n(z - d_0) \cdots (z - d_n)\big| &\leq |\varepsilon_n|\big(n + |d_0|\big) \cdots \big(n + |d_n|\big) \\ &\leq |\varepsilon_n| n(1 + |d_0|) \cdots n(1 + |d_n|) \\ &= |\varepsilon_n| n^n \big(1 + |d_0|\big) \cdots \big(1 + |d_n|\big) < \frac{1}{2^n} \end{aligned}$$

so the series uniformly converges on every disc. [P. Erdős: An interpolation problem associated with the continuum hypothesis, *Michigan Math. Journ.* **11**(1964), 9–10]

20. (a) Every first category set is included in a first-category F_σ set. The number of the latter sets is $c = \aleph_1$. Let $\{A_\alpha : \alpha < \omega_1\}$ be a list of first-category F_σ sets. Notice that for every $\alpha < \omega_1$ the set $\bigcup\{A_\beta : \beta < \alpha\}$ is of first category. If we pick $x_\alpha \in \mathbf{R} \setminus (\bigcup\{A_\beta : \beta < \alpha\})$, then $A = \{x_\alpha : \alpha < \omega_1\}$ is a Lusin set for $A \cap A_\alpha$ is included in the countable set $\{x_\beta : \beta \leq \alpha\}$. [P. Mahlo: Über Teilmengen des Kontinuums von dessen Machtigkeit, *Sitzungberichte der Sachsischen Akademie der Wissenschaften zu Leipzig, Mathematisch-Naturwissenschaftliche Klasse*, **65**(1913), 283–315. N. Lusin: Sur un probléme de M. Baire, *Comptes Rendus Hebdomadaires Siences Acad. Sci. Paris*, **158**(1914), 1258–1261]

(b) There is a decomposition $\mathbf{R} = X \cup Y$ where X is of the first category and Y is of measure zero. Indeed, for every n we can cover the rational numbers by open intervals of total length $< 1/2^n$, and the intersection of all these covering sets is of measure 0, while its complement is of the first category (since it is the union of countably many nowhere dense sets).

Now if A is a Lusin set, then $A \cap X$ is countable, so all but countably many elements of A are in Y, so A is of measure zero.

21. One direction is clear by Problem 20 and by the fact that if CH holds then every set of cardinality $< \mathbf{c}$ is countable.

Conversely, suppose that A is a Lusin set and every subset of $\mathbf{R}$ of cardinality $< \mathbf{c}$ is of first category. Let us enumerate the reals into a sequence r_α, $\alpha < \mathbf{c}$, and consider the sets

$$A_\alpha = A \bigcap \{r_\beta \ : \ \beta < \alpha\}, \qquad \alpha < \mathbf{c}.$$

By the assumption the set in the bracket is of first category, and hence by the Lusin property of A each set A_α is countable. But $\cup_{\alpha<\mathbf{c}} A_\alpha = A$, hence we have a representation of a set of power continuum as the union of an increasing chain of countable sets. Now apply Problem 7.

22. Every set of measure zero is included in a G_δ set of measure zero. The number of the latter sets is $\mathbf{c} = \aleph_1$. Let $\{A_\alpha : \alpha < \omega_1\}$ be a list of the G_δ sets of measure zero. Notice that for every $\alpha < \omega_1$ the set $\bigcup\{A_\beta : \beta < \alpha\}$ is of measure zero. If we pick $x_\alpha \in \mathbf{R} \setminus (\bigcup\{A_\beta : \beta < \alpha\})$, then $A = \{x_\alpha : \alpha < \omega_1\}$ is a Sierpiński set for $A \cap A_\alpha$ is included in the countable $\{x_\beta : \beta \leq \alpha\}$. [W. Sierpiński: Sur l'hypothèse du continu ($2^{\aleph_0} = \aleph_1$), *Fund. Math.*, **5**(1924), 177–187]

(b) There is a decomposition $\mathbf{R} = X \cup Y$ where X is of first categoiy and Y is of measure zero (see the solution to 20(b)). If A is a Sierpiński set, then $A \cap Y$ is countable, so all but countably many elements of A are in X, so A is of first category.

23. The proof is similar to that of Problem 21. One direction is clear by Problem 22 and by the fact that if CH holds, then every set of cardinality $< \mathbf{c}$ is countable.

Conversely, suppose that A is a Sierpiński set and every subset of $\mathbf{R}$ of cardinality $< \mathbf{c}$ is of zero measure. Let us enumerate the reals into a sequence r_α, $\alpha < \mathbf{c}$, and consider the sets

$$A_\alpha = A \bigcap \{r_\beta \ : \ \beta < \alpha\}, \qquad \alpha < \mathbf{c}.$$

By the assumption the set in the bracket is of measure zero, and hence by the Sierpiński property of A, each set A_α is countable. But $\cup_{\alpha<\mathbf{c}} A_\alpha = A$, hence we have a representation of a set of power continuum as the union of an increasing chain of countable sets. Again apply Problem 7.

24. A set $B \subseteq [0,1]$ is of outer measure 1 if it intersects every compact set $K \subseteq [0,1]$ of positive measure. Let us assume CH and let K_α, $\alpha < \omega_1$, be an enumeration of the compact subsets of $[0,1]$ of positive measure. We define by induction the increasing sequence of sets B_α, C_α, $\alpha < \omega_1$, in such a way that for all α both B_α and C_α are countable, $B_\alpha \cap K_\beta \neq \emptyset$, $C_\alpha \cap K_\beta \neq \emptyset$ for $\beta < \alpha$ and $B_\alpha \times C_\alpha \subset A$. Then clearly $B = \cup_{\alpha<\omega_1} B_\alpha$ and $C = \cup_{\alpha<\omega_1} C_\alpha$ are suitable. In order that the induction run through we also require that for any $b \in B_\alpha$ the set $\{y \ : \ (b, y) \in A\}$ is of linear measure 1 and for any $c \in C_\alpha$ the set $\{x \ : \ (x, c) \in A\}$ is of linear measure 1.

For limit ordinal $\alpha < \omega_1$ just set $B_\alpha = \cup_{\beta<\alpha} B_\beta$, $C_\alpha = \cup_{\beta<\alpha} C_\beta$. Now let B_α and C_α be defined and we define the next sets $B_{\alpha+1}$ and $C_{\alpha+1}$ by adding one–one points to B_α and C_α. By the hypothesis for each $b \in B_\alpha$

the set $\{y : (b,y) \in A\}$ is of linear measure 1, therefore the same is true of $\bigcap_{b\in B_\alpha}\{y : (b,y) \in A\}$. Thus, this set intersects K_α in a set of positive measure, and for almost all points c of the intersection $K_\alpha \cap (\bigcap_{b\in B_\alpha}\{y : (b,y) \in A\})$ the set $\{x : (x,c) \in A\}$ is of linear measure 1. Pick such a $c = c_\alpha$ and let $C_{\alpha+1} = C_\alpha \cup \{c_\alpha\}$. By the choice of c_α we have $B_\alpha \times \{c_\alpha\} \subset A$, hence $B_\alpha \times C_{\alpha+1} \subset A$. Select in an analogous way a point b_α in $K_\alpha \cap (\bigcap_{c\in C_{\alpha+1}}\{x : (x,c) \in A\})$ and let $B_{\alpha+1} = B_\alpha \cup \{b_\alpha\}$. The construction gives that these sets satisfy all the requirements.

25. Enumerate the rationals as $\mathbf{Q} = \{q_i : i < \omega\}$. By CH we can enumerate the sequences of positive reals as $\{(\varepsilon_i^\alpha : i < \omega) : \alpha < \omega_1\}$. The set

$$G_\alpha = \bigcup_{i<\omega}(q_i - \varepsilon_i^\alpha, q_i + \varepsilon_i^\alpha)$$

is dense and open. By the Baire category theorem the set $X_\alpha = \bigcap\{G_\beta : \beta < \alpha\}$ is a dense G_δ set of cardinality continuum. We can therefore inductively select $a_\alpha \in X_\alpha$ different from every a_β, $\beta < \alpha$. Now the set $A = \{a_\alpha : \alpha < \omega_1\}$ is as required. [A. S. Besicovitch: Concentrated and rarified sets, *Annals of Mathematics*, **62**(1934), 289–300]

26. A Lusin set (Problem 20) A has this property: if $B \subset A$ is not dense in any interval, then it is nowhere dense in $\mathbf{R}$, hence $B = A \cap B$ must be countable.

27. If L is a Lusin set (Problem 20) then $A = L \cup \mathbf{Q}$ has this property: if $B \subset A$ is nowhere dense in the interval topology, then it is nowhere dense in $\mathbf{R}$, hence $B = A \cap B$ must be countable.

28. Let A be a set as constructed in Problem 20. Assume that we are given the positive reals $\varepsilon_0, \varepsilon_1, \ldots$. There are intervals $I_0, I_2, I_4, \ldots$ (one around each rational point) of length $\varepsilon_0, \varepsilon_2, \varepsilon_4, \ldots$ such that $A \setminus (I_0 \cup I_2 \cup I_4 \cup \cdots)$ is countable . This countable set can be covered by some intervals of the respective lengths $\varepsilon_1, \varepsilon_3, \ldots$.

29. Assume CH, and enumerate the first-category F_σ sets in $\mathbf{R}$ into a sequence I_α, $\alpha < \omega_1$, and the G_δ sets of measure zero into a sequence O_α, $\alpha < \omega_1$. We may also assume that $I_0 = O_0 = \emptyset$ and that these sequences of sets are increasing. It is easy to verify that the complement of a first-category set includes a first-category set of cardinality continuum, and likewise the complement of a set of measure zero includes a set of measure zero and of cardinality continuum. Thus, for every $\alpha > 0$ there is an index γ_α such that both sets $I_{\gamma_\alpha} \setminus I_\alpha$ and $O_{\gamma_\alpha} \setminus O_\alpha$ are of cardinality continuum. Define now the sequence τ_α, $\alpha < \omega_1$, as $\tau_0 = 0$, $\tau_\alpha = \sup_{\beta<\alpha} \tau_\alpha$ if α is a limit ordinal, and $\tau_{\alpha+1} = \gamma_{\tau_\alpha}$ otherwise. Then

$$\bigcup_{\alpha<\omega_1} (I_{\tau_{\alpha+1}} \setminus I_{\tau_\alpha}) = \bigcup_{\alpha<\omega_1} I_{\tau_{\alpha+1}} = \bigcup_{\alpha<\omega_1} I_\alpha = \mathbf{R}$$

and

$$\bigcup_{\alpha<\omega_1} (O_{\tau_{\alpha+1}} \setminus O_{\tau_\alpha}) = \bigcup_{\alpha<\omega_1} O_{\tau_{\alpha+1}} = \bigcup_{\alpha<\omega_1} O_\alpha = \mathbf{R}$$

are decompositions of $\mathbf{R}$ into disjoint subsets of power continuum. Thus, any one-to-one correspondences between the sets $I_{\tau_{\alpha+1}} \setminus I_{\tau_\alpha}$ and $O_{\tau_{\alpha+1}} \setminus O_{\tau_\alpha}$ induce a permutation π of $\mathbf{R}$. If A is of first category, then $A \subset I_\alpha$ for some α, hence, as $\pi[A] \subset O_\alpha$, the set $\pi[A]$ is of measure zero. Similarly, it follows that if B is of measure zero, then $\pi^{-1}[B]$ is of first category.

17

Ultrafilters on ω

1. Let $\mathcal{F}$ be a maximal filter and $A \subseteq \omega$, $A \notin \mathcal{F}$. If the intersection of A with any member of $\mathcal{F}$ is nonempty, then $\mathcal{F} \cup \{A\}$ generates a filter that is larger than $\mathcal{F}$, but this is not possible. Thus, there is an $F \in \mathcal{F}$ with $A \cap F$, but then $F \subseteq \omega \setminus A$, hence $\omega \setminus A \in \mathcal{F}$.

Conversely, if for every $A \subseteq \omega$ either $A \in \mathcal{F}$ or $\omega \setminus A \in \mathcal{F}$, then for every $A \notin \mathcal{F}$ there is an $F \in \mathcal{F}$, namely $F = \omega \setminus A$, with $A \cap F = \emptyset$. Thus, there cannot be a filter that would include $\mathcal{F}$ as its proper subset, hence $\mathcal{F}$ is an ultrafilter.

2. See Problem 14.6(c).

3. By Problem 4.43 there is an independent family $\mathcal{F}$ of cardinality continuum of subsets of ω i.e., $\mathcal{F}$ is such that if $F_1, \ldots, F_n \in \mathcal{F}$ are different elements of $\mathcal{F}$ and $F_i^* = F_i$ or $\omega \setminus F_i$ independently of each other, then $\cap_{i=1}^n F_i^* \neq \emptyset$. This means that if $g : \mathcal{F} \to \{0, 1\}$ is an arbitrary mapping and $\mathcal{F}_g$ is the family that contains $F \in \mathcal{F}$ if $g(F) = 1$ and contains $\omega \setminus F$ if $g(F) = 0$, then $\mathcal{F}_g$ has the property that any finite subset of $\mathcal{F}_g$ has nonempty intersection. But then $\mathcal{F}_g$ generates a filter which is included in a $\mathcal{U}_g$ ultrafilter, and it is clear that if $g \neq h$ then $\mathcal{U}_g \neq \mathcal{U}_h$ because there is an $F \in \mathcal{F}$ with $g(F) \neq h(F)$, and then F is contained in one of $\mathcal{U}_g$ and $\mathcal{U}_h$ and $\omega \setminus F$ is contained in the other one. Since there are $2^{\mathfrak{c}}$ possibilities for g, this shows that there are at least $2^{\mathfrak{c}}$ ultrafilters on ω. But the total number of systems of subsets of ω is $2^{\mathfrak{c}}$, so there cannot be more than $2^{\mathfrak{c}}$ ultrafilters either.

4. Partition ω into $n+1$ infinite sets: $\omega = A_0 \cup \cdots \cup A_n$. For each $1 \leq i \leq n$ there is some $0 \leq k_i \leq n$ that $A_{k_i} \in \mathcal{U}_i$. The union of these sets is in every $\mathcal{U}_i$, and is coinfinite, as is disjoint from the A_j for which $j \notin \{0, 1, \ldots, n\} \setminus \{k_1, \ldots, k_n\}$.

5. Note that of the two sets $A = \cup_i [n_{2i}, n_{2i+1})$ and $B = \cup_i [n_{2i+1}, n_{2i+2})$ exactly one of them is in $\mathcal{U}$. If, say, $A \in \mathcal{U}$, then is appropriate for $A \cap [n_{2i+1}, n_{2i+2}) = \emptyset$ for all i.

6. This follows from the preceding problem if $n_{i+1}/n_i \to \infty$ as $i \to \infty$.

7. The set $\mathcal{F}$ of all subsets of ω of density 1 is a filter. Let $\mathcal{U}$ be an ultrafilter including $\mathcal{F}$. Then no $A \in \mathcal{U}$ can have zero density, for then $\omega \setminus A$ would be of density 1, hence it would also belong to $\mathcal{U}$.

8. There is no translation-invariant ultrafilter on ω as exactly one of the sets of the odd, resp. even numbers is in any ultrafilter.

Assume now that $\mathcal{U}$ is a translation-invariant ultrafilter on $\mathbf{Q}$. Then exactly one of $\mathbf{Q} \cap (-\infty, 0)$, $\mathbf{Q} \cap [0, \infty)$ is in $\mathcal{U}$, say the latter. Now exactly one of

$$\mathbf{Q} \cap ([0,1) \cup [2,3) \cup \cdots)$$

and

$$\mathbf{Q} \cap ([1,2) \cup [3,4) \cup \cdots)$$

is in $\mathcal{U}$ and that contradicts translation invariance. So there is no translation-invariant ultrafilter on $\mathbf{Q}$.

9. Let us suppose that the second player II has a winning strategy σ. Let player I select first $n_0 = 1$, for which player II responds with some number n_1. Now from this point on if until the kth step the game proceeds as $n_0 < n_1 < n_2 < \cdots < n_{2k-1}$, then let player I respond with the number n_{2k}, which would be the second player's response (under σ) for the play $n_1 < n_2 < \ldots < n_{2k-1}$ (in other words, I plays the strategy σ as if n_0 has not been played). Since σ is a winning strategy for player II, and player I is playing the σ strategy, the set $[0, n_1) \cup [n_2, n_3) \cup \cdots$ does not belong to $\mathcal{U}$, hence the set $[0, n_0) \cup [n_1, n_2) \cup \cdots$ must belong to $\mathcal{U}$. Thus, with this strategy of player I he/she wins, so σ cannot be a winning strategy for player II.

The same consideration shows that player I cannot have a winning strategy.

10. By recursion on $\alpha < \omega_1$ we build the increasing, continuous sequence $\{\mathcal{G}_\alpha : \alpha < \omega_1\}$ of countable centered subfamilies of $\mathcal{P}(\omega)$ (i.e., each $\mathcal{G}$ has the property that any finite subset of $\mathcal{G}$ has infinite intersection, and the continuity means that if α is a limit ordinal then $\mathcal{G}_\alpha = \cup_{\beta<\alpha}\mathcal{G}_\beta$). At every step we perform one of two possibilities. Either regard some $A \subseteq \omega$ and add A or $\omega - A$ to $\mathcal{G}_\alpha$ to get $\mathcal{G}_{\alpha+1}$ or make sure that for a given sequence $A_0 \supseteq A_1 \supseteq A_2 \supseteq \cdots$ from $\mathcal{G}_\alpha$ there is a $B \in \mathcal{G}_{\alpha+1}$ with $B \setminus A_n$ finite for all $n < \omega$. This will clearly work, as by CH there are $\aleph_1$ many such "tasks" so it suffices to treat one at a time.

There is no problem with the first type, given $A \subseteq \omega$ and the centered $\mathcal{G}_\alpha$, either A or its complement can be added to $\mathcal{G}_\alpha$ and still keep it centered. Assume, therefore, that we are given $\mathcal{G}_\alpha$ and the decreasing sequence $A_0 \supseteq A_1 \supseteq A_2 \supseteq \cdots$ from $\mathcal{G}_\alpha$. Enumerate $\mathcal{G}_\alpha$ as $\mathcal{G}_\alpha = \{C_0, C_1, \ldots\}$. Pick

$$a_n \in A_n \cap (C_0 \cap C_1 \cap \cdots \cap C_n)$$

and set $B = \{a_0, a_1, \ldots\}$. Now $\mathcal{G}_{\alpha+1} = \mathcal{G}_\alpha \cup \{B\}$ will be fine.

Finally, $\bigcup\{\mathcal{G}_\alpha : \alpha < \omega_1\}$ is clearly appropriate.

11. Enumerate the triplets $\langle r, n, f\rangle$ where $1 \le r < \omega$, $f : [\omega]^r \to \{1, \ldots, n\}$ as $\{\langle r_\alpha, n_\alpha, f_\alpha\rangle : \alpha < \omega_1\}$ (this is possible since the number of such triplets is $\mathbf{c}$ and we have assumed CH, i.e., $\mathbf{c} = \omega_1$). We construct by transfinite recursion infinite sets $A_\alpha \subset \omega$, $\alpha < \omega_1$ such that each A_α is monochromatic with respect to f_α and $\{A_\alpha : \alpha < \omega_1\}$ is centered, i.e., any finite subset of it has infinite intersection. In fact, we choose A_α so that $A_\alpha \setminus A_\beta$ is finite for $\beta < \alpha$, which clearly implies that $\{A_\alpha : \alpha < \omega_1\}$ is centered. Assuming that at step $\alpha < \omega_1$ we have the set $\{A_\beta : \beta < \alpha\}$ with this property. Select an infinite set B such that $B \setminus A_\beta$ is finite for all $\beta < \alpha$ (enumerate the sets $\{A_\beta : \beta < \alpha\}$ into a sequence $A_0^*, A_1^*, \ldots$, and select one–one different points from the sets $A_0^* \cap A_1^* \cap \cdots \cap A_n^*$, $n = 0, 1, \ldots$). By Ramsey's theorem (Problem 24.1) there is an infinite $B' \subset B$ on which f_α is monochromatic, and let $A_\alpha = B'$.

Since $\{A_\alpha : \alpha < \infty\}$ is centered, it can be extended to an ultrafilter $\mathcal{U}$. Now if $f : [\omega]^r \to \{1, 2, \ldots, n\}$ is a coloring of all r-element subsets of ω with finitely many colors, then $\langle r, n, f\rangle = \langle r_\alpha, n_\alpha, f_\alpha\rangle$ for some $\alpha < \omega_1$, and then $A_\alpha \in \mathcal{U}$ is monochromatic with respect to f.

12. Let $f : A \to \omega$ be a bijection. This induces an ordering $\prec_f$ on A: $x \prec_f y$ if and only if $f(x) < f(y)$. Color pairs of A by 2 colors as follows. For $x \prec y$ in A let $g(x, y) = 0$ if and only if $x \prec_f y$, otherwise set $g(x, y) = 1$. As $\mathcal{U}$ is a Ramsey ultrafilter, there is a monochromatic $B \in \mathcal{U}$ with respect to g. If the pairs in B have color 0, then on B the two orderings $\prec$ and $\prec_f$ coincide, hence $\langle B, \prec\rangle$ is of type ω (note that with respect to $\prec_f$ the type of A is ω). If, however, the pairs in B have color 1, then on B the two orderings $\prec$ and $\prec_f$ are each other's reverses, so in this case $\langle B, \prec\rangle$ has type ω^*.

13. Color $[\omega]^2$ as follows: let $g(x, y) = 0$ if $f(x) = f(y)$, and otherwise set $g(x, y) = 1$. Let $A \in \mathcal{U}$ be a monochromatic subset with respect to g. If the pairs in A have color 0, then f is constant on A, and if they have color 1, then f is one-to-one on A.

14. Color $[\omega]^2$ by two colors: let $g(x, y) = 0$ if x and y belong to the same interval $[n_i, n_{i+1})$, $i = 1, 2, \ldots$, and otherwise let $g(x, y) = 1$. A monochromatic infinite set $B \subset \omega$ can only be of color 1, in which case it intersects every interval of the above type in at most one element. Add to B elements so that every intersection has exactly one element.

15. Apply the previous problem by making $n_0 = 0$ and with n_i so large that $a_j \le \epsilon/2^{i+1}$ is true for $j \ge n_i$. If $B \in \mathcal{U}$ is such that it intersects every interval $[n_i, n_{i+1})$ in exactly one element, then for $A = B \setminus [0, n_0) \in \mathcal{U}$ we clearly have $\sum_{i \in A} a_i < \sum_i \epsilon/2^{i+1} = \epsilon$.

16. **First solution**. Instead of ω, work with $S = \bigcup\{S_n : n < \omega\}$ as the underlying set, where the S_n's are disjoint, finite sets, $|S_n| = n^2$. For $X \subseteq S$,

set $X \in \mathcal{F}$ if and only if $|X \cap S_n| \geq n^2 - cn$ for some constant $c > 0$ and all n. Then $\mathcal{F}$ is a filter. Extend it to an ultrafilter $\mathcal{U}$. Now define $a_i = 1/n$ for $i \in S_n$. Then, if $X \in \mathcal{U}$, $|X \cap S_n| \geq n$ holds for infinitely many n (otherwise the complement of X would belong to $\mathcal{F}$), so

$$\sum_{i \in X} a_i = \infty.$$

[I. Juhász]

Second solution. Let $a_i = 1/(i+1)$ for $i = 0, 1, \ldots$, and consider the family $\mathcal{I}$ of those subsets $H \subset \omega$ for which

$$\sum_{i \in H} a_i < \infty.$$

This $\mathcal{I}$ is clearly an ideal, so the family

$$\mathcal{F} = \{K \subset \omega \, : \, \omega \setminus K \in \mathcal{I}\}$$

is a filter, which can be extended to an ultrafilter $\mathcal{U}$. Since no $H \in \mathcal{I}$ can belong to $\mathcal{F}$, we are done.

17. Let $\mathcal{H} = \{H_\alpha \, : \, \alpha < \mathbf{c}\}$ be an independent family of subsets of ω (cf. the solution to Problem 3 above). For $S \in [\mathcal{H}]^\omega$ set

$$X(S) = \omega \setminus (\cap S) = \omega \setminus (\cap_{H \in S} H).$$

We claim that the set

$$\mathcal{F} = \mathcal{H} \bigcup \{X(S) \, : \, S \in [\mathcal{H}]^\omega\}$$

is centered, i.e. if $\alpha_1, \ldots, \alpha_n < \mathbf{c}$ and $S_1, \ldots, S_m \in [\mathcal{H}]^\omega$, then

$$Z = H_{\alpha_1} \cap \cdots \cap H_{\alpha_n} \cap X(S_1) \cap \cdots \cap X(S_m)$$

is infinite. Indeed, if $H_{\beta_i} \in S_i \setminus \{H_{\alpha_1}, \ldots, H_{\alpha_n}\}$ for $i = 1, \ldots, m$, then every element of the infinite set

$$H_{\alpha_1} \cap \cdots \cap H_{\alpha_n} \cap (\omega \setminus H_{\beta_1}) \cap \cdots \cap (\omega \setminus H_{\beta_m})$$

is in Z. Extend $\mathcal{F}$ to an ultrafilter $\mathcal{U}$. If $\mathcal{U}$ was generated by $\mathcal{U}' \subseteq \mathcal{U}$, $|\mathcal{U}'| < \mathbf{c}$, then $\mathcal{U}'$ would generate every member of $\mathcal{H}$ as well. So, as $|\mathcal{U}'| < |\mathcal{H}|$, there would be $\alpha_1, \alpha_2, \ldots$ such that $H_{\alpha_1}, H_{\alpha_2}, \ldots$ are generated by the same element T of $\mathcal{U}'$, i.e., $T \subset H_{\alpha_1} \cap H_{\alpha_2} \cap \cdots$. But this is impossible as

$$T \cap X(\{H_{\alpha_1}, H_{\alpha_2}, \ldots\}) = \emptyset,$$

though these are two elements of $\mathcal{U}$.

18. Suppose to the contrary that $X_{\mathcal{U}}$ is Lebesgue measurable. If $x \in (0,1)$ is not diadically rational, then it has a unique binary expansion, thus if $x = x_A$ then $1 - x = x_{\omega\setminus A}$ and these are the only representations for the numbers x and $1-x$ as an x_B, $B \subset \omega$. Since exactly one of A and $\omega \setminus A$ belongs to $\mathcal{U}$, we get that exactly one of x and $1-x$ belongs to $X_{\mathcal{U}}$. Thus, the mapping $x \to 1 - x$ maps $X_{\mathcal{U}}$ into $[0,1] \setminus X_{\mathcal{U}}$ with the exception of countably many points, so $X_{\mathcal{U}}$ must have measure $1/2$.

Note also that the nonprincipality of $\mathcal{U}$ implies that adding to or deleting from A finitely many elements does not change the fact if $A \in \mathcal{U}$ or not. This means that $X_{\mathcal{U}}$ is periodic (mod 1) with period a for any diadically rational a. Now let $x \in X_{\mathcal{U}} \cap (0,1)$ and $y \in (0,1) \setminus X_{\mathcal{U}}$ be two points of density 1 for the sets $X_{\mathcal{U}}$ and $(0,1) \setminus X_{\mathcal{U}}$, respectively, and let $\delta > 0$ be so small that

$$|(x-\delta, x+\delta) \cap X_{\mathcal{U}}| > 3\delta/2 \qquad \text{and} \qquad |(y-\delta, y+\delta) \cap ((0,1) \setminus X_{\mathcal{U}})| > 3\delta/2.$$

If a is a diadically rational number such that $|y - (x+a)| < \delta/8$, then

$$|(X_{\mathcal{U}} + a) \cap (y-\delta, y+\delta)| \geq 3\delta/2 - 2\delta/8 > \delta,$$

hence

$$(X_{\mathcal{U}} + a) \cap ([0,1] \setminus X_{\mathcal{U}}) \neq \emptyset,$$

which is impossible, since $X_{\mathcal{U}} + a = X_{\mathcal{U}}$ (mod 1) and $X_{\mathcal{U}}$ and $[0,1] \setminus X_{\mathcal{U}}$ are disjoint.

19. (a) Let K be so large that $-K < x_n < K$ is true for every n. Set $x \in A$ if $\{n : x \leq x_n\} \in D$, and $y \in B$ if $\{n : x_n < y\} \in D$. Then $A \cup B = \mathbf{R}$, $-K \in A$, $K \in B$ and if $x \in A$, $z < x$, then $z \in A$ while if $y \in B$ and $z > y$ then $z \in B$. There is therefore a unique real number $-K \leq r \leq K$ such that $r - \epsilon \in A$ while $r + \epsilon \in B$ for all $\epsilon > 0$. This means that the set $\{n : r - \epsilon \leq x_n < r + \epsilon\}$, being the intersection of two elements in D, lies in D, and so $\lim_D x_n = r$. This shows the existence of the D-limit.

Since any two real numbers can be separated by disjoint neighborhoods, the unicity of the D-limit is clear.

(b) This immediately follows from the definition of ordinary and D-limits.

(c) For $c > 0$ we have $p < x_n < q$ if and only if $cp < cx_n < cq$. For $c < 0$ we have $p < x_n < q$ if and only if $cq < cx_n < cp$, and finally $\{cx_n\}$ is the constant sequence for $c = 0$. Now just apply the definition of D-limit.

(d) Let $a = \lim_D x_n$, $b = \lim_D y_n$ and $c = a + b$. For $\epsilon > 0$ the sets $\{n : a - \epsilon/2 < x_n < a + \epsilon/2\}$ and $\{n : b - \epsilon/2 < y_n < b + \epsilon/2\}$ are in D, hence so is their intersection, which is included in the set $\{n : c - \epsilon < x_n + y_n < c + \epsilon\}$. Thus, this latter set is in D for any $\epsilon > 0$, which means that $\lim_D(x_n + y_n) = c$.

(e) This is immediate from the definition of D-limit.

(f) This is a consequence of parts (b) and (d) (but also immediately follows from the definition of D-limit).

(g) Let $a = \lim_D x_n$ and $\epsilon > 0$. There is a $\delta_\epsilon > 0$ such that for $x \in (a-\delta_\epsilon, a+\delta_\epsilon)$ we have $f(x) \in (f(a) - \epsilon, f(a) + \epsilon)$. Thus,

$$\{n \ : \ a - \delta_\epsilon < x_n < a + \delta_\epsilon\} \subseteq \{n \ : \ f(a) - \epsilon < f(x_n) < f(a) + \epsilon\}.$$

Since here the set on the left-hand side belongs to D for all $\epsilon > 0$, the set on the right-hand side also has to belong to D, which proves part (d).

(h) Let $A \subset \omega$ be an infinite set such that the sequence $\{x_n\}_{n\in A}$ converges to r. Now for any nonprincipal ultrafilter D with $A \in D$ we have $\lim_D x_n = A$.

(i) For a sequence $\{x_n\}$ let $y_n = \arctan(x_n)$ and let us also set $\arctan(\pm\infty) = \pm\pi/2$. If we copy the proof of part (g) with the monotone and continuous functions $f(x) = \arctan x$, $x \in [-\infty, \infty]$ and $f^{-1}(x)$, $x \in [-\pi/2, \pi/2]$ we can easily see that $\lim_D x_n = r$ exists if and only if $\lim_D y_n = \arctan(r)$ exists. But $\{y_n\}$ is already a bounded sequence, hence we can apply part (a).

20. Let D be a nonprincipal ultrafilter on ω and let $f(A) = \lim_D(x_n)$, where

$$x_n = \frac{|A \cap n|}{n}$$

for $n > 0$. The properties (b) and (d) from the previous problem show that this f is suitable.

21. Formally, we consider functions $f : \omega \to \{0,1,2\}$ and an operator Φ assigning to every such f a value in $\{0,1,2\}$. The property is that if f_0, f_1 differ everywhere, then $\Phi(f_0) \neq \Phi(f_1)$. We have also assumed that if g_i is the identically constant function $g_i(j) = i$, $j = 0,1,2,\ldots$, then $\Phi(g_i) = i$.

First, assume that $A \subseteq \omega$, $B = \omega \setminus A$. If $f : A \to \{0,1,2\}$, $g : B \to \{0,1,2\}$ we simply write fg for the union of the functions f and g, and we also use the notation $(c)_A$ for the function that is identically c on A. Then $\Phi((0)_A(0)_B) = 0$, $\Phi((1)_A(1)_B) = 1$, hence we must have $\Phi((0)_A(0)_B) \neq \Phi((1)_A(0)_B)$ or $\Phi((1)_A(0)_B) \neq \Phi((1)_A(1)_B)$. By interchanging the sets A, B we may assume that the first of these holds, i.e., $\Phi((1)_A(0)_B) \neq 0$. Also, $\Phi((1)_A(0)_B) \neq \Phi((2)_A(2)_B) = 2$, so we must have $\Phi((1)_A(0)_B) = 1$. If $g : B \to \{1,2\}$, then $(2)_A g$ is pointwise different from both $(0)_A(0)_B$ and $(1)_A(0)_B$ so necessarily $\Phi((2)_A g) = 2$. This we denote by $\Phi((2)_A(1-2)_B) = 2$. If now $g : B \to \{1,2\}$ and $\overline{g}(i) = 3 - g(i)$ for every $i \in B$, then the functions $(0)_A(0)_B, (1)_A g, (2)_A \overline{g}$ assume 3 different values everywhere, so we get $\Phi((1)_A(1-2)_B) = 1$. Similarly, $\Phi((0)_A(1-2)_B) = 0$.

From this we get that $\Phi((0-1)_A g)$ is either 0 or 1, and similarly for 0, 2 and 1, 2. The first of these gives $\Phi((2)_A \overline{g}) = 2$, and similarly we get from the other ones that $\Phi((i)_A \overline{g}) = i$ for all $i \in \{0,1,2\}$.

Assume finally that for some functions $f : A \to \{0,1,2\}$ and $g_0, g_1 : B \to \{0,1,2\}$ we have $\Phi(fg_0) \neq \Phi(fg_1)$, say, $\Phi(fg_0) = i_0$, $\Phi(fg_1) = i_1$. Then there is a function $\overline{f} : A \to \{i_0, i_1\}$ which is everywhere different from f, and a function $h : B \to \{0,1,2\}$ which is everywhere different from g_0, g_1, then $\Phi(\overline{f}h) \neq i_0, i_1$ but must be in $\{i_0, i_1\}$ by the above, a contradiction.

What we showed is that if $\omega = A \cup B$ is a decomposition, then one and just one of A, B has the property that $\Phi(f)$ depends on $f|A$ (say). Let $\mathcal{U}$ be the system of those sets with this property. We get that exactly one of A, B is in $\mathcal{U}$. Clearly, $\emptyset \notin \mathcal{U}$, moreover if $A \in \mathcal{U}$ and $A \subset B$, then $B \in \mathcal{U}$. It is also immediate that $\mathcal{U}$ is closed under intersection. This implies that $\mathcal{U}$ is an ultrafilter. To conclude the proof let $C_i(f) = \{j \ : \ f(j) = i\}$, $i = 0,1,2$. These are disjoint sets with union ω, hence exactly one of them, say C_{i_0}, belongs to $\mathcal{U}$. Since on C_{i_0} the function f coincides with the constant function $g_{i_0}(j) = i_0$, and $\Phi(f)$ depends only on $f|_{C_0}$, we have $\Phi(f) = \Phi(g_{i_0}) = i_0$ as was claimed. [D. Greenwell–L. Lovász: Applications of product colouring, *Acta Math. Acad. Sci. Hung*, **25** (1974), 335–340]

22. First let us consider the case when there are only finitely many voters (I is finite). We call a voter dominant if the outcome of the vote is always her list.

First we show that if there are two voters A and B, then one of them is dominant. Let us agree in the following notation: the fact that candidates a, b, c are listed in A's list in some order like $\dots, a, \dots, c, \dots, b, \dots$ and in B's list in another order like $\dots, c, \dots, a, \dots, b, \dots$, and then in the outcome their order is like $\dots, b, \dots, c, \dots, a, \dots$ will be denoted by

A :	*acb*
B :	*cab*
	—
outcome :	*bca*

Suppose now that A is not dominant. Then there are some candidates ab on his list in this order such that in the outcome their order is ba. Then necessarily on B's list their order is ba (otherwise a and b would be listed in both lists in the order ab, which should be the outcome as well). We show that B is dominant. Since the order in the outcome is the result of the order of the pairs of the candidates, it is sufficient to show that B is dominant for each pair of candidates. Let c be a third candidate. Each column in the following table implies the next one:

A :	*ab*	*acb*	*ac*	*abc*	*bc*	*bac*	*ba*
B :	*ba*	*cba*	*ca*	*cab*	*cb*	*acb*	*ab*
	—	—	—	—	—	—	—
outcome :	*ba*	*cba*	*ca*	*cab*	*cb*	*acb*	*ab*

This proves that B is dominant for the pair a, b. But applying what we have obtained to column 3 resp. 5 we can see that B is also dominant for the pairs

a, c resp. b, c. Thus, the dominance of B for the pair a, b has been established, and here a and b can be replaced by any other $c \neq a, b$. By at most two such replacements we can get to any pair of candidates, and the dominance of B has been established.

Next we show that if there are 4 voters A,B,C,D, then one of them is dominant. In fact, suppose first that A and B form a block, i.e., they always vote the same way and C and D also form a block. Then we have two block voters, hence one of them is dominant, say the AB block. We claim that if A and B vote the same way, then they are dominant. If this is not the case, then there are candidates p, q such that A and B vote them in the order pq, but in the outcome the order is qp. In the following table $p'q'$ and $p''q''$ denote permutations of p, q, and again each column implies the next one:

$$\begin{array}{rlll} \mathrm{A}: & pq & paq & aq \\ \mathrm{B}: & pq & paq & aq \\ \mathrm{C}: & p'q' & p'q'a & qa \\ \mathrm{D}: & p''q'' & p''q''a & qa \\ & \text{---} & \text{---} & \text{---} \\ \text{outcome}: & qp & qpa & qa \end{array} \tag{17.1}$$

contradicting the dominance of the block AB over CD. Now fix the votes of CD in some order $\pi(c_1), \ldots, \pi(c_n)$ (for both of them), where $\pi : \{c_1, \ldots, c_n\} \to \{c_1, \ldots, c_n\}$ is some permutation, but A and B vote as they wish. Then we get a two-member voting scheme, hence either A or B is dominant, say A. We claim that A is dominant in the original 4 voter scheme. Suppose this is not the case. Then there are some candidates p, q such that A votes them in the order pq, but their order in the outcome is qp. Since the block AB was dominant, this is possible only if B votes in the order qp. Then, if the last element of the fixed order is b, each column in the following table implies the next one:

$$\begin{array}{rlll} \mathrm{A}: & pq & pbq & bq \\ \mathrm{B}: & qp & qpb & qb \\ \mathrm{C}: & p'q' & p'q'b & qb \\ \mathrm{D}: & p''q'' & p''q''b & qb \\ & \text{---} & \text{---} & \text{---} \\ \text{outcome}: & qp & qpb & qb \end{array}$$

contradicting the dominance of A when C and D vote in the fixed order $\pi(c_1), \ldots, \pi(c_n)$ (if one of p, q is the largest element b, then work symmetrically with the smallest element in the fixed order, and if p and q agree with the largest and smallest elements, then first replace one of them in the indicated manner by a third element and then we are back to the previously considered cases). With this the claim that in a four-member voting scheme there is always a dominant voter has been verified.

The same argument shows the same claim if there are three voters.

Now let I be an arbitrary set of voters. We call a subset $F \subseteq I$ dominant if it is true that if all members of F vote the same way then this is always the outcome. An argument similar to that in Table (17.1) shows that if F is dominant in the two-block voting scheme consisting of the blocks F and $I \setminus F$, then F is dominant. Let $\mathcal{F}$ be the set of dominant subsets of I. We show that it is an ultrafilter on I. It is clear that $\emptyset \notin \mathcal{F}$ (that would mean a fixed outcome irrespective of the votes) if $F \in \mathcal{F}$, $F \subset F'$, then $F' \in \mathcal{F}$, and out of F and $I \setminus F$ only one can belong to $\mathcal{F}$. That one of them is actually in $\mathcal{F}$ follows from the dominance in the two-member voting schemes. Thus, to show that $\mathcal{F}$ is an ultrafilter, it is sufficient to show that if $F_1, F_2 \in \mathcal{F}$ then $F_1 \cap F_2 \in \mathcal{F}$. Consider the 4-member block voting scheme when the blocks are $F_1 \cap F_2$, $I \setminus (F_1 \cup F_2)$, $F_1 \setminus F_2$, and $F_2 \setminus F_1$ (i.e., the voters in each block vote the same way, and if one of these sets is empty then the appropriate block voter is missing). We know that one of them is dominant (we have verified dominancy if there are at most four voters). Since both F_1 and F_2 are dominant, this dominant block cannot be any of the last three ones, so it must be $F_1 \cap F_2$, hence $F_1 \cap F_2 \in \mathcal{F}$.

Now let us consider an arbitrary voting, and for a permutation π of the candidates consider the set F_π of those voters $i \in I$ who voted in the order given by π. Since $I = \cup_\pi F_\pi$ is a disjoint decomposition, exactly one of the F_π belongs to $\mathcal{F}$, say $F_{\pi_0} \in \mathcal{F}$. Then F_{π_0} is dominant, so the outcome must be π_0. [K. J. Arrow]

18

Families of sets

1. For each $\alpha < \kappa^+$ let $f_\alpha : \kappa \to \alpha + 1$ be a surjection, and let

$$A_{\xi,\eta} = \{\alpha \;:\; f_\alpha(\xi) = \eta\}.$$

Since f_α is single-valued, the elements in each row are disjoint. If for $\eta < \kappa^+$ and $\alpha < \kappa^+$ there is no $\xi < \kappa$ with $f_\alpha(\xi) = \eta$, then $\alpha < \eta$; therefore, the union of the sets in the ηth column is $\kappa^+ \setminus \eta$, and we are done.

2. **First solution.** We replace the ground set κ with $\kappa \times \kappa$. Our sets will be $\kappa \to \kappa$ functions, so it is enough to construct a sequence $\{f_\alpha : \alpha < \kappa^+\}$ of $\kappa \to \kappa$ functions, such that any two differ from a certain point onward. We construct the functions by transfinite recursion on α. If $\{f_\beta : \beta < \alpha\}$ are already given, enumerate α as $\alpha = \{\gamma_\xi(\alpha) : \xi < \kappa\}$. Then select the value of $f_\alpha(\xi)$ to be different from every $f_{\gamma_\zeta(\alpha)}(\xi)$, $\zeta < \xi$ ("the first ξ values"). Having defined the functions, if we have $\beta < \alpha < \kappa^+$, then $\beta = \gamma_\zeta(\alpha)$ for some $\zeta < \kappa$ and then $f_\beta(\xi) \neq f_\alpha(\xi)$ for $\xi > \zeta$. [Erdős, 1934]

This is a condensed form of the solution for Problem 12.5.

Second solution. Let λ be the smallest cardinal with the property $\kappa^\lambda > \kappa$. By Cantor's theorem (Problem 10.16) $\lambda \leq \kappa$. Let X be the set of all transfinite sequences of length $< \lambda$ of ordinals $\xi < \kappa$. For each $\rho < \lambda$ there are at most $\kappa^\rho = \kappa$ such sequences of length ρ, hence X is of cardinality κ. Furthermore, let $\mathcal{H}^*$ be the set of all transfinite sequences $\{\alpha_\xi\}_{\xi<\lambda}$ of type λ of ordinals $\xi < \kappa$. Then, by the definition of λ, $\mathcal{H}^*$ is of cardinality bigger than κ. For every $s = \{\alpha_\xi\}_{\xi<\lambda} \in \mathcal{H}^*$ let H_s be the set of initial segment subsequences of s, i.e., the set $\{\{\alpha_\xi\}_{\xi<\eta}\}_{\eta<\lambda}$. Then $H_s \subseteq X$, and if $s' = \{\alpha'_\xi\}_{\xi<\lambda} \in \mathcal{H}$ is a different sequence in $\mathcal{H}^*$, then there is a $\tau < \lambda$ such that $\alpha_\tau \neq \alpha'_\tau$, hence the elements in the subsequences $\{\alpha_\xi\}_{\xi<\eta}$ and $\{\alpha'_\xi\}_{\xi<\eta}$ are different for all $\tau < \eta$. This shows that $H_s \cap H_{s'}$ is of cardinality smaller than $\lambda \leq \kappa$, hence the set $\mathcal{H} = \{H_s\}_{s\in\mathcal{H}^*}$ satisfies all the requirements. [W. Sierpiński, Mathematica,

14(1938), p. 15, Tarski, Func. Math., **12**(1928), 188–205 and **14**(1949), 205–215]

3. Let us decompose X as a disjoint union of the sets X_0, X_1, and X_2 each of cardinality κ, let $f : X_1 \to X_2$ be a 1-to-1 correspondence between the elements of X_1 and X_2, and let us also decompose X_1 into a disjoint family of sets $X_{1,\alpha}$, $\alpha < \kappa$ of cardinality κ. Now for any subset A of κ consider the set

$$H_A = X_0 \bigcup (\cup_{\alpha \in A} X_{1,\alpha}) \bigcup (X_2 \setminus (\cup_{\alpha \in A} f[X_{1,\alpha}]))\,.$$

It is clear that if A and B are different subsets of κ, say $\alpha \in A \setminus B$, then $X_{1,\alpha} \subset H_A \setminus H_B$, and $f[X_{1,\alpha}] \subset H_B \setminus H_A$, and since X_0 is part of every H_A, the family of the 2^κ sets H_A, $A \subseteq \kappa$ satisfies the properties set forth in the problem. [cf. W. Sierpiński, Cardinal and Ordinal Numbers, Polish Sci. Publ., Warszawa, 1965, XVII.4. Theorem 1]

4. This is a special case of Problem 6.90, since out of two initial segments one of them includes the other one.

5. The statement follows from Problem 3.

6. We show by transfinite induction on α that there is a mapping $\varphi_\alpha : [\alpha]^\kappa \to 2^\kappa$ such that for every $\xi < 2^\kappa$ the set $\varphi^{-1}(\xi)$ is an antichain. For $\alpha = \kappa$ we can take a bijection. Assume that $\mathrm{cf}\,(\alpha) > \kappa$ and φ_β exists for every $\beta < \alpha$. For $X \in [\alpha]^\kappa$ set $\varphi^*_\alpha(X) = \big(\mathrm{tp}\,(X), \varphi_{\beta(X)}(X)\big)$ where $\mathrm{tp}\,(X)$ is the order type of X and $\beta(X) = \sup(X)$. If $X \subseteq Y$ and $\varphi^*_\alpha(X) = \varphi^*_\alpha(Y)$, then specifically $\mathrm{tp}\,(X) = \mathrm{tp}\,(Y)$ so by $X \subseteq Y$ we have $\beta(X) = \beta(Y)$ and then the inductive assumption on $\varphi_{\beta(X)}$ gives $X = Y$. Thus, $X \subset Y$, $X \neq Y$ implies $\varphi^*_\alpha(X) \neq \varphi^*_\alpha(Y)$, i.e., the inverse image under φ^*_α of any set is an antichain. This φ^*_α is a mapping to $\kappa^+ \times 2^\kappa$, which can easily be transformed into a mapping φ_α to 2^κ.

If $\mathrm{cf}\,(\alpha) \leq \kappa$, then α can be decomposed into the union of at most κ disjoint intervals $\{I_j : j \in J\}$ where the order type of each I_j is smaller than α (this covers both the successor and the limit cases). By the inductive assumption for each $j \in J$ there is a $\psi_j : [I_j]^\kappa \to 2^\kappa$ such that $\psi_j^{-1}(\xi)$ is always an antichain. Define $\psi : [\alpha]^\kappa \to (2^\kappa)^\kappa$ by $\psi(X) = \langle \psi_j(X \cap I_j) : j \in J\rangle$. We show that ψ is a mapping with the required property (as it maps into a set of cardinality $(2^\kappa)^\kappa = 2^\kappa$ we are done). Assume that $X \subseteq Y$ and $\psi(X) = \psi(Y)$. Then $X \cap I_j \subseteq Y \cap I_j$ and $\psi_j(X \cap I_j) = \psi_j(Y \cap I_j)$ hold for every $j \in J$, so by hypothesis $X \cap I_j = Y \cap I_j$ for all $j \in J$, and therefore $X = Y$. [Erdős–Milner]

7. Let $H = \kappa \times \{0, 1\}$, and for an $f : \kappa \to \{0, 1\}$ set $A_f = \{(\alpha, f(\alpha))\}_{\alpha<\kappa}$, $B_f = H \setminus A_f$. There are 2^κ such pairs of sets, and if $f \neq g$ then for some $\alpha < \kappa$ we have $f(\alpha) \neq g(\alpha)$, say, $f(\alpha) = 0$ and $g(\alpha) = 1$, and then $(\alpha, 0) \in A_f \cap B_g$, i.e., $A_f \cap B_g \neq \emptyset$.

8. See Problem 24.31.

9. Let us enumerate the infinite subsets of $\mathbf{N}$ into a transfinite sequence A_ξ, $\xi < \mathbf{c}$. By transfinite induction we define different infinite subsets $B_\xi, C_\xi \subset A_\xi$, $\xi < \mathbf{c}$ such that B_ξ and C_ξ are different from every B_η, C_η, $\eta < \xi$. Since A_ξ has $\mathbf{c}$ infinite subsets, at each step $\xi < \mathbf{c}$ we can select $B_\xi \neq C_\xi$ with this property. Now if $\mathcal{F}$ is the set of the B_ξ's and $\mathcal{G}$ is the set of the C_ξ's, then these are clearly suitable, since every infinite subset of $\mathbf{N}$ is one of the A_ξ's, and this includes B_ξ and C_ξ.

10. Let $\mathcal{H}$ be a maximal set of almost disjoint countably infinite subsets of X, i.e., if $H_1, H_2 \in \mathcal{H}$, then $|H_1| = |H_2| = \aleph_0$ but $H_1 \cap H_2$ is finite, and there is no family with this property that properly includes $\mathcal{H}$ (the existence of $\mathcal{H}$ follows from Zorn's lemma; see Chapter 14). Thus, if $A \subset X$ is countably infinite, then $A \cap H$ is infinite for some $H \in \mathcal{H}$.

For each $H \in \mathcal{H}$ fix families $\mathcal{F}_H, \mathcal{G}_H \subset \mathcal{P}(H)$ with the properties from the preceding problem, and let $\mathcal{F}$ be the union of all the $\mathcal{F}_H$'s and $\mathcal{G}$ the union of all the $\mathcal{G}_H$'s for all $H \in \mathcal{H}$. These are disjoint families. In fact, if we had $S \in \mathcal{F} \cap \mathcal{G}$, then S would belong to some $\mathcal{F}_{H_1}$ and also to some $\mathcal{G}_{H_2}$. Here $H_1 = H_2$ is not possible since $\mathcal{F}_{H_1}$ and $\mathcal{G}_{H_1}$ are disjoint. If, however, $H_1 \neq H_2$, then S would be a common subset of both H_1 and of H_2, which cannot be the case because $H_1 \cap H_2$ is finite.

Finally, if $A \subseteq X$ is infinite, then A has a countably infinite subset A_1 which intersects one of the H's in an infinite set. By the choice of $\mathcal{F}_H$ there is an $F \in \mathcal{F}_H$ such that $F \subseteq A_1 \cap H$, i.e., this $F \in \mathcal{F}$ is a subset of A. A similar argument shows that A contains an element of $\mathcal{G}$, and the proof is over. [A. Hajnal]

11. **First solution.** Obviously, the existence of an appropriate family depends only on the cardinality of the ground set. Rather than working on κ we work on the set $X = \{(s,h) : s \in [\kappa]^{<\omega}, h \subseteq \mathcal{P}(s)\}$. As there are κ finite subsets of κ each carrying finitely many families of subsets, we have $|X| = \kappa$. If $A \subseteq \kappa$, then we associate the set $Y(A) = \{(s,h) : s \in [\kappa]^{<\omega}, A \cap s \in h\} \subseteq X$ with A. This way we have created the family $\mathcal{F} = \{Y(A) : A \subseteq \kappa\}$. If we show that it is independent, then, in particular, we find that the elements of $\mathcal{F}$ are distinct and so $|\mathcal{F}| = 2^\kappa$. Toward showing independence, assume that we are given the different sets $A_1, \ldots, A_n \subseteq \kappa$ and $\varepsilon_1, \ldots, \varepsilon_n < 2$. To any two sets there is a point which is in one but not in the other, so there is a finite set $s \subseteq \kappa$ such that the intersections $A_i \cap s$ are different. Now set $h = \{A_i \cap s : 1 \le i \le n, \varepsilon_i = 1\}$. Clearly, $(s,h) \in Y(A_1)^{\varepsilon_1} \cap \cdots \cap Y(A_n)^{\varepsilon_n}$.

Second solution. For every $\xi < \kappa$ and natural number l let us choose l sets $A_{\xi,l,j}$, $j < l$ that form an independent family over a finite set $A_{\xi,l}$ (e.g., if $A_j \subset {}^l\{0,1\}$, $j < l$ is the set of those 0–1 sequences of length l which have 1 at the jth position, then A_j, $j < l$ is an independent family of sets over

${}^{l}\{0,1\}$). Let us also assume that these sets are selected in such a way that the ground sets $A_{\xi,l}$ are disjoint for different (ξ, l)'s. Now with the functions f from Problem 13 consider the sets

$$H_f = \bigcup_{\xi<\kappa,\ l\in\mathbf{N}} A_{\xi,l,\min(l-1,f(\xi))}, \qquad f \in \mathcal{F}.$$

If $f_1, \dots, f_n$ are different, then there is a $\xi^* < \kappa$ such that all the values $f_i(\xi^*)$ are different, and let us choose l^* so large that all these values are less than l^*. Now for $\epsilon_i = 0$ or 1 the intersection

$$H_{f_1}^{\epsilon_1} \cap \ldots \cap H_{f_n}^{\epsilon_n}$$

includes the nonempty set

$$A^{\epsilon_1}_{\xi^*,l^*,f_1(\xi^*)} \cap \cdots \cap A^{\epsilon_n}_{\xi^*,l^*,f_n(\xi^*)}$$

(where in the latter case complements are taken with respect to the set A_{ξ^*,l^*}). [F. Hausdorff, Studia Math., **6**(1936), 18–19, A. Tarski, Fund. Math., **32**(1939), 45–63]

12. (See also the solution to Problem 17.3.) By Problem 11 there is an independent family $\mathcal{F}$ of cardinality 2^κ of subsets of κ. This means that if $g : \mathcal{F} \to \{0,1\}$ is an arbitrary mapping and $\mathcal{F}_g$ is the family that contains $F \in \mathcal{F}$ if $g(F) = 1$ and contains $\kappa \setminus F$ if $g(F) = 0$, then $\mathcal{F}_g$ has the property that any finite subset of $\mathcal{F}_g$ has nonempty intersection. But then $\mathcal{F}_g$ generates a filter which is included in a U_g ultrafilter, and if $g \neq h$, then $U_g \neq U_h$, so we get this way $2^{|\mathcal{F}|} = 2^{2^\kappa}$ different ultrafilters. In fact, if $g \neq h$, then there is an $F \in \mathcal{F}$ with $g(F) \neq h(F)$, and then this F is contained in one of U_g and U_h and $\kappa \setminus F$ is contained in the other one.

13. Without loss of generality, we may assume that A is the set of the nonempty finite subsets of κ, and for a function $g : \{\xi_0, \xi_1, \dots, \xi_m\} \to \{0,1\}$, $\xi_0 < \xi_1 < \xi_2 < \ldots < \xi_m$ let

$$t(g) = g(\xi_0) + 2g(\xi_1) + \cdots + 2^n g(\xi_m).$$

Note that if $g' : \{\xi_0, \xi_1, \dots, \xi_m\} \to \{0,1\}$ is another function and $t(g) = t(g')$, then we must have $g(\xi_i) = g'(\xi_i)$ for all $i = 0, \dots, m$.

Now if $f : \kappa \to \{0,1\}$ is arbitrary, then associate with it that function F on A for which $F(I) = t(f|_I)$ for every $I \in A$, and let $\mathcal{F}$ be the set of all these functions F. If $F_1, \dots, F_n$ are different functions that correspond to $f_1, \dots, f_n$, then these are also pairwise different, hence for each $1 \le i < j \le n$ there is a $\xi_{i,j} < \kappa$ with $f_i(\xi_{i,j}) \neq f_j(\xi_{i,j})$. Thus, for the set I consisting of all these $\xi_{i,j}$ we have $t(f_i|_I) \neq t(f_j|_I)$ for all $1 \le i < j \le n$, which means that the values $F_i(I)$ are all different.

14. We may assume that $A = {}^{\kappa}\{0,1\}$ is the set of infinite 0–1 sequences of length κ, and for a finite subset I of κ let A_I be the set of all functions $h : I \to \{0,1\}$ that map the given finite set I into $\{0,1\}$. Any mapping $g : A_I \to \kappa$ generates a mapping $f_g : A \to \kappa$ defined as

$$f_g(h) = g(h|_I).$$

Now there are κ many ways to map A_I into κ, and the set of finite subsets I of κ is also of cardinality κ, so if $\mathcal{F}$ is the set of all f_g's with all possible $g : A_I \to \kappa$ and all possible $I \subset \kappa$, $|I| < \omega$, then $\mathcal{F}$ is of cardinality κ. If $F : A \to \kappa$ is any given function and $h_1, \ldots, h_n$ are different elements of A, then there is a finite subset I of κ such that $h_i|_I$ are all different. Now if we define the function g as $g(h_i|_I) = F(h_i)$ for $i = 1, \ldots, n$ and $g(h)$ is arbitrary for other $h : I \to \{0,1\}$, then $F(h_i) = f_g(h_i)$ for all $1 \le i \le n$, so $\mathcal{F}$ satisfies the requirements. (See also the solution to Problem 4.27.)

15. See the solution to Problem 4.28, and apply Problem 14 instead of 4.27.

16. Set $\mathcal{F} = \{A_0, A_1, \ldots\}$. By induction on $n < \omega$ we build the finite sets $X_0 \subseteq X_1 \subseteq \cdots$ with the property that $1 \le |X_n \cap A_i| \le 2$ holds for $i \le n$ and $|X_n \cap A_i| \le 2$ holds for $i > n$. If we can do this, then $X = \bigcup\{X_n : n < \omega\}$ will be good. Assume therefore that we have reached step number n and we have the finite set X_n. The choice $X_{n+1} = X_n$ is good unless $X_n \cap A_n = \emptyset$. In this latter case we have to choose some $x \in A_n$ so that $X_{n+1} = X_n \cup \{x\}$ is good. This requires that $|(X_n \cup \{x\}) \cap A_i| \le 2$ should hold for every $i \ne n$, that is, $x \notin A_i$ for every $i < \omega$ for which $|X_n \cap A_i| = 2$ holds. We argue that only finitely many elements $x \in A_n$ are disqualified by this requirement (and this concludes the proof). Indeed, for any pair $Y \subseteq X_n$ there can be only one A_i with $Y \subseteq A_i$ (by our intersection condition on the family) and for that i there is only one $x \in A_i \cap A_n$ (again by the intersection condition).

17. Let $\mathcal{F} = \{A_0, A_1, \ldots\}$. For each $k \ge 3$ we construct some finite disjoint sets E_k and F_k so that

$$\text{for } i \le k \text{ either } |A_i \cap E_k| = 1 \text{ or } |A_i \cap F_k| = 1 \tag{18.1}$$

and

$$\text{for } i > k \text{ either } |A_i \cap E_k| \le 1 \text{ or } |A_i \cap F_k| \le 1 \tag{18.2}$$

is true. It will also be true that $E_k \subseteq E_{k+1}$ and $F_k \subseteq F_{k+1}$ and so the sets $X = \cup_k E_k$ $Y = \cup_k F_k$ will have the desired property.

For $k = 3$ it is easy to construct such sets E_3, F_3. Let us suppose that E_k, F_k have already been constructed with the above properties. First let us assume that A_{k+1} intersects both E_k and F_k. In view of (18.2) this is possible if, say, $|A_{k+1} \cap E_k| = 1$, in which case $E_{k+1} = E_k$ and $F_{k+1} = F_k$ is suitable for $k + 1$.

Now suppose that A_{k+1} does not intersect one of the sets E_k, F_k, say $A_{k+1} \cap E_k = \emptyset$. Consider all the sets A_i that intersect $E_k \cup F_k$ in at least three elements. Since three given elements can be only in one A_i (recall that $|A_i \cap A_j| \le 2$ for $i \ne j$), and $E_k \cup F_k$ is finite, there are only finitely many such i, let these be $i_0, i_1, \ldots, i_m$. Let e_{k+1} be an arbitrary element of the set

$$A_{k+1} \setminus \left\{ (\bigcup_{r=0}^{m} A_{i_r}) \bigcup (\bigcup_{i=0}^{k} A_i) \cup F_k \right\},$$

and let $E_{k+1} = E_k \cup \{e_{k+1}\}$, $F_{k+1} = F_k$. These sets obviously satisfy (18.1) with k replaced by $k+1$, and if $i > k+1$ and $e_{k+1} \notin A_i$ then (18.2) is true, as well. Finally, if $e_{k+1} \in A_i$, then A_i differs from the sets $A_{i_0}, \ldots, A_{i_m}$ by the choice of e_{k+1}, and so $|A_i \cap (E_k \cup F_k)| \le 2$. Now this yields that if $|A_i \cap F_k| = 2$ then $|A_i \cap E_{k+1}| = |\{e_{k+1}\}| = 1$, while if $|A_i \cap F_k| \le 1$ then $|A_i \cap F_{k+1}| \le 1$ is true because $F_{k+1} = F_k$. Thus, in any case we have (18.2) for $k+1$, and this completes the proof.

18. We prove by induction on $\aleph_1 \le \kappa < \aleph_\omega$ the result for κ. For $\kappa = \aleph_1$ we can take the set $\{\beta : \omega \le \beta < \alpha\}$, $\alpha < \omega_1$, of all countably infinite initial segments of ω_1. Assume we have the result for some κ. As the existence of such a system is a property of the cardinality of the ground set, there is an appropriate system $\mathcal{F}_\alpha$ on every set α for $\kappa \le \alpha < \kappa^+$. We claim that $\mathcal{F} = \bigcup\{\mathcal{F}_\alpha : \kappa \le \alpha < \kappa^+\}$ is a good system for κ^+. The cardinality of $\mathcal{F}$ is $\kappa\kappa^+ = \kappa^+$. Assume that $X \subseteq \kappa^+$ is countable. Then, as $\kappa^+ > \aleph_0$ is regular, $X \subseteq \alpha$ for some $\kappa \le \alpha < \kappa^+$. Then some $Y \in \mathcal{F}_\alpha$ covers X, as required.

To show that the same is not true for $\aleph_\omega$, let A be a set of cardinality $\aleph_\omega$, and let $\mathcal{F}$ be an arbitrary system of cardinality $\aleph_\omega$ of countable subsets of A. We represent it as $\mathcal{F} = \mathcal{F}_0 \cup \mathcal{F}_1 \cup \cdots \cup \mathcal{F}_n \cup \cdots$ where the $\mathcal{F}_n$'s are increasing subsets of $\mathcal{F}$ of cardinality $\aleph_n$. For each n the set $\mathcal{F}_n$ is of cardinality $\aleph_n$, hence there is an element $a_{n+1} \in A$ outside this set. Consider $B = \{a_n\}_{n=1}^{\infty} \subset A$. If $F \in \mathcal{F}$, then $F \in \mathcal{F}_n$ for some n, hence $a_{n+1} \in B \setminus F$, i.e., B is not covered by any one of the sets in $\mathcal{F}$. This shows that for $\aleph_\omega$ there is no system $\mathcal{F}$ with the prescribed properties.

19. We can assume that $\kappa \ge \mu$. Let V be an enumeration of all functions $f : \alpha \to \kappa$, $\alpha < \kappa^+$, say $V = \{v(f) : f : \alpha \to \kappa,\ \alpha < \kappa^+\}$. Clearly, $|V| = 2^\kappa$. Let $f : \alpha \to \kappa$, $\alpha < \kappa^+$ be one of these functions. If $\mathrm{cf}\,(\alpha) = \mathrm{cf}\,(\mu)$, and there is some cofinal set in α of order type μ which is monocolored by f, then let $B \subseteq \alpha$ be one such set, and define $H(f) = \{v(f|_\beta) : \beta \in B\}$. If no such set exists, then leave $H(f)$ undefined. Let $\mathcal{H}$ be the collection of all these sets $H(f)$.

We claim that $\mathcal{H}$ is as required. Assume first that $|H(f) \cap H(f')| = \mu$, with $f : \alpha \to \kappa$, $f' : \alpha' \to \kappa$. If $v(f|_\beta) = v(f'|_{\beta'})$ is a common element, then $\beta = \beta'$ and $f|_\beta = f'|_\beta'$. As the common elements are necessarily cofinal in

α, α', we get that $\alpha = \alpha'$ and $f = f'$. This shows that $\mathcal{H}$ is an almost disjoint family.

Assume now that $F : V \to \kappa$ is a coloring. By transfinite recursion on $\alpha < \kappa^+$ we construct the increasing sequence of functions $\{f_\alpha : \alpha < \kappa^+\}$ with $f_\alpha : \alpha \to \kappa$. $f_0 = \emptyset$. If α is a limit ordinal, then let $f_\alpha = \bigcup\{f_\beta : \beta < \alpha\}$. Finally, set $f_{\alpha+1}(\alpha) = F(v(f_\alpha))$. There is some $f : \kappa^+ \to \kappa$ such that $f_\alpha = f|_\alpha$ holds for every $\alpha < \kappa^+$.

By the pigeon hole principle there is a value ξ assumed by f on a set of cardinality κ^+. Let α be the supremum of the first μ elements of $f^{-1}(\xi)$. Then for this α it is true that there is a cofinal set $B \subseteq \alpha$ of order type μ that is monocolored by $f|_\alpha = f_\alpha$. Therefore, $H(f_\alpha)$ is defined, using some set B' (possibly different from B) on which the color of f_α is some ξ'. But then $H(f_\alpha)$ is monocolored by F:

$$F\left(v(f_\alpha|_\beta)\right) = F(v(f_\beta)) = f_{\beta+1}(\beta) = f_\alpha(\beta) = \xi'$$

for every $\beta \in B'$. [G. Elekes, G. Hoffmann: On the chromatic number of almost disjoint families of countable sets, *Coll. Math. Soc. J. Bolyai*, **10** *Infinite and Finite Sets*, Keszthely (Hungary), 1973, 397–402]

19

The Banach–Tarski paradox

1. $A \sim A$ is obvious using the identity. If $A \sim B$ then there are partitions $A = A_1 \cup \cdots \cup A_t$ and $B = B_1 \cup \cdots \cup B_t$ with $B_i = f_i[A_i]$ for some $f_i \in \Phi$. Then $B \sim A$ holds using the same partitions as $A_i = f_i^{-1}[B_i]$. What remains to be proved is that $A \sim B$ and $B \sim C$ imply $A \sim C$. As $A \sim B$, there are decompositions $A = A_1 \cup \cdots \cup A_n$ and $B = B_1 \cup \cdots \cup B_n$ such that $B_i = f_i[A_i]$ for some $f_i \in \Phi$. Similarly, by $B \sim C$ there are decompositions $B = B^1 \cup \cdots \cup B^m$ and $C = C^1 \cup \cdots \cup C^m$ such that $C^j = g_j[B^j]$ for some $g_j \in \Phi$. Set $B_{ij} = B_i \cap B^j$ for $1 \le i \le n$, $1 \le j \le m$. If now $A_{ij} = f_i^{-1}[B_{ij}]$, $C_{ij} = g_j[B_{ij}]$, then

$$A = \bigcup \{A_{ij} : 1 \le i \le n, 1 \le j \le m\},$$

$$C = \bigcup \{C_{ij} : 1 \le i \le n, 1 \le j \le m\}$$

are decompositions of A and C, respectively, and $C_{ij} = h_{ij}[A_{ij}]$ where $h_{ij} = g_j \circ f_i \in \Phi$ (some of the pieces may be empty but that does not invalidate the argument).

2. As A is equidecomposable to a subset of B, there is a decomposition $A = A_1 \cup \cdots \cup A_n$, and there are functions $f_1, \ldots, f_n \in \Phi$ such that $f_1[A_1] \cup \cdots \cup f_n[A_n]$ is a disjoint decomposition of a subset of B. Similarly, there are a decomposition $B = B_1 \cup \cdots \cup B_m$ and functions $g_1, \ldots, g_m \in \Phi$ such that $g_1[B_1] \cup \cdots \cup g_m[B_m]$ is a disjoint decomposition of a subset of A. Now define $f : A \to B$, as well as $g : B \to A$ the following way. $f(x) = f_i(x)$ for $x \in A_i$ and $g(x) = g_j(x)$ for $x \in B_j$. As f and g are both injective, by Problem 3.1 there are decompositions $A = A' \cup A''$, $B = B' \cup B''$ such that $B' = f[A']$, $A'' = g[B'']$. As $A' = (A' \cap A_1) \cup \cdots \cup (A' \cap A_n)$, $B' = (B' \cap f_1[A_1]) \cup \cdots \cup (B' \cap f_n[A_n])$ are decomposition of A', B', respectively, we get that $A' \sim B'$. Likewise, $A'' \sim B''$ and these two together give $A \sim B$.

3. As $pA \preceq qB$ there is a p-cover $A_1 \cup \cdots \cup A_n$ of A such that $f_1[A_1] \cup \cdots \cup f_n[A_n]$ is a $\leq q$ cover of B for some elements $f_1, \ldots, f_n \in \Phi$. Similarly, $qB \preceq rC$ is witnessed by a q-cover $B_1 \cup \cdots \cup B_m$ of B such that $g_1[B_1] \cup \cdots \cup g_m[B_m]$ is a $\leq r$-cover of C with some elements $g_1, \ldots, g_m \in \Phi$. For $1 \leq i \leq n$, $1 \leq j \leq m$, $1 \leq s \leq q$ we define the set $A_{ijs} \subseteq A$ as follows. $x \in A_{ijs}$ if and only if $x \in A_i$, $y = f_i(x) \in B_j$ and $\{1 \leq u \leq i : y \in f_u[A_u]\}$ and $\{1 \leq v \leq j : y \in B_v\}$ both have exactly s elements. Set $h_{ijs} = g_j \circ f_i$. We claim that

$$\bigcup \{A_{ijs} : 1 \leq i \leq n, 1 \leq j \leq m, 1 \leq s \leq q\}$$

is a p-cover of A and

$$\bigcup \{h_{ijs}[A_{ijs}] : 1 \leq i \leq n, 1 \leq j \leq m, 1 \leq s \leq q\}$$

is a $\leq r$-cover of C (and so they witness $pA \preceq rC$). Pick some $x \in A$. For every i with $x \in A_i$ set $y = f_i(x)$ and $s = |\{1 \leq u \leq i : y \in f_u[A_u]\}|$. For these y, s there is a unique j such that $y \in B_j$ and $\{1 \leq v \leq j : y \in B_v\}$ has exactly s elements. That is, $x \in A_{ijs}$, and this shows that the system of A_{ijs}'s is a p-covering of A (as so is the system of A_i's). In a similar manner, if $z \in C$, for every j with $z \in g_j[B_j]$ (and there are $\leq r$ of them) there are unique s and i such that $z \in h_{ijs}[A_{ijs}]$, and so the system of $h_{ijs}[A_{ijs}]$'s is a $\leq r$-cover, as claimed.

4. Assume that $A = A_1 \cup \cdots \cup A_n$ is a p-cover, $f_1[A_1] \cup \cdots \cup f_n[A_n]$ is a $\leq q$-cover of B, $B = B_1 \cup \cdots \cup B_m$ is a q-cover, $g_1[B_1] \cup \cdots \cup g_m[B_m]$ is a $\leq p$-cover of A. Refining the decomposition, if needed, we can assume that every $x \in A_i$ is in the same number of sets among $A_1, \ldots, A_i$, say, in $a(i)$ of them, for every $x \in A_i$, $f_i(x)$ is in the same number of sets among $f_1[A_1], \ldots, f_i[A_i]$, say, in $b(i)$ of them, and similarly, we assume that every $x \in B_j$ is in the same number of sets among $B_1, \ldots, B_j$, say, in $c(j)$ of them, and, finally, for every $x \in B_j$, $g_j(x)$ is in the same number of sets among $g_1[B_1], \ldots, g_j[B_j]$, say, in $d(j)$ of them.

Set $A^* = A \times \{1, \ldots, p\}$, $B^* = B \times \{1, \ldots, q\}$. Notice that for every $x \in A$, $r \in \{1, \ldots, p\}$ there is a unique i such that $x \in A_i$, $r = a(i)$. Define, for $\langle x, r\rangle \in A^*$, $F(\langle x, r\rangle) = \langle y, s\rangle$, $y = f_i(x)$, $s = b(i)$ where i is such that $x \in A_i$, $r = a(i)$. Likewise, for $\langle y, s\rangle \in A^*$, define $G(\langle y, s\rangle) = \langle x, r\rangle$ where $y \in B_j$, $s = c(j)$, $x = g_j(y)$, $r = d(j)$.

Notice that $F : A^* \to B^*$, $G : B^* \to A^*$ are injective. In fact, let x, r, y, s, and i be as above, and suppose that $F(\langle x', r'\rangle) = \langle y, s\rangle$ is also true. Then $y = f_j(x)$ for some j. Here $i < j$ is not possible, for then $b(j) \geq b(i) + 1$ and hence $b(j) = s = b(i)$ cannot hold. For the same reason neither is $j < i$, hence $i = j$ and then the bijective character of f_i gives $x = x'$. By Problem 3.1 there exist decompositions $A^* = A_0^* \cup A_1^*$, $B^* = B_0^* \cup B_1^*$, such that F is bijective between A_0^* and B_0^* and G is bijective between B_1^* and A_1^*.

Now define

$$A_i' = \{x \in A_i : \langle x, a(i)\rangle \in A_0^*\} \quad (1 \le i \le n),$$

$$A_j'' = \{g_j(y) : \langle y, c(j)\rangle \in B_1^*\} \quad (1 \le j \le m).$$

If we now apply f_i on A_i' and g_j^{-1} on A_j'' then we get $pA \sim qB$. In fact, for any $x \in A$ each of the points $\langle x, r\rangle$, $r = 1, \dots, p$ lie either in A_0^* or in A_1^*, so $\{A_1', \dots, A_n', A_1'', \dots, A_m''\}$ forms a p-cover of A. In a similar way we find that their images form a q-cover of B.

5. This is an immediate consequence of the preceding two problems.

6. Assume that $kpA \preceq kqB$ holds. There are, therefore, subsets $A_1, \dots, A_t \subseteq A$ such that every $x \in A$ is in exactly kp of them, $f_i \in \Phi$, $B_i = f_i[A_i] \subseteq B$, and every $y \in B$ is in at most kq of the B_i's. We construct a bipartite graph with the bipartition classes A, B, as follows. We join every $x \in A$ with an edge to each $f_i(x)$ (in case $f_i(x) = y$ for say $s \ge 1$ of the i's, we keep only one edge between x and y; but the number s appears below as $f(x, y)$). This way, we defined a locally finite graph (that is, every vertex has finite degree). By hypothesis, there is a function f from the edge set into the natural numbers (namely the one that associates with an edge $e = (x, y)$ the number of i for which $y = f_i(x)$) such that

$$\sum_{x \in e} f(e) = kp \quad (x \in A)$$

and

$$\sum_{y \in e} f(e) \le kq \quad (y \in B).$$

This implies, by simple counting, that for every finite $A' \subseteq A$ the set $\Gamma[A']$ of points in B joined into A' has

$$|\Gamma[A']| \ge \frac{kp}{kq}|A'| = \frac{p}{q}|A'|$$

elements. Using Problem 23.15 we find that there is a function g from the edge set into the natural numbers such that

$$\sum_{x \in e} g(e) = p \quad (x \in A)$$

and

$$\sum_{y \in e} g(e) \le q \quad (y \in B).$$

We define the sets $A_1^*, \dots, A_t^*$ as follows. $x \in A_i^*$ if and only if $x \in A_i$ and

$$|\{1 \le j \le i : f_j(x) = f_i(x)\}| \le g\left(x, f_i(x)\right).$$

If $x \in A$, then for every $y \in B$, x is contained in exactly $g(x,y)$ of the A_i^*'s (namely, in the first $g(x,y)$ of those A_i^*'s for which $f_i(x) = y$), so altogether it is in p of them. Similarly, if $y \in B$, the number of i's for which $y \in f_i[A_i^*]$ holds, is $\sum\{g(x,y) : x \in A\}$, which is at most q by condition.

7. (d) $\to$ (c) $\to$ (b) $\to$ (a) and (e) $\to$ (b) are obvious.

(b) $\to$ (c) Assume that $A \sim 2A$, i.e., there is a partition $A = A_1 \cup \cdots \cup A_n$ and there are $f_1, \ldots, f_n \in \Phi$ such that $f_1[A_1], \ldots, f_n[A_n] \subseteq A$ and every point in A is covered exactly twice. Set

$$A_i' = \{x \in A_i : f_i(x) \notin f_1[A_1] \cup \cdots \cup f_{i-1}[A_{i-1}]\},$$

$$A_i'' = \{x \in A_i : f_i(x) \in f_1[A_1] \cup \cdots \cup f_{i-1}[A_{i-1}]\}.$$

Consider $A' = A_1' \cup \cdots \cup A_n'$, $A'' = A_1'' \cup \cdots \cup A_n''$. Then, $A'' = A \setminus A'$, and as

$$A = f_1[A_1'] \cup \cdots \cup f_n[A_n'] = f_1[A_1''] \cup \cdots \cup f_n[A_n''],$$

we have $A' \sim A \sim A''$.

(c) $\to$ (d) By induction on k. The case $k = 2$ is just (c). To proceed from k to $k+1$ let $A = A_1 \cup \cdots \cup A_k$ be a partition appropriate for k. As $A_k \sim A$ there are partitions $\{B_1, \ldots, B_t\}$ and $\{C_1, \ldots, C_t\}$ of A_k, A respectively such that $C_i = f_i[B_i]$ hold for appropriate $f_i \in \Phi$. By (c) there is a decomposition $A = A' \cup A''$ with $A' \sim A'' \sim A$. Set $C_i' = C_i \cap A'$, $C_i'' = C_i \cap A''$, $B_i' = f_i^{-1}[C_i']$, $B_i'' = f_i^{-1}[C_i'']$. If we put $A_k' = B_1' \cup \cdots \cup B_t'$, $A_k'' = B_1'' \cup \cdots \cup B_t''$, then we get $A_k' \sim A' \sim A$ and $A_k'' \sim A'' \sim A$. Hence $A = A_1 \cup \cdots \cup A_{k-1} \cup A_k' \cup A_k''$ is a partition appropriate for $k+1$.

(a) $\to$ (e) Assume that $(n+1)A \preceq nA$. Then by Problem 4 we have $nA \sim (n+1)A$. If we add one–one copy of A to the covers on the two sides we can see that this implies $(n+1)A \sim (n+2)A$, and iteration gives

$$nA \sim (n+1)A \sim (n+2)A \sim \cdots \sim npA,$$

and by Problem 5 we find that $pA \sim A \sim qA$ for any $p, q \ge 1$.

8. Using Problem 7 we find that

$$2B \preceq 2nA \preceq A \preceq B$$

and that suffices again by Problem 7.

9. (a) We consider the complex plane. Let $c \in \mathbf{C}$ be a transcendental number with $|c| = 1$. Let $A \subseteq \mathbf{C}$ be the set of all complex numbers of the form $a_n c^n + \cdots + a_0$ with $a_n, \ldots, a_0$ nonnegative integer. Notice that A is countable and every element in A has a unique representation of the above form. Now the congruences $z \mapsto z+1$, $z \mapsto cz$ (a rotation, as $|c| = 1$) map A onto

disjoint subsets of itself so $2A \sim A$. [S. Mazurkiewicz, W. Sierpiński: Sur un ensemble superposable avec chacune des ses deux parties, *Comptes Rendus Acad. Sci. Paris* **158**(1914) 618–619]

(b) Let $D = \{z \in \mathbf{C} : |z| \leq 1\}$ be the unit disc, $\epsilon = (-1 + i\sqrt{3})/2$, $c \in \mathbf{C}$ again some transcendental with $|c| = 1$. For every $z \in D$ one of $z+1$, $z+\epsilon$, $z+\epsilon^2$ is in D so there is a function $f : D \to D$ such that $f(z) \in \{z+1, z+\epsilon, z+\epsilon^2\}$ holds for $z \in D$. Let A be the smallest set containing 0 and having the property that if $z \in A$ then $cz, f(cz) \in A$. Again, A is countable and each element can uniquely be represented as a polynomial of c with coefficients $0, 1, \epsilon$, or ϵ^2. As $A \subseteq D$, it is bounded. Set $A_1 = cA$,

$$A_2 = \{z \in cA : f(z) = z + 1\}, A_3 = \{z \in cA : f(z) = z + \epsilon\},$$

$$A_4 = \{z \in cA : f(z) = z + \epsilon^2\}.$$

Now $A = A_1 \cup (A_2 + 1) \cup (A_3 + \epsilon) \cup (A_4 + \epsilon^2)$ is a partition, $A_1 = cA \sim A$ and $A_2 \cup A_3 \cup A_4 = cA$ so $2A \sim A$. [W. Just: A bounded paradoxical subset of the plane, *Bull. Polish Acad. Sci. Math* **36**(1988), 1–3]

10. Let φ be some rotation with an angle incommensurable to 2π, such that no $x \in A$ is a fixed point of φ, and $\varphi^n(x) \neq x'$ holds for $x, x' \in A$, $n = 1, 2, \ldots$. Such a φ exists as the second and the third conditions disqualify only $< \mathbf{c}$ rotations, once we fix the angle of it. Now define $B = \{\varphi^n(x) : x \in A, n < \omega\}$. Notice that every $y \in B$ can uniquely be written in the form $\varphi^n(x)$ with $x \in A, n < \omega$. φ now transforms B into $B \setminus A$. As $\mathbf{S}^2 = (\mathbf{S}^2 \setminus B) \cup B$, applying the identity on the first set and φ on the second we find that $\mathbf{S}^2 \sim \mathbf{S}^2 \setminus A$.

11. Let $0 < \alpha < 1$ be an irrational number. Let x_n be the fractional part of αn. Notice that $x_0 = 0$ and $x_n \neq x_m$ holds for $n \neq m$. Also,

$$x_{n+1} = \begin{cases} x_n + \alpha & \text{if } 0 \leq x_n < 1 - \alpha, \\ x_n + \alpha - 1 & \text{if } 1 - \alpha \leq x_n < 1. \end{cases}$$

Set $X = \{x_n : n = 0, 1, \ldots\}$, $Y = [0, 1] \setminus X$, $X' = X \cap [0, 1 - \alpha)$, $X'' = X \cap [1 - \alpha, 1]$.

Then, by according to what was said above the set $X \setminus \{0\}$ decomposes as $(X' + \alpha) \cup (X'' - (1 - \alpha))$ so $[0, 1] = X' \cup X'' \cup Y$ and $(0, 1] = (X' + \alpha) \cup (X'' - (1 - \alpha)) \cup Y$. [W. Sierpiński: *On the congruence of sets and their equivalence by finite decomposition*. Lucknow, 1954. Reprinted by Chelsea, 1967]

12. Notice that $Q = [0, 1] \times [0, 1]$. We know from Problem 10 that there are a decomposition $[0, 1] = A_1 \cup \cdots \cup A_t$ and real numbers $\alpha_1, \ldots, \alpha_t$ that the translates $(A_1 + \alpha_1), \ldots, (A_t + \alpha_t)$ give a decomposition of $(0, 1]$. If we multiply these sets by $[0, 1]$ we get the decomposition

$$[0, 1] \times [0, 1] = (A_1 \times [0, 1]) \cup \cdots \cup (A_t \times [0, 1]),$$

which can be transformed by translations in the direction of the x-axis into

$$((A_1+\alpha_1)\times[0,1])\cup\cdots\cup((A_t+\alpha_t)\times[0,1]),$$

a decomposition of $(0,1]\times[0,1]$.

13. We notice that a proof virtually identical to the one given to Problem 12 shows that if S is a parallelogram that contains arbitrarily the points of its boundary, A is a subset of one of the sides of S, then $S\sim S\setminus A$.

Given P, a planar polygon, decompose its boundary as $A_1\cup\cdots\cup A_{2n}$ where each A_i is a half of one of the sides. Then use the above argument to show that

$$P\sim P\setminus A_1\sim P\setminus(A_1\cup A_2)\sim\cdots\sim P\setminus(A_1\cup\cdots\cup A_{2n})$$

(for every j we can find a small enough parallelogram in $P\setminus(A_1\cup\cdots\cup A_j)$ that includes A_{j+1} on its boundary).

14. What we suppose is that P, Q can geometrically be decomposed into the subpolygons $P_1,\ldots,P_t$, and $Q_1,\ldots,Q_t$ such that Q_i is the f_i-map of P_i via some congruences f_i. The problem is with the boundary points. Each and every one of them can be multiply covered or not covered at all. First we assign the boundary points arbitrarily to one of the sets, so we have the partitions $P=P_1\cup\cdots\cup P_t$ and $Q=Q_1\cup\cdots\cup Q_t$. (Notice that $Q_i=f_i[P_i]$ does not necessarily hold.) Using Problem 13 we can equidecompose each P_i into its interior, $P_i\sim\operatorname{int}(P_i)$ and similarly treat each Q_i. Now we are done as clearly each f_i maps the interior of P_i onto the interior of Q_i.

15. We show that $E\sim\mathbf{Z}$ if and only if $E=\mathbf{Z}\setminus A$ for some finite A. For one direction, let A be finite. If $A=\{a_1,\ldots,a_n\}$ is its increasing enumeration, then $\mathbf{Z}\setminus A=(-\infty,a_1-1]\cup[a_1+1,a_2-1]\cup\cdots\cup[a_n+1,\infty)$. To get a decomposition of $\mathbf{Z}$ it suffices to shift the second interval by 1 to the left, the next, by 2, etc., the last interval, $[a_n+1,\infty)$, will be shifted by n to the left.

For the other direction, we show that if $\mathbf{Z}\sim\mathbf{Z}\setminus A$ then A is finite. Assume, therefore, that $\mathbf{Z}=B_1\cup\cdots\cup B_k$ is a decomposition of $\mathbf{Z}$ and $\mathbf{Z}\setminus A=f_1[B_1]\cup\cdots\cup f_k[B_k]$ where $f_i(x)=x+c_i$ for $i=1,\ldots,k$. Put $c=\max(|c_1|,\ldots,|c_k|)$. Let N be a large natural number. Notice that $f_i[B_i]\cap[-N,N]$ must include $(B_i\cap[-N+c,N-c])+c_i$ so it has at least $|B_i\cap[-N,N]|-2c$ elements. Adding up, we get that $(\mathbf{Z}\setminus A)\cap[-N,N]$ has at least

$$|(B_1\cup\cdots\cup B_k)\cap[-N,N]|-2kc=2N+1-2kc$$

elements. As N can be arbitrarily large, $|A|\le 2kc$ follows.

16. (a) Assume that some nonempty $H\subseteq\mathbf{Z}^n$ is paradoxical. By translating H we can assume that $\mathbf{0}\in H$ holds. By assumption, there is a decomposition

$H = H_1 \cup \cdots \cup H_t$ and there are vectors $\mathbf{c}_1, \ldots, \mathbf{c}_t \in \mathbf{Z}^n$ such that $(H_1 + \mathbf{c}_1) \cup \cdots \cup (H_t + \mathbf{c}_t)$ covers every point of H twice and no other points. If we set $f(\mathbf{x}) = \mathbf{x} + \mathbf{c}_i$ for $\mathbf{x} \in H_i$, then $f : H \to H$, $f(\mathbf{x}) - \mathbf{x} \in \{\mathbf{c}_1, \ldots, \mathbf{c}_t\}$ for every $\mathbf{x} \in H$, and every element of H is assumed twice by f. If $g(N) = |H_N|$ where $H_N = H \cap [-N, N]^n$ for $N = 0, 1, \ldots$ then $1 = g(0) \leq g(1) \leq \cdots$. Let the natural number c be larger than all the coordinates of all the vectors $\mathbf{c}_1, \ldots, \mathbf{c}_t$. As $f^{-1}[H_N] \subseteq H_{N+c}$ we get $g(N + c) \geq 2g(N)$, which, by induction, gives $g(Nc) \geq 2^N$. But that contradicts the obvious inequality $g(Nc) \leq (2Nc+1)^n$ for N large.

(b) Assume that the nonempty subset H of the Abelian group G is paradoxical. This means, that there is a partition $H = H_1 \cup \cdots \cup H_t$ and there are elements $g_1, \ldots, g_t \in G$ that the sets $(H_1 + g_1), \ldots, (H_t + g_t)$ cover every element in H exactly twice (and no other element). Set $f(x) = x + g_i$ for $x \in H_i$. Then $f : H \to H$, $f(x) - x \in \{g_1, \ldots, g_t\}$ for every $x \in H$, and every element of H is assumed exactly twice by f. We can assume that $0 \in H$. Let A be the subgroup of G generated by $g_1, \ldots, g_t$. Then f maps $H \cap A$ to $H \cap A$ and has exactly the same properties as f; therefore, the nonempty $H \cap A$ is paradoxical, as well. We reduced, therefore, the problem to the finitely generated case.

By the fundamental theorem of finitely generated Abelian groups, A is the direct product of finitely many cyclic groups, that is, isomorphic to $B \times \mathbf{Z}^n$ where n is a natural number and B is a finite Abelian group. $n \geq 1$ as a finite set obviously cannot be paradoxical.

Set $g(N) = |H_N|$ where $H_N = H \cap (B \times [-N, N]^n)$ for $N = 0, 1, \ldots$. Again, $1 = g(0) \leq g(1) \leq \cdots$. Let the natural number c be larger than all the coordinates in the $\mathbf{Z}^n$ part of all the vectors $g_1, \ldots, g_t$. As $f^{-1}[H_N] \subseteq H_{N+c}$ we get $g(N+c) \geq 2g(N)$, which as above gives $g(Nc) \geq 2^N$ and that contradicts the obvious inequality $g(Nc) \leq |B|(2Nc + 1)^n$ for N large.

(c) Assume that the nonempty $H \subseteq \mathbf{R}$ is paradoxical with congruences. This means that there is a decomposition $H = H_1 \cup \cdots \cup H_t$ and there exist functions $f_i : H_i \to H$ of the form $f_i(x) = x + c_i$ or $f_i(x) = -x + c_i$ such that $f_1[H_1], \ldots, f_t[H_t]$ cover every element of H exactly twice. Pick $a \in H$. Let A be the additive subgroup of $\mathbf{R}$ generated by $a, c_1, \ldots, c_t$. A is isomorphic to $\mathbf{Z}^n$ for some $n \geq 1$ and as A is closed under the functions f_i, $A \cap H$ is a similarly paradoxical set in $\mathbf{Z}^n$ where now in $\mathbf{Z}^n$ we consider the bijections generated by translations and the reflection $x \mapsto -x$. Now, as above, if the coordinates of the elements $c_1, \ldots, c_t$ in $\mathbf{Z}^n$ are bounded by c, and $g(N) = |H \cap [-N, N]^n|$ then $g(N) > 0$ for $N \geq N_0$, and $g(N + c) \geq 2g(N)$, which gives rise to an exponential growth of g, an impossibility.

17. (a) Let the generators of F_2 be x and y and let $A \subseteq F_2$ consist of those words that start with a power (positive or negative) of x. Then, the words in $yA, y^2A, \ldots$ start with $y, y^2, \ldots$ respectively, so they are disjoint, $\aleph_0 A \preceq F_2$. On the other hand, xA contains 1 and every word that starts with a power of y, so $A \cup xA = F_2$.

(b) Consider φ, the rotation around axis z with angle $\cos^{-1}\frac{1}{3}$ and ψ, the rotation around axis x with angle $\cos^{-1}\frac{1}{3}$. That is,

$$\varphi^{\pm 1} = \frac{1}{3}\begin{bmatrix} 1 & \mp 2\sqrt{2} & 0 \\ \pm 2\sqrt{2} & 1 & 0 \\ 0 & 0 & 3 \end{bmatrix} \quad \text{and} \quad \psi^{\pm 1} = \frac{1}{3}\begin{bmatrix} 3 & 0 & 0 \\ 0 & 1 & \mp 2\sqrt{2} \\ 0 & \pm 2\sqrt{2} & 1 \end{bmatrix}.$$

We show that no nontrivial product of powers of φ and ψ is the identity. Assume that $w = g_n g_{n-1} \cdots g_1$ is such a product. Suppose first that only ψ occurs in it. We can assume that $w = \psi^n$ with $n > 0$. By induction on n we get that $\psi^n(0,0,1) = \frac{1}{3^n}(0, b_n\sqrt{2}, c_n)$ where b_n, c_n are integers, in fact $b_0 = 0$, $c_0 = 1$, $b_{n+1} = b_n - 2c_n$, and $c_{n+1} = 4b_n + c_n$. Again, induction shows that if $n > 0$ is odd, then $b_n \equiv c_n \equiv 1 \pmod 3$, if $n > 0$ is even, then $b_n \equiv c_n \equiv 2 \pmod 3$, so whenever $n > 0$ then $\psi^n(0,0,1) \neq (0,0,1)$ holds.

We can assume, therefore, that $\varphi^{\pm 1}$ properly occurs in w. We show that $w(1,0,0) \neq (1,0,0)$. As $\psi(1,0,0) = (1,0,0)$ we can assume that $g_1 = \varphi^{\pm 1}$. Set $v_0 = (1,0,0)$ and $v_{i+1} = g_{i+1}(v_i)$ for $0 \leq i < n$. Induction gives that $v_i = (a_i, b_i\sqrt{2}, c_i)/3^i$ where a_i, b_i, c_i are integers, and in fact, if $v_{i+1} = \varphi^{\pm 1}(v_i)$, then

$$\begin{aligned} a_{i+1} &= a_i \mp 4b_i, \\ b_{i+1} &= \pm 2a_i + b_i, \\ c_{i+1} &= 3c_i, \end{aligned}$$

If, however, $v_{i+1} = \psi^{\pm 1}(v_i)$, then

$$\begin{aligned} a_{i+1} &= 3a_i, \\ b_{i+1} &= b_i \mp 2c_i, \\ c_{i+1} &= c_i \pm 4b_i. \end{aligned}$$

We prove that b_i is not divisible by 3 for $i > 0$ (and therefore it cannot be 0). As $g_1 = \varphi^{\pm 1}$, $(a_1, b_1, c_1) = (1, \pm 2, 0)$, we have this for $i = 1$. We complete the induction by considering cases.

- if $g_i = \varphi^{\pm 1}$, $g_{i-1} = \psi^{\pm 1}$, then $b_{i+1} = \pm 2a_i + b_i = \pm 2 \cdot 3a_{i-1} + b_i \equiv b_i \pmod 3$,
- if $g_i = \psi^{\pm 1}$, $g_{i-1} = \varphi^{\pm 1}$, then $b_{i+1} = b_i \mp 2c_i = b_i \mp 6c_{i-1} \equiv b_i \pmod 3$,
- if $g_i = g_{i-1} = \varphi^{\pm 1}$, then $b_{i+1} = \pm 2a_i + b_i = \pm 2(a_{i-1} \mp 4b_{i-1}) + b_i = -8b_{i-1} \pm 2a_{i-1} + b_i \equiv (b_{i-1} \pm 2a_{i-1}) + b_i = 2b_i \pmod 3$,
- if $g_i = g_{i-1} = \psi^{\pm 1}$, then $b_{i+1} = b_i \mp 2c_i = b_i \mp 2(c_{i-1} \pm 4b_{i-1}) = b_i \mp 2c_{i-1} - 8b_{i-1} \equiv b_i \mp 2c_{i-1} + b_{i-1} = 2b_i \pmod 3$.

(c) Let φ, ψ be two independent rotations around the center of $\mathbf{S}^2$ as in part (b). Let x, y be the generators of F_2. If $w = x^{n_0}y^{m_0} \cdots x^{n_t}y^{m_t}$ is some element of F_2, set $g(w) = \varphi^{n_0}\psi^{m_0} \cdots \varphi^{n_t}\psi^{m_t}$, this gives an isomorphic embedding of F_2 into the group of rotations of $\mathbf{S}^2$. Notice that $A = \{x \in \mathbf{S}^2 : g(w)(x) = x, \text{some } 1 \neq w \in F_2\}$ is countable, therefore $\mathbf{S}^2 \sim \mathbf{S}^2 \setminus A$ by Problem 10. Set $B = \mathbf{S}^2 \setminus A$. It suffices to prove that B is paradoxical. Define the following equivalence relation on B: $x \in B$ is equivalent to $y \in B$ if and only if $g(w)(x) = y$ for some $w \in F_2$. Then B decomposes into (countable) equivalence classes; $B = \bigcup\{B_j : j \in J\}$, and pick an element $b_j \in B_j$ from every class (this is the point where we use the axiom of choice).

As F_2 is paradoxical by part (a) and Problem 8, there are a decomposition $F_2 = A_1 \cup \cdots \cup A_t$ and elements $w_1, \ldots, w_t \in F_2$ such that the sets $w_1 A_1, \ldots, w_t A_t$ cover every element of F_2 exactly twice.

Set

$$B^i = \bigcup\{g(w)(b_j) : w \in A_i, j \in J\}$$

for $i = 1, \ldots, t$. Then $B = B^1 \cup \cdots \cup B^t$ is such a decomposition that the rotated sets $g(w_1)[B^1], \ldots, g(w_t)[B^t]$ cover every point of B exactly twice, i.e., $B \sim 2B$.

(d) It suffices to show that $A \sim \mathbf{S}^2$. As A has inner points, it includes a small open set, so finitely many, say n rotated copies of it cover $\mathbf{S}^2$. Thus $\mathbf{S}^2 \preceq nA$. As $\mathbf{S}^2$ is paradoxical, $n\mathbf{S}^2 \preceq \mathbf{S}^2$, so we get $n\mathbf{S}^2 \preceq nA$. By Problem 6 this gives $\mathbf{S}^2 \preceq A$ and as obviously $A \preceq \mathbf{S}^2$, we get, using Problem 4, that $\mathbf{S}^2 \sim A$.

(e) First we show that the centerless unit ball,

$$\mathbf{B}' = \{(x, y, z) : 0 < x^2 + y^2 + z^2 \leq 1\},$$

is paradoxical. By part (c), if $A = \{(x, y, z) : x^2 + y^2 + z^2 = 1\}$, then A can be partitioned as $A = A_1 \cup \cdots \cup A_n$ and there are rotations (around the origin) $\varphi_1, \ldots, \varphi_n$ such that in $\varphi_1[A_1] \cup \cdots \cup \varphi_1[A_n]$ every point of A is covered exactly twice. We set

$$\mathbf{B_i'} = \{(rx, ry, rz) : 0 < r \leq 1, (x, y, z) \in A_i\}.$$

Then $\mathbf{B}'$ is partitioned as $\mathbf{B}' = \mathbf{B_1'} \cup \cdots \cup \mathbf{B_n'}$ and in $\varphi_1[\mathbf{B_1'}] \cup \cdots \cup \varphi_n[\mathbf{B_n'}]$ every point of $\mathbf{B}'$ is covered exactly twice, that is, $\mathbf{B}' \sim 2\mathbf{B}'$.

Finally, as clearly $\mathbf{B}' \preceq \mathbf{B}^3 \preceq 2\mathbf{B}'$ we get that $\mathbf{B}^3$ is paradoxical, by Problem 8. (Alternatively, we can get $\mathbf{B}' \sim \mathbf{B}^3$, by considering a segment inside $\mathbf{B}^3$ one of whose endpoints is $(0, 0, 0)$, and applying Problem 11.) [S. Banach, A. Tarski: Sur la decomposition des ensembles de points en parties respectivement congruents, *Fund. Math*, **6**(1924), 244–277]

(f) Let D be a ball small enough such that both A and B include a translated copy of D. Let n be a natural number large enough that both A and B can be covered by n copies of D. Then $D \preceq A \preceq nD$ and, as D is paradoxical by part (d), $nD \sim D$, so $A \sim D$ holds by Problems 2 and 3. Similarly $B \sim D$, so by Problem 1, $A \sim B$.

18. We can assume that $\epsilon < \frac{1}{2}$. Let A, B be subsets of the $\{\langle x, y, z\rangle : z = 0\}$ plane. Let $r > 0$ be large enough that the disc $D = \{\langle x, y\rangle : x^2 + y^2 \leq r^2\}$ covers both A and B. Let E be the upper half-sphere above D, that is,

$$E = \{\langle x, y, z\rangle : x^2 + y^2 + z^2 = r^2, z \geq 0\}.$$

Notice that the projection $\pi(x, y, z) = \langle x, y\rangle$ is a bijection between E and D. Let $\delta > 0$ be a small number ($\delta = \frac{\epsilon}{10}$ suffices). The mapping $g_\delta(x, y) = \langle \delta x, \delta y\rangle$ is a δ-contraction from D to $D_\delta = \{\langle x, y\rangle : x^2 + y^2 \leq \delta^2 r^2\}$, the disc of radius δr around the origin. Set $F = \pi^{-1}[D_\delta]$, a small set around $\langle 0, 0, r\rangle$, the North Pole of E. The connecting line of any two points of F has angle $< \frac{\pi}{4}$ with our original plane. Therefore, π^{-1} on D_δ can multiply distances by at most $\cos^{-1}(\frac{\pi}{4}) = \sqrt{2}$. This implies that the composed mapping $\pi^{-1} \circ g_\delta$ is still an ϵ-contract on A.

The sets $A^* = \pi^{-1} \circ g_\delta[A]$ and $B^* = \pi^{-1}[B]$ are subsets of E with inner points. Therefore, by Problem 17 part (d) $A^* \sim B^*$, that is, there are decompositions $A^* = A_1^* \cup \cdots \cup A_n^*$ and $B^* = B_1^* \cup \cdots \cup B_n^*$ and congruences $f_i : A_i^* \to B_i^*$. If we set $A_i = g_\delta^{-1} \circ \pi[A_i^*]$ and $B_i = \pi[B_i^*]$, then $h_i : A_i \to B_i$ is a bijection, where $h_i = \pi \circ f_i \circ \pi^{-1} \circ g_\delta$. Also, h_i is an ϵ-contract, as $\pi^{-1} \circ g_\delta$ is an ϵ-contract, f_i and π are 1-contracts. [W. Sierpiński: Sur un paradoxe de M. J. von Neumann, *Fundamenta Mathematicae*, **35**(1948), 203–207]

20

Stationary sets in ω_1

1. If $A \subseteq \omega_1$ is finite then $\omega_1 \setminus A$ is a club if and only if A contains only 0 and successor ordinals. Indeed, as the limit of any sequence (of distinct elements) is a limit ordinal, if A excludes limits then $\omega_1 \setminus A$ is closed. On the other hand, if $\alpha \in A$ is limit, then every sequence $\alpha_n \to \alpha$ has a tail in $\omega_1 \setminus A$ so $\omega_1 \setminus A$ is not closed.

2.

(a) Yes. Indeed, if $A \cap C \neq \emptyset$ holds for every club set C, then it holds for B as well.

(b) No. Set $A = (\omega, \omega_1)$ and $B = [0, \omega) \cup (\omega, \omega_1)$. A is a club, but B is not even closed.

(c) Yes. This is just the contrapositive form of (a).

3. Assume that $C_0, C_1, \ldots$ are club sets, we are to show that $C = C_0 \cap C_1 \cap \cdots$ is a club set, too.

Closure is immediate: if $\alpha_n \to \alpha$, $\alpha_n \in C$, then $\alpha \in C_i$ holds for $i = 0, 1, \ldots$, so $\alpha \in C$, too.

For unboundedness assume that $\beta < \omega_1$. Recursively choose $\beta < \alpha_1 < \alpha_2 < \cdots$ such that if $n = 2^i(2j+1)$ then $\alpha_n \in C_i$. This is possible as each C_i is unbounded. Set $\alpha = \lim_n \alpha_n$. For every $i < \omega$ there is an infinite subsequence of $\{\alpha_1, \alpha_2, \ldots\}$ from C_i, so as $\alpha_n \to \alpha$, we get $\alpha \in C_i$, that is, $\alpha \in C$.

4. Assume that $N_0, N_1, \ldots$ are nonstationary sets. By definition, there exist club sets $C_0, C_1, \ldots$ such that $N_i \cap C_i = \emptyset$ $(i < \omega)$. Set $N = N_0 \cup N_1 \cup \cdots$, $C = C_0 \cap C_1 \cap \cdots$. By Problem 3, C is a club, clearly $C \cap N = \emptyset$, so N is nonstationary, as claimed.

5. One has to show that $(S \cap C) \cap D \neq \emptyset$ if D is a club. Indeed, $C \cap D$ is a club, so $S \cap (C \cap D) \neq \emptyset$.

6. Closure is easy: assume that $\alpha_n \to \alpha$ and $\alpha_n \in C = \bigtriangledown\{C_\alpha : \alpha < \omega_1\}$. For every $\beta < \alpha$ an end segment of $\{\alpha_0, \alpha_1, \ldots\}$ is in C_β (namely all terms that are greater than β) so their limit, α is in C_β, as well. As this holds for every $\beta < \alpha$, we get $\alpha \in C$.

For unboundedness let $\beta < \omega_1$ be given. Select, by recursion, the elements $\beta = \alpha_0 < \alpha_1 < \cdots$ such that $\alpha_{n+1} \in \bigcap\{C_\gamma : \gamma < \alpha_n\}$ (possible as by Problem 3 the intersection of countably many club sets is unbounded again). Let $\alpha = \lim_n \alpha_n$. We claim that $\alpha \in C$. Indeed, if $\gamma < \alpha$ then, for some n, $\gamma < \alpha_n < \alpha_{n+1} < \cdots < \alpha$ holds, so $\alpha_{n+1}, \alpha_{n+2}, \ldots$ are all in C_γ, therefore $\alpha \in C_\gamma$. As this holds for all $\gamma < \alpha$, $\alpha \in C$ holds.

7. For closure, assume that $\alpha_0 < \alpha_1 < \cdots$ are from $C(f)$ and $\alpha = \lim_n \alpha_n$. in order to show that $\alpha \in C(f)$ assume that $\beta_1, \ldots, \beta_k < \alpha$. There is some n that $\beta_1, \ldots, \beta_k < \alpha_n$ holds, so $f(\beta_1, \ldots, \beta_k) < \alpha_n < \alpha$ and we are done.

For unboundedness, let $\beta < \omega_1$ be given. Select $\beta = \alpha_0 < \alpha_1 < \cdots$ in such a fashion that α_{k+1} is a strict upper bound for the countable set

$$\{f(\beta_1, \ldots, \beta_n) : \beta_1, \ldots, \beta_n < \alpha_k\}.$$

If $\alpha = \lim_k \alpha_k$ then whenever $\beta_1, \ldots, \beta_n < \alpha$ then for some $k < \omega$ we will have $\beta_1, \ldots, \beta_n < \alpha_k$ so $f(\beta_1, \ldots, \beta_n) < \alpha_{k+1} < \alpha$, so $\alpha \in C(f)$ holds.

8. For $\alpha < \omega_1$ let $f(\alpha)$ be the least element of C, strictly above α (and define $f(\alpha_1, \ldots, \alpha_n)$ arbitrarily for $n \geq 2$). We show that $C(f) \setminus \{0\} \subseteq C$. Assume that $\gamma \in C(f) \setminus \{0\}$. γ cannot be successor, as if $\gamma = \beta + 1$ then $\beta < \gamma$, so $f(\beta) < \gamma$, an impossibility. So γ is limit, and select a sequence $\gamma_0, \gamma_1, \ldots$, converging to γ. For every $n < \omega$, $\gamma_n < f(\gamma_n) < \gamma$, that is, γ is the limit of $f(\gamma_0), f(\gamma_1), \ldots$, and, as these ordinals are elements of C, so is γ.

9. Assume that $f : \omega_1 \to \omega_1$ is strictly increasing and C is its range. As $f(\alpha) \geq \alpha$ for every $\alpha < \omega_1$, C is unbounded. Assume that $\alpha_0 < \alpha_1 < \ldots$ are from C and $\alpha = \lim_n \alpha_n$. Then $\alpha_n = f(\beta_n)$ for some β_n, and $\beta_0 < \beta_1 < \cdots$ as f is strictly increasing. Set $\beta = \lim_n \beta_n$. Then $f(\beta) = \alpha$ holds by continuity, and we are done.

For the other direction let $C \subseteq \omega_1$ be a club set. Define $f : \omega_1 \to \omega_1$ the following way: $f(\alpha) =$ the αth element of C. Clearly, C is the range of f and f is strictly increasing. For continuity, assume that $\beta_0 < \beta_1 < \cdots$ and $\beta_n \to \beta$. Set $\alpha_n = f(\beta_n)$. Then $\alpha_0 < \alpha_1 < \cdots$ and if $\alpha_n \to \alpha$, then $\alpha \in C$, so it is the βth element of C, therefore $\alpha = f(\beta)$.

10. The continuity of f and g guarantees that $\{\alpha : f(\alpha) = g(\alpha)\}$ is closed (if $\alpha_n \to \alpha$ then $f(\alpha) = \lim_n f(\alpha_n) = \lim_n g(\alpha_n) = g(\alpha)$). Toward showing unboundedness let $\beta < \omega_1$ be given. Define the sequence $\alpha_0 < \alpha_1 < \cdots$

the following way. $\alpha_0 = \beta$ and for $n < \omega$, α_{n+1} is greater than $f(\alpha_n)$ and $g(\alpha_n)$. Set $\alpha = \lim_n \alpha_n$. Then, by monotonicity, $\alpha \le f(\alpha), g(\alpha)$, and also $f(\alpha) = \lim_n f(\alpha_n) \le \alpha$, $g(\alpha) = \lim_n g(\alpha_n) \le \alpha$ hold. Therefore, $\beta < \alpha$ and $f(\alpha) = g(\alpha)$.

11. Assume that $\alpha_0 < \alpha_1 < \cdots$ are epsilon numbers, $\alpha = \lim_n \alpha_n$. Then

$$\alpha = \lim_n \alpha_n = \lim_n \omega^{\alpha_n} = \omega^\alpha,$$

and this gives closure.

Toward proving unboundedness assume that $\beta < \omega_1$. Define the sequence $\beta = \alpha_0 \le \alpha_1 \le \cdots$ by taking $\alpha_{n+1} = \omega^{\alpha_n}$. Unless β is an epsilon number (in which case we are done), this sequence is strictly increasing. Therefore, for $\alpha = \lim_n \alpha_n$ we have

$$\omega^\alpha = \lim_n \omega^{\alpha_n} = \lim_n \alpha_{n+1} = \alpha.$$

12. Assume the statement is false. Then for every ordinal $\alpha < \omega_1$ there is a bound $g(\alpha) < \omega_1$ such that if ξ is greater than $g(\alpha)$ then $f(\xi) \ne \alpha$. Define the sequence $0 = \alpha_0 < \alpha_1 < \cdots$ as follows. For every $n < \omega$, the ordinal α_{n+1} is greater than $g(\beta)$ for every $\beta < \alpha_n$. This is possible, as every countable set of countable ordinals is bounded below ω_1. Set $\alpha = \lim_n \alpha_n$. Now we are in trouble with $f(\alpha)$: if $\beta = f(\alpha)$, then $\beta < \alpha$ by condition, so $\beta < \alpha_n$ for some n, but then α, an element of $f^{-1}(\beta)$, must be smaller than $\alpha_{n+1} < \alpha$, a contradiction.

13. If, on the contrary, every value is assumed only countably many times, then there is a function g such that for $\alpha < \omega_1$, $g(\alpha)$ is an upper bound for the countably many elements of $f^{-1}(\alpha)$. By Problem 7 there is a club set $C(g)$ such that if $\alpha < \beta \in C(g)$ then $g(\alpha) < \beta$ holds. Pick $\alpha \in S \cap C(g)$, $\alpha > 0$, and let $\beta = f(\alpha)$. Then $\beta < \alpha$ (as f is regressive) and $\alpha \le g(\beta)$ but that contradicts $\alpha \in C(g)$. [W. Neumer: Verallgemeinerung eines Satzes von Alexandrov und Urysohn, *Math. Z.*, **54**(1951), 254–261]

14. As N is nonstationary, there is a club set C, disjoint from N. For $\alpha \in N$, $\alpha > 0$ let $f(\alpha) = \sup(C \cap \alpha)$. Clearly, f is regressive. Notice that $f(\alpha) = 0$ if α is smaller than the least element of C. To show the property required assume that $\beta < \omega_1$. Choose some element $\gamma \in C$, with $\gamma > \beta$. Then $f(\alpha) \ge \gamma > \beta$ holds for $\alpha \in N$, $\alpha > \gamma$, that is, all elements of $f^{-1}(\beta)$ are below γ, and so $f^{-1}(\beta)$ is countable.

15. We first show that for every $\alpha < \omega_1$ there is a regressive function g_α : $(0, \alpha] \to [0, \alpha)$ that assumes every value at most twice. In order to prove this by transfinite induction, we will only consider ordinals of the form $\alpha = \omega \cdot \beta$,

indeed, if $\alpha < \omega \cdot \beta$, then the restriction of $g_{\omega\cdot\beta}$ to $(0,\alpha]$ is appropriate. Also, we require that infinitely many values in $[0,\alpha)$ be taken at most once (this will be our inductive side condition).

For $\beta = 1$ we can take $f_\omega(n+1) = n$, $f_\omega(\omega) = 0$.

To proceed from β to $\beta+1$, if $g_{\omega\cdot\beta}$ is given, we define

$$g_{\omega\cdot\beta+\omega}(\xi) = \begin{cases} g_{\omega\cdot\beta}(\xi), & \text{for } \xi \le \omega\cdot\beta; \\ \omega\cdot\beta+n, & \text{for } \xi = \omega\cdot\beta+n+1; \\ \omega\cdot\beta, & \text{for } \xi = \omega\cdot\beta+\omega. \end{cases}$$

If β is limit, we can present it as a sum $\beta = \beta_0+\beta_1+\cdots$ of nonzero smaller ordinals. Now set, for $\xi \le \omega\cdot\beta_i$,

$$g_{\omega\cdot\beta}\big(\omega\cdot(\beta_0+\cdots+\beta_{i-1})+\xi\big) = \omega\cdot(\beta_0+\cdots+\beta_{i-1}) + g_{\omega\cdot\beta_i}(\xi),$$

and let $g_{\omega\cdot\beta}(\omega\cdot\beta)$ be any of the infinitely many values $< \omega\cdot\beta$ that are taken at most once.

Turning to the solution of the problem, let N be a nonstationary set. We can as well assume that $0 \notin N$, and so $0 \in C$, where C is a closed, unbounded set, disjoint from N. Let $C = \{\gamma_\xi : \xi < \omega_1\}$ be the increasing enumeration of C. For every $\xi < \omega_1$ let δ_ξ be the unique ordinal such that $\gamma_{\xi+1} = \gamma_\xi + \delta_\xi$. If we now define

$$f(\gamma_\xi + \alpha) = \gamma_\xi + g_{\delta_\xi}(\alpha)$$

for $\alpha < \delta_\xi$, then f is a regressive function on $\omega_1 \setminus C$, and as it maps $(\gamma_\xi, \gamma_{\xi+1})$ into $[\gamma_\xi, \gamma_{\xi+1})$, and these intervals are disjoint, it will have the property that every value is assumed at most twice. [G. Fodor, A. Máté: Some results concerning regressive functions, *Acta Sci. Math.*, **30**(1969), 247–254]

16. **First solution.** Let f be a putative counterexample on the stationary set S. Then for every $\alpha < \omega_1$ there is a club set C_α such that $f^{-1}(\alpha) \cap C_\alpha = \emptyset$. Set $C = \triangledown\{C_\alpha : \alpha < \omega_1\}$, the diagonal intersection. By Problem 6 this C is a club set. Pick $\alpha \in C \cap S$, $\alpha > 0$. Let $\beta = f(\alpha)$. This gives a contradiction: $\beta < \alpha$, $\alpha \notin C_\beta$, so $\alpha \notin C$, either.

Second solution. We use the characterization of stationary sets given in Problems 13 and 14: a set $A \subseteq \omega_1$ is stationary if and only if every regressive function on A assumes some value on an uncountable set. For a proof by contradiction let $S \subseteq \omega_1$ be a stationary set and $f : S \to \omega_1$ a regressive function, such that every $f^{-1}(\alpha)$ is nonstationary, so let $f_\alpha : f^{-1}(\alpha) \to \omega_1$ be a regressive function that assumes every value countably many times. Then $g(\xi) = \max\big(f(\xi), f_{f(\xi)}(\xi)\big)$ is a regressive function on S, so there is a value, say γ, which is assumed on an uncountable set X. For $\xi \in X$, $f(\xi) \le \gamma$ holds, so by the pigeon hole principle there is an uncountable $Y \subseteq X$, such that $f(\xi) = \delta$ for $\xi \in Y$. For $\xi \in Y$ we have $f_\delta(\xi) \le \gamma$, so $f_\delta(\xi) = \epsilon$ for $\xi \in Z$, with Z uncountable, a contradiction.

17. **First solution.** Decompose S as

$$S = S_0 \cup S_1 \cup \cdots \qquad \text{with} \qquad S_n = \{\alpha \in S : |F(\alpha)| = n\}.$$

As the union of countably many nonstationary sets is nonstationary, we can consider an S_n that is stationary. For every $\alpha \in S_n$, let $F(\alpha) = \{f_1(\alpha), \dots, f_n(\alpha)\}$ be the increasing enumeration. With n successive applications of Fodor's theorem (Problem 16) we get a stationary set $S^* \subseteq S_n$ and ordinals $\gamma_1 < \cdots < \gamma_n$ such that for $\alpha \in S^*$, $f_1(\alpha) = \gamma_1, \dots, f_n(\alpha) = \gamma_n$ hold. That is, $F(\alpha) = s$ for $\alpha \in S^*$, where $s = \{\gamma_1, \dots, \gamma_n\}$.

Second solution. Let g be a function that codes, in a one-to-one fashion, the finite sets in ω_1 into countable ordinals (identifying finite sets with increasing sequences). For example, $g(0) = 0$, $g(\gamma_1, \dots, \gamma_n) = \omega^{\gamma_n} + \cdots + \omega^{\gamma_1}$ is one possibility. By Problem 7 there is a club set C that is closed under g, that is, if $\gamma_1 < \cdots < \gamma_n < \alpha \in C$ then $g(\gamma_1, \dots, \gamma_n) < \alpha$ holds. $S' = S \cap C$ is stationary, and on S' we consider the regressive $f(\alpha) = g\big(F(\alpha)\big)$. By Fodor's theorem $f(\alpha) = \gamma$ on a stationary subset S^* of S', and clearly F assumes the finite set $g^{-1}(\gamma)$ on S^*.

18. Suppose the player can play through ω_1 steps. The coin that she inserts at step $0 < \alpha < \omega_1$ must have been obtained at some step $f(\alpha) < \alpha$. By Problem 13 there are a value τ and uncountably many α such that $f(\alpha) = \tau$. But that means that at step τ the machine returned uncountably many quarters, a contradiction.

19. **First solution.** If not, then every subset of ω_1 is either nonstationary or includes a closed, unbounded subset. If $\alpha < \omega_1$, let $f_n(\alpha) \to \alpha$ be a sequence converging to α. By Fodor's theorem (Problem 16) for every $n < \omega$ there is a γ_n such that $X_n = f_n^{-1}(\gamma_n)$ is stationary. By our hypothesis, every X_n includes a club subset, hence so does $X = \bigcap\{X_n : n < \omega\}$ (Problem 3). But then the elements of X are all the limits of the same convergent sequence $(\gamma_n)_{n<\omega}$, an impossibility.

Second solution. If not, then every subset of ω_1 is either nonstationary or includes a closed, unbounded subset. Let $\alpha \mapsto f(\alpha) = \langle f_0(\alpha), f_1(\alpha), \dots \rangle$ be an injection of ω_1 into ${}^{\omega}\{0,1\}$. For every $i < \omega$ there is, by our indirect assumption, a unique $\epsilon_i = 0$ or 1, such that $A_i = \{\alpha : f_i(\alpha) = \epsilon_i\}$ includes a club subset. But then, as the intersection of countably many club sets is closed, unbounded again, for club many α we have $f(\alpha) = \langle \epsilon_0, \epsilon_1, \dots \rangle$ which is impossible, as there is at most one ordinal α with $f(\alpha)$ assuming this fixed value.

20. We can assume that f maps into $(0,1)$ (by composing the original function with a monotonic $\mathbf{R} \to (0,1)$ mapping). For every limit $\alpha < \omega_1$, $1 \le n < \omega$, there is a $g_n(\alpha) < \alpha$ such that the oscillation of f in $(g_n(\alpha), \alpha)$ is at most

$\frac{1}{n}$. By Problem 12 there are γ_n and uncountable X_n such that $g_n(\alpha) = \gamma_n$ for $\alpha \in X_n$. As X_n is unbounded, f oscillates at most $\frac{1}{n}$ in (γ_n, ω_1), and if $\gamma = \sup_n \gamma_n$ then the oscillation of f in (γ, ω_1) is 0, i.e., it is constant there.

21. See the previous proof. See also Problem 8.43.

22. Assume that d is a metric on the ordered ω_1. For $1 \le n < \omega$, if $\alpha < \omega_1$ is limit, there is some $f_n(\alpha) < \alpha$, such that $d\big(f_n(\alpha), \alpha\big) \le \frac{1}{n}$. By Problem 12 there exist γ_n and uncountable sets X_n such that $f_n(\alpha) = \gamma_n$ for $\alpha \in X_n$. Let C_n be the closure of X_n. C_n is closed and as $X_n \subseteq C_n$ it is uncountable, so it is closed, unbounded. As $d(\gamma_n, \alpha) \le \frac{1}{n}$ for $\alpha \in X_n$, the diameter of X_n, and therefore of C_n, is at most $\frac{2}{n}$. So the diameter of $C = C_1 \cap C_2 \cap \cdots$ is 0, and this contradicts that C, a club set, has more than one point.

23. (a) We start by noticing that $\alpha \times \beta$ is normal for α, β countable (it can be embedded into $\mathbf{R} \times \mathbf{R}$). Assume that $F_0, F_1 \subseteq \alpha \times \omega_1$ are disjoint, closed sets. For $\beta < \alpha$, $i < 2$ set $K_i(\beta) = \big\{\gamma < \omega_1 : \langle \beta, \gamma\rangle \in F_i\big\}$, and set $\beta \in H_i$ if $K_i(\beta)$ is uncountable. Notice that $K_i(\beta)$ is always closed and so if $\beta \in H_i$ then $K_i(\beta)$ is a club subset of ω_1. As F_0, F_1 are disjoint, so are $K_0(\beta)$, $K_1(\beta)$ for $\beta < \alpha$, therefore $H_0 \cap H_1 = \emptyset$. We further claim that H_0, H_1 are closed. Indeed, let $x_n \to x$, $x_n \in H_i$ $(n < \omega)$. If $\gamma \in K_i(x_0) \cap K_i(x_1) \cap \cdots$, then $\langle x_n, \gamma\rangle \in F_i$ for all n, and since F_i is closed, it follows that $\langle x, \gamma\rangle \in F_i$. Thus, $C \subseteq K_i(x)$ where C is the closed, unbounded, therefore uncountable set of the limit points of $K_i(x_0) \cap K_i(x_1) \cap \cdots$, so $x \in H_i$. Let $\gamma < \omega_1$ be large enough to bound every bounded $K_i(\beta)$ $(i < 2,\ \beta < \alpha)$. Now $\alpha \times \omega_1$ splits into the open components $\alpha \times (\gamma + 1)$ and $\alpha \times [\gamma + 1, \omega_1)$. It suffices to separate F_0, F_1 in both of them, separately. The first space is normal, as we have seen.

Let π denote the projection to the first coordinate in $\alpha \times [\gamma + 1, \omega_1)$: $\pi\big(\langle x, y\rangle\big) = x$. Then $H_0 = \pi[F_0]$, $H_1 = \pi[F_1]$ are disjoint, closed subsets of α. They can, therefore, be separated by disjoint open sets: $F_0 \subseteq G_0$, $F_1 \subseteq G_1$, $G_0 \cap G_1 = \emptyset$, and then the disjoint, open $\pi^{-1}[G_0]$, $\pi^{-1}[G_1]$ will separate F_0, F_1.

(b) Let F_0, F_1 be disjoint closed sets in $\omega_1 \times \omega_1$. Assume first that for every $\alpha < \omega_1$, both sets have points in $[\alpha, \omega_1) \times [\alpha, \omega_1)$. Then, we can select by induction the points $\langle x_0, y_0\rangle, \langle x_1, y_1\rangle, \ldots$ such that $\max(x_n, y_n) < \min(x_{n+1}, y_{n+1})$ for $n = 0, 1, \ldots$, $\langle x_{2n}, y_{2n}\rangle \in F_0$ and $\langle x_{2n+1}, y_{2n+1}\rangle \in F_1$ $(n = 0, 1, \ldots)$. Then the two increasing sequences $x_0, x_1, \ldots$ and $y_0, y_1, \ldots$ converge to the same ordinal α and $\langle \alpha, \alpha\rangle \in F_0 \cap F_1$ a contradiction. We have, therefore, that for some $\alpha < \omega_1$, either F_0 or F_1 has no elements in $[\alpha, \omega_1) \times [\alpha, \omega_1)$. Then we have to separate F_0, F_1 in the disjoint components $[\alpha + 1, \omega_1) \times [\alpha + 1, \omega_1)$, $(\alpha + 1) \times \omega_1$, $[\alpha + 1, \omega_1) \times (\alpha + 1)$. In the first set this is trivial (one of them is the empty set), the other two are treated in part (a).

24. Set

$$F_0 = \{\omega_1\} \times \omega_1 = \big\{\langle \omega_1, \alpha\rangle : \alpha < \omega_1\big\},$$

$$F_1 = \{\langle \alpha, \alpha \rangle : \alpha < \omega_1\}.$$

We show that they are disjoint, closed sets that cannot be separated. It is obvious that $F_0 \cap F_1 = \emptyset$. F_0 is closed as its complement is the union of the open sets of the form $\alpha \times \omega_1$ ($\alpha < \omega_1$). F_1 is closed as its complement is the union of the open sets of the forms $\alpha \times (\alpha, \omega_1)$ or $(\alpha, \omega_1] \times \alpha$ ($\alpha < \omega_1$).

Assume that $G_0 \supseteq F_0$, $G_1 \supseteq F_1$ are disjoint open sets. For every limit $\alpha < \omega_1$ there is some $f(\alpha) < \alpha$ such that

$$(f(\alpha), \alpha] \times (f(\alpha), \alpha] \subseteq G_1.$$

By Problem 13 there are γ and an uncountable X such that $f(\alpha) = \gamma$ holds for $\alpha \in X$. Then, G_1 includes $(\gamma, \omega_1) \times (\gamma, \omega_1)$, the union of the sets $(\gamma, \alpha] \times (\gamma, \alpha]$ for $\alpha \in X$. But every point in $\{\omega_1\} \times (\gamma, \omega_1)$ is a limit point of this latter set, so $G_0 \supseteq F_0$ cannot be disjoint from G_1. This contradiction proves the claim.

25. Enumerate the sets as $\{A_\alpha : \alpha < \omega_1\}$ with $\min(A_\alpha)$ strictly increasing. Clearly, $\min(A_\alpha) \geq \alpha$. We claim that the union

$$A = \bigcup \{A_{\alpha+1} : \alpha < \omega_1\}$$

is nonstationary. If it was stationary, then the regressive function $f(x) = \alpha$ for $x \in A_{\alpha+1}$ would assume a value on a stationary set by Problem 16, but this contradicts the assumption that every $A_{\alpha+1}$ is nonstationary. [G. Elekes]

26. Fix, prior to the game, the distinct reals $\{r_\alpha : \alpha < \omega_1\} \subseteq [0,1]$. What II has to do in her nth step is to force $\{r_\alpha : \alpha \in A_{2n+1}\}$ into an interval of length $1/2^{n+1}$. We show by induction that this can be done. Assume that $\{r_\alpha : \alpha \in A_{2n}\}$ is in some interval $[x, y]$ of length $1/2^n$. Then one of $\{\alpha \in A_{2n} : r_\alpha \in [x, \frac{x+y}{2}]\}$ and $\{\alpha \in A_{2n} : r_\alpha \in [\frac{x+y}{2}, y]\}$ is stationary so can be chosen as A_{2n+1}. Now $A_0 \cap A_1 \cap \cdots$ can only contain one point.

27. Assume that the stationary sets $\{A_\alpha : \alpha < \omega_2\}$ have pairwise nonstationary intersections. For a given $\alpha < \omega_2$ enumerate α as $\alpha = \{\gamma_\xi(\alpha) : \xi < \omega_1\}$. Define

$$B_\alpha = A_\alpha \cap \bigcup \{A_{\gamma_\xi(\alpha)} \setminus (\xi + 1) : \xi < \omega_1\}.$$

This set is nonstationary for the following reason. If it was not, then for every $x \in B_\alpha$ there was a $\xi < x$ such that $x \in A_{\gamma_\xi(\alpha)}$. The regressive function $x \mapsto \xi$ assumes—by Fodor's theorem—a constant value ξ on some stationary set, but then that stationary set would be a subset of the nonstationary $A_\alpha \cap A_{\gamma_\xi(\alpha)}$, a contradiction.

We can now consider the system $\{A_\alpha \setminus B_\alpha : \alpha < \omega_2\}$ of stationary sets. We show that the pairwise intersections are countable: if $\beta < \alpha$, say $\beta = \gamma_\xi(\alpha)$, then

$$(A_\beta \setminus B_\beta) \cap (A_\alpha \setminus B_\alpha) \subseteq A_\beta \cap (A_\alpha \setminus B_\alpha) \subseteq \xi + 1.$$

28. Assume that $\{C_\alpha : \alpha < \omega_2\}$ are closed, unbounded sets in ω_1. First, we claim that there is some $\alpha < \omega_2$ such that for every $\xi < \omega_1$ there are $\aleph_2$ indices β that $C_\beta \cap \xi = C_\alpha \cap \xi$ holds. Indeed, otherwise, for every α there is some $\xi < \omega_1$ such that only for at most $\aleph_1$ values of β does $C_\beta \cap \xi = C_\alpha \cap \xi$ hold. By the pigeon hole principle, for $\aleph_2$ many $\alpha < \omega_2$ the same $\xi < \omega_1$ applies. As CH holds, for $\aleph_2$ many $\alpha < \omega_2$ the sets $C_\alpha \cap \xi$ are identical (there are only $\aleph_1$ subsets of ξ). But this is a contradiction to the stated property of ξ.

Let $\alpha < \omega_2$ have the above property. Choose, by transfinite recursion, distinct ordinals $\{\beta_\xi : \xi < \omega_1\}$ as follows. If $\{\beta_\zeta : \zeta < \xi\}$ are already chosen, then let $\beta_\xi \notin \{\beta_\zeta : \zeta < \xi\}$ be such that $C_{\beta_\xi} \cap (\xi + 1) = C_\alpha \cap (\xi + 1)$. Set $A = \bigcap\{C_{\beta_\xi} : \xi < \omega_1\}$. As it is the intersection of closed sets, A is closed. In order to show that A is also unbounded, we prove that $C_\alpha \setminus A$ is nonstationary. Indeed, if $x \in C_\alpha$, but $x \notin A$ then there is some ξ that $x \notin C_{\beta_\xi}$. Here we cannot have $x \leq \xi$ for then $x \in C_\alpha \cap (\xi+1) = C_{\beta_\xi} \cap (\xi+1) \subseteq C_{\beta_\xi}$ would hold, so $\xi < x$. That is, the mapping $x \mapsto \xi$ is regressive, and as clearly it assumes every value on a nonstationary set (ξ on a subset of $\omega_1 \setminus C_{\beta_\xi}$), its domain must be nonstationary, as well by Problem 16. [F. Galvin, cf. J. E. Baumgartner, A. Hajnal, A. Máté: Weak saturation properties of ideals, *Coll. Math. Soc. J. Bolyai* **10**, *Infinite and Finite Sets*, Keszthely, 1973, 137–158]

29. Assume that $\{f_\alpha : \alpha < \omega_2\}$ are $\omega_1 \to \omega$ functions such that f_β and f_α differ on the closed, unbounded $D_{\beta\alpha}$ for $\beta < \alpha < \omega_2$. By taking diagonal intersection (see Problem 6) of the sets $\{D_{\beta\alpha} : \beta < \alpha\}$ we can get a closed, unbounded C_α such that for $\beta < \alpha$ $f_\beta(\xi) \neq f_\alpha(\xi)$ on an end segment of C_α. Assume that for $\nu < \omega_1$ the function $h_\nu : \nu \cup \{\nu\} \to \omega$ is an injection. For $\alpha < \omega_2$, $\xi < \omega_1$ let $\delta = \sup(C_\alpha \cap \xi)$ (with $\sup(\emptyset) = 0$), $g_\alpha(\xi) = \langle h_\xi(\delta), f_\alpha(\delta)\rangle$. That is, $g_\alpha : \omega_1 \to \omega \times \omega$. We show that if $\beta < \alpha < \omega_2$ then g_α, g_β are eventually different. Assume that $g_\beta(\xi) = g_\alpha(\xi)$. Then, the corresponding δ values are the same and also $f_\beta(\delta) = f_\alpha(\delta)$ holds for this common value δ. We get, therefore, that $\xi < \delta'$, where $\delta' \in C_\alpha$ is so large that $f_\beta(\gamma) \neq f_\alpha(\gamma)$ holds for $\gamma \in C_\alpha$, $\gamma \geq \delta'$. [R. Jensen]

30. Define f with the following property: if $\gamma < \omega_1$ is 0 or a limit ordinal, then let f restricted to $[\gamma, \gamma + \omega)$ assume every value below $\gamma + \omega$ (and of course, we must make f regressive). To show that f is, indeed, as required, assume that $\alpha < \omega_1$ is a limit ordinal. Then it is of the form $\alpha = \omega \cdot \beta$, and if here β is a successor ordinal, say $\beta = \gamma + 1$, then $\alpha = (\omega \cdot \gamma) + \omega$ where $\omega \cdot \gamma$ is 0 or a limit ordinal. In this case simply select $\alpha_0 = \omega \cdot \gamma$, then inductively find $\alpha_n \alpha_{n+1} < \omega \cdot \gamma + \omega$ with the property that $f(\alpha_{n+1}) = \alpha_n$. Clearly, this sequence converges to α. If β is not a successor ordinal then it is a limit ordinal, and then first select an increasing sequence $\beta_n \to \beta$. Then $\omega \cdot \beta_n$ converges to $\omega \cdot \beta$. Set $\alpha_0 = \omega \cdot \beta_0$, and inductively let α_{n+1} be the unique ordinal in $[\omega \cdot \beta_n, \omega \cdot \beta_n + \omega)$ for which $f(\alpha_{n+1}) = \alpha_n$ holds. This can be done, and the sequence $\{\alpha_n\}$ must clearly converge to α. [J. E. Baumgartner]

21

Stationary sets in larger cardinals

1. Assume that $\mu < \kappa$ and $\{C_\xi : \xi < \mu\}$ are club sets, we are to show that $C = \bigcap\{C_\xi : \xi < \mu\}$ is a club set, too.

Closure is immediate, in fact, the intersection of an arbitrary number of closed sets is closed as well.

In order to show unboundedness, assume that $\beta < \kappa$. Recursively define $\beta = \alpha_0 < \alpha_1 < \cdots$ such that every C_ξ has a point in each interval $[\alpha_i, \alpha_{i+1})$. This is possible as the C_ξ's are unbounded and every set of at most μ points is bounded. Set $\alpha = \sup\{\alpha_i : i < \omega\}$. Then, for each $\xi < \mu$, α is the limit of points in C_ξ, therefore $\alpha \in C_\xi$, so $\alpha \in C$ as well.

2. Let B be the set of ordinals specified in the problem. It is immediate that B is closed. For unboundedness, let $\alpha < \kappa$ be arbitrary, we are going to find $\beta \in B, \beta \geq \alpha$. If $\alpha \in B$, we are done. Otherwise, select inductively the ordinals $\alpha = \alpha_0 < \alpha_1 < \cdots$ in such a way that the order type of $C \cap \alpha_{n+1}$ is exactly α_n. As the order type of C is κ, this is possible, and induction gives that this sequence is strictly increasing, and so $C \cap (\alpha_{n+1} \setminus \alpha_n) \neq \emptyset$ for all n. Hence, if β is the limit of the sequence, then $C \cap \beta$ is cofinal in β, and the order type of $C \cap \beta$ is the limit of the order types of $C \cap \alpha_n$, i.e., it is β. That is, $\beta \in B$, and we are done.

3. For the forward direction, assume that $f : [\kappa]^{<\omega} \to [\kappa]^{<\kappa}$. To show that $C(f)$ is closed, assume that $\alpha_\tau \in C(f)$ for $\tau < \mu$, $\alpha_\tau \to \alpha < \kappa$. If $\beta_1, \ldots, \beta_n < \alpha$, then $\beta_1, \ldots, \beta_n < \alpha_\tau$ holds for some $\tau < \mu$, and as $\alpha_\tau \in C(f)$, $f(\beta_1, \ldots, \beta_n) \subseteq \alpha_\tau \subseteq \alpha$ holds.

To show that $C(f)$ is unbounded, let $\beta < \kappa$ be given. We define the increasing sequence $\beta = \alpha_0 < \alpha_1 < \cdots$ where $\alpha_{i+1} > \alpha_i$ is a bound for every $f(s)$, $s \in [\alpha_i]^{<\omega}$: $f(s) \subseteq \alpha_{i+1}$. Such an α_{i+1} exists as the union of $< \kappa$ sets each of cardinality $< \kappa$ is bounded below κ. If we set $\alpha = \lim\{\alpha_i : i < \omega\}$ then $\alpha \in C(f)$, as every finite set $s \subseteq \alpha$ is in some α_i, therefore $f(s) \subseteq \alpha_{i+1} \subseteq \alpha$.

For the other direction, let $C \subseteq \kappa$ be a club. Set $f(\alpha) = \min(C \setminus (\alpha+1))$, that is, the least element of C which is strictly larger than α. In order to show that $C(f) \setminus \{0\} \subseteq C$, pick $\gamma \in C(f)$, $\gamma > 0$. Clearly, $\gamma \geq f(0)$. If $\gamma \notin C$, then $\gamma > f(0) \in C$, so $\delta = \sup(C \cap \gamma)$ exists and clearly is in C. But then $\delta < \gamma < f(\delta)$, and so $\gamma \notin C(f)$, a contradiction. Thus, we must have $\gamma \in C$.

4. By Problems 1 and 3 for almost all α the set $\{a_\gamma : \gamma < \alpha\}$ is closed under all operations of $\mathcal{A}$ hence it is a substructure.

5. Set $C = \{\alpha < \kappa : \beta < \alpha \longrightarrow \alpha \in C_\beta\}$. For closure, assume that $\alpha_\tau \to \alpha$ for $\tau < \mu$, $\alpha_\tau \in C$. If $\beta < \alpha$, then there is an $\eta < \mu$ such that $\beta < \alpha_\tau < \alpha$ holds for $\eta < \tau < \mu$, therefore $\alpha_\tau \in C_\beta$, so (as C_β is closed) $\alpha \in C_\beta$.

For unboundedness, let $\beta < \kappa$. Define, by induction, the sequence $\beta = \alpha_0 < \alpha_1 < \cdots$ where $\alpha_{i+1} > \alpha_i$ is in $\bigcap\{C_\xi : \xi < \alpha_i\}$ (possible, by Problem 1). Set $\alpha = \sup\{\alpha_i : i < \omega\}$. Then, if $\gamma < \alpha$, there is some $i < \omega$ such that $\gamma < \alpha_i$. Now, $\alpha_{i+1}, \alpha_{i+2}, \ldots \in C_\gamma$ by construction, so $\alpha \in C_\gamma$.

6. Assume that $\{N_\alpha : \alpha < \mu\}$ are nonstationary sets in κ ($\mu < \kappa$). By definition, there exist club sets C_α such that $N_\alpha \cap C_\alpha = \emptyset$ ($\alpha < \mu$). Set $N = \bigcup\{N_\alpha : \alpha < \mu\}$, $C = \bigcap\{C_\alpha : \alpha < \mu\}$. By Problem 1, C is a club, clearly $C \cap N = \emptyset$, so N is nonstationary, as claimed.

7. One has to show that $(S \cap C) \cap D \neq \emptyset$ if D is a club. Indeed, $C \cap D$ is a club, so $S \cap (C \cap D) \neq \emptyset$.

8. We have to show that every club set $C \subseteq \kappa$ contains an element with cofinality μ. Indeed, choose a strictly increasing sequence $\{\alpha_\tau : \tau < \mu\}$ of elements of C (possible, as C is unbounded in κ), then $\alpha = \sup\{\alpha_\tau : \tau < \mu\}$ is in C by closure, and obviously $\mathrm{cf}(\alpha) = \mu$. The set in question will be a club set if and only if μ is the only regular cardinal below κ, that is, if $\mu = \omega$ and $\kappa = \omega_1$.

The set $\{\alpha < \kappa : \mathrm{cf}(\alpha) \leq \mu\}$ is a club set exactly when μ is the largest regular cardinal that is less than κ, i.e., when $\kappa = \mu^+$. Finally, $\{\alpha < \kappa : \mathrm{cf}(\alpha) \geq \mu\}$ is a club set exactly when μ is the least (infinite) regular cardinal below κ, that is, when $\mu = \omega$ and $\kappa > \omega$ is arbitrary.

9. Assume the stationary $S \subseteq \kappa$ and the regressive $f : S \to \kappa$ contradict the statement. Then, for every $\alpha < \kappa$ there is a club $C_\alpha \subseteq \kappa$ such that $C_\alpha \cap f^{-1}(\alpha) = \emptyset$. Set $C = \triangledown\{C_\alpha : \alpha < \kappa\}$, the diagonal intersection of the C_α's. As C is a club set (Problem 5), there is an ordinal $\alpha > 0$, $\alpha \in S \cap C$. If $\beta = f(\alpha)$, then $\beta < \alpha$ (as f is regressive), and $\alpha \in C$ implies $\alpha \in C_\beta$ which in turn implies that $\alpha \notin f^{-1}(\beta)$, a contradiction. [G. Fodor: Eine Bemerkung zur Theorie der regressiven Funktionen, *Acta Sci. Math. (Szeged)*, **17**(1956), 139–142]

10. Set $g(\alpha) = \sup(f(\alpha))$ for $\alpha \in S$. Then, $g(\alpha) < \alpha$, as $\mathrm{cf}(\alpha) = \mu^+$ and $|f(\alpha)| \le \mu$. By Fodor's theorem, there is a stationary set $S' \subseteq S$, such that $g(\alpha) = \gamma$ for $\alpha \in S'$. There are $|\gamma|^\mu < \kappa$ distinct subsets of γ with cardinality at most μ, this splits S' into $|\gamma|^\mu$ subsets, if we consider those $\alpha \in S'$ which have a given image under f. One of them must be stationary (Problem 6), and so f is constant on that set.

11. For each $\alpha < \kappa$ there is a club set C_α disjoint from A_α. Let C be their diagonal intersection (Problem 5). If $\xi \in B = \bigcup\{A_\alpha \setminus (\alpha+1) : \alpha < \kappa\}$, then $\xi \in A_\alpha$ for some $\xi > \alpha$. Hence $\xi \notin C_\alpha$ and therefore $\xi \notin C$. Thus, C is disjoint from B and so B is nonstationary.

12. Clearly, if B is stationary then so is $A \supseteq B$.

Assume A is stationary. Set $f(x) = \min(A_\alpha)$ where $x \in A_\alpha$. Plainly, $f(x) \le x$ for $x \in A$. If $f(x) = x$ on a stationary set, then the range of f, that is, B, must be stationary. In the other case, $f(x) < x$ on a stationary set, so by Fodor's theorem, $f(x) = \alpha$ on a stationary set, but this is impossible, as A_α is nonstationary.

13. Immediate from the preceding problem.

14. Assume first that there are κ sets; $A_\alpha \subseteq \kappa$ for $\alpha < \kappa$. Set

$$A = \{\alpha < \kappa : \text{there is } \beta < \alpha, \alpha \in A_\beta\}$$

(the diagonal union). We claim that A is the least upper bound for $\{A_\alpha : \alpha < \kappa\}$. First, for every $\alpha < \kappa$, $A_\alpha \setminus A \subseteq \alpha+1$, a bounded, therefore nonstationary set. Next, assume that $B < A$. Then $A \setminus B$ is stationary, and for $\alpha \in A \setminus B$ let $f(\alpha)$ be the least $\beta < \alpha$ such that $\alpha \in A_\beta$. By the Fodor theorem, for a stationary $A' \subseteq A \setminus B$, $f(\alpha) = \beta$ holds, that is, $A' \subseteq A_\beta \setminus B$, so $A_\beta \not\le B$.

The case when there are less than κ sets is easier, and in fact it is covered by the above case if we repeat one of the sets κ times.

15. (a) As it must intersect every end segment.

(b) Set $\kappa = \mathrm{cf}(\alpha)$, a regular cardinal. By the definition of cofinality, no unbounded subset of α may have order type or cardinality less than κ. For the other direction, let $X \subseteq \alpha$ be an arbitrary cofinal set or order type κ. Let C be its closure in α. C is a closed, unbounded subset of α and its order type is still κ as the following mapping $f : C \to X$ is order preserving. For $y \in C$, $f(y)$ is the least $x \in X$, $x > y$. Indeed, if $y_0 < y_1$ are in C, then $y_0 < f(y_0) \le y_1 < f(y_1)$.

(c) Let $\alpha_0 < \alpha_1 < \cdots$ be a sequence converging to α. Then $C_0 = \{\alpha_0, \alpha_2, \ldots\}$ and $C_1 = \{\alpha_1, \alpha_3, \ldots\}$ are disjoint closed, unbounded sets.

(d) Identical with the classical case (Problem 1).

(e) See (c).

16. (a) As C is unbounded, every subset unbounded in C is unbounded in α. As C is closed, every subset closed in C is closed in α.

(b) Set $E = \{\gamma : c_\gamma \in D\}$. As $C \cap D$ is unbounded (Problem 15(d)), E is unbounded in κ. Let $\mu < \mathrm{cf}\,(\kappa)$ be a limit ordinal, and assume that $\gamma_\tau \in E$ $(\tau < \mu)$, and $\lim\{\gamma_\tau : \tau < \mu\} = \gamma$. As C is closed, $\lim\{c_{\gamma_\tau} : \tau < \mu\} = c_\gamma$ holds. As D is closed, $c_{\gamma_\tau} \in D$ for $\tau < \mu$ implies $c_\gamma \in D$, so $\gamma \in E$.

(c) If $S \subseteq \alpha$ is stationary then, by part (a), it intersects the closed, unbounded $\{c_\gamma : \gamma \in D\}$ for every closed, unbounded $D \subseteq \kappa$, so $Y = \{\gamma : c_\gamma \in S\}$ is stationary. Conversely, if $Y \subseteq \kappa$ is stationary, then, by part (b), $S = \{c_\gamma : \gamma \in Y\}$ meets every closed, unbounded subset of C, so it meets every closed, unbounded subset, therefore it is stationary.

17 (a) Let $C \subseteq \alpha$ be a closed, unbounded set of order type $\mathrm{cf}\,(\alpha)$. We can as well assume that the first two elements of C are 0 and $\mathrm{cf}\,(\alpha)$. Let $g : C \setminus \{0\} \to \mathrm{cf}\,(\alpha) \setminus \{0\}$ be a bijection. Define the regressive function $f : \alpha \setminus \{0\} \to \alpha$ as follows. If $\gamma \in C$, then let $f(\gamma) = g(\gamma)$. If, however, $\gamma \notin C$, then set $f(\gamma) = \max(C \cap \gamma)$. It is easy to see, that f is well defined and regressive. If $\gamma < \alpha$, then either $0 < \gamma < \mathrm{cf}\,(\alpha)$ and $f^{-1}(\gamma)$ consists of one point, or $\gamma \in C$ and then $f^{-1}(\gamma)$ is the open interval (γ, γ') where γ' is the next element of C, or else $f^{-1}(\gamma)$ is the empty set. In all three cases, $f^{-1}(\gamma)$ is bounded.

(b) Recall that if $\mathrm{cf}\,(\alpha) = \omega$ then there are no stationary subsets of α. Set $\kappa = \mathrm{cf}\,(\alpha)$, an uncountable regular cardinal. Let $C \subseteq \alpha$ be a closed, unbounded set of order type κ, let $C = \{c_\gamma : \gamma < \kappa\}$ be its increasing enumeration. As $S \subseteq \alpha$ is stationary, then so is $\{\gamma < \kappa : c_\gamma \in S\}$. Even $T = \{\gamma < \kappa : c_\gamma \in S, \gamma\,\mathrm{limit}\}$ is stationary, as we only remove the successor elements, a nonstationary set. For $\gamma \in T$, let $g(\gamma)$ be the least β such that $f(\gamma) < c_\beta$. As c_γ is the supremum of $\{c_\xi : \xi < \gamma\}$, we have $g(\gamma) < \gamma$ by the regressivity of f. That is, we have a regressive function (g) on a stationary set (T), so we can invoke Fodor's theorem, and get a stationary $T' \subseteq T$ such that $g(\gamma) = \beta$ holds for $\gamma \in T'$, for some $\beta < \kappa$. Then, $S' = \{c_\gamma : \gamma \in T'\}$ is stationary by Problem 16, and f is bounded by c_β on S'.

18 let B be the set of those limit ordinals $\alpha < \kappa$ for which $C \cap \alpha$ is a club set in α. Since C is closed, B is just the set of limit ordinals $\alpha < \kappa$ for which $C \cap \alpha$ is cofinal in α. It is immediate that B is closed, and that it is unbounded, follows from Problem 2.

19. (a) Assume that $S < S$. Then, there is a closed, unbounded set C such that if $\alpha \in C \cap S$, then $S \cap \alpha$ is stationary in α. Let C' be the set of limit points of C. Set $\gamma = \min(C' \cap S)$. Then, in γ, $S \cap \gamma$ is stationary, so, in particular, $\mathrm{cf}\,(\gamma) > \omega$. As $\gamma \in C'$, γ is a limit point of C, and by $\mathrm{cf}\,(\gamma) > \omega$ it is also a

limit point of C'. So C' is closed, unbounded in γ, and also $S\cap\gamma$ is stationary, therefore $C'\cap S\cap\gamma$ is nonempty, but this contradicts the minimality of γ.

(b) Assume that $S < T < U$. There is, therefore, a closed, unbounded set C, such that if $\alpha \in C\cap T$ then $S\cap\alpha$ is stationary in α and if $\alpha\in C\cap U$ then $T\cap\alpha$ is stationary in α (and so $\mathrm{cf}\,(\alpha)>\omega$). Let C' be the closed, unbounded set of the limit points of C. We show that if $\alpha\in C'\cap U$ then $S\cap\alpha$ is stationary in α and this will prove that $S<U$. Assume that for some $\alpha\in C'\cap U$, $S\cap\alpha$ is nonstationary in α. There is, therefore, a club set $D\subseteq\alpha$ such that $D\cap S=\emptyset$. As $\mathrm{cf}\,(\alpha)>\omega$, the set $C'\cap D'$ is a club set in α (here D' is the set of limit points of D), and as $T\cap\alpha$ is stationary in α, there is $\beta\in C'\cap D'\cap T$. Then, on the one hand $S\cap\beta$ is stationary in β, on the other hand, D is closed, unbounded in β, and disjoint from S, so $S\cap\beta$ is nonstationary in β, a contradiction.

(c) Assume that $S_{n+1} < S_n$ for $n=0,1,\ldots$ and this is witnessed by the closed, unbounded sets C_n, that is, for $\alpha\in C_n\cap S_n$, $S_{n+1}\cap\alpha$ is stationary in α. Set $C=\bigcap\{C_n : n<\omega\}$, a closed, unbounded set. For every n, let $\gamma_n=\min(C'\cap S_n)$ where C' is the set of limit points of C. By definition, S_{n+1} is stationary in γ_n, so in particular, $\mathrm{cf}\,(\gamma_n)>\omega$ and therefore $C'\cap\gamma_n$ is closed and unbounded in γ_n. So $C'\cap S_{n+1}\cap\gamma_n\neq\emptyset$, that is, $\gamma_{n+1}<\gamma_n$, so $\gamma_0>\gamma_1>\cdots$ is a decreasing sequence of ordinals, a contradiction. [T. Jech: Stationary subsets of inaccessible cardinals, in: *Axiomatic Set Theory* (J. E. Baumgartner, D. A. Martin, S. Shelah, eds), Boulder, Co. 1983, Contemporary Math., **31**, Amer. Math. Soc., Providence, R.I., 1984, 115–142]

20. (a) If some stationary $S'\subseteq S$ is the union of κ disjoint stationary sets then so is S, by adding the difference $S\setminus S'$ to any of the components.

(b) Let the regressive $f: S\to\kappa$ be a counterexample. Then, for any $\gamma<\kappa$, the set $\{\alpha\in S:\gamma<f(\alpha)\}$ is stationary. We now construct by transfinite recursion on $\xi<\kappa$ an increasing sequence $\{\gamma_\xi:\xi<\kappa\}$ of ordinals. If γ_ζ is defined for $\zeta<\xi$, then by the above property the set

$$S_\xi=\{\alpha\in S: f(\alpha)>\sup\{\gamma_\zeta:\zeta<\xi\}\}$$

is stationary. As f is regressive on S_ξ, by Fodor's theorem (Problem 9) there are a γ_ξ and a stationary $S'_\xi\subseteq S_\xi$ such that $f(\alpha)=\gamma_\xi$ holds for $\alpha\in S'_\xi$. As obviously $\gamma_\xi>\gamma_\zeta$ holds for $\zeta<\xi$, the stationary sets $\{S'_\xi:\xi<\kappa\}$ are pairwise disjoint, contrary to our hypothesis.

(c) Assume indirectly that $S'=\{\alpha\in S:\mathrm{cf}\,(\alpha)<\alpha\}$ is stationary. Then, as the function cf is regressive on S', using parts (a) and (b) we get that there is some $\mu<\kappa$ such that $\mathrm{cf}\,(\alpha)\le\mu$ holds for the elements of a stationary $S''\subseteq S'$. For $\alpha\in S''$ let $\{f_\xi(\alpha):\xi<\mathrm{cf}\,(\alpha)\}$ be a set cofinal in α. Again by (b), there are club sets C_ξ and values $\gamma_\xi<\kappa$ such that if $\alpha\in C_\xi\cap S''$ then $f_\xi(\alpha)\le\gamma_\xi$ $(\xi<\mu)$. Define $C=\bigcap\{C_\xi:\xi<\mu\}$, a club set. Notice that $S^*=C\cap S''$ is stationary. But if $\alpha\in S^*$, then

$$\alpha \le \sup\{f_\xi(\alpha) : \xi < \mu\} \le \sup\{\gamma_\xi : \xi < \mu\},$$

that is, S^* is bounded in κ, a contradiction.

(d) Assume indirectly that there is a stationary $S' \subseteq S$ consisting of regular cardinals such that for $\alpha \in S'$ there is a closed, unbounded $C_\alpha \subseteq \alpha$, such that $C_\alpha \cap S = \emptyset$. Set, for $\xi < \kappa$, $f_\xi(\alpha) = \min(C_\alpha \setminus \xi)$ (the least element of C_α that $\ge \xi$). This is a regressive function for $\alpha \in S'$, $\alpha > \xi$, so by part (b), there are a closed, unbounded $D_\xi \subseteq \kappa$, and a $\gamma_\xi < \kappa$ such that $f_\xi(\alpha) < \gamma_\xi$ holds for $\alpha \in D_\xi \cap S'$. Set $D = \triangledown\{D_\xi : \xi < \kappa\}$, the diagonal intersection (Problem 5). Let $E \subseteq \kappa$ be a closed, unbounded set, consisting of limit ordinals, that are closed under γ_ξ, that is, if $\xi < \delta \in D$, then $\gamma_\xi < \delta$ (cf. Problem 3). Pick $\alpha \in S' \cap D$. If $\delta \in \alpha \cap E$, then for $\xi < \delta$ we have $f_\xi(\alpha) < \gamma_\xi < \delta$, therefore C_α has an element in the interval $[\xi, \delta)$. As this holds for every $\xi < \delta$, δ is a limit point of C_α, so $\delta \in C_\alpha$. That is, if $\alpha \in S' \cap D$, then $E \cap \alpha \subseteq C_\alpha$, so $(E \cap S) \cap \alpha = \emptyset$. As $S' \cap D$ has arbitrarily large elements below κ, we conclude that $E \cap S = \emptyset$, a contradiction, as E is a closed, unbounded set.

(e) Assume that there is a club $D \subseteq \kappa$ as in (d). Let D' be the club set of limit points of D. Set $\alpha = \min(D' \cap S)$. Then, α is a regular, uncountable cardinal and $S \cap \alpha$ is stationary in α. $D \cap \alpha$ is a club set in α, but then so is $D' \cap \alpha$. But then, $(D' \cap \alpha) \cap (S \cap \alpha) \ne \emptyset$, so $D' \cap S$ has an element smaller than α, a contradiction. [R. M. Solovay: Real-valued measurable cardinals, in: *Axiomatic Set Theory*, Proc. Symp. Pure Math. **XIII**, Amer. Math. Soc., Providence, R.I., 1971, 397–428]

21. By Problem 20 there are pairwise disjoint stationary sets $\{S_\alpha : \alpha < \kappa\}$. By increasing S_0, if needed, we can assume that $\bigcup\{S_\alpha : \alpha < \kappa\} = \kappa$. Set $f(\xi) = \alpha$ if and only if $\xi \in S_\alpha$. Assume now that $X \subseteq \kappa$ includes a club subset C. Then for every $\alpha < \kappa$ there is some $x \in X$ such that $f(x) = \alpha$, namely, any element of (the nonempty) $C \cap S_\alpha$.

22. By the previous problem S can be decomposed into the disjoint union of κ stationary sets, $S = \bigcup\{S_\alpha : \alpha < \kappa\}$. Let $\mathcal{H}$ be a family of 2^κ subsets of κ, none being a subset of any other (see Problem 18.5). Set, for $A \in \mathcal{H}$, $X(A) = \bigcup\{S_\alpha : \alpha \in A\}$. Then $\{X(A) : A \in \mathcal{H}\}$ is as required. Indeed, if $A \ne B$ are in $\mathcal{H}$, then $\alpha \in A \setminus B$ for some $\alpha < \kappa$ and then $S_\alpha \subseteq X(A) \setminus X(B)$.

23. Fix an arbitrary closed, unbounded subset $C_\alpha \subseteq \alpha$ of order type μ, for every $\alpha < \kappa$, $\mathrm{cf}(\alpha) = \mu$. For any closed, unbounded $E \subseteq \kappa$ consider the system $\mathcal{C}(E) = \{C_\alpha \cap E : \alpha < \kappa, \mathrm{cf}(\alpha) = \mu\}$. We claim that for some closed, unbounded E^* the system $\mathcal{C}(E^*)$ is as required in the sense that for every closed, unbounded $E \subseteq \kappa$ there is some $\alpha < \kappa$ such that $C_\alpha \cap E^* \subseteq E$ and $C_\alpha \cap E^*$ is of order type μ. Assume that it is not the case. Then for every closed, unbounded set E^* there is a closed, unbounded E such that for every α ($\alpha < \kappa, \mathrm{cf}(\alpha) = \mu$) either $|C_\alpha \cap E^*| < \mu$ or $C_\alpha \cap E^* \not\subseteq E$. By replacing E by $E^* \cap E$ if needed, we can assume that $E \subseteq E^*$ holds.

Define the decreasing sequence $\{E_\gamma : \gamma < \mu^+\}$ of closed, unbounded sets in κ as follows. Set $E_0 = \kappa$. If $\gamma < \mu^+$ is limit, set $E_\gamma = \bigcap\{E_\xi : \xi < \gamma\}$. Finally, let $E_{\gamma+1} \subseteq E_\gamma$ be a set, as above which shows that $\mathcal{C}(E_\gamma)$ is not good.

Let α be the μth element of the closed, unbounded set $E = \bigcap\{E_\xi : \xi < \mu^+\}$. For every $\gamma < \mu^+$ the intersection $C_\alpha \cap E_\gamma$ is of order type μ (this holds even for E). Thus, necessarily $C_\alpha \cap E_\gamma \not\subseteq E_{\gamma+1}$ holds, but then $\{C_\alpha \cap E_\gamma : \gamma < \mu^+\}$ is a properly descending sequence of sets, with the first set of cardinality μ, and this is impossible. [S. Shelah: *Cardinal Arithmetic*, Oxford Logic Guides **34**, Oxford Science Publications, Clarendon Press, Oxford, 1994]

24. Fix, for every ordinal $\alpha < \kappa$ with cf $(\alpha) = \omega$ a sequence $0 = x_0^\alpha < x_1^\alpha < \cdots$ converging to it. If $E \subseteq \kappa$ is closed, unbounded, α is as above, then set $n \in T(E, \alpha)$ if and only if $E \cap (x_n^\alpha, x_{n+1}^\alpha]$ is nonempty, and then let $y_n^\alpha(E) = \max(E \cap (x_n^\alpha, x_{n+1}^\alpha])$. If we set $X_\alpha(E) = \{y_n^\alpha(E) : n \in T(E, \alpha)\}$ then our claim is that for some E the system

$$\mathcal{H}(E) = \{X_\alpha(E) : |T(E, \alpha)| = \omega\}$$

is as required. Suppose to the contrary that this is not true. If the closed, unbounded set D witnesses that $\mathcal{H}(E)$ is not good, i.e., $X_\alpha(E) \subseteq D$ never holds, then $E \cap D$ also witnesses this, so we can assume that $D \subseteq E$.

Construct the closed unbounded sets $\{E_\gamma : \gamma < \omega_1\}$ as follows. E_0 is arbitrary. If $\gamma < \omega_1$ is limit, then let $E_\gamma = \bigcap\{E_\beta : \beta < \gamma\}$. And finally, if E_γ is given, let $E_{\gamma+1} \subseteq E_\gamma$ be a closed unbounded set witnessing that $\mathcal{H}(E_\gamma)$ is not good, that is, there is no α such that $T(E_\gamma, \alpha)$ is infinite and $X_\alpha(E_\gamma) \subseteq E_{\gamma+1}$ holds. Let α be the ωth element of (the closed unbounded) $\bigcap\{E_\gamma : \gamma < \omega_1\}$. In α every E_γ is unbounded, so every $T(E_\gamma, \alpha)$ is infinite. Moreover, $T(E_\gamma, \alpha) \supseteq T(E_{\gamma'}, \alpha)$ holds for $\gamma < \gamma'$, so there is some γ^* that $T(E_\gamma, \alpha) = T$ holds for $\gamma \geq \gamma^*$. We have

$$\{y_n^\alpha(E_\gamma) : n \in T\} \not\subseteq E_{\gamma+1} \quad (\gamma \geq \gamma^*)$$

so for some $n \in T$ we have $y_n^\alpha(E_\gamma) \notin E_{\gamma+1}$ and hence $y_n^\alpha(E_\gamma) > y_n^\alpha(E_{\gamma+1})$. By the pigeon hole principle for infinitely many γ, say for $\gamma_0 < \gamma_1 < \cdots$ the same n applies here, which is impossible, as then $y_n^\alpha(E_{\gamma_0}) > y_n^\alpha(E_{\gamma_1}) > \cdots$, a decreasing sequence of ordinals. [S. Shelah: *Cardinal Arithmetic*, Oxford Logic Guides **34**, Oxford Science Publications, Clarendon Press, Oxford, 1994]

22

Canonical functions

1. Induction gives that $h_\alpha(\gamma) = \alpha$ for $\alpha < \kappa$. Next we get that $h_\kappa(\gamma) = \gamma$ for almost every γ, namely for all γ with the property $\gamma = \sup_{\tau<\gamma} \alpha_\tau$. Further, if $0 < \alpha < \kappa$ then $h_{\kappa+\alpha}(\gamma) = \gamma + \alpha$ holds for a.e. γ, and $h_{\kappa\cdot 2}(\gamma) = \gamma \cdot 2$ again on a closed, unbounded set of γ.

2. This can be proved by induction on α. For $\alpha = \beta + 1$ clearly $h_\alpha(\gamma) = h_\beta(\gamma) + 1 > h_\beta(\gamma)$ holds for all γ. If $\alpha = \alpha' + 1$ with $\alpha' > \beta$, then $h_\alpha(\gamma) = h_{\alpha'}(\gamma) + 1 > h_{\alpha'}(\gamma)$, and by the inductive hypothesis, the last term $> h_\beta(\gamma)$ for a.e. γ. If α is limit with $\mu = \mathrm{cf}\,(\alpha) < \kappa$, $\alpha = \sup\{\alpha_\tau : \tau < \mu\}$, then $\beta < \alpha_\tau$ for some $\tau < \mu$, and so by the induction hypothesis $h_\beta(\gamma) < h_{\alpha_\tau}(\gamma) \le h_\alpha(\gamma)$ holds for almost every γ. If, finally, α is limit with $\mathrm{cf}\,(\alpha) = \kappa$, $\alpha = \sup\{\alpha_\tau : \tau < \kappa\}$, then again, $\beta < \alpha_\tau$ for some $\tau < \kappa$, and in this case if $\gamma > \tau$ and if γ is in a closed, unbounded set, then $h_\beta(\gamma) < h_{\alpha_\tau}(\gamma) \le h_\alpha(\gamma)$ holds.

3. By induction on α. The step $\alpha \mapsto \alpha + 1$ is obvious: if $f_{\alpha+1}(\gamma) > f_\alpha(\gamma)$, then

$$f_{\alpha+1}(\gamma) \ge f_\alpha(\gamma) + 1 \ge h_\alpha(\gamma) + 1 = h_{\alpha+1}(\gamma)$$

and these are true for a.e. γ. Assume that α is limit, $\mu = \mathrm{cf}\,(\alpha) < \kappa$, $\alpha = \sup\{\alpha_\tau : \tau < \mu\}$. By condition, there is a closed, unbounded set $C_\tau \subseteq \kappa$, such that for $\gamma \in C_\tau$ $f_\alpha(\gamma) > f_{\alpha_\tau}(\gamma)$ holds. Also, by the inductive hypothesis, there is a closed, unbounded set $D_\tau \subseteq \kappa$, such that $f_{\alpha_\tau}(\gamma) \ge h_{\alpha_\tau}(\gamma)$ holds for $\gamma \in D_\tau$. If C is the intersection of all the C_τ's and D_τ's, then for $\gamma \in C$ we have $f_\alpha(\gamma) > f_{\alpha_\tau}(\gamma) \ge h_{\alpha_\tau}(\gamma)$ for every τ, that is,

$$f_\alpha(\gamma) \ge \sup\{h_{\alpha_\tau}(\gamma) : \tau < \mu\} = h_\alpha(\gamma).$$

Assume finally, that $\mathrm{cf}\,(\alpha) = \kappa$, $\alpha = \sup\{\alpha_\tau : \tau < \kappa\}$. Let $C_\tau \subseteq \kappa$ be a closed, unbounded set such that if $\gamma \in C_\tau$ then $f_\alpha(\gamma) > f_{\alpha_\tau}(\gamma) \ge h_{\alpha_\tau}(\gamma)$ holds. Set $C = \triangledown\{C_\tau : \tau < \kappa\}$, their diagonal intersection (see Problem 21.5). Then for $\gamma \in C$, if $\tau < \gamma$, then $f_\alpha(\gamma) > f_{\alpha_\tau}(\gamma) \ge h_{\alpha_\tau}(\gamma)$, that is,

$$f_\alpha(\gamma) \geq \sup\{h_{\alpha_\tau}(\gamma) : \tau < \gamma\} = h_\alpha(\gamma)$$

holds.

4. The statement is obvious if α is 0 or successor. Assume that it is a limit ordinal. If $\mu = \mathrm{cf}(\alpha) < \kappa$, $\alpha = \sup\{\alpha_\tau : \tau < \mu\}$, then, as $f(\gamma) < h_\alpha(\gamma)$ for $\gamma \in S$ (S a stationary set), for every $\gamma \in S$, we have $f(\gamma) \leq h_{\alpha_{\tau(\gamma)}}(\gamma)$ for some $\tau(\gamma) < \mu$. For a stationary set $S' \subseteq S$, $\tau(\gamma) = \tau$ for some $\tau < \mu$ (see Problem 21.6) and we are done.

If, however, $\mathrm{cf}(\alpha) = \kappa$, $\alpha = \sup\{\alpha_\tau : \tau < \kappa\}$, and for $\gamma \in S$ (S a stationary set), we have

$$f(\gamma) < h_\alpha(\gamma) = \sup\{h_{\alpha_\tau}(\gamma) : \tau < \gamma\},$$

then for every $\gamma \in S$, there is $\tau(\gamma) < \gamma$ that $f(\gamma) \leq h_{\alpha_{\tau(\gamma)}}(\gamma)$, and then by Fodor's theorem (Problem 21.9) $\tau(\gamma) = \tau$ with some τ and stationary many γ and we are done again.

5. Let $\beta < \alpha$ be the least ordinal such that $f(\gamma) \leq h_\beta(\gamma)$ holds on a stationary set (say, for $\gamma \in S$). By Problem 4, $\{\gamma \in S : f(\gamma) < h_\beta(\gamma)\}$ is nonstationary, so f and h_β indeed agree on a stationary set, namely, at a.e. point of S.

6. By Problem 3, $f_\alpha(\gamma) \geq h_\alpha(\gamma)$ holds for a.e. γ. If the conclusion is not true then there is a least ordinal α such that $f_\alpha(\gamma) > h_\alpha(\gamma)$ holds for $\gamma \in S$, where S is some stationary set. Set

$$f(\gamma) = \begin{cases} h_\alpha(\gamma), & \text{for } \gamma \in S; \\ f_\alpha(\gamma), & \text{for } \gamma \notin S. \end{cases}$$

Then $f : \kappa \to \kappa$ contradicts property (c) of the Problem. In fact, if $\beta < \alpha$ then $f_\beta(\gamma) < f_\alpha(\gamma)$ holds on a club set D_1 (by property (b)), $f_\beta(\gamma) \leq h_\beta(\gamma)$ holds on a club set D_2 (by the minimality of α) and $h_\beta(\gamma) < h_\alpha(\gamma)$ holds for $\gamma \in D_3$ by Problem 2. Thus, $f_\beta(\gamma) < \min\bigl(f_\alpha(\gamma), h_\alpha(\gamma)\bigr) \leq f(\gamma)$ for $\gamma \in D_1 \cap D_2 \cap D_3$, hence $f(\gamma) \leq f_\beta(\gamma)$ cannot hold for stationarily many γ.

7. Suppose that the conclusion is false. Let α be the least ordinal, such that $h_\alpha(\gamma) \geq |\gamma|^+$ holds on a stationary set, say, for $\gamma \in S$. Clearly, α is limit. Assume first that $\mu = \mathrm{cf}(\alpha) < \kappa$, and $\alpha = \sup\{\alpha_\tau : \tau < \mu\}$. If $\gamma \in S$, $\gamma > \mu$, then there is some $\tau(\gamma) < \mu$ such that $h_{\alpha_{\tau(\gamma)}}(\gamma) \geq \gamma^+$. For a stationary subset $S' \subseteq S$, $\tau(\gamma) = \tau$ holds for some τ (Problem 21.9), hence $h_{\alpha_\tau}(\gamma) \geq |\gamma|^+$ for $\gamma \in S'$, a contradiction to the minimality of h_α.

Assume finally that $\mathrm{cf}(\alpha) = \kappa$, $\alpha = \sup\{\alpha_\tau : \tau < \kappa\}$, and for a stationary set S, if $\gamma \in S$, then

$$|\gamma|^+ \leq h_\alpha(\gamma) = \sup\{h_{\alpha_\tau}(\gamma) : \tau < \gamma\}$$

holds. As $|\gamma|^+$ is regular, it is not the supremum of a γ-sequence of smaller ordinals, so for every $\gamma \in S$ there is some $\tau(\gamma) < \gamma$ such that $h_{\alpha_{\tau(\gamma)}}(\gamma) \geq |\gamma|^+$.

By Fodor's theorem (Problem 21.9) $\tau(\gamma) = \tau$ holds for the elements of a stationary subset $S' \subseteq S$, and then again we get that $h_{\alpha_\tau}(\gamma) \geq |\gamma|^+$ for $\gamma \in S$ and here $\alpha_\tau < \alpha$, a contradiction.

8. For every $\delta < \kappa$ there are δ', δ'' such that $g'_\alpha(\delta') = g_\alpha(\delta)$, $g_\alpha(\delta'') = g'_\alpha(\delta)$. By Problem 21.3, there is a closed, unbounded set C, such that if $\gamma \in C, \delta < \gamma$, then the corresponding δ' and δ'' are also below γ. For $\gamma \in C$, $g_\alpha[\gamma] = g'_\alpha[\gamma]$, so $f_\alpha(\gamma) = f'_\alpha(\gamma)$ holds as well.

9. For every $\delta < \kappa$ there are δ', δ'' such that $g_\alpha(\delta') = g_\beta(\delta)$ and if $g_\alpha(\delta) < \beta$ then $g_\beta(\delta'') = g_\alpha(\delta)$. By Problem 21.3, there is a closed, unbounded set C, such that if $\gamma \in C$, $\delta < \gamma$, then the corresponding δ' and δ'' are also below γ. If, now, $\gamma \in C$, then $g_\beta[\gamma] = g_\alpha[\gamma] \cap \beta$ holds.

10. We can assume $\beta > 0$. Let C be a closed, unbounded set such that for $\gamma \in C$, $g_\beta[\gamma] = g_\alpha[\gamma] \cap \beta$ holds (see the previous problem). If $\delta < \kappa$ is such that $g_\alpha(\delta) = \beta$, then for the closed, unbounded set $C^* = C \setminus (\delta + 1)$ we have $g_\alpha[\gamma] \supseteq g_\beta[\gamma] \cup \{\beta\}$, so surely $f_\alpha(\gamma) \geq f_\beta(\gamma) + 1$.

11. Assume that $f(\gamma) < f_\alpha(\gamma)$ for $\gamma \in S$, S stationary. For $\gamma \in S$, there is $\tau(\gamma) < \gamma$ such that $g_\alpha(\tau(\gamma))$ is the $f(\gamma)$th element of $g_\alpha[\gamma]$ (which has type $f_\alpha(\gamma)$). By Fodor's theorem (Problem 21.9) there are a stationary $S^* \subseteq S$ and a τ such that $\tau(\gamma) = \tau$ holds for every $\gamma \in S^*$. Set $\beta = g_\alpha(\tau) < \alpha$. By Problem 9, there is a closed, unbounded set C, such that $g_\beta[\gamma] = g_\alpha[\gamma] \cap \beta$ holds for $\gamma \in C$. If now γ is in the stationary set $S^* \cap C$, then $f(\gamma)$ equals to the order type of

$$g_\alpha[\gamma] \cap g_\alpha(\tau(\gamma)) = g_\alpha[\gamma] \cap \beta = g_\beta[\gamma],$$

i.e., $f(\gamma) = f_\beta(\gamma)$.

12. By induction on α. The statement is obvious for $\alpha = 0$. Assume that $\alpha > 0$ and the statement holds for every $\beta < \alpha$. Then, $F = h_\alpha$ has the property that $F(\gamma) > h_\beta(\gamma)$ for almost every γ ($\beta < \alpha$), and if $F^*(\gamma) < F(\gamma)$ holds for stationarily many γ, then $F^*(\gamma) = h_\beta(\gamma)$ for stationarily many γ, for some $\beta < \alpha$. The same properties apply to f_α, by Problems 7 and 8 and by the induction hypothesis. But if F_1, F_2 have the above properties, then $F_1 = F_2$ almost everywhere. Indeed, should, e.g., $F_1(\gamma) < F_2(\gamma)$ hold for $\gamma \in S$, where S is a stationary set, we could define the function F^* equal to F_1 on S, and to F_2 otherwise, we would get a contradiction: $F^*(\gamma) = h_\beta(\gamma)$ for some $\beta < \alpha$ and $\gamma \in S'$ (S' is some stationary set), but $h_\beta(\gamma) < \min(F_1(\gamma), F_2(\gamma)) \leq F^*(\gamma)$ a.e.

13. $f_\alpha(\gamma)$ is the order type of a well-ordered set of cardinality $|\gamma|$, so it is $< |\gamma|^+$.

23

Infinite graphs

1. Let V be the vertex set. We consider cases.

First Case. Whenever $W \subseteq V$ is infinite then there is a vertex $v \in W$ that is joined to infinitely many vertices in W. Choose $v_0 \in V$, which is joined to the infinite set V_0 of elements. Then, applying the condition to V_0, pick $v_1 \in V_0$, which is joined to the infinite set $V_1 \subseteq V_0$. Continuing the induction we get the vertices $v_0, v_1, \ldots$ and infinite subsets $V_0 \supseteq V_1 \supseteq \cdots$ and as any two of those vertices are joined, we are done.

Second Case. There is an infinite $W \subseteq V$ such that every $v \in W$ is joined to finitely many vertices in W.

In this case inductively choose the vertices $v_0, v_1 \ldots \in W$ such that they form an independent set. This can be carried out, as when $v_0, \ldots, v_{n-1}$ are already given, each of them is joined to finitely many elements of W, therefore all but finitely many elements of W are not joined to any of them.

2. Assume that $f : [\omega]^2 \to k$. Let $\mathcal{U}$ be a nonprincipal ultrafilter on ω. For $x < \omega$ set $g(x) = i$ if and only if $\{y : f(x, y) = i\} \in \mathcal{U}$. Clearly, $g : \omega \to k$ is well defined.

We are going to construct the vertex disjoint paths step by step. At step j we will have the vertex disjoint finite sets $A_0^j, \ldots, A_{k-1}^j$ covering at least $\{0, \ldots, j-1\}$ such that A_i^j is the vertex set of a path in color i, and if it is nonempty, we specify an end-vertex y_i^j with $g(y_i^j) = i$.

To proceed from step j to step $j+1$ assume that $j \notin A_0^j \cup \cdots \cup A_{k-1}^j$ (otherwise we do nothing). Set $i = g(j)$. If $A_i^j = \emptyset$ simply make $A_i^{j+1} = \{j\}$, $y_i^{j+1} = j$, and $A_l^{j+1} = A_l^j$ for all other l. Otherwise, pick $z \notin A_0^j \cup \cdots \cup A_{k-1}^j$ with

$$z \in \{t : f(j, t) = f(y_i^j, t) = i\}$$

(remark that this latter set is in $\mathcal{U}$, so it is infinite). We can now extend the path A_i^j at its end at y_i^j with the vertices z and j and make $y_i^{j+1} = j$. [R. Rado]

3. Let X be a graph on V with $|V| = \kappa$. Let $\mathcal{U}$ be a uniform ultrafilter on V, i.e., with $|A| = \kappa$ for every $A \in \mathcal{U}$. [Such a $\mathcal{U}$ can be obtained by applying Problem 14.6,(c) to the filter $\mathcal{F} = \{V \setminus X : X \in [V]^{<\kappa}\}$.] Either for $Y = X$ or for $Y = \overline{X}$ (the complement of X) the following holds. There is a set $A \in \mathcal{U}$ such that for every $x \in A$ the $\Gamma(x) = \{y : \{x, y\} \in Y\}$ is in $\mathcal{U}$. We show that Y includes a topological K_κ. Notice that by uniformity $|A| = \kappa$ and $|\Gamma(x)| = \kappa$ holds for every $x \in A$. We recursively choose the nodes $\{v(\alpha) : \alpha < \kappa\} \subseteq A$ of the topological K_κ and the vertices $\{w(\beta, \alpha) : \beta < \alpha < \kappa\}$ such that $w(\beta, \alpha)$ is joined to $v(\alpha)$ and $v(\beta)$ (and so there are disjoint paths of length 2 between the $v(\alpha)$'s). We need to maintain, of course, that the vertices of the form $v(\alpha)$ and $w(\beta, \alpha)$ be all distinct. At step α we first choose $v(\alpha) \in A$ which is not in $\{v(\beta) : \beta < \alpha\} \cup \{w(\gamma, \beta) : \gamma < \beta < \alpha\}$ (possible, as the first set is of cardinality κ, the second is smaller), then similarly by recursion on $\beta < \alpha$ we choose an element of $\Gamma(v(\alpha)) \cap \Gamma(v(\beta))$ (a κ-sized set) which differs from all earlier elements. [P. Erdős, A. Hajnal]

4. Well-order the vertex set of the graph as $V = \{v_\alpha : \alpha < \varphi\}$ for some ordinal φ. Define by transfinite recursion the coloring $f : V \to \{0, 1, \ldots, n\}$ so that each v_α gets a color different from any of its already colored neighbors v_β, $\beta < \alpha$. Since there are at most n such neighbors, this is possible.

5. For vertices u and v set $u \sim v$ if $u = v$ or they are connected by a path. This is clearly an equivalence relation, its classes are the connected components. The number of vertices reachable from a specified vertex by paths is at most $1 + \kappa + \kappa^2 + \cdots = \kappa$. Therefore, each class has cardinality at most κ and so it can be colored by κ colors. As there are no edges between classes, this suffices.

6. The proof is the same as that of Problem 4.

7. If, for every vertex v, $f(v)$ is the set of smaller vertices joined to v, then f is a set mapping with $|f(v)| < \kappa$ for every v. By Problem 26.10 the vertex set is the union of κ free sets and a free set is obviously an independent set in the graph.

8. **First solution.** The vertex set of the graph can be enumerated as $V = \{v_\alpha : \alpha < \varphi\}$ for some ordinal φ. Using transfinite recursion we construct $f_\alpha : \{v_\beta : \beta < \alpha\} \to \{1, \ldots, n\}$ such that f_α is a good n-coloring of $\{v_\beta : \beta < \alpha\}$, and if $\beta < \alpha$ then f_α extends f_β. If we succeed with this then f_φ will witness that X is n-colorable. Our inductive hypothesis is somewhat stronger; we will require not just that $f_\alpha : \alpha \to \{1, \ldots, n\}$ is a good coloring but that it can be extended to a good coloring on every finite subset of $\{v_\gamma : \alpha \leq \gamma < \varphi\}$.

For $\alpha = 0$, f_0 can (and must) be chosen to be the empty function. This function is good—this is exactly the assumption of the theorem.

Assume that α is limit, and f_β exists for every $\beta < \alpha$. We show that $f_\alpha = \bigcup\{f_\beta : \beta < \alpha\}$ is good for our purposes. If $A \subseteq \{v_\gamma : \alpha \leq \gamma < \varphi\}$ is a finite subset, by hypothesis, for every $\beta < \alpha$ there is some $g : A \to \{1, \ldots, n\}$ such that $f_\beta \cup g$ is a good coloring. As there are only finitely many n-colorings of A, there is a g that occurs for a cofinal set of the β's. Then this g gives a good extension of f_α to A.

Finally, assume we have f_α, and let us show the existence of $f_{\alpha+1}$. For every $1 \le i \le n$ we try to define the function $f^i_{\alpha+1}$ by extending f_α to v_α with $f^i_{\alpha+1}(v_\alpha) = i$. Assume indirectly that $f^i_{\alpha+1}$ is not good. Then, there is some finite $A_i \subseteq \{v_\gamma : \alpha < \gamma < \varphi\}$ such that $f^i_{\alpha+1}$ cannot be extended to a good coloring of A_i. Take $A = \{v_\alpha\} \cup \bigcup\{A_i : 1 \le i \le n\}$. Then there is no good extension of f_α to the finite set A, a contradiction. [P. Erdős, N. G. de Bruijn: A color problem for infinite graphs and a problem in the theory of relations, *Proceedings of the American Mathematical Society* **54**(1951), 371–373]

Second solution. Assume that X is a graph on the vertex set V such that every finite subgraph of X is n-colorable. We consider the following partially ordered set $\langle \mathcal{P}, \le \rangle$. $Y \in \mathcal{P}$ if Y is a graph on V with $X \subseteq Y$ and every finite subgraph of Y is n-colorable. Order $\mathcal{P}$ the obvious way: $Y_0 \le Y_1$ if $Y_0 \subseteq Y_1$, that is, Y_0 is a subgraph of Y_1. As $X \in \mathcal{P}$, our set is nonempty.

We show that $\langle \mathcal{P}, \le \rangle$ satisfies the condition of Zorn's lemma. Indeed, assume that $\{Y_i : i \in I\}$ is an ordered family of elements of $\mathcal{P}$. We have to show that every finite subgraph of $Y = \bigcup\{Y_i : i \in I\}$ is n-colorable. If Z is such a subgraph, then every edge of Z appears in some Y_i, so, among those finitely many indices i there is a largest one, and the corresponding Y_i shows that Z is n-colorable.

We can, therefore, apply Zorn's lemma (Chapter 14), and get a maximal element $Y \in \mathcal{P}$. That is, we extended X to Y, a graph saturated to our condition. We show that the relation "not joined in Y" is an equivalence relation on V. Of the three properties of equivalence only transitivity is not obvious. Assume that x is not joined to y, y is not joined to z. As Y is maximal and x and y are not joined there is a finite set A that will not be n-colorable, once we join x and y. Phrased differently, in every n-coloration of the graph Y on A the vertices x and y get the same color. Similarly, as we cannot extend Y by the edge $\{y, z\}$, there is a finite set B such that in every n-coloration of the graph Y on B the vertices y and z get the same color. But then, in every n-coloration of the of the graph Y on $A \cup B$ (and by assumption, there is such a coloring) the vertices x and z get the same color, so, x and z can not be joined.

So, we proved that there is some decomposition $V = \bigcup\{V_i : i \in I\}$ such that two points are joined if and only if they are in distinct classes. But there cannot be more than n classes, as that would mean a subgraph of type K_{n+1}. That is, we have at most n classes, therefore Y can be colored by n colors, and so can be X. [G. Dirac, L. Pósa]

9. In order to show the nontrivial direction let X be a graph which is not finitely chromatic. Then, by Problem 8, for every $k < \omega$ there is a finite sub-

graph G_k which cannot be colored with k colors. The union of these subgraphs is a countable subgraph that is not finitely chromatic.

10. Let X be an infinitely chromatic graph on the well-ordered set $\langle V, < \rangle$. We can assume that for every $a \in V$ the graph on $V_a = \{x \in V : x < a\}$ is finitely chromatic (otherwise replace X by X restricted to V_a where $a \in V$ is the least element such that X on V_a is infinitely chromatic). Clearly, X on $V^a = \{x \in V : a < x\}$ is then infinitely chromatic. Now choose by mathematical induction the increasing sequence of elements $a_0 < a_1 < \cdots$ from V and (using the de Bruijn–Erdős theorem, Problem 8) the finite subgraphs F_n with elements between a_n and a_{n+1} such that X restricted to F_n is at least n-chromatic. Then, the union of the sets $F_0, F_1, \ldots$ will give an ω-type subset that is infinitely chromatic. [L. Babai: Végtelen gráfok színezéséről, *Matematikai Lapok*, **20**(1969), 141–143.]

11. Replace the ground set with $A = \bigcup\{A_\alpha : \alpha < \omega_1\}$ where $A_\alpha = [\omega_1 \cdot \alpha + \alpha, \omega_1 \cdot (\alpha + 1))$ (ordinal interval). The order type of A is still ω_1^2 so it suffices to construct the graph on A. Join $\omega_1 \cdot \alpha + \beta$ and $\omega_1 \cdot \alpha' + \beta'$ if and only if $\alpha < \alpha'$ and $\beta > \beta'$ (notice that then $\alpha < \alpha' \leq \beta' < \beta$). If $B \subseteq A$ is some subset of order type ω_1 then either all but countably many elements of B are in one A_α or else it has a countable intersection with every A_α. In the former case it is a countable set plus an independent set. In the latter case every vertex has countable degree: $\omega_1 \cdot \alpha + \beta \in B$ is certainly not joined to vertices in $\bigcup\{A_\xi : \xi > \beta\}$. These and Problem 5 show that X on B is countably chromatic.

In order to show that X is uncountably chromatic assume that $A = B_0 \cup B_1 \cup \cdots$ is a decomposition into independent sets (note that points with the same color form an independent set). Observe that if for some $\alpha < \omega_1$ and $i < \omega$ the intersection $A_\alpha \cap B_i$ is uncountable then B_i has no elements in any $A_{\alpha'}$, $\alpha' > \alpha$. As for every A_α $(\alpha < \omega_1)$ there is some $i < \omega$ such that $A_\alpha \cap B_i$ is uncountable, and to different α's we get different i's, we get the desired contradiction. [P. Erdős–A. Hajnal]

12. The chromatic number of $X \times Y$ is at most k. Indeed, if $f : V \to \{1, 2, \ldots, k\}$ is a good coloring of X, then $F(\langle x, y \rangle) = f(x)$ will be a good coloring of $X \times Y$.

For the other direction, in order to get a contradiction, assume that $F : V \times W \to \{1, 2, \ldots, k-1\}$ is a good coloring of $X \times Y$. On W, let $\mathcal{F}$ be the family of those subsets A of W for which $W \setminus A$ is independent (in Y). As W is not the union of finitely many independent sets, $\mathcal{F}$ has the finite intersection property, that is, the intersection of finitely many elements of $\mathcal{F}$ is always nonempty. We can therefore extend $\mathcal{F}$ to an ultrafilter $\mathcal{U}$ on W (see Problem 14.6(c)). The ultrafilter property gives that for every $x \in V$ there is a unique $i(x)$ such that $\{y \in W : F(\langle x, y \rangle) = i(x)\} \in \mathcal{U}$. The mapping $x \mapsto i(x)$ cannot be a good coloring of X, so there are $x, x' \in V$ with $i = i(x) = i(x')$ and

$\{x, x'\} \in X$. The set $A = \{y : F(\langle x, y\rangle) = F(\langle x', y\rangle) = i\}$ is in $\mathcal{U}$, therefore it is not independent. Now if $y, y' \in A$ are joined in Y, then $\langle x, y\rangle$ and $\langle x', y'\rangle$ are joined in $X \times Y$, and they get the same color, viz. i, a contradiction. [A.Hajnal: The chromatic number of the product of two $\aleph_1$-chromatic graphs can be countable, *Combinatorica*, **5**(1985), 137–139]

13.

(a) If $f_i : V_i \to C_i$ is a good coloring of (V_i, X_i), and the color sets $\{C_i : i \in I\}$ are disjoint, then the union of the colorings is a good coloring to the union of the C_i's.

(b) Let $f_i : V \to \mathrm{Chr}(X_i)$ be a good coloring of (V, X_i). Then f is a good coloring, where $f(x) = \langle f_i(x) : i \in I\rangle$.

14. The condition is obviously necessary. It is known that for finite graphs it is also sufficient (Hall's theorem). We get, therefore, that every finite subset of A has a matching into B. Let $A = \{x_\alpha : \alpha < \varphi\}$ be a well-ordering of the elements of A. We define, by transfinite recursion on $\alpha \le \varphi$ a function $f_\alpha : \{x_\beta : \beta < \alpha\} \to B$ which is an extendable partial matching, that is, it is injective, $\{x_\beta, f_\alpha(x_\beta)\}$ is always an edge of X and every finite subset of $A \setminus \mathrm{Dom}(f_\alpha)$ is matchable into $B \setminus \mathrm{Ran}(f_\alpha)$. Further, if $\beta < \alpha$ then f_α extends f_β. If we can reach f_φ then we will be done. f_0 clearly exists, the empty function is good for our purposes. If α is a limit ordinal and f_β exists for $\beta < \alpha$ then $f_\alpha = \bigcup\{f_\beta : \beta < \alpha\}$ is as required: if A' is a finite subset of $A \setminus \mathrm{Dom}(f_\alpha)$ then by the condition of the finiteness of the elements of A it has finitely many matchings into B, and for every f_β some of them are good. Therefore, there must be one that is good for unboundedly many $\beta < \alpha$. Then it is good for α.

To cover the successor case, suppose to the contrary that we succeeded in selecting f_α but we cannot extend it to $f_{\alpha+1}$. This means that every finite subset of $A' = A \setminus \mathrm{Dom}(f_\alpha)$ can be matched into $B' = B \setminus \mathrm{Ran}(f_\alpha)$ but for every $y \in \Gamma(x_\alpha)$ there is a finite set $A_y \subseteq A' \setminus \{x_\alpha\}$ such that A_y has no matching into $B' \setminus \{y\}$. Let A^* be the union of all these sets A_y plus $\{x_\alpha\}$. By the condition on f_α, the finite A^* has a matching into B' but if now x_α is matched into y then we reach a contradiction by observing that this matching gives a matching of A_y into $B' \setminus \{y\}$. [M. Hall, Jr.: Distinct representatives of subsets, *Bull. Amer. Math. Soc.* **54**(1948), 922–926]

15. We reduce the statement to the previous problem. Given p, q and the graph X on A and B, replace every vertex in A by p copies and every vertex in B by q copies with the copies joined if and only if the original vertices are joined. Call the so obtained graph X' with its corresponding sides A' and B'. It is clear that in X' side A' has a matching if and only if the original graph X has a function as described. We have to show that the condition in the problem holds if and only if the Hall condition holds for X'. One direction is

obvious: if we pick k vertices of A in X that are joined to m vertices in B then we get pk vertices of A' that are joined to qm vertices in B', so the Hall condition means that $qm \geq pk$, which is indeed $m \geq pk/q$. Assume now that the above condition holds for X and try to establish the Hall condition for X'. Assume that we are given a finite subset F of A'. F splits as $F = F_1 \cup \cdots \cup F_p$ where F_i is obtained by replacing every vertex in some set $T_i \subseteq A$ by i copies. Notice that $|F| = |T_1| + 2|T_2| + \cdots + p|T_p|$. In X the vertices of $T = T_1 \cup \cdots \cup T_p$ are joined, by condition, to at least $\frac{p}{q}|T|$ vertices of B. These give in X' at least $q \cdot \frac{p}{q}|T| = p|T| \geq |T_1| + 2|T_2| + \cdots + p|T_p|$ vertices and we are done.

16. It is obvious that (c) is necessary. (a) is also necessary by (the easy direction of) Kuratowski's theorem on planar graphs. To show the necessity of (b) assume that some graph X is planar yet it contains uncountably many vertices p_i with degree ≥ 3. Let p_i be joined to the distinct vertices a_i, b_i, c_i. Let A_i, B_i, C_i be rational discs, that is, whose radii are rational and centers have rational coordinates, such that $a_i \in A_i$, $b_i \in B_i$, $c_i \in C_i$, they are disjoint and exclude p_i. As there are just countably many choices for A_i, B_i, C_i, there are i_0, i_1, i_2 such that $A_{i_0} = A_{i_1} = A_{i_2} = A$, and similarly $B_{i_0} = B_{i_1} = B_{i_2} = B$, $C_{i_0} = C_{i_1} = C_{i_2} = C$. Let a, b, c be the centers of A, B, C. p_{i_0} is joined with an edge in X to a_{i_0}, which is in fact a curve K between them. Consider k, the first point of K common with A, and replace the part of K after k with the radius between k and a. Perform the same operation with all the other edges (= curves) between the p_i's and a_i's, b_i's, c_i's, then we get a $K_{3,3}$ drawn on the plane, an impossibility.

For the other direction we first notice that if X is a finite graph not including a topological K_5 or $K_{3,3}$, then it is planar by Kuratowski's theorem. We first extend this to countable graphs.

Let X be a graph on the vertices $v_0, v_1, \ldots$ that has no topological K_5 or $K_{3,3}$ subgraphs. By Kuratowski's theorem, for every n, there is a drawing φ_n of X_n, X restricted to $\{v_0, \ldots, v_n\}$ on the plane. It is easy to see that there are just finitely many nonhomeomorphic ways of drawing X_n on the plane. In fact, an easy induction on n shows that there are just finitely many nonhomeomorphic n-vertex graphs drawn on the plane. Using the König infinity lemma (Problem 27.1), we get that there is a sequence $\varphi_0, \varphi_1, \ldots$ such that every φ_n is homeomorphic to φ_{n+1}'s restriction to X_n. We can modify φ_{n+1} such that it actually extends φ_n, and then the union of them draws X on the plane. Actually, this process can be carried out in such a way that each edge e is represented by a C^∞ curve l_e and to each edge $e = \{x, y\}$ we can associate a neighborhood U_e of l_e such that the closures of any two U_e and $U_{e'}$ are disjoint except possibly for common endpoints of l_e and $l_{e'}$. This latter property can easily be preserved in the previous induction of creating the φ_n's.

Assume finally that X is a graph satisfying (a), (b), and (c). Then X has a countable part X', spanned by the vertices with degree at least 3, and it has additionally at most continuum many paths, circuits, and isolated vertices. Let X^* be X' augmented with a simple path $\sigma(p, q)$ of length 2 for every pair of nodes p, q in X' ($p = q$ is possible) that are connected in X by at least one path of length ≥ 2. Then we can reconstruct X from X^* by replacing each $\sigma(p,q)$ by (at most countably many) appropriate paths by adding circuits, finite or infinite paths, unconnected to X^* and to each other, and by adding finite or infinite paths emanating from some points of X^*.

By the previous step, X^* can be drawn on the plane in a manner specified above. We can easily add to this representation the required objects to get a planar representation of X.

17. It suffices to show that there are continuum many points in the 3-space such that the connecting segments are pairwise disjoint (except, possibly, at their extremities). For this, see Problem 13.3.

18. First we show that K_{κ^+} has a decomposition into κ forests. Without loss of generality, we can assume that the graph is the complete graph on κ^+. Decompose the edges into κ classes in such a way that for every $\alpha < \kappa$ the edges going down from α, i.e., those of the form $\{\beta, \alpha\}$ with $\beta < \alpha$ are put into distinct classes. This is possible as the number of those edges is $|\alpha| \leq \kappa$. No circuit occurs with edges in the same class; indeed, if the vertices of a putative circuit are $v_1, \dots, v_n$, then if v_i is the largest of them under the ordering of the ordinals, then it is joined to two vertices (namely, to v_{i-1} and v_{i+1}) with edges going down, a contradiction.

For the other direction it suffices to note that if the edges of the complete graph on $(\kappa^+)^+$ vertices, and even if the edges of the complete bipartite graph on classes of cardinalities κ^+ and $(\kappa^+)^+$ are colored with κ colors, then there is a monochromatic circuit of length 4, by Problem 24.27. [P. Erdős, S. Kakutani: On non-denumerable graphs, *Bull. Amer. Math. Soc.*, **49**(1943), 457–461]

19. One direction is a special case of Problem 13(b): if some graph is the edge-union of countably many bipartite graphs, then its chromatic number is at most $2^{\aleph_0} = \mathbf{c}$.

For the other direction, assume that the chromatic number of some graph (V, X) is at most continuum. There is, therefore, a good coloring $f : V \to \mathbf{R}$ with the reals as the colors. Fix an enumeration $q_0, q_1, \dots$ of the rational numbers. If $\{x, y\} \in X$ is an edge, put it into Y_i if and only if i is the least number such that q_i is strictly between $f(x)$ and $f(y)$. This works: all edges of Y_i go between $A_i = \{x \in V : f(x) < q_i\}$ and $B_i = \{x \in V : f(x) \geq q_i\}$.

20. Fix, for every $n < \omega$, an enumeration $\{H_n(i) : 1 \leq i \leq 2^n\}$ of the subsets of $\{0, \dots, n-1\}$. The vertex set of our strongly universal graph will be the union of the disjoint finite sets $V_0, V_1, \dots$ where V_n consists of the vertices $v(i_0, \dots, i_n)$ indexed with the natural numbers $i_0, \dots, i_n$ on the condition $1 \leq i_k \leq 2^k$ for $0 \leq k \leq n$, so

$$|V_n| = 2^0 \cdot 2^1 \cdot \dots \cdot 2^n = 2^{\frac{n(n+1)}{2}}.$$

If $v = v(i_0, \ldots, i_n)$ then we join v for every $j \in H_n(i_n)$ to $v(i_0, \ldots, i_j)$ and to no other vertices in $V_0 \cup \cdots \cup V_n$. This defines a graph X on V and we show that it is strongly universal, i.e., if (W, Y) is a countable graph then (W, Y) is isomorphic to an induced subgraph of (V, X).

Enumerate W as $w_0, w_1, \ldots$. We find $f(w_n) \in V_n$ by induction on n. Set $f(w_0) = v(1)$, the only element of V_0. If we have already found $f(w_{n-1}) = v(i_0, \ldots, i_{n-1})$ then set $f(w_n) = v(i_0, \ldots, i_{n-1}, i_n)$ where $1 \leq i_n \leq 2^n$ is the only number that

$$H_n(i_n) = \{0 \leq j < n : \{w_j, w_n\} \in Y\}$$

holds. These steps can be executed and it is clear that for $0 \leq j < n$ $\{w_j, w_n\} \in Y$ holds if and only if $\{f(w_j), f(w_n)\} \in X$ holds, that is, f isomorphically embeds (W, Y) into (V, X). [R. Rado: Universal graphs and universal functions, *Acta Arithmetica* **9** (1964), 331–340]

21. Assume indirectly that the graph X on the countable vertex set V is universal for the countable K_ω-free graphs. Let $v \notin V$ be a further vertex and join v to every element of V. The graph X' so obtained is still K_ω-free, so by hypothesis there is $f : V \cup \{v\} \to V$, an embedding of X' into X. Set $v_0 = f(v)$, and inductively $v_{n+1} = f(v_n)$. As v is joined (in X') to every element of V, v_0 will be joined in X to $v_1, v_2, \ldots$. As f preserves adjacency, v_1 is joined to $v_2, \ldots$. Carrying out the induction we get that $v_0, v_1, \ldots$ are pairwise joined in X, and therefore they are distinct, so they form a K_ω, a contradiction.

22. Let (V, X) be a putative universal, countable, locally finite graph. For $v \in V$, $1 \leq i < \omega$, let $f_i^X(v)$ be the number of vertices reachable from v in X in at most i steps. As (V, X) is locally finite, $f_i^X(v)$ is a natural number for every $v \in V$, $1 \leq i < \omega$. Enumerate V as $V = \{v_1, v_2, \ldots\}$. Construct a countable, locally finite graph (W, Y) with a vertex $w \in W$ such that $f_i^Y(w) > f_i^X(v_i)$ holds for $i = 1, 2, \ldots$, where f_i^Y is the analogous function for (W, Y). This can be done easily; for example, we can take as (W, Y) a tree with large enough successive levels. Now it is impossible to isomorphically embed (W, Y) into (V, X): for every $1 \leq i < \omega$ the condition $f_i^Y(w) > f_i^X(v_i)$ excludes that w be mapped into v_n. [N. G. de Bruijn]

23. Suppose that (V, X) is a $K_{\aleph_1}$-free graph of cardinality $\leq \mathbf{c}$. Let W be the set of functions $f : \alpha \to V$ injecting a countable ordinal α into V in such a way that its range spans a complete subgraph in X. Join two such functions if one extends the other. This way we get a graph (W, Y), and we are going to show that $|W| \leq \mathbf{c}$, (W, Y) is $K_{\aleph_1}$-free, and it cannot be embedded into (V, X). This proves that (V, X) is not universal.

As for any $\alpha < \omega_1$ there are at most $\mathbf{c}^{\aleph_0} = \mathbf{c}$ functions from α into V, we have $|W| \leq \aleph_1 \cdot \mathbf{c} = \mathbf{c}$. Next, assume that $\{f_\alpha : \alpha < \omega_1\}$ spans a complete subgraph in (W, Y). Then, they are defined on different ordinals, and the one on larger ordinal extends the one on smaller ordinal. But this gives a complete $K_{\aleph_1}$ in (V, X), a contradiction.

Finally, assume to the contrary that $F : W \to V$ embeds (W, Y) into (V, X). By transfinite recursion on $\alpha < \omega_1$ we define functions $f_\alpha : \alpha \to V$, $f_\alpha \in W$ and vertices $v_\alpha = F(f_\alpha) \in V$ in such a way that f_α extends f_β for $\beta < \alpha$. To start, set $f_0 = \emptyset$, $v_0 = F(f_0)$. Assume that $\alpha > 0$ and f_β is determined for $\beta < \alpha$ with the above properties. Then the $\{f_\beta : \beta < \alpha\}$ forms a complete subgraph in (W, Y); therefore, $\{v_\beta : \beta < \alpha\}$ forms a complete subgraph in (V, X). This implies that if $f_\alpha(\beta) = v_\beta$ for $\beta < \alpha$ then $f_\alpha \in W$. If all f_β were defined analogously, then f_α extends every f_β $(\beta < \alpha)$. Thus, we can construct f_α, v_α as required for $\alpha < \omega_1$, but then $\{v_\alpha : \alpha < \omega_1\}$ forms a complete subgraph in (V, X), a contradiction. [R. Laver]

24. First we remark that it suffices to prove the result for κ a successor cardinal. Indeed, if κ is a limit cardinal, and for every successor $\tau < \kappa$ there is a triangle-free graph with chromatic number τ then the vertex disjoint union of them will be a triangle-free graph with chromatic number κ (this does not work for $\kappa = \aleph_0$, but this case follows along the same lines if we notice that the $\kappa = \aleph_1$ case and the de Bruijn–Erdős theorem (Problem 8) easily imply the statement in the problem for all finite cardinals).

Given a successor cardinal $\kappa = \mu^+$ we define the graph as follows. The vertex set is $[\mu^+]^3$, the set of 3-element subsets of κ. Join $\{x, y, z\}$ and $\{x', y', z'\}$ if and only if $x < y < x' < z < y' < z'$ holds (or vice versa). It is immediate that there is no triangle.

In order to show that the chromatic number is μ^+ assume to the contrary that $f : [\mu^+]^3 \to \mu$ is a good coloring. For $x < y < \mu^+$ we define the set $A(x, y) \subseteq \mu$ of colors as follows. Set $\alpha \in A(x, y)$ if and only if there are arbitrarily large $z < \mu^+$ such that $f(x, y, z) = \alpha$. We argue that $A(x, y) \neq \emptyset$ for every $x < y < \mu^+$. Indeed, if $\alpha \notin A(x, y)$ then there is $\gamma_\alpha < \mu^+$ such that $f(x, y, z) \neq \alpha$ holds for $\gamma_\alpha < z < \mu^+$. But then, if $z < \mu^+$ is larger than the supremum of the γ_α's, then $\{x, y, z\}$ can get no color at all. Next, we define, for every $x < \mu^+$ the set $B(x) \subseteq \mu$ as follows. $\alpha \in B(x)$ if and only if there are arbitrarily large $y < \mu^+$ with $\alpha \in A(x, y)$. An argument similar to the above one gives $B(x) \neq \emptyset$ for every $x < \mu^+$. Finally, set $\alpha \in C$ if and only if α occurs in $B(x)$ for cofinally many x. Again, we get that $C \neq \emptyset$.

Pick $\alpha \in C$. Choose an $x < \mu^+$ with $\alpha \in B(x)$, then select $y > x$ with $\alpha \in A(x, y)$. Then choose $y < x' < \mu^+$ such that $\alpha \in B(x')$ (possible, as $\alpha \in C$). Next choose $x' < z < \mu^+$ such that $f(x, y, z) = \alpha$ (again, such a z exists, as $\alpha \in A(x, y)$). Then choose $y' < \mu^+$ such that $y' > z$ and $\alpha \in A(x', y')$ (this is possible, as $\alpha \in B(x')$. Finally, as $\alpha \in A(x', y')$, we can select $y' < z'$ such that $f(x', y', z') = \alpha$. Now we are done: $\{x, y, z\}$ and $\{x', y', z'\}$ are joined, and they get the same color (α), a contradiction. [P. Erdős, R. Rado: A construction of graphs without triangles having preassigned order and chromatic number, *Journal London Math. Soc.*, **35**(1960), 445–448]

25. We can disregard all triples (α,β,γ) with not $\alpha<\beta<\gamma$ since these are isolated points in the graph. Note also that $A\subset\omega_1^3$ is of order type ω_1^3 if and only if for uncountably many α there are uncountably many β with the property that for uncountably many γ we have $(\alpha,\beta,\gamma)\in A$.

The same graph appeared for $\mu=\omega$ in Problem 24, and the proof given there shows that if $A\subset\omega_1^3$ is of type ω_1^3, then it spans an uncountable chromatic subgraph.

Suppose now that A is of type $<\omega_1^3$. Then only for countably many α can the set $\{(\beta,\gamma)\ :\ (\alpha,\beta,\gamma)\in A\}$ be of order type ω_1^2, and for all such α we can color any $(\alpha,\beta,\gamma)\in A$ by α (note that two triples with the same α are not connected). Let the rest of the points in A form the set A_1, and we have to show that A_1 is also countable chromatic. For every α there is an $f(\alpha)<\omega_1$ such that if $\beta>f(\alpha)$, then there are only countably many $(\alpha,\beta,\gamma)\in A_1$, i.e., there is an $f(\alpha,\beta)<\omega_1$ such that $(\alpha,\beta,\gamma)\notin A_1$ if $\gamma>f(\alpha,\beta)$. By Problem 20.7 there is an increasing sequence δ_ξ, $\xi<\omega_1$ such that $\delta_0=0$, $\delta_\xi=\sup_{\eta<\xi}\delta_\eta$ if ξ is a limit ordinal, and for any $\alpha<\delta_\xi$ we have $f(\alpha)<\delta_\xi$ and for any $\alpha<\beta<\delta_\xi$ we have $f(\alpha,\beta)<\delta_\xi$. Then $D_\xi=\{\alpha\ :\ \delta_\xi\le\alpha<\delta_{\xi+1}\}$, $\xi<\omega_1$ is a partition of ω_1 into disjoint sets, and for $(\alpha,\beta,\gamma)\in A_1$ the ordinals α, β and γ cannot belong to three different D_ξ: if $\alpha\in D_\xi$, $\beta\in D_\eta$ and $\gamma\in D_\theta$ with $\xi<\eta<\theta$, then $\beta>f(\alpha)$ and $\gamma>f(\alpha,\beta)$, hence $(\alpha,\beta,\gamma)\notin A_1$.

Thus, $A_1=A_2\cup A_3\cup A_4$, where

- in A_2 we have $\alpha,\beta,\gamma\in D_\xi$ for some ξ,
- in A_3 we have $\alpha\in D_\xi$ and $\beta,\gamma\in D_\eta$ for some $\xi<\eta$, while
- in A_4 we have $\alpha,\beta\in D_\xi$ and $\gamma\in D_\eta$ for some $\xi<\eta$,

and it is enough to color each of these sets by countably many colors.

Every $A_2\cap D_\xi$, $\xi<\omega_1$ is countable, and we can simply color the elements of this set by different colors $0,1,\ldots$ (note that no vertex from $A_2\cap D_\xi$ is connected to any vertex in $A_2\cap D_\eta$ if $\xi\ne\eta$).

Let $F:\omega_1^2\to\omega$ be a function such that $F(\xi,\eta)\ne F(\eta,\xi')$ for any $\xi<\eta<\xi'$. Such a function/coloring was constructed in Problem 24.8. If $(\alpha,\beta,\gamma)\in A_3$ and $\alpha\in D_\xi$, $\beta,\gamma\in D_\eta$, then let the color of (α,β,γ) be $F(\xi,\eta)$. This is clearly a good coloring on A_3: if $(\alpha,\beta,\gamma)\in A_3$ with $\alpha\in D_\xi$, $\beta,\gamma\in D_\eta$ is connected to $(\alpha',\beta',\gamma')\in A_3$ with $\alpha'\in D_{\xi'}$, $\beta',\gamma'\in D_{\eta'}$, then, because of, say, $\beta<\alpha'<\gamma$, we have $\eta=\xi'$, hence $F(\xi,\eta)\ne F(\xi',\eta')$.

Finally, one can similarly define a good coloring of A_4 with the aid of F: if $(\alpha,\beta,\gamma)\in A_4$ and $\alpha,\beta\in D_\xi$, $\gamma\in D_\eta$, then let the color of (α,β,γ) be $F(\xi,\eta)$.

26. (a) For one direction, if $f:V'\to\kappa$ is a good coloring of (V',X') then we can set

$$F(x)=\bigl\{f(\{y,x\}) : \{y,x\}\in X, y<x\bigr\}$$

for $x\in V$, that is, we color $x\in V$ with the set of colors of the edges going down from x. This is a good coloring, as otherwise there are $y<x$ with $\{y,x\}\in X$

and $F(y) = F(x)$ but then there is a $z < y < x$ with $f(\{z,y\}) = f(\{y,x\})$ and this contradicts to the hypothesis that f is a good coloring of (V', X').

For the other direction, assume that $\mathrm{Chr}(V, X) \le 2^\kappa$. Let $F : V \to {}^\kappa 2$ be a good coloring (i.e., we color with the $\kappa \to \{0, 1\}$ functions). If $\{y, x\} \in X$, $y < x$, then there is a least $\alpha < \kappa$ with $F(y)(\alpha) \ne F(x)(\alpha)$. We let $f(\{y,x\}) = \langle \alpha, 0\rangle$ if $F(y)(\alpha) = 0$, $F(x)(\alpha) = 1$, and, dually, let $f(\{y,x\}) = \langle \alpha, 1\rangle$ if $F(y)(\alpha) = 1$, $F(x)(\alpha) = 0$. We cannot have $f(\{z,y\}) = f(\{y,x\})$ for some values $z < y < x$, for, if the common value is say, $\langle \alpha, 0\rangle$ then $0 = F(y)(\alpha) = 1$ and we get a similar contradiction in the other case, too. [F. Galvin: Chromatic numbers of subgraphs, *Periodica Mathematica Hungarica*, **4**(1973), 117–119]

(b) Assume that there is a circuit C in (V', X') of some odd length $2t + 1 \le 2n + 1$. The vertices of C are edges of (V, X), $e_1, \dots, e_{2t+1}$, and there are vertices $v_1, \dots, v_{2t+1}$ such that v_i is the larger vertex of e_i and the smaller vertex of e_{i+1} or vice versa (and $e_{2t+2} = e_1$). So C forms a cycle in (V, X) (circuit with possibly repeated vertices). Choose $1 \le i \le 2t + 1$ such that $v_{i-1} \ne v_i$ (with $v_0 = v_{2t+1}$) and there is no value $v_j > v_i$ (this is possible as the v_i's cannot be all equal). Then v_i is the larger endpoint of e_i, the smaller of e_{i+1}, and again the larger endpoint of e_{i+2}, so $v_{i+1} = v_i$. We can, therefore, remove e_i from C, and likewise we can remove one edge corresponding to the smallest element among the v_i's. This way, we get an odd cycle of length $2t - 1 \le 2n - 1$ in (V, X) and that includes an odd circuit.

(c) By repeated applications of (a), (b) and for $n = 1$ by starting from some large enough complete graph. [P. Erdős, A. Hajnal: Some remarks on set theory, IX, *Michigan Math. Journal*, **11**(1964), 107–127]

27. Let (V, X) be the complete graph on $\mathbf{c}^+$, and let (V', X') be the graph defined in Problem 26. Using (a) of that problem, as $\mathrm{Chr}(X) > \mathbf{c}$, $\mathrm{Chr}(X') > \aleph_0$ holds. Every subgraph of X' of cardinality at most $\mathbf{c}$ is the subgraph of Y' for some induced subgraph Y of X with $|Y| \le \mathbf{c}$. As then $\mathrm{Chr}(Y) \le \mathbf{c}$, we must have $\mathrm{Chr}(Y') \le \aleph_0$ again by Problem 26.

28. Let (V, X) be the complete graph on ω_3, and let (V', X') be the graph derived from it in Problem 26. As $2^{\aleph_1} < 2^{\aleph_2} = \aleph_3$, $\mathrm{Chr}(X') = \aleph_2$. Every induced subgraph of X' is of the form Y' for some (not necessarily induced) subgraph Y of X. Now, if $\mathrm{Chr}(Y) = \aleph_3$ then $\mathrm{Chr}(Y') = \aleph_2$, and if $\mathrm{Chr}(Y) \le \aleph_2$ then $\mathrm{Chr}(Y') \le \aleph_0$, by Problem 26(a), and by the cardinal arithmetic hypothesis. That is, $\mathrm{Chr}(Y') \ne \aleph_1$ for every such graph. [F. Galvin: Chromatic numbers of subgraphs, *Periodica Mathematica Hungarica*, **4**(1973), 117–119]

29. Assume the contrary and let X be an uncountably chromatic graph which does not include $K_{n,\aleph_1}$ as a subgraph. By passing to a subgraph, if needed, we can assume that its vertex set V has cardinality κ and every subgraph of cardinality less than κ is countably chromatic. Obviously, $\kappa > \aleph_0$.

We first show that every vertex set $U \subseteq V$ has a "closure", a unique minimal set $F(U) \supseteq U$ with the property that if $x \in V$ is joined to at least n

elements of $F(U)$ then $x \in F(U)$. For this, set $F_0(U) = U$ and for $k = 0, 1, \dots$ let $F_{k+1}(U)$ consist of the elements of $F_k(U)$ plus all the vertices which are joined to at least n vertices in $F_k(U)$. Then take $F(U) = F_0(U) \cup F_1(U) \cup \cdots$.

We further have, by the condition imposed on the graph, that if U is finite, then $F_0(U), F_1(U), \dots$ and so $F(U)$ are countable, and if U is infinite, then $|F(U)| = |U|$.

Enumerate V as $\{v_\alpha : \alpha < \kappa\}$. For every $\alpha < \kappa$ set $V_\alpha = F(\{v_\beta : \beta < \alpha\})$. Then $V = \bigcup\{V_\alpha : \alpha < \kappa\}$, an increasing, continuous union. Also, by our above remark, each V_α is a set of cardinality $< \kappa$. If we now set $W_\alpha = V_{\alpha+1} \setminus V_\alpha$, then $\{W_\alpha : \alpha < \kappa\}$ is a partition of V into smaller sets.

Decompose X, the set of edges, as $X = Y \cup Z$ where Y is the set of crossing edges, that is, between points in different W_α's, and Z is the set of edges going between vertices in the same W_α. Z is the vertex disjoint union of—by the selection of κ—countably chromatic graphs, so itself is countably chromatic. Further, by Problem 6 Y is $n+1$-colorable, so we get that $X = Y \cup Z$ is countably chromatic, a contradiction. [P. Erdős, A. Hajnal: On chromatic number of graphs and set-systems, *Acta Math. Acad. Sci. Hung.*, **17**(1966), 61–99]

30. Let X be an uncountably chromatic graph. Decompose X as $X = Y \cup Z$ where an edge is put into Y if and only if for every n it is an edge of a complete bipartite graph $K_{n,n}$. Then there is an n such that Z does not include $K_{n,n}$ so by Problem 29, Z is countably chromatic. Y is therefore uncountably chromatic, so it includes an odd circuit C of length $2m+1$ for some m. We claim that every odd number $> 2m+1$ occurs as the length of a circuit in X. Let e be an edge of C. As e is in Y, for every n there is a $K_{n,n}$ containing e, so for every n there is a $K_{n,n}$ containing e and meeting C only in the end vertices of e. Now it is easy to choose a circuit of length $2(m+n)-1$ by adding to C a circuit of length $2n$ and by removing the edge e. [P. Erdős, A. Hajnal, S. Shelah: On some general properties of chromatic numbers, *Topics in topology (Proc. Colloq. Keszthely, 1972), Colloq. Math. Soc. J. Bolyai, Vol. 8.* North Holland, Amsterdam, 1974, 243–255, C. Thomassen: Cycles in graphs of uncountable chromatic number, *Combinatorica* **3** (1983), 133–134.]

31. Let (V, X) be an uncountably chromatic graph. If there is a nonempty subset $W \subseteq V$ that induces a graph in which every vertex has infinite degree, then we can easily choose by induction the vertices of an infinite path. We can therefore assume that no such subset of V exists, that is, if $W \subseteq V$ is nonempty, then there is a vertex $F(W) \in W$ joined to only finitely many vertices in W. Using this, determine recursively the elements $\{v_\alpha : \alpha < \varphi\}$ for some ordinal φ by making $v_\alpha = F\big(V \setminus \{v_\beta : \beta < \alpha\}\big)$. This process must terminate for some $\varphi < |V|^+$ and that can only happen when $V = \{v_\alpha : \alpha < \varphi\}$. If we now order V by $v_\alpha < v_\beta$ if $\beta > \alpha$, then Problem 7 gives that (V, X) is countably chromatic, a contradiction.

32. We can assume that $V = \omega_1$.

Recall that there exists an Ulam matrix, i.e., $\{U_{n,\alpha} : n < \omega, \alpha < \omega_1\}$ with $U_{n,\alpha} \subseteq \omega_1$, $U_{n,\alpha} \cap U_{n,\beta} = \emptyset$ for $\alpha \neq \beta$, and for a fixed α, $\bigcup\{U_{n,\alpha} : n < \omega\}$ is a co-countable subset of ω_1 (see Problem 18.1). The latter condition implies that for every $\alpha < \omega_1$ there is $n(\alpha) < \omega$ such that $U_{n(\alpha),\alpha}$ induces an $\aleph_1$-chromatic subgraph of X. For uncountably many α, $n(\alpha) = n$ for some n, and then these $U_{n,\alpha}$'s give $\aleph_1$ disjoint sets spanning $\aleph_1$-chromatic subgraphs.

33. Assume that the (first) statement fails and X is some uncountably chromatic graph that does not split into two uncountably chromatic induced subgraphs. Let $\{A_\alpha : \alpha < \lambda\}$ be a least family (with respect to the cardinality λ) of disjoint subsets such that each A_α induces a countably chromatic graph while $A = \bigcup\{A_\alpha : \alpha < \lambda\}$ does not. For $B \subseteq \lambda$ set $B \in I$ if and only if X on $\bigcup\{A_\alpha : \alpha \in B\}$ is countably chromatic. I is a proper, σ-complete ideal on λ, and by our minimal choice of λ, it contains every subset of cardinality less than λ. Furthermore, by our hypothesis, it is a prime ideal (i.e., for every $B \subseteq \lambda$ either $B \in I$ or $\lambda \setminus B \in I$). Let $f_\alpha : \bigcup\{A_\beta : \beta < \alpha\} \to \omega$ be a good coloring. Define $F : A \to \omega$ as follows. Let $F(x) = i$ if and only if $\{\alpha < \lambda : f_\alpha(x) = i\} \notin I$. As I is σ-complete and prime, this is well defined and is a good coloring of X on A with countably many colors: if $\{x, y\} \in X$, say $x, y \in \bigcup\{A_\beta : \beta < \alpha_0\}$, then $f_\alpha(x) \neq f_\alpha(y)$ for all $\alpha \geq \alpha_0$, hence $\{\alpha : f_\alpha(x) \neq f_\alpha(y)\} \in I$ and so $F(x) \neq F(y)$. This contradiction proves the claim.

The stronger statement follows by recursively splitting the vertex set into more and more subsets inducing uncountably chromatic graphs. [A. Hajnal: On some combinatorial problems involving large cardinals, *Fundamenta Mathematicae*, **LXIX**(1970), 39–53]

34. **First Solution.** Assume indirectly that $F : V \to \omega$ is a good coloring. Define by transfinite recursion on $\alpha < \omega_1$ the following function $f(\alpha) = F(f|_\alpha)$. It is clear that f is a function from ω_1 to ω. We show that it is injective and that gives the desired contradiction. Indeed, let $\alpha < \omega_1$ be the least ordinal such that $f(\alpha) = f(\beta)$ holds for some $\beta < \alpha$. Then, $f|_\beta$ and $f|_\alpha$ are injective functions, so they are elements of V, and they are joined in X. But as F is a good coloring of X, $f(\beta) = F(f|_\beta)$ and $f(\alpha) = F(f|_\alpha)$ are distinct, a contradiction.

Second Solution. Assume indirectly that $F : V \to \omega$ is a good coloring. Set $A_0 = \{0\}$, $\alpha_0 = 0$, $f_0 = \emptyset$. Suppose that at step n we are given the finite set $A_n \subseteq \omega$, the ordinal $\alpha_n < \omega_1$, and the function $f_n : \alpha_n \to \omega$. Set $A_{n+1} = A_n \cup \{i_n\}$ where i_n is the least element of $\omega \setminus \mathrm{Ran}(f_n)$ above $\max(A_n)$. If there exists some $f \supseteq f_n$ with $\omega \setminus \mathrm{Ran}(f)$ infinite, $A_{n+1} \cap \mathrm{Ran}(f) = \emptyset$, $F(f) = n$ then let $f_{n+1} : \alpha_{n+1} \to \omega$ be one such f. Otherwise let f_{n+1} be an arbitrary proper extension of f_n to a one–one function $f_{n+1} : \alpha_{n+1} \to \omega$ with co-infinite range that is disjoint from A_{n+1}. This way we get a strictly increasing sequence $f_0 \subseteq f_1 \subseteq \cdots$ of one–one functions. Their union $f_\omega = \bigcup\{f_n : n < \omega\}$ is also a one–one function that properly extends each. Assume that $F(f_\omega) = n$. Notice that $\omega \setminus \mathrm{Ran}(f_\omega)$ is infinite (it includes $\bigcup\{A_k : k < \omega\}$) and $\mathrm{Ran}(f_\omega)$ is disjoint from A_{n+1}. Therefore, we had the first case in the definition of f_{n+1} and selected f_{n+1} with $F(f_{n+1}) = n$. But now f_{n+1} and f_ω are distinct functions which are joined and get the same color, a contradiction. [F. Galvin, R. Laver]

35. We can assume that the ground set of the set system is some cardinal κ. We show by transfinite recursion that there is a good 2-coloring f of κ, and to do that we define $f|_\alpha : \alpha \to \{0,1\}$ inductively on α, where the inductive hypothesis that $f|_\alpha : \alpha \to \{0,1\}$ is a partial good coloring, i.e., there is no monocolored $H \in \mathcal{H}$, $H \subset \alpha$. If $f|_\beta$ is a partial good coloring for every $\beta < \alpha$ and α is a limit ordinal, then (recall that the sets in $\mathcal{H}$ are finite) clearly so is $f|_\alpha = \cup_{\beta<\alpha} f|_\beta$. Suppose now that $\alpha = \beta + 1$ is a successor ordinal, and $f|_\beta$ is already given. If there is no extension of it to α, then there is an $A \in \mathcal{H}$ such that $A \subseteq \alpha$, $\beta \in A$, and $A \setminus \{\beta\} \subseteq f^{-1}(0)$. Similarly, there is a $B \in \mathcal{H}$ such that $B \subseteq \alpha$, $\beta \in B$, and $B \setminus \{\alpha\} \subseteq f^{-1}(1)$. But then $A \cap B = \{\beta\}$ and exactly this configuration is excluded.

36. Let $\{A_i : i \in I\}$ be a maximal subfamily of $\mathcal{H}$ of pairwise disjoint sets (exists by Zorn's lemma). Devise a function $f : \bigcup\{A_i : i \in I\} \to \omega$ which is one-to-one when restricted to any particular A_i. Extend f arbitrarily to the remaining points. We show that f is a good ω-coloring of $\mathcal{H}$. Pick $H \in \mathcal{H}$. By condition, there is some $i \in I$ that $A_i \cap H \neq \emptyset$ and also by condition, $|A_i \cap H| \geq 2$. But then f assumes at least two different values on H and this is what we wanted to show.

For the other part, let $\mathcal{H}$ be a nontrivial ultrafilter on ω. It is not finitely chromatic, as in any finite coloring one of the color classes is in the ultrafilter, and no intersection is a singleton, actually, the intersection of any two members is infinite. [R. Aharoni, P. Komjáth]

37. Let the underlying set of $\mathcal{H}$ be V. We first claim that for every $U \subseteq V$ there is a "closure" of U, a unique minimal set $F(U) \supseteq U$ with the property that if $|H \cap F(U)| \geq 2$ holds for some $H \in \mathcal{H}$ then $H \subseteq F(U)$. Indeed, let $F(U) = F_0(U) \cup F_1(U) \cup \cdots$ where $F_0(U) = U$ and for $n = 0, 1, \ldots$ we set

$$F_{n+1}(U) = F_n(U) \cup \bigcup\{H \in \mathcal{H} : |H \cap F_n(U)| \geq 2\}.$$

Notice that as $\mathcal{H}$ satisfies the condition mentioned in the problem, $F(U)$ is countable whenever U is.

Enumerate V as $\{v_\alpha : \alpha < \omega_1\}$ and set $V_\alpha = F(\{v_\beta : \beta < \alpha\})$. Now each V_α is countable and $V = \bigcup\{V_\alpha : \alpha < \omega_1\}$ is an increasing, continuous decomposition. Moreover, every V_α is "closed", that is, no $H \in \mathcal{H}$ can intersect it in exactly 2 points. This gives that for every $H \in \mathcal{H}$ there is an $\alpha < \omega_1$ such that H has 2 or 3 points in $W_\alpha = V_{\alpha+1} \setminus V_\alpha$ and at most one point in

V_α. As $\{W_\alpha : \alpha < \omega_1\}$ is a system of pairwise disjoint, countable sets, there is an injection $f_\alpha : W_\alpha \to \omega$ and then the union of the f_α's will give a coloring of V with ω such that no $H \in \mathcal{H}$ is monocolored. [P. Erdős, A. Hajnal: On chromatic number of graphs and set-systems, *Acta Math. Acad. Sci. Hung.*, **17**(1966), 61–99]

38. Assume that S^n is colored by $n+1$ colors, and V_i is the set of points of color i. With $\text{dist}(x,y) = \|x-y\|$ the Euclidean distance on $\mathbf{R}^{n+1}$, the functions

$$g_i(x) = \inf_{y \in V_i} \text{dist}(x,y)$$

are continuous functions of $x \in S^n$; therefore,

$$F(x) = (g_1(x), \dots, g_n(x))$$

is a continuous mapping of S^n into $\mathbf{R}^n$. By Borsuk's antipodal theorem there is an $x \in S^n$ with $F(x) = F(-x)$. If for some $1 \le i \le n$ we have $g_i(x) = 0$, then $g_i(-x) = 0$ as well, and so there are points arbitrarily close to x and $-x$ of color i. On the other hand, if for all $1 \le i \le n$ we have $g_i(x) = g_i(-x) > 0$, then necessarily x and $-x$ are of color $n+1$. In any case, under any $(n+1)$-coloring we obtain points with distance arbitrarily close to 2 that have the same color, hence the chromatic number of $G_{n,\alpha}$ must be at least $n+2$.

To see that $G_{n,\alpha}$ can be colored by $n+2$ colors for $\alpha < 2$ close to 2 do this: take a regular $(n+1)$-simplex with vertices on S^n, project from the origin each face of the simplex onto S^n, and let the points of these projected sets have the same color.

39. To show $\text{Chr}(G) \le \aleph_0$, choose $\epsilon < \alpha/2$, and let the color set $\mathcal{F}$ be the set of those $F \subset [0,1]$ which consist of finitely many intervals with rational endpoints. This is a countable set. Let the color of a vertex E be $F \in \mathcal{F}$ if $\text{meas}(E \Delta F) < \epsilon$. Since E contains compact subsets E' with measure arbitrarily close to $\text{meas}(E)$, and for each such E' there is an $F \in \mathcal{F}$ with $E' \subset F$ and $\text{meas}(F \setminus E') < \epsilon/2$, each E gets at least one color from $\mathcal{F}$ (of course, each E gets more than one colors, just keep one of them). Now if both E_1 and E_2 get the same color F, then $E_1 \cap E_2 \ne \emptyset$, so they are not connected in G. This shows that the above coloring is appropriate, and hence $\text{Chr}(G) \le \aleph_0$.

In the other direction we have to show that $\text{Chr}(G) > n$ for all $n = 1, 2, \dots$. Let S_n be a sphere in $\mathbf{R}^n$ with surface measure equal to 1, and let r_n be the radius of S_n. It is known (see e.g., P. Halmos and J. v. Neumann, Ann. Math., 43(1942), 332–350) that there is a measure-preserving bijective mapping $T_n : [0,1] \to S_n$. For $X \in S_n$ consider the (closed) spherical cap U_X with center at X and of surface measure equal to α, and let $E_X = T^{-1}(U_X)$ be the inverse image of U_X. Note that there is a $\beta_{n,\alpha} < 2r_n$ such that $E_{X_1} \cap E_{X_2} = \emptyset$ (which is the same as $U_{X_1} \cap U_{X_2} = \emptyset$) precisely if the distance of X_1 and X_2 is bigger than $\beta_{\alpha,n}$. Hence the chromatic number of the subgraph spanned by $\{E_X : X \in S_n\}$ is at least $n+1$ by the previous problem. [P. Erdős and A. Hajnal, Matematikai Lapok, **18**(1967), 1–4]

24

Partition relations

1. For $k = 2$ this is just a reformulation of Problem 23.1. Suppose the statement is known for some k, and let $f : [\omega]^2 \to \{0, 1, \ldots, k\}$ be a coloring with $k+1$ colors. Unite color classes 0 and 1 into a new color class -1. This way we obtain a coloring of the pairs of ω with k colors: $-1, 2, \ldots, k$. By the inductive hypothesis there is an infinite monochromatic subset V' for the latter coloring. If its color is one of $2, \ldots, k$, we are done, V' is monochromatic in the original coloring. In the remaining case, V' is colored by -1; therefore, it was originally colored by 0 and 1. The case $k = 2$, applied to V', gives an infinite monochromatic set of color 0 or 1, in the original coloring.

2. We prove the statement by induction on r. The case $r = 1$ is obvious: if we decompose an infinite set into finitely many parts, then one of the parts is infinite. Suppose the statement has been verified for r. Let $f : [\omega]^{r+1} \to k$ be a coloring. We argue that there is an infinite set A such that the following is true. If $a_1 < \cdots < a_r < a < b$ are from A, then $f(a_1, \ldots, a_r, a) = f(a_1, \ldots, a_r, b)$ holds (that is, A is endhomogeneous). Accepting the existence of A we conclude the proof as follows. Color the r-tuples of A by putting $g(a_1, \ldots, a_r)$ the common value of $f(a_1, \ldots, a_r, a)$ where $a \in A$, $a > a_r$. By the induction hypothesis there are an infinite $B \subseteq A$ and a color i such that all r-tuples from B get color i under g. But then clearly B is monochromatic in color i for f as well.

To obtain A we inductively select the decreasing sequence of infinite sets $Y_0 \supseteq Y_1 \supseteq \cdots$ and the elements $x_0 < x_1 < \cdots$ as follows. Set $Y_0 = \omega$. If Y_i is determined, let x_i be its least element. After this, for every $z \in Y_i \setminus \{x_i\}$, z determines a coloring g_z of the r-tuples of $\{x_0, \ldots, x_i\}$ by making $g_z(x_{j_1}, \ldots, x_{j_r}) = f(x_{j_1}, \ldots, x_{j_r}, z)$. As $[\{x_0, \ldots, x_i\}]^r$ is finite (possibly empty), there are finitely many possibilities of coloring it with k colors. There is, therefore, an infinite $Y_{i+1} \subseteq Y_i \setminus \{x_i\}$ such that the g_z functions are identical for $z \in Y_{i+1}$, and so the definition of Y_{i+1} is complete. We get, therefore, an infinite set $\{x_0, x_1, \ldots\}$ such that the color of an $(r+1)$-tupie does not depend on the last element. [F. P. Ramsey: On a problem of formal logic, *Proc. London Math. Soc.* (2), **30**(1930), 264–286]

3. Color the pairs of elements as follows. A pair gets color 0 if it consists of comparable elements, and color 1 otherwise. By Problem 1 there is an infinite monochromatic set and it can only be a chain or antichain, according to its color.

4. Let $a_0, a_1, \dots$ be infinitely many elements of the ordered set $\langle A, \prec \rangle$. Color $\{a_i, a_j\}$ with $i < j$ zero if $a_i \prec a_j$ and with one otherwise. By Problem 1 there is an infinite monochromatic set and it is an increasing or decreasing sequence, according to its color.

5. **First solution.** An easy geometry argument gives that out of 5 planar points some 4 form a convex quadruple. Color every 4-element subset of X by 0 or 1 accordingly if they form a convex quadruple or not. By the above remark there is no monochromatic 5-element subset of color 1, so, by Problem 2 there is an infinite monochromatic set of color 0, which is exactly a convex set. [P. Erdős, G. Szekeres: A combinatorial problem in geometry, *Compositio Math.*, **2**(1935), 463–470]

Second solution. Working on the plane with x-, y-axes we can assume that the points of X are $\langle a_0, b_0 \rangle, \langle a_1, b_1 \rangle, \dots$. We can equally assume (by shrinking X, and rotating the coordinate system, if needed) that $a_i \neq a_j$ for $i \neq j$. Given a triple $\{i, j, k\}$ of natural numbers there can be two cases: of the points (a_i, b_i), (a_j, b_j), and (a_k, b_k), the one whose x-coordinate is between those of the other two, can be above or below the segment determined by the other two points. If we color the point triple by 0 or 1 according to which case holds, we get a coloring of $[\omega]^3$ by two colors. An application of Problem 2 gives a subset as required. [N. Tarsi, cf. M. Lewin: A new proof of a theorem of Erdős and Szekeres, *Math. Gaz.*, **60**(1976), 136–138]

6. If we are given a tournament on ω, for $u < v < \omega$ color the edge $\{u, v\}$ green, if $\overrightarrow{uv}$, and blue otherwise. By Ramsey's theorem, there is an infinite monochromatic set, and it is obviously a transitive subtournament.

Another possibility is to observe that a tournament is transitive if and only if every triangle in it is transitive, and every tournament on 4 nodes includes a transitive triangle. Then we can apply the relation $\omega \to (\omega, 4)^3$. [P.Erdős–R.Rado]

7. As in the first solution of Problem 6, we assume that the graph is on ω, and color the pair $\{u, v\}$ $(u < v < \omega)$ with 0, if u and v are not joined in X, with 1, if $\overrightarrow{uv}$, and with 2, if $\overleftarrow{uv}$. By Ramsey's theorem there is an infinite monochromatic set. If its color is 0, then it is an independent set, if it is 1 or 2, it is a transitively directed subset.

8. Let the vertices be the functions $f : \omega \to \{0,1\}$, and if f, g are two such functions and n is the smallest number with $f(n) \neq g(n)$, then let the color of (f,g) be $(n,0)$ if $f(n) < g(n)$, and otherwise let it be $(n,1)$. It is easy to see that this is an appropriate coloring.

9. The proof is identical with the corresponding part of the solution of Problem 2.

10. For a triple $\{x,y,z\} \in [\omega]^3$ with $x < y < z$ there are 5 possibilities if we consider which of $f(x,y), f(x,z), f(y,z)$ are equal. Similarly, given a quadruple $\{x,y,z,t\} \in [\omega]^4$, with $x < y < z < t$, there are a finite number, say s possibilities, on equalities of the values of f on $\left[\{x,y,z,t\}\right]^2$. Accordingly, we get colorings $g : [\omega]^3 \to 5$ and $h : [\omega]^4 \to s$, which give the types of the triples and quadruples in the above sense. By Ramsey's theorem (Problem 2) there is an infinite set $H \subseteq \omega$ homogeneous to both g and h. We claim that H is as required. Assume that there are $s, t \in [H]^2$ with $f(s) = f(t)$ (otherwise we land in case (d)). As H is homogeneous for g, h, $f(s') = f(t')$ holds every time the relative (ordered) position of $s', t' \in [H]^2$ is the same as that of s, t. One can find $s', t', t'' \in [H]^2$ such that s', t' and s', t'' both are similar to s, t (in the above sense) and either $\min(t') = \min(t'')$ or $\max(t') = \max(t'')$. For simplicity, assume the former case. We get, therefore, one occurrence of $f(s) = f(t)$ in $[H]^2$ with $\min(s) = \min(t)$, and, as H is homogeneous for g, this must always hold in this situation. We get (b), unless there are $s, t \in [H]^2$ with $\min(s) \neq \min(t)$ yet $f(s) = f(t)$. Then, using the properties of H again, we get that to any $x < y$ in H there are $s, t \in [H]^2$ with $\min(s) = x$, $\min(t) = y$ and $f(s) = f(t)$, and eventually we get that H is homogeneous. [P. Erdős, R. Rado: A combinatorial theorem, *Jour. Lond. Math. Soc.*, **25**(1950), 249–255]

11. Select the sequence $1 = r_0 < r_1 < \cdots$ in such a way that if $r \geq r_t$ then $f(r) \geq 2^t$. Let A_1 be an infinite subset of ω that is homogeneous for every H_r, $r < r_1$ (exists by Ramsey's theorem, Problem 2). Choose $x_1 = \min(A_1)$. By induction on t, if we have found $\{x_1, \ldots, x_t\}$ and A_t, choose an infinite $A_{t+1} \subseteq A_t \setminus \{x_t\}$ such that if $r_t \leq r < r_{t+1}$, $B \subseteq \{x_1, \ldots, x_t\}$ and $C \subseteq A_{t+1}$, $|B| + |C| = r$, then $H_r(B \cup C)$ depends only on B. Such a set can be found; it only requires a(n enormous but) finite number of applications of Ramsey's theorem. Having finished the inductive construction, set $X = \{x_1, \ldots\}$. If $r_t \leq r < r_{t+1}$, then H_r assumes at most $2^t \leq f(r)$ values on X and we are done.

For the other direction if $s = \{x_1, \ldots, x_r\} \subseteq \omega$ make $H_r(s) = i$ if there are precisely i indices $1 \leq j < r$ for which $x_{j+1} - x_j < r$. We claim that if $X \subseteq \omega$ is infinite and i is given then for r sufficiently large there is an $s \in [X]^r$ with $H_r(s) = i$, thus the number of colors occurring in $[X]^r$ tends to infinity as $r \to \infty$. In fact, let $y_1 < y_2 < \cdots < y_{i+1}$ be the first $i+1$ elements of X, choose $r > y_{i+1}$, further let $x_j = y_j$ for $1 \leq j \leq i+1$ and inductively choose

$x_{i+2}, \ldots, x_r \in X$ in such a way that $x_{j+1} - x_j > r$ for $j = i+1, \ldots, r-1$. Then $\{x_1, \ldots, x_r\}$ has color i. [J. E. Baumgartner, P. Erdős, A. Hajnal, R. Rado]

12. From finite Ramsey theory we know that there is a natural number d such that $d^n \to (3n)^2_3$ holds for every n. Set $c = d + 1$. Assume that we are given a coloring $f : [\omega]^2 \to 3$. By induction we select the finite sets $A_0, A_1, \ldots$ as follows. If $A_0, \ldots, A_t$ have already been selected, set $p = |A_0| + \cdots + |A_t|$, $q = \max(A_t)$. There is a number $n = n_{t+1} > q$ so large that $q + 3^p d^n < c^n$. By the pigeon hole principle, there are at least d^n elements in the interval $[q+1, q+3^p d^n]$ that are joined to $A_0 \cup \cdots \cup A_t$ the same way, i.e., $f(x, y)$ depends only on x. Using the above-mentioned Ramsey property, there is a $3n_{t+1}$-element subset, which is homogeneous to f, this will be our A_{t+1}.

Applying Problem 9, we get an infinite subset $X \subseteq \omega$ such that for $i < j$ in X if $x \in A_i$, $y \in A_j$, then $f(x, y) = g(x)$, that is, the color does not depend on y or even on j. This g 3-colors A_i, so there is a $B_i \subseteq A_i$, $|B_i| = n_i$, for which $g(x)$ only depends on i.

For an infinite $Y \subseteq X$ this value is the same (say e_0), and also the color of pairs in B_i is the same (say e_1).

The set $\bigcup\{B_i : i \in Y\}$ uses only the colors $\{e_0, e_1\}$, the index of the largest element of B_i is at least n_i and its value is at most c^{n_i} for $i \in Y$. [P. Erdős, cf. P. Erdős, F. Galvin: Some Ramsey-type theorems, *Discrete Mathematics*, **87**(1991), 261–269]

13. (a) Let $f : [\kappa]^2 \to \{0, 1\}$. Assume first that for every $x < \kappa$, the set $\{y < \kappa : f(x, y) = 1\}$ is of cardinality less than κ, that is, if we consider the graph of those pairs $\{x, y\}$ for which $f(x, y) = 1$, then every vertex has degree $< \kappa$. Then, by transfinite recursion, we can choose the vertices $\{x_\alpha : \alpha < \kappa\}$ such that $f(x_\beta, x_\alpha) = 0$ holds for $\beta < \alpha < \kappa$. Indeed, if at step α, the vertices $\{x_\beta : \beta < \alpha\}$ have already been selected, then each of them disqualifies (by hypothesis) a set of cardinality $< \kappa$ as possible x_α, and as κ is regular, the union of these $< \kappa$ sets each with cardinality $< \kappa$ is still a set of cardinality $< \kappa$ so it is possible to choose x_α. Now observe that $\{x_\alpha : \alpha < \kappa\}$ is a set of cardinality κ monochromatic in color 0.

We have proved that if there is no monochromatic set of size κ in color 0, then there must be some vertex v_0 such that if $A_0 = \{y < \kappa : f(v_0, y) = 1\}$, then A_0 is of cardinality κ. Repeating the previous argument inside A_0 we get that there must be some vertex $v_1 \in A_0$ such that the set $A_1 = \{y \in A_0 : f(v_1, y) = 1\}$ if of cardinality κ. Continuing, we get the vertices $v_0, v_1, \ldots$ and sets $A_0, A_1, \ldots$ and the set $\{v_0, v_1, \ldots\}$ is an infinite set monochromatic in color 1. [B. Dushnik, E. W. Miller: Partially ordered sets, *American Journal of Mathematics*, **63**(1941), 600–610]

(b) Using the argument in part (a) it suffices to show the following. If X is a graph on κ with no infinite complete subgraph and in which every degree is less than κ, then there is an independent set in X of cardinality κ. Let $\{\kappa_\alpha : \alpha < \mu\}$ be a strictly increasing sequence of cardinals cofinal in κ where

$\mu = \mathrm{cf}(\kappa)$, with $\mu < \kappa_0$. Decompose κ into the union $\kappa = \bigcup\{S_\alpha : \alpha < \mu\}$ with $|S_\alpha| = \kappa_\alpha^+$. Using part (a) we can shrink each S_α to an independent set $S'_\alpha \subseteq S_\alpha$, $|S'_\alpha| = \kappa_\alpha^+$. For each $x \in S'_\alpha$ there is a least $\beta = \beta(x) < \mu$ such that the degree of x is $\leq \kappa_{\beta(x)}$. The mapping $x \mapsto \beta(x)$ decomposes S'_α into at most μ parts (taking the inverse images of the elements). As $\mathrm{cf}(\kappa_\alpha) > \mu$, some of them must have cardinality κ_α^+, that is, there is $S''_\alpha \subseteq S'_\alpha$, $|S''_\alpha| = \kappa_\alpha^+$ and there is $g(\alpha) < \mu$ such that if $x \in S''_\alpha$ then the degree of x is at most $\kappa_{g(\alpha)}$.

Select, by transfinite recursion, an increasing sequence $\{\alpha_i : i < \mu\}$ of ordinals smaller than μ such that $\sup\{g(\alpha_j) : j < i\} \leq \alpha_i$ holds for every $i < \mu$. This is possible as μ is regular and at every step we must choose an ordinal that is greater than the supremum of some $< \mu$ ordinals below μ. We finally choose the sets $\{T_i : i < \mu\}$ by transfinite recursion on $i < \mu$ with the properties $|T_i| = \kappa_{\alpha_i}^+$, $T_i \subseteq S''_{\alpha_i}$ so that the set $\bigcup\{T_i : i < \mu\}$ will be independent. Assume we are at step i and the sets $\{T_j : j < i\}$ have already been constructed. In order to get T_i we remove from S''_{α_i} all vertices that are joined to some element of $T = \bigcup\{T_j : j < i\}$. The number of these removed elements can be estimated as

$$\sum_{j<i} |T_j| \kappa_{g(\alpha_j)} \leq \kappa_{\alpha_i} \sum_{j<i} \kappa_{g(\alpha_j)} \leq \kappa_{\alpha_i} \cdot \kappa_{\alpha_i} \cdot i = \kappa_{\alpha_i}.$$

As $|S''_{\alpha_i}| = \kappa_{\alpha_i}^+$, there remain $\kappa_{\alpha_i}^+$ elements, so T_i can be chosen. As $\bigcup\{T_i : i < \mu\}$ is an independent set of cardinality κ, we are done. [P. Erdős]

14. Assume that $\{f_\alpha : \alpha < \kappa^+\}$ is a lexicographically decreasing sequence. Then, $\{f_\alpha(0) : \alpha < \kappa^+\}$ is a nonincreasing sequence of ordinals; therefore, it stabilizes, that is, $f_\alpha(0) = g(0)$ holds for $\alpha > \alpha_0$ for some $\alpha_0 < \kappa^+$. Restricting to those values of α, $\{f_\alpha(1) : \alpha < \kappa^+\}$ is a nonincreasing sequence of ordinals, so again, $f_\alpha(1) = g(1)$ holds for $\alpha > \alpha_1$ for some $\alpha_1 < \kappa^+$. Continuing, we get the ordinals $\alpha_i < \kappa^+$ for $i < \kappa$, and the values $g(i) < \lambda$ that $f_\alpha(i) = g(i)$ holds for $\alpha > \alpha_i$. But then, all functions f_α with $\alpha > \sup\{\alpha_i : i < \kappa\}$ are identical, a contradiction.

Assume that $\{f_\alpha : \alpha < \mu^+\}$ is a lexicographically increasing sequence for $\mu = \max(\kappa, \lambda)$. $\{f_\alpha(0) : \alpha < \mu^+\}$ is a nondecreasing sequence of ordinals $< \lambda$, only at λ places can they properly increase. So it stabilizes, that is, $f_\alpha(0) = g(0)$ holds for $\alpha > \alpha_0$ for some $\alpha_0 < \mu^+$. Restricting to those values of α, $\{f_\alpha(1) : \alpha < \mu^+\}$ is a nondecreasing sequence of ordinals, so again, $f_\alpha(1) = g(1)$ holds for $\alpha > \alpha_1$ for some $\alpha_1 < \mu^+$. Continuing, we find the ordinals $\alpha_i < \mu^+$ for $i < \kappa$, and the values $g(i) < \lambda$ that $f_\alpha(i) = g(i)$ holds whenever $\alpha > \alpha_i$. As before, all functions f_α with $\alpha > \sup\{\alpha_i : i < \kappa\}$ will be identical, a contradiction. See also Problems 6.93–94.

15. As $|A| \leq 2^\kappa$ we have an injection $\Phi : A \to {}^\kappa 2$. For $x < y$ in A there is a least $\alpha < \kappa$ that $\Phi(x)(\alpha) \neq \Phi(y)(\alpha)$. Set $f(x, y) = \langle \alpha, 0 \rangle$ if $\Phi(x)(\alpha) = 0$ and $\Phi(y)(\alpha) = 1$, and set $f(x, y) = \langle \alpha, 1 \rangle$ when $\Phi(x)(\alpha) = 1$ and $\Phi(y)(\alpha) = 0$. If,

for $x < y < z$, $f(x, y) = f(y, z) = \langle \alpha, 0 \rangle$, say, then $\Phi(y)(\alpha)$ would be 0 and 1 in the same time, a contradiction.

16. Let $\langle A, \prec \rangle$ be an ordered set whose order type is a Specker type (see 27.15). Enumerate A as $A = \{a(\alpha) : \alpha < \omega_1\}$ and let $\{x(\alpha) : \alpha < \omega_1\}$ be a set of distinct reals in $[0, 1]$. We construct the tournament on ω_1: if $\alpha < \beta < \omega_1$, set $\overrightarrow{\alpha\beta}$, i.e., direct the edge $\{\alpha, \beta\}$ from α to β if and only if either $x(\alpha) < x(\beta)$ and $a(\alpha) \prec a(\beta)$ or $x(\beta) < x(\alpha)$ and $a(\beta) \prec a(\alpha)$.

First we observe that if $B \subseteq A$ is uncountable then there is an $a(\alpha) \in B$ such that for uncountably many $\beta > \alpha$ the relations $a(\beta) \in B$ and $a(\alpha) \prec a(\beta)$ hold. Indeed, otherwise, we could inductively select a sequence from B of order type ω_1^*, which contradicts the properties of $\langle A, \prec \rangle$.

Assume that $X \subseteq \omega_1$ is uncountable. We claim that there is $\alpha \in X$ such that the set $\{\beta \in X : a(\alpha) \prec a(\beta),\ x(\alpha) < x(\beta)\}$ is uncountable. In fact, for $\alpha \in X$ let $f(a(\alpha))$ be the least $t \in [0, 1]$ such that $x(\beta) < t$ holds for all but countably many $\beta \in X$ with $a(\alpha) \prec a(\beta)$. Since f is a nonincreasing real-valued function on a subset of A, it can only have countably many different values; otherwise, there would be an uncountable subset of A similar to an uncountable subset of the reals, an impossibility. Hence f is constant, say t_0 on an uncountable set. Set $X_0 = \{\alpha \in X : f(a(\alpha)) = t_0\}$. As we have remarked above, there is an $\alpha_0 \in X_0$ such that $\{\beta \in X_0 : a(\alpha_0) \prec a(\beta)\}$ is uncountable, and then in this set there is an $\alpha \in X_0$ such that $a(\alpha_0) \prec a(\alpha)$ and $x(\alpha) < t_0$ (by the choice of $t_0 = f(a(\alpha_0))$). Since $f(a(\alpha)) = t_0$ also holds and $x(\alpha) < t_0$, there are uncountably many $\beta \in X$ such that $a(\alpha) \prec a(\beta)$ and $x(\alpha) < x(\beta)$, and the claim has been proved.

A similar argument shows (by reversing $\prec$ and $<$) that there is an α with $\{\beta \in X : a(\beta) \prec a(\alpha), x(\beta) < x(\alpha)\}$ uncountable.

We next claim that there are uncountable $X_0, X_1 \subseteq X$ such that if $\alpha \in X_0$, $\beta \in X_1$, then $a(\alpha) \prec a(\beta)$ and $x(\alpha) < x(\beta)$. Toward proving this, let U be the set of those $\alpha \in X$ such that $\{\beta \in X : a(\alpha) \prec a(\beta), x(\alpha) < x(\beta)\}$ is countable, and let L be the set of those $\alpha \in X$ such that $\{\beta \in X : a(\beta) \prec a(\alpha), x(\beta) < x(\alpha)\}$ is countable. Both U and L are countable. Indeed, should, say, U be uncountable, then, by our first claim, it would contain an α with $\{\beta \in U : a(\alpha) \prec a(\beta), x(\alpha) < x(\beta)\}$ uncountable, but this is nonsense since then α cannot belong to U. Thus U and L are countable, and so we can pick an $\alpha \in X \setminus (U \cup L)$. Then the sets

$$X_0 = \{\beta \in X : a(\beta) \prec a(\alpha) \quad \text{and} \quad x(\beta) < x(\alpha)\}$$

and

$$X_1 = \{\beta \in X : a(\alpha) \prec a(\beta) \quad \text{and} \quad x(\alpha) < x(\beta)\}$$

establish our second claim.

Fix now X_0 and X_1 as in the second claim. A further application of the same claim to X_0 and to the reversely ordered $\langle A, \succ \rangle$ we get uncountable $Y_0, Y_1 \subseteq X_0$ such that $\alpha \in Y_0$, $\gamma \in Y_1$ satisfy $a(\alpha) \prec a(\gamma)$ and $x(\alpha) > x(\gamma)$. As

Y_0, Y_1, X_1 are uncountable subsets of ω_1, we can choose $\alpha < \beta < \gamma$, $\alpha \in Y_0$, $\beta \in X_1$, and $\gamma \in Y_1$. Then $\overrightarrow{\alpha\beta}$, $\overrightarrow{\beta\gamma}$, $\overrightarrow{\gamma\alpha}$ are edges in our tournament, so it is not transitive on X. [R. Laver, see F. Galvin, S. Shelah: Some counterexamples in the partition calculus, *Jour. Comb. Th.*, **15**(1973), 167–174]

17. It suffices to give a function $F : [\omega_1]^2 \to \omega_1$ such that the range of F on any uncountable $X \subseteq \omega_1$ includes a closed, unbounded set. Indeed, if $f : \omega_1 \to \omega_1$ is a function as described in Problem 21.21, then their composition $f \circ F$ is as required.

Select the distinct functions $r_\alpha : \omega \to 2$ for $\alpha < \omega_1$. For $\alpha \neq \beta < \omega_1$ let $d(\alpha, \beta)$ be the least n with $r_\alpha(n) \neq r_\beta(n)$. Fix, for every $0 < \alpha < \omega_1$ a (possibly repetitive) enumeration $\alpha = \{x_n^\alpha : n < \omega\}$. For $\alpha < \beta < \omega_1$ set $A(\alpha, \beta) = \{x_n^\beta : n \leq d(\alpha, \beta)\}$, $F(\alpha, \beta) = \min(A(\alpha, \beta) \setminus \alpha))$.

Assume that $X \subseteq \omega_1$ is uncountable. If $g : n \to 2$ for some $n < \omega$, set $T(g) = \{\alpha \in X : g \subseteq r_\alpha\}$. Set $\gamma \in C$ if γ is a limit ordinal and the following is true. For every $g : n \to 2$, $(n < \omega)$, if $T(g)$ is countable, then $\gamma > \sup(T(g))$, if $T(g)$ is uncountable, then $T(g) \cap \gamma$ is cofinal in γ. C is closed, unbounded in ω_1 by Problems 21.2 and 21.1. We claim, and that suffices, that every element of C is in the range of F on $[X]^2$.

Assume that $\gamma \in C$. Pick $\beta \in X$, $\beta > \gamma$ (possible, as X is uncountable). For $n < \omega$ set $g_n = r_\beta|_{(n+1)}$. Notice that $\gamma < \beta \in T(g_n)$, therefore $T(g_n)$ is uncountable for $n < \omega$. For $n < \omega$ let $g_n^* : (n+1) \to 2$ be the (unique) function that agrees with g_n at all but the last place: $g_n^*|_n = g_n|_n$, $g_n^*(n) \neq g_n(n)$. Clearly, $T(g_n) \setminus \{\beta\} = T(g_{n+1}^*) \cup T(g_{n+2}^*) \cup \cdots$, so for every $n < \omega$ there is $N \geq n$ with $T(g_N^*)$ uncountable.

As $\beta > \gamma$, $\gamma = x_k^\beta$ holds for some $k < \omega$. Choose $N \geq k$ with $T(g_N^*)$ uncountable. Notice that for $\alpha \in T(g_N^*)$, $d(\alpha, \beta) = N$, hence $A = A(\alpha, \beta) = \{x_n^\beta : n \leq N\}$ is the same finite set containing γ. Recalling the definition of F, we get that for $\alpha \in T(g_N^*) \cap \beta$, $F(\alpha, \beta) = \min(A \setminus \alpha)$. As $A \cap \gamma$ is finite and $T(g_N^*) \cap \gamma$ is cofinal in γ, we can choose an $\alpha \in T(g_N^*) \cap \gamma$ so large that the least element of A which is $\geq \alpha$ is γ. For this α, we have $F(\alpha, \beta) = \gamma$, as desired. [S. Todorcevic: Partitioning pairs of countable ordinals, *Acta. Math.*, **159**(1987), 261–294]

18. Set $S = \left\{\alpha < (2^\kappa)^+ : \mathrm{cf}\,(\alpha) = \kappa^+\right\}$, a stationary set in $(2^\kappa)^+$ by Problem 21.8. For every $\alpha \in S$ start building the endhomogeneous set $\{x_\xi^\alpha : \xi < \kappa^+\} \subseteq \alpha$ in the sense that we require that

$$f(x_{\xi_1}^\alpha, \ldots, x_{\xi_r}^\alpha, x_\eta^\alpha) = f(x_{\xi_1}^\alpha, \ldots, x_{\xi_r}^\alpha, \alpha)$$

hold for $\xi_1 < \cdots < \xi_r < \eta < \kappa^+$. For every given α we can either continue for κ^+ steps or get stuck somewhere. If there is some α for which the first case holds, lovely, we have the sought-for endhomogeneous set: $X = \{x_\xi^\alpha : \xi < \kappa^+\}$. We can therefore assume that for every $\alpha \in S$ there is a point where we get stuck: for some ordinal $\gamma(\alpha) < \kappa^+$ we cannot extend the set $\{x_\xi^\alpha : \xi < \gamma(\alpha)\}$.

Notice that as $\mathrm{cf}(\alpha)=\kappa^+$, $\{x_\xi^\alpha : \xi<\gamma(\alpha)\}$ is a bounded subset of α. Applying Problem 21.10, we get that there is a stationary $S'\subseteq S$ such that these values are constant: for $\alpha\in S'$ we have $\gamma(\alpha)=\gamma$ and for $\xi<\gamma$, $x_\xi^\alpha=x_\xi$. The number of $h:[\gamma]^r\to\kappa$ functions is $\kappa^\kappa=2^\kappa$, and for each $\alpha\in S'$

$$\{\xi_1,\dots,\xi_r\}\mapsto f(x_{\xi_1},\dots,x_{\xi_r},\alpha)$$

is such a function, so there must be $\alpha<\beta$ in S' such that $f(x_{\xi_1},\dots,x_{\xi_r},\alpha)=f(x_{\xi_1},\dots,x_{\xi_r},\beta)$ holds for $\xi_1<\dots<\xi_r<\gamma$. But then we reached a contradiction; α can be added to the set $\{x_\xi^\beta:\xi<\gamma(\beta)\}$ and still keep it endhomogeneous.

19. Assume that $f:\left[(2^\kappa)^+\right]^2\to\kappa$. Set $S=\left\{\alpha<(2^\kappa)^+ : \mathrm{cf}(\alpha)=\kappa^+\right\}$, a stationary set in $(2^\kappa)^+$ (see Problem 21.8). For every $\alpha\in S$ and every color $i<\kappa$ we start building the increasing sequence $Z(\alpha,i)=\{x_\xi^{\alpha,i}:\xi<\kappa^+\}\subset\alpha$ such that for $\xi<\zeta$ we have

$$f(x_\xi^{\alpha,i},x_\zeta^{\alpha,i})=f(x_\xi^{\alpha,i},\alpha)=i,$$

that is, $Z(\alpha,i)\cup\{\alpha\}$ is homogeneous in color i. If, for some $\alpha\in S$ and some $i<\kappa$ we can proceed through κ^+ steps, we get a homogeneous set of cardinality κ^+ in color i. We can assume, therefore, that for every $\alpha\in S$, $i<\kappa$ we have the nonextendable set $Z(\alpha,i)=\{x_\xi^{\alpha,i}:\xi<\gamma(\alpha,i)\}$ with some $\gamma(\alpha,i)<\kappa^+$. As the mapping $\alpha\mapsto\langle\gamma(\alpha,i):i<\kappa\rangle$ has a range of cardinality at most $\left(\kappa^+\right)^\kappa=2^\kappa$, there is, by Problem 21.6 a stationary $S'\subseteq S$ such that $\gamma(\alpha,i)=\gamma(i)$ with some $\gamma(i)$ for every $\alpha\in S'$. On S' we have a system of κ regressive functions, for every $i<\kappa$ and $\xi<\gamma(i)$, the mapping $\alpha\mapsto x_\xi^{\alpha,i}$. By Problem 21.10, there is a stationary set $S''\subseteq S'$ where they all are constant, that is, on S'' the sets $Z(\alpha,i)$ are identical, $Z(\alpha,i)=Z(i)$. Now pick $\alpha<\beta$ in S'', let $i=f(\alpha,\beta)$. Then, as $\sup(Z(i))<\alpha$, and $f(\alpha,\beta)=i$, α is a good continuation of $Z(i)=Z(\beta,i)$, and this contradicts the maximality of the latter set. [P. Erdős: Some set-theoretical properties of graphs, *Revista de la Univ. Nac. de Tucumán, Ser. A. Mat. y Fis. Teór.* **3**(1942), 363–367. For an alternative proof, see the solution to Problem 25.]

20. Assume that $f:\left[(2^\kappa)^+\right]^2\to\kappa$. We repeat the argument in the previous problem for the colors $0<i<\kappa$. That is, for every $\alpha<(2^\kappa)^+$, $\mathrm{cf}(\alpha)=\kappa^+$, $0<i<\kappa$, we build the set $Z(\alpha,i)\subseteq\alpha$ such that $Z(\alpha,i)\cup\{\alpha\}$ is homogeneous in color i. If there are some α and i such that we can proceed through κ^+ steps, then we are finished; we have found a homogeneous set of cardinality κ^+ in one of the colors $0<i<\kappa$. In the other case, for each α and each $0<i<\kappa$ there is a nonextendable $Z(\alpha,i)$ as above, of cardinality $\le\kappa$. By the above argument, there is a stationary set S'', such that we get a contradiction if for some $\alpha<\beta$ in S'', the color $f(\alpha,\beta)$ is any of the values $0<i<\kappa$. This

exactly means that S'' is a homogeneous set in color 0, and, as it is stationary, it has cardinality $(2^\kappa)^+$.

21. Define the coloring $F : (2^\kappa)^+ \to \kappa + 1$ as follows. For $\alpha < \beta < (2^\kappa)^+$ set $F(\alpha, \beta) = \kappa$ if $f_\alpha(\xi) \le f_\beta(\xi)$ for every $\xi < \kappa$, otherwise let $F(\alpha, \beta)$ be the least ξ such that $f_\alpha(\xi) > f_\beta(\xi)$ holds. By Problem 20 either there is a homogeneous subset in color κ of cardinality $(2^\kappa)^+$, in which case we are done, or there is a homogeneous subset of cardinality κ^+ in color ξ for some $\xi < \kappa$. But in the latter case, if Z is the homogeneous set, then $\{f_\alpha(\xi) : \alpha \in Z\}$ is a decreasing sequence of ordinals of length κ^+, an impossibility.

22. Suppose first that $|X| > \mathbf{c}$ and $d : X \times X \to [0, \infty)$ is a symmetric mapping with $d(x, y) = 0$ if and only if $x = y$. Color the pair $\{x, y\}$ with color $k \in \mathbf{Z}$ if $2^k \le d(x, y) < 2^{k+1}$. By Problem 19 there is a homogeneous triangle, $\{x, y, z\}$, in some color, say, in color k. Now if $\{x', y', z'\}$ is any permutation of $\{x, y, z\}$, then $d(x', z') < 2^{k+1} = 2^k + 2^k \le d(x', y') + d(y', z')$ so d is not an antimetric.

For the other direction notice that if $X \subseteq \mathbf{R}$ then $d(x, y) = (x - y)^2$ is an antimetric on X. [V. Totik]

23. Consider two orderings on the same set ${}^\kappa 2$, the set of all $\kappa \to \{0, 1\}$ functions. One is the lexicographic ordering, denoted by $<$. The other is an arbitrary well-ordering, denoted by $<_w$. For $f, g \in {}^\kappa 2$ color the pair $\{f, g\}$ by 0 if the orders agree on the pair, that is either $f < g$ and $f <_w g$ hold, or else $g < f$ and $g <_w f$ hold. In the other case color the pair $\{f, g\}$ by 1.

Assume that X is some homogeneous set in color 0 with $|X| = \kappa^+$. Then the orderings agree on X. As one of them is a well-ordering, so is the other; therefore, X is a set on which $<$ is a well order. But this is impossible as by Problem 14 there is no subset of $\langle {}^\kappa 2, < \rangle$ of order type $\ge \kappa^+$.

A similar argument works for a homogeneous set in color 1. [W. Sierpiński: Sur un problème de la théorie des relations, *Ann. Scuola Norm. Sup. Pisa, Sci. Fis. Matem.*, **2**(1933), 285–287]

24. Consider the set of all $\kappa \to \{0, 1\}$ functions as S. Color a pair $\{g, h\} \in [S]^2$ with color $i < \kappa$ if and only if i is the least coordinate that $g(i) \ne h(i)$ holds. There is no monochromatic triangle as that would mean three functions g_0, g_1, g_2 with $g_0(i), g_1(i), g_2(i)$ being three distinct elements of $\{0, 1\}$, an impossibility. [K. Gödel]

25. By induction on r. The case $r = 0$ is trivial: if κ^+ is colored with κ colors, then (as κ^+ is regular) there are κ^+ points with the same color.

Assume the statement for r and let $f : [\exp_{r+1}(\kappa)^+]^{r+2} \to \kappa$. By Problem 18 there is an endhomogeneous set X with $|X| = \exp_r(\kappa)^+$, that is, for $x_1 < \cdots < x_{r+1} < y$, $f(x_1, \ldots, x_{r+1}, y)$ does not depend on y, say $f(x_1, \ldots, x_{r+1}, y) = g(x_1, \ldots, x_{r+1})$ holds on X. Applying the case for r to g we get that there is a set of cardinality κ^+ that is homogeneous for g and so it is homogeneous for f, as well. [P.Erdős, R. Rado: A partition calculus in set theory, *Bull. Amer. Math. Soc.*, **62**(1956), 427–489]

26 We show the existence of the required function by induction on r. For $r = 0$ the function $f(x) = x$ $(x < \kappa)$ is good. Assume that we have the statement for r and want to prove it for $r + 1$. Given the infinite cardinal κ, let F be a function on $\left[\exp_r(2^\kappa)\right]^{r+1}$ with the required properties. We can assume that F maps into ${}^\kappa 2$, the set of all $\kappa \to \{0, 1\}$ functions. Define f on the $r + 2$-tuples of $\exp_r(2^\kappa) = \exp_{r+1}(\kappa)$ as follows. If $x_0 < x_1 < \cdots < x_{r+1}$ are given, then $g = F(x_0, \cdots, x_r)$ and $h = F(x_1, \cdots, x_{r+1})$ are two distinct $\kappa \to 2$ functions. Let $\alpha < \kappa$ be the point of first difference. If $g(\alpha) = 0$, $h(\alpha) = 1$, then set $f(x_0, x_1, \cdots, x_{r+1}) = \langle \alpha, 0\rangle$, if it is the other way around, set $f(x_0, x_1, \cdots, x_{r+1}) = \langle \alpha, 1\rangle$. This f is a coloring as required: if $x_0 < x_1 < \cdots < x_{r+2}$ and $f(x_0, \cdots, x_{r+1}) = f(x_1, \cdots, x_{r+1}) = \langle \alpha, 0\rangle$, say, then $F(x_1, \cdots, x_{r+1})$ must be a function which assumes at place α the values 0 and 1 in the same time. [P. Erdős, A. Hajnal: On chromatic number of infinite graphs, in: *Theory of graphs, Proc. of the Coll. held at Tihany 1966, Hungary* (ed. P. Erdős, G. Katona), Akadémiai Kiadó, Budapest, Academic Press, New York, 1968, 83–89]

27. Assume that $f : A \times B \to \kappa$ is a counterexample. For $S \in [A]^k$, $i < \kappa$, set

$$T_i(S) = \{y \in B : f(x, y) = i \text{ for all } x \in S\}.$$

By our indirect assumptions, $|T_i(S)| < k$ holds for all $S \in [A]^k$, $i < \kappa$. Their union, $T = \bigcup\{T_i(S) : i < \kappa, S \in [A]^k\}$ has cardinality at most $\kappa \cdot \kappa^+ = \kappa^+$. We can therefore pick some $y \in B \setminus T$. This y has the property that for every $i < \kappa$, the set $\{x \in A : f(x, y) = i\}$ has at most $k - 1$ elements, which is a contradiction, as they must cover the set A of cardinality κ^+.

28. Assume that A, B, k, and f are as in the problem. Let $\mathcal{U}$ be a nonprincipal ultrafilter on B. For $x \in A$, $i < k$, set $B_x^i = \left\{y \in B : f(x, y) = i\right\}$. For every $x \in A$, $(B_x^0, \ldots, B_x^{k-1})$ is a partition of B into k parts, there is, therefore, a unique $i(x) < k$ such that $B_x^{i(x)} \in \mathcal{U}$. By the pigeon hole principle there are an $i < k$ and an uncountable $A'' \subseteq A$ such that $i(x) = i$ holds for $x \in A''$. We can now apply Problem 4.36 to the system $\{B_x^i : x \in A''\}$ to get $A' \subset A''$, $B' = \cap\{B_x^i : x \in A'\}$ infinite. Hence f is homogeneous of color i on $A' \times B'$.

29. Select the increasing sequence $\{\lambda_\alpha : \alpha < \mu\}$ of regular cardinals, cofinal in λ, with $\lambda_0 > \kappa^\mu$ and $\lambda_{\alpha+1} \geq \left(2^{\lambda_\alpha}\right)^+$. Thinning the sequence $\{S_\alpha : \alpha < \mu\}$ we can achieve that $|S_\alpha| \geq \lambda_{\alpha+1}$ holds for every $\alpha < \mu$. Next, by shrinking the individual sets S_α we can assume that actually $|S_\alpha| = \left(2^{\lambda_\alpha}\right)^+$ holds for $\alpha < \mu$. For $\beta < \alpha$, we have $|S_\beta| \leq \lambda_{\beta+1} \leq \lambda_\alpha$, so $|\bigcup\{S_\beta : \beta < \alpha\}| \leq \lambda_\alpha$. There are at most 2^{λ_α} different $\bigcup\{S_\beta : \beta < \alpha\} \to \kappa$ functions so there are sets S'_α with $|S'_\alpha| = \left(2^{\lambda_\alpha}\right)^+$ such that if $\beta < \alpha$, $x \in S_\beta$, y, $y' \in S'_\alpha$, then $f(x, y) = f(x, y')$.

This can be reformulated in the following way. For every $x \in S'_\alpha$ there is some function $g_x : (\alpha, \mu) \to \kappa$ such that $f(x, y) = g_x(\beta)$ holds for $y \in S'_\beta$, $\alpha < \beta < \mu$ (here (α, μ) denotes the ordinal interval). As the number of different such functions is at most $\kappa^\mu < \lambda_\alpha$, there is some $S''_\alpha \subseteq S'_\alpha$ with $|S''_\alpha| = \left(2^{\lambda_\alpha}\right)^+$ that is homogeneous in this sense, that is, $g_x = g_{x'}$ holds for $x, x' \in S''_\alpha$. Another formulation of this is that there is some function h such that for $\alpha < \beta < \mu$, if $x \in S''_\alpha$, $y \in S''_\beta$, then $f(x, y) = h(\alpha, \beta)$.

We are almost finished, we only have to apply Problem 19 and shrink S''_α to a homogeneous S'''_α with $|S'''_\alpha| = \lambda_\alpha^+$. Of course, the homogeneous color of S'''_α may depend on α. [P. Erdős]

30. Assume that $f : [\lambda]^2 \to \{1, 2, \ldots, k\}$. Problem 29 gives that there are disjoint sets $S_0, S_1, \ldots$ with $|S_n| \to \lambda$ and there are functions $g : [\omega]^2 \to \{1, 2, \ldots, k\}$, $h : \omega \to \{1, \ldots, k\}$ such that if $i < j$, $x \in S_i$, $y \in S_j$ then $f(x, y) = g(i, j)$ and likewise if $x, y \in S_i$ then $f(x, y) = h(i)$. Applying Ramsey's theorem (Problem 1) and the pigeon hole principle we get an infinite set $A \subseteq \omega$ such that if $i, j \in A$ then $g(i, j) = c$ for some $c \in \{1, 2, \ldots, k\}$ and if $i \in A$ then $h(i) = d$ for some $d \in \{1, 2, \ldots, k\}$. Now $\bigcup\{S_i : i \in A\}$ is a set of cardinality λ in which the pairs only get colors c and d. [P. Erdős]

31. Enumerate every A_i as $A_i = \{a^i_\alpha : \alpha < \kappa\}$ and every B_i as $B_i = \{b^i_\alpha : \alpha < \kappa\}$. Let $<$ order I. For $i < j$ in I color the pair $\{i, j\}$ with the ordered pair $\langle \alpha, \beta \rangle$ where $a^i_\alpha = b^j_\beta$ is some element of the nonempty $A_i \cap B_j$. If $|I| > 2^\kappa$ then by Problem 19 there are $i < j < k$ forming a monocolored triangle, and if the color is $\langle \alpha, \beta \rangle$, then

$$b^j_\beta = a^i_\alpha = b^k_\beta = a^j_\alpha$$

an element of $A_j \cap B_j$, a contradiction. [R. Engelking, M. Karlowicz: Some theorems of set theory and their topological consequences, *Fundamenta Mathematicae* **57** (1965), 275–285]

32. For every limit ordinal $\alpha < \kappa$ select the ordinals $x^\alpha_0 < x^\alpha_1 < \cdots < \alpha$, as long as possible, with $f(x^\alpha_n, x^\alpha_m) = f(x^\alpha_n, \alpha) = 1$. If, for some α, we can choose infinitely many such ordinals, we are done: $\{x^\alpha_n : n < \omega\} \cup \{\alpha\}$ is a set of type $\omega + 1$, homogeneous in color 1. In the other case, for every limit $\alpha < \kappa$ there is some nonextendable $\{x^\alpha_n : n \leq N(\alpha)\}$. The mapping $\alpha \mapsto N(\alpha)$ decomposes the stationary set of all limit ordinals below κ into countably many parts, so by Problem 21.6 there is some $N < \omega$ that $\{\alpha : N(\alpha) = N\}$ is stationary. On this set, all functions $\alpha \mapsto x^\alpha_n$ are regressive ($n \leq N$), so repeated applications of Fodor's lemma (Problem 21.9) give a stationary subset S on which they are constant; $x^\alpha_n = x_n$. Then, S is homogeneous in color 0. Indeed, if $f(\beta, \alpha) = 1$ held for some $\beta < \alpha$ in S then β would be a good extension of the set $\{x^\alpha_0, \ldots, x^\alpha_N\}$, i.e., it would be a possible choice for x^α_{N+1} contradicting nonextandability. [P. Erdős, R. Rado: A partition calculus in set theory, *Bull. Amer. Math. Soc.* **62** (1956), 427–489]

33. By induction on k. The case $k = 2$ follows from Problem 32. If the case for k is established, and $f : [\omega_1]^2 \to k+1$ then, again by Problem 32, either there is a monocolored set of order type $\omega+1$ in color k, or there is an uncountable set using only colors $0, 1, \ldots, k-1$. In the former case we are done, in the latter case we use the case for k.

34. Assume that $f : [\mathbf{R}]^2 \to k$ for some $k < \omega$. Let $X \subseteq \mathbf{R}$ be a nonempty, countable set which has the following property. For every choice of $x_1, \ldots x_t \in X$, $j_1, \ldots, j_t < k$ if the set $Y = Y(x_1, \ldots x_t; j_1, \ldots, j_t)$ of those $y > \max(x_1, \ldots x_t)$ with $f(x_1, y) = j_1, \ldots, f(x_t, y) = j_t$ is nonempty, then if $\min(Y)$ exists, then $\min(Y) \in X$, if $\min(Y)$ does not exist, then there are $y_n \in X \cap Y$ with $y_n \to \inf(Y)$. [Such an X can be obtained by putting $X = X_0 \cup X_1 \cup \cdots$ where $X_0 \subseteq \mathbf{R}$ is an arbitrary countably infinite set, and

$$X_{n+1} = X_n \cup \bigcup \{Z(x_1, \ldots x_t; j_1, \ldots, j_t) : x_1, \ldots x_t \in X_n, j_1, \ldots, j_t < k\}$$

where $Z(x_1, \ldots x_t; j_1, \ldots, j_t) \subseteq Y(x_1, \ldots x_t; j_1, \ldots, j_t)$ is a countable, coinitial subset.]

Pick some $y \in \mathbf{R} \setminus X$, bigger than $\inf(X)$. Let $x_0 \in X$, $x_0 < y$ be arbitrary. If $x_0, \ldots, x_n$ are already selected, let $x_{n+1} \in X$ be chosen subject to the conditions $x_n < x_{n+1} < y$ and $f(x_i, x_{n+1}) = f(x_i, y)$ $(i \le n)$. This is possible as $Y\big(x_0, \ldots, x_n; f(x_0, y), \ldots, f(x_n, y)\big)$ is nonempty, y is not its least element (note that $y \notin X$) and there is an element of X in it which is smaller than y.

Now the set $\{x_0, x_1, \ldots, y\}$ is endhomogeneous for f: for $i < j < \omega$, $f(x_i, x_j) = f(x_i, y) = \gamma_i$, say. As $k < \omega$, for an infinite set $Z \subseteq \omega$ and for some $\gamma < k$ we have $\gamma_i = \gamma$ $(i \in Z)$, and then $\{x_i : i \in Z\} \cup \{y\}$ is a homogeneous set in color γ of order type $\omega + 1$. [P. Erdős–R. Rado]

35. (a) Suppose to the contrary that κ is singular. Define $f : [\kappa]^2 \to \{0, 1\}$ with no homogeneous set of cardinality κ as follows. Decompose κ as a disjoint union $\kappa = \bigcup\{S_\alpha : \alpha < \mu\}$ where $\mu < \kappa$ and each S_α has cardinality less than κ. Now set $f(x, y) = 0$ if x and y are in the same S_α, otherwise $f(x, y) = 1$. Every homogeneous set of color one intersects every S_α in at most one point, so it is of cardinality at most μ. Every homogeneous set of color zero is a subset of some S_α so it is of cardinality $< \kappa$.

The problem also follows from Problems 27.44(c) and 27.42.

(b) If $\lambda < \kappa$ and $2^\lambda \ge \kappa$ then, by Problem 23, $2^\lambda \not\to (\lambda^+)^2_2$ holds, so certainly $\kappa \not\to (\kappa)^2_2$.

(c) This is an immediate consequence of 27.44(c) and 27.43.

36. We prove the equivalent statement that for $\beta_0, \ldots, \beta_{k-1} < \omega_1$ there is some $G(\beta_0, \ldots, \beta_{k-1}) < \omega_1$ that if $\alpha = G(\beta_0, \ldots, \beta_{k-1})$ and f is a semihomogeneous coloring of the pairs of ω^α then for some $j < k$ there is a homogeneous set of type β_j in color j. Set $\langle \beta'_0, \ldots, \beta'_{k-1} \rangle \prec \langle \beta_0, \ldots, \beta_{k-1} \rangle$

if and only if $\beta'_j \le \beta_j$ holds for every $j < k$ and there is strict inequality at least once. This gives a well-founded partial ordering on the sequences of countable ordinals of length k. Assume there is some $\langle\beta_0, \dots, \beta_{k-1}\rangle$ for which the statement fails. Then there is a $\prec$-minimal such sequence, $\langle\beta_0, \dots, \beta_{k-1}\rangle$. Notice that $\beta_j \ge 2$ holds for every j. As there are countably many $\prec$-smaller sequences, there is some $\alpha < \omega_1$ such that $\alpha \ge G(\beta'_0, \dots, \beta'_{k-1})$ holds for every $\langle\beta'_0, \dots, \beta'_{k-1}\rangle \prec \langle\beta_0, \dots, \beta_{k-1}\rangle$. We claim that $\alpha + 1$ is a good choice for $G(\beta_0, \dots, \beta_{k-1})$ (and that concludes the indirect argument). Assume that $f : [\omega^{\alpha+1}]^2 \to \{0, 1, \dots, k-1\}$ is semihomogeneous. The ground set of type $\omega^{\alpha+1}$ decomposes into the ordered union of the sets $S_0, S_1, \dots$ each of type ω^α. Assume that the edges between different S_i's get color j. Decompose β_j into the ordered sum of smaller ordinals: $\beta_j = \gamma_0 + \gamma_1 + \cdots$. As $G(\beta_0, \dots, \gamma_i, \dots, \beta_{k-1}) \le \alpha$ holds for every i, we have that for every $i < \omega$ either there is a homogeneous set of type β_r for some $r \ne j$ or there is one of type γ_i in color j. If the first clause holds even for one i, then we are done, we get a homogeneous set of the required type. If the second clause holds for every i, then we have homogeneous sets of order types $\gamma_0, \gamma_1, \dots$ in color j and as the crossing edges all get color j as well, together they form a homogeneous set of type $\gamma_0 + \gamma_1 + \cdots = \beta_j$ in color j, as was required. [F. Galvin: On a partition theorem of Baumgartner and Hajnal, Colloquia Mathematica Societatis János Bolyai, **10.**, *Infinite and Finite Sets*, Keszthely, Hungary, 1973, 711–729]

37. As $\aleph_1 + \aleph_2 = \aleph_2$, there are linearly independent vectors $\{a_\alpha : \alpha < \omega_1\} \cup \{b_\beta : \beta < \omega_2\}$ in V. If V is colored with countably many colors, specifically all vectors of the form $a_\alpha + b_\beta$ are colored, so we get a derived coloring of $\omega_1 \times \omega_2$. In this latter coloring, by Problem 27 there is a monochromatic $\{\alpha, \alpha'\} \times \{\beta, \beta'\}$, that is, $x = a_\alpha + b_\beta$, $z = a_{\alpha'} + b_\beta$, $u = a_\alpha + b_{\beta'}$, $y = a_{\alpha'} + b_{\beta'}$ get the same color, and clearly $x + y = z + u$.

38. If $\{v_\alpha : \alpha < \mathbf{c}^+\}$ is a set of linearly independent vectors, a coloring of V colors in particular the vectors of the form $v_\alpha - v_\beta$ $(\alpha < \beta < \mathbf{c}^+)$. This gives a derived coloring of the pairs $\{\alpha, \beta\} \in [\mathbf{c}^+]^2$, and so, by Problem 19, there are $\alpha < \beta < \gamma$ such that $\{\alpha, \beta\}$, $\{\beta, \gamma\}$, $\{\alpha, \gamma\}$ get the same color. That is, in the original coloring, $x = v_\alpha - v_\beta$, $y = v_\beta - v_\gamma$, $z = v_\alpha - v_\gamma$ have identical colors, and obviously, $x + y = z$.

For the other direction, it suffices to color any vector space of cardinality $\mathbf{c}$, let our choice be $\mathbf{R}$. Let the color classes be $[1, 2)$, $[2, 4)$, $[4, 8), \dots$, and downward $[\frac{1}{2}, 1)$,$[\frac{1}{4}, \frac{1}{2}), \dots$. We define similar color classes on the negative numbers, and let 0 form a color class alone. Now obviously, there is no nontrivial solution of $x + y = z$ in one color class.

39. Assume that S is dense with $|S| = \kappa$. For $x \in X$ let $f(x)$ be the set of those sets $G \cap S$ where G is an open set containing x. We show that $f : X \to \mathcal{P}(\mathcal{P}(S))$ is injective, and so $|X| \le 2^{2^\kappa}$. Assume that x, $y \in X$,

$x \neq y$. As the space is Hausdorff, there are disjoint open sets $x \in U$, $y \in V$. Then $S' = S \cap U \in f(x)$. But $S' \notin f(y)$. Indeed, if $y \in G$ is open, then $G \cap S$ contains elements from $G \cap V \cap S$ and this latter set is disjoint from S'. [B. Pospíšil: Sur la puissance d'un espace contenant une partie dense de puissance donnée, *Časopis Pro Pěstování Matematiky a Fysiky*, **67**(1937), 89–96]

40. Let for $x \neq y \in X$, $x \in U(x,y)$, $y \in V(x,y)$ be disjoint open sets. Assume that $\left\{y(\alpha) : \alpha < \left(2^{2^\kappa}\right)^+\right\}$ are distinct points in X. Color the triplets of this set in such a way that for $\alpha < \beta < \gamma$ the color of $\{\alpha, \beta, \gamma\}$ gives the information if $y(\alpha) \in U\left(y(\beta), y(\gamma)\right)$, or $y(\alpha) \in V\left(y(\beta), y(\gamma)\right)$ or neither holds, and similarly for the other combinations of α, β, γ. This can be done with $3^3 = 27$ colors. By the Erdős–Rado theorem (Problem 25) there is a homogeneous set of cardinality κ^+. Let $\{x(\alpha) : \alpha < \kappa^+\}$ be the corresponding set of points. By homogeneity, either $x(\gamma) \notin U\left(x(\alpha), x(\beta)\right)$ holds whenever $\gamma < \alpha < \beta$ or $x(\gamma) \notin V\left(x(\alpha), x(\beta)\right)$ holds whenever $\gamma < \alpha < \beta$. A similar statement holds for all $\alpha < \gamma < \beta$ and for all $\alpha < \beta < \gamma$. This implies that if

$$W_\alpha = U\left(x(\alpha+1), x(\alpha+2)\right) \cap V\left(x(\alpha), x(\alpha+1)\right)$$

for $\alpha < \kappa^+$, then $x(\alpha+1) \in W_\alpha$ holds for every α and W_α does not contain any other $x(\gamma)$, that is, if $\gamma < \alpha$ or $\gamma > \alpha+2$. Since $x(\alpha+1)$ can be separated from $x(\alpha)$ and $x(\alpha+2)$ by a neighborhood $x(\alpha+1) \in W'_\alpha \subseteq W_\alpha$, it follows that the subspace $\{x(\alpha+1) : \alpha < \kappa^+\}$ is discrete. [A. Hajnal, I. Juhász: On discrete subspaces of topological spaces, *Indag. Math.*, **29**(1967), 343–356]

41. As $\langle X, \mathcal{T}\rangle$ is a Hausdorff space, for $x \neq y \in X$ there are disjoint open sets $x \in U(x,y)$, $y \in V(x,y)$. Assume that every subspace is Lindelöf and $\{y(\alpha) : \alpha < \mathbf{c}^+\}$ is a set of $\mathbf{c}^+$ distinct points. For $\alpha < \beta < \gamma < \mathbf{c}^+$ color the triplet $\{\alpha, \beta, \gamma\}$ with 3 colors depending on if $y(\gamma) \in U(y(\alpha), y(\beta))$ or $y(\gamma) \in V(y(\alpha), y(\beta))$ or neither holds. By Problem 18 there is a set of cardinality $\aleph_1$ which is endhomogeneous, that is, we have the set $\{x(\alpha) : \alpha < \omega_1\}$ of distinct elements such that for $\alpha < \beta < \omega_1$ either $x(\gamma) \in U(x(\alpha), x(\beta))$ holds for every $\beta < \gamma < \omega_1$, or $x(\gamma) \in V(x(\alpha), x(\beta))$ holds for every $\beta < \gamma < \omega_1$, or neither. For a fixed α the open sets $\{V(x(\alpha), x(\beta)) : \alpha < \beta < \omega_1\}$ surely cover $\{x(\beta) : \alpha < \beta < \omega_1\}$, so by Lindelöfness, countably many of them cover as well. Therefore, there is some $\beta(\alpha)$ such that $V(x(\alpha), x(\beta(\alpha)))$ covers uncountably many $x(\gamma)$, so by endhomogeneity it covers every $x(\gamma)$ with $\gamma > \beta(\alpha)$. Now consider the set $U(x(\alpha), x(\beta(\alpha)))$. By disjointness, it does not contain any $x(\gamma)$ with $\gamma > \beta(\alpha)$. So the sets $\{U(x(\alpha), x(\beta(\alpha))) : \alpha < \omega_1\}$ cover the subspace $\{x(\alpha) : \alpha < \omega_1\}$, but each of them only covers countably many elements, so there is no countable subcover, a contradiction. [A. Hajnal–I. Juhász]

42. For $x \in X$ let $\{V_n(x) : n < \omega\}$, $V_0(x) \supseteq V_1(x) \supseteq \cdots$ be a neighborhood base of x. As $\langle X, \mathcal{T}\rangle$ is a Hausdorff space, for $x \neq y$ there is some $n < \omega$ such

that $V_n(x) \cap V_n(y) = \emptyset$ holds. This is a coloring of the pairs with countably many colors, so if $|X| > \mathbf{c}$ then, by Problem 19, there is some n and there are uncountably many points $\{x_\alpha : \alpha < \omega_1\}$ such that this n works for any two points. But then $\{V_n(x_\alpha) : \alpha < \omega_1\}$ is an uncountable system of pairwise disjoint nonempty open sets. [A. Hajnal, I. Juhász: On discrete subspaces of topological spaces, *Indag. Math.* **29**(1967) 343–356]

43. Assume indirectly that $f : \mathcal{P}(\omega) \to \omega$ is a coloring with no distinct X, Y, Z such that $Z = X \cup Y$ and $f(X) = f(Y) = f(Z)$.

We claim that given $i < \omega$, A, B with $A \subseteq B$, $|B \setminus A| = \omega$, there are A', B', with $A \subseteq A' \subseteq B' \subseteq B$, $|B' \setminus A'| = \omega$, such that for no X with $A' \subseteq X \subseteq B'$ does $f(X) = i$ hold. Indeed, otherwise choose some C with $A \subseteq C \subseteq B$, $|C \setminus A| = \omega$, $f(C) = i$ (if no such C exists the choice $A' = A \cup S$, $B' = B$ works, where $S \subseteq B \setminus A$ is such that $|S| = \big|B \setminus (A \cup S)\big| = \omega$). Partition the infinite $C \setminus A$ into two infinite parts: $C \setminus A = U \cup V$. It cannot happen that there is an i-colored X between $A \cup U$ and C, and another i-colored Y between $A \cup V$ and C, for then we would have the monocolored set X, Y and $C = X \cup Y$. If the first case fails then we can choose $A' = A \cup U$, $B' = C$, if the second case fails, $A' = A \cup V$, $B' = C$,

Repeatedly using the claim we choose the sets $\emptyset = A_0 \subseteq A_1 \subseteq \cdots$ and $\omega = B_0 \supseteq B_1 \supseteq \cdots$ with $B_i \supseteq A_i$, $|B_i \setminus A_i| = \omega$ and no X between A_{i+1} and B_{i+1} gets color i. But then $X = A_0 \cup A_1 \cup \cdots$ can get no color at all. [G. Elekes: On a partition property of infinite subsets of a set, *Periodica Math. Hung.* **5** (1974), 215–218]

44. Let S be any set of cardinality $\mathbf{c}^+$. Let $<$ be any ordering on it. If $f : \mathcal{P}(S) \to \omega$, then let $g : [S]^2 \to \omega$ be the following coloring: if $x < y$ then $g(x, y) = f([x, y))$ where $[x, y) = \{z \in S : x \leq z < y\}$. By Problem 19 there are $x < y < z$ with $g(x, y) = g(x, z) = g(y, z)$ and so $[x, y)$ and $[y, z)$ are two disjoint sets such that they, as well as their union, get the same color. [P. Komjáth]

45. The subsets of S with symmetric difference as addition form a vector space over the two-element field. Notice that for disjoint subsets symmetric difference is the same as union. Fix a basis $B = \{b_i : i \in I\}$ and color the subsets of S according to the number of basis elements in their representation. Notice that

$$\sum_{i \in J} b_i + \sum_{i \in J'} b_i = \sum_{i \in J \triangle J'} b_i$$

as the characteristic is 2. The required property in this form reduces to showing that there are no infinitely many n-element sets such that the symmetric difference of any finitely many of them is still an n-element set. Indeed, there is a 3-element Δ-subsystem of it, $A_0 = S \cup B_0$, $A_1 = S \cup B_1$, $A_2 = S \cup B_2$ with S, B_0, B_1, B_2 pairwise disjoint (see Problem 25.1). Clearly, $|B_i| = |B_j|$, and since $|A_i \triangle A_j| = |B_i| + |B_j|$, it follows that $|S| = |B_0| = |B_1| = |B_2| = \frac{n}{2}$

but then $|A_0 \triangle A_1 \triangle A_2| = |S \cup B_0 \cup B_1 \cup B_2| = 2n$, a contradiction. [G. Elekes, A. Hajnal, P. Komjáth: Partition theorems for the power set, *Coll. Math. Soc. János Bolyai* **60**, Sets, graphs, and numbers, Budapest (Hungary), 1991, 211–217]

46. We first treat the case $S = \omega$. Enumerate $[\omega]^{\aleph_0}$ as $\{A_\alpha : \alpha < \mathbf{c}\}$. Choose, by transfinite recursion, the sets X_α, Y_α such that $X_\alpha, Y_\alpha \subseteq A_\alpha$ and the sets $\{X_\alpha, Y_\alpha : \alpha < \mathbf{c}\}$ are all different. This is possible, as at step α we have $2|\alpha| < \mathbf{c}$ sets already chosen and $\left|[A_\alpha]^{\aleph_0}\right| = \mathbf{c}$ possibilities to choose X_α, Y_α. If now f satisfies $f(X_\alpha) = 0$, $f(Y_\alpha) = 1$ for every $\alpha < \mathbf{c}$, then f has no homogeneous infinite subset.

Passing to the general case, let S be uncountable. Let $\mathcal{H}$ be a maximal almost disjoint subfamily of $[S]^{\aleph_0}$, that is, if $A \neq B \in \mathcal{H}$, then $|A \cap B| < \omega$, and no proper extension of $\mathcal{H}$ has this property. Such an $\mathcal{H}$ exists by Zorn's lemma. For $H \in \mathcal{H}$ let f_H be a function on the infinite subsets of H, as constructed in the previous paragraph. Let $f : [S]^{\aleph_0} \to \{0, 1\}$ be an arbitrary function that extends every f_H. It is possible to find such an f, as the f_H's operate on disjoint domains. We claim that f is as required for S. Indeed, if $A \in [S]^{\aleph_0}$, then for some $H \in \mathcal{H}$ the intersection $B = A \cap H$ is infinite (otherwise $\{A\} \cup \mathcal{H}$ would properly extend $\mathcal{H}$). By the construction, there are $X, Y \subseteq B \subseteq A$ such that $f(X) = f_H(X) = 0$, $f(Y) = f_H(Y) = 1$. [P. Erdős, R. Rado: Combinatorial theorems on classifications of subsets of a given set, *Proc. London Math. Soc.*, **3**(1952), 417–432]

25

Δ-systems

1. We prove the claim by induction on n. The result is obvious for $n = 1$. Assume we have it for n, and we have $\mathcal{F}$, an infinite family of $(n+1)$-element sets. If there is some element x that is contained in infinitely many members of $\mathcal{F}$, then we can consider the infinite family of n-element sets $\mathcal{F}' = \{A \setminus \{x\} : x \in A \in \mathcal{F}\}$. If the sets $\{A_0 \setminus \{x\}, \ldots\}$ form a Δ-subfamily of $\mathcal{F}'$, then the corresponding members $\{A_0, \ldots\}$ of $\mathcal{F}$ will give a Δ-subfamily of $\mathcal{F}$. We can therefore assume that every point is contained in only finitely many members of $\mathcal{F}$. We select, by induction, infinitely many pairwise disjoint sets. If we have $A_0, \ldots, A_t$ then there are just finitely many sets in $\mathcal{F}$ containing elements from $A_0 \cup \cdots \cup A_t$, so we can choose a further element of $\mathcal{F}$, and that will be disjoint from each of $A_0, \ldots, A_t$.

2. By the pigeon hole principle we can assume that every member of the family $\mathcal{F}$ has n elements for some natural number n. We show, by induction on n, that $\mathcal{F}$ has an uncountable Δ-subfamily. This is obvious for $n = 1$. For the inductive step let $\mathcal{F}$ be an uncountable system of $(n+1)$-element sets. If some element x is contained in uncountably many members of $\mathcal{F}$ then we apply the statement for the system $\mathcal{F}' = \{A \setminus \{x\} : x \in A \in \mathcal{F}\}$. We get an uncountable Δ-subsystem of $\mathcal{F}'$ and by adding x to the common part of it, we arrive at a Δ-subsystem of $\mathcal{F}$.

We can, therefore, assume, that every $x \in \bigcup \mathcal{F}$ is contained in only countably many members of $\mathcal{F}$. In this case, we can select an uncountable disjoint subsystem of $\mathcal{F}$ as follows. Let $A_0 \in \mathcal{F}$ be arbitrary. If $\{A_\beta : \beta < \alpha\}$ are selected for some $\alpha < \omega_1$, then, by hypothesis, only countably many $A \in \mathcal{F}$ contain one or more elements of $X = \bigcup\{A_\beta : \beta < \alpha\}$, so we can choose $A_\alpha \in \mathcal{F}$, disjoint from X (and so continue constructing the disjoint subsystem).

3. If $\kappa = |\mathcal{F}|$ is a regular cardinal, then we can repeat the argument in the previous solution by replacing "uncountable" with "of cardinality κ" and "countable" with "of cardinality $< \kappa$".

If κ is singular then there is $\mu < \kappa$ and there are cardinals $\{\kappa_\alpha : \alpha < \mu\}$ below κ that sum up to κ. Consider the distinct points $\{x_\alpha, y_{\alpha,\xi} : \alpha < \mu,\ \xi < \kappa_\alpha\}$. Let $\mathcal{F}$ consist of the pairs of the form $\{x_\alpha, y_{\alpha,\xi}\}$. Clearly, $|\mathcal{F}| = \kappa$. As no point is covered κ times ($y_{\alpha,\xi}$ is in one set, x_α is in κ_α sets) the only possibility for a Δ-subsystem of cardinality κ if there is a disjoint subsystem with κ members. But there are no more than μ pairwise disjoint elements in $\mathcal{F}$; after all, every member meets the set $\{x_\alpha : \alpha < \mu\}$ of cardinality μ.

4. No. Let S be a set of cardinality $\aleph_1$. Our counterexample is $[S]^2$, the set of all pairs from S. Assume indirectly, that $[S]^2 = \mathcal{F}_0 \cup \mathcal{F}_1 \cup \cdots$ where $\mathcal{F}_0, \mathcal{F}_2, \ldots$ are systems of disjoint sets while $\mathcal{F}_1, \mathcal{F}_3, \ldots$ are Δ-systems with kernels $\{x_0\}, \{x_1\}, \ldots$. Pick $y \neq x_0, x_1, \ldots$ (possible, as S is uncountable). There are $z, t \neq x_0, x_1, \ldots, y$ such that $\{y, z\}$, $\{y, t\}$ are in the same $\mathcal{F}_{2i}$, but this is a contradiction.

5. Assume that $f_\alpha \in F(A, B)$ $(\alpha < \omega_1)$. By Problem 2 there is an uncountable subfamily $\{f_\alpha : \alpha \in X\}$ such that $\{\mathrm{Dom}(f_\alpha) : \alpha \in X\}$ is a Δ-system; $\mathrm{Dom}(f_\alpha) = s \cup s_\alpha$ for $\alpha \in X$ where the sets $\{s, s_\alpha : \alpha \in X\}$ are pairwise disjoint. As B is countable there are just countably many $s \to B$ functions, so, with a further trim we get an uncountable subfamily $\{f_\alpha : \alpha \in Y\}$ such that $f_\alpha|_S = f$ for every $\alpha \in Y$ with some $f : S \to B$. Now, any f_α, f_β with $\alpha, \beta \in Y$ have a common extension, namely $g : s \cup s_\alpha \cup s_\beta \to B$ where

$$g(x) = \begin{cases} f(x) & \text{if } x \in s, \\ f_\alpha(x) & \text{if } x \in s_\alpha, \\ f_\beta(x) & \text{if } x \in s_\beta. \end{cases}$$

6. Assume that $\{G_\alpha : \alpha < \omega_1\}$ are disjoint, nonempty open sets in ${}^S\mathbf{R}$, for some set S. Pick a basic open subset I_α of G_α, that is, for a finite set $s_\alpha \subseteq S$ there are open intervals $K_x^\alpha \subseteq \mathbf{R}$ $(x \in s_\alpha)$ such that $f : S \to \mathbf{R}$ is in I_α if and only if $f(x) \in K_x^\alpha$ holds for all $x \in s_\alpha$. Further shrinking I_α we can arrange that every K_x^α be an interval with rational endpoints. By Problem 2 there is an uncountable Δ-subfamily of $\{s_\alpha : \alpha < \omega_1\}$. That is, there is an uncountable $X \subseteq \omega_1$, and there are pairwise disjoint $\{s, t_\alpha : \alpha \in X\}$ such that $s_\alpha = s \cup t_\alpha$ holds for $\alpha \in X$. As we restricted the K_x^α to rational intervals, for each $x \in s$ there are countably many possibilities for K_x^α. As s is finite, there can be just countably many different systems $\langle K_x^\alpha : x \in s\rangle$, so there are $\alpha \neq \beta$ in X with $K_x^\alpha = K_x^\beta$ for every $x \in s$. But then $I_\alpha \cap I_\beta$ is nonempty, indeed, the following function f is in the intersection: $f(x) \in K_x^\alpha$ for $x \in s_\alpha$, $f(x) \in K_x^\beta$ for $x \in t_\beta$, and $f(x) \in \mathbf{R}$ arbitrary for $x \in S \setminus (s_\alpha \cup s_\beta)$.

The same proof shows that in any product of topological spaces with a countable base there is no uncountable system of pairwise disjoint open sets.

7. **First solution.** We prove the stronger statement that if $S \subseteq \omega_1$ is stationary, $\{A_\alpha : \alpha \in S\}$ is a system of finite sets then $\{A_\alpha : \alpha \in S'\}$ is a Δ-system

for some stationary $S' \subseteq S$. As the union of countably many nonstationary sets is nonstationary (see Problem 20.4), we can, by shrinking S, assume that $|A_\alpha| = n$ holds for every $\alpha \in S$, for some natural number n. We prove the statement, which now has parameter n, by induction on n. The case $n = 0$ is obvious. If $n > 0$ and there is some point x which is contained in stationary many A_α, that is, $x \in A_\alpha$ for $\alpha \in S' \subseteq S$ stationary, then we consider the system $\{A'_\alpha : \alpha \in S'\}$ of $(n-1)$-element sets, where $A'_\alpha = A_\alpha \setminus \{x\}$ and so we can apply the statement for $n-1$. If $\{A'_\alpha : \alpha \in S''\}$ is a Δ-subsystem of the latter system, then $\{A'_\alpha : \alpha \in S''\}$ is a Δ-subsystem of the original system (just it contains one more element in the common core: x). We can therefore assume that no element is contained in stationary many A_α. Split S as $S = S_0 \cup S_1$ where $\alpha \in S_0$ if and only if there is some $\beta < \alpha$ with $A_\beta \cap A_\alpha \neq \emptyset$. For $\alpha \in S_0$ we let $f(\alpha) =$ the least such β. This is a regressive function, so if S_0 is stationary, then by Fodor's theorem (Problem 20.16) there is some β that $\{\alpha \in S_0 : f(\alpha) = \beta\}$ is stationary. That is, stationary many sets A_α intersect the same A_β. Further refining we get stationary many A_α such that they contain the same element $x \in A_\beta$, and that is impossible, as we have assumed that no element is contained in stationary many A_α.

We proved that S_0 is nonstationary; therefore, S_1 is stationary. But then obviously $\{A_\alpha : \alpha \in S_1\}$ is a system of pairwise disjoint sets, so it is a Δ-system.

Second solution. As the union of $\aleph_1$ finite sets has cardinality at most $\aleph_1$ we can assume that each A_α is a subset of ω_1. There is a closed, unbounded set $C \subseteq \omega_1$ such that if $\beta < \alpha \in C$ then $\max(A_\beta) < \alpha$. This easily follows from Problem 20.6 but we can argue as follows. If the set of those α's is stationary for which there is some $\beta < \alpha$ with $\max(A_\beta) \geq \alpha$, then the function f that selects such a β for every such α is regressive, so by Fodor's theorem it assumes some fixed value β on a stationary set which is impossible, as clearly $f(\alpha) \neq \beta$ if $\alpha > \max(A_\beta)$.

As the union of countably many nonstationary sets is nonstationary, there are a natural number n and a stationary set $S \subseteq C$ such that $|A_\alpha \cap \alpha| = n$ holds for every $\alpha \in S$ ($n = 0$ is possible). Let $\{x_1^\alpha, \ldots, x_n^\alpha\}$ be the increasing enumeration of $A_\alpha \cap \alpha$ ($\alpha \in S$). As for every $1 \leq i \leq n$ the mapping $\alpha \mapsto x_i^\alpha$ is a regressive function on S, with n successive applications of Fodor's theorem we get that there is a stationary set $S' \subseteq S$ such that $x_i^\alpha = y_i$ holds for $\alpha \in S'$ for some $y_1 < \cdots < y_n$. Then, $\{A_\alpha : \alpha \in S'\}$ is a Δ-system with pairwise intersection $Y = \{y_1, \ldots, y_n\}$. Indeed, $Y \subseteq A_\alpha$ obviously holds for all $\alpha \in S'$, and if $\beta < \alpha$ are in S', $A_\beta \neq Y$, $A_\alpha \neq Y$, then

$$\max(A_\beta \setminus Y) = \max(A_\beta) < \alpha \leq \min(A_\alpha \setminus \alpha) = \min(A_\alpha \setminus Y).$$

8. (a) As $\bigcup \mathcal{F}$ has cardinality at most $\mathbf{c}^+$, we can assume that our system is $\mathcal{F} = \{A_\alpha : \alpha < \mathbf{c}^+\}$ with $A_\alpha \subseteq \mathbf{c}^+$ for $\alpha < \mathbf{c}^+$. From Problem 21.5 it follows that there is a closed, unbounded set $C \subseteq \mathbf{c}^+$ such that if $\alpha < \beta$ and $\beta \in C$

then $\sup(A_\alpha) < \beta$. By Problem 21.8 the set $S = \{\alpha \in C : \mathrm{cf}(\alpha) = \omega_1\}$ is stationary. As A_α is countable, $A_\alpha \cap \alpha$ is a bounded subset of α, therefore $f(\alpha) = \sup(A_\alpha \cap \alpha)$ is a regressive function on S. By Fodor's theorem (Problem 21.9) there is some $\gamma < \mathbf{c}^+$ such that $S' = \{\alpha \in S : f(\alpha) = \gamma\}$ is stationary. For each $\alpha \in S'$, $A_\alpha \cap \alpha$ is a countable subset of $\gamma + 1$. $\gamma + 1$, being a set of cardinality $\leq \mathbf{c}$, has $\mathbf{c}$ countable subsets, or less. As the union of $\mathbf{c}$ nonstationary sets (in $\mathbf{c}^+$) is nonstationary, there is some countable set $B \subseteq \mathbf{c}^+$ such that $S'' = \{\alpha \in S' : A_\alpha \cap \alpha = B\}$ is stationary. Now $\mathcal{F}' = \{A_\alpha : \alpha \in S''\}$ is a Δ-system. Indeed, if $\alpha < \beta$ are in S'', then certainly $B \subseteq A_\alpha \cap A_\beta$, and $A_\alpha \setminus B$ and $A_\beta \setminus B$ are disjoint, as every element of the first set $< \beta \leq$ every element of the second set.

(b) The proof is identical to the one given in part (a), just replace $\mathbf{c}^+$ by λ, "countable" by "of cardinality $\leq \mu$", and let $S = \{\alpha < \lambda : \mathrm{cf}(\alpha) = \mu^+\}$.

9. For a set S of cardinality μ let $T = (S \times \{0\}) \cup (S \times \{1\})$, that is, we consider "two copies" of S. For $A \subseteq S$ set $H(A) = (A \times \{0\}) \cup ((S \setminus A) \times \{1\})$. We claim that $\mathcal{F} = \{H(A) : A \subseteq S\}$ is as required. Obviously, $\mathcal{F}$ is a system of 2^μ sets, each of cardinality μ. To conclude, assume that $H(A), H(B), H(C)$ form a Δ-system. This means that $\{A, B, C\}$ as well as $\{S \setminus A, S \setminus B, S \setminus C\}$ both are Δ-systems. The first assumption implies that if some $x \in S$ is in two of A, B, C then it is in the third one. The second assumption implies that if some $x \in S$ is in one of A, B, C then it is in another. Putting together, we get that $A = B = C$.

10. Assume to the contrary that $|I| \geq (2^\mu)^+$. By Problem 8(b) we can shrink I in two steps to a $J \subseteq I$ with $|J| = (2^\mu)^+$ such that $\{A_i : i \in J\}$ and $\{B_i : i \in J\}$ are both Δ-systems, that is, $A_i = A \cup A^i$, $B_i = B \cup B^i$ with $\{A, A^i : i \in J\}$ as well as $\{B, B^i : i \in J\}$ systems of pairwise disjoint sets. $A \cap B = \emptyset$ as otherwise we had $A_i \cap B_i \neq \emptyset$ whenever $i \in J$. The set $J_0 = \{i \in J : A^i \cap B \neq \emptyset\}$ has cardinality at most μ as the corresponding sets $A^i \cap B$ are disjoint nonempty subsets of B, which is of cardinality $\leq \mu$. Fix $i' \in J \setminus J_0$. As before, the set $J_1 = \{i \in J : B^i \cap A_{i'} \neq \emptyset\}$ has cardinality $\leq \mu$. Choose $i'' \in J \setminus (J_0 \cup J_1 \cup \{i'\})$. Then $i' \neq i''$ and $A \cup A^{i'}$ and $B \cup B^{i''}$, that is, $A_{i'}$ and $B_{i''}$ are disjoint, contradiction.

For an alternative proof see Problem 24.31.

11. (a) We can assume that the members of $\mathcal{F}$ are indexed by the elements of λ: $\mathcal{F} = \{A_\alpha : \alpha < \lambda\}$. Also, without loss of generality, $A_\alpha \subseteq \lambda$ for every $\alpha < \lambda$. Assume first that κ is regular. Then if S is the set of ordinals smaller λ with cofinality κ then for $\alpha \in S$ we have $f(\alpha) = \sup(A_\alpha \cap \alpha) < \alpha$. By Fodor's theorem, for a stationary set $S' \subseteq S$ we have $f(\alpha) = \gamma$ $(\alpha \in S')$. There is a closed, unbounded set $C \subseteq \lambda$ such that if $\alpha < \beta \in C$ then $\sup(A_\alpha) < \beta$ holds (cf. Problem 21.3). Now $\mathcal{F}' = \{A_\alpha : \alpha \in S' \cap C\}$ is as required; if α, β are in $S' \cap C$, and $\alpha < \beta$, then $A_\alpha \cap A_\beta \subseteq \gamma$, namely $A_\alpha \cap \alpha$, $A_\beta \cap \beta$ are subsets of γ and $A_\alpha \setminus \alpha$, $A_\beta \setminus \beta$ are subsets of the disjoint intervals $[\alpha, \beta)$, $[\beta, \lambda)$.

Assume now that κ is singular, $\mathrm{cf}(\kappa) = \mu$, $\kappa = \sup\{\kappa_\xi : \xi < \mu\}$ where $\kappa_\xi < \kappa$ is regular. Then $S = \bigcup\{S_\xi : \xi < \mu\}$ where $S_\xi = \{\alpha : |A_\alpha| < \kappa_\xi\}$. For some $\xi < \mu$ the set S_ξ must be stationary (as the union of μ nonstationary sets is nonstationary) and then we can repeat the above argument for S_ξ.

(b) By part (a) we can assume that λ is singular. Set $\lambda = \sup\{\lambda_\xi : \xi < \mu\}$, where $\mu = \mathrm{cf}(\lambda)$, and $\lambda_\xi > \sup\{\lambda_\zeta : \zeta < \xi\}$ is regular, with $\lambda_0 > \kappa^+$. Accordingly, $\mathcal{F}$ is decomposed into the union of the subfamilies $\mathcal{F}_\xi$ with $|\mathcal{F}_\xi| = \lambda_\xi$. Using Problem 8(b) we get a Δ-subsystem $\mathcal{F}'_\xi \subseteq \mathcal{F}_\xi$ of the form (say) $\mathcal{F}'_\xi = \{A_\xi \cup A_{\xi,\alpha} : \alpha < \lambda_\xi\}$ where the sets $\{A_\xi, A_{\xi,\alpha} : \alpha < \lambda_\xi\}$ are pairwise disjoint. If we remove from $\mathcal{F}'_\xi$ every set for which $A_{\xi,\alpha}$ has a nonempty intersection with some $A_{\zeta,\beta}$ ($\zeta < \xi$, $\beta < \lambda_\zeta$) then we remove at most $\sup\{\lambda_\zeta : \zeta < \xi\} < \lambda_\xi$ sets, so the remaining system $\mathcal{F}''_\xi$ still has cardinality λ_ξ. Now the system $\bigcup\{\mathcal{F}''_\xi : \xi < \mu\}$ is as required; the intersection of two elements is a subset of $\bigcup\{A_\xi : \xi < \mu\}$, a set of cardinal at most $\mu\kappa < \lambda$. [G. Fodor: Some results concerning a problem in set theory, *Acta Sci. Math.*, **16**(1955), 232–240., W. W. Stothers, M. J. Thomkinson: On infinite families of sets, *Bull. of the London Math. Soc.*, **11**(1979), 23–26]

26

Set mappings

1. For every real x there is an open interval I with rational endpoints such that $x \in I$ and $f(x) \cap I = \emptyset$. If we consider the set of those x that are associated with a given I, then we get a decomposition of $\mathbf{R}$ into countably many classes. One of them, say A, associated with I, is of the second category, one of them, say A', associated with I', is of cardinality $\mathbf{c}$ (see Problem 4.15). $A \cap f[A] = \emptyset$ as A and $f[A]$ are separated by I, and likewise for A'.

2. Let n be the least natural number which is greater than $|x|$. Set $f(x) = (-n, n) \setminus \{x\}$. If $x \neq y$ are reals and $|x| \leq |y|$ then $x \in f(y)$.

3. There is an enumeration $\mathbf{R} = \{r_\alpha : \alpha < \mathbf{c}\}$ of $\mathbf{R}$. Set $f(r_\alpha) = \{r_\beta : \beta < \alpha\}$. Then $|f(r_\alpha)| = |\alpha| < \mathbf{c}$ and whenever $r_\alpha \neq r_\beta$ are reals then either $r_\alpha \in f(r_\beta)$ or $r_\beta \in f(r_\alpha)$ holds (according to whether $\alpha < \beta$ or $\beta < \alpha$ holds).

4. Enumerate the intervals with rational endpoints as $I_0, I_1, \ldots$. Our plan is to select $a_i \in I_i$ such that the set $\{a_0, a_1, \ldots\}$ is free. The trick is that we keep the side condition that the set $A_i = \{x : a_0, \ldots, a_{i-1} \notin f(x)\}$ is everywhere (i.e., in every interval) of the second category. As $A_0 = \mathbf{R}$ we can start the inductive construction. Assume we have reached the ith stage. Set $A_i^* = A_i \setminus (f(a_0) \cup \cdots \cup f(a_{n-1}))$; it is also a set everywhere of the second category. If we choose $a_i \in A_i^* \cap I_i$, then the freeness of $\{a_0, \ldots, a_i\}$ is kept so the only problem can be that for every $b \in I_i \cap A_i^*$ there is an interval J such that $B_b = \{x \in A_i^* : b \notin f(x)\}$ is of the first category in J. We are going to show that this is impossible. Indeed, if this is not the case, then there is such a rational interval J and as there are countably many rational intervals there is some J such that for a second-category set of $b \in I_i \cap A_i^*$ the above B_b is of the first category in that same J. Select a set $\{b_0, b_1, \ldots\} \subseteq I_i \cap A_i^*$ that is dense in some subinterval K. For every n the set B_{b_n} is of the first category in J, hence there is an $x \in A_i^* \cap J \setminus (B_{b_0} \cup B_{b_1} \cup \cdots)$. For this x we have $\{b_0, b_1, \ldots\} \subseteq f(x)$ so $f(x)$ is dense in K, a contradiction. [F. Bagemihl:

The existence of an everywhere dense independent set, *Michigan Math. J.*, **20**(1973), 1–2]

5. Let $\{a_n : n < \omega\}$ be an everywhere-dense set. By Problem 4.15 (König's theorem) $\mathrm{cf}(\mathbf{c}) > \omega$. Therefore, $f(a_0) \cup f(a_1) \cup \cdots$, the union of countably many sets with cardinality less than continuum, is a set of cardinality less than continuum, we can choose $b \in \mathbf{R} \setminus (f(a_0) \cup f(a_1) \cup \cdots)$. As $f(b)$ is not everywhere dense, there is some $a_n \notin f(b)$ and then $\{a_n, b\}$ is a free set.

There is not necessarily a 3-element free set. To show this, let $<_w$ be a well-ordering of $\mathbf{R}$ into order type $\mathbf{c}$. For $x > 0$ set $f(x) = \{y > 0 : y <_w x\}$ and similarly, for $x \leq 0$ set $f(x) = \{y \leq 0 : y <_w x\}$. Then neither $(0, \infty)$ nor $(-\infty, 0]$ includes a 2-element free set.

6. Let λ^* denote the outer Lebesgue measure on $\mathbf{R}$. Choose a set $A_1 \subseteq \mathbf{R}$ with $\lambda^*(A_1) > n - 1$ such that for $x \in A_1$ we have $f(x) \subseteq [-k_1, k_1]$ for an appropriate k_1. Such a choice is possible, as $\bigcup_k \{x : f(x) \subseteq [-k, k]\} = \mathbf{R}$. Next choose an $A_2 \subseteq (k_1, \infty)$ with $\lambda^*(A_2) > n - 2$ such that for $x \in A_2$ we have $f(x) \subseteq [-k_2, k_2]$ for an appropriate k_2. Keep going. We finally select some $A_{n-1} \subseteq (k_{n-2}, \infty)$ with $\lambda^*(A_{n-1}) > 1$ such that for $x \in A_{n-1}$ we have $f(x) \subseteq [-k_{n-1}, k_{n-1}]$ for an appropriate k_{n-1}. Then inductively choose the elements

$$x_n > k_{n-1}, x_{n-1} \in A_{n-1} \setminus f(x_n), \ldots, x_1 \in A_1 \setminus (f(x_2) \cup \cdots \cup f(x_n)),$$

the set $\{x_1, \ldots, x_n\}$ will be the required free set. [P. Erdős, A. Hajnal: Some remarks onset theory, VIII, *Michigan Math. J.*, **7**(1960), 187–191]

7. As CH holds, we can enumerate $\mathbf{R}$ as $\mathbf{R} = \{r_\alpha : \alpha < \omega_1\}$ and the collection of the countable sets with uncountable closure as $\{H_\alpha : \alpha < \omega_1\}$. Define $f(r_\alpha)$ in such a way that it is a sequence converging to r_α, and for every $\beta < \alpha$ if r_α is a limit point of H_β, then $f(r_\alpha)$ contains a point from H_β. This is possible as there are countably many such sets H_β, so reordering them into an ω-sequence we can select the appropriate points closer and closer to r_α.

Assume now that $X \subseteq \mathbf{R}$ is an uncountable set. Let $H \subseteq X$ be countable and dense in X. Then $H = H_\alpha$ for some $\alpha < \omega_1$. If $r_\beta \in X \setminus H_\alpha$ with $\beta > \alpha$ (all but countably many elements of X satisfy this), then $f(r_\beta)$ contains an element of $H \subseteq X$, so X is not free. [S. Hechler]

8. (a) Assume to the contrary that there is no free set of cardinality κ. Using Zorn's lemma we can inductively select the maximal free sets $A_0 \subseteq \kappa$, $A_1 \subseteq \kappa \setminus A_0$, $A_2 \subseteq \kappa \setminus (A_0 \cup A_1)$, $A_\xi \subseteq \kappa \setminus \bigcup_{\eta<\xi} A_\eta$, for $\xi < \mu$. By our indirect hypothesis each A_ξ has cardinality $< \kappa$, so every set A_ξ is nonempty, and even $A = \bigcup\{A_\xi : \xi < \mu\}$ has cardinality less than κ. Also, $|f[A]| < \kappa$. Select $x \in \kappa \setminus (A \cup f[A])$. For every $\xi < \mu$ the set $A_\xi \cup \{x\}$ is not free, which, as $x \notin f[A_\xi]$, can only mean that $A_\xi \cap f(x) \neq \emptyset$. As the sets A_ξ are disjoint, this gives $|f(x)| \geq \mu$, a contradiction. [Sophie Piccard: Sur un problème de M. Ruziewicz

de la théorie des relations, *Fundamenta Mathematicae*, **29**(1937), 5–9. This proof was given by Dezső Lázár, see P. Erdős: Some remarks on set theory, *Proc Amer. Math. Soc*, **1**(1950), 127–141]

(b) Increasing μ if needed, we can assume that $\mu > \mathrm{cf}(\kappa)$. Let $\{\kappa_\xi : \xi < \mathrm{cf}(\kappa)\}$ be a strictly increasing sequence of cardinals, cofinal in κ, with $\kappa_0 > \mu^+$.

Decompose the ground set κ into the disjoint union of the sets $\{S_\xi : \xi < \mathrm{cf}(\kappa)\}$ where $|S_\xi| = \kappa_\xi^+$. Using the result of part (a) we can assume, by shrinking it, if necessary, that every S_ξ is free. As the cardinality of $\bigcup\{f[S_\eta] : \eta < \xi\}$ is at most κ_ξ, by a further reduction we can achieve that if $x \in S_\eta$ and $\eta < \xi$ then $f(x) \cap S_\xi = \emptyset$. We have to show, therefore, that we can select subsets $A_\xi \subseteq S_\xi$ with $|A_\xi| = \kappa_\xi$ such that for no $x \in A_\xi$, $\eta < \xi$ does $f(x) \cap A_\eta \neq \emptyset$ hold.

Assume that this cannot be done. By transfinite recursion on $\alpha < \mu^+$ define the ordinal $\eta(\alpha) < \mathrm{cf}(\kappa)$ and construct the sets A_ξ^α again by transfinite recursion on $\xi < \eta(\alpha)$ as follows: if $\{A_\xi^\beta : \xi < \eta(\beta), \beta < \alpha\}$ are all constructed, choose, as long as possible, a subset $A_\xi^\alpha \subseteq S_\xi \setminus \left(\bigcup\{A_\xi^\beta : \beta < \alpha\}\right)$ of cardinality κ_ξ such that the union of these sets is free. Our construction must stop at some point $\eta(\alpha) < \mathrm{cf}\,(\kappa)$, as otherwise we would get a free set of cardinality κ and the proof was over.

As μ^+ is a regular cardinal greater than $\mathrm{cf}(\mu)$, there are μ^+ many values α, say $\alpha \in T$ such that $\eta(\alpha) = \eta$, the same value. For these ordinals α we are unable to select an appropriate $A_\eta^\alpha \subseteq S_\eta$, that is, at the given point of the construction only $< \kappa_\eta$ many points in $S_\eta \setminus \left(\bigcup\{A_\eta^\beta : \beta < \alpha\}\right)$ were free from $\bigcup\{A_\xi^\alpha : \xi < \eta\}$. For $\alpha \in T$ let B_α be the set of those points, then $|B_\alpha| < \kappa_\eta$. Let $C_\alpha = B_\alpha \bigcup(\bigcup\{A_\xi^\beta : \xi < \eta\})$, $\alpha \in T$. The union of the C_α's has cardinality at most $\mu^+(\kappa_\eta \mathrm{cf}\,(\kappa) + \kappa_\eta) = \kappa_\eta < |S_\eta|$, so there is a point $x \in S_\eta$ not in any of the C_α's. By our conditions, $f(x)$ intersects every $\bigcup\{A_\xi^\alpha : \xi < \eta\}$, $\alpha \in T$, and these μ^+ sets are disjoint, so $|f(x)| \geq \mu^+$, a contradiction. [A. Hajnal: Proof of a conjecture of S. Ruziewicz, *Fund. Math.*, **50**(1961), 123–128. This proof is from S. Shelah: *Classification theory and the number of non-isomorphic models*, North-Holland, 1978]

9. Join two points $x, y \in S$ if $x \in f(y)$ or $y \in f(x)$ holds. This gives a graph and the claim in the problem is equivalent to the fact that this graph can be colored by $2k+1$ colors, i.e., we have to show that the chromatic number is at most $2k+1$. By the de Bruijn–Erdős theorem (Problem 23.8) it suffices to show this for the finite subsets of S, in other words, it suffices to show the statement for finite S. This we prove by induction on $n = |S|$. The result is obvious for $n \leq 2k+1$ (we can color the vertices with different colors). Assume that $n > 2k+1$. The number e of edges is at most kn so the sum of the degrees is $2e \leq 2kn$. There is, therefore, a vertex x with degree at most $2k$. Remove x. By the inductive hypothesis $S \setminus \{x\}$ has a good coloring with $2k+1$ colors. As the degree of x is at most $2k$, this coloring can be extended to x, and we are done.

10. As in the previous problem, join two points $x, y \in S$ if either $x \in f(y)$ or $y \in f(x)$ holds and get graph X on S. Again we have to show that X can be colored with μ colors. To this end we prove that S has a well-ordering $\prec$ such that every point is joined to less than μ points that precede it. With this, we can color X with a straightforward transfinite recursion along $\prec$ (see Problem 23.3).

Enumerate S as $S = \{s_\alpha : \alpha < \kappa\}$. Set $x \in A_\alpha$ if there is a sequence $x_0 = s_\alpha, \ldots, x_n = x$ with $x_{i+1} \in f(x_i)$ for $i = 0, \ldots, n-1$. Notice that $s_\alpha \in A_\alpha$ and $|A_\alpha| \leq \aleph_0\mu = \mu$. Set $B_\alpha = A_\alpha \setminus (\bigcup\{A_\beta : \beta < \alpha\})$ for $\alpha < \kappa$. Notice that $S = \bigcup\{B_\alpha : \alpha < \kappa\}$ is a partition of S. Set $x \prec y$ if $x \in B_\alpha$, $y \in B_\beta$ for some $\alpha < \beta < \kappa$, inside a B_α let $\prec$ be a well order into order type $\leq \mu$.

Fix an element $x \in S$. We show that it is joined into less than μ elements that precede it. There is some α such that $x \in B_\alpha$. The number of elements in B_α that precede x is less than μ, anyway. And if $y \in B_\beta$ for some $\beta < \alpha$ and x is joined to y, then $x \in f(y)$ is impossible (as that would imply $x \in A_\beta$ so $x \in B_\gamma$ for some $\gamma \leq \beta$), so $y \in f(x)$ and there are less than μ elements like this. [Géza Fodor: Proof of a conjecture of P. Erdős, *Acta Sci. Math*, **14**(1952), 219–227]

11. If $f(\alpha)$ is uncountable for some $\alpha < \omega_1$ then, when restricted to $f(\alpha)$, f is a set mapping with finite images, and we have an uncountable free subset by Problem 8(a). So, we can assume that $f(\alpha)$ is countable for every $\alpha < \omega_1$.

By closing, we can get a closed, unbounded set $C \subseteq \omega_1$ such that if $\gamma \in C$, and

(A) $x < \gamma$, then $f(x) \subseteq \gamma$;

(B) if $s \subseteq z < \gamma$, s is finite, and $\{x : f(x) \cap s = \emptyset\}$ is countable, then $\sup(\{x : f(x) \cap s = \emptyset\}) < \gamma$;

(C) if $s \subseteq z < \gamma$, s is finite, and $\{x : f(x) \cap s = \emptyset\}$ is uncountable, then there is a $z < z' < \gamma$ such that (z, z') contains infinitely many elements of $\{x : f(x) \cap s = \emptyset\}$;

(D) if $s \subseteq w < z < \gamma$, s is finite, and there is a finite t with $\min(t) > z$, such that for all x with $w < x < z$ either $f(x) \cap s \neq \emptyset$ or $x \in f[t]$ holds, then there is such a t' with $|t'| = |t|$, $t' \subseteq \gamma$.

This can be achieved, as all conditions are of the form "if $x_1, \ldots, x_n < \gamma$ then some countable ordinal depending on $x_1, \ldots, x_n$ is $< \gamma$" so we can apply Problem 20.7.

Let $\alpha < \omega_1$. We produce a free subset of order type α. Let $\gamma_0 < \cdots < \gamma_\alpha$ be the first $\alpha + 1$ elements of C. Let $y \geq \gamma_\alpha$ be arbitrary. Enumerate α as $\alpha = \{z_i : i < \omega\}$. By induction on $i < \omega$ we are going to choose $\gamma_{z_i} < x_i < \gamma_{z_i+1}$ such that $\{y, x_0, x_1, \ldots\}$ is free. If we succeed, we are done, as this latter set has order type $\alpha + 1$. Assume that $0 \leq i < \omega$ and we have already selected $\{x_0, \ldots, x_{i-1}\}$. If we cannot choose x_i, then setting $s = \{x_j : j < i, z_j < z_i\}$, $t = \{x_j : j < i, z_j > z_i\} \cup \{y\}$, we get that there is no $\gamma_{z_i} < x < \gamma_{z_i+1}$ such that $x \notin f[t]$ and $s \cap f(x) = \emptyset$. As $y > \gamma_{z_i+1}$ has $f(y) \cap s = \emptyset$, by

(B) the set $\{x : f(x) \cap s = \emptyset\}$ is uncountable, therefore, by (C), there is a z with $\gamma_{z_i} < z < \gamma_{z_i+1}$ such that $X = \{\gamma_{z_i} < x < z : f(x) \cap s = \emptyset\}$ is infinite. Our hypothesis gives that $X \subseteq f[t]$ and so by (D) there is another finite $t' \subseteq [z, \gamma_{z_i+1})$ with $X \subseteq f[t']$. We obtained that t, t' are disjoint finite sets and $f[t] \cap f[t']$ is infinite, which is impossible, as it is the union of finitely many sets of the form $f(\beta) \cap f(\gamma)$, and those sets are finite.

Virtually the same proof gives that, under GCH, if κ is regular, f is a set mapping on κ^+ with $|f(x) \cap f(y)| < \kappa$ for $x \neq y$, then there are free sets of arbitrary large ordinal below κ^+. [S. Shelah: Notes on combinatorial set theory, *Israel Journal of Mathematics*, **14**(1973), 262–277]

12. First we verify the last statement by induction on k. If $k = 1$ and $|S| \leq \aleph_0$, enumerate S as $S = \{s_0, s_1, s_2, \ldots\}$ and define $F(s_n) = \{s_0, s_1, s_2, \ldots, s_{n-1}\}$. Clearly, there is no 2-element free set. Assume we have the result for k. Then for every $\alpha < \omega_k$ there is a set mapping $F_\alpha : [\alpha]^k \to [\alpha]^{<\omega}$ with no free set of cardinality $k+1$. Define $F : [\omega_k]^{k+1} \to [\omega_k]^{<\omega}$ as follows: for $x_0 < \cdots < x_k$ set $F(x_0, \ldots, x_k) = F_{x_k}(x_0, \ldots, x_{k-1})$. If now $\{y_0, \ldots, y_{k+1}\}$ was a free subset with $y_0 < \cdots < y_{k+1}$ then $\{y_0, \ldots, y_k\}$ would be free for $F_{y_{k+1}}$ which is impossible.

For the other direction assume that $|S| \geq \aleph_k$ and $F : [S]^k \to [S]^{<\omega}$ is a set mapping. Choose disjoint subsets $A_0, \ldots, A_k$ with $|A_0| = \aleph_0, \ldots, |A_k| = \aleph_k$. The set $A_0 \times \cdots \times A_{k-1}$ and together with it the set $F[A_0 \times \cdots \times A_{k-1}]$ has cardinality $\aleph_{k-1}$; therefore, we can select $y_k \in A_k \setminus F[A_0 \times \cdots \times A_{k-1}]$. As $|A_0 \times \cdots \times A_{k-2}| \leq \aleph_{k-2}$ we can select $y_{k-1} \in A_k \setminus F[A_0 \times \cdots \times A_{k-2} \times \{y_k\}]$. Continuing this way we define $y_{k-2} \in A_{k-1} \setminus F[A_0 \times \cdots \times A_{k-1} \times \{y_{k-1}\} \times \{y_k\}]$, etc., finally picking $y_0 \in A_0 \setminus F(y_1, \ldots, y_k)$, which is again possible as we subtract a finite set from a set of cardinality $\aleph_0$. The set $\{y_0, \ldots, y_k\}$ is a free set of cardinality $k+1$.

13. Fix the natural number $n \geq 3$. Set $S_1 = S$. By induction on $1 \leq i \leq n$ we make the following construction. If already we have S_i with $|S_i| = \aleph_2$ then first choose an arbitrary countably infinite subset $A_i \subseteq S_i$. Then define the following set mapping F on $S_i \setminus A_i$.

$$F(x) = \{f(x, y) : y \in A_i\} \cap (S_i \setminus A_i)$$

for $x \in S_i \setminus A_i$. By Problem 8(a) there is a free set of cardinal $\aleph_2$; let S_{i+1} be one of those free sets.

This way, we get the countably infinite sets $A_1, \ldots, A_n$ with the property that if $1 \leq i < j < k \leq n$, then for $x \in A_i$, $y \in A_j$, $z \in A_k$ neither $y \in F(x, z)$ nor $z \in F(x, y)$ holds.

Now select $x_n \in A_n$, $x_{n-1} \in A_{n-1}$ arbitrarily. Then by reverse induction on $1 \leq i \leq n-2$ pick

$$x_i \in A_i \setminus \left[\bigcup \{F(x_j, x_k) : i < j < k \leq n\}\right].$$

The set $\{x_1, \ldots, x_n\}$ will be free. [A. Hajnal–A. Máté: Set mappings, partitions, and chromatic numbers, in: *Logic Colloquium '73*, Bristol, North-Holland, 1975, 347–379]

27

Trees

1. Let $\langle T, \prec \rangle$ be an ω-tree, that is, an infinite tree with levels $T_0, T_1, \ldots$, which are all finite. $T = \bigcup\{T_{\geq x} : x \in T_0\}$; therefore, for one of them (at least), say for x_0, $T_{\geq x_0}$ is infinite. x_0 has finitely many immediate successors on level 1, repeating the previous argument, for (at least) one of them, say for x_1, $T_{\geq x_1}$ is infinite. Repeating the argument we get an infinite branch, $\{x_0, x_1, \ldots\}$.

Another possibility is to argue that, as $|T| = \aleph_0$, there is a nonprincipal ultrafilter, D, on T. For every $n < \omega$, $T_{\geq n}$ is partitioned into the finitely many sets $T_{\geq x}$ $(x \in T_n)$. Exactly one of them, say $T_{\geq x_n}$ is in D. Clearly, $x_0 \prec x_1 \prec \cdots$ as otherwise we would get disjoint sets in D.

2. Let T be the union of the disjoint branches b_n $(n = 1, 2, \ldots,)$ of height n. Then T has no infinite branch and T_i consist of the $(i + 1)$th elements of $b_{i+1}, \ldots$.

3. Pick a vertex v, let it be the sole element of T_0. By induction on $n = 0, 1, \ldots$ add x to T_{n+1} if and only if it is joined to some $y \in T_n$ and $x \notin T_0 \cup \cdots \cup T_n$. Choose one such y and make $y \prec x$. This gives T, a spanning tree of the graph. T is infinite, as by connectivity it contains all vertices, and by local finiteness each T_n is finite. Therefore, König's theorem applies, and there is an infinite branch, which is an infinite path in the graph.

Another possibility is to fix again a vertex, and let T consist of the finite paths from v, that is, $t = \langle v_0, \ldots, v_n \rangle \in T$ if and only if $v_0 = v$, each v_i is joined to v_{i+1} and $v_0, \ldots, v_n$ are different. $t \prec t'$ if and only if t' end-extends t. An ω-branch in T gives rise to an infinite path in the graph.

4. Build a tree with vertex set $\mathcal{H}$ and let $S \prec R$ in $\mathcal{H}$ if and only if R properly extends S. By König's lemma there is an infinite branch, which is a collection of finite strings, between any two of them one being an initial segment of the other. Then their union is an infinite 0–1 sequence all whose initial segments belong to the branch and therefore to $\mathcal{H}$.

5. Construct a tree T whose nth level consists of those functions $F \in \prod_{i<n} A_i$ for which there is an f_k with $F = f_k\big|_{\{0,\ldots,n-1\}}$. Set $F \prec G$ if $G \in \prod_{i<m} A_i$ with $n < m$ and G extends F. By König's lemma there is an infinite branch $F_0 \prec F_1 \prec \cdots$, and if we let F be the union of the functions $F_0, F_1, \ldots$, then $F \in \prod_{i<\omega} A_i$ is as required: if $S \subseteq \omega$ is finite, then $S \subseteq \{0, 1, \ldots, n-1\}$ holds for some n, and if f_k is a function that extends F_n, then $F|_S = f_k|_S$.

6. Assume that $A \subseteq [0,1]$ is infinite. Define an ω-tree $\langle T, \prec \rangle$ as follows. $I \in T$ if and only if I is a dyadic interval of the form $I = [\frac{p}{2^n}, \frac{p+1}{2^n}]$ for some natural number p and $I \cap A$ is infinite. We make $I \prec I'$ if I' is a subinterval of I. It is easy to check that if $I = [\frac{p}{2^n}, \frac{p+1}{2^n}] \in T$, then there are n intervals below I, therefore $I \in T_n$. Clearly, $|T_n| \leq 2^n$, and, as A is infinite, $T_n \neq \emptyset$. By the König lemma, there is an infinite branch, $I_0 \prec I_1 \prec \cdots$, and these intervals shrink to a real number, which is a limit point of A.

7. Assume to the contrary that for some r, k, and s no number n as described exists. That is, for every $n < \omega$ there is some coloring of the r-tuples of $\{0, 1, \ldots, n-1\}$ with k colors with no homogeneous subset as indicated. Define an ω-tree T as follows. T_n contains the above colorings of $\{0, 1, \ldots, n-1\}$. $s \prec t$ if t extends s. By König's lemma, there is an infinite branch $t_0 \prec t_1 \prec \cdots$ and the union of these colorings gives a coloring F of the k-tuples of ω with no homogeneous subsets as described. By Ramsey's theorem (Problem 24.1), there is an infinite homogeneous set for F, say $a_1 < a_2 < \cdots$. Choose p such that $p \geq s$, $p \geq a_1$. Then for $n \geq a_p$, $\{a_1, \ldots, a_p\}$ is a homogeneous subset for the restriction of F to $\{0, 1, \ldots, n-1\}$ of the forbidden type, a contradiction. [As we have just shown, the statement in the problem is true. The proof used infinity, and this is inevitable, because Jeff Paris and Leo Harrington proved that the statement is unprovable in the axiom system Peano Arithmetic, that is, number theory. So this is a true but unprovable statement of arithmetic. J. Paris, L. Harrington: A mathematical incompleteness in Peano Arithmetic, in: *Handbook of Mathematical Logic*, (Jon Barwise, ed.), Studies in Logic, **90**, North-Holland, 1977, 1133–1142]

8. (a) For $n = 0, 1, \ldots$ let T_n be the set of tilings of the $\{-n, \ldots, n\} \times \{-n, \ldots, n\}$ square. For $p \in T_n$, $q \in T_m$, $n < m$, set $p \prec q$ if and only if q extends p. As there are finitely many different color types of the dominoes, every T_n is finite. Also, they are nonempty by our condition. Applying König's lemma we get an ω-branch $p_0 \prec p_1 \prec \cdots$ the union of which is a tiling of the plane.

(b) Using part (a), it suffices to show that for every n there is a tiling of an n-by-n square using dominoes from D'. Indeed, consider a tiling of the plane with D. In this tiling, all dominoes from $D \subseteq D'$ form a finite, therefore bounded part of the plane. Beyond that part, one can find arbitrarily large squares, necessarily using dominoes only from D'. [H. Wang: Proving theorems by pattern recognition, *Bell System Tech. Journal*, **40**(1961), 1–42]

9. It suffices to show the statement for connected graphs, so we can assume that X is either finite or countably infinite.

If X is finite, choose the decomposition $V = A \cup B$ of the vertex set V for which the number $e(A, B)$ of edges between A and B is maximal. If $v \in A$, then for the choice $A' = A \setminus \{v\}$, $B' = B \cup \{v\}$ we have

$$e(A', B') = e(A, B) + d_A(v) - d_B(v) \leq e(A, B),$$

so $d_A(v) \leq d_B(v)$. A similar argument applies if $v \in B$.

Assume now that X is countably infinite. Enumerate its vertices as $\{v_0, v_1, \ldots\}$. By the previous argument there is an appropriate decomposition $A_n \cup B_n = \{v_0, \ldots, v_n\}$ for the induced graph on $\{v_0, \ldots, v_n\}$. By Problem 4 there is a decomposition $A \cup B$ of $\{v_0, v_1, \ldots\}$ such that for every m the sets $A \cap \{v_0, \ldots, v_m\}$, $B \cap \{v_0, \ldots, v_m\}$ are the restrictions of A_n, B_n for some $n \geq m$. If now $v \in A$ then choose m so large that v as well as all vertices neighboring v are among $v_0, \ldots, v_m$ (here we use local finiteness). Then, by the above claim $d_A(v) \leq d_B(v)$ and similarly for $v \in B$.

10. (a) For an index sequence $0 = k(0) < k(1) < \cdots < k(r) = n$ let $Q_i = a_{k(i-1)+1}^2 + \cdots + a_{k(i)}^2$. As there are finitely many index sequences as above, we can consider one with the sum

$$Z = (S_1^2 + \cdots + S_r^2) + 2(Q_1 + 2Q_2 + \cdots + rQ_r)$$

minimal. We claim that this sequence is as required. Let $a = a_{k(i)+1}$ be the first term of S_{i+1}. If we remove it from S_{i+1} and add it to S_i, then in Z, in the first sum $S_i^2 + S_{i+1}^2$ will be changed to $(S_i + a)^2 + (S_{i+1} - a)^2$, while the second sum will be decremented by $2a^2$, so

$$S_i^2 + S_{i+1}^2 \leq (S_i + a)^2 + (S_{i+1} - a)^2 - 2a^2,$$

and this implies $S_i \geq S_{i+1}$.

In order to show the other property, let j be the unique index with

$$a_1 + \cdots + a_{j-1} < \frac{S_1}{2} \leq a_1 + \cdots + a_j.$$

Split S_1 into the subsums $S_1' = a_1 + \cdots + a_j$ and $S_1'' = a_{j+1} + \cdots + a_{k(1)}$. There is some $d \geq 0$ such that $S_1' = \frac{S_1}{2} + d$ and $S_1'' = \frac{S_1}{2} - d$ and clearly $d < a_j$ holds. Again comparing the old and the new values of Z we get that

$$S_1^2 \leq \left(\frac{S_1}{2} + d\right)^2 + \left(\frac{S_1}{2} - d\right)^2 + 2(a_{j+1}^2 + \cdots + a_n^2)$$

and so

$$S_1^2 \leq 4(d^2 + a_{j+1}^2 + \cdots + a_n^2) < 4(a_j^2 + \cdots + a_n^2).$$

(b) As the series diverges, there are natural numbers $M_1 < M_2 < \cdots$ such that $\sum_{M_s+1}^{M_{s+1}} a_j > 2\sqrt{a_1^2 + a_2^2 + \cdots}$. Let T be the following tree. $\langle k(1), \ldots, k(s) \rangle \in T$ if there are n and $0 = k(0) < k(1) < \cdots < k(r) = n$, $r \geq s$, as in part (a). $t \prec t'$ in T if and only if t' is an end-extension of t. Obviously, $t \in T_s$ if and only if t is of length s. As by conditions $k(i) \leq M_i$ always holds, each T_s is finite. It is also nonempty, so König's lemma gives an infinite branch, and that produces a decomposition as claimed. [M. Szegedy, G. Tardos: On the decomposition of infinite series into monotone decreasing parts, *Studia Sci. Math. Hung.*, **23**(1998), 81–83.]

11. In our construction of an Aronszajn tree T, every element of the tree will be some increasing function $f : \alpha \to \mathbf{Q}$ $(\alpha < \omega_1)$ with $f \prec g$ if g extends f. As all these functions are necessarily injective, in a putative ω_1-branch the union of the elements would give an injective function $\omega_1 \to \mathbf{Q}$, which is impossible.

We require that if $f : \alpha \to \mathbf{Q}$ is in T, $\beta < \alpha$, then $f|_\beta \in T$ and also that for every $f \in T$, the supremum of the range of f, denoted by $s(f)$, is finite. Notice that, under these conditions, if $f : \alpha \to \mathbf{Q}$ is in T, then $f \in T_\alpha$.

We construct T_α by transfinite recursion on α.

For $\beta < \alpha$ we add the following stipulation, which we call $P(\beta, \alpha)$:

if $f \in T_\beta$, $1 \leq k < \omega$, *then there is* $f \prec g \in T_\alpha$, $s(g) < s(f) + \frac{1}{k}$.

Notice that $P(\beta, \alpha)$ and $P(\alpha, \alpha + 1)$ imply $P(\beta, \alpha + 1)$.

For $\alpha = 0$ we take $f = \emptyset$, the empty function as the sole element of T_0, and formally set $s(\emptyset) = 0$.

If T_α is determined then for every $f \in T_\alpha$ and for every $1 \leq k < \omega$ we define a one-point extension f_k of f with $s(f_k) < s(f) + \frac{1}{k}$. For this we only have to choose a rational number $s(f) < q_k < s(f) + \frac{1}{k}$ and make $f_k(\alpha) = q_k$. This assures $P(\alpha, \alpha + 1)$ and that suffices by the above remark.

Assume that α is limit and we are to construct T_α. For every choice of $f \in T_\beta$, $(\beta < \alpha)$, and $1 \leq k < \omega$ we are going to build an α-branch b through f such that for $g = \cup b$ we have that $g : \alpha \to \mathbf{Q}$ with $s(g) < s(f) + \frac{1}{k}$. As there are countably many choices for f, k, if all these functions g form T_α, the latter set will be countable.

Given f, β, and k as above, select a sequence of ordinals $\beta = \alpha_0 < \alpha_1 < \cdots$ converging to α. Using $P(\beta, \alpha_1)$,$P(\alpha_1, \alpha_2)$, etc., get the elements $f = f_0 \prec f_1 \prec \cdots$ with $f_i \in T_{\alpha_i}$,

$$s\left(f_{i+1}\right) < s\left(f_i\right) + \frac{1}{k 2^{i+1}}.$$

Then $g = f_0 \cup f_1 \cup \cdots$ is as required. [N. Aronszajn, cf. DJ. Kurepa: Ensembles ordonnés et ramifiés, *Publ. Math. Univ. Belgrade*, **4**(1935), 1–138]

12. We slightly modify the construction of an Aronszajn tree in Problem 11 by requiring that $s(f) \in \mathbf{Q}$ for every $f \in T$. In this case, if $\mathbf{Q} = \{q_0, q_1, \ldots\}$, then $T = A_0 \cup A_1 \cup \cdots$ is a decomposition into antichains, where $A_i = \{f \in$

$T : s(f) = q_i\}$. The condition $P(\beta, \alpha)$ is now changed to the following: if $f \in T_\beta$, $q > s(f)$ is rational, then there is some $f \prec g \in T_\alpha$, $s(g) = q$. If α is limit (the only problematic case), proceed as follows. Given $f \in T_\beta$, $\beta < \alpha$, $p = s(f) < q$, select the sequence of ordinals $\beta = \alpha_0 < \alpha_1 < \cdots$ converging to α, and also the sequence $p = p_0 < p_1 < \cdots$ of rational numbers, converging to q. Then inductively choose the elements $f = f_0 \prec f_1 \prec \cdots$ such that $f_i \in T_{\alpha_i}$, $s(f_i) = p_i$, and then add $g = f_0 \cup f_1 \cup \cdots$ to T_α.

13. As an antichain can contain only one element of a branch, a special ω_1-tree may not have an uncountable branch.

14. Only the transitivity of $<_{\text{lex}}$ is not immediately clear. Assume therefore that $x <_{\text{lex}} y <_{\text{lex}} z$ and try to show that $x <_{\text{lex}} z$. We consider cases.

If $x \prec y \prec z$, then $x \prec z$ and we are done.

Assume that $x \prec y$ and $p_\alpha(y) <_\alpha p_\alpha(z)$ with α least. If now $\alpha > o(x)$, then $x \prec z$; otherwise $p_\alpha(x) = p_\alpha(y) <_\alpha p_\alpha(z)$ and $p_\beta(x) = p_\beta(y) = p_\beta(z)$ for $\beta < \alpha$ so $x <_{\text{lex}} z$.

Assume that $p_\alpha(x) <_\alpha p_\alpha(y)$ for some least α and $y \prec z$. Then $p_\alpha(x) <_\alpha p_\alpha(z)$ holds and so again $x <_{\text{lex}} z$.

Finally, assume that there is a minimal $\alpha \leq o(x), o(y)$ such that $p_\alpha(x) <_\alpha p_\alpha(y)$ and there is a minimal $\beta \leq o(y), o(z)$ such that $p_\beta(y) <_\beta p_\beta(z)$. If $\alpha < \beta$, then $p_\alpha(x) <_\alpha p_\alpha(y) = p_\alpha(z)$. If $\alpha = \beta$, then $p_\alpha(x) <_\alpha p_\alpha(y) <_\alpha p_\alpha(z)$. If $\alpha > \beta$, then $p_\beta(x) = p_\beta(y) <_\beta p_\beta(z)$. In each case we are done, for α, resp. β is minimal with the given property.

An alternative possibility for the proof is to add a new element, say $\star$, to every T_α, make it precede all elements of T_α by $<_\alpha$, and then identify $x \in T$ with the following function f_x: for $\alpha \leq o(x)$, set $f_x(\alpha) = p_\alpha(x)$, while for $o(x) < \alpha < h(T)$, set $f_x(\alpha) = \star$. The functions f_x are now functions defined on the same well-ordered set, so we can use the usual lexicographic ordering and note that $x <_{\text{lex}} y$ if and only if $f_x <^{\text{lex}} f_y$ where $<^{\text{lex}}$ is the lexicographic ordering on the set $\{f_x : x \in T\}$.

15. Toward an indirect proof, assume first that $\{x_\xi : \xi < \omega_1\}$ is increasing or decreasing by $<_{\text{lex}}$. The elements $\{p_0(x_\xi) : \xi < \omega_1\}$ form a weakly increasing (or weakly decreasing) sequence by $<_0$ and as T_0 is countable, this sequence eventually stabilizes: for $\gamma_0 < \xi < \omega_1$ we have $p_0(x_\xi) = s_0$. Repeating the argument we get ordinals $\gamma_\alpha < \omega_1$ and elements $s_\alpha \in T_\alpha$ for every $\alpha < \omega_1$ such that for $\xi > \gamma_\alpha$ we have $o(x_\xi) \geq \alpha$ and $p_\alpha(x_\xi) = s_\alpha$. Now $\{s_\alpha : \alpha < \omega_1\}$ is an ω_1-branch of $\langle T, \prec \rangle$.

Assume, finally, that $X \subseteq T$ is uncountable and $Y \subseteq T$ is countable. There is some $\alpha < \omega_1$ such that $Y \subseteq T_{<\alpha}$. As X is uncountable, there are uncountable many elements of it with height $> \alpha$, there are two of them, say x and y, with $p_\alpha(x) = p_\alpha(y)$. But then, no element with $o(z) < \alpha$ can be between them. Thus, for X there is no countable Y that separates its elements, hence $\langle X, <_{\text{lex}}\rangle$ cannot be similar to a subset of $\mathbf{R}$. [E. Specker: Sur un problème de Sikorski, *Coll. Math.*, **2**(1949), 9–12]

16. We construct e_α by transfinite recursion on $\alpha < \omega_1$ with the added assumption that $A_\alpha = \text{Ran}(e_\alpha)$ is a coinfinite subset of ω. e_0 is the empty function. If e_α is determined then we let $e_{\alpha+1}$ be a one point extension of f_α with $e_{\alpha+1}(\alpha) = x$ for some $x \in \omega \setminus A_\alpha$. Assume finally that $\alpha < \omega_1$ is limit and e_β is given for every $\beta < \alpha$. Let $\alpha_0 < \alpha_1 < \cdots$ be a sequence converging to α. By induction, we are going to determine $e_\alpha|_{\alpha_n}$, a finite modification of e_{α_n} and we also pick an element $x_n < \omega$. To start, set $e_\alpha|_{\alpha_0} = e_{\alpha_0}$, and let $x_0 \in \omega \setminus A_{\alpha_0}$. Assume that $e_\alpha|_{\alpha_n}$ and $x_0, \ldots, x_n$ are determined. We need to determine $e_\alpha|_{[\alpha_n, \alpha_{n+1})}$. As $e_{\alpha_{n+1}}$ is injective and almost extends e_{α_n}, there are only finitely many points in the range of both $e_\alpha|_{\alpha_n}$ and $e_{\alpha_{n+1}}|_{[\alpha_n, \alpha_{n+1})}$. By reassigning values from $\omega \setminus A_{\alpha_{n+1}}$ we can achieve, by modifying $e_{\alpha_{n+1}}$ at finitely many places, that the range of $e_{\alpha_{n+1}}|_{[\alpha_n, \alpha_{n+1})}$ is disjoint from the range of $e_\alpha|_{\alpha_n}$ and also from $\{x_0, \ldots, x_n\}$. This modified function will be $e_\alpha|_{[\alpha_n, \alpha_{n+1})}$ and finally we let x_{n+1} be any element of ω not in $\{x_0, \ldots, x_n\}$ or the range of $e_\alpha|_{\alpha_{n+1}}$.

This induction defines $e_\alpha : \alpha \to \omega$. It is injective, as it is the union of injective functions. Its range is disjoint from $\{x_0, x_1, \ldots\}$. And finally, for every n, the functions $e_\alpha|_{\alpha_n}$ and e_{α_n} differ only at finitely many places. Now, if $\beta < \alpha$, then there is some n such that $\beta < \alpha_n < \alpha$, and e_β, $e_{\alpha_n}|_\beta$, and $e_\alpha|_\beta$ also differ only at finitely many places. [K. Kunen: Combinatorics, in: *Handbook of Mathematical Logic*, (Jon Barwise, ed.), Studies in Logic, **90**, North-Holland, 1977, 371–401]

17. $\langle T, \prec\rangle$ is an ω_1-tree and T_α is a set of $\alpha \to \omega$ injections. If $g \in T_\alpha$ then $g = e_\beta|_\alpha$ for some $\beta \geq \alpha$, so g differs from e_α at finitely many places. As this is possible only countably many ways, T_α is countable. If $b = \{g_\alpha : \alpha < \omega_1\}$ with $g_\alpha \in T_\alpha$ was an ω_1-branch, then $\bigcup b$ would be an injection $\omega_1 \to \omega$, an impossibility.

18. It suffices to decompose the pairs $\langle e_\beta, e_\alpha\rangle$ $(\beta < \alpha)$ into countably many chains (use symmetry and notice that $\{\langle e_\alpha, e_\alpha\rangle : \alpha < \omega_1\}$ is a chain). Given such a pair $\langle e_\beta, e_\alpha\rangle$ let $n < \omega$ be so large that $e_\alpha(\beta) \leq n$ and, if $e_\alpha(\gamma) \neq e_\beta(\gamma)$ holds for some $\gamma < \beta$, then $e_\alpha(\gamma)$, $e_\beta(\gamma) \leq n$. This is possible by the condition imposed on our functions. Set $\Gamma = \{\gamma < \alpha : e_\alpha(\gamma) \leq n\}$, a finite set. Enumerate Γ increasingly as $\{\gamma_1, \ldots, \gamma_k\}$, let t be that number with $\gamma_t = \beta$. Finally, let $a_i = e_\alpha(\gamma_i)$ $(1 \leq i \leq k)$, $b_i = e_\beta(\gamma_i)$ $(1 \leq i < t)$. Classify $\langle e_\beta, e_\alpha\rangle$ according to the corresponding ordered sequence $s = \langle n, k, t, a_1, \ldots, a_k, b_1, \ldots, b_{t-1}\rangle$.

We show that if $\langle e_\beta, e_\alpha \rangle$, $\langle e_{\beta'}, e_{\alpha'} \rangle$ get the same sequence then they are comparable and this will conclude the proof. Let Γ' be the corresponding set for $\langle e_{\beta'}, e_{\alpha'} \rangle$.

Assume first that $e_{\alpha'}$ extends e_α. Then $\Gamma' \supseteq \Gamma$, so $\Gamma' = \Gamma$, as both have k elements, but then $\beta = \beta'$ and therefore $\langle e_\beta, e_\alpha \rangle \preceq \langle e_{\beta'}, e_{\alpha'} \rangle$.

Assume now that the first difference between e_α and $e_{\alpha'}$ occurs at $\delta < \alpha$: $e_\alpha(\delta) < e_{\alpha'}(\delta)$. Since n is the same for both pairs $\langle e_\beta, e_\alpha \rangle$ and $\langle e_{\beta'}, e_{\alpha'} \rangle$, we have $\Gamma \cap \delta = \Gamma' \cap \delta$. Therefore, if $\beta \leq \delta$ or $\beta' \leq \delta$, then necessarily $\beta = \beta'$ (recall also that $e_\alpha(\beta) \leq n$, $e_{\alpha'}(\beta') \leq n$), in which case these pairs are comparable. So assume from now on that $\beta, \beta' > \delta$. Since s is the same for both pairs, it follows from $\Gamma \cap \delta = \Gamma' \cap \delta$ that $e_\beta|_\delta = e_{\beta'}|_\delta$.

If $\delta \notin \Gamma$, $\delta \notin \Gamma'$, then $e_\beta(\delta) = e_\alpha(\delta) < e_{\alpha'}(\delta) = e_{\beta'}(\delta)$, so $\langle e_\beta, e_\alpha \rangle \preceq \langle e_{\beta'}, e_{\alpha'} \rangle$.

The possibility $\delta \notin \Gamma$, $\delta \in \Gamma'$ is ruled out as then $e_\alpha(\delta) > n \geq e_{\alpha'}(\delta)$.

Finally, if $\delta \in \Gamma$, $\delta \notin \Gamma'$, then $e_\alpha(\delta), e_\beta(\delta) \leq n < e_{\alpha'}(\delta) = e_{\beta'}(\delta)$, so again $\langle e_\beta, e_\alpha \rangle \preceq \langle e_{\beta'}, e_{\alpha'} \rangle$ holds. [S. Shelah: Decomposing uncountable squares to countably many chains, *J. Comb. Theory* (A), **21**(1976), 110–114. This proof is due to S. Todorcevic]

19. In order to prove that a Countryman type may not include a subtype of order type ω_1, ω_1^*, or the type of an uncountable subset of the reals it suffices to show that neither $\omega_1 \times \omega_1$ nor $B \times B$ (with $B \subseteq \mathbf{R}$, uncountable) is the union of countable many chains.

Assume first indirectly that $\omega_1 \times \omega_1 = A_0 \cup A_1 \cup \cdots$ where every A_i is a chain. By the pigeon hole principle, for every $\alpha < \omega_1$ there is an $i(\alpha) < \omega$ such that $\{\beta : \langle \alpha, \beta \rangle \in A_{i(\alpha)}\}$ is uncountable. There must be $\alpha < \alpha'$ with $i(\alpha) = i(\alpha') = i$. Choose β' such that $\langle \alpha', \beta' \rangle \in A_i$, then choose $\beta > \beta'$ with $\langle \alpha, \beta \rangle \in A_i$. Now $\langle \alpha, \beta \rangle$, $\langle \alpha', \beta' \rangle$ are incomparable, a contradiction.

If we assume that B is an uncountable set of reals and $B \times B = A_0 \cup A_1 \cup \cdots$, then for every $x \in B$ there exist two elements of B, $y'(x) < y''(x)$, such that $\langle x, y'(x) \rangle, \langle x, y''(x) \rangle \in A_{i(x)}$ with some $i(x) < \omega$. There is a rational number $p(x)$ with $y'(x) < p(x) < y''(x)$. As there are only countably many possibilities for $\langle i(x), p(x) \rangle$, there are $x' < x''$ in B such that $i(x') = i(x'') = i$, $p(x') = p(x'') = p$. But then $\langle x', y''(x') \rangle, \langle x'', y'(x'') \rangle \in A_i$ and they are incomparable: $x' < x''$ and $y''(x') > p > y'(x'')$.

20. One direction is obvious: if $f : \langle T, \prec \rangle \to \langle \mathbf{Q}, < \rangle$ is order-preserving, then T is the union of the countable many antichains of the form $f^{-1}(x)$ where $x \in \mathbf{Q}$.

For the other direction, assume that $T = A_0 \cup A_1 \cup \cdots$ where $A_0, A_1, \ldots$ are antichains. If $t \in A_n$, set

$$f(t) = \sum_{i<n} \frac{\epsilon_i}{2^i},$$

where, for $i < n$, $\epsilon_i = 1$, if there exists an $s \prec t$ with $s \in A_i$, $\epsilon_i = -1$ if there exists an $s \succ t$ with $s \in A_i$, and $\epsilon_i = 0$ if neither holds. Clearly, f is a mapping from T into $\mathbf{Q}$. For order preservation, assume that $t \prec t'$, $t' \in A_m$, and $f(t') = \sum_{i<m} \epsilon'_i 2^{-i}$. Assume that $n < m$. Case analysis shows that for $i < n$, $\epsilon'_i \geq \epsilon_i$ holds, moreover $\epsilon'_n = 1$ while there is no corresponding ϵ_n. No matter what the later terms of $f(t')$ are, this implies that $f(t') > f(t)$. A similar argument works if $n > m$, just then $\epsilon_m = -1$ and there is no ϵ'_m.

21. For every $s \in T$ let $g_s : f^{-1}(s) \to \omega$ be a decomposition of $f^{-1}(s)$ into countably many antichains, i.e., $u \in f^{-1}(s)$ belongs to the ith antichain if and only if $g_s(u) = i$. Given $t \in T$ construct the sequence $t = t_0 \succ t_1 \succ$ with $t_{i+1} = f(t_i)$. As there is no infinite decreasing sequence of ordinals, there is some finite n such that $t_n \in T_0$. Set $\xi_i = g_{t_{i+1}}(t_i)$ and decompose the elements of T according to the string $F(t) = \langle \xi_0, \ldots, \xi_{n-1} \rangle$. This decomposition shows that $\langle T, \prec \rangle$ is special: assume that $F(s) = F(t)$ and $s \prec t$. Let $s = s_0 \succ s_1 \succ \cdots \succ s_n \in T_0$ and $t = t_0 \succ t_1 \succ \cdots \succ t_n \in T_0$ be the corresponding sequences. As $s \prec t$, we have that $s_n = t_n$, and the elements $s_0, \ldots, s_n, t_0, \ldots, t_n$ are all comparable. But then, as $\xi_{n-1} = g_{t_n}(t_{n-1}) = g_{s_n}(s_{n-1})$, $t_{n-1} = s_{n-1}$ must hold (as they are comparable elements of an antichain), and repeating this we inductively get $t_{n-1} = s_{n-1}, \ldots, t_0 = s_0$, a contradiction.

22. We have to show that if the normal ω_1-tree $\langle T, \prec \rangle$ has an ω_1-branch then it includes an uncountable antichain. Let $b = \{t_\alpha : \alpha < \omega_1\}$ be an ω_1-branch. Let $x_\alpha \in T_{\alpha+1}$ be an element such that $t_\alpha \prec x_\alpha$, but $x_\alpha \neq t_{\alpha+1}$. x_α exists by normality. $A = \{x_\alpha : \alpha < \omega_1\}$ is an antichain, as for $\beta < \alpha$ we have $o(x_\beta) < o(x_\alpha)$ and x_α's predecessor on level $\beta + 1$ is $t_{\beta+1} \neq x_\beta$. Obviously, A is uncountable.

23. Otherwise we can recursively select the elements $A = \{x_\alpha : \alpha < \omega_1\}$ with $T_{\geq x_\alpha}$ countable such that if $\beta < \alpha$, then $o(x_\alpha)$ is larger than the height of any element above x_β. This is possible as every element excludes only countably many elements. But then A is an antichain: if $\beta < \alpha$ then, considering height, only $x_\beta \prec x_\alpha$ is possible, but that cannot happen by the construction.

24. If $\langle T, \prec \rangle$ is a Suslin tree, then by Problem 23 the set A is countable where $x \in A$ if and only if $T_{>x}$ is countable. Let α be a countable ordinal with $A \subseteq T_{<\alpha}$ and remove $T_{\leq\alpha}$ from T. This way, we get a Suslin tree that satisfies property (A) in normality.

Assume that the Suslin tree $\langle T, \prec \rangle$ satisfies (A) of the definition of normality. For every $x \in T$ the set $T_{>x}$ is uncountable. It cannot consist of pairwise comparable elements, as that would give rise to an ω_1-branch. There are, therefore, incomparable $y, z \succ x$. If we increase them they stay incomparable, so there are incomparable elements with identical height, and actually, for every $\alpha < \omega_1$ there is $\beta(\alpha) > \alpha$ such that any $x \in T_\alpha$ has incomparable successors on $T_{\beta(\alpha)}$. If we select the increasing sequence $0 = \alpha_0 < \alpha_1 < \cdots < \alpha_\xi < \cdots$

($\xi < \omega_1$) such that $\beta(\alpha_\eta) < \alpha_\xi$ holds for $\eta < \xi$, then the tree restricted to $\bigcup\{T_{\alpha_\xi} : \xi < \omega_1\}$ satisfies properties (A) and (B).

Assume finally that the Suslin tree $\langle T, \prec\rangle$ satisfies (A)+(B) of normality. Set $b(x) = T_{\leq x}$ if $o(x) = 0$ or a successor ordinal, and $b(x) = T_{<x}$ if $o(x)$ is limit. Let $U = \{b(x) : x \in T\}$. Set $b(x) \prec b(y)$ if $b(y)$ end-extends $b(x)$. Notice that if $x \prec y$, then $b(x) \prec b(y)$ (but not the other way around) and in the tree $\langle U, \prec\rangle$ $b(x)$ has rank $o(x)$. It is also clear that $\langle U, \prec\rangle$ satisfies (A) and (B). As for property (C), if $\alpha < \omega_1$ is a limit ordinal and $b(x) \neq b(y)$ are at level α of $\langle U, \prec\rangle$ then there is a $\beta < \alpha$ such that if x_β resp. y_β are the elements of $b(x)$ resp. $b(y)$ on T_β then $x_\beta \neq y_\beta$. But then $b(x_{\beta+1}) \prec b(x)$, $b(y_{\beta+1}) \prec b(y)$, and $b(x_{\beta+1}) \neq b(y_{\beta+1})$. Finally, if $\{b(x_\alpha) : \alpha < \omega_1\}$ was an uncountable antichain, then $\{x_\alpha : \alpha < \omega_1\}$ would be an uncountable antichain in $\langle T, \prec\rangle$ by the above remark, so $\langle U, \prec\rangle$ has no uncountable antichains. But in Problem 22 we showed that these properties imply that $\langle U, \prec\rangle$ has no ω_1-branch, either. Therefore, $\langle U, \prec\rangle$ is a normal Suslin tree.

25. For the forward direction, if there exists a Suslin tree, then by Problem 24 there is a normal Suslin tree $\langle T, \prec\rangle$. Then $\bigcup\{T_\alpha : \alpha < \omega_1,\ \alpha \text{ limit}\}$ is again a normal Suslin tree such that every element has infinitely many immediate successors, so we may assume that T has this property. Let $<_{\alpha+1}$ be an ordering of $T_{\alpha+1}$ that orders the immediate successors of any element of T_α into a dense set with no first or last element, then use these orderings to define the ordered set $\langle T, <_{\text{lex}}\rangle$. We claim that $\langle T, <_{\text{lex}}\rangle$ is a Suslin line.

Assume that $I_\alpha = (a_\alpha, b_\alpha)$ ($\alpha < \omega_1$) are intervals. Case analysis shows that for every α there is some element x_α such that $T_{\geq x_\alpha} \subseteq (a_\alpha, b_\alpha)$. But then some two x_α's are comparable (or identical) and then the corresponding intervals intersect. This proves that $\langle T, <_{\text{lex}}\rangle$ has property ccc.

Assume now that $X \subseteq T$ is countable. There is some $\alpha < \omega_1$ such that $X \subseteq T_{<\alpha}$. But if $x \in T_\alpha$, then the elements of $T_{\geq x}$ (an uncountable set) cannot be separated by X, i.e., if u, $v \in T_{\geq x}$, then no element of X lies in between u and v. This proves that $\langle T, <_{\text{lex}}\rangle$ has no countable dense subset.

For the other direction assume that $\langle S, <\rangle$ is a Suslin line. We call a subset $R \subseteq S$ separable if there is a countable set $P \subseteq S$ such that if $a < b$ are in R, and there is at least one element between them then there is an element of P between them.

We notice that if $C \subseteq S$ is convex, i.e., if $x, y \in C$, $x < z < y$, then $z \in C$, then C can be decomposed into the disjoint union of a countable set and disjoint nonempty open intervals of the form $(a, b) = \{z \in S : a < z < b\}$. Indeed, if C has no largest element, there is, by Hausdorff's theorem (Problem 6.44), a cofinal sequence $a_0 < a_1 < \cdots$ of some length, which cannot be $\geq\omega_1$, as $\langle S, <\rangle$ is ccc. As the length $<\omega_1$, we can select a subsequence, denoted again by $a_0 < a_1 < \cdots$ of length ω. Similarly, unless there is a minimal element, there is a decreasing, co-initial sequence $a_0 > a_{-1} > \cdots$ and then C is split into the intervals (a_i, a_{i+1}) and the countable set $\{a_i : i \in \mathbf{Z}\}$.

We construct a Suslin tree $\langle T, \prec\rangle$ consisting of open, nonempty intervals of the form (a, b) of $\langle S, <\rangle$ in such a fashion that for $\alpha < \omega_1$ the elements of T_α will be pairwise disjoint intervals that cover S save a separable part. Moreover, $I \prec I'$ holds if and only if I' is a proper subinterval of I, and if I, I' are incomparable, then $I \cap I' = \emptyset$. If we can construct the tree with these requirements, then, as $\langle S, <\rangle$ is not separable, $T_\alpha \neq \emptyset$ will hold, and by the ccc property T_α is countable for every $\alpha < \omega_1$ and there is no uncountable antichain in $\langle T, \prec\rangle$.

Let T_0 be a decomposition, as described above, of S. If we have $(a, b) = I \in T_\alpha$ and $|I| \geq 3$, say $d < c < e$ are in I, then split I into $I_0 = (a, c)$ and $I_1 = (c, b)$, and make I_0, I_1 the immediate successors of I. If, however, $|I| \leq 2$ then I will have no successors. Notice that by this construction, if some $I \in T$ has successors, then it has two immediate successors, therefore we can use the argument of Problem 22 to show that if $\langle T, \prec\rangle$ has an ω_1-branch, then it has an uncountable antichain, as well.

Let $\alpha < \omega_1$ be a limit ordinal. If we consider the nonempty convex sets of the form $\bigcap b$ where b is an α-branch of $T_{<\alpha}$, then they constitute a partition of S minus a separable set (the union of the countably many exceptional separable sets on lower levels) into convex sets: $\bigcup\{C_j : j \in J\}$. Set $J' = \{j \in J : |C_j| = 1\}$, $J'' = J \setminus J'$. If, for $j \in J'$, we have $C_j = \{x_j\}$, then the set $\{x_j : j \in J'\}$ is separable. Indeed, if $x_i < x_j$ for i, $j \in J'$, then there is some $I \in T_{<\alpha}$ where the branches corresponding to i, j split, so x_i and x_j are separated by one of the endpoints occurring. As there are countably many endpoints in $T_{<\alpha}$, the statement is proved. As $\langle S, <\rangle$ is a Suslin line, J'' is countable, so it suffices to apply the treatment described at the beginning of the proof to every C_j ($j \in J''$), and we get the elements of T_α.

26. Let $A \subseteq D$ be a maximal subset of pairwise incomparable elements. Such a set exists by Zorn's lemma. A is countable as T is Suslin. There is an $\alpha < \omega_1$ such that $o(x) < \alpha$ holds for every $x \in A$. We claim that $T_{\geq\alpha} \subseteq D$ (and that suffices). For this, let $x \in T_{\geq\alpha}$ be arbitrary. As D is dense, there is $y \in D$, $y \succ x$. The set $A \cup \{y\}$, a proper extension of A, cannot consist of incomparable elements, so there is some $z \in A$, such that y and z are comparable. As $o(z) < \alpha \leq o(y)$, the only possibility is that $z \prec y$, and then necessarily $z \prec x \prec y$. As D is open, this implies that $x \in D$, as claimed.

27. By Problem 26 for every D_n there is some $\alpha_n < \omega_1$ such that $T_{\geq\alpha_n} \subseteq D_n$. If $\alpha = \sup_n \alpha_n$, then $T_{\geq\alpha} \subseteq D_0 \cap D_1 \cap \cdots$ and $T_{\geq\alpha}$ is dense by normality. It is also clear that the intersection of open sets is open and we are done.

28. Define the set D as follows. $a \in D$ if and only if there is no element x of A such that $x \succeq a$. If the statement of the problem fails, then D is dense. As D is clearly open, we get by Problem 26 that D is a co-countable subset of T, but this is a contradiction as then $A \cap D \neq \emptyset$ and this is impossible.

29. We can assume that f maps into $[0,1]$. For $n = 1, 2, \ldots$ set $x \in D_n$ if and only if $f(y) < f(x) + \frac{1}{n}$ holds for every $y \succeq x$.

We claim that every D_n is dense. Indeed, if $x \in T$ and $x \notin D_n$, then there is an $x_1 \succ x$ such that $f(x_1) \geq f(x) + \frac{1}{n}$. If $x_1 \notin D_n$ then there is an $x_2 \succ x_1$ such that $f(x_2) \geq f(x_1) + \frac{1}{n} \geq f(x) + \frac{2}{n}$, etc. As this procedure must stop, we end up with some $y \succeq x$, $y \in D_n$.

By Problem 27 there is an $\alpha < \omega_1$ such that $T_{\geq\alpha} \subseteq D_1 \cap D_2 \cap \cdots$, but then, if $x \in T_\alpha$ and $y \succ x$, then $f(x) \leq f(y) < f(x) + \frac{1}{n}$ holds for every n, that is, $f(y) = f(x)$. This implies that all values of f are attained on $T_{\leq\alpha}$, a countable set, so f has countable range.

30. In this solution "dense" and "open" are used as in the introduction to this chapter, but the continuity of f is meant in the topology on trees defined before Problem 30.

If $p < q$ are rational numbers, set $t \in D_{p,q}$ if either $f(y) \geq p$ holds for every $y \succeq t$ or $f(y) \leq q$ holds for every $y \succeq t$. We show that each $D_{p,q}$ is dense.

Indeed, assume that some $D_{p,q}$ has no elements above some $a \in T$. Passing to $T_{\geq a}$ we can assume that $D_{p,q} = \emptyset$. Set $\alpha_0 = 0$ and define $\alpha_0 < \alpha_1 < \cdots$, a sequence of countable ordinals, and A_i, a maximal set of incomparable elements x in $T \setminus T_{\leq\alpha_i}$ and either with $f(x) \leq p$ (if i is even) or with $f(x) \geq q$ (if i is odd). By Zorn's lemma, such a set A_i exists, and it is countable, as $\langle T, \prec\rangle$ is Suslin. Choose α_{i+1} so that $A_i \subseteq T_{\leq\alpha_{i+1}}$. Let α be the limit of $\alpha_0, \alpha_1, \ldots$. If $x \in T_\alpha$, then by $D_{p,q} = \emptyset$ there are y_0, $y_1 \succ x$ such that $f(y_0) < p$ and $f(y_1) > q$. By the maximality of A_{2i} and A_{2i+1} there are some $x_{2i} \in A_{2i}$ comparable with y_0 and $x_{2i+1} \in A_{2i+1}$ comparable with y_1. As $o(x_{2i}), o(x_{2i+1}) < \alpha < o(y_0), o(y_1)$ we must have $x_{2i} \prec y_0$, $x_{2i+1} \prec y_1$, and then necessarily x_{2i}, $x_{2i+1} \prec x$. Hence $x_0 \prec x_1 \prec \cdots$ is a sequence converging to x, with $f(x_{2i}) \leq p$, $f(x_{2i+1}) \geq q$, so f is not continuous.

Having proved that each $D_{p,q}$ is dense, one can easily observe that they are open. There is, therefore, by Problems 26 and 27 an $\alpha < \omega_1$ such that $T_{\geq\alpha}$ is in the intersection of all of them. If now $x \in T_\alpha$, $x \prec y$, and $p < q < f(x)$ are rationals, then we get that, as $x \in D_{p,q}$, $f(y) \geq p$ holds. Since $p < q < f(x)$ were arbitrary, $f(y) \geq f(x)$ follows. Selecting $f(x) < p < q$ an identical argument gives $f(y) \leq f(x)$, i.e., actually $f(y) = f(x)$. Therefore, f attains all its values on the countable set $T_{\leq\alpha}$, so it has countable range.

31. We have to show that if F_0, F_1 are disjoint, closed sets then they can be separated by disjoint open sets.

We first consider the case when F_0, F_1 are both countable. Enumerate them as $F_0 = \{u_0, u_1, \ldots\}$, $F_1 = \{v_0, v_1, \ldots\}$. By induction on $n = 0, 1, \ldots$ we construct the closed and open sets $U_n \supseteq \{u_0, \ldots, u_{n-1}\}$, $V_n \supseteq \{v_0, \ldots, v_{n-1}\}$, such that $\emptyset = U_0 \subseteq U_1 \cdots$, $\emptyset = V_0 \subseteq V_1 \cdots$, and $U_n \cap V_n = U_n \cap F_1 = V_n \cap F_0 = \emptyset$. If we can do this, then the open sets $U_0 \cup U_1 \cup \cdots$ and $V_0 \cup V_1 \cup \cdots$ separate F_0 and F_1. Assume that we have reached step n. If u_n is isolated (i.e., it is on

T_0 or a successor level) then we can let $U_{n+1} = U_n \cup \{u_n\}$. If $o(u_n)$ is limit, then we choose a closed and open neighborhood of u_n of the form $[y, u_n]$ which is disjoint from $F_1 \cup V_n$. This is possible, as the latter is a closed set excluding u_n, we only have to choose $y \prec u_n$ with a large enough successor $o(y)$. Then we set $U_{n+1} = U_n \cup [y, u_n]$. Argue similarly for V_{n+1} by selecting it to be disjoint from $F_0 \cup U_{n+1}$.

We now consider the general case. Set $a \in A$ if there are $x \in F_0$, $y \in F_1$, such that $a \prec x, y$. If A is uncountable, then by Problem 28 there is some $d \in T$ such that A is dense over d. As $\langle T, \prec\rangle$ is normal, $T_{\geq d}$ is uncountable, and so is a Suslin tree. In $T_{\geq d}$ both D_0 and D_1 are dense, open (in the sense that is defined for trees in the introduction to this chapter), where $a \in D_i$ if and only if there is $d \prec x \prec a$, $x \in F_i$. There is, by Problem 26 some α_0 such that the α_0th level of $T_{\geq d}$ is in $D_0 \cap D_1$, that is, if $x \in T_{\alpha_0}$, $x \succ d$, then there are $y_0 \in F_0$, $z_0 \in F_1$ such that $d \prec y_0, z_0 \prec x$. Repeating this argument we get ordinals $\alpha_0 < \alpha_1 < \cdots$ such that if $x \succ d$, $x \in T_{\alpha_i}$, then there exist y_i, z_i with y_i, $z_i \prec x$, $y_i \in F_0$, $z_i \in F_1$, $\alpha_{i-1} < o(y_i), o(z_i)$. If now α is the limit of the sequence $\alpha_0, \alpha_1, \ldots$, $x \in T_\alpha$, $x \succ d$, then x is an element of F_0, as well as of F_1, a contradiction.

We proved, therefore, that A is countable, so there is some $\alpha < \omega_1$ such that $A \subseteq T_{<\alpha}$. Our space splits into the disjoint union of the closed and open sets $T_{\leq\alpha}$, $T_{>\alpha}$. It suffices to separate F_0 and F_1 in these components, separately. In the former we can separate F_0 and F_1 by the argument at the beginning of the proof. The closed and open set $T_{>\alpha}$ splits into the disjoint union of the closed and open sets of the form $T_{>x}$ ($x \in T_\alpha$), and, as $A \subseteq T_{<\alpha}$, none of them contains points from both F_0 and F_1. In this situation it is easy to separate $F_0 \cap T_{>\alpha}$ and $F_1 \cap T_{>\alpha}$; include $F_0 \cap T_{>\alpha}$ into the union of those sets $T_{>x}$ which intersect it, and similarly for F_1.

32. (a) Let σ be a putative winning strategy for I. By closure (see Problem 20.7), there is a limit ordinal $\alpha < \omega_1$ with the property, that if $t_0, \ldots, t_{2n-1}$ are in $T_{<\alpha}$ then so is t_{2n}, I's response according to σ. Let $\alpha_0 < \alpha_1 < \cdots$ converge to α. Enumerate T_α as $T_\alpha = \{p_0, p_1, \ldots\}$. We make II play as follows. Given t_{2n}, she chooses a_n, an immediate successor of t_{2n} such that $a_n \not\prec p_n$. Then let her response be some $t_{2n+1} \succ a_n$, with $\alpha_n < o(t_{2n+1}) < \alpha$. This is possible, as $\langle T, \prec\rangle$ is normal. This way, a play $t_0, t_1, \ldots$ is determined, with $o(t_n)$ converging to α, but no element of T_α can extend the sequence, so I loses, although he played according to his winning strategy σ. This contradiction shows that σ does not exist.

(b) If $\langle T, \prec\rangle$ is special, then $T = A_0 \cup A_1 \cup \cdots$ with A_n an antichain. II can have the following strategy. Given t_{2n}, if there is some $t \succ t_{2n}$ with $t \in A_n$, then let t_{2n+1} be such an element, otherwise let $t_{2n+1} \succ t_{2n}$ be arbitrary. This way, if $t_0 \prec t_1 \prec \cdots \prec t$, then t can be in no A_n, so such a t cannot exist.

(c) Assume to the contrary that σ is a winning strategy for II. We exhibit a play in which II responds by σ yet she loses. For every $a_0 \in T$, if a_0 is the opening move by I, II answers by $\sigma(a_0) \succ a_0$. Set $t \in D_0$ if there is some a_0

with $\sigma(a_0) \prec t$. D_0 is obviously open, and it is dense, as for every $a_0 \in T$ there is an element of D_0 above a_0, namely any element of $T_{>\sigma(a_0)}$. By Problem 26, there is some $\alpha_0 < \omega_1$ such that $T_{\alpha_0} \subseteq D_0$.

We notice that for every $x \in T_{\alpha_0}$ there is a partial play $P_0(x)$ (consisting of one round of moves), $a_0^x, \sigma(a_0^x)$, which can be continued by I by saying x or any element of $T_{\geq x}$.

We repeat the above argument for every $x \in T_{\alpha_0}$ separately on $T_{\geq x}$ by using the second round of σ, continuing the play $P_0(x)$. We get an ordinal $\alpha_1 > \alpha_0$, for every $x \in T_{\alpha_1}$ a partial (2-round) play $P_1(x)$, which, on the one hand, continues the appropriate $P_0(y)$ (where y is the predecessor of x on level α_0), on the other hand, it can be continued to any element of $T_{\geq x}$.

Continuing this way we get $\alpha_0 < \alpha_1 < \cdots$. Let α be the limit of these ordinals. If $x \in T_\alpha$, then, letting t_i be the predecessor of x on level α_n, we get that the union of the partial plays $P_i(t_i)$ is a play in which II plays according to σ, and the element of the tree played by I and II remain below x, therefore II loses, a contradiction.

33. We first consider the case when λ is regular. We claim that we can assume property (C) of normality. Indeed, using the argument in the solution of Problem 24 given a κ-tree $\langle T, \prec\rangle$ is as in the problem, we can consider $\langle U, \prec\rangle$ where for $\alpha = 0$ or successor $b \in U_\alpha$ if and only if b is an $\alpha + 1$-branch of $\langle T, \prec\rangle$, for α limit $b \in U_\alpha$ if and only if b is an α-branch of $\langle T, \prec\rangle$ that extends to T_α. Set $b \prec b'$ if b' extends b. Then $\langle U, \prec\rangle$ is a κ-tree, $1 \leq |U_\alpha| \leq |T_\alpha| < \lambda$, and if $B = \{b_\alpha : \alpha < \kappa\}$ is a κ-branch of $\langle U, \prec\rangle$, then $\bigcup B$ is a κ-branch of $\langle T, \prec\rangle$.

We therefore assume that $\langle T, \prec\rangle$ of the problem satisfies (C) of normality. Set $S = \{\alpha < \kappa : \mathrm{cf}\,(\alpha) = \lambda\}$, a stationary set by Problem 21.8. For $\alpha \in S$, $x \neq y \in T_\alpha$, there is some $\beta < \alpha$ such that $T_{<x}$ and $T_{<y}$ differ from level β. As $|T_\alpha| < \lambda = \mathrm{cf}\,(\alpha)$, there is some $f(\alpha) < \alpha$ such that the elements of T_α have distinct predecessors in $T_{f(\alpha)}$. As f is a regressive function on a stationary set, we can apply Fodor's lemma (Problem 21.9) and get a stationary $S' \subseteq S$ and some $\gamma < \kappa$ such that $f(\alpha) = \gamma$ holds for $\alpha \in S'$. Pick $x_\alpha \in T_\alpha$ for $\alpha \in S'$. Let y_α be the predecessor of x_α on level γ. As κ is regular and $|T_\gamma| < \kappa$, $y_\alpha = y$ holds for $\alpha \in S''$, $|S''| = \kappa$. We claim that $Z = \{x_\alpha : \alpha \in S''\}$ is totally ordered. Indeed, if $\alpha < \beta$ are in S'' and z is x_β's predecessor on level α, then $y \prec x_\alpha, z$ so $x_\alpha = z$, i.e., $x_\alpha \prec x_\beta$. Finally, $B = \bigcup\{T_{\leq x} : x \in Z\}$ is a κ-branch.

If λ is singular and $\langle T, \prec\rangle$ is as in the problem, then for every $\alpha < \kappa$ there is some regular cardinal $\mu_\alpha < \lambda$ such that $|T_\alpha| < \mu_\alpha$ holds. As κ is regular and there are at most λ regular cardinals below λ, there is a set $Z \subseteq \kappa$, $|Z| = \kappa$ such that $\mu_\alpha = \mu$ holds for $\alpha \in Z$. We can now apply the already covered case for the tree on $\bigcup\{T_\alpha : \alpha \in Z\}$ and get a κ-branch.

If κ is singular, let $\mu = \mathrm{cf}\,(\kappa)$, $\kappa = \sup\{\kappa_\xi : \xi < \mu\}$. Let $\langle T, \prec\rangle$ be the disjoint union of the branches b_ξ with b_ξ of height κ_ξ. Then $\langle T, \prec\rangle$ has no

κ-branch and $|T_\alpha| < \mu^+ < \kappa$ holds for every $\alpha < \kappa$. [DJ. Kurepa: Ensembles ordonnés et ramifiés, *Publ. Math. Univ. Belgrade*, **4**(1935), 1–138]

34. Let $\langle T, \prec\rangle$ be a κ-Aronszajn tree. Set $x \in T'$ if and only if $T_{>x}$ contains elements of arbitrarily large height $< \kappa$. Notice that if $y \prec x \in T'$, then $y \in T'$ so for $x \in T'$, $T'_{<x} = T_{<x}$ holds and so the height of x in T' is the same as the height of x in T.

We claim that T' has elements of arbitrarily large height $< \kappa$. Indeed, if no $x \in T_\alpha$ is in T' for some $\alpha < \kappa$, then for every $x \in T_\alpha$ there is a $\beta(x) < \kappa$ such that for no $y \succ x$ does $o(y) \geq \beta(x)$ hold. As $|T_\alpha| < \kappa$, the set $\{\beta(x) : x \in T_\alpha\}$ has a bound $\beta < \kappa$, but then T_β can have no element.

Assume that $x \in T'$, $o(x) < \alpha < \kappa$. By the definition of T', there are κ elements $y \in T$, with $x \prec y$. Some κ of them has $o(y) > \alpha$. Then, let $p_\alpha(y) \in T_\alpha$ be y's predecessor at level α. For κ many y, $p_\alpha(y) = p$ holds for the same $p \in T_\alpha$ and so $p \in T'$. We proved therefore property (A) of normality for T' and so from now we will assume (A) for T.

Now assume that $\langle T, \prec\rangle$ is a κ-Aronszajn tree satisfying (A). If $x \in T$, then the set $T_{>x}$ must contain incomparable elements as otherwise it would be a branch of cardinality κ. If $x \prec y, z$ and y, z are incomparable then there are $y' \succ y$, $z' \succ z$ with $o(y') = o(z')$ and of course, y', z' are also incomparable. We get, therefore, that if $x \in T_\alpha$ then some $T_{\beta(x)}$ $(\beta(x) > \alpha)$ contains incomparable successors of x. As κ is regular and $|T_\alpha| < \kappa$, some $\beta(\alpha) > \alpha$ applies for all $x \in T_\alpha$.

We can then choose, by transfinite recursion, the increasing sequence $\{\alpha_\xi : \xi < \kappa\}$ such that every $x \in T_{\alpha_\xi}$ has incomparable successors in $T_{\alpha_{\xi+1}}$, so $T' = \bigcup\{T_{\alpha_\xi} : \xi < \kappa\}$ satisfies (A)+(B) of the definition of normality.

Assume finally that $\langle T, \prec\rangle$ is a κ-Aronszajn tree satisfying (A)+(B) of the definition of normality. Define the tree T' as follows. If $\alpha = 0$ or a successor ordinal, then let $b \in T'$ if and only if b is an $\alpha+1$-branch of T. If α is limit then let $b \in T'$ if and only if b is an α-branch of T that has an extension on level T_α. Set $b \prec b'$ if and only if b' extends b. Now $|T'_\alpha| = |T_\alpha|$ if $\alpha = 0$ or a successor, and $1 \leq |T'_\alpha| \leq |T_\alpha|$ if α is a limit ordinal. Moreover, if $\{b_\alpha : \alpha < \kappa\}$ was a κ-branch in $\langle T', \prec\rangle$ then $\bigcup\{b_\alpha : \alpha < \kappa\}$ would be a κ-branch in $\langle T, \prec\rangle$. $\langle T', \prec\rangle$ is therefore a κ-Aronszajn tree and it is easy to see that it satisfies the definition of normality.

35. One direction is obvious; if b is a κ-branch, then b is a subset of order type κ in $\langle T, <_{\text{lex}}\rangle$.

For the other direction assume that $\{x_\xi : \xi < \kappa\}$ is a subset of $\langle T, <_{\text{lex}}\rangle$ of order type either κ or κ^*. As κ is regular and every level of $\langle T, \prec\rangle$ has cardinality less than κ, we have that $o(x_\xi) \to \kappa$ Therefore, for any given $\alpha < \kappa$, $p_\alpha(x_\xi)$, the predecessor of x_ξ on level α, is defined for all large ξ.

As by the definition of $<_{\text{lex}}$ the sequence $p_0(x_\xi)$ $(\xi < \kappa)$ is weakly increasing (or decreasing) in the ordered set $\langle T_0, <_0\rangle$ of cardinality $< \kappa$, we have that $p_0(x_\xi) = a_0$ for some a_0 and for $\xi \geq \gamma_0$ with an appropriate $\gamma_0 < \kappa$. Repeating the argument for level one, but only using the above "tail" of the sequence, we get that $p_1(x_\xi) = a_1$ holds for $\xi \geq \gamma_1$, etc. By recursion we get the elements $\{a_\alpha : \alpha < \kappa\}$ and increasing ordinals $\{\gamma_\alpha : \alpha < \kappa\}$ such that $p_\alpha(x_\xi) = a_\alpha$ for $\xi \geq \gamma_\alpha$. But then $\{a_\alpha : \alpha < \kappa\}$ is a κ-branch.

36. Let S be the set of finite sequences of elements of κ. Clearly, $|S| = \kappa$. For $\alpha < \kappa^+$ we are going to construct an injection $f_\alpha : \alpha \to S$ by transfinite recursion on α. f_0 can only be the empty function. If f_α is given, set $f_{\alpha+1}(\beta) = 0$ for $\beta = \alpha$ and $f_{\alpha+1}(\beta) = 1f_\alpha(\beta)$, that is, if $f_\alpha(\beta) = \gamma_1 \cdots \gamma_n$ then we let $f_{\alpha+1}(\beta) = 1\gamma_1 \cdots \gamma_n$ (concatenation). If α is a limit ordinal, enumerate C_α increasingly as $\{x(\alpha, \xi) : \xi < \epsilon_\alpha\}$. We may assume that $x(\alpha, 0) = 0$. If $\beta < \alpha$, let $\gamma = x(\alpha, \xi)$ be the least element of C_α greater than β. Set $f_\alpha(\beta) = \xi f_\gamma(\beta)$, where again, the right-hand side denotes the string starting with ξ and then continuing with the sequence $f_\gamma(\beta)$.

We claim that f_α is an injection of α into S. This we prove by induction on α. The induction step is obvious, if α is zero or a successor ordinal. Assume that α is limit and $f_\alpha(\beta) = f_\alpha(\beta')$. Then $\xi f_\gamma(\beta) = \xi' f_{\gamma'}(\beta')$ with the ordinals ξ', γ' corresponding to β'. As the two strings are equal, we must have $\xi = \xi'$ but then $\gamma = \gamma'$. We then get $f_\gamma(\beta) = f_\gamma(\beta')$ and so $\beta = \beta'$ by the inductive hypothesis.

We define the κ^+-Aronszajn tree $\langle T, \prec \rangle$ as follows. The nodes on T_β are the functions $f_\alpha|_\beta$ for $\beta \leq \alpha < \kappa^+$. $t \prec t'$ if and only if t' extends t. It is obvious that T has no κ^+-branches as its elements are injective functions so a κ^+-branch would give rise to an injection of κ^+ into S, a set of cardinality κ.

We show that $|T_\beta| \leq \kappa$ holds for $\beta < \kappa^+$. Assume we are given $f_\alpha|_\beta \in T_\beta$. We prove that there are finitely many ordinals $0 = \gamma_0 < \cdots < \gamma_t = \beta$ and corresponding strings $s_1, \ldots, s_t$ such that if $\gamma_{i-1} \leq \delta < \gamma_i$ then $f_\alpha(\delta) = s_i f_{\gamma_i}(\delta)$ holds. This suffices for our claim as there are at most κ ways of selecting $\gamma_0, \ldots \gamma_t, s_1, \ldots, s_t$.

We prove this claim by induction on α. It is obvious for $\alpha = \beta$ and the inductive step from α to $\alpha + 1$ is equally clear.

Assume finally that α is limit. There exist successive elements $\gamma_0 < \gamma_1 < \cdots \gamma_n$ of C_α such that $\gamma_{n-1} \leq \beta < \gamma_n$ and $\gamma_0 = x(\alpha, \delta)$ with either $\delta = 0$ or δ a limit ordinal. Inspection of the definition of f_α shows that on the intervals $[0, \gamma_0), [\gamma_0, \gamma_1), \ldots, [\gamma_{n-1}, \beta)$ f_α equals to $f_{\gamma_0}, (\delta+1)f_{\gamma_1}, (\delta+2)f_{\gamma_2}, \ldots, (\delta+n)f_{\gamma_n}$, respectively. (Here we use that $C_{\gamma_0} = C_\alpha \cap \gamma_0$.) Each of these terms gives restriction of the required type except the last one. In that case, however, as $\gamma_n < \alpha$, we can refer to the inductive hypothesis, and argue again, that f_{γ_n} restricted to $[\gamma_{n-1}, \beta)$ splits into finitely many functions of the required type. [This proof is due to S. Todorcevic.]

37. We slightly modify the solution of the previous problem. Fix a system $\{C_\alpha : \alpha < \kappa^+\}$ where, for every limit ordinal $\alpha < \kappa^+$, C_α is a closed, unbounded subset of α, of order type cf (α) (which is always $\leq \kappa$). We assume that $0 \in C_\alpha$,

and $C_\alpha = \{x(\alpha,\xi) : \xi < \epsilon_\alpha\}$ is its increasing enumeration. Let S be the set of finite sequences of elements of κ, clearly, $|S| = \kappa$. For $\alpha < \kappa^+$ we are going to construct an injection $f_\alpha : \alpha \to S$, by transfinite recursion on α. f_0 is the empty function. If f_α is given, set $f_{\alpha+1}(\beta) = 0$ for $\beta = \alpha$ and $f_{\alpha+1}(\beta) = 1f_\alpha(\beta)$, (concatenation). If α is limit, $\beta < \alpha$, then let $\gamma = x(\alpha,\xi)$ be the least element of C_α, greater than β and set $f_\alpha(\beta) = \xi f_\gamma(\beta)$, again, concatenating ξ and the finite string $f_\gamma(\beta)$.

Just as in the preceding problem, f_α is an injection of α into S.

We again define the κ^+-Aronszajn tree $\langle T, \prec\rangle$ as follows. The nodes on T_β are the functions $f_\alpha|_\beta$ for $\beta \le \alpha < \kappa^+$, and $t \prec t'$ if and only if t' extends t. As in the preceding problem, T has no κ^+-branches.

In order to show that $|T_\beta| \le \kappa$ holds for every $\beta < \kappa^+$ we claim that if $\beta \le \alpha < \kappa^+$ then $f_\alpha|_\beta$ has the following specific form. β, that is, $[0,\beta)$ splits into the disjoint union of fewer than κ disjoint intervals of the form $I = [\gamma,\delta)$ and on each of them, f_α restricts to a function of the form $x \mapsto sf_\delta(x)$ with some $s \in S$. This proves that $|T_\beta| \le \kappa$, as by the hypothesis on cardinal exponentiation, there are at most κ functions of the required type. We prove the above statement by transfinite induction on α. It is obvious if $\alpha = \beta$, and the inductive step from α to $\alpha+1$ is equally clear. Assume now that $\alpha > \beta$ is a limit ordinal. Then C_α splits α into $\mathrm{cf}(\alpha) \le \kappa$ many intervals of the form $[\gamma,\delta)$, only $< \kappa$ of those having nonempty intersection with β. On only one of them it is not clear that $f_\alpha|_\beta$ is of the required form: the one for which $\gamma \le \beta < \delta$ holds. In this interval, f_α's restriction is equal to (the restriction of) ξf_δ where $\xi < \kappa$ is the index of δ in the increasing enumeration of C_α. But now we can refer to the inductive hypothesis which says that $[\gamma,\beta)$ splits into $< \kappa$ intervals, with each of them f_δ restricting to some function of the required form, and so the statement holds for α. [E. Specker: Sur un problème de Sikorski, *Coll. Math.*, **2**(1949), 9–12. The present proof is due to S. Todorcevic.]

38. Notice that κ is regular and $\kappa > \omega_1$ by Problems 28.3 and 28.5. Let $\langle T, \prec\rangle$ be a κ-tree. Let μ be a κ-additive measure on T. By additivity, $\mu(T_{\le\alpha}) = 0$ for every $\alpha < \kappa$. For $t \in T$ set $f(t) = \mu(T_{\ge t})$. For every $\alpha < \kappa$, the set $U_\alpha = \{t \in T_\alpha : f(t) > 0\}$ is countable and nonempty, and if $s \prec t \in U_\alpha$, $s \in T_\beta$, then $s \in U_\beta$. If $U = \bigcup\{U_\alpha : \alpha < \kappa\}$, then $\langle U, \prec\rangle$ is a tree of height κ, with countable levels, so by Problem 33 it has a κ-branch which is a κ-branch of $\langle T, \prec\rangle$, as well. [J. Silver: Some applications of model theory in set theory, *Ann. Math. Logic*, **3**(1970), 45–110]

39. We are going to use the condition in the form that on every set of cardinality κ^+ there is an ultrafilter consisting of sets of cardinality κ^+ such that if we decompose the ground set into λ parts then exactly one of them is in the ultrafilter.

Let T be a κ^+-tree. Set $\mu = \mathrm{cf}(\kappa)$ and choose an increasing continuous sequence of cardinals $\langle \kappa_\xi : \xi < \mu\rangle$ with $\kappa_0 = 0$ and $\kappa = \sup\{\kappa_\xi : \xi < \mu\}$.

Enumerate each T_α as $T_\alpha = \{t^\alpha_\xi : \xi < \kappa\}$. As $|T| = \kappa^+$, there is an ultrafilter D_μ on T, as described above. For every $\alpha < \kappa^+$,

$$T_{>\alpha} = \bigcup\{B^\alpha_\beta : \beta < \mu\},$$

where

$$B^\alpha_\beta = \bigcup\{T_{>x} : x = t^\alpha_\xi, \kappa_\beta \le \xi < \kappa_{\beta+1}\}.$$

As $|T_{\le\alpha}| \le \kappa$, $T_{>\alpha} \in D_\mu$. By the property prescribed for D_μ, for every $\alpha < \kappa^+$ there is a unique $\beta(\alpha) < \mu$ with $B^\alpha_{\beta(\alpha)} \in D_\mu$. For a set $Z \subseteq \kappa^+$ of cardinality κ^+ we have $\beta(\alpha) = \beta$ for $\alpha \in Z$ with some common $\beta < \mu$.

We notice that

(∗) if $\alpha < \alpha'$ are in Z then there are $x = t^\alpha_\xi \in T_\alpha$, $x' = t^{\alpha'}_{\xi'} \in T_{\alpha'}$ with $\kappa_\beta \le \xi, \xi' < \kappa_{\beta+1}$, such that $x \prec x'$.

Indeed, if $z \in B^\alpha_\beta \cap B^{\alpha'}_\beta \in D_\mu$, then there are $x \in T_\alpha$, $x' \in T_{\alpha'}$, $x \prec x' \prec z$.

Now consider an ultrafilter $D_{\kappa_{\beta+1}}$ on the set Z (of cardinality κ^+) and put the elements of Z into the (not necessarily disjoint) classes $Z(\xi,\eta)$ for $\kappa_\beta \le \xi, \eta < \kappa_{\beta+1}$ by putting $\alpha \in Z(\xi,\eta)$ provided $\{\gamma : t^\alpha_\xi \prec t^\gamma_\eta\} \in D_{\kappa_{\beta+1}}$. Note that every α belongs to some $Z(\xi,\eta)$; therefore, there is a class $Z(\xi,\eta) \in D_{\kappa_{\beta+1}}$. We claim that $\{t^\alpha_\xi : \alpha \in Z(\xi,\eta)\}$ generates a κ^+-branch. Indeed, if $\alpha, \alpha' \in Z(\xi,\eta)$, $\alpha < \alpha'$ then there is some $\gamma > \alpha'$ such that $t^\alpha_\xi, t^{\alpha'}_\xi \prec t^\gamma_\eta$, so $t^\alpha_\xi \prec t^{\alpha'}_\xi$. [M. Magidor, S. Shelah: The tree property at successors of singular cardinals, *Arch. for Math. Logic*, **35**(1996), 385–404]

40. Let $\langle A, <\rangle$ be an ordered set of cardinality κ and let $\prec$ be a well-ordering of A in type κ. Color a pair $\{a,b\} \in [A]^2$, $a < b$, green if $a \prec b$, and let it be blue if $b \prec a$. As $\kappa \to (\kappa)^2_2$, there is a monochromatic $B \subseteq A$ of size κ. If the color of B is green, then on B the orders $<$ and $\prec$ are the same, if the color is blue, the order $<$ on B agrees with $\prec^*$, the reverse of the well order. In the former case $\langle B, <\rangle$ is of order type κ, in the latter case it is of order type κ^*.

41. Assume to the contrary first that κ is singular with $\mu = \mathrm{cf}(\kappa) < \kappa$ and $\kappa = \sup\{\kappa_\alpha : \alpha < \mu\}$ where $\kappa_\alpha < \kappa$. Let $\langle A, \prec\rangle$ be the ordered union with respect to $\alpha < \mu$ of the disjoint ordered sets $\langle A_\alpha, <_\alpha\rangle$ of order type κ^*_α. That is, $x \prec y$ holds if $x \in A_\alpha$, $y \in A_\beta$ with $\alpha < \beta$. If $B \subseteq A$ is well ordered, then for each α the intersection $B \cap A_\alpha$ is finite (otherwise $B \cap A_\alpha$ would include an infinite decreasing sequence), so $|B| \le \mu$. If, however, B is reversely well ordered, then it can meet only finitely many A_α (otherwise the set of elements in different A_α would include an infinite increasing sequence) hence its cardinality is again smaller than κ. This contradicts the hypothesis, hence κ must be regular.

Now let $\lambda < \kappa \le 2^\lambda$. By Problem 6.93 the lexicographically ordered set ${}^\lambda\{0,1\}$ (of size 2^λ) does not include increasing or decreasing sequences of length $\lambda^+ \leq \kappa$. Therefore, the assumption in the problem implies $2^\lambda < \kappa$, and so κ is strongly inaccessible.

42. Assume κ is singular, $\mathrm{cf}(\kappa) = \mu$, $\kappa = \sup\{\kappa_\alpha : \alpha < \mu\}$ with $\kappa_\alpha < \kappa$ for $\alpha < \mu$. Let the tree $\langle T, \prec\rangle$ be the disjoint union of the branches b_α with height κ_α. Formally, $b_\alpha = \kappa_\alpha \times \{\alpha\}$ and $\langle \xi, \alpha\rangle \prec \langle \xi', \alpha'\rangle$ if and only if $\alpha = \alpha'$ and $\xi < \xi'$. Then $\langle T, \prec\rangle$ is of height κ, every level is of size $\mu < \kappa$, and there is no κ-branch. Hence, if κ has the tree property, then it is regular.

43. Let κ be the smallest strong limit regular cardinal bigger than ω, and let $C = \{c_\alpha : \alpha < \kappa\}$ be a closed, unbounded set in κ consisting of infinite cardinals plus $c_0 = 0$, satisfying $2^{c_\alpha} < c_{\alpha+1}$ for $\alpha < \kappa$. We claim that for every $\xi < \kappa$ there is a function f_ξ defined on $C \cap \xi$ which is regressive and assumes every value finitely many times. We prove this by induction on ξ. It suffices to prove the result for ordinals of the form c_α, $c_\alpha + 1$. This latter case is easy: if f_{c_α} is given, we simply extend it to $f_{c_\alpha+1}$ by associating an arbitrary image for c_α. The same argument works for $f_{c_{\alpha+1}}$. We are done unless $\xi = c_\alpha$ where α is a limit ordinal. By our conditions on κ and on C, ξ must be a singular ordinal, let $\mu < \xi$ be its cofinality. Let $D = \{d_\beta : \beta < \mu\} \subseteq C$ be a closed, unbounded subset in ξ, $d_0 = 0 < \mu < d_1$. We define f_ξ as follows. If $x \in C \cap \xi$ and $x \notin D$, then there is a unique $\beta < \mu$ such that $d_\beta < x < d_{\beta+1}$. Now set $f_\xi(x) = d_\beta + f_{d_{\beta+1}}(x)$ (ordinal addition). Notice that $f_\xi(x) < x$ as x is an infinite cardinal by condition and so the sum of two smaller ordinals is a smaller ordinal. Finally, let f_ξ be an injection between $D \setminus \{0\}$ and the interval $(0, d_1)$. It is clear that f_ξ is as required.

Let T be the set of all functions defined on some $C \cap \lambda$, $\lambda \in C$, which are regressive and take every value finitely many times. Put $f \prec g$ if g is an extension of f. This way we get a tree, the αth level of which is formed by all functions in question with domain $C \cap \lambda_\alpha$. Thus, these levels are not empty for all $\alpha < \kappa$, so the tree is of height κ. The number of functions $f : \lambda_\alpha \to \lambda_\alpha$ is at most $\lambda_\alpha^{\lambda_\alpha} = 2^{\lambda_\alpha} < \kappa$, so every level is of cardinality smaller than κ. Finally, in this tree there is no branch of length κ, for if there was such a branch then its union would be a regressive function on C that takes all values finitely many times, which is impossible by Problem 21.9.

44. **(b)**$\Rightarrow$**(a)** is trivial, **(a)**$\Rightarrow$**(d)** holds by Problem 40, **(d)**$\Rightarrow$**(c)** by Problems 41 and 35. All that remains to show is that **(c)** implies **(b)**.

Assume **(c)**. κ is a strongly inaccessible cardinal; therefore, the product of fewer than κ cardinals, each smaller than κ, is less than κ. For $n = 1$ the statement of **(b)** holds by the regularity of κ. We prove it for larger n by mathematical induction. Suppose that we have it for n and consider a coloring $f : [\kappa]^{n+1} \to \sigma$ of the $(n+1)$-tuples. We construct an endhomogeneous set $A \subseteq \kappa$ of cardinality κ, i.e., if $x_1 < \cdots < x_n < y < y'$ are $n+2$ elements of A, then $f(x_1, \ldots, x_n, y) = f(x_1, \ldots, x_n, y')$. This suffices, as then we can define a coloring of the n-tuples by assigning $g(x_1, \ldots, x_n) = f(x_1, \ldots, x_n, y)$ where

$y > x_n$ is an arbitrary element of A. By the inductive hypothesis, there is a κ-sized set $B \subseteq A$ that is homogeneous for g, and so it is homogeneous for f, as well.

The construction of A will be done via a construction of a κ-tree T (a ramification tree). Every node of T will be an element of κ and to each $x \in T$ we associate a set $H_x \subseteq \kappa$. It will have the property that if $x \neq y$ are elements of the same level then H_x and H_y are disjoint, in fact, the sets $\{H_x : x \in T_\alpha\}$ constitute a partition of $\kappa \setminus T_{\leq\alpha}$. To start, let 0 be the sole element of T_0. Accordingly, $H_0 = \kappa \setminus \{0\}$.

Assume that the tree below level α is constructed and α is a limit ordinal. For every α-branch B of the tree $T_{<\alpha}$ if the set $K_B = \bigcap\{H_x : x \in B\}$ is nonempty, then we let $t(B) = \min(K_B)$ be the only successor of B on level α and $H_{t(B)} = K_B \setminus \{t(B)\}$. We notice that for $a \in T_{<\alpha}$, $a \prec t(B)$ if and only if $a \in B$. $|T_\alpha|$ is at most as large as the number of α-branches in $T_{<\alpha}$, which, a product of $< \kappa$ cardinals each smaller than κ, by the induction hypothesis is less than κ itself.

We notice that $\{H_x : x \in T_\alpha\}$ forms a partition of $\kappa \setminus T_{\leq\alpha}$.

Assume now that $\alpha = \beta + 1$ is a successor ordinal. Consider an element $x \in T_\beta$. Define an equivalence relation $\sim$ on H_x as follows. For $c, d \in H_x$, $c \sim d$ if and only if for any $a_1 < \cdots < a_n$ from $T_{\leq x}$ $f(a_1, \ldots, a_n, c) = f(a_1, \ldots, a_n, d)$ holds. This is clearly an equivalence relation on H_x, and the number of equivalence classes is at most

$$\sigma^{|[T_{\leq x}]^n|} \leq \sigma^{|T_{\leq x}|+\omega} \leq 2^{\sigma+|\alpha|+\omega} < \kappa.$$

For each nonempty equivalence class C let y_C be its least element. We make these elements y_C the immediate successors of x, and set $H_{y_C} = C \setminus \{y_C\}$ for them. Once again as some H_x were split to more classes plus some elements put into T we will have the required condition that $\{H_x : x \in T_\alpha\}$ partitions $\kappa \setminus T_{\leq\alpha}$. In particular, $T_\alpha \neq \emptyset$ for every $\alpha < \kappa$, so the height of the tree is κ.

Note also that if $x \prec y$ in the tree, and x^* is the element for which $x^* \preceq y$ and $o(x^*) = o(x)+1$, then either $y = x^*$ or $y \in H_{x^*}$. Hence if $a_1 < \cdots < a_n$ are elements of $T_{\leq x}$, then $f(a_1, \ldots, a_n, y) = f(a_1, \ldots, a_n, x^*)$ holds. In particular, any branch in $\langle T, \prec\rangle$ is endhomogeneous.

By the tree property of κ there is a κ-branch A in T, which, as we have just remarked, is endhomogeneous. This completes the proof of **(c)⇒(b)**.

28

The measure problem

1. Let X be an infinite set and $\mathcal{F}_0$ the set of those $F \subset X$ for which $X \setminus F$ is finite. This $\mathcal{F}_0$ is a filter, and it can be extended to an ultrafilter $\mathcal{F}$ (see Problem 14.6(c)). Now for $A \subset X$ set $\mu(A) = 1$ if and only if $A \in \mathcal{F}$. The properties of ultrafilters show that μ is a finitely additive nontrivial measure on X.

2. It is clear that μ is a κ-additive 0–1-valued measure on X if and only if the set of measure 0 sets is a κ-complete prime ideal. Furthermore, $\mathcal{I}$ is a κ-complete prime ideal if and only if $\mathcal{F} = \{X \setminus F \,:\, F \in \mathcal{I}\}$ is a κ-complete ultrafilter.

3. Suppose the contrary, and assume that μ is a real-valued measure on all subsets of ω_1. Let $\{U_{n,\alpha} \,:\, n < \omega,\ \alpha < \omega_1\}$ be an Ulam matrix (see Problem 18.1). For every $\alpha < \omega_1$, as $\bigcup\{U_{n,\alpha} : n < \omega\}$ has countable complement and every countable set is of measure 0, there are some $n = n(\alpha) < \omega$ and $k = k(\alpha) < \omega$ such that $\mu(U_{n,\alpha}) > 1/k$. There is a pair $\langle n, k\rangle$ that occurs as $\langle n(\alpha), k(\alpha)\rangle$ for uncountably many α, say for $\alpha \in S$, $|S| = \aleph_1$, which is absurd as then $U_{n,\alpha}$, $\alpha \in S$, would be disjoint sets each having measure greater than $1/k$ (there cannot be even k such sets).

4. It is enough to show that the union of some of the sets is nonmeasurable in $[0,1]$; therefore, instead of $\mathbf{R}$ consider $[0,1]$ and instead of the sets their intersection with $[0,1]$. Thus, if the sets are A_α, $\alpha < \omega_1$, then they are of measure 0 and $A_\alpha \subset [0,1]$. Now if all unions were measurable then for $Y \subset \omega_1$ we could set

$$\mu(Y) = m\left(\cup_{\alpha \in Y} A_\alpha\right)$$

with m standing for Lebesgue measure, and this way we would get a nontrivial $[0,1]$-valued σ-additive measure on ω_1, which is not possible by the preceding problem.

5. Let μ be a $[0,1]$-valued measure on κ. First of all, any set $Y \subset \kappa$ of cardinality $< \kappa$ is of measure 0 (by nontriviality and κ-additivity), hence κ must be regular, for otherwise it is the union of fewer than κ sets of cardinality smaller than κ, hence it would have measure 0. That κ cannot be a successor cardinal, say $\kappa = \lambda^+$, can be proven the same way as Problem 3 was solved, just use a $\lambda \times \lambda^+$-Ulam matrix instead of an $\omega \times \omega_1$ one.

6. Assume that μ is a $[0,1]$-valued measure on the subsets of $[0,1]$. For $0 \le x \le 1$ define $f(x) = \mu([0,x])$. This is a nondecreasing continuous (by $\mu(\{x\}) = 0$ and the σ-additivity of μ) function with $f(0) = 0$, $f(1) = 1$ (not necessarily strictly monotone). For $A \subseteq [0,1]$ set $\overline{\mu}(A) = \mu(f^{-1}[A])$. It is easy to verify that $\overline{\mu}$ is a κ-additive (at this moment possibly trivial) $[0,1]$-valued measure on $[0,1]$. We show that $\overline{\mu}([0,x]) = x$ for $0 \le x \le 1$. By σ-additivity then $\overline{\mu}(A)$ equals the Lebesgue measure of A for every Borel set A (see Problem 12.23 and recall that the intervals $[0,x]$, $x \in [0,1]$ generate the Borel sets of $[0,1]$), and by completeness it also follows that every set that is a subset of the set of a Borel set of measure 0 is also of measure 0. Hence $\overline{\mu}$ extends the Lebesgue measure (and in particular, it is a nontrivial measure, i.e., $\overline{\mu}(\{x\}) = 0$ for all $x \in [0,1]$).

To prove $\overline{\mu}([0,x]) = x$ notice that the set $\{y : 0 \le f(y) \le x\}$ is of the form $[0,u]$ with some u satisfying $f(u) = x$. Therefore, $\overline{\mu}([0,x]) = \mu([0,u]) = x$.

7. Let μ be a $[0,1]$-valued measure on $\kappa > \mathbf{c}$, and let $\mathcal{I}$ be the set of measure 0 subsets of κ. Then $\mathcal{I}$ is a κ-complete ideal (not necessarily a prime ideal), in particular every $A \notin \mathcal{I}$ is of cardinality κ. If there is an $A \subset \kappa$, $A \notin \mathcal{I}$ such that for all disjoint decomposition $A = A^0 \cup A^1$ one of the A^j belongs to $\mathcal{I}$, then $\mathcal{I}' = \{A \cap I \ : \ I \in \mathcal{I}\}$ is a κ-complete prime ideal on A, hence $\kappa = |A|$ is measurable.

Therefore, if we assume to the contrary that κ is not a measurable cardinal, then for all $A \subset \kappa$ there is a disjoint decomposition $A = A^0 \cup A^1$ such that if $A \notin \mathcal{I}$ then $A^0, A^1 \notin \mathcal{I}$. By transfinite induction on $\alpha < \omega_1$ for every function $f : \alpha \to \{0,1\}$ we define a set A_f in the following way. Set $A_\emptyset = \kappa$. Suppose $\alpha < \omega_1$, and that A_g have already been defined for all $g : \beta \to \{0,1\}$, $\beta < \alpha$. Let $f : \alpha \to \{0,1\}$. If α is a limit ordinal, then set $A_f = \cap_{\beta<\alpha} A_{f|\beta}$. On the other hand, if $\alpha = \beta + 1$, then set $A_f = (A_{f|\beta})^{f(\beta)}$, and this completes the definition of the sets A_f. Extend this definition to $f : \omega_1 \to \{0,1\}$ by setting $A_f = \cap_{\alpha<\omega_1} A_{f|\alpha}$. It is clear that if $\beta < \alpha < \omega_1$ and $f : \alpha \to \{0,1\}$ then $A_f \subseteq A_{f|\beta}$, and if $f \ne g$ are both mapping α into $\{0,1\}$, then $A_f \cap A_g = \emptyset$. Transfinite induction on α gives that for each $\alpha < \omega_1$

$$\bigcup_{f \in {}^\alpha\{0,1\}} A_f = \kappa, \tag{28.1}$$

and then actually this is also true for $\alpha = \omega_1$ as well. Note also that the union on the left is a disjoint union.

Now let

$$B = \bigcup \{A_f \,:\, f \in {}^{\alpha}\{0,1\},\ \alpha < \omega_1,\ A_f \in \mathcal{I}\}.$$

There are at most $\mathbf{c}$ many terms in the union on the right, hence by the κ-completeness of $\mathcal{I}$, we have $B \in \mathcal{I}$ (recall that $\kappa > \mathbf{c}$). Let $\gamma \notin B$. On applying (28.1) for $\alpha = \omega_1$ we can see that there is an $f : \omega_1 \to \{0,1\}$ such that $\gamma \in A_f$. But then $\gamma \in A_{f|_\alpha}$ for all $\alpha < \omega_1$, and hence, by the definition of B, $A_{f|_\alpha} \notin \mathcal{I}$ for all $\alpha < \omega_1$. Therefore, the sets $A_{f|_\alpha} \setminus A_{f|_{\alpha+1}} = (A_{f|_\alpha})^{1-f(\alpha)}$ do not belong to $\mathcal{I}$, hence we get the ω_1 disjoint sets

$$A_{f|_\alpha} \setminus A_{f|_{\alpha+1}}, \qquad \alpha < \omega_1,$$

of positive measure, which is absurd. This contradiction proves that κ is measurable.

8. We know from Problem 3 that $\kappa > \aleph_1$ (actually, Problem 5 shows that $\kappa > \aleph_\omega$). Let $\mu : \mathcal{P}(\kappa) \to [0,1]$ be a σ-additive measure on κ. We claim that it is κ-additive. If this is not the case, then there is a $\omega < \lambda < \kappa$ and disjoint sets A_α, $\alpha < \lambda$ such that for $A = \cup_{\alpha<\lambda} A_\alpha$ we have $\mu(A) \neq \sum_{\alpha<\lambda} \mu(A_\alpha)$. Since the sum on the right is the same as the supremum of its finite partial sums, the σ-additivity of μ gives that there are only countably many A_α's with $\mu(A_\alpha) > 0$, and necessarily $\mu(A) > \sum_{\alpha<\lambda} \mu(A_\alpha)$. Using again the σ-additivity we may exclude all A_α with $\mu(A_\alpha) > 0$, i.e., we may assume that $\mu(A_\alpha) = 0$ for all α but $A = \cup_{\alpha<\lambda} A_\alpha$ is of positive measure. For $B \subset \lambda$ define

$$\nu(B) = \frac{1}{\mu(A)} \mu\left(\cup_{\alpha \in B} A_\alpha)\right).$$

Because of the disjointness of the A_α's this is a σ-additive measure on λ with the property that $\nu(\lambda) = \mu(A)/\mu(A) = 1$ and $\nu(\{\alpha\}) = \mu(A_\alpha)/\mu(A) = 0$ for all $\alpha < \lambda$, i.e., ν is a nontrivial σ-additive measure on $\lambda < \kappa$. But this contradicts the minimality of κ, and this contradiction proves that κ is real measurable.

9. The solution to Problem 8 can be repeated word for word.

10. Instead of $\mathbf{R}$ we work with ${}^{\omega}\{0,1\}$, and suppose there is a σ-additive 0–1-valued measure on all subsets. Fcr each $n < \omega$ one of the sets $A_n^0 = \{f : f(n) = 0\}$ or $A_n^1 = \{f : f(n) = 1\}$ is of measure 1, and the one with this property is denoted by $A_n^{g(n)}$. Then $g \in {}^{\omega}\{0,1\}$ and $\{g\}$ is the intersection of the measure 1 sets $A_n^{g(n)}$, $n = 0, 1, \ldots$, hence it must have measure 1 (the complement is the union of the countably many ${}^{\omega}\{0,1\} \setminus A_n^{g(n)}$ sets of measure 0). But then μ is trivial, and this proves that there is no nontrivial σ-additive 0–1-valued measure on $\mathbf{R}$.

11. Regularity follows from Problem 5. Now let $\lambda < \kappa$, and suppose that $2^\lambda \geq \kappa$. Let ν be a κ-additive 0–1-valued measure on κ. Extend it to a κ-additive 0–1-valued measure μ on 2^λ by stipulating $\mu(A) = \nu(A \cap \kappa)$. Now we can repeat the proof of the preceding problem on ${}^\lambda\{0,1\}$. For each $\alpha < \lambda$ one of the sets $A_\alpha^0 = \{f : f(\alpha) = 0\}$ or $A_\alpha^1 = \{f : f(\alpha) = 1\}$ is of measure 1, and the one with this property is denoted by $A_\alpha^{g(\alpha)}$. Then $g \in {}^\lambda\{0,1\}$ and $\{g\}$ is the intersection of the measure 1 sets $A_\alpha^{g(\alpha)}$, $\alpha < \lambda$, hence it must have measure 1 (the complement is the union of the $\lambda < \kappa$ many ${}^\lambda\{0,1\} \setminus A_\alpha^{g(\alpha)}$ sets of measure 0). But then μ is trivial, and so ν must be trivial, and this proves that we must have $2^\lambda < \kappa$ if κ is measurable. Therefore, κ is a strong limit cardinal.

12. Suppose first that $\mathcal{F}$ is not closed for diagonal intersection, i.e., there are $F_\alpha \in \mathcal{F}$ such that if $F = \{\alpha : \alpha \in F_\beta \text{ for all } \beta < \alpha\}$ is their diagonal intersection, then $F \notin \mathcal{F}$. Since $\mathcal{F}$ is an ultrafilter, $\kappa \setminus F \in \mathcal{F}$. For each $\alpha \in \kappa \setminus F$ there is a $\beta_\alpha < \alpha$ such that $\alpha \notin F_{\beta_\alpha}$. The mapping f defined by $\alpha \to \beta_\alpha$ is regressive on $\kappa \setminus F \in \mathcal{F}$, but $f^{-1}(\beta) \subseteq (\kappa \setminus F) \setminus F_\beta$ is not in $\mathcal{F}$ for any $\beta < \kappa$. Thus, $\mathcal{F}$ is not a normal filter.

Conversely, suppose that $\mathcal{F}$ is an ultrafilter but it is not normal, i.e., there is an $F \in \mathcal{F}$ and a regressive $f : F \to \kappa$ such that $F_\alpha = f^{-1}(\alpha) \notin \mathcal{F}$ for all $\alpha < \kappa$. Then $\kappa \setminus F_\alpha$ is in $\mathcal{F}$, and let G be their diagonal intersection. Then $\gamma \in G \cap F$ would mean $\gamma \in \kappa \setminus F_{f(\gamma)}$ (note that $f(\gamma) < \gamma$), which is not the case (because $\gamma \in f^{-1}(f(\gamma))$). Therefore, $G \cap F = \emptyset$, and since $F \in \mathcal{F}$, the set G cannot be a member of $\mathcal{F}$, and this shows that $\mathcal{F}$ is not closed for diagonal intersection.

13. It is clear that if $\mathcal{F}$ is κ-complete and nontrivial (i.e., all $\kappa \setminus \{\alpha\} \in \mathcal{F}$) then it does not contain a subset of cardinality smaller than κ. Now suppose that $\mathcal{F}$ is a normal ultrafilter on κ for which every element is of cardinality κ, hence if $A \subset \kappa$, $|A| < \kappa$, then $\kappa \setminus A \in \kappa$. We want to show that if $\lambda < \kappa$ and A_α, $\alpha < \lambda$ are fewer than κ sets from $\mathcal{F}$, then their intersection is also in $\mathcal{F}$. Set $A_\alpha = \kappa$ for $\lambda \leq \alpha < \kappa$, and form the diagonal intersection B of all these A_α. By Problem 12 this B belongs to $\mathcal{F}$. But it is clear that $(\bigcap_{\alpha<\lambda} A_\alpha) \setminus \lambda = B \setminus \lambda = B \cap (\kappa \setminus \lambda)$, and here the right-hand side is the intersection of two elements of $\mathcal{F}$; therefore, it belongs to $\mathcal{F}$. Hence $\bigcap_{\alpha<\lambda} A_\alpha$ also belongs to $\mathcal{F}$.

14. **(a)** Let $\mathcal{G}$ be the set of measure 1 sets. Then $\mathcal{G}$ is a κ-complete ultrafilter on κ from which it easily follows that $\equiv$ is an equivalence relation on ${}^\kappa\kappa$, and $\prec$ is irreflexive and transitive on the set of equivalence classes. As for trichotomy, if $f, g \in {}^\kappa\kappa$, then the union of the sets $\{\alpha : f(\alpha) < g(\alpha)\}$, $\{\alpha : f(\alpha) = g(\alpha)\}$, and $\{\alpha : g(\alpha) < f(\alpha)\}$ is κ, so one (and only one) of them belongs to $\mathcal{G}$. If it is the first one then $\overline{f} \prec \overline{g}$, if it is the second then $\overline{f} = \overline{g}$, and if it is the third then $\overline{g} \prec \overline{f}$. This proves that $\prec$ is an ordering. To show that it is a well-ordering it is sufficient to show that there is no infinite

decreasing sequence $\cdots \prec \overline{f}_2 \prec \overline{f}_1 \prec \overline{f}_0$. In fact, if such a sequence existed then all the sets $A_n = \{\alpha < \kappa \ : \ f_{n+1}(\alpha) < f_n(\alpha)\}$ would belong to $\mathcal{G}$, and, by κ-completeness, so would do their intersection, i.e., $\cap_{n<\omega} A_n \neq \emptyset$. But this would lead to nonsense, for then $\{f_n(\alpha)\}_{n<\omega}$ would be a decreasing sequence of ordinals for any $\alpha \in \cap_{n<\omega} A_n$.

(b) First of all, no $\{\alpha\}$ belongs to $\mathcal{F}$, since $f_0^{-1}[\{\alpha\}] = f_0^{-1}(\alpha)$ is of measure 0. Next, it is clear that if $F \in \mathcal{F}$ and $F \subseteq F'$ then F' also belongs to $\mathcal{F}$. Finally, if $f_0^{-1}[F_\gamma]$, $\gamma < \lambda$ with $\lambda < \kappa$ are of measure 1, then so is

$$f_0^{-1}[\cap_{\gamma<\lambda} F_\gamma] = \cap_{\gamma<\lambda} f_0^{-1}[F_\gamma],$$

and this shows that $\mathcal{F}$ is closed for fewer than κ intersections. Therefore, $\mathcal{F}$ is a κ-complete (nontrivial) filter on κ. But it is an ultrafilter, since either $f_0^{-1}[Y]$ or its complement $\kappa \setminus f_0^{-1}[Y] = f_0^{-1}[\kappa \setminus Y]$ is of measure 1 for all $Y \subseteq \kappa$.

It is left to show the normality. Let $F \in \mathcal{F}$ and let $f : F \to \kappa$ be a regressive function. We may assume that $0, 1 \notin F$, and $f(\alpha) \geq 1$ for all $\alpha \in F$ (otherwise consider $\max\{f, 1\}$). Extend f to a κ by setting it equal to 0 outside F. For the function $f(f_0)$ we have for all $\alpha \notin f_0^{-1}(0)$ the inequality $f(f_0(\alpha)) < f_0(\alpha)$, and since $f_0^{-1}(0)$ is of measure 0, this means that $\overline{f(f_0)} \prec \overline{f}_0$. By the minimality of f_0 this is possible only if $f(f_0) \notin Y$, i.e., $(f(f_0))^{-1}(\alpha) = f_0^{-1}[f^{-1}(\alpha)]$ is of measure 1 for some $\alpha < \kappa$. Therefore, by the definition of $\mathcal{F}$, we have $f^{-1}(\alpha) \in \mathcal{F}$. Here $\alpha = 0$ is not possible because $f^{-1}(0) = \kappa \setminus F$ is not in $\mathcal{F}$, hence $f^{-1}(\alpha) \subset F$ is an inverse image of the original f belonging to $\mathcal{F}$. This proves that $\mathcal{F}$ is a normal ultrafilter.

15. Let $\mathcal{F}$ be a κ-complete normal ultrafilter on κ (see Problem 14). We prove the stronger statement that if $g : [\kappa]^r \to \sigma$ is an arbitrary coloring, then there is an $F \in \mathcal{F}$ homogeneous for f.

For $r = 1$ this is clear by the κ-completeness of $\mathcal{F}$, and from here we use induction on r. So let us suppose that the claim has already been proven for some r, and let $g : [\kappa]^{r+1} \to \sigma$ be a coloring of the $(r+1)$-tuples. For each $\alpha < \kappa$ define the coloring g_α on $[\kappa \setminus (\alpha+1)]^r$ by setting $g_\alpha(V) = g(\{\alpha\} \cup V)$ for any $V \in [\kappa \setminus (\alpha+1)]^r$. By the induction hypothesis there is an $F_\alpha \in \mathcal{F}$ homogeneous for g_α in some color, say in color $\tau_\alpha < \sigma$. Let F' be the diagonal intersection of the F_α's. By Problem 12 this also belongs to $\mathcal{F}$. To each $\alpha \in F'$ there is an associated color τ_α, therefore, by the κ-completeness of $\mathcal{F}$, there is an $F \subset F'$, $F \in \mathcal{F}$ and a $\tau < \sigma$ such that for $\alpha \in F$ we have $\tau_\alpha = \tau$. We claim that F is homogeneous in color τ for g. In fact, if $V \subset F$ has $r+1$ elements and α is its smallest element, then $V \setminus \{\alpha\} \subset F_\alpha$ by the definition of the diagonal intersection, hence $g(V) = g_\alpha(V \setminus \{\alpha\}) = \tau_\alpha = \tau$.

16 Let $\mathcal{F}$ be a κ-complete normal ultrafilter on κ (see Problem 14). If $g : [\kappa]^{<\omega} \to \sigma$ is a coloring, then the restriction g_r of g to $[\kappa]^r$ is a coloring on the set of r-tuples. By the preceding problem for each r there is an $F_r \in \mathcal{F}$

homogeneous with respect to g_r. Then $\cap F_r \in \mathcal{F}$ is clearly a set of cardinality κ such that all r-tuples of it for any fixed $r < \omega$ have the same color.

17. **(a)** Such a linear functional I was constructed in Problem 17.19. Note that the functional I from Problem 17.19 has the property that if $f, g \in \mathcal{B}_{\mathbf{N}}$ are such that $f(n) - g(n) \to 0$ as $n \to \infty$, then $I(f) = I(g)$.

(b) Let I_0 be the functional from part (a), and for an $f \in \mathcal{B}_{\mathbf{N}}$ let

$$F(n) = \frac{f(0) + \cdots + f(n)}{n+1}.$$

Now $I(f) = I_0(F)$ is clearly linear, normed, and translation invariant, for if $g(n) = f(n+1)$ and

$$G(n) = \frac{g(0) + \cdots + g(n)}{n+1},$$

then we have $G(n) - F(n) \to 0$ as $n \to \infty$, and hence $I_0(F) = I_0(G)$.

(c) Let I_0 be the functional from part (b). Note that such a functional is necessarily independent of finitely many values of $f \in \mathcal{B}_{\mathbf{N}}$. In fact, if f and g differ only in finitely many values, then there are translations of them (in the sense of part (b)) which are identical. Now for an $f \in \mathcal{B}_{\mathbf{Z}}$ let $f^+(n) = f(n)$ and $f^-(n) = f(-n-1)$ for all $n \in \mathbf{N}$. Then $f^{\pm} \in \mathcal{B}_{\mathbf{N}}$, hence we can set $I(f) = (I_0(f^+) + I_0(f^-))/2$. This is clearly nontrivial, linear and normed. Its translation invariance follows from the fact that if $F(n) = f(n+1)$, then $F^+(n)$ differs from $g_+(n) := f^+(n+1)$ only in finitely many values and $F^-(n)$ differs from $g_-(n) := f^-(n-1)$ only in finitely many values, hence

$$\begin{aligned} I(F) &= (I_0(F^+) + I_0(F^-))/2 = (I_0(g_+) + I_0(g_-))/2 \\ &= (I_0(f^+) + I_0(f^-))/2 = I(f). \end{aligned}$$

(d) We prove that there is a translation invariant normed linear functional I_n on $\mathcal{B}_{\mathbf{Z}^n}$ by induction on n. For $n = 1$ this was done in part (c), and suppose now that I_n is already known to exist. Let $f_{\mathbf{a}}(\mathbf{y}) = f(\mathbf{y} + \mathbf{a})$ denote the translate of f by the vector $\mathbf{a}$. If $f \in \mathcal{B}_{\mathbf{Z}^{n+1}}$, then for each fixed $x \in \mathbf{Z}$ the function $f^x(y_1, \ldots, y_n) = f(x, y_1, \ldots, y_n)$ belongs to $\mathcal{B}_{\mathbf{Z}^n}$, hence $x \to I_n(f^x)$ is well defined and belongs to $\mathcal{B}_{\mathbf{N}}$, and we can set $I_{n+1}(f) = I_1(I_n(f^x))$. This I_{n+1} is clearly nontrivial, linear, and normed. If $\mathbf{a} = (a_0, \ldots, a_n) \in \mathbf{Z}^{n+1}$, then we have $(f_{\mathbf{a}})^x = (f^{x+a_0})_{\mathbf{b}}$ where $\mathbf{b} = (a_1, \ldots, a_n)$. Therefore, by the translation invariance of I_n we have $I_n\big((f_{\mathbf{a}})^x\big) = I_n(f^{x+a_0})$, and then by the translation invariance of I_1 it follows that $I_1\Big(I_n\big((f_{\mathbf{a}})^x\big)\Big) = I_1\big(I_n(f^x)\big)$, which proves the translation invariance of I_{n+1}.

(e) Consider the I_n from part (d). For $f \in \mathcal{B}_A$ the function $F(y_1, \ldots, y_n) = f(y_1 s_1 + \ldots + y_n s_n)$, $y_i \in \mathbf{N}$, is in $\mathcal{B}_{\mathbf{Z}^n}$, hence we can set $I(f) = I_n(F)$. This clearly satisfies all the requirements.

(f) Let $\mathbf{B} = \{f \in \mathcal{B}_A : \|f\| \le 1\} = {}^A[-1,1]$ be the unit ball of $\mathcal{B}_A$. Consider $\mathbf{C} = \{I \mid I : \mathbf{B} \to [-1,1]\} = {}^{\mathbf{B}}[-1,1]$ equipped with the product topology on ${}^{\mathbf{B}}[-1,1]$. Being the product of compact spaces, this is compact. For a finite subset S of A let $\mathbf{C}_S$ be the set of all normed linear functionals I from $\mathbf{C}$ that are invariant for translation with any $s \in S$, where by linearity we mean that if $f_1, f_2, c_1f_1 + c_2f_2 \in \mathbf{B}$, then $I(c_1f_1 + c_2f_2) = c_1I(f_1) + c_2I(f_2)$. We claim that this is a closed subset of $\mathbf{C}$, and to this end it is sufficient to show that its complement relative to $\mathbf{C}$ is open. If $I \in {}^{\mathbf{B}}[-1,1]$ is not in $\mathbf{C}_S$, then either

- $I(1) \ne 1$, or
- there is an $f \in \mathbf{B}$ with $|I(f)| > \|f\|$, or
- there are $f_1, f_2, f_1 + f_2 \in \mathbf{B}$, $c_1, c_2 \in \mathbf{R}$ with $I(c_1f_1 + c_2f_2) \ne c_1I(f_1) + c_2I(f_2)$, or
- there is an $s \in S$ and an $f \in \mathbf{B}$ such that if f_s is the translate of f with s then $I(f_s) \ne I(f)$.

In each case the corresponding property depends only on finitely many coordinates in ${}^{\mathbf{B}}[-1,1]$, hence it holds in a neighborhood of I, and this proves that the complement of $\mathbf{C}_S$ is open (relative to $\mathbf{C}$).

Thus, each $\mathbf{C}_S$ is compact and nonempty by part (e). Since

$$\mathbf{C}_{S_1} \cap \cdots \cap \mathbf{C}_{S_m} = \mathbf{C}_{S_1 \cup \cdots \cup S_m},$$

we can conclude that the intersection of all $\mathbf{C}_S$ with $S \subset A$, $|S| < \infty$ is nonempty, and any I^* in this intersection is a translation-invariant normed linear functional on $\mathbf{B}$. Thus, all we need to do is to extend I^* from $\mathbf{B}$ to all of $\mathcal{B}_A$ while preserving its properties.

Let $f \in \mathcal{B}_A$ and select a natural number N with $N > \|f\|$. Then $f/N \in \mathbf{B}$ and we can set $I(f) = NI^*(f/N)$. This is a good definition: if $M > \|f\|$ is another integer, then by the additivity of I^* on $\mathbf{B}$ we have $NI^*(f/N) = N(MI^*(f/NM)) = MI^*(f/M)$, and similar argument gives that I is an extension of I^*, and that it is a translation-invariant normed linear functional on $\mathcal{B}_A$.

(g) Let I be the linear functional from part (f), and for a subset H of A set $\mu(H) = I(\chi_H)$ where χ_H is the characteristic function of H (i.e., it is 1 on H and 0 on $A \setminus H$). This clearly satisfies the requirements.

(h) Consider the I_0 from part (f) for the Abelian group $\mathbf{R}$ (with addition as operation). The isometries of $\mathbf{R}$ are translations $(x \to x + y)$ and reflection $(x \to -x)$ coupled with translations. Now set $I(f) = (I_0(f) + I_0(f^-))/2$, where $f^-(x) = f(-x)$. This is clearly invariant for reflection. But it is also invariant for translation, for if the translate of f by y is $f_y = f(\cdot + y)$, then $I(f_y) = (I_0(f_y) + I_0((f_y)^-))/2$, and since $(f_y)^- = (f^-)_{-y}$, the translation invariance of I_0 gives that this is the same as $I(f)$.

(i) This follows from (h) the same way as we deduced (g) from (f).

(j) We identify $\mathbf{R}^2$ with the complex plane $\mathbf{C}$. The isometries of $\mathbf{R}^2$ are of the form $T_x R_t$ and $T_x R_t S$, where S is the reflection onto the real line (complex conjugation: $z \to \overline{z}$), R_t for $|t| = 1$ is rotation about the origin by angle $\arg(t)$ (multiplication by t: $z \to tz$) and T_x is translation by x (adding x: $z \to z+x$). Let I_0 be a translation-invariant normed linear functional from part (f) for the Abelian group $\mathbf{R}^2$ (where $\mathbf{R}^2$ is equipped with the addition operation). Exactly as in part (i) (by considering $(I_0(f)+I_0(f^-))/2$ where $f^-(z) = f(\overline{z})$) this gives rise to a translation-invariant normed linear functional, which is also invariant with respect to reflection S, so we may assume that already I_0 has this property. Let $\mathbf{T}$ be the unit circle with multiplication as operation. It is an Abelian group, and let I_1 be a rotation-invariant normed linear functional on $\mathcal{B}_{\mathbf{T}}$ (see part (f)). Now for an $f \in \mathcal{B}_{\mathbf{R}^2}$ and $t \in \mathbf{T}$ we set $f^t(z) = f(tz)$, and define $I(f) = I_1(I_0(f^t))$. This is clearly linear, normed, and rotation invariant (for rotations about the origin, which is enough). It is also translation invariant: if $g_x(z) = g(z + x)$ is the translate of a function $g \in \mathcal{B}_{\mathbf{R}^2}$ by x, then $(f_x)^t = (f^t)_{x/t}$, hence by the translation invariance of I_0 we get $I_0((f_x)^t) = I_0(f^t)$, and so $I(f_x) = I(f)$. The same argument shows that I is invariant with respect to reflection (S), hence I is invariant with respect to all isometries of $\mathbf{R}^2$.

(k) This again follows from (j) by considering characteristic functions of sets (see the proof of (g)).

(l) Let I_0 be the translation-invariant functional from (f) for $\mathcal{B}_R$, and let f be a bounded function on $\mathbf{R}$ with bounded support. Note that I_0 is a positive linear functional. For an integer n let f^n be the periodic extension of the restriction $f|_{[n, n+1)}$ with period 1, and set

$$I(f) = \sum_n I_0(f^n).$$

Note that all but finitely many terms in this sum are zero. We claim that this I satisfies all the requirements. That I is a positive linear functional is clear.

First we prove translation invariance. Let $a = k + b$ with k an integer and $0 \le b < 1$, and let $f_a(x) = f(x + a)$ be the a-translate of f. Making use that (recall that χ_E denotes the characteristic function of E)

$$(f_a)^n = \left((f\chi_{[n+a,n+k+1)})^{n+k}\right)_b + \left((f\chi_{[n+k+1,n+a+1)})^{n+k+1}\right)_b,$$

we obtain from the additivity and translation invariance of I_0:

$$\begin{aligned} I(f_a) &= \sum_n \left\{I_0\left((f\chi_{[n+a,n+k+1)})^{n+k}\right) + I_0\left((f\chi_{[n+k+1,n+a+1)})^{n+k+1}\right)\right\} \\ &= \sum_n \left\{I_0\left((f\chi_{[n+a,n+k+1)})^{n+k}\right) + I_0\left((f\chi_{[n+k,n+a)})^{n+k}\right)\right\} \\ &= \sum_n I_0\left((f\chi_{[n+k,n+k+1)})^{n+k}\right) = I(f), \end{aligned}$$

i.e., translation invariance holds.

Next observe that $I(\chi_{[0,1)}) = 1$, and as a consequence $I(\chi_{[0,1/m)}) = 1/m$, because m translates of $\chi_{[0,1/m)}$ add up to $\chi_{[0,1)}$. Then for any $k \geq 1$ we have $I(\chi_{[0,k/m)}) = k/m$, and hence the positivity of I gives

$$\frac{k_1}{m_1} \leq I(\chi_{[0,x)}) \leq \frac{k_2}{m_2}$$

whenever $k_1/m_1 \leq x \leq k_2/m_2$. But here k_2/m_2 can be arbitrarily close to k_1/m_1, and $I(\chi_{[0,x)}) = x$ follow for all $x > 0$. This implies again by translation invariance $I(\chi_{[a,b))}) = b - a$ for all $a < b$, and a similar equality is true for all intervals (open, semi-closed, or closed) by the monotonicity of I. Therefore, if g is a step function with bounded support and finitely many steps, then $I(g) = \int g$, where $\int$ denotes Riemann integration. Finally, if f is a Riemann integrable function with bounded support, then for every $\epsilon > 0$ there are step functions $g_1 \leq f \leq g_2$ such that

$$\int g_2 - \int g_1 < \epsilon,$$

and since we also have

$$\int g_1 = I(g_1) \leq I(f) \leq I(g_2) = \int g_2,$$

the equality

$$I(f) = \int f$$

follows if we let $\epsilon \to 0$.

(m) We prove the statement by induction on n. For $n = 1$ this was done in part (l), and suppose now that I_{n-1} is already known to exist. Exactly as in part (d) let $f_{\mathbf{a}}(\mathbf{y}) = f(\mathbf{y} + \mathbf{a})$ denote the translate of f by the vector $\mathbf{a}$, and if $f \in \mathcal{B}_{\mathbf{R}^n}$ is of bounded support, then set $I_n(f) = I_1(I_{n-1}(f^x))$, where $f^x(y_1, \ldots, y_{n-1}) := f(x, y_1, \ldots, y_{n-1})$. This I_n is clearly linear and positive, and the proof used in part (d) shows that it is translation invariant (just repeat that proof with $\mathbf{R}^n$ in place of $\mathbf{Z}^n$). We have by induction $I_n(\chi_{[0,1)^n}) = 1$, and then as in part (l) one can see that I_n agrees with the Riemann integral for the functions $\chi_{[0,1/m)^n}$, from which just as in part (l) one can deduce that I_n agrees with the Riemann integral for all finite linear combinations of characteristic functions of sets of the form $\prod_{j=1}^n [a_j, b_j)$ as well as their open and closed variants. Since for every Riemann integrable function f and for every $\epsilon > 0$ one can find two such linear combinations g_1 and g_2 with $g_1 \leq f \leq g_2$ and $\int g_2 - \int g_1 < \epsilon$, $I(f) = \int f$ follows just as in part (l).

(n) Let I be the functional from part (m) for $\mathbf{R}^n$. If E is a bounded subset of $\mathbf{R}^n$ then set $\mu(E) = I(\chi_E)$. This is clearly finitely additive and translation invariant, and extends Jordan measure. For unbounded E define

$$\mu(E) = \lim_{r\to\infty} I(\chi_{E\cap B_r}),$$

where B_r denotes the closed ball around the origin of radius r. By monotonicity the limit on the right-hand side exists, and for bounded sets E we get back the $\mu(E) = I(\chi_E)$ definition. Using monotonicity and translation invariance of I it is easy to verify the translation invariance of μ.

(o) Use the reflection-rotation technique of parts (h), (j) to generate isometry invariant functionals I from the translation invariant I_0 defined in part (m). E.g., in $\mathbf{R}^2$ considering $(I_0(f)+I_0(f^-))/2$ where $f^-(z) = f(\overline{z})$ gives rise to a translation invariant positive linear functional, which is also invariant with respect to reflection $z \to \overline{z}$ and extends Riemann integral (note that if f is Riemann integrable, then so is f^-, and $\int f = \int f^-$), so we may assume that already I_0 has this property. Let $\mathbf{T}$ be the unit circle with multiplication as operation, and let I_1 be a rotation-invariant normed linear functional on $\mathcal{B}_{\mathbf{T}}$ (see part (f)). Now for an $f \in \mathcal{B}_{\mathbf{R}^2}$ with bounded support and $t \in \mathbf{T}$ we set $f^t(z) = f(tz)$, and define $I(f) = I_1(I_0(f^t))$. This is clearly linear, positive, rotation invariant, and the same proof that was given in part (j) shows that it is also translation invariant. Hence I is invariant with respect to all isometries of $\mathbf{R}^2$. Finally, if $f \in \mathcal{B}_{\mathbf{R}^2}$ is Riemann integrable, then so is every f^t with the same integral as f, hence

$$I(f) = I_1(I_0(f^t)) = I_1\left(\int f^t\right) = I_1\left(\int f\right) = \int f,$$

i.e., I extends the Riemann integral.

(p) Set as in part (n)

$$\mu(E) = \lim_{r\to\infty} I(\chi_{E\cap B_r}),$$

where I is the functional from part (o) for $\mathbf{R}^n$, $n = 1, 2$. The same argument that we gave in part (n) shows that this is an isometry-invariant measure that extends Jordan measure.

29

Stationary sets in $[\lambda]^{<\kappa}$

1. $[\lambda]^{<\kappa} = \bigcup\{X_\alpha : \alpha < \kappa\}$ where $X_\alpha = \{P \in [\lambda]^{<\kappa} : \alpha = \min(\kappa \setminus P)\}$.

2. Assume that $\gamma < \kappa$ and $\{X_\alpha : \alpha < \gamma\}$ are bounded sets, $X = \bigcup\{X_\alpha : \alpha < \gamma\}$. For $\alpha < \gamma$ choose $P_\alpha \in [\lambda]^{<\kappa}$ with the following property: no $Q \supseteq P_\alpha$ is in X_α. If now $P = \bigcup\{P_\alpha : \alpha < \gamma\}$, then $P \in [\lambda]^{<\kappa}$ (as κ is regular) and no $Q \supseteq P$ is in X.

3. It is obvious that the increasing union of sets each containing α is again a set containing α. For unboundedness, for every $P \in [\lambda]^{<\kappa}$, $P \cup \{\alpha\}$ will be an element of the set in question. The other claim is similar.

4. Assume that S is stationary and $Q \in [\lambda]^{<\kappa}$ is given. In order to show that there exists a $P \in S$ with $P \supseteq Q$ we remark that the set $\{P \in [\lambda]^{<\kappa} : P \supseteq Q\}$ is closed, unbounded by Problem 3. Hence it must intersect S.

5. If $A \subseteq \kappa$ is unbounded, then it is unbounded in $[\kappa]^{<\kappa}$ as well: if $P \in [\kappa]^{<\kappa}$, then $P \subseteq \alpha$ where $\alpha \in A$ is any element with $\sup(P) \leq \alpha$.

If $A \subseteq \kappa$ is unbounded in $[\kappa]^{<\kappa}$ then it is unbounded in κ, as well: if $\beta < \kappa$ then, by unboundedness, some $P \in A$ has $\beta \subseteq P$. As $A \subseteq \kappa$, $P = \alpha$ for some $\alpha < \kappa$, so $\beta \leq \alpha \in A$, as claimed.

If $A \subseteq \kappa$ is closed, then it is closed in $[\kappa]^{<\kappa}$ as well: if $\{\alpha_\tau : \tau < \mu\}$ is some increasing sequence from A then it is an increasing sequence of ordinals less than κ. There is, therefore, a supremum α of them, which is in A, and then $\bigcup\{\alpha_\tau : \tau < \mu\} = \alpha$ as is required for closure of A in $[\kappa]^{<\kappa}$.

If $A \subseteq \kappa$ is closed in $[\kappa]^{<\kappa}$ then it is closed in κ as well: indeed, assume that $\{\alpha_\tau : \tau < \mu\}$ is an increasing sequence of elements of A. Then $\{\alpha_\tau : \tau < \mu\}$ is $\subseteq$-increasing in $[\kappa]^{<\kappa}$, so by hypothesis $P = \bigcup\{\alpha_\tau : \tau < \mu\} \in A$. But P is an ordinal, that is, an initial segment of κ, and it can only be the supremum of our sequence.

From what was just said, it follows that if $A \subseteq \kappa$ and it is a stationary subset of $[\kappa]^{<\kappa}$, then it is also stationary in κ. For the other implication it suffices to prove that if $C \subseteq [\kappa]^{<\kappa}$ is closed, unbounded, then so is $C \cap \kappa$, that is, the set of those elements of C that are initial segments of κ. Closure is immediate. For unboundedness, pick $\beta < \kappa$. Then select the increasing sequence $P_0, P_1, \ldots$ of elements of C with $\beta \subseteq P_0$ and then $\sup(P_n) \subseteq P_{n+1}$. Then $P = P_0 \cup P_1 \cup \cdots$ will be in C, and by construction it is an initial segment in κ, i.e., $P = \alpha$ for some ordinal $\alpha < \kappa$, and clearly $\alpha \geq \beta$.

6. One direction is obvious as every increasing sequence is manifestly a directed system.

For the other direction assume that $\gamma < \kappa$ is an infinite cardinal and $Y = \{P_\alpha : \alpha < \gamma\}$ is a directed subsystem of a system X closed under increasing unions of length $< \kappa$.

We show $\bigcup Y \in X$ by induction on γ. For $\gamma = \omega$ select $n_0 < n_1 < \cdots$ such that $n_0 = 0$ and $P_{n_{i+1}} \supseteq P_i \cup P_{n_i}$. Clearly, $P_{n_0} \subseteq P_{n_1} \subseteq \cdots$ is an increasing sequence with union $P_0 \cup P_1 \cup \cdots$.

For $\gamma > \omega$ we use the fact that if Y is a directed system and $Z \subseteq Y$, then there is a directed subsystem $Z \subseteq Z' \subseteq Y$ with $|Z'| \leq |Z| + \omega$. By this, we can decompose Y as an increasing, continuous union $Y = \bigcup\{Y_\alpha : \alpha < \gamma\}$ of directed systems Y_α of smaller cardinality. By our inductive hypothesis we get $\bigcup Y_\alpha \in X$ for every $\alpha < \gamma$, so finally this holds for Y, as $\{Y_\alpha : \alpha < \gamma\}$ is an increasing family of sets.

7. (a) It is obvious that $C(f)$ is closed under increasing unions, as the increasing union of sets, each closed under f, is again a set closed under f. To show unboundedness, assume that $P \in [\lambda]^{<\kappa}$. Set $P_0 = P$ and for $n = 0, 1, 2, \ldots$ let

$$P_{n+1} = P_n \cup \bigcup \{f(s) : s \in [P_n]^{<\omega}\}.$$

Induction gives, as $\kappa > \omega$ is regular, that $|P_n| < \kappa$. Now $P_0 \cup P_1 \cup P_2 \cup \cdots$ is an f-closed set of cardinality $< \kappa$, containing P.

(b) Assume that C is closed, unbounded. Define $f(s)$ for every $s \in [\lambda]^{<\omega}$ by recursion on $|s|$ as follows. Let $f(\emptyset) \in C$ be arbitrary. For $|s| > 0$ let $f(s) \in C$ be such that $s \subseteq f(s)$ and also $f(t) \subseteq f(s)$ holds for every $t \subseteq s$, $t \neq s$ (these values have been determined before $f(s)$). Assume that $P \in C(f)$, $P \neq \emptyset$. Then $P = \bigcup\{f(s) : s \in [P]^{<\omega}\}$ and as this is the union of a directed subsystem of C, it is in C by Problem 6.

8. Assume that $\{C_\alpha : \alpha < \gamma\}$ are closed, unbounded sets, $\gamma < \kappa$. It is obvious that $C = \bigcap\{C_\alpha : \alpha < \gamma\}$ is closed. By Problem 7 for every $\alpha < \gamma$ there is some $f_\alpha : [\lambda]^{<\omega} \to [\lambda]^{<\kappa}$ such that $C(f_\alpha) \setminus \{\emptyset\} \subseteq C_\alpha$. If we now set $f(s) = \bigcup\{f_\alpha(s) : \alpha < \gamma\}$ for every $s \in [\lambda]^{<\omega}$, then $f : [\lambda]^{<\omega} \to [\lambda]^{<\kappa}$ and clearly $C(f) \subseteq \bigcap\{C(f_\alpha) : \alpha < \gamma\}$ so by Problem 7(a) this latter set, therefore C, is an unbounded set.

9. Let $f : [\lambda]^{<\omega} \to [\lambda]^{<\kappa}$ be an arbitrary function with $f(\{\alpha\}) = \alpha$ for $\alpha < \kappa$ (that is, f assigns the set $\alpha \in [\kappa]^{<\kappa}$ to the point $\alpha < \kappa$). By Problem 7 almost every P is in $C(f)$. If $P \in C(f)$ then $P \cap \kappa$ has the property that if $\alpha \in P \cap \kappa$ then $\alpha \subseteq P \cap \kappa$ so $P \cap \kappa$ is an initial segment.

10. Given an algebra on λ with the operations $f_i : [\lambda]^{n_i} \to \lambda$ for $i = 0, 1, \ldots$, set $f(s) = f_0(s) \cup f_1(s) \cup \cdots$ and apply Problem 7(a).

11. We consider various properties of $P \in [\lambda]^{<\kappa}$ and notice that they hold for $P \in C(f)$ for certain functions $f : [\lambda]^{<\omega} \to [\lambda]^{<\kappa}$. Then, if we take the pointwise union of these functions, then all the properties hold for the elements of the appropriate $C(f)$. First, if $f(\{\alpha\}) = \alpha$ for $\alpha < \kappa$ then $P \cap \kappa < \kappa$ holds for every $P \in C(f)$. Second, assume that $f(\kappa \cdot \alpha + \beta) \ni \kappa \cdot \alpha, \beta$ (for $\beta < \kappa$) and $\kappa \cdot \alpha + \beta \in f(\{\kappa \cdot \alpha, \beta\})$. Then, if $P \cap [\kappa \cdot \alpha, \kappa \cdot (\alpha + 1)) \neq \emptyset$ holds for some α, then $P \cap [\kappa \cdot \alpha, \kappa \cdot (\alpha + 1))$ is the left translation of the interval $\kappa \cap P$ by $\kappa \cdot \alpha$, so we are done.

12. Set $C = \nabla\{C_\alpha : \alpha < \lambda\}$. In order to show that C is closed and unbounded, assume that $\gamma < \kappa$ and $\{P_\xi : \xi < \gamma\}$ is an increasing sequence of elements of C, $P = \bigcup\{P_\xi : \xi < \gamma\}$. If $\alpha \in P$, then $\alpha \in P_\xi$ for some $\xi < \gamma$, so $\alpha \in P_\zeta$ holds for every $\xi < \zeta < \gamma$, therefore $P_\zeta \in C_\alpha$, and then $P = \bigcup\{P_\zeta : \xi < \zeta < \gamma\} \in C_\alpha$ as C_α is closed.

In order to show that C is unbounded, assume that $P \in [\lambda]^{<\kappa}$ is arbitrary. Set $P_0 = P$ and then choose $P_1, P_2, \ldots$ as follows. Let $P_{n+1} \supseteq P_n$ be an element of $\bigcap\{C_\alpha : \alpha \in P_n\}$ (the latter set is closed, unbounded by Problem 8). Set $P' = P_0 \cup P_1 \cup \cdots$. Then $P' \in C$ as if $\alpha \in P'$ then $\alpha \in P_n \subseteq P_{n+1} \subseteq \cdots$ for some n and then $P_n, P_{n+1}, \ldots \in C_\alpha$, so $P' \in C_\alpha$.

13. Assume that the statement fails, i.e., for every $\alpha < \lambda$ there is some closed, unbounded set C_α such that $f(P) \neq \alpha$ holds for $P \in C_\alpha$. By the previous problem, the diagonal intersection C of the closed, unbounded sets $\{C_\alpha : \alpha < \lambda\}$ is closed, unbounded again. But then, if $P \in S \cap C$, then $f(P) \neq \alpha$ holds for every $\alpha \in P$, a contradiction.

14. Let $G : [\lambda]^{<\omega} \to \lambda$ be a bijection. By Problem 7, almost every P is "closed" under G, G^{-1}, that is, the following are true: if $s \in [P]^{<\omega}$, then $G(s) \in P$ and if $\alpha \in P$, then $G^{-1}(\alpha) \subseteq P$. Given f as in the problem, for a.e. $P \in S$ we have $g(P) = G(f(P)) \in P$, so by Problem 13 there are a stationary $S' \subseteq S$ and a $\gamma < \lambda$ such that for $P \in S'$ we have $g(P) = \gamma$. But then, $f(P) = G^{-1}(\gamma)$ is true for $P \in S'$.

15. Assume that $X \subseteq [\lambda]^{<\kappa}$ is nonstationary. Then $X \cap C = \emptyset$ holds for some closed, unbounded set C. By Problem 7(b) $C(g) \setminus \{\emptyset\} \subseteq C$ for some $g : [\lambda]^{<\omega} \to [\lambda]^{<\kappa}$, so $C(g) \cap X = \emptyset$ or $\{\emptyset\}$, i.e., no $\emptyset \neq P \in X$ is closed under g. Let $f(P)$ be a finite subset $s \subseteq P$ such that $g(s) \not\subseteq P$. Now clearly $f^{-1}(s)$ is bounded: it contains no $P \supseteq g(s)$.

16. If $\{P_\alpha : \alpha < \gamma\}$ is an increasing sequence with $\gamma < \kappa$ and $\kappa \cap P_\alpha \in C$ holds for every P_α then, as C is closed, it holds for $\bigcup\{P_\alpha : \alpha < \gamma\}$. For unboundedness, assume that $P \in [\lambda]^{<\kappa}$. Pick $\delta \in C$, $\delta > \sup(P \cap \kappa)$. Then $P \cup \delta$ is above P and has the required property.

17. To show that A is closed, assume that $\{P_\alpha : \alpha < \gamma\}$ is an increasing sequence of elements of A, $\gamma < \kappa$, $P = \bigcup\{P_\alpha : \alpha < \gamma\}$. Set $\xi_\alpha = \sup(P_\alpha) \in C$. Now if $\xi = \sup\{\xi_\alpha : \alpha < \gamma\}$, then $\xi = \sup(P)$ and this is in C as C is closed.

To show that A is unbounded, let $P \in [\lambda]^{<\kappa}$ be arbitrary. Let ξ be the least element of C above $\sup(P)$. Clearly, $P \cup \{\xi\} \in A$.

18. Set $B = \{P \in [\lambda]^{<\kappa} : \kappa(P) \in S\}$. We have to show that B is stationary, that is, $B \cap C \neq \emptyset$ holds for every closed, unbounded set C. Assume that C is closed, unbounded. Without loss of generality, $\kappa \cap P < \kappa$ holds for every $P \in C$ (Problem 9). For $\alpha < \kappa$ we define the increasing, continuous sequence of elements of C as follows. Let $P_0 \in C$ be arbitrary. If $0 < \alpha < \kappa$ is a limit ordinal, then, of course, $P_\alpha = \bigcup\{P_\beta : \beta < \alpha\}$. And for successor ordinals, let $P_{\alpha+1}$ be some element of C with $P_{\alpha+1} \supseteq P_\alpha$ and $\kappa(P_{\alpha+1}) > \kappa(P_\alpha)$. This done, we observe that $\{\kappa(P_\alpha) : \alpha < \kappa\}$ is closed, unbounded, so there exists an element of it in S and then we are done.

19. For $\omega_1 \leq \alpha < \omega_2$ let φ_α be a bijection between ω_1 and α. Set, for $\omega_1 \leq \alpha < \omega_2$, $\gamma < \omega_1$, $P_{\alpha\gamma} = \varphi_\alpha[\gamma] = \{\varphi_\alpha(\xi) : \xi < \gamma\}$. We claim (and this suffices) that $S = \{P_{\alpha\gamma} : \gamma < \omega_1 \leq \alpha < \omega_2\}$ is stationary. To this end, by Problem 7(b), it suffices to show that $S \cap C(f) \neq \emptyset$ holds for every $f : [\lambda]^{<\omega} \to [\lambda]^{<\aleph_1}$. Indeed, given f, there is some $\omega_1 \leq \alpha < \omega_2$ such that α is closed under f, that is, $f(s) \subseteq \alpha$ holds for every $s \in [\alpha]^{<\omega}$. Repeating this argument for the underlying set α, we get that there is some $\gamma < \omega_1$ such that $\varphi_\alpha[\gamma] = P_{\alpha\gamma}$ is closed under f. Indeed, set $\gamma_0 = 1$ and inductively select $\gamma_{n+1} < \omega_1$ such that $\varphi_\alpha[\gamma_{n+1}] \supseteq f[\varphi_\alpha[\gamma_n]]$. Then $\gamma = \sup\{\gamma_n : n < \omega\}$ is as required.

20. As $\aleph_2^{\aleph_0} = \max(\aleph_2, 2^{\aleph_0})$, see Problem 10.27(b), it suffices to show that every closed, unbounded set C has cardinality at least $2^{\aleph_0}$. By Problem 7(b) this can further be reduced to the case when C is of the form $C(f)$ for some $f : [\lambda]^{<\omega} \to [\lambda]^{<\kappa}$. Assume therefore that we are given such an f. Set $T = \{\alpha < \omega_2 : \operatorname{cf}(\alpha) = \omega\}$, a stationary set in ω_2. For $\alpha \in T$ let $A_\alpha = \{g_n(\alpha) : n < \omega\}$ be an ω-sequence of ordinals less than α converging to α. Let B_α be the f-closure of A_α. Notice that $B_\alpha \supseteq A_\alpha$ is a countable set.

We argue that the following statement suffices:

(+) If $T' \subseteq T$ is stationary, then there exist x_0, x_1 and disjoint stationary $T_0, T_1 \subseteq T'$ such that for $\alpha \in T_0$, $x_0 \in A_\alpha$ and $x_1 \notin B_\alpha$ hold and for $\alpha \in T_1$, $x_1 \in A_\alpha$, and $x_0 \notin B_\alpha$ hold.

Indeed, assuming (+) we can recursively construct $x(s) < \omega_2$ and a stationary $T(s) \subseteq T$ for every finite 0–1 sequence s such that $x(s) \in A_\alpha$ $(\alpha \in T(s))$

and $x(s0) \notin B_\alpha$ $(\alpha \in T(s1))$, $x(s1) \notin B_\alpha$ $(\alpha \in T(s0))$. This implies that if we set

$$U_g = \bigcap\{B_\alpha : \alpha \in T(g|_n), n < \omega\}$$

for $g : \omega \to \{0,1\}$, that is for the continuum many infinite 0–1 sequences, then $\{x(g|_n) : n < \omega\} \subseteq U_g$ and if $s \not\subseteq g$ then $x(s) \notin U_g$, so the f-closed sets $\{U_g : g \in {}^\omega\{0,1\}\}$ are distinct.

In order to show (+) we first reduce it to

(++) If $T_0, T_1 \subseteq T$ are stationary, then there are some $x < \omega_2$ and stationary $T_0' \subseteq T_0$ and $T_1' \subseteq T_1$ such that $x \in A_\alpha$ $(\alpha \in T_0')$, $x \notin B_\alpha$ $(\alpha \in T_1')$.

Clearly, two applications of (++) give (+).

To show (++) we first argue that there are $\aleph_2$ ordinals, $\{x_\beta : \beta < \omega_2\}$ such that for every x_β there are stationarily many $\alpha \in T_0$ that $x_\beta \in A_\alpha$. [By transfinite recursion. If $\{x_\gamma : \gamma < \beta\}$ is already constructed, $\xi = \sup\{x_\gamma : \gamma < \beta\}$ then for $\alpha \in T_0$, $\alpha > \xi$ there is some $f(\alpha) \in A_\alpha$, $f(\alpha) > \xi$, and for stationary many α, $f(\alpha)$ is the same by Fodor's theorem (Problem 21.3). Now this value can be taken as x_β.] Now for every $\alpha \in T_1$, $\alpha > \sup\{x_\beta : \beta < \omega_1\}$, as B_α is countable, there exists some $\beta < \omega_1$ that $x_\beta \notin B_\alpha$. Again, by Fodor's theorem, for stationary many $\alpha \in T_1$ this x_β is the same and this can be chosen as the x in (++). [J. E. Baumgartner: On the size of closed unbounded sets, *Annals of Pure and Applied Logic* **54**(1991), 195–227]

21. (a) Assume that $C \subseteq [\lambda]^{<\kappa}$ is a closed, unbounded set. Choose $P_0 = P \in C$ arbitrarily. For $n = 0, 1, \ldots$ choose $P_{n+1} \in C$ such that $P_{n+1} \supseteq P_n$, $\kappa \cap P_{n+1} > |P_n|$. If $P = P_0 \cup P_1 \cup \cdots$, then $P \in C \cap Z$ will hold.

(b) First we remark that it suffices to show that if $S \subseteq Z$ is stationary and $\kappa < \mu \leq \lambda$ is regular, then S is the disjoint union of μ stationary sets. Indeed, if this holds, and λ is singular (otherwise we are done) then we can write λ as $\lambda = \sup\{\lambda_\alpha : \alpha < \mathrm{cf}(\lambda)\}$ the supremum of *regular* cardinals. Decompose first S as the union of $\mathrm{cf}(\lambda)$ disjoint stationary sets, then split the αth set into the union of λ_α disjoint stationary sets.

In order to prove the claim, let $f_P : P \to \kappa(P)$ be injective for $P \in S$. For $\alpha < \mu$ set $g_\alpha(P) = f_P(\alpha)$ for $\alpha \in P$. Notice that any given $\alpha < \mu$ is contained in almost every $P \in S$ (Problem 3). By Problem 13 there are a γ_α and a stationary $S_\alpha \subseteq S$ such that $g_\alpha(P) = \gamma_\alpha < \kappa$ holds for $P \in S_\alpha$. As $\mu > \kappa$ is regular, there is a set $B \subseteq \mu$ of cardinality μ that $\gamma_\alpha = \gamma$ holds for $\alpha \in B$. Now $\{S_\alpha : \alpha \in B\}$ are disjoint stationary subsets of S, indeed, if $P \in S_\alpha \cap S_\beta$ then $\gamma = g_\alpha(P) = g_\beta(P)$, so $f_P(\alpha) = f_P(\beta) = \gamma$ would hold, contradicting the injectivity of f_P.

22. (a) If some S' decomposed as $S' = \bigcup\{S_\alpha : \alpha < \kappa\}$ then we would get $S = ((S \setminus S') \cup S_0) \cup \bigcup\{S_\alpha : 1 \leq \alpha < \kappa\}$, a decomposition into κ stationary sets.

(b) Indeed, if $S \cap Z$ is stationary, then, by Problem 21(b), it decomposes into κ stationary sets, and this contradicts part (a).

(c) Assume that the statement fails. Then by Problem 13 we can select by transfinite recursion on $\xi < \kappa$ the distinct elements x_ξ and stationary sets $S_\xi \subseteq S$ such that for $P \in S_\xi$, $f(P) = x_\xi$ holds. But then, the κ stationary sets $\{S_\xi : \xi < \kappa\}$ are disjoint.

(d) If not, then $\kappa = \mu^+$ for some cardinal μ. For $P \in S$ we let $f_P : \mu \to P$ be surjective. By part (c) for every $\alpha < \mu$ there is a closed, unbounded set C_α such that for $P \in C_\alpha \cap S$, $f_P(\alpha) \in Q_\alpha$ holds, where $Q_\alpha \in [\lambda]^{<\kappa}$. If we set $C = \bigcap\{C_\alpha : \alpha < \mu\}$, $Q = \bigcup\{Q_\alpha : \alpha < \mu\}$, then $Q \in [\lambda]^{<\kappa}$ and for the elements P of the stationary, and so unbounded $S \cap C$ we will have $P \subseteq Q$, a contradiction.

(e) For almost every $P \in S'$, $\mu_P = |f(P)| < \kappa(P)$, so $\mu_P \in P$. By part (c), there is a $\mu < \kappa$ such that for almost every $P \in S'$, $\mu_P \leq \mu < \kappa(P)$ holds. For these P we can set $f(P) \subseteq \{x^P_\alpha : \alpha < \mu\}$. By part (c) again, for a closed, unbounded set C_α we have that $x^P_\alpha \in Q_\alpha$ for $P \in C_\alpha \cap S$ with $|Q_\alpha| < \kappa$. Set $C = \bigcap\{C_\alpha : \alpha < \mu\}$, a closed, unbounded set and $Q = \bigcup\{Q_\alpha : \alpha < \mu\}$. Then for $P \in S' \cap C$, that is, for almost every $P \in S'$, we have $f(P) \subseteq Q$.

(f) First we show that $\kappa(P)$ is regular, in particular a cardinal, for a.e. $P \in S$. Indeed, if not, then for a stationary $S' \subseteq S$ there are a $\mu(P) < \kappa(P)$ and some $f_P : \mu(P) \to \kappa(P)$ with cofinal range. By part (c), for almost every $P \in S'$, we have $\mu(P) \leq \mu$ with some $\mu < \kappa$, and by part (e) for every $\alpha < \mu$ for almost every $P \in S'$ we have $f_P(\alpha) \in Q_\alpha$, for some $Q_\alpha \in [\lambda]^{<\kappa}$. If $Q = \bigcup\{Q_\alpha : \alpha < \mu\}$ then $Q \in [\lambda]^{<\kappa}$ and for almost every $P \in S'$ we have $\kappa(P) \subseteq Q$, which is impossible by Problem 4.

If $\kappa(P) = \mu(P)^+$ held for stationary many $P \in S$, then, as $\mu(P) < \kappa(P)$, we had, by part (c), that $\mu(P) = \mu$ for stationary many $P \in S$ with some μ, but that is impossible for then this stationary set would not meet the closed and unbounded set $\{P : \mu^+ \leq \kappa(P)\}$ (see Problems 3, 9).

(g) Assume otherwise, that is, there is a stationary $S' \subseteq S$ such that for $P \in S'$ there is a closed, unbounded set $C_P \subseteq [P]^{<\kappa(P)}$ such that $S \cap C_P = \emptyset$. By part (e) for every $Q \in [\lambda]^{<\kappa}$ there exists some $Q' \supseteq Q$, $Q' \in [\lambda]^{<\kappa}$ such that the following holds: for a. e. $P \in S'$ there is some $R \in C_P$ with $Q \subseteq R \subseteq Q'$. If we now set $Q^* = Q \cup Q' \cup Q'' \cup \cdots$, then for a. e. $P \in S'$, $Q^* \in C_P$ holds. We have, therefore, that the set $D = \{Q \in [\lambda]^{<\kappa} : \text{for a.e. } P \in S', Q \in C_P\}$ is unbounded. It is obviously closed (see Problem 8), and as $D \cap S = \emptyset$ we get a contradiction.

(h) By the previous parts we find that there is a closed, unbounded $C \subseteq [\lambda]^{<\kappa}$ such that for $P \in S \cap C$ we have that $\kappa(P)$ is inaccessible and $S \cap [P]^{<\kappa(P)}$ is stationary in $[P]^{<\kappa(P)}$. By Problem 7(b) there is some $f : [\lambda]^{<\omega} \to [\lambda]^{<\kappa}$ such that $C(f)\backslash\{\emptyset\} \subseteq C$. For $s \in [\lambda]^{<\omega}$ let $g(s) \supseteq f(s)$ be such that $|g(s)|+1 \subseteq g(s)$ (add every $\gamma \leq |f(s)| + 1$ to $f(s)$). As $C(g) \subseteq C(f)$, the above things hold for $C(g)$ as well. By Problem 7(a), $C(g) \cap S$ is unbounded, so we can choose $P \in C(g) \cap S$, $P \neq \emptyset$, with $\kappa(P)$ minimal. Notice that for $s \in [P]^{<\omega}$ we have $|g(s)| + 1 \subseteq g(s) \subseteq P$, hence $|g(s)| < \kappa(P)$ so the restriction of g to $[P]^{<\omega}$

is a function $[P]^{<\omega} \to [P]^{<\kappa(P)}$. If now $D \subseteq [P]^{<\kappa(P)}$ is the set of elements closed under this function, then D is closed, unbounded in $[P]^{<\kappa(P)}$. As S is stationary there, there is $Q \in D \cap S$, but then $\kappa(Q) < \kappa(P)$ and $Q \in C(g) \cap S$, a contradiction. [A. Hajnal., M. Gitik, Nonsplitting subset of $\mathcal{P}_\kappa(\kappa^+)$, *Journal of Symbolic Logic*, **50**(1985), 881–894]

23. By Problem 10.20 we have $\left|[\lambda]^{\aleph_0}\right| > \lambda$ hence GCH gives $\left|[\lambda]^{\aleph_0}\right| = \lambda^+$. Enumerate $[\lambda]^{\aleph_0}$ as $\{A_\alpha : \alpha < \lambda^+\}$ in such a way that $A_\alpha \neq A_\beta$ holds for $\alpha \neq \beta$, and similarly enumerate the functions from $[\lambda]^{<\omega}$ to $[\lambda]^{\aleph_1}$ as $\{f_\alpha : \alpha < \lambda^+\}$. We define S the following way. For $Y \in [\lambda]^{\aleph_1}$ we let Y be an element of S if and only if the following holds: $A_\alpha \subseteq Y$ implies that Y is closed under f_α. To show that S is stationary it suffices to show (by Problem 7(b)) that $S \cap C(f) \neq \emptyset$ holds for every $f : [\lambda]^{<\omega} \to [\lambda]^{\aleph_1}$. Define the increasing, continuous sequence $\{Y_\xi : \xi \leq \omega_1\}$ of elements of $[\lambda]^{\aleph_1}$ with $Y_0 \in [\lambda]^{\aleph_1}$ arbitrary and such that for every $\xi < \omega_1$ the set $Y_{\xi+1}$ includes $f\left[[Y_\xi]^{<\omega}\right]$ as well as $f_\alpha\left[[Y_\xi]^{<\omega}\right]$ for every α with $A_\alpha \subseteq Y_\xi$. Then clearly $Y_{\omega_1} \in S \cap C(f)$. For the other property assume that $U \subseteq S$ is an unbounded subset that is not stationary. By Problem 7(b) again, there is some function $f : [\lambda]^{<\omega} \to [\lambda]^{\aleph_1}$ that $U \cap C(f) = \emptyset$. Let α be the ordinal that $f = f_\alpha$. As U is unbounded, $A_\alpha \subseteq Y$ holds for some $Y \in U$. As $Y \in S$ this implies that Y is closed under f_α, which shows $U \cap C(f) \neq \emptyset$, a contradiction. [J. E. Baumgartner]

24. (a) A is obviously closed under any $f : [A]^{<\omega} \to [A]^{\leq\aleph_0}$.

(b) If $x \in A$, then let $f : [A]^{<\omega} \to [A]^{\leq\aleph_0}$ be a function such that $x \in f(s)$ holds for every $s \in [A]^{\leq\aleph_0}$. Then, if $B \in S$ is closed under f, then $x \in B$ holds, so $x \in \bigcup S$.

(c) By Problem 7 a set $S \subseteq [\lambda]^{<\aleph_1}$ is stationary if and only if it intersects every set of form $C(f)$, where $f : [\lambda]^{<\omega} \to [\lambda]^{\leq\aleph_0}$, and that is λ-stationarity in the new sense.

(d) Let $f : [B]^{<\omega} \to [B]^{\leq\aleph_0}$ be a function. Fix $x \in B$. Let $f' : [A]^{<\omega} \to [A]^{\leq\aleph_0}$ defined by $f(s) \cup \{x\}$ for $s \in [B]^{<\omega}$ and $\{x\}$ otherwise. If $P \in S$, $P \neq \emptyset$ is closed under f', then $P \cap B$ is closed under f, and obviously $x \in P \cap B$ so it is nonempty.

(e) Assume that $f : [B]^{<\omega} \to [B]^{\leq\aleph_0}$. For $X \subseteq B$ let $f^*(X)$ be the closure of X, i.e., $f^*(X) = f_0(X) \cup f_1(X) \cup \cdots$, where $f_0(X) = X$, and $f_{n+1}(X) = \bigcup\{f(s) : s \in [f_n(X)]^{<\omega}\}$ for $n = 0, 1, 2, \ldots$. Notice that $f^*(X)$ is countable for X countable and $f^*(X) = \bigcup\{f^*(s) : s \in [X]^{<\omega}\}$. Set $g(s) = f^*(s) \cap A$ for $s \in [A]^{<\omega}$. Choose $P \in S$ such that P is closed under g. Set $Q = f^*(P)$. Clearly, Q is closed under f. We claim that $P = Q \cap A$, so $Q \in T$. Indeed,

$$Q \cap A = \bigcup\{f^*(s) \cap A : s \in [P]^{<\omega}\} = \bigcup\{g(s) : s \in [P]^{<\omega}\} = P.$$

(f) If the statement fails, then for every $x \in A$ there is some $f_x : [A]^{<\omega} \to [A]^{\leq\aleph_0}$ such that $F(a) \neq x$ holds for every a which is closed under f_x. Set

$$f(s) = \bigcup\{f_x(t) : \{x\} \cup t \subseteq s\}$$

for every $s \in [A]^{<\omega}$. Notice that $f(s)$ is countable, as it is the union of finitely many countable sets. As S is stationary, there is some $a \in S$ which is closed under f. If now $x = F(a) \in a$, then for every $t \in [a]^{<\omega}$ we have $f_x(t) \subseteq f(\{x\} \cup t) \subseteq a$, as a is f-closed, that is, a is closed under f_x, so $F(a) \neq x$, contradiction. [S. Shelah, W. H. Woodin]

30

The axiom of choice

1. By Cantor's theorem $\kappa < 2^\kappa = \aleph_0$ (see, e.g., Problem 10.21), that is, κ is finite, and then so is 2^κ.

2. First we construct the functions f_{ω^α}. For $\omega^\alpha = \omega$ set $f_\omega = g$ where $g(n,m) = 2^n(2m+1)-1$. Notice that $g(0,0) = 0$. Our intention is to construct f_{ω^α} by transfinite recursion on α in such a way that for $\beta < \alpha$ the function f_{ω^α} extends f_{ω^β}. Given f_{ω^α} define $f_{\omega^{\alpha+1}}$ as follows.

$$f_{\omega^{\alpha+1}}(\omega^\alpha n + \xi, \omega^\alpha m + \zeta) = \omega^\alpha g(n,m) + f_{\omega^\alpha}(\xi,\zeta).$$

If α is limit then, using the above extension property, we can take

$$f_{\omega^\alpha} = \bigcup\{f_{\omega^\beta} : \beta < \alpha\}.$$

Next we define $f_{\omega^\alpha \cdot n}$ for $1 < n < \omega$ by using f_{ω^α} and composing it with the following bijection $h : \omega^\alpha \cdot n \to \omega^\alpha$ (and its inverse); $h(\omega^\alpha \cdot m + \xi) = n \cdot \xi + m$ where $m < n$ and $\xi < \omega^\alpha$.

Finally, to construct $f_{\omega^\alpha \cdot n+\gamma}$ from $f_{\omega^\alpha \cdot n}$ for $\gamma < \omega^\alpha$ it suffices to give a bijection h between $\omega^\alpha \cdot n + \gamma$ and $\omega^\alpha \cdot n$ (and then we can compose h, $f_{\omega^\alpha \cdot n}$, and h^{-1}). Let h be the following function. $h(\omega^\alpha \cdot n + \xi) = \xi$ for $\xi < \gamma$, $h(\xi) = \gamma + \xi$ for $\xi < \omega^\alpha$, and finally, $h(\xi) = \xi$ for $\omega^\alpha \le \xi < \omega^\alpha n$.

3. It suffices to show that there is a surjection $\mathbf{R} \to \omega_1$ as for every $0 < \alpha < \omega_2$ there is a surjection from ω_1 onto α. For this, if $x \in \mathbf{R}$ codes some ordinal $\beta < \omega_1$ we map it to β, otherwise map it to 0. x codes β for example, if $\langle \omega, < \rangle$ is a well-ordered set of order type β, where $i < j$ if and only if the $2^i 3^j$-th digit of x is 1. We map all reals to 0 that do not code an ordered set or they do code, but the ordered set is not well ordered.

4. (a) We consider cases. Assume first that $\aleph_1 = \mathbf{c}$. We claim that there are exactly $\mathbf{c}$ perfect sets. For this, it suffices to show that there are at most $\mathbf{c}$

closed sets, which is equivalent to showing that there are at most $\mathbf{c}$ open sets. Every open set is the union of open intervals of rational endpoints, so, if $\mathcal{I}$ is the set of open intervals of rational endpoints, then the power set of $\mathcal{I}$, a set of cardinality $\mathbf{c}$ can be mapped onto the set of open sets. The latter set, being the surjective image of a set of cardinality $\mathbf{c} = \aleph_1$, itself is of cardinality at most $\aleph_1$, as claimed. Given this, one can define an uncountable set with no perfect subsets via diagonalization.

If $\aleph_1 < \mathbf{c}$, then any subset of the reals of cardinality $\aleph_1$ is obviously an uncountable set with no perfect subsets.

Assume, finally, that $\aleph_1 \not\leq \mathbf{c}$. By Problem 3, $\mathbf{R}$ has a surjection onto ω_1.

$\mathbf{c} + \mathbf{c} = \mathbf{c}$, as can be seen from the decomposition of any interval into two subintervals. Therefore, $\mathbf{R}$ has a surjection onto a set of cardinality $\mathbf{c} + \aleph_1$. But $\mathbf{c} + \aleph_1 > \mathbf{c}$ as $\mathbf{c} + \aleph_1 \geq \mathbf{c}$ and $\mathbf{c} + \aleph_1 = \mathbf{c}$ would give $\aleph_1 \leq \mathbf{c}$.

(b) Consider the Vitali decomposition of $\mathbf{R}$, i.e., $\mathcal{P} = \mathbf{R}/\sim$ where $x \sim y$ if and only if $x - y \in \mathbf{Q}$. If $f : \mathbf{R} \to \mathcal{P}$ is the mapping that sends $x \in \mathbf{R}$ into its class in $\mathcal{P}$, then f is an onto mapping. It is easy to give continuum many reals with pairwise irrational difference, so $|\mathbf{R}| \leq |\mathcal{P}|$. Assume that $|\mathbf{R}| < |\mathcal{P}|$ does not hold, i.e., $|\mathbf{R}| = |\mathcal{P}|$. Then, as $\mathbf{R}$ can be ordered, the set $\mathcal{P}$ can also be ordered, let $<$ be an ordering of it. Now let

$$A = \{x \in \mathbf{R} \setminus \mathbf{Q} : x + \mathbf{Q} < (-x) + \mathbf{Q}\}.$$

Then A cannot be measurable, as the mapping $x \mapsto r - x$ bijects A onto its relative complement in $\mathbf{R} \setminus \mathbf{Q}$ for every rational number r, and therefore A cannot have relative density greater than half in any rational interval, and the same also holds for its complement. [W. Sierpiński: Sur une proposition qui entraîne l'existence des ensembles non measurables, *Fund. Math.*, **34**(1947), 157–162]

(c) If there are no two disjoint stationary sets in ω_1, then by the second solution of Problem 20.19, there is no subset of $\mathbf{R}$ with cardinality $\aleph_1$. The argument there requires to prove that if $A_0, A_1, \ldots \subseteq \omega_1$ all include closed, unbounded subsets, then $A_0 \cap A_1 \cap \cdots$ is nonempty. For this, we need to fix club sets witnessing this, and this requires AC_ω. Now, if $\aleph_1 \not\leq \mathbf{c}$ then we can conclude as in part (a).

5. Assume that $m = kn$. In order to prove C_n let $\{A_i : i \in I\}$ be a collection of n-element sets. Let S be some set with k elements. As $\{A_i \times S : i \in I\}$ is a set of m-element sets, we can apply C_m to get a choice function g. Finally, just project g to get the required function f: let $f(i) = x$, where $g(i) = \langle x, y \rangle$ for some $y \in S$.

6. Let $\mathcal{F} = \{A_i : i \in I\}$ be a system of 4-element sets. Let $\mathcal{G}$ be the family of all 2-piece, 2-element set partitions of the elements of $\mathcal{F}$, that is, $\{X, Y\} \in \mathcal{G}$ if and only if $X, Y \in [A_i]^2$ for some $i \in I$ and $X \cap Y = \emptyset$.

Let $\mathcal{H}$ be the set of all two-element subsets of all A_i-s, that is,

$$\mathcal{H} = \bigcup\{[A_i]^2 : i \in I\}.$$

By hypothesis, there is a choice function g for $\mathcal{G}$, and another, h for $\mathcal{H}$. We are going to describe, in terms of g, h, a choice function f for $\mathcal{F}$. Given $i \in I$, there are 3 partitions in $\mathcal{G}$ corresponding to A_i. Given one of them $\{X, Y\}$, evaluate $h\big(g(\{X,Y\})\big)$. This is an element of A_i so we select 3 times some element of A_i.

We now consider cases. If the same point is selected 3 times then let it be $f(i)$. If a point is selected twice and another once, then let $f(i)$ be the point chosen twice. Finally, if three different points are chosen, then let $f(i)$ be the remaining point. [A. Tarski]

7. Let $\mathcal{F} = \{A_i : i \in I\}$ be a system of 6-element sets.

Let $\mathcal{G}, \mathcal{H}$ be the set of all two-element, respectively all three-element subsets of all A_i-s, that is,

$$\mathcal{G} = \bigcup\{[A_i]^2 : i \in I\},$$

$$\mathcal{H} = \bigcup\{[A_i]^3 : i \in I\}.$$

Let g, h be choice functions for $\mathcal{G}, \mathcal{H}$.

We first argue that it suffices to find somehow a function F such that $F(i)$ is a nonempty, proper subset of A_i. Indeed, from F, g, h we can construct a choice function for $\mathcal{F}$ as follows. If $F(i)$ is a singleton, let its only element be $f(i)$. If $|F(i)| = 2$, apply g to select an element of it. If $|F(i)| = 3$, apply h to select an element of it. If $|F(i)| = 4$, apply g to select an element of its complement. If $|F(i)| = 5$, let $f(i)$ be the only element of its complement.

To find a function F as described in the previous paragraph let G be a choice function on those 3-element sets that occur as 3-piece partitions of some A_i. For every such partition $\{X, Y, Z\}$ of some A_i we can canonically choose an element as follows. Set $e\big(\{X,Y,Z\}\big) = g\big(G(\{X,Y,Z\})\big)$. This way we associate with every such partition an element from A_i. As there are 15 such partitions we select 15 times an element of A_i. Let $F(i)$ be the set of those elements chosen at least 3 times. Then clearly $1 \leq |F(i)| \leq 5$. [A.Mostowski: Axiom of choice for finite sets, *Fundamenta Mathematicae*, **33**(1945), 137–168]

8. Let $\{A_i : i \in I\}$ be a system of nonempty, finite sets. By assumption, the set $\bigcup\{A_i : i \in I\}$ can be ordered. Now let, for $i \in I$, $f(i)$ be equal to the least (by the presumed ordering) element of A_i. Clearly, f is a choice function.

9. By induction, we can assume that $n = 1$. Assume that $|A| = \kappa$, $|B| = \lambda$, and $F : A \cup \{x\} \to B \cup \{y\}$ is a bijection. We seek for a bijection between A and B. If $F(x) = y$ we are done, $F|_A$ is a bijection from A to B. Otherwise, if $F(x) = y'$, $F(x') = y$, let F' be equal to F on $A - \{x'\}$, and $F'(x') = y'$.

10. As $\aleph_0 \leq \kappa$ holds κ can be written as $\kappa = \lambda + \aleph_0$ for some cardinal λ. We can then write $\kappa + \aleph_0 = (\lambda + \aleph_0) + \aleph_0 = \lambda + (\aleph_0 + \aleph_0) = \lambda + \aleph_0 = \kappa$.

11. Clearly, $\kappa + 1 \leq 2^\kappa$ holds for every κ. [For every set S the power set $\mathcal{P}(S)$ contains all one-element subsets of S plus the empty set.] Assume we have equality for some $\kappa > 1$. Then κ must be infinite. By assumption, for some set A of cardinality κ and for some element $p \notin A$ we have a bijection $F : \mathcal{P}(A) \to A \cup \{p\}$ and we can assume $F(A) = p$. Define the elements $a_0, a_1, \dots$ as follows. $a_0 = F(\emptyset)$, $a_{n+1} = F(\{a_0, \dots, a_n\})$. Induction shows that the elements $a_0, a_1, \dots$ are distinct. That is, $\aleph_0 \leq \kappa$, so by Problem 10 we have $\kappa+1 \leq \kappa+\aleph_0 = \kappa < 2^\kappa$, a contradiction. [E. Specker: Verallgemeinerte Kontinuumshypothese und Auswahlaxiom, *Archiv der Mathematik*, **5** (1954), 332–337]

12. Using Problem 10 and the fact that $\kappa \leq 2^\kappa$, we get

$$\kappa + 2^\kappa \leq 2^\kappa + 2^\kappa = 2^{1+\kappa} = 2^\kappa.$$

13. For transitivity, assume that $a \ll b \ll c$. Then $a + c = a + (b + c) = (a + b) + c = b + c = c$, so $a \ll c$ holds as well.

As for the second statement, for one direction, if $\aleph_0\kappa \leq \lambda$ then $\lambda = \aleph_0\kappa + \mu$ for some cardinal μ. Then,

$$\kappa + \lambda = \kappa + (\aleph_0\kappa + \mu) = (\kappa + \aleph_0\kappa) + \mu = (1 + \aleph_0)\kappa + \mu = \aleph_0\kappa + \mu = \lambda.$$

For the other direction, let A, B be disjoint sets with $|A| = \kappa$, $|B| = \lambda$, and assume that $f : A \cup B \to B$ is an injection. Then, for $j > 0$ we have $A \cap f^j[A] \subseteq A \cap B = \emptyset$, so $f^i[A] \cap f^{i+j}[A] = \emptyset$, that is, the sets $f[A], f^2[A], \dots$ are disjoint subsets of B of cardinal κ. This shows that $\aleph_0\kappa \leq \lambda$.

14. In the former case $\kappa + \kappa = 2\kappa = 2(\aleph_0\lambda) = (2\aleph_0)\lambda = \aleph_0\lambda = \kappa$. In the latter case we have $1+\lambda = \lambda$ by Problem 10, and so $\kappa+\kappa = 2^\lambda+2^\lambda = 2^{1+\lambda} = 2^\lambda = \kappa$.

15. We have to show that if A, B are sets and $A \times \{0,1\} \sim B \times \{0,1\}$, then $A \sim B$. We can assume that A, B are disjoint. Let $f : A \times \{0,1\} \to B \times \{0,1\}$ be a bijection.

We construct an edge-colored, directed graph as follows. The vertices are the sets of the form $\{\langle x, i\rangle, \langle y, j\rangle\}$, where $f(\langle x, i\rangle) = (\langle y, j\rangle)$, that is, we *identify* $\langle x, i\rangle$ and $\langle y, j\rangle$. Draw an edge from $\langle x, 0\rangle$ to $\langle x, 1\rangle$ and color it red (for $x \in A$), and draw an edge from $\langle y, 0\rangle$ to $\langle y, 1\rangle$ and color it blue (for $y \in B$). (So every point in either A or B is represented as an edge of this graph.) We have a directed graph in which every vertex is on exactly one red and one blue edge. Therefore, its connected components are finite cycles of even lengths (possibly of length 2) and 2-way infinite paths. Our task is to determine a bijection between the red and the blue edges.

We do this individually for the components. There is no problem (actually, no choice) in the case of a cycle of length 2: map the edges to each other. Also, if the edges of a cycle or an infinite path are consecutively directed, i.e., it is a directed cycle or path, we define the bijection, if the red edge $\overrightarrow{\langle x,0\rangle\langle x,1\rangle}$ is followed by the blue $\overrightarrow{\langle y,0\rangle\langle y,1\rangle}$, then map x to y.

Otherwise, there are some pairs of edges with the same vertex as the ends of the arrow, i.e., of the type $\overrightarrow{uv}$ and $\overleftarrow{vw}$. Pair all these edges to each other, and cut them out from the cycle/path in question. This means that, using the above notation, we remove v and identify u and w (and of course, the point represented by $\overrightarrow{uv}$ is mapped to the point represented by $\overleftarrow{vw}$). Notice that, as we remove pairs of consecutive edges, in the remaining part of the cycle or path, the edges again come interchangingly as red, blue, red, etc. Repeat this operation inductively.

In the case of finite cycles in finitely many steps, we either pair all the edges or eventually we get a fully directed cycle, and this case is handled as above.

In the case of infinite paths, if we repeat this argument infinitely many times, either all the edges get eventually paired up, in which case we are done, or there remains a finite, or infinite part. If the remaining part is infinite, the edges are necessarily consecutively directed and we are done with the above argument.

If the remaining part has an even number of edges, we can pair them up starting with either end.

If, however, the remaining part has an odd number of edges, then it has a medium edge, so we identified an edge in the path, and we can use that to define a pairing of the edges (in the original) path, for example if it is a red edge $\overrightarrow{uv}$ then map it to the next edge (that is, to $\overleftarrow{vw}$ or $\overrightarrow{vw}$), and continue this bijection both ways. [F. Bernstein: *Untersuchungen aus der Mengenlehre*, Inaugural Dissertation, Halle, 1901. This proof is from W. Sierpiński: Sur l'égalité $2m = 2n$ pour les nombres cardinaux, *Fund. Math.* **3**(1922), 1–6.]

16. Let A be a set of cardinality κ. As A is infinite for every natural number n there are subsets of A of cardinality n. Let $f(n) = \{X \subseteq A : |X| = n\}$. Then $f(n)$ is a nonempty subset of $\mathcal{P}(A)$ so f is an embedding of the set of natural numbers into $\mathcal{P}(\mathcal{P}(A))$.

17. If α is a countable ordinal, then there are subsets of $(\mathbf{Q},<)$ of order type α. [As we are not assuming the axiom of choice this is a little delicate. Let $\langle A,\prec\rangle$ be an ordered set of order type α. Enumerate A as $A = \{a_i : i < \omega\}$ and $\mathbf{Q}$ as $\mathbf{Q} = \{q_i : i < \omega\}$. Define the order preserving $f : A \to \mathbf{Q}$ as follows. $f(a_0) = q_0$, and then by induction let $f(a_{i+1})$ be q_j, where j is minimal with respect to the condition that this choice be consistent with order preservation.] Let F be the function that maps every $\alpha < \omega_1$ to the set of all subsets of $\mathbf{Q}$ with order type α. By the above argument, F is injective. We have that $F(\alpha) \subseteq \mathcal{P}(\mathbf{Q})$ that is, F maps into $\mathcal{P}(\mathcal{P}(\mathbf{Q}))$.

18. If A is some set and $\langle x, y\rangle \in A \times A$, then by the definition of ordered pairs $\langle x, y\rangle = \{\{x\}, \{x, y\}\} \in \mathcal{P}(\mathcal{P}(A))$. This implies that $|A \times A| \leq 2^{2^{|A|}}$

19. If $|A| = \kappa$ and β is an ordinal with $|\beta| \leq \kappa$, then there exist binary relations $R \subseteq B \times B$ for some $B \subseteq A$ that well-order B into type β. We can recover β from R; therefore, the mapping

$$\beta \mapsto \{R \subseteq A \times A : R \text{ orders some subset of } A \text{ into type } \beta\}$$

is an injective mapping of those ordinals into $\mathcal{P}(A \times A)$. We cannot inject all ordinals (a proper class) into a set, because then by the axiom of comprehension (which states that if X is a set and φ is a formula with one free variable, then the elements of X that satisfy φ form again a set) the image is a set, and then the inverse is a mapping from that set to the class of ordinals which contradicts the axiom of replacement. This shows that $H(\kappa)$ exists with $|H(\kappa)| \not\leq \kappa$. By the above argument we have an injection of the ordinals below $H(\kappa)$ into $\mathcal{P}(A \times A)$, and by the previous problem $|H(\kappa)| \leq 2^{\kappa \cdot \kappa} \leq 2^{2^{2^{\kappa}}}$. [F.Hartogs: Über das Problem der Wohlordnung, *Matematische Annalen*, **76**(1915), 436–443]

20. Let A be an infinite set, we show it has a well-ordering. By Hartogs' lemma (Problem 19) there is a well-ordered set $\langle B, \prec\rangle$ such that $|B| \not\leq |A|$. We show that $|A| \leq |B|$ and this suffices, for the ordering on B can be pulled back to A. We can assume that A, B are disjoint. By assumption, there is an injection $F : (A \cup B) \times (A \cup B) \to A \cup B$. If $x \in A$, then the mapping $y \mapsto F(x, y)$ cannot map into A, as there is no injection from B into A. There are, therefore, elements $y \in B$ with $F(x, y) \in B$. Let y_x be the least (by $\prec$, the well-ordering of B) such element. Then $x \mapsto F(x, y_x)$ is an injection from A into B.

21. It suffices, by Problem 20, to show that GCH implies $\kappa^2 = \kappa$ for every infinite cardinal κ. Assume κ is infinite. As $\kappa \leq \kappa + 1 < 2^\kappa$ holds by Problem 11 we have $\kappa + 1 = \kappa$, so $\aleph_0 \leq \kappa$ (Problem 13). Next, $\kappa \leq \kappa + \kappa \leq 2^\kappa + 2^\kappa = 2^{\kappa+1} = 2^\kappa$ as $\kappa + 1 = \kappa$. We have, therefore, either $\kappa + \kappa = \kappa$ or $\kappa + \kappa = 2^\kappa$. In the latter case, we had $2\kappa = 2^\kappa = 2^{1+\kappa} = 2 \cdot 2^\kappa$ so we could deduce, using Problem 15, that $\kappa = 2^\kappa$, a contradiction to Problem 11. We have, therefore, $\kappa + \kappa = \kappa$. Also, $\kappa \leq \kappa^2 \leq 2^\kappa 2^\kappa = 2^{\kappa+\kappa} = 2^\kappa$ so either $\kappa^2 = \kappa$ or $\kappa^2 = 2^\kappa$, and we can assume the latter. Let S be a set of cardinality κ. We show that it has a well-ordering (and so by Problem 2 $\kappa^2 = \kappa$ holds). Fix some bijection $F : \mathcal{P}(S) \to S \times S$. As $\aleph_0 \leq \kappa$ there are infinite well-orderable sets $X \subseteq S$. If we give a method to select an element from $S \setminus X$ for every infinite well-ordered $\langle X, \prec\rangle$ with $X \subseteq S$, $X \neq S$, we can copy the proof of the well-ordering theorem. Here is what we do for such an X. Let $f : X \times X \to X$ be the injection from Problem 2. For $x \in X$, set $x \in Y$ if and only if $f^{-1}(x)$ is defined, and $x \notin F^{-1}(f^{-1}(x))$. $F(Y)$ cannot be an element of $X \times X$ as

if $F(Y) = f^{-1}(y)$ then $y \in Y$ holds if and only if $y \notin Y$. We get, therefore, that $F(Y) = \langle u, v\rangle$ where either $u \in S \setminus X$, in which case we select u, or else $v \in S \setminus X$ and then we select v. [A. Lindenbaum, A. Tarski: Communication sur les recherches de la Théorie des Ensembles, *Comptes Rendus des Séances de la Société des Sciences et des Lettres de Varsovie*, **19** (1926), 299–330. First published proof given in W. Sierpiński: L'hypothèse généralisée du continu et l'axiome du choix, *Fund. Math.* **34** (1947), 1–5.]

22. By the axiom of foundation every set is a subset of some V_α of the cumulative hierarchy. It suffices, therefore, to show that every V_α can be well ordered under the stated condition. Given V_α, let g be a function defined on the nonempty subsets of V_α such that $g(X)$ is always a finite, nonempty subset of X. From g, we will construct a well-ordering $<_\alpha$ of V_α. Actually, we construct by transfinite recursion on $\gamma \leq \alpha$, a well-ordering $<_\gamma$ of V_γ. Let $<_0$ be the only ordering of the one-element V_0. If $\gamma \leq \alpha$ is limit and $<_\delta$ is already defined for all $\delta < \gamma$ we let, for $x, y \in V_\gamma$, $x <_\gamma y$ if and only if either $\mathrm{rk}(x) < \mathrm{rk}(y)$ or else $\delta = \mathrm{rk}(x) = \mathrm{rk}(y)$ and $x <_\delta y$. That is, we endow each $V_\delta \setminus \bigcup\{V_\xi : \xi < \delta\}$ with the ordering $<_\delta$ and place them one after the other. Assume now that $\gamma = \delta + 1$ and we have the well-ordering $<_\delta$ on V_δ. Now $V_\gamma = \mathcal{P}(V_\delta)$. The proof of the well-ordering theorem gives a well-ordering of V_γ once we give a choice function f on all nonempty subsets of V_γ. We define f as follows. If $X \subseteq V_\gamma$, $X \neq \emptyset$, our g gives a finite, nonempty subset $g(X)$ of X, say $\{Y_1, \dots, Y_n\}$. Notice that each Y_i is a subset of V_δ, which is already well ordered by $<_\delta$. We can now select the lexicographically least Y_i as $f(X)$. [H. Rubin, J. Rubin: *Equivalents of the Axiom of Choice*, North-Holland, 1963]

23. Let $\{A_i : i \in I\}$ be a system of nonempty sets; it suffices (by Problem 22) to show that there is a function selecting a nonempty finite subset of each. We can assume, without loss of generality, that the A_i's are disjoint. Let k be an arbitrary field, and adjoin all elements of $X = \bigcup\{A_i : i \in I\}$ as indeterminates to k. We get the field $k(X)$ of rational functions of X. Call a polynomial $p \in k[X]$ *i-homogeneous of degree d* if the sum of the exponents of elements of A_i is d in every monomial in p. Call a rational function $\frac{p}{q} \in k(X)$ *i-homogeneous of degree d* if there is some n that p is i-homogeneous of degree $n + d$ and q is i-homogeneous of degree n. Let K be the subfield of $k(X)$ generated by k and all elements of the form y/x where $x, y \in A_i$ $(i \in I)$. Clearly, every element of K is i-homogeneous of degree 0 for every i. Let V be the vector space over K generated by X. By assumption, V has a basis B. For $i \in I$, $x \in A_i$, x can uniquely be written as

$$x = \sum_{b \in B(x)} \alpha_b(x) b,$$

where $B(x)$ is a finite subset of B and $\alpha_b(x)$ is a nonzero element of K. If $y \in A_i$ is another element, then

$$y = \sum_{b \in B(y)} \alpha_b(y)b = \sum_{b \in B(x)} \frac{y}{x}\alpha_b(x)b,$$

so we get that $B(x) = B(y)$ and $\alpha_b(y) = \frac{y}{x}\alpha_b(x)$. Thus, the ratio $\alpha_b(x)/x$ depends only on i, not on $x \in A_i$. As $\alpha_b(x) \in K$, the rational function $\alpha_b(x)/x$ has homogeneous i-degree -1. Therefore, some variables from A_i must occur in the denominator. Let B_i be the set of those variables for all $b \in B(x)$. Then, for every $i \in I$, B_i is a nonempty finite subset of A_i and we are done. [A. Blass: Existence of bases implies the axiom of choice, *Axiomatic Set Theory* (J.E. Baumgartner, D.A. Martin, S. Shelah, eds), Contemporary Mathematics, **31**, 1984, 31–33]

24. One direction is easy: if the axiom of choice is assumed, X is some graph on a set $V \neq \emptyset$, then the set of those cardinals, for which there is a good coloring of X, is a well-ordered set of cardinals, with $|V|$ as the largest element. It has, therefore, a smallest element, and that is the chromatic number of X.

For the other implication assume that there is a cardinal κ that cannot be well ordered. There is, by Hartogs' lemma (Problem 19) an ordinal φ such that $|\varphi| \not\leq \kappa$. Notice that $\kappa \not\leq |\varphi|$ also holds (as otherwise κ would be well orderable). Let A be some set of cardinality κ. Let the vertex set V of our graph X be $A \times \varphi$. Join two vertices $\langle x, y\rangle$ and $\langle x', y'\rangle$ if and only if $x \neq x'$ and $y \neq y'$ both hold. Notice that for this graph both projections $\langle x, y\rangle \mapsto x$ and $\langle x, y\rangle \mapsto y$ are good colorings, therefore if μ, the chromatic number of X exists, then $\mu \leq \kappa, |\varphi|$. As $|\varphi|$, κ are incomparable, equality cannot hold, so $\mu < \kappa, |\varphi|$. As $\mu < |\varphi|$, μ is a well-orderable cardinal. By the definition of chromatic number there is a decomposition $A \times \varphi = \bigcup\{A_i : i \in I\}$ into independent vertex sets with $|I| = \mu$.

Consider first the case when for every $x \in A$ there is some $i \in I$ with A_i intersecting $\{x\} \times \varphi$ in more than one element. Let $I(x)$ be the set of these indices i, so $I(x) \subseteq I$, nonempty. Now $I(x) \cap I(x') = \emptyset$ holds for $x \neq x'$, indeed, otherwise we could find $\langle x, y\rangle, \langle x', y'\rangle \in A_i$ for some i with $x \neq x'$, $y \neq y'$, an impossibility. As I can be well ordered, we can choose the least element (by that ordering) of each $I(x)$, let this be $f(x)$. Then $f : A \to I$ is an injection, contradicting $\mu < \kappa$.

Finally, we consider the case that there is some $x \in A$ such that $A_i \cap (\{x\} \times \varphi)$ has at most one element for every $i \in I$. Then the mapping $\alpha \mapsto i(\alpha)$ for $\alpha < \varphi$, where $\langle x, \alpha\rangle \in A_{i(\alpha)}$, will be an injection $\varphi \to I$ which contradicts $\mu < |\varphi|$. [F.Galvin, P.Komjáth: Graph colorings and the axiom of choice, *Periodica Math. Hung.* **22** (1991), 71–75]

25. Assume, toward a contradiction, that A is a set that cannot be well ordered. Let $\kappa = H(A)$ be its Hartog's ordinal (see Problem 19). We notice that κ is a cardinal and $|A|$ and κ are both smaller than their product, $|A|\kappa$. Indeed, $|A| \leq |A|\kappa$ and $\kappa \leq |A|\kappa$ are obvious, and equality in either case would give either $|A| \leq \kappa$ or $\kappa \leq |A|$, which are ruled out by the non-well orderability of A and by the Hartog's property of κ, respectively.

On the set $A \times \kappa$ define the following set mapping. For $\langle x, \alpha\rangle \in A \times \kappa$, let $f(x, \alpha) = \{x\} \times \alpha$. Notice that $|f(x, \alpha)| = |\alpha| < \kappa < |A \times \kappa|$ by the above remark. Assume by Hajnal's theorem that $X \subseteq A \times \kappa$ is a free set of cardinality $|A \times \kappa|$. For every $x \in A$, X intersects $\{x\} \times \kappa$ in at most one point, so the projection to the first coordinate shows $|X| \leq |A|$, which contradicts $|A| < |A \times \kappa|$. [Norbert Brunner: Set-mappings on Dedekind sets, *Notre Dame Journal of Formal Logic*, **30**(1989), 268–270]

26 Let $\mathbf{R} = A_0 \cup A_1 \cup \cdots$ be a decomposition into countable sets. By Problem 3 there is a surjection $f : \mathbf{R} \to \omega_1$. If $B_i = f[A_i]$ $(i < \omega)$ then $\omega_1 = B_0 \cup B_1 \cup \cdots$ is a decomposition of ω_1 into the union of countably many countable sets.

For the second claim we first observe that there is a countable subset B with $\sup(B) = \omega_1$. Indeed, if no B_i satisfies this, then $\beta_i = \sup(B_i) < \omega_1$ for each of them, but then $B = \{\beta_0, \beta_1, \ldots\}$ is as required. As by Hausdorff's theorem there is a cofinal ω-sequence in B, we are done.

27. Assume that $\omega_2 = A_1 \cup A_2 \cup \cdots$ where every A_i is countable. We may as well assume that the sets are disjoint (otherwise replace A_i by $A_i \setminus (A_1 \cup \cdots \cup A_{i-1})$). Every A_i inherits a well-ordering from ω_2, let its order type be δ_i. Clearly, each δ_i is countable. We can map A_i onto

$$[\delta_0 + \cdots + \delta_{i-1}, \delta_0 + \cdots + \delta_i)$$

(where $\delta_0 = 0$) by mapping the αth element of A_i to $\delta_0 + \cdots + \delta_{i-1} + \alpha$. This will map $\omega_2 = A_1 \cup A_2 \cup \cdots$ to $\delta_1 + \delta_2 + \cdots = \lim\{\delta_1 + \cdots + \delta_i : i < \omega\}$. The latter ordinal is the increasing limit of countable ordinals; it is therefore at most ω_1. So we reached an embedding of ω_2 into ω_1, a contradiction. [T. Jech: On hereditarily countable sets, *Journ. Symb. Logic*, **47**(1982), 43–47]

31

Well-founded sets and the axiom of foundation

1. (a) → (b) If we are given a partially ordered set $\langle P, <\rangle$, then define $R(x, y)$ iff $y < x$. Then the condition for DC holds, so we get that there are $a_0, a_1, \ldots$ with $R(a_0, a_1), R(a_1, a_2), \ldots$, i.e., $\cdots < a_1 < a_0$.

(b) → (a) Assume that we are given the binary relation R on the nonempty set A such that for every $x \in A$ there is some $y \in A$ with $R(x, y)$. Let P be the set of all finite sequences $\langle a_0, \ldots, a_n\rangle$ where $a_0, \ldots, a_n$ are elements of A and $R(a_0, a_1), \ldots, R(a_{n-1}, a_n)$ all hold. Partially order P by making $\langle b_0, \ldots, b_m\rangle < \langle a_0, \ldots, a_n\rangle$ if and only if $\langle b_0, \ldots, b_m\rangle$ is a proper end-extension of $\langle a_0, \ldots, a_n\rangle$, i.e., $m > n$ and $b_i = a_i$ holds for $0 \le i \le n$. P is clearly nonempty (it contains the one-element sequences) and it has no minimal element, as $\langle a_0, \ldots, a_n\rangle$ has proper end-extensions, for example, $\langle a_0, \ldots, a_n, a_{n+1}\rangle$ where a_{n+1} any element for which $R(a_n, a_{n+1})$ holds. There is, by (b), an infinite decreasing chain in $\langle P, <\rangle$ and this gives a sequence $a_0, a_1, \ldots$ of elements of A such that $R(a_i, a_{i+1})$ holds for $i = 0, 1, \ldots$.

(b) → (c) Assume that a partially ordered set $\langle Q, <\rangle$ is ill founded. Then there is a nonempty $P \subseteq Q$ with no minimal element. By (b) there is an infinite descending chain in $\langle P, <\rangle$, therefore in $\langle Q, <\rangle$.

(c) → (b) If a partially ordered set $\langle P, <\rangle$ has no minimal element, then it is certainly not well founded so by (c) there is an infinite descending chain in $\langle P, <\rangle$.

2. One direction is obvious: assume that there is a monotonic ordinal-valued function f on P and $Q \subseteq P$ is nonempty. Pick $p \in Q$ with $f(p)$ minimal. Then p is a minimal element in Q: should $q < p$ hold for some $q \in Q$ we would get $f(q) < f(p)$, contradicting the minimality of $f(p)$.

Assume now that $\langle P, <\rangle$ is well founded. By transfinite recursion on α we select the subsets $P_\alpha \subseteq P$ as follows. P_0 is the set of minimal elements of $\langle P, <\rangle$. In general, P_α is the set of minimal elements of

$$P \setminus \bigcup_{\beta<\alpha} P_\beta.$$

By well foundedness, P_α is nonempty, so long as the above set is nonempty, and obviously these sets are disjoint. So eventually we decompose P as $P = \bigcup\{P_\alpha : \alpha < \varphi\}$ for some ordinal φ. Assume that $p < q$ are in P and $q \in P_\alpha$. Then q is a minimal element in the corresponding set, so p cannot be in that set, hence $p \in P_\beta$ for some $\beta < \alpha$. We can, therefore, define $f(p) = \alpha$ iff $p \in P_\alpha$.

3. By the well-ordering theorem we can enumerate P as $P = \{p_\alpha : \alpha < \varphi\}$ for some ordinal φ. Put p_α into Q iff there is no $\beta < \alpha$ with $p_\alpha < p_\beta$.

We show that $Q \subseteq P$ is as required.

$\langle Q, <\rangle$ is well founded: if there is a decreasing chain $\cdots < q_1 < q_0$ in Q, that is, $\cdots < p_{\alpha_1} < p_{\alpha_0}$, then, by the well-ordering property of ordinals, we have $\alpha_n < \alpha_{n+1}$ for some n, that is, p_{α_n} is greater than the later $p_{\alpha_{n+1}}$, a contradiction.

Q is cofinal: assume that $p \in P$. Choose $p_\alpha \geq p$ with α minimal. Then $p_\alpha \in Q$, indeed, otherwise, there is some $p \leq p_\alpha < p_\beta$ with $\beta < \alpha$, but that contradicts the minimal choice of α.

4. The counterexample will be built on the Cartesian product $\omega_1 \times \omega_1$. We make $\langle \alpha, \beta\rangle \prec \langle \alpha', \beta'\rangle$ if and only if $\alpha < \alpha'$ and $\beta > \beta'$. In a supposed infinite decreasing/increasing sequence the first/second coordinates would give an infinite decreasing sequence of ordinals, which is impossible. Assume, toward a contradiction, that $\omega_1 \times \omega_1 = A_0 \cup A_1 \cup A_2 \cup \cdots$ is a decomposition into countable many antichains. For every $\alpha < \omega_1$ there is some natural number $i(\alpha)$ such that for uncountably many β we have $\langle \alpha, \beta\rangle \in A_{i(\alpha)}$. By the pigeon hole principle there are ordinals $\alpha < \alpha'$ and some number i that $i = i(\alpha) = i(\alpha')$ holds. Pick an $\langle \alpha', \beta'\rangle \in A_i$. As there are arbitrarily large β with $\langle \alpha, \beta\rangle \in A_i$ we can select with $\beta > \beta'$ and then we get $\langle \alpha, \beta\rangle, \langle \alpha', \beta'\rangle \in A_i$ that is, $\langle \alpha, \beta\rangle \prec \langle \alpha', \beta'\rangle$, a contradiction.

5. Indeed, the function f constructed in the solution of Problem 2 has this property; if $f(p) = \alpha$ and $\beta < \alpha$ then p is an element of $P_\alpha \subseteq \bigcup\{P_\gamma : \gamma \geq \beta\}$ and by the well foundedness of $\langle P, <\rangle$ there is a minimal element q of this latter set *below* p so $q \leq p$ and $q \in P_\beta$. Clearly, $q < p$ as $f(q) \neq f(p)$.

Toward unicity, assume that r_0 and r_1 both have the properties described in the problem and $r_0 \neq r_1$. Then $\{x \in P : r_0(x) \neq r_1(x)\}$ is nonempty, so there is a minimal element p in it. But then both $r_0(p)$ and $r_1(p)$ are the least strict upper bound for $\{r_0(x) : x < p\} = \{r_1(x) : x < p\}$. We get therefore $r_0(p) = r_1(p)$ and this gives a contradiction to the choice of p.

6. If $T \subseteq \mathrm{FS}(\kappa)$ is a tree, $R(T) = \alpha$, and r is a rank function witnessing this, then by $|\mathrm{FS}(\kappa)| = \kappa$ implies $|T| \leq \kappa$, and as r assumes every value $\leq \alpha$ we have $|\alpha + 1| \leq \kappa$, i.e., $\alpha < \kappa^+$.

We prove the other statement by transfinite induction on α.

For $\alpha = 0$ we can (and must) take the one-element tree, that is, the one consisting of $\emptyset$, the empty sequence.

Assume that $\alpha < \kappa^+$ and $T \subseteq \mathrm{FS}(\kappa)$ is a tree with $R(T) = \alpha$. Let T' be the tree consisting of the empty string plus all strings of the form $0\hat{\ }s$ for $s \in T$. Clearly, T' is also well founded. If r, r' are the rank functions assigned to T, T', respectively, then by induction on $r(s)$ we get that $r'(0\hat{\ }s) = r(s)$ holds for every $s \in T$, so $r'(0) = \alpha$, and finally $r'(\emptyset) = \alpha + 1$.

Assume finally that $0 < \alpha < \kappa^+$ is a limit ordinal, and we have the construction for every ordinal less than α. Enumerate α, that is, the ordinals below α, as $\alpha = \{\beta(i) : i < \kappa\}$ (with possible repetitions). For $i < \kappa$ let T_i be a tree with $R(T_i) = \beta(i)$. Let the tree T consist of $\emptyset$, the empty string, and of the strings of the form $i\hat{\ }s$ for $s \in T_i$. T is obviously well founded (all but the first elements of a putative infinite decreasing sequence would be in some T_i). Let r be the rank function of T and r_i that of T_i. We again get that $r(i\hat{\ }s) = r_i(s)$, so $r(i) = \beta(i)$ and hence

$$R(T) = r(\emptyset) = \sup\{\beta(i) + 1 : i < \kappa\} = \alpha.$$

7. Let r, r' be the rank functions for T, T'. We construct the appropriate $f : T \to T'$ by recursion on the level, and during the recursion we keep the property $r(x) \leq r'(f(x))$. First map the root of T to the root of T'. Extend f from level n to level $n+1$ by keeping the condition $r(x) \leq r'(f(x))$ for every x. This is possible as if we have $r(x) \leq r'(f(x))$ for some $x \in T$, then the left-hand side is the strict smallest upper bound of all values $r(y)$ for $x \lhd y$, the right-hand side is the strict smallest upper bound of similar values $r'(z)$ for $f(x) \lhd z$, so for each y in the former set we can choose an appropriate $f(y)$ in the latter set with $r(y) \leq r'(f(y))$.

8. Using the previous problem it suffices to show that if T' is ill founded, then $T \preceq T'$ holds for any tree T. Indeed, if $\emptyset = y_0 \lhd y_1 \lhd \cdots$ is an infinite branch in T', then we can set $f : T \to T'$ where $f(x) = y_n$ whenever x is on level n in T.

9. Irreflexivity and trichotomy are clear. For transitivity assume that $s <_{\mathrm{KB}} t <_{\mathrm{KB}} u$ where

$$s = s(0)s(1)\cdots s(n), \quad t = t(0)t(1)\cdots t(m), \quad \text{and} \quad u = u(0)u(1)\cdots u(k).$$

We have to show that $s <_{\mathrm{KB}} u$ holds. There are several cases to consider. If s extends t, t extends u, then obviously s extends u and we are done. If s extends t and $t(i) < u(i)$ at the first difference, then clearly $s(i) < u(i)$ also holds, and that is where the first difference occurs. Next, assume that $s(i) < t(i)$ holds at the first difference and t extends u. Then either s extends u (if u is so short that $u(i)$ does not exist) or else $s(i) < u(i) = t(i)$ and this is the least difference, we are done in either case. Assume finally that $s(i) < t(i)$

holds at the least difference of s, t and $t(j) < u(j)$ holds at the least difference of t, u. If $i = j$ then $s(i) < t(i) < u(i)$ hold and i is the place of the first difference of s and u. If $i < j$ then $s(i) < t(i) = u(i)$ hold, and j is the place of the first difference, if $j < i$ then $s(j) = t(j) < u(j)$ hold, and it is the place of the first difference.

Concerning the other statement first notice that if T is not well founded then there is a infinite chain $s_0 \lhd s_1 \lhd s_2 \lhd \cdots$ and this itself constitutes a $<_{KB}$-descending sequence. For the other direction assume that $s_0, s_1, s_2, \ldots$ is a $<_{KB}$-descending sequence. Only s_0 can be the empty sequence. Therefore, $s_1(0), s_2(0), \ldots$ all exist. As we have a $<_{KB}$-descending sequence, we must have $s_1(0) \geq s_2(0) \geq \cdots$, so $s_i(0)$ stabilizes from some point on: $s_i(0) = t(0)$ for $i \geq n_0$. Only the first of these elements, s_{n_0} can possibly be of length one, for the rest we have that

$$s_{n_0+1}(1) \geq s_{n_0+2}(1) \geq \cdots$$

holds, and that must stabilize again from some point: $s_i(1) = t(1)$ for $i \geq n_1$. Repeating this argument we get an infinite string $t = t(0)t(1)\cdots$ whose every finite initial segment is the initial segment of some s_i. Therefore, these segments are elements of T, and they form an infinite decreasing sequence in T as was needed.

10. (a) Assume that W has no winning strategy. That is, at the starting position, W has no winning strategy. He cannot make a step after which he will possess a winning strategy as this would mean that he had one at the beginning. After W's first step, B can always answer that W still won't have a winning strategy. Indeed, if for every answer of B, W could produce a winning strategy, by combining them into one strategy, he could get a winning strategy outright. This argument gives that B can forever prolong the situation that W has no winning strategy. But this strategy must be a winning strategy for B, as the game certainly ends in finitely many steps (the trees are well founded), and otherwise if the play was a win for W then the last move is obviously made by W and he, therefore, has a winning strategy at the very last moment (before making the final, and winning, move).

(b) In virtue of (a) it suffices to derive a contradiction from the assumption that B has a winning stategy. Let σ be such a strategy. Let $T_0, T_1, \ldots$ be trees, isomorphic to the tree on which the original game is played. We place a pawn on the root of every T_i. At every step, one of the pawns is moved one step up. We also have players $p_0, p_1, p_2, \ldots$. p_0 is a moron, he makes a move on T_0, whenever asked for. Each p_i for $i \geq 1$ sees only T_{i-1} and T_i, p_i believes that she is B, she thinks that T_{i-1} is T_W and T_i is T_B and she playes according to σ. We also have some function $f(\alpha)$ that tells us where the game is played at moment α.

First $f(0) = 0$ and p_0 makes an arbitrary move on T_0. Next $f(1) = 1$. In general, if $f(\alpha) = i > 0$, then player p_i wakes up and investigates T_{i-1}

and T_i. If she observes that one of the pawns has been moved up one step since her last action, then she answers according to σ. If she moves on T_i then we let $f(\alpha+1) = i+1$, otherwise (if she moves on T_{i-1} or passes) we let $f(\alpha+1) = i-1$. If, however, she observes that there was no movement then she passes and we let $f(\alpha+1) = i-1$. When $f(\alpha) = 0$, p_0 makes a step on T_0, and we define $f(\alpha+1) = 1$. Observe that if there is a pass by a p_i then everybody will pass until p_0 makes a move.

Notice that f cannot attain the same value infinitely many times as T_i is well founded and if the pawn on it reaches a terminal node, then p_{i+1} would observe in the next step that she lost, although she played according to σ. We get, therefore, that $f(\alpha)$, that is, the center of action, must tend to infinity. Then we write $\alpha = \omega$, $f(\omega) = 0$, and again have p_0 make a move. This way we can continue the game so long as $\alpha < \omega^2$. But this is impossible, for then at some step $\alpha < \omega^2$ the pawn on T_0 must reach a terminal node, which is a contradition, as we have seen. [Fred Galvin]

11. Let $\langle P, <\rangle$ consist of an increasing sequence $x_0 < x_1 < \cdots$. Let $\langle Q, <\rangle$ contain one largest element, y, plus a chain L_n of length n, for every positive natural number n. We make the chains L_n incomparable, but smaller, of course, than y. It is obvious that both $\langle P, <\rangle$ and $\langle Q, <\rangle$ are well founded (every element in $\langle P, <\rangle$ and all but one elements in $\langle Q, <\rangle$ have finitely many elements below).

Assume that f is an order-preserving mapping from $\langle P, <\rangle$ into $\langle Q, <\rangle$. If $f(x_0) \in L_n$, then $f(x_{n+1})$ would be greater than y, an impossibility. Thus, such an f does not exist.

Assume that f is an order-preserving mapping from $\langle Q, <\rangle$ into $\langle P, <\rangle$. If $f(y) = x_n$ then we are in trouble in finding room for the image of the chain L_{n+1} of size $n+1$. Thus, such an f does not exist, either.

12. Assume $x \in x$. Set $A = \{x\}$. By the axiom of foundation there is $y \in A$ with $y \cap A = \emptyset$. $y = x$ is the only possibility, but as $x \in x \cap A$ we have a contradiction.

13. Assume that $x \in y$ and $y \in x$. Set $A = \{x, y\}$. Applying the axiom of foundation we get that there is some $z \in A$ with $z \cap A = \emptyset$, which is nonsense, as if $z = x$ then y is a common element in z and A, and if $z = y$ then x is.

14. For $n = 0$ the empty set (and only it) will be good. If some set A is good for n, then $A \cup \{A\}$ is good for $n+1$.

We now prove by induction on n that there is only one good set for n. This clearly holds for $n = 0$. Assume we have this for some n and A, B are good sets for $n+1$. That is, A, B are both $n+1$-element transitive sets, ordered by $\in$. Let a, b be the largest elements. Then a, b are good sets for n, so $a = b$. Moreover, $A = a \cup \{a\} = b \cup \{b\} = B$, and we are done.

15. Assume that $\{x\}$ is transitive. We have that $x \in \{x\}$, so every element of x (if there are any) is an element of $\{x\}$, i.e., it can only be x. So either $x = \emptyset$ or $x = \{x\}$. The latter case is impossible by Problem 12, so the only solution is $\{\emptyset\}$.

16. Assume that $\{A_i : i \in I\}$ is a nonempty set of transitive sets. If $x \in \bigcap\{A_i : i \in I\}$ and $y \in x$ then, by transitivity, $y \in A_i$ holds for all $i \in I$ so $y \in \bigcap\{A_i : i \in I\}$. If $x \in \bigcup\{A_i : i \in I\}$, then $x \in A_i$ for some A_i, so, if $y \in x$ then $y \in A_i$ (as A_i is transitive), so $y \in \bigcup\{A_i : i \in I\}$.

17. Assume first that $y \in x \in \mathrm{TC}(A)$. Then $x \in A_n$ for some n and then $y \in A_{n+1}$ so surely $y \in \mathrm{TC}(A)$.

For the other statement, if $A \in B$ and B is transitive, then we get by induction for every n that $A_n \subseteq B$. $A_0 \subseteq B$ is just a reformulation of $A \in B$, and if $A_n \subseteq B$ holds then, by transitivity, all elements of elements of A_n, i.e., all the elements of A_{n+1} should be in B. But then, $\mathrm{TC}(A) = A_0 \cup A_1 \cup \cdots \subseteq B$.

18. (a) We show by transfinite induction on α that V_α is transitive. This is obvious for $V_0 = \emptyset$. If α is limit, then we can use the inductive assumption and argue that V_α, the union of transitive sets, is itself transitive (see Problem 16). To finish the proof we have to show that if V_α is transitive then so is $V_{\alpha+1}$. Assume that $y \in x \in V_{\alpha+1}$. That is, $x \subseteq V_\alpha$, so $y \in V_\alpha$, so by the assumption, $y \subseteq V_\alpha$ which means that $y \in V_{\alpha+1}$ and that was to be proved.

(b) We show by transfinite induction on $\alpha \geq \beta$ that $V_\beta \subseteq V_\alpha$ holds. This is obvious for $\beta = \alpha$, the base case. If $\alpha > \beta$ is limit then again it is obvious (V_α is defined as a union with V_β in it). To cover the successor case it suffices to show that $V_\alpha \subseteq V_{\alpha+1}$. That is, $x \in V_\alpha$ implies $x \in V_{\alpha+1}$, i.e., $x \in V_\alpha$ implies $x \subseteq V_\alpha$, i.e., that V_α is transitive, which is just part (a).

(c) $\mathrm{rk}(x) = 0$ is impossible, as $V_0 = \emptyset$ has no elements. $\mathrm{rk}(x)$ cannot be some limit ordinal α either, as V_α is the union of the sets V_β for $\beta < \alpha$ and so every element of it appears earlier.

(d) Assume that $y \in x$ and $\mathrm{rk}(x) = \alpha + 1$. Thus, $x \in V_{\alpha+1}$, or, equally, $x \subseteq V_\alpha$, from which we get $y \in V_\alpha$, that is, $\mathrm{rk}(y) \leq \alpha$.

(e) Assume that every element of x is ranked. By the axiom of replacement there is some ordinal α such that $\mathrm{rk}(y) \leq \alpha$ holds for every $y \in x$. Thus, $x \subseteq V_\alpha$, so $x \in V_{\alpha+1}$, x is indeed ranked.

(f) For one direction assume that every set is ranked and A is a nonempty set. Select $x \in A$ with $\mathrm{rk}(x)$ minimal. Such an x exists, by the well-ordering property of ordinals. We claim that $x \cap A = \emptyset$. Indeed, if $y \in x \cap A$ then by (d) we have $\mathrm{rk}(y) < \mathrm{rk}(x)$ and, as $y \in A$, this would contradict the minimal choice of x.

For the other direction assume that the set A is not ranked. Let B be the transitive closure of A (see Problem 17). Set $C = \{x \in B : x \text{ is not ranked}\}$. C is not empty (as, for example, $A \in C$). We claim that C contradicts the axiom of foundation. Indeed, if $x \in C$ then x is an element of the transitively closed B and as x is not ranked, by (e) there is some $y \in x$ which is not ranked. But then $y \in B$ as well, so $y \in x \cap C$.

19. $X = \emptyset$ is clearly a solution. We prove that there is no other solution. Assume that $X \times Y = X$ and X is nonempty. Pick $x \in X$ with $\mathrm{rk}(x)$ minimal. Then, as $X \times Y = X$, $x = \langle u, v \rangle$ for some $u \in X$, $v \in Y$, we get that as $u \in \{u\} \in \{\{u\}, \{u, v\}\} = \langle u, v \rangle = x \in X$ holds, we have $\mathrm{rk}(u) < \mathrm{rk}(x)$, a contradiction.

20. (a) Define first $\mathcal{F}(x) = \mathrm{rk}(x)$ for $x \in \mathcal{C}$. Then $\mathcal{F}$ is an operation from $\mathcal{C}$ into the class of ordinals. If α is an ordinal, then $\mathcal{F}^{-1}(\alpha)$ is necessarily a set, as it is a subset of V_α. We do not know if the range $\mathcal{H}$ of $\mathcal{F}$ is all the ordinals, but it is certainly a proper class, as otherwise, by the axiom of replacement, we would get that

$$\mathcal{C} = \bigcup \{\mathcal{F}^{-1}(\alpha) : \alpha \in \mathcal{H}\}$$

is a set.

To eliminate the gaps, let $\mathcal{G}$ map the αth element of $\mathcal{H}$ to α. $\mathcal{G}$ maps $\mathcal{H}$ onto an initial segment of the ordinals, which, being a proper class, can only be the class of all ordinals. So we are finished by taking the composition of $\mathcal{F}$ and $\mathcal{G}$.

(b) Using (a), it suffices to give a mapping $\mathcal{F}$ from the ordinals to the ordinals such that $\mathcal{F}^{-1}(\alpha)$ is a proper class for every ordinal α. For this we let $\mathcal{F}(\kappa + \alpha) = \alpha$ where κ is an infinite cardinal and $\alpha < \kappa$, on the other places we let $\mathcal{F}$ be defined arbitrarily. Notice that this definition is unambigious, as κ can be calculated from $\kappa + \alpha$ by considering its cardinality, and α can be determined from $\kappa + \alpha$ and κ by left subtraction. Every value α is attained on a proper class; namely, on the ordinals of the form $\kappa + \alpha$ where $\kappa > \alpha$ is a cardinal.

21. For every $x \in \mathcal{C}$ let $\alpha(x)$ be the least ordinal that occurs as the rank of some $y \sim x$. Such an $\alpha(x)$ exists as every set is ranked and it is uniquely determined by the minimality property of ordinals. Then set

$$\mathcal{F}(x) = \left\{ y \in V_{\alpha(x)} : y \sim x \right\}.$$

This is always a set, as is a subset of $V_{\alpha(x)}$. $\mathcal{F}$ is, therefore, an operation. Notice that $\mathcal{F}(x)$ is always a nonempty set. Now, if $x \sim y$ then $\alpha(x) = \alpha(y)$ and so $\mathcal{F}(x) = \mathcal{F}(y)$. On the other hand, if $\mathcal{F}(x) = \mathcal{F}(y)$ then any $z \in \mathcal{F}(x) = \mathcal{F}(y)$ witnesses $x \sim z \sim y$.

22. Assume the statement holds. If A is any set, it has an embedding into the class of ordinals. Then, we can get a well-ordering of A by pulling back the well ordering of the ordinals.

Assume now that the axiom of choice holds, $\mathcal{C}$ is a proper class, κ a cardinal. We have to show that $\mathcal{C}$ has a subset of cardinality precisely κ. By AC, κ is well orderable, we can simply assume that it is an ordinal. The class $\{\mathrm{rk}(x) : x \in \mathcal{C}\}$ is a proper class of ordinals, so κ can be embedded (actually, there is an initial segment of order type κ). We have, therefore, found a subset B of $\mathcal{C}$, such that B has a surjective image of cardinality κ. Using the axiom of choice again, we get that B has a subset of cardinality κ. [John von Neumann: Die Axiomatisierung der Mengenlehre, *Mathematische Zeitschrift* **27**(1928), 669–752]

23.

(c) $\to$ (b) $\to$ (a) is obvious.

(a) $\to$ (c). Given a global choice operation $\mathcal{F}$ we well-order the universe as follows. Let $<_\alpha$ be a well-ordering of $V_{\alpha+1} \setminus V_\alpha$ determined by the proof of the well-ordering theorem using $\mathcal{F}$ (restricted to the nonempty subsets of $V_{\alpha+1} \setminus V_\alpha$). Then set $x < y$ iff either $\mathrm{rk}(x) < \mathrm{rk}(y)$ or else $\mathrm{rk}(x) = \mathrm{rk}(y) = \alpha+1$ for some ordinal α and $x <_\alpha y$. In this case, the predecessors of x are included into the set $V_{\alpha+1}$ where $\mathrm{rk}(x) = \alpha + 1$.

(e) $\to$ (d) is obvious.

(d) $\to$ (c). Apply (d) to the universe and the class of ordinals, which obviously has a setlike well order.

(c) $\to$ (e). Assume that $\mathcal{A}$ is a proper class with $<$, its inherited setlike well-ordering. For any ordinal α there is exactly one element of $\mathcal{A}$ which is the αth (whose set of predecessors form a well-ordered set of ordinal α), and this gives a bijection between $\mathcal{A}$ and the class of ordinals.

24. Increasing κ if needed, we can assume that κ is uncountable, regular. It suffices to show that $H_\kappa \subseteq V_\kappa$. Assume that $|\mathrm{TC}(x)| < \kappa$. We show by transfinite induction on the rank of $y \in \mathrm{TC}(x)$ that $y \in V_\kappa$. As $x \in \mathrm{TC}(x)$ this will give the result. Assume that we reached some $y \in \mathrm{TC}(x)$. We know that $|y| < \kappa$ (by condition) and that $\mathrm{rk}(z) < \kappa$ holds for every $z \in y$ (by the inductive hypothesis). As κ is regular, there is some $\alpha < \kappa$, such that $\mathrm{rk}(z) < \alpha$ holds for every $z \in y$, that is, $y \subseteq V_\alpha$, so $y \in V_{\alpha+1} \subseteq V_\kappa$.

25. We define the following subclasses M_α of M by transfinite recursion on α for every ordinal α. If M_β is defined for $\beta < \alpha$ then let M_α consist of those elements x of M that are not in any of the M_β's but every yEx is.

We claim that every element of M is in some M_α. Assume first that some $x \in M$ is not in any of the M_α's, but every yEx is. Then

$$y \mapsto \min\{\beta : y \in M_\beta\}$$

is an operation defined on $\{y : yEx\}$ that is a set, as M was supposed to be setlike. By the axiom of replacement the range of this set under the operation is a set of ordinals, it is therefore bounded by some ordinal α. Then x will be an element of $M_{\alpha+1}$ at the latest. We proved, therefore, that if $x \in M$ is such that it is not in any of the M_α's then necessarily some $x_1 E x$ has the exact same property. Repeating, we get a decreasing sequence $\ldots x_2 E x_1 E x$, contradicting the well foundedness of E.

We now define $\pi(x)$ for $x \in M_\alpha$ by transfinite recursion on α:

$$\pi(x) = \{\pi(y) : yEx\}\,.$$

π is injective: if $x \neq y$ then there is z, zEx, $z \not\!\!E\ y$ (or vice versa) and then $\pi(z) \in \pi(x)$, $\pi(z) \notin \pi(y)$, so $\pi(x) \neq \pi(y)$. π is an isomorphism: yEx if and only if $\pi(y) \in \pi(x)$. We set N as the range of π.

For unicity, assume that $\pi_1 : (M, E) \to (N_1, \in)$, $\pi_2 : (M, E) \to (N_2, \in)$ are isomorphisms. By transfinite induction on α we get that $\pi_1(x) = \pi_2(x)$ holds for $x \in M_\alpha$, that is, $\pi_1 = \pi_2$ and therefore $N_1 = N_2$. [A. Mostowski: An undecidable arithmetical statement, *Fund. Math.*, **36**(1949), 143–164]

Part III

Appendix

1

Glossary of Concepts

Abelian group is a group with commutative operation.

algebraic number is a complex number z that satisfies an equation of the form $a_n z^n + \cdots + a_0 = 0$ where $n > 0$, $a_n \neq 0$ and all coefficients a_i are integers.

algebraically closed field is a field F such that if $a_0, \ldots, a_n \in F$, $a_n \neq 0$, $n > 0$, then there is an $x \in F$ with $a_0 + a_1 x + \cdots + a_n x^n = 0$.

analytic set in $\mathbf{R}^n$ is a set that is the continuous image of a Borel set.

antichain in a partially ordered set is a subset no two elements of which are comparable.

antilexicographic ordering in a product of ordered sets is the ordering in which the last difference in the coordinates is decisive (i.e., if $\langle A_i, <_i \rangle$, $i \in I$, where $\langle I, \prec \rangle$ is an ordered set, then for $f, g \in \prod_{i \in I} A_i$ the element f is smaller in the antilexicographic ordering than g if there is an $i_0 \in I$ such that $f(i_0) <_{i_0} g(i_0)$, but for all $i \in I$ with $i_0 \prec i$ the equality $f(i) = g(i)$ holds).

antisymmetric relation is a binary relation ρ such that $(a, b) \in \rho$ and $(b, a) \in \rho$ implies $a = b$ ($a\rho b$ and $b\rho a$ implies $a = b$).

Aronszajn tree is a tree of height ω_1 with all levels and branches countable. In general, a κ-Aronszajn tree is a tree of height κ such that each level and each branch is of cardinality smaller than κ.

associative operation is a binary operation h such that $h(a, h(b, c)) = h(h(a, b), c)$ holds for all elements.

automorphism of an algebraic structure is a 1-to-1 mapping of the ground set onto itself that preserves operations and relations.

axiom of choice (AC) is the statement that for any family of nonempty sets there is a choice function, i.e., if $\{A_i\}_{i\in I}$ is a family of nonempty sets then there is a mapping $f : I \to \cup_{i\in I} A_i$ with $f(i) \in A_i$ for all $i \in I$.

axiom of comprehension states that if A is a set then the elements of A with a given property again form a set. Formally, if $\varphi(x_1, x_2, \ldots, x_{n+1})$ is a formula in the first order language of set theory and $A, a_1, \ldots, a_n$ are sets, then $\{x \in X \ : \ \varphi(x, a_1, \ldots, a_n)\}$ is a set.

axiom of replacement claims that if $\mathcal{F}(x)$ is an operation and A is a set, then $\{\mathcal{F}(x) : x \in A\}$ is a set.

Baire function is an element of the smallest family of functions (say on an interval) that contains all continuous functions and that is closed for pointwise limits.

basis in a vector space V over a field F is a set B such that every element of V can be uniquely written in a form $\lambda_1 v_1 + \cdots + \lambda_n v_n$ with $v_i \in B$ and $\lambda_i \in F$.

bijective mapping is an injective and surjective mapping (the same as a "1–1 and onto" mapping).

binary relation on a set A is a subset of $A \times A$.

bipartite graph is a graph in which the vertex set has a decomposition $V = V_1 \cup V_2$, $V_1 \cap V_2 = \emptyset$ such that all edges go between points of V_1 and V_2.

Boolean algebra is an algebraic structure $(A, +, \cdot, ', 0, 1)$, such that the structure $(A, +, \cdot, 0)$ is a commutative ring with multiplicative unit 1 in which $+$ and $\cdot$ are idempotent operations, $'$ is a unary operation such that $(a')' = a$ for all a, and for all a we have $a \cdot a' = 0$, $a + a' = 1$.

Borel function is a real-valued function f (defined on a topological space) such that $f^{-1}(-\infty, a)$ is a Borel set for all $a \in \mathbf{R}$. Complex-valued Borel function has real-valued Borel functions as its real and imaginary parts.

Borel set is an element of the smallest σ-algebra containing the open sets.

branch in a tree $\langle T, \prec\rangle$ is an ordered subset B that intersects every level of the tree. An α-branch of a tree $\langle T, \prec\rangle$ is an ordered subset $b \subseteq T_{<\alpha}$ which intersects every level T_β $(\beta < \alpha)$.

Cantor set is $\cap_{n=0}^{\infty} I_n$, where $I_0 = [0,1]$ and I_{n+1} is obtained from I_n by removing from every subinterval $[a,b]$ of I_n the middle third $(a+(b-a)/3, b-(b-a)/3)$. It is a perfect set of measure zero and of cardinality **c**. The Cantor set is precisely the set of those $x \in [0,1]$ which have a ternary expansion (i.e., expansion in base 3) that does not contain the digit 1.

cardinal is an ordinal α such that for $\beta < \alpha$ we have $\beta \not\sim \alpha$.

cardinal exponentiation: κ^λ is the cardinality of ${}^B A$ where A has cardinality κ and B has cardinality λ.

cardinality of a set is its size: two sets have the same cardinality if and only if they are equivalent. The cardinality of the set A is the smallest ordinal α with $A \sim \alpha$.

Cartesian product $\prod_{i\in I} A_i$ of a family A_i, $i \in I$ of sets is the set of all choice functions $f : I \to \cup_{i\in I} A_i$, $f(i) \in A_i$ for all $i \in I$. When I is finite, say $I = 1, 2, \ldots, n$, then this is often identified with the set

$$\{(a_1, \ldots, a_n) \;:\; a_i \in A_i\}$$

of n-tuples with ith coordinate from A_i, and then we write for it $A_1 \times \cdots \times A_n$.

ccc (countable chain condition) property holds in an ordered set (topological space) if every family of pairwise disjoint nonempty open intervals (sets) is countable.

chain in a partially ordered set is an ordered subset.

choice function for a family A_i, $i \in I$, of sets is a function $f : I \to \cup_{i\in I} A_i$ such that $f(i) \in A_i$ for all $i \in I$.

chromatic number of a graph G is the smallest κ such that G has a coloring with κ colors.

circuit in a graph is a is a sequence of distinct vertices $v_1, v_2, \ldots, v_n$ $(n \geq 3)$ such that v_i is joined to v_{i+1} and v_n is joined to v_1.

class is a well-determined part of the universe (of sets) that is not necessarily a set. Formally, if $\varphi(x_1, \ldots, x_{n+1})$ is a formula in the first-order language of set theory with free variables $x_1, \ldots, x_{n+1}$ and $a_1, \ldots, a_n$ are sets, then the collection of sets x which satisfy $\varphi(x, a_1, \ldots, a_n)$ forms a class.

closed unbounded set in an ordinal α is a set $C \subset \alpha$ that is closed in the order topology on α and that is cofinal with α.

club set in an ordinal is the same as closed and unbounded set.

cofinality: the ordered set $\langle A, \prec\rangle$ is cofinal with its subset B if for every $a \in A$ there is a $b \in B$ such that $a \preceq b$. There is always a well-ordered B with this property.

cofinality $\mathrm{cf}(\langle A, \prec\rangle)$ of an ordered set $\langle A, \prec\rangle$ is the smallest ordinal α such that there is a cofinal well-ordered $B \subset A$ for which the order type of $\langle B, <\rangle$ is α. It is always true that $\mathrm{cf}\langle A, \prec\rangle \le |A|$.

coloring of an infinite graph (V, E) with colors I is a mapping $f : V \to I$ such that $f(x) \neq f(y)$ whenever $(x, y) \in E$ (neighboring points have different colors). In this case we say that (V, E) is $|I|$-colorable.

commutative operation is a binary operation h such that $h(a, b) = h(b, a)$ holds for all elements.

compact space is a topological space in which every open cover includes a finite subcover.

complete metric space is a metric space $\langle X, d\rangle$, for which it is true that if $x_n \in X$, $n = 0, 1, \ldots$ is a Cauchy sequence (i.e., $d(x_n, x_m) \to 0$ as $n, m \to \infty$), then there is an element $x \in X$ such that $d(x_n, x) \to 0$ as $n \to \infty$.

connected graph is a graph such that any two vertices are connected by a path.

continuously ordered set is an ordered set $\langle A, \prec\rangle$ such that for any disjoint decomposition $A = B \cup C$, where $B \neq A$ is a nonempty initial segment, either B has a largest element or C has a smallest element (but not both).

Continuum hypothesis (CH) is the assumption that $\mathbf{c} = \aleph_1$, i.e., that every infinite subset of $\mathbf{R}$ is equivalent either with $\mathbf{R}$ or with $\mathbf{N}$. It is a statement neither provable nor disprovable in the Zermelo–Fraenkel axiom system.

Countryman type is the order type of an ordered set $\langle S, \prec\rangle$ if $S \times S$ is the union of countably many chains under the partial order "$\langle x, y\rangle \preceq \langle x', y'\rangle$ if and only if $x \preceq x'$ and $y \preceq y'$"

dense set

- in a topological space: A is dense in the topological space $\mathcal{T}$ if every open set contains a point of A.

- in an ordered set: A is dense in the ordered set $\langle B, \prec \rangle$ if for every $x, y \in B$ there is an element $a \in A$ with $x \preceq a \preceq y$, where $x \preceq y \Leftrightarrow x \prec y$ or $x = y$ (this is the same definition as density in topological spaces if one uses the order topology).

densely ordered set is an ordered set $\langle A, \prec \rangle$ such that for any $a, b \in A$ with $a \prec b$ there is a $c \in A$ with $a \prec c \prec b$.

density of a set $A \subset \mathbf{N}$ is defined as

$$\lim_{n \to \infty} \frac{|A \cap \{0, 1, \ldots, n-1\}|}{n}$$

provided this limit exists.

dichotomous relation on A is a binary relation ρ on A such that either $a\rho b$ or $b\rho a$ holds for all $a, b \in A$.

discrete set in a topological space is a set A such that each point in A has a neighborhood that does not contain any other point of A.

distributivity: a binary operation h is called left (right) distributive with respect to the binary operation g if $h(a, g(b, c)) = g(h(a, b), h(a, c))$ $(h(g(b, c), a) = g(h(b, a), h(c, a)))$ holds for all elements.

divisible group is an Abelian group $(G, +)$ such that for all $x \in G$ and for all $n \geq 1$ there is an y such that $x = \overbrace{y + \cdots + y}^{n-\text{times}}$.

domain of a function $f : A \to B$ is the set A.

edge coloring (also called good edge coloring) of a graph (V, X) (with edge set X) with colors I is a mapping $f : X \to I$ such that $f(e) \neq f(e')$ whenever e and e' have common endpoints.

end segment in an ordered set $\langle A, \prec \rangle$ is a subset $B \subseteq A$ such that $b \in B$, $b \prec c$ imply $c \in B$. It is called proper if it is not the whole set.

equivalence of sets: $A \sim B$ if and only there is a bijection between A and B.

equivalence relation is a reflexive, transitive, and symmetric relation.

field is a commutative ring $(F, +, \cdot)$ such that $(F \setminus \{0\}, \cdot)$ is an Abelian group (here 0 is the additive unit), i.e., if there is a multiplicative unit and every nonzero element is multiplicatively invertible.

filter is a family $\mathcal{F}$ of subsets of a ground set X such that $\emptyset \notin \mathcal{F}$, if $A, B \in \mathcal{F}$ then $A \cap B \in \mathcal{F}$, and if $A \in \mathcal{F}$ and $A \subseteq B \subseteq X$, then $B \in \mathcal{F}$.

first-category set in a topological space is a set that is the countable union of nowhere-dense sets.

function is a set f consisting of ordered pairs (x, y) such that $(x, y), (x, y') \in f$ imply $y = y'$. The set

$$A = \{x \ : \ (x, y) \in f \text{ for some } y\}$$

is called the domain (Dom(f)) of f, and

$$C = \{y \ : \ (x, y) \in f \text{ for some } x\}$$

is called its range (Ran(f)). If $C \subset B$, then we write $f : A \to B$. It is also customary to write $f(x)$ for y when $(x, y) \in f$.

generating set in a vector space is a set B such that every element in the space is a linear combination of elements of B.

Generalized continuum hypothesis (GCH) is the assumption that $2^\kappa = \kappa^+$ for all infinite cardinals κ. It is a statement neither provable nor disprovable in the Zermelo-Fraenkel axiom system.

G**-free graph** is a graph that does not include the graph G as a subgraph.

graph is a pair (V, X) where V is a set (the vertex set) and X is a set of two element subsets $\{x, y\}$ of V. Think of V as the set of vertices (points), X as the set of edges, $\{x, y\}$ the edge connecting x and y.

group is an algebraic structure $(G, \cdot)$ where $\cdot$ is an associative binary operation on G with unit element ($e \in G$ such that $e \cdot g = g \cdot e = g$ for all g) such that every element $g \in G$ has an inverse (an h such that $g \cdot h = h \cdot g = e$).

Hausdorff topological space is a topological space in which any two (different) elements have disjoint neighborhoods. Same as T_2 space.

height

- of an element x in a tree is the order type of the set of the elements smaller than x.
- of a tree is the smallest ordinal α for which the αth level of the tree is empty.

homogeneous set (monochromatic set) in a coloring is a set with constant color.

ideal

- in a ring R is a subring I such that for all. $a \in I$ and $b \in R$ the products ab and ba belong to I
- of sets is a set $\mathcal{I}$ of subsets of some ground set X such that $X \notin \mathcal{I}$, if $I \in \mathcal{I}$ and $J \subseteq I$ then $J \in \mathcal{I}$ and if $I, J \in \mathcal{I}$ then $I \cup J \in \mathcal{I}$.

idempotent operation is a binary operation h with $h(a,a) = a$ for all elements.

independent set in a graph is any set of vertices such that no two are connected by an edge.

initial segment in an ordered set $\langle A, \prec \rangle$ is a subset $B \subseteq A$ such that $b \in B$, $c \prec b$ imply $c \in B$. It is called proper if it is not the whole set.

injective mapping is the same as a one-to-one mapping ($f(x) \neq f(y)$ if $x \neq y$).

interval in an ordered set $\langle A, \prec \rangle$ is a subset $B \subseteq A$ such that $a, b \in B$ and $a \prec c \prec b$ imply $c \in B$.

interval topology (order topology) on an ordered set is the topology generated by the intervals of the set.

irreflexive relation is a binary relation ρ such that no element is in relation with itself: $(a, a) \notin \rho$.

$\boldsymbol{K_\kappa}$ is the full graph with κ vertices.

$\boldsymbol{K_{\kappa,\lambda}}$ is the full bipartite graph with bipartition classes of cardinality κ and λ (i.e., the vertex set is $\{x_\xi\}_{\xi<\kappa} \cup \{y_\eta\}_{\eta<\lambda}$ with $x_\xi \neq y_\eta$ and every element x_ξ is joined to every y_η).

lattice is an algebraic structure $(A, \wedge, \vee)$, in which $\wedge, \vee$ are commutative, associative, and idempotent operations such that $a \wedge (a \vee b) = a$ and $a \vee (a \wedge b) = a$ hold for all elements. A lattice is called distributive if $\wedge$ and $\vee$ are distributive with respect to each other.

level (level set) in a tree is the set of elements with the same "height", i.e., the αth level set of $\langle T, \prec \rangle$ is the set of those elements $x \in T$ for which the order type of $\{y : y \prec x\}$ is α.

lexicographic ordering in a product of ordered sets is the ordering in which the first difference in the coordinates is decisive (i.e., if $\langle A_i, <_i\rangle$, $i \in I$, where $\langle I, \prec\rangle$ is an ordered set, then for $f, g \in \prod_{i\in I} A_i$ the element f is smaller in the lexicographic ordering than g if there is an $i_0 \in I$ such that $f(i_0) <_{i_0} g(i_0)$, but for all $i \in I$ with $i \prec i_0$ the equality $f(i) = g(i)$ holds).

limit cardinal is an uncountable cardinal κ such that $\lambda < \kappa$ implies $\lambda^+ < \kappa$.

limit ordinal is a non-successor, nonzero ordinal α (i.e., $\beta < \alpha$ implies $\beta + 1 < \alpha$).

linearly independent system B in a vector space means that if $v_1, \dots, v_n \in B$ are different elements then and $\lambda_1 v_1 + \cdots + \lambda_n v_n = 0$ if and only if $\lambda_1 = \cdots = \lambda_n = 0$ (nontrivial linear combinations cannot be zero).

lower density of a set $A \subset \mathbf{N}$ is defined as

$$\liminf_{n\to\infty} \frac{|A \cap \{0, 1, \dots, n-1\}|}{n}.$$

matching in a graph (V, E) is a set M of disjoint edges such that every $v \in V$ is the endpoint of an edge in M.

maximal element in a partially ordered set is an x such that no element is larger than x.

measurable cardinal is a cardinal κ for which there is a κ-additive nontrivial 0–1-valued measure on all subsets of a set of cardinality κ (i.e., if $|X| = \kappa$, then there is a $\mu : \mathcal{P}(X) \to \{0, 1\}$ such that $\mu(X) = 1$, $\mu(\{x\}) = 0$ for all $x \in X$, and if Y_i, $i \in I$, $|I| < \kappa$ is a disjoint family of fewer than κ sets then $\mu(\cup_i Y_i) = \sum_i \mu(Y_i)$).

metric on a set X is a mapping $d : X \to [0, \infty)$ with the properties that $d(x, y) = 0$ if and only if $x = y$, $d(x, y) = d(y, x)$ and (triangle inequality) $d(x, y) \le d(x, z) + d(z, y)$.

metric space is a set X with a metric d on it, in which the topology is generated by balls, i.e., sets of the form $\{y : d(x, y) < r\}$, $x \in X$, $r > 0$.

metrizable topology is a topology equivalent to the topology of a metric (i.e., there is a metric on the space such that the open sets in the topology and in the metric are the same).

minimal element in a partially ordered set is an x such that no element is smaller than x.

monochromatic set (homogeneous set) in a coloring is a set with constant color.

monotone mapping between two ordered sets $\langle A, \prec \rangle$ and $\langle B, < \rangle$ is a mapping $f : A \to B$ such that $a \prec b$ implies $f(a) < f(b)$.

nonstationary set in an ordinal α is a set that is disjoint from some closed and unbounded set.

normal topological space is where any two disjoint closed sets can be separated by disjoint open sets (i.e., if $F_1 \cap F_2 = \emptyset$ are closed sets then there are disjoint open sets $F_1 \subseteq G_1$ and $F_2 \subseteq G_2$).

nowhere dense set in a topological space is a set such that its closure has no inner point.

one-to-one correspondence (1-to-1 correspondence) between sets A and B is a one-to-one mapping of A *onto* B. Such mappings are often called bijections. This is nothing else than an equivalence between A and B.

one-to-one mapping (1-to-1 mapping) is the same as an injective mapping, i.e., it maps different elements into different elements ($f(x) \neq f(y)$ if $x \neq y$).

operation (in set theory) is a mapping $x \mapsto \mathcal{F}(x)$ that is not necessarily a function (i.e., its domain or range may not be sets). Formally, it is the correspondence $x \mapsto y$ given by $\varphi(x, y, a_1, \ldots, a_n)$, where $\varphi(x_1, \ldots, x_{n+2})$ is a formula in the first-order language of set theory with the property that for every x there is at most one y for which φ holds, and $a_1, \ldots, a_n$ are given sets.

ordered pair is a set of the form $(a, b) = \{\{a\}, \{a, b\}\}$.

ordered set is a pair $\langle A, \prec \rangle$ where A is a set and $\prec$ is an irreflexive, transitive and trichotomous relation on A.

ordered sum of order types θ_i with respect to $\langle I, \prec \rangle$ is the order type of the ordered union of ordered sets $\langle A_i, <_i \rangle$ with respect to $\langle I, \prec \rangle$, where $\langle A_i, <_i \rangle$ has order type θ_i (denoted by $\sum_{i \in I\ (\prec)} \theta_i$).

ordered union of the ordered sets $\langle A_i, <_i \rangle$, $i \in I$ with respect to the ordered set $\langle I, < \rangle$ (where the A_i's are disjoint sets) is the ordered set $\langle B, < \rangle$ in which $B = \cup_{i \in I} A_i$, and for $a \in A_i$ and $b \in A_j$ the relation $a \prec b$ holds if and only if $i < j$ or $i = j$ and $a <_i b$.

ordering is a binary relation that is irreflexive, transitive, and trichotomous (on a ground set).

order topology (interval topology) in an ordered set $\langle A, <\rangle$ is the topology generated by the intervals. It is also the topology generated by sets of the form A, $\{x : x < a\}$, $\{x : a < x\}$ with $a, b \in A$.

order type: two ordered sets are said to have the same order types if they are similar.

ordinal is the order types of a well-ordered set. We identify every ordinal α with the set of ordinals smaller than α, i.e., $\alpha = \{\beta : \beta < \alpha\}$. The von Neumann definition of ordinals: ordinals are transitive sets A (i.e., $a \in A$ implies $a \subseteq A$) that are well ordered by the $\in$ relation.

partial ordering is a binary relation that is irreflexive and transitive.

partially ordered set is a pair $\langle A, \prec\rangle$ where A is a set and $\prec$ is an irreflexive and transitive relation on A.

partition relation $\kappa \to (\lambda)^r_\rho$ means that if we color the r-element subsets of a set X of cardinality κ with ρ colors, then there is a homogeneous set (monochromatic set) $Y \subset X$ of cardinality λ (i.e., every r-element subset of Y has the same color).

path in a graph is a sequence $v_1, \ldots, v_n$ of distinct vertices such that each v_i is connected to v_{i+1} by an edge.

perfect set is a nonempty closed set (in a topological space) that is dense in itself, i.e., any neighborhood of any point x contains a point different from x.

permutation of a set is a one-to-one mapping of the set onto itself.

planar graph is a graph that can be represented in the plane with noncrossing (curved) edges.

predecessor (immediate predecessor) to $a \in A$ in a partially ordered set $\langle A, \prec\rangle$ is an element b such that $b \prec a$ but there is no $c \in A$ with $b \prec c \prec a$.

prime field of a field $\mathcal{F}$ is the subfield generated by 1. It is isomorphic to either Q or to one of Z_p (the field of integers mod p with a prime number p). In fact, if the characteristic of $\mathcal{F}$ is $p > 0$ (i.e., if $p \cdot 1 = 0$) then the prime field is Z_p, and if the characteristic is 0 (i.e., $m \cdot 1 \neq 0$ for any $m > 0$) then the prime field is isomorphic to $(Q, +, \cdot)$.

prime ideal is a maximal ideal $\mathcal{I}$ over a ground set X (alternatively, for every $Y \subset X$ either Y or $X \setminus Y$ belongs to $\mathcal{I}$). It is trivial if $\mathcal{I} = \{Y : x \notin Y\}$ for some $x \in X$.

product of cardinals κ_i, $i \in I$ is the cardinality of the product set $\prod_{i\in I} A_i$ where the A_i's are sets of cardinality κ_i (denoted by $\prod_{i\in I} \kappa_i$).

product of order types θ and ρ is the order type of the antilexicographically ordered product of two ordered sets of order type θ and ρ, respectively (denoted by $\theta \cdot \rho$).

product of sets A_i, $i \in I$ is the set of all choice functions $f : I \to \cup_i A_i$, $f(i) \in A_i$, $i \in I$ for the sets A_i.

proper class is a class that is not a set.

proper initial segment is an initial segment of an ordered set that is not the whole set.

Pythagorean triplet: positive integers a, b, c with the property $c^2 = a^2+b^2$. If a, b, c do not have a common divisor, then one of a and b, say b, is even, and then they are of the form $a = n^2 - m^2$, $b = 2mn$, and $c = n^2 + m^2$ where m, n are relatively prime natural numbers of different parity.

range of a function $f : A \to B$ is the set of all elements $y = f(x)$, $x \in A$.

real-valued measurable cardinal is a cardinal κ for which there is a κ-additive nontrivial $[0, 1]$-valued measure on all subsets of a set of cardinality κ (i.e., if $|X| = \kappa$, then there is a $\mu : \mathcal{P}(X) \to [0, 1]$ from the power set of X into $[0, 1]$ such that $\mu(X) = 1$, $\mu(\{x\}) = 0$ for all $x \in X$, and if Y_i, $i \in I$, $|I| < \kappa$ is a disjoint family of fewer than κ sets then $\mu(\cup_i Y_i) = \sum_i \mu(Y_i)$).

reflexive relation is a binary relation ρ for which $(a, a) \in \rho$ for all a ($a\rho a$ for all a).

regressive function f on a subset A of an ordinal α is a function $f : A \to \alpha$ with the property that $f(\xi) < \xi$ for all $\xi \in A$, $\xi \neq 0$.

regular cardinal is an infinite cardinal κ that coincides with its cofinality ($\text{cf}(\kappa) = \kappa$). Equivalently, κ is not the sum of fewer than κ cardinals each smaller than κ.

relation: A subset of the Cartesian product $\overbrace{A \times A \times \cdots \times A}^{k-\text{times}}$ is called a k-ary relation on A. It is called a binary relation when $k = 2$. For easier notation $(a, b) \in \rho$ is often denoted $a\rho b$.

reverse order type θ^* to an order type θ is the order type of $\langle A, \prec^* \rangle$ where θ is the order type of $\langle A, \prec \rangle$, and $\prec^*$ is the reverse ordering on A, i.e., $a \prec^* b \iff b \prec a$.

ring is an algebraic structure $(A, +, \cdot, 0)$, in which $(A, +, 0)$ is a commutative group (i.e., $+$ is a commutative and associative operation, $a + 0 = a$ for all a, and for all a there is an element a^* such that $a + a^* = 0$) and $\cdot$ is an associative operator that is distributive with respect to $+$ from both sides (i.e., $a \cdot (b + c) = a \cdot b + a \cdot c$ and $(b + c) \cdot a = b \cdot a + c \cdot a$).

σ-algebra is a family $\mathcal{A} \subset X$ of subset of a ground set X such that $\emptyset \in \mathcal{A}$, if $A_i \in \mathcal{A}$, $i = 0, 1, 2, \ldots$ then $\cup_{i=0}^{\infty} A_i \in \mathcal{A}$, and if $A \in \mathcal{A}$ then $X \setminus A \in \mathcal{A}$.

second-category set in a topological space is a set that is not the countable union of nowhere-dense sets.

separable metric/topological space is a metric/topological space including a countable dense subset.

similarity mapping between two ordered sets $\langle A, \prec \rangle$ and $\langle B, < \rangle$ is a monotone and surjective mapping $f : A \to B$.

similarity of ordered sets $\langle A, \prec \rangle$ and $\langle B, < \rangle$ means that there is a similarity mapping between them.

singular cardinal is a non-regular infinite cardinal.

spanning tree in a graph G is a subgraph T which is a tree that contains all points of G.

spanned subgraph (induced subgraph) $G' = (V', X')$ of a graph $G = (V, X)$ is a graph with $V' \subseteq V$ and $X' = X \cap (V' \times V')$.

Specker type is the order type of an uncountable ordered set that does not embed ω_1, ω_1^* (the reverse of ω_1), or an uncountable subset of the reals.

stationary set is a set that intersects every closed and unbounded sets (in an ordinal α).

strong limit cardinal is an uncountable cardinal κ such that $\lambda < \kappa$ implies $2^\lambda < \kappa$.

strongly inaccessible cardinal is a strong limit regular cardinal, i.e., a regular κ such that $\lambda < \kappa$ implies $2^\lambda < \kappa$.

subbase in a topological space $\langle X, \mathcal{T} \rangle$ is a set $\mathcal{B} \subset \mathcal{T}$ such that for every $x \in X$ and for every $V \in \mathcal{T}$ with $x \in V$ (i.e., for every neighborhood of x) there are finitely many $U_1, \ldots, U_m \in \mathcal{B}$ with $x \in \cap_{j=1}^m U_j \subseteq V$.

subgraph $G' = (V', X')$ of a graph $G = (V, X)$ is a graph with $V' \subseteq V$ and $X' \subseteq X$.

successor

- to $a \in A$ in a partially ordered set $\langle A, \prec \rangle$ is an element b such that $a \prec b$ but there is no $c \in A$ with $a \prec c \prec b$.
- to a cardinal κ is the smallest cardinal λ that is bigger than κ (it is denoted by κ^+).
- to an ordinal θ is the smallest ordinal ξ that is bigger than θ (it is actually $\theta + 1$).

successor cardinal is an infinite cardinal of the form κ^+.

successor ordinal is an ordinal of the form $\beta + 1$.

sum of cardinals κ_i, $i \in I$ is the cardinality of the set $\cup_{i \in I} A_i$ where the A_i's are disjoint sets of cardinality κ_i (denoted by $\sum_{i \in I} \kappa_i$).

surjective mapping: $f : A \to B$ such that every $b \in B$ has a pre-image (i.e., $f[A] = B$). It is also said that f maps A onto B.

Suslin line is a nonseparable ordered set which is ccc, that is, it does not include a countable dense set and every family of pairwise disjoint nonempty open intervals is countable .

Suslin tree is an ω_1-tree with no ω_1-branch or uncountable antichain in it.

symmetric relation is a binary relation ρ for which $(a, b) \in \rho$ implies $(b, a) \in \rho$ ($a\rho b$ implies $b\rho a$).

$\boldsymbol{T_2}$ space is a Hausdorff topological space.

topological space $\langle X, \mathcal{T} \rangle$, where $\mathcal{T} \subseteq \mathcal{P}(X)$ is a set of subsets of X (the set of open sets in the space) with $X \in \mathcal{T}$ and closed under finite intersection and arbitrary union.

tournament is a complete directed graph (i.e., all edges of a complete undirected graph is directed in exactly one way). It is called transitive if whenever $\overrightarrow{uv}$ and $\overrightarrow{vw}$ are edges, then so is $\overrightarrow{uw}$.

transcendence basis in a field F is a set B such that the elements of B are algebraically independent over the prime field F_1 of F (i.e., if $p(x_1, \ldots, x_n)$ is a nonzero polynomial over F_1, i.e., with coefficients in F_1, and $a_1, \ldots, a_n \in B$ are different elements, then $p(a_1, \ldots, a_n) \neq 0$), but for every $a \in F$ there is a nonzero polynomial $p(x_1, \ldots, x_n, x_{n+1})$ over F_1 and different elements $a_1, \ldots, a_n \in B$ so that $p(a_1, \ldots, a_n, a) = 0$.

transcendental number is a non-algebraic number (in **C** or **R**).

transitive relation is a binary relation ρ for which $(a,b) \in \rho$ and $(b,c) \in \rho$ implies $(a,c) \in \rho$ ($a\rho b$ and $b\rho c$ implies $a\rho c$).

tree

- as a graph: is a connected graph without circuit.
- as a partially ordered set: is a partially ordered set $\langle X, \prec \rangle$ such that for every $x \in X$ the set $\{y : y \prec x\}$ is well ordered. It is called a κ-tree if its height is κ and every level is of cardinality smaller than κ.

tree property of a cardinal κ means that every tree of height κ the levels of which are of cardinality smaller than κ includes a branch of length κ.

trichotomous relation on A is a binary relation ρ on A such that one of $(a,b) \in \rho$, $(b,a) \in$ and $a = b$ holds for all $a, b \in A$ (one of $a\rho b$, $b\rho a$ and $a = b$ holds for all $a, b \in A$).

ultrafilter is a maximal filter $\mathcal{F}$ over a ground set X (alternatively, for every $Y \subseteq X$ either Y or $X \setminus Y$ belongs to $\mathcal{F}$). It is called trivial if it is generated by an element (i.e., there is an x such that the elements in the ultrafilter are those subsets of the ground set that contain x).

upper density of a set $A \subset \mathbf{N}$ is defined as

$$\limsup_{n\to\infty} \frac{|A \cap \{0, 1, \ldots, n-1\}|}{n}.$$

vector space over a field F is an Abelian group $(V, +)$ such that for every $\lambda \in F$ and $v \in V$ the product $\lambda v \in V$ is also defined and is an element of V, if 1 is the multiplicative unit of F then $1v = v$ for all $v \in V$, and the following identities hold: $\lambda(u + v) = \lambda u + \lambda v$, $(\lambda_1\lambda_2)u = \lambda_1(\lambda_2 u)$ and $(\lambda_1 + \lambda_2)u = \lambda_1 u + \lambda_2 u$.

weakly compact cardinal is a cardinal $\kappa > \omega$ for which $\kappa \to (\kappa)^2_2$ holds.

well-founded partially ordered set is a partially ordered set in which every nonempty subset contains a minimal element.

well-ordered set is an ordered set in which every nonempty subset contains a smallest element.

well-ordering theorem is the statement that on every set there is a well-ordering (i.e., for every set X there is a binary relation $\prec$ on X such that $\langle X, \prec \rangle$ is a well-ordered set).

2

Glossary of Symbols

$\aleph_\alpha$ is the αth infinite cardinal (same as ω_α).

${}^AB = \{f : f : A \to B\}$ is the set of all mappings from A to B.

$[A]^\kappa$ is the set of subsets of A of cardinality κ.

$[A]^{<\kappa}$ is the set of subsets of A of cardinality $< \kappa$.

$A \sim B$, equivalence of A and B.

$A \Delta B = (A \setminus B) \cup (B \setminus A)$, symmetric difference.

∇C_α, diagonal intersection.

$A \times B = \{(a,b) : a \in A, \ b \in B\}$, Cartesian product.

$A^c = X \setminus A$, complement with respect to the ground set X.

$\text{cf}(\langle A, \prec \rangle)$ is the cofinality of the ordered set $\langle A, \prec \rangle$.

$\text{cf}(\alpha)$ is the cofinality of the ordinal α.

CH stands for the continuum hypothesis.

$\mathbf{c}$, the cardinality continuum, i.e., the cardinality of $\mathbf{R}$.

χ_A is the characteristic function of the set A.

$\text{Dom}(f)$ is the domain of the function f.

η is the order type of $\langle \mathbf{Q}, < \rangle$.

$f[A] = \{f(a) : a \in A\}$ is the set of the images of elements of A under f.

$f^{-1}(y) = \{x : f(x) = y\}$ is the inverse image of the element A under f.

$f^{-1}[A] = \{x : f(x) \in A\} = \cup_{y \in A} f^{-1}(y)$ is the inverse image of the set A under f.

$\text{FS}(\kappa)$ is the set of finite sequences of ordinals smaller than κ.

GCH stands for the generalized continuum hypothesis.

κ^λ is the λ power of κ.

κ^+ is the successor cardinal to the cardinal κ.

K_λ is the complete graph with vertex set of cardinality λ.

$K_{\lambda,\rho}$ is the complete bipartite graph with bipartition classes of size λ and ρ.

λ is the order type of $\langle \mathbf{R}, < \rangle$.

$\mathbf{N} = \omega = \{0, 1, 2, \ldots\}$ is the set of natural numbers.

$\omega = \{0, 1, 2, \ldots\}$ is the set of natural numbers and also its order type.

ω_1 is the first uncountable ordinal.

ω_α is the αth infinite cardinal, same as $\aleph_\alpha$ ($\omega_0 = \aleph_0 = \omega$, i.e., counting is started at 0).

$\mathcal{P}(A) = \{B : B \subset A\}$ is the power set of A (set of all subsets of A).

$\prod_{i \in I} A_i = \{f : I \to \cup_{i \in I} A_i : f(i) \in A_i \text{ for all } i \in I\}$ is the set of all choice functions for the family $\{A_i\}_{i \in I}$.

$\prod_{i \in I} \kappa_i$ is the product of the cardinals κ_i.

$\mathbf{Q}$ is the set of rational numbers.

$\mathrm{Ran}(f)$ is the range of the function f.

$\mathbf{R}$ is the set of real numbers.

$\mathbf{R}^n$ is the n-dimensional Euclidean space.

$\mathbf{R}^\infty$ is the set of infinite sequences of real numbers.

$\sum_{i \in I\ (\prec)} \theta_i$ is the ordered sum of the order types θ_i with respect to $\langle I, \prec \rangle$.

$\sum_{i \in I} \kappa_i$ is the sum of the cardinals κ_i.

$\theta_0 + \theta_1$ is the sum of the order types θ_0 and θ_1 in this order (with respect to the ordered set $\langle \{0, 1\}, < \rangle$.

$\theta_0 \cdot \theta_1$ is the product of the order types θ_0 and θ_1 in this order.

θ^*, reverse order type to θ.

$\mathbf{Z}$ is the set of integers.

3

Index